T0189998

Lecture Notes in Artificial Intelligence 12465

Subseries of Lecture Notes in Computer Science

More information about this series at http://www.springer.com/series/1244

De-Shuang Huang · Prashan Premaratne (Eds.)

Intelligent Computing Methodologies

16th International Conference, ICIC 2020
Bari, Italy, October 2–5, 2020
Proceedings, Part III

Editors
De-Shuang Huang
Machine Learning and Systems Biology
Tongji University
Shanghai, China

Prashan Premaratne
School of Electrical, Computer
and Telecommunications Engineering
University of Wollongong
North Wollongong, NSW, Australia

ISSN 0302-9743 ISSN 1611-3349 (electronic)
Lecture Notes in Artificial Intelligence
ISBN 978-3-030-60795-1 ISBN 978-3-030-60796-8 (eBook)
https://doi.org/10.1007/978-3-030-60796-8

LNCS Sublibrary: SL7 – Artificial Intelligence

This Springer imprint is published by the registered company Springer Nature Switzerland AG
The registered company address is: Gewerbestrasse 11, 6330 Cham, Switzerland

Preface

The International Conference on Intelligent Computing (ICIC) was started to provide an annual forum dedicated to the emerging and challenging topics in artificial intelligence, machine learning, pattern recognition, bioinformatics, and computational biology. It aims to bring together researchers and practitioners from both academia and industry to share ideas, problems, and solutions related to the multifaceted aspects of intelligent computing.

ICIC 2020, held in Bari, Italy, during October 2–5, 2020, constituted the 16th edition of this conference series. It built upon the success of ICIC 2019 (Nanchang, China), ICIC 2018 (Wuhan, China), ICIC 2017 (Liverpool, UK), ICIC 2016 (Lanzhou, China), ICIC 2015 (Fuzhou, China), ICIC 2014 (Taiyuan, China), ICIC 2013 (Nanning, China), ICIC 2012 (Huangshan, China), ICIC 2011 (Zhengzhou, China), ICIC 2010 (Changsha, China), ICIC 2009 (Ulsan, South Korea), ICIC 2008 (Shanghai, China), ICIC 2007 (Qingdao, China), ICIC 2006 (Kunming, China), and ICIC 2005 (Hefei, China).

This year, the conference concentrated mainly on the theories and methodologies as well as the emerging applications of intelligent computing. Its aim was to unify the picture of contemporary intelligent computing techniques as an integral concept that highlights the trends in advanced computational intelligence and bridges theoretical research with applications. Therefore, the theme for this conference was "Advanced Intelligent Computing Technology and Applications." Papers that focused on this theme were solicited, addressing theories, methodologies, and applications in science and technology.

ICIC 2020 received 457 submissions from 21 countries and regions. All papers went through a rigorous peer-review procedure and each paper received at least three review reports. Based on the review reports, the Program Committee finally selected 162 high-quality papers for presentation at ICIC 2020, included in three volumes of proceedings published by Springer: two volumes of *Lecture Notes in Computer Science* (LNCS), and one volume of *Lecture Notes in Artificial Intelligence* (LNAI).

This volume of LNAI includes 54 papers.

The organizers of ICIC 2020, including Tongji University, China, and Polytechnic University of Bari, Italy, made an enormous effort to ensure the success of the conference. We hereby would like to thank the members of the Program Committee and the referees for their collective effort in reviewing and soliciting the papers. We would like to thank Alfred Hofmann, executive editor from Springer, for his frank and helpful advice and guidance throughout as well as his continuous support in publishing the proceedings. In particular, we would like to thank all the authors for contributing their papers. Without the high-quality submissions from the authors, the success of the conference would not have been possible. Finally, we are especially grateful to the International Neural Network Society and the National Science Foundation of China for their sponsorship.

August 2020

De-Shuang Huang
Prashan Premaratne

Organization

General Co-chairs

De-Shuang Huang, China
Vitoantonio Bevilacqua, Italy

Program Committee Co-chairs

Eugenio Di Sciascio, Italy
Kanghyun Jo, South Korea

Organizing Committee Co-chairs

Ling Wang, China
Phalguni Gupta, India
Vincenzo Piuri, Italy
Antonio Frisoli, Italy
Eugenio Guglielmelli, Italy
Silvestro Micera, Italy
Loreto Gesualdo, Italy

Organizing Committee Members

Andrea Guerriero, Italy
Nicholas Caporusso, USA
Francesco Fontanella, Italy
Vincenzo Randazzo, Italy
Giacomo Donato Cascarano, Italy
Irio De Feudis, Italy
Cristian Camardella, Italy
Nicola Altini, Italy

Award Committee Co-chairs

Kyungsook Han, South Korea
Jair Cervantes Canales, Mexico
Leonarda Carnimeo, Italy

Tutorial Co-chairs

M. Michael Gromiha, India
Giovanni Dimauro, Italy

Publication Co-chairs

Valeriya Gribova, Russia
Antonino Staiano, Italy

Special Session Co-chairs

Abir Hussain, UK
Antonio Brunetti, Italy

Special Issue Co-chairs

Mario Cesarelli, Italy
Eros Pasero, Italy

International Liaison Co-chairs

Prashan Premaratne, Australia
Marco Gori, Italy

Workshop Co-chairs

Laurent Heutte, France
Domenico Buongiorno, Italy

Publicity Co-chairs

Giansalvo Cirrincione, France
Chun-Hou Zheng, China
Salvatore Vitabile, Italy

Exhibition Contact Co-Chairs

Michal Choras, Poland
Stefano Cagnoni, Italy

Program Committee Members

Daqi Zhu
Xinhong Hei
Yuan-Nong Ye
Abir Hussain
Khalid Aamir
Kang-Hyun Jo
Andrea Guerriero
Angelo Ciaramella
Antonino Staiano
Antonio Brunetti
Wenzheng Bao
Binhua Tang
Bin Qian
Bingqiang Liu
Bo Liu
Bin Liu
Chin-Chih Chang
Wen-Sheng Chen
Michal Choras
Xiyuan Chen
Chunmei Liu
Cristian Camardella
Zhihua Cui
Defu Zhang
Dah-Jing Jwo
Dong-Joong Kang
Domenico Buongiorno
Domenico Chiaradia
Ben Niu
Shaoyi Du
Eros Pasero
Fengfeng Zhou
Haodi Feng
Fei Guo
Francesco Fontanella
Chuleerat Jaruskulchai
Fabio Stroppa
Gai-Ge Wang
Giacomo Donato Cascarano
Giovanni Dimauro
L. J. Gong
Guoquan Liu
Wei Chen
Valeriya Gribova
Michael Gromiha
Maria Siluvay
Guoliang Li
Huiyu Zhou
Tianyong Hao
Mohd Helmy Abd Wahab
Honghuang Lin
Jian Huang
Hao Lin
Hongmin Cai
Xinguo Lu
Ho-Jin Choi
Hongjie Wu
Irio De Feudis
Dong Wang
Insoo Koo
Daowen Qiu
Jiansheng Wu
Jianbo Fan
Jair Cervantes
Junfeng Xia
Junhui Gao
Juan Carlos
Juan Carlos Figueroa-García
Gangyi Jiang
Jiangning Song
Jing-Yan Wang
Yuhua Qian
Joaquín Torres-Sospedra
Ju Liu
Jinwen Ma
Ji Xiang Du
Junzhong Gu
Ka-Chun Wong
Kyungsook Han
K. R. Seeja
Yoshinori Kuno
Weiwei Kong
Laurent Heutte
Leonarda Carnimeo
Bo Li
Junqing Li
Juan Liu
Yunxia Liu
Zhendong Liu
Jungang Lou
Fei Luo
Jiawei Luo
Haiying Ma
Marzio Pennisi
Nicholas Caporusso
Nicola Altini
Giansalvo Cirrincione
Gaoxiang Ouyang
Pu-Feng Du
Shaoliang Peng
Phalguni Gupta
Ping Guo
Prashan Premaratne
Qinghua Jiang
Qingfeng Chen
Roman Neruda
Rui Wang
Stefano Squartini
Salvatore Vitabile
Wei-Chiang Hong
Jin-Xing Liu
Shen Yin
Shiliang Sun
Saiful Islam
Shulin Wang
Xiaodi Li
Zhihuan Song
Shunren Xia
Sungshin Kim
Stefano Cagnoni
Stefano Mazzoleni
Surya Prakash
Tar Veli Mumcu
Xu-Qing Tang

Vasily Aristarkhov
Vincenzo Randazzo
Vito Monaco
Vitoantonio Bevilacqua
Waqas Bangyal
Bing Wang
Wenbin Liu
Weidong Chen
Weijia Jia
Wei Jiang
Shanwen Zhang
Takashi Kuremoto

Reviewers

Wan Hussain Wan Ishak
Nureize Arbaiy
Shingo Mabu
Lianming Zhang
Xiao Yu
Shaohua Li
Yuntao Wei
Jinglong Wu
Wei-Chiang Hong
Sungshin Kim
Tianhua Guan
Shutao Mei
Yuelin Sun
Hai-Cheng Yi
Zhan-Heng Chen
Suwen Zhao
Medha Pandey
Mike Dyall-Smith
Xin Hong
Ziyi Chen
Xiwei Tang
Khanh Le
Shulin Wang
Di Zhang
Sijia Zhang
Na Cheng
Menglu Li
Zhenhao Guo
Limin Jiang
Kun Zhan
Cheng-Hsiung Chiang
Yuqi Wang
Anna Esposito
Salvatore Vitabile
Bahattin Karakaya
Tejaswini Mallavarapu
Sheng Yang
Heutte Laurent
Seeja
Pu-Feng Du
Wei Chen
Jonggeun Kim
Eun Kyeong Kim
Hansoo Lee
Yiqiao Cai
Wuritu Yang
Weitao Sun
Shou-Tao Xu
Min-You Chen
Yajuan Zhang
Guihua Tao
Jinzhong Zhang
Wenjie Yi
Miguel Gomez
Lingyun Huang
Chao Chen
Jiangping He
Jin Ma
Xiao Yang
Sotanto Sotanto
Liang Xu
Chaomin Iuo
Rohitash Chandra
Hui Ma
Lei Deng
Di Liu
María I. Giménez
Ansgar Poetsch
Dimitry Y. Sorokin
Jill F. Banfield
Can Alkan
Ji-Xiang Du
Xiao-Feng Wang
Zhong-Qiu Zhao
Bo Li
Zhong rui Zhang
Yanyun Qu
Shunlin Wang
Jin-Xing Liu
Shravan Sukumar
Long Gao
Yifei Wu
Qi Yan
Tianhua Jiang
Fangping Wan
Lixiang Hong
Sai Zhang
Tingzhong Tian
Qi Zhao
Leyi Wei
Lianrong Pu
Chong Shen
Junwei Wang
Zhe Yan
Rui Song
Xin Shao
Xinhua Tang
Claudia Guldimann
Saad Abdullah Khan Bangyal
Giansalvo Cirrincione
Bing Wang
Xiao Xiancui
X. Zheng
Vincenzo Randazzo
Huijuan Zhu
DongYuan Li
Jingbo Xia
Boya Ji
Manilo Monaco
Xiao-Hua Yu
Pierre Leblond
Zu-Guo Yu
Jun Yuan

Shenggen Zheng
Xiong Chunhe
Punam Kumari
Li Shang
Sandy Sgorlon
Bo Wei Zhao
X. J. Chen
Fang Yu
Takashi Kurmeoto
Huakuang Li
Pallavi Pandey
Yan Zhou
Mascot Wang
Chenhui Qiu
Haizhou Wu
Lulu Zuo
Jiangning Song
Rafal Kozik
Wenyan Gu
Shiyin Tan
Yaping Fang
Xiuxiu Ren
Antonino Staiano
Aniello Castiglione
Qiong Wu
Atif Mehmood
Wang Guangzhong
Zheng Tian
Junyi Chen
Meineng Wang
Xiaorui Su
Jianping Yu
Jair Cervantes
Lizhi Liu
Junwei Luo
Yuanyuan Wang
Jiayin Zhou
Mingyi Wang
Xiaolei Zhu
Jiafan Zhu
Yongle Li
Hao Lin
Xiaoyin Xu
Shiwei Sun
Hongxuan Hua
Shiping Zhang
Yuxiang Tian
Zhenjia Wang
Shuqin Zhang
Angelo Riccio
Francesco Camastra
Xiong Yuanpeng
Jing Xu
Zou Zeyu
Y. H. Tsai
Chien-Yuan Lai
Guo-Feng Fan
Shaoming Pan
De-Xuan Zou
Zheng Chen
Renzhi Cao
Ronggen Yang
Azis Azis
Shelli Shelli
Zhongming Zhao
Yongna Yuan
Kamal Al Nasr
Chuanxing Liu
Panpan Song
Joao Sousa
Min Li
Wenying He
Kaikai Xu
Ming Chen
Laura Dominguez Jalili
Vivek Kanhangad
Zhang Ziqi
Davide Nardone
Liangxu Liu
Huijian Han
Qingjun Zhu
Hongluan Zhao
Chyuan-Huei Thomas Yang
R. S. Lin
N. Nezu
Chin-Chih Chang
Hung-Chi Su
Antonio Brunetti
Xie conghua
Caitong Yue
Li Yan
Tuozhong Yao
Xuzhao Chai
Zhenhu Liang
Yu Lu
Hua Tang
Liang Cheng
Jiang Hui
Puneet Rawat
Kulandaisamy Akila
Niu Xiaohui
Zhang Guoliang
Egidio Falotico
Peng Chen
Cheng Wang
He Chen
Giacomo Donato Cascarano
Vitoantonio Bevilacqua
Shaohua Wan
Jaya Sudha J. S.
Sameena Naaz
Cheng Chen
Jie Li
Ruxin Zhao
Jiazhou Chen
Abeer Alsadhan
Guoliang Xu
Fangli Yang
Congxu Zhu
Deng Li
Piyush Joshi
Syed Sadaf Ali
Qin Wei
Kuan Li
Teng Wan
Hao Liu
Yexian Zhang
Xu Qiao
Ce Li
Lingchong Zhong
Wenyan Wang
Xiaoyu Ji
Weifeng Guo
Yuchen Jiang
Yuanyuan Huang
Zaixing Sun

Honglin Zhang
Yu Jie He
Benjamin Soibam
Sungroh Yoon
Mohamed Chaabane
Rong Hu
Youjie Yao
NaiKang Yu
Carlo Bianca
Giulia Russo
Dian Liu
Cheng Liang
Iyyakutti Iyappan Ganapathi
Mingon Kang
Zhang Chuanchao
Hao Dai
Geethan
Brendan Halloran
Yue Li
Qianqian Shi
Zhiqiang Tian
Yang Yang
Jalilah Arijah Mohd Kamarudin
Jun Wang
Ke Yan
Hang Wei
David A. Hendrix
Ka-Chun Wong
Yuyan Han
Hisato Fukuda
Yaning Yang
Lixiang Xu
Yuanke Zhou
Shihui Ying
Wenqiang Fan
Zhao Li
Zhe Zhang
Xiaoying Guo
Yiqi Jiang
Zhuoqun Xia
Jing Sun
Na Geng
Chen Li
Xin Ding
Balachandran Manavalan
Bingqiang Liu
Lianrong Pu
Di Wang
Fangping Wan
Guosheng Han
Renmeng Liu
Yinan Guo
Lujie Fang
Ying Zhang
Yinghao Cao
Xhize Wu
Le Zou
G. Brian Golding
Viktoriya Coneva
Alexandre Rossi Paschoal
Ambuj Srivastava
Prabakaran R.
Xingquan Zuo
Jiabin Huang
Jingwen Yang
Liu Qianying
Markus J. Ankenbrand
Jianghong Meng
Tongchi Zhou
Zhi-Ping Liu
Xinyan Liang
Xiaopeng Jin
Jun Zhang
Yumeng Liu
Junliang Shang
L. M. Xiao
Shang-han Li
Jianhua Zhang
Han-Jing Jiang
Daniele Nardi
Kunikazu
Shenglin Mu
Jing Liang
Jialing Li
Yu-Wen-Tian Sun
Zhe Sun
Wentao Fan
Wei Lan
Jiancheng Zhong
Josue Espejel Cabrera
José Sergio Ruiz Castilla
Juan de Jesus Amador
Nanxun Wang
Rencai Zhou
Moli Huang
Yong Zhang
Daniele Loiacono
Grzegorz Dudek
Joaquín Torres-Sospedra
Xingjian Chen
Saifur Rahaman
Olutomilayo Petinrin
Xiaoming Liu
Xin Xu
Zi-Qi Zhu
Punam Kumari
Pallavy Pandey
Najme Zehra
Zhenqing Ye
Hao Zhang
Zijing Wang
Lida Zhu
Lvzhou Li
Junfeng Xia
Jianguo Liu
Jia-Xiang Wang
Gongxin Peng
Junbo Liang
Linjing Liu
Xian Geng
Sheng Ding
Jun Li
Laksono Kurnianggoro
Minxia Cheng
Meiyi Li
Qizhi Zhu
Peng Chao Li
Ming Xiao
Guangdi Liu
Jing Meng
Kang Xu
Cong Feng
Arturo Yee
Yi Xiong
Fei Luo
Xionghui Zhou

Kazunori Onoguchi
Hotaka Takizawa
Suhang Gu
Zhang Yu
Bin Qin
Yang Gu
Zhibin Jiang
Chuanyan Wu
Wahyono Wahyono
Van-Dung Hoang
My-Ha Le
Kaushik Deb
Danilo Caceres
Alexander Filonenko
Van-Thanh Hoang
Ning Guo
Deng Chao
Soniya Balram
Jian Liu
Angelo Ciaramella
Yijie Ding
Ramakrishnan
Nagarajan Raju
Kumar Yugandhar
Anoosha Paruchuri
Dhanusa
Jino Blessy
Agata Gie
Lei Che
Yujia Xi
Ma Haiying
Huanqiang Zeng
Hong-Bo Zhang
Yewang Chen
Farheen Sidiqqui
Sama Ukyo
Parul Agarwal
Akash Tayal
Ru Yang
Junning Gao
Jianqing Zhu
Joel Ayala
Haizhou Liu
Nobutaka Shimada
Yuan Xu
Ping Yang
Chunfeng Shi
Shuo Jiang
Xiaoke Hao
Lei Wang
Minghua Zhao
Cheng Shi
Jiulong Zhang
Shui-Hua Wang
Xuefeng Cui
Sandesh Gupta
Nadia Siddiqui
Syeda Shira Moin
Sajjad Ahmed
Ruidong Li
Mauro Castelli
Leonardo Bocchi
Leonardo Vanneschi
Ivanoe De Falco
Antonio Della Cioppa
Kamlesh Tiwari
Puneet Gupta
Zuliang Wang
Luca Tiseni
Francesco Porcini
Ruizhi Fan
Grigorios Skaltsas
Mario Selvaggio
Xiang Yu
Abdurrahman Eray Baran
Alessandra Rossi
Jacky Liang
Robin Strudel
Stefan Stevsic
Ariyan M. Kabir
Lin Shao
Parker Owan
Rafael Papallas
Alina Kloss
Muhammad Suhail
Saleem
Neel Doshi
Masaki Murooka
Huitan Mao
Christos K. Verginis
Joon Hyub Lee
Gennaro Notomista
Donghyeon Lee
Mohamed Hasan
ChangHwan Kim
Vivek Thangavelu
Alvaro Costa-Garcia
David Parent
Oskar Ljungqvist
Long Cheng
Huajuan Huang
Vasily Aristarkhov
Zhonghao Liu
Lichuan Pan
Yongquan Zhou
Zhongying Zhao
Kunikazu Kobayashi
Masato Nagayoshi
Atsushi Yamashita
Wei Peng
Haodi Feng
Jin Zhao
Shunheng Zhou
Xinguo Lu
Xiangwen Wang
Zhe Liu
Pi-Jing Wei
Bin Liu
Haozhen Situ
Meng Zhou
Muhammad Ikram Ullah
Hui Tang
Sakthivel Ramasamy
Akio Nakamura
Antony Lam
Weilin Deng
Haiyan Qiao
Xu Zhou
Shuyuan Wang
Rabia Shakir
Shixiong Zhang
Xuanfan Fei
Fatih Ad
Aysel Ersoy Yilmaz
Haotian Xu
Zekang Bian
Shuguang Ge
Dhiya Al-Jumeily

Thar Baker
Haoqian Huang
Siguo Wang
Huan Liu
Jianqing Chen
Chunhui Wang
Xiaoshu Zhu
Wen Zhang
Yongchun Zuo
Dariusz Pazderski
Elif Hocaoglu
Hyunsoo Kim
Park Singu
Saeed Ahmed
Youngdoo Lee
Nathan D. Kent
Areesha Anjum
Sanjay Sharma
Shaojin Geng
Andrea Mannini
Van-Dung Hoang
He Yongqiang
Kyungsook Han
Long Chen
Jialin Lyu
Zhenyang Li
Tian Rui
Khan Alcan
Alperen Acemoglu
Duygun Erol Barkana
Juan Manuel Jacinto Villegas
Zhenishbek Zhakypov
Domenico Chiaradia
Huiyu Zhou
Yichuan Wang
Sang-Goo Jeong
Nicolò Navarin
Eray A. Baran
Jiakai Ding
Dehua Zhang
Giuseppe Pirlo
Alberto Morea
Giuseppe Mastronardi
Insoo Koo
Dah-Jing Jwo
Yudong Zhang
Zafaryab Haider
Mahreen Saleem
Quang Do
Vladimir Shakhov
Daniele Leonardis
Simona Crea
Byungkyu Park
Pau Rodr´
Alper Gün
Mehmet Fatih Demirel
Elena Battini
Radzi Ambar
Mohamad Farhan
Mohamad Mohsin
Nur Azzah Abu Bakar
Noraziah ChePa
Sasalak Tongkaw
Kumar Jana
Hafizul Fahri Hanafi
Liu Jinxing
Alex Moopenn
Liang Liang
Ling-Yun Dai
Raffaele Montella
Maratea Antonio
Xiongtao Zhang
Sobia Pervaiz Iqbal
Fang Yang
Si Liu
Natsa Kleanthous
Zhen Shen
Jing Jiang
Shamrie Sainin
Suraya Alias
Mohd Hanafi Ahmad Hijazi
Mohd Razali Tomari
Chunyan Fan
Jie Zhao
Yuchen Zhang Casimiro
Dong-Jun Yu
Jianwei Yang
Wenrui Zhao
Di Wu
Chao Wang
Alex Akinbi
Fuyi Li
Fan Xu
Guangsheng Wu
Yuchong Gong
Weitai Yang
Mohammed Aledhari
Yanan Wang
Bo Chen
Binbin Pan
Chunhou Zheng
Abir Hussain
Chen Yan
Dhanjay Singh
Bowen Song
Guojing
Weiping Liu
Yeguo Liao
Laura Jalili
Quan Zou
Xing Chen
Xiujuan Lei
Marek Pawlicki
Haiying Ma
Hao Zhu
Wang Zhanjun
Mohamed Alloghani
Yu Hu
Haya Alaskar
Baohua Wang
Hanfu Wang
Hongle Xie
Guangming Wang
Yongmei Liu
Fuchun Liu
Farid Garcia-Lamont
Yang Li
Hengyue Shi
Gao Kun
Wen Zheng Ma
Jin Sun
Xing Ruiwen
Zhong Lianxin
Zhang Hongyuan
Han Xupeng
Mon Hian Chew

Jianxun Mi
Michele Scarpiniti
Hugo Morais
Alamgir Hossain
Felipe Saraiva
Xuyang Xuyang
Yasushi Mae
Haoran Mo
Pengfei Cui
Yoshinori Kobayashi
Qing Yu Cui
Kongtao Chen
Feng Feng
Wenli Yan
Zhibo Wang
Ying Qiao
Qiyue Lu
Geethan Mendiz
Dong Li
Liu Di
Feilin Zhang
Haibin Li
Heqi Wang
Wei Wang
Tony Hao
Yingxia Pan
Chenglong Wei
My Ha Le
Yu Chen
Eren Aydemir
Naida Fetic
Bing Sun
Zhenzhong Chu
Meijing Li
Wentao Chen
Mingpeng Zheng
Zhihao Tang
Li keng Liang
Alberto Mazzoni
Domenico Buongiorno
Zhang Lifeng
Chi Yuhong
Meng-Meng Yin
Yannan Bin
Wasiq Khan
Yong Wu
Qinhu Zhang
Jiang Liu
Yuzhen Han
Pengcheng Xiao
Harry Haoxiang Wang
Fengqiang Li
Chenggang Lai
Dong Li
Shuai Liu
Cuiling Huang
Lian-Yong Qi
Qi Zhu
Wenqiang Gu
Haitao Du
Bingbo Cui
Qinghua Li
Xin Juan
Emanuele Principi
Xiaohan Sun
Inas Kadhim
Jing Feng
Xin Juan
Hongguo Zhao
Masoomeh Mirrashid
Jialiang Li
Yaping Hu
Xiangzhen Kong
Mi-Xiao Hou
Zhen Cui
Juan Wang
Na Yu
Meiyu Duan
Pavel Osinenko
Chengdong Li
Stefano Rovetta
Mingjun Zhong
Baoping Yuan
Akhilesh Mohan Srivastatva
Vivek Baghel
Umarani Jayaraman
Somnath Dey
Guanghui Li
Lihong Peng
Wei Zhang
Hailin Chen
Fabio Bellavia
Giosue' Lo Bosco
Giuseppe Salvi
Giovanni Acampora
Zhen Chen
Enrico De Santis
Xing Lining
Wu Guohua
Dong Nanjiang
Jhony Heriberto Giraldo Zuluaga
Waqas Haider Bangyal
Cong Feng
Autilia Vitiello
TingTing Dan
Haiyan Wang
Angelo Casolaro
Dandan Lu
Bin Zhang
Raul Montoliu
Sergio Trilles
Xu Yang
Fan Jiao
Li Kaiwen
Wenhua Li
Ming Mengjun
Ma Wubin
Cuco Cristanno
Chao Wu
Ghada Abdelmoumin
Han-Zhou Wu
Antonio Junior Spoleto
Zhenghao Shi
Ya Wang
Tao Li
Shuyi Zhang
Xiaoqing Li
Yajun Zou
Chuanlei Zhang
Berardino Prencipe
Feng Liu
Yongsheng Dong
Yatong Zhou
Carlo Croce
Rong Fei
Zhen Wang

Huai-Ping Jin
Mingzhe She
Sen Zhang
Yifan Zheng
Christophe Guyeux
Jun Sang
Huang Wenzhun
Jun Wu
Jing Luo
Wei Lu
Heungkyu Lee
Yinlong Qian
Hong wang
Daniele Malitesta
Fenqiang Zhao
Xinghuo Ye
Hongyi Zhang
Xuexin Yu
Guanshuo Xu
Mehdi Yedroudj
Xujun Duan
Xing-Ming Zhao
Jiayan Han
Yan Xiao
Weizhong Lu
Weiguo Shen
Hongzhen Shi
Zeng Shangyou
Zhou Yue
TaeMoon Seo
Sergio Cannata
Weiqi Luo
Feng Yanyan
Pan Bing
Jiwen Dong
Yong-Wan Kwon
Heng Chen
S. T. Veena
J. Anita Christaline
R. Ramesh
Shadrokh Samavi
Amin Khatami
Min Chen
He Huang
Qing Lei
Shuang Ye
Francesco Fontanella
Kang Jijia
Rahul Kumar
Alessandra Scotto Freca
Nicole Cilia
Alessandro Aliberti
Gabriele Ciravegna
Jacopo Ferretti
Jing Yang
Zheheng Jiang
Dan Yang
Dongxue Peng
Wenting Cui
Francescomaria Marino
Wenhao Chi
Ruobing Liang
Feixiang Zhou
Jijia Kang
Xinshao Wang
Huawei Huang
Zhi Zhou
Yanrui Ding
Peng Li
Yunfeng Zhao
Guohong Qi
Xiaoyan Hu
Li Guo
Xia-an Bi
Xiuquan Du
Ping Zhu
Young-Seob Jeong
Han-Gyu Kim
Dongkun Lee
Jonghwan Hyeon
Chae-Gyun Lim
Nicola Altini
Claudio Gallicchio
Dingna Duan
Shiqiang Ma
Mingliang Dou
Jansen Woo
Shanshan
ShanShan Hu
Hai-tao Li
Francescomaria Marino
Jiayi Ji
Jun Peng
Jie Hu
Jipeng Wu
Shirley Meng
Prashan Premaratne
Lucia Ballerini
Haifeng Hu
JianXin Zhang
Xiaoxiao Sun
Shaomin Mu
Yongyu Xu
Jingyu Hou
Zhixian Liu

Contents – Part III

Intelligent Computing in Robotics

Intelligent Computing in Computer Vision

Intelligent Fault Diagnosis

Fuzzy Theory and Algorithms

Kernel Methods and Supporting Vector Machines

Machine Learning

Knowledge Discovery and Data Mining

Natural Language Processing and Computational Linguistics

Recent Advances in Swarm Intelligence: Computing and Applications

Information Security

Intelligent Computing in Robotics

Automatic Pose Estimation of Micro Unmanned Aerial Vehicle for Autonomous Landing

Manish Shrestha[1], Sanjeeb Prasad Panday[2(✉)], Basanta Joshi[2(✉)], Aman Shakya[2], and Rom Kant Pandey[3]

[1] Nepal College of Information Technology, Pokhara University, Lalitpur, Nepal
[2] Pulchowk Campus, Institute of Engineering, Tribhuvan University, Lalitpur, Nepal
{sanjeeb,basanta,aman.shakya}@ioe.edu.np
[3] Sanothimi Campus, Tribhuvan University, Bhaktapur, Nepal

Abstract. The guided navigation has enabled users with minimal amount of training to navigate and perform flight mission of micro unmanned aerial vehicle (MAV). In non-urban areas, where there are no other aerial traffic and congestion, MAV take-off & travel does not need much Global Positioning System (GPS) accuracy. The critical part seems to be during the landing of the MAV, where slight GPS inaccuracy can lead to landing of the vehicle in the dangerous spot, causing damage to the MAV. This paper aims to propose a low cost portable solution for the Autonomous landing of the MAV, using object detection and machine learning techniques. In this work, You Only Look Once (YOLO) has been used for object detection and corner detection algorithm along with projective transformation equation has been used for getting the position of MAV with respect to the landing spot has been devised. The experiments were carried with Raspberry Pi and the estimation shows up to 4% of error in height and 12.5% error in X, Y position.

Keywords: Micro unmanned aerial vehicle · UAV · GPS · Autonomous landing · Object detection · CNN · YOLO

1 Introduction

Micro Unmanned Aerial Vehicles (MAV) or drones has been using Global Positioning System (GPS) to execute flight missions easily. Even though there are some fluctuations in GPS readings from time to time even for the same spot, they are commonly being used in such missions. Instead of GPS, the landing at specified spot can also be done with the help of other sensors, like camera. Takeoff, hovering, moving forward and landing are some of the basic phases for autonomous flight of MAV. Among them, landing visually on a specified target is especially complex because it requires robust recognition of the landing pad and precise position control; and a slight offset of few meters can also cause crash landing of the vehicle.

Vision based approach was also used by Yang et al. [1] presented an on- board vision system that can detect a landing pad consisting of the letter "H" surrounded by a circle, from images captured by a monocular camera on a MAV and determine the 6

D.-S. Huang and P. Premaratne (Eds.): ICIC 2020, LNAI 12465, pp. 3–15, 2020.
https://doi.org/10.1007/978-3-030-60796-8_1

DOF pose of the MAV relative to the landing pad using projective geometry. The 5 DOF pose is estimated from the elliptic projection of the circle. The remaining geometric ambiguity is resolved by incorporating the gravity vector estimated by the inertial measurement unit (IMU). The last degree of freedom pose, yaw angle of the MAV, is estimated from the ellipse fitted from the letter "H". A neural network was used to initially detect the landing pad. The structure of the neural network is a multilayer perceptrons with 196 input units (one per pixel of patterns resized to 14×14), only one hidden layer consisting of 20 hidden units and three output units.

In another paper, Yang et al. [2] presented a solution for micro aerial vehicles (MAVs) to autonomously search for and land on an arbitrary landing site using real-time monocular vision. The autonomous MAV is provided with only one single reference image of the landing site with an unknown size before initiating this task. The autonomous navigation of MAV was achieved by implementing a constant-time monocular visual SLAM framework, while simultaneously detecting an arbitrarily textured landing site using ORB features, and estimating its global pose.

Daniel et al. [3] employed visual odometry techniques with feature-based methods to compute the aircraft motion and thereby allowing the position estimation in GPS denied environments. With regards to GPS inaccuracy, Stark et al. [4] showed that almost half (49.6%) of all ≈68,000 GPS points recorded with the Qstarz Q1000XT GPS units fell within 2.5 m of the expected location, 78.7% fell within 10 m and the median error was 2.9 m.

Traditional object detection systems are variants of the following pipeline: Firstly, find potential objects and their bounding boxes, then do feature ex- traction, and finally classify using a good classifier. Selective Search (SS) [5] enjoyed being the state-of-the-art for detection on PASCAL VOC etc. competitions. HOG [6] and SIFT [7] are the popular choices for feature extractions. A classifier is applied on image pyramid to overcome problems with scale.

The current state-of-the-art object detectors such as Fast R-CNN [8], YOLO [9], SSD [10] etc. are based on convolutional neural networks (CNN) and have outperformed the traditional techniques. The key to the success of CNNs is their ability to extract/learn generic features. Furthermore, the advancement in computational resources such as high-performance GPUs and its easy availability through the use of high-performance cloud computing platforms played an important role in the recent success of neural networks.

In this work, monocular vision based system has been proposed to localize the MAV position with respect to the landing spot. Detection of the landing spot has been carried out with more advanced and recent classifiers known as You Only Look Once (YOLO) [9]. A simpler projective transform with 3-DOF variables based on rectangular feature points of a simple landing spot has been used. The proposed system also aims to develop effective system using a simple camera (Raspberry Pi camera) instead of advanced camera (with global shutter).

2 Methodology

This work is divided into two phases: Learning Phase and Implementation phase.

2.1 Learning Phase

As shown in Fig. 1, the learning phase involves data collection, pre-processing and training and evaluating two classifiers, namely YOLO v3 and YOLO Tiny v3.

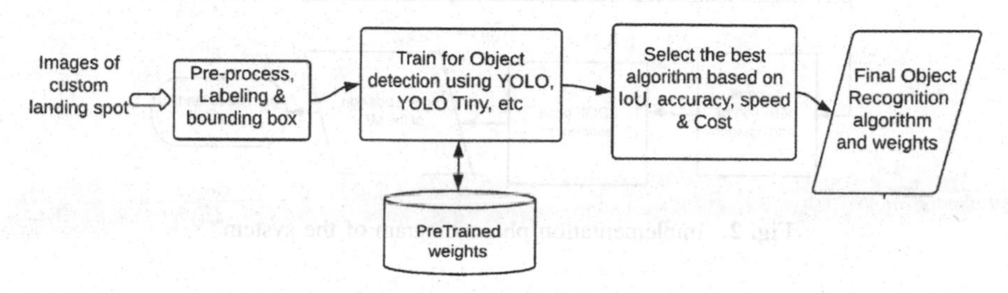

Fig. 1. Learning phase diagram.

A custom landing pad of dimension 142 cm × 112 cm with 4 rectangular regions of color red, blue, white and black of equal size was designed. Images of the landing spot in various background and orientation is captured from different height, using simple web camera in a flying MAV, for object detection training. For pose estimation, images will be captured along with roll, pitch and yaw angle using a handheld MAV. Data augmentation technique like rotation, etc. is done to increase the samples of our captured data set for verifying the corner detection phase. For training the object detection classifier, the images has been classified to contain the landing spot and those images are also tagged with the bounding box that indicates the location of the landing spot within the image.

Using the pre-trained available weights and collected datasets, the neural networks YOLO v3 and YOLO tiny v3 have been trained to detect our custom landing spot. The YOLO Tiny v3 has been made to detect the landing spot in Raspberry Pi 3 hardware too.

2.2 Implementation Phase

During the real time application phase, the object detection of the landing spots will be followed by corner detection phase and then the pose estimation phase as shown in Fig. 2.

2.3 Object Detection

Whenever the MAV arrives near the final landing spot as reported by GPS, the task for object detection comes into action. The live images from the camera installed in the MAV are feed into the object detection classifier (YOLO). The classifier, using the

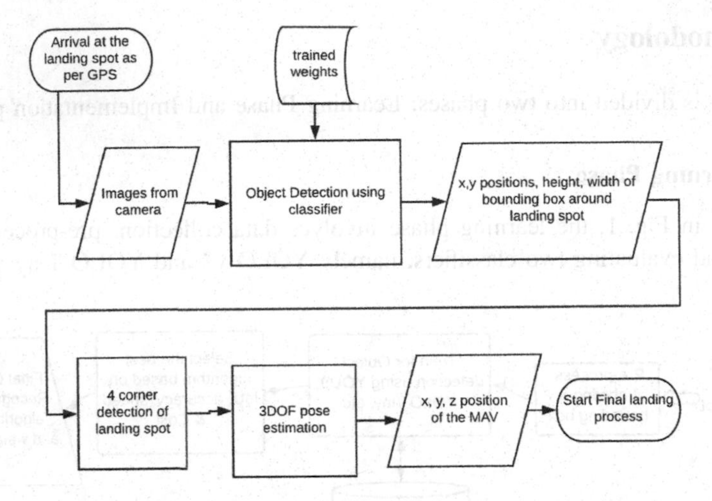

Fig. 2. Implementation phase diagram of the system

already trained weights, calculates the bounding box position of the landing spot from the image. The bounding box position constitutes a rough approximate of x, y position of the landing spot in the image and the height and the width of the bounding box.

2.4 Corner Detection

The information about the bounding box and the image from the first phase would be feed into the Corner estimation algorithm. In this phase, the top left, top right, bottom left and bottom right corners of the landing spot will be identified in the image. The section of the image, which is slightly larger than the detected bounding box area and which encompasses the detected bounding box, should be chosen for the corner detection. Here, the width and height of the selected section of the image can be 1.5 times that of the detected bounding box. The steps for acquiring the corners from the selection section of the image is listed below:

1. Converting to gray scale image.
2. Application of Canny edge detection to get the edges of the landing spot.
3. Perform Hough lines detection to find the lines of the landing spot.
4. Augment the image with the detected lines. The lines to be augmented are chosen such that they are among the top 14 lengthiest lines among the detected lines.
5. Perform Harris corner detection on the augmented image.
6. Take the top leftmost, top rightmost, bottom leftmost, bottom rightmost as the four corners in the images as the corresponding points of the landing spot.

In order to calculate the effectiveness of both methods, the difference between the ground truth value and detected positions of each of the four corners would be calculated. Then mean error distance for each corner detected will be calculated, which is the average of the distance from ground truth calculated for each of the four points.

2.5 Pose Estimation

The position of the four corners detected in the image using the previous phase is used, to get the pose of the camera with respect to the landing spot. The pose information contains the x, y, z position and the rotation around x, y & z axis. Estimating of the pose of a calibrated camera, given a set of n 3D points in the world and their corresponding 2D projections in the image, is a Perspective- n-Point problem. Since the camera position is fixed with respect to the MAV frame, the pose of the camera also gives the pose of MAV with respect to the landing spot.

After solving the Perspective-n-Point problem using the 4 point correspondence between the detected four corners and the actual four corners in the world coordinate, more accurate estimation of the x, y, z positions of the MAV with respect to the landing spot is obtained. This position information would finally be used by the landing mechanism for landing the MAV into the landing spot. The equation governing this projective transformation is shown below:

$$s\begin{bmatrix} u \\ v \\ 1 \end{bmatrix} = \begin{bmatrix} f_x & \gamma & u_0 \\ 0 & f_y & v_0 \\ 0 & 0 & 1 \end{bmatrix} \begin{bmatrix} r_{11} & r_{12} & r_{13} & t_1 \\ r_{21} & r_{22} & r_{23} & t_2 \\ r_{31} & r_{32} & r_{33} & t_3 \end{bmatrix} \begin{bmatrix} x \\ y \\ z \\ 1 \end{bmatrix}.$$

Here, s is the scaling factor. fx and fy are the x and y focal length in pixels. x, y, z are the world coordinate of a given point and u, v are the x and y locations of the point in an image. The transformation in x, y, z direction is given by t1, t2, and t3.

Since obtaining the ground truth data for the flying MAV is difficult in absence of state-of-the-art object tracking laboratory, the ground truth data of the x, y, z position of the MAV with respect to the landing spot needs to be obtained using hand-held MAV. The difference in the x, y and z position of the calculated observed value and the ground truth value needs to be calculated. The error percentage in x, y and z positions would be given by the formula below:

$$\textit{Error percent in X position} = \frac{(Xg - Xc) * 100\%}{Xg}$$

$$\textit{Error percent in Y position} = \frac{(Yg - Yc) * 100\%}{Yg}$$

$$\textit{Error percent in Z position} = \frac{(Zg - Zc) * 100\%}{Zg}$$

Here, Xg, Yg, Zg denotes the ground truth value of the measured distance between the MAV and landing spot. Xc, Yc, Zc, represents the calculated distance from the pose estimation phase.

3 Experimental Setup

3.1 Development of Experimental MAV for Handheld Experiment

Before making actual outdoor flights, the indoor testing was carried with an experimental setup up consisting of MAV, Raspberry Pi3, Raspberry Pi Camera as shown in the Fig. 3a. An existing MAV was fitted with Raspberry Pi3, and its camera. The existing MAV had 4 motors, 4ESC units, flight controller based on PixHawk, GPS, telemetry unit and LiPo battery. Also a landing pad of 1/4 th of the original size was also made for indoor and handheld experiments as shown in Fig. 3b.

(a) MAV setup for experiments

(b) Mini Landing pad

Fig. 3. MAV and landing pad for indoor and handheld experiments.

3.2 Software Setup

The Raspberry Pi3 was installed with the operating system Raspbian Stretch with desktop (release date 20180418). OpenCV 3.3 was compiled and installed into it. Also picamera module was also installed for accessing the Raspberry Pi Camera.

Since the object detection is based on YOLO, darknet system had to be compiled. It was compiled in Linux, Windows and Raspberry Pi. Also in order to take advantage of GPU based calculation, NVIDIA CUDA Deep Neural Network library (CuDNN) had to be installed in both Windows and Linux machine.

3.3 Camera Calibration

Before sending the recorded image through a set of image processing pipeline, a process to correct the image from deformations resulting from camera geometry, image plane placement, etc. needs to be done. For this camera calibration is done to determine extrinsic and intrinsic parameters [11]. The Raspberry Pi Camera was calibrated using a simple checkerboard pattern and the API provided by OpenCV.

3.4 Experiment Parameters

The experiments were performed in three separate environments with the specified parameters as shown in Table 1. The training was done in Environment 1 and

Environment 2. Due to low hardware resources, Raspberry Pi3 has been only used for testing purpose with trained weights from Environment 2. Since YOLO v3 network requires minimum of 4 GB of RAM, it could not be tested in Raspberry Pi3. The YOLO was only tested in Environment 1 with the Windows 10 machine with GTX 1060 Nvidia graphics card.

Table 1. Experiment environment and parameters

Parameters	Environment 1	Environment 2	Environment 3
Hardware	Laptop with 16 GB RAM	Alienware Ddsktop	Raspberry Pi3
Operating System	Windows 10	Ubuntu 16.04	Ubuntu 16.04
Graphics card	Nvidia GTX 1060	Nvidia GTX 1080 Ti	NA
Used for	Training/Testing	Training	Testing
Detection Type	YOLO v3	YOLO v3 Tiny	YOLO v3 Tiny
1. S. Yang	S. A. Scherer and A. Zell	"An onboard monocular vision system for autonomous takeoff	hovering and landing of a aerial vehicle
No of training Image	351	320	N/A
Network input size	416 by 416	448 by 448	448 by 448
Batch Iteration	3100	21000	N/A
Batch size	64	64	N/A
Training Time	9 h	4 h	N/A

4 Dataset Collection and Pre-processing

4.1 Images of Landing Spot for Object Detection

In order to train the neural network for recognizing the landing spot, images of the landing spot from different height was needed. First, video in mp4 format using GoPro Hero 3 camera was taken at the resolution of 1920 * 1080. Then the mp4 format video was converted to still images using YOLO mark tools and using OpenCV API. Then some images in the original size were used, while some of the images were down sized to 448 * 448, using OpenCV API. Then each of those images were labeled with class name and bounding box. Also in order to test the output of our pose estimation algorithm, images were captured from different height using a simple web camera at resolution of 680 $\times$ 480.

Labeling of Landing Spot for Yolo Training: The labeling of the landing spot in the captured images was done using an open source tool called Yolo mark. The YOLO mark was tool was obtained from and was compiled in Windows 10 machine.

4.2 Pre-trained Weights

In order to make the training with few amount of custom images, we use pre- trained weights, that had been created using thousands of images from standard image dataset. The pre-trained weights for YOLO v3 and YOLO tiny v3 were **darknet53.conv.74** and **yolov3-tiny.weights** and were taken from official site of darknet.

4.3 Images of Landing Spot for Pose Estimation

Since it is not possible to obtain the ground truth value of the x, y, z position of a flying MAV in a simple lab setup, the handheld MAV was used to capture the image of the mini landing spot from different height and angle. During the capture, the roll, pitch and yaw angles were also noted. The position, from where the images were taken, was also measured with the help of measuring tape. The images were taken from around 6, 10 and 16 m of height from a building. These images and the measured distances are used to validate the results of the object detection phase.

Marking the Landing Spot Corners in Images: In order to generate the ground truth data for the corner detection phase, all the images used for the verification the corner detection phase, was one by one marked with the x, y position in the image. A Python program was written in order to display the image, on which four corners can be clicked by mouse and then the clicked positions would be recorded. For each image, four points representing the top left, top right, bottom right and bottom left were to be clicked sequentially.

5 Results and Analysis

5.1 Training on Environment 1

YOLO version 3 was trained on Environment 1 with parameters as mentioned in Table 1 with pre-trained weight obtained from official site of YOLO. Images that were directly converted from the video of 1080p, with the resolution of 1920 * 1080 were used. It took around 9 h of training in the Windows machine. The average loss in the network was around 0.08 to 0.07 for about an hour, and hence the training was stopped.

While testing against the test image set, the Intersection over Union (IoU) was calculated. For different IoU detection thresholds, the resulting Average IoU% F1 score are tabulated as shown in Table 2. It can be observed that even for high IoU threshold like 0.95, the results are quite satisfactory with F1-score of 0.99 and False negative of only 1.

Table 2. Comparison of validation results while mapping the YOLO v3 trained network with the test set using different IoU thresholds.

S No	Threshold %	Average IoU %	True positive	False positive	False negative	F1-score
1	0.25	88.85	38	0	0	1
2	0.5	88.85	38	0	0	1
3	0.75	88.85	38	0	0	1
4	0.85	88.85	38	0	0	1
5	0.9	89.14	37	0	1	0.99
6	0.95	89.14	37	0	1	0.99
7	0.99	89.4	33	0	5	0.93

5.2 Training on Environment 2

YOLO Tiny version 3 was trained on Environment 2 with parameters as mentioned in Table 1. The final average loss was also around 0.08. In the training hardware, the trained network of YOLO v3 tiny was able to detect low resolution images (640 pixels * 480 pixels) taken from the web-camera at different height.

5.3 Object Detection

A comparison of the object detection using YOLO version 3 and YOLO Tiny version 3 is tabulated in Table 3.

Table 3. Object Detection Comparison of YOLO v3 versus YOLO Tiny v3.

S No	Detector type	Environment on	Detection time in seconds	Frames per seconds
1	Yolo v3	Environment 1	0.0451	22.172949
2	Yolo v3 tiny	Environment 3	10.901	0.0917347

With Environment 1, Yolo V3 was able to detect the landing spot within 0.0451 s, resulting about 22 Frames per Seconds. A slight modification in the original YOLO v3 code was done in order to get the position of the detected bounding box and the size of the bounding box. Here 0.707 and 0.858 are the relative x and y position of the detected bounding box with respect to original image size. And similarly 0.258 and 0.271 are the width and height of the bounding box detected. An image depicting the bounding box is detected by the trained YOLO v3 in Environment 1.

With Environment 3(Raspberry Pi3), Yolo Tiny version took 10.9 s to detect the landing spot resulting in 0.09 frames per seconds of speed. An image depicting the bounding box is detected by the YOLO tiny v3.

5.4 Corner Detection

After the object detection phase, the x, y position of the landing spot with its height and width is obtained. This gives rough location of the landing spot in the image. Then for the area that is 1.5 times of the indicated dimensions (by the object detection) is considered for the corner detection.

For estimating corners, Canny edge detection, Hough Transform and Harris Corner detection is used. The result of Canny edge detection is shown in Fig. 4a. After the lines are detected using Hough transform, those lines are super imposed on the edge detected image, as shown in Fig. 4b. Then applying Harris corner on this superimposed image, results in the 16 corners as shown in Fig. 4c. Finally, 4 corners obtained after this step is shown in Fig. 4d. The results gave mean error distance of 4 pixels and standard deviation of 2 pixels.

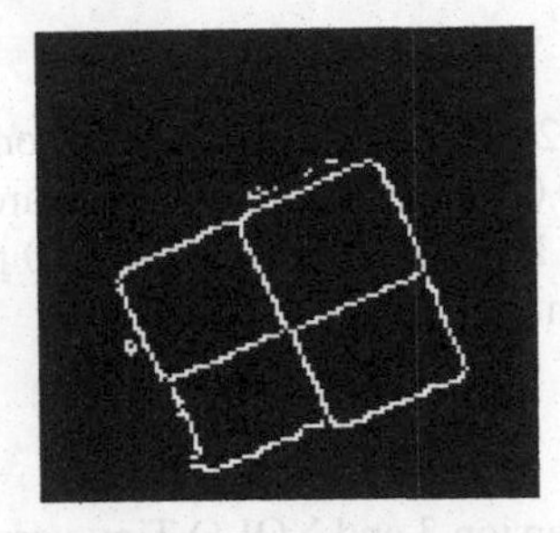

(a) Canny edge detection

(b) Lines detected from Hough transform super imposed on the probable area.

(c) Harris Corner detection in the super imposed image.

(d) Four Corner detection

Fig. 4. Corner detection for landing spot

5.5 Pose Estimation

Camera Calibration. The calibration of Raspberry Pi camera v2.1 was done by taking various images in 640 x 480 pixels and then detecting the corners in the checker board pattern using OpenCV API. The focal length in x & y direction are at 499 and 501. The optical center position in x & y are at 323 and 234 pixels, which sounds reasonable. The radial distortion parameters k1, k2 and k3 are 0.17, −0.27, −0.20 respectively. And tangential distortion coefficients are −0.0043 and 0.0006 respectively.

Pose Estimation Calculation. After the 4 corners in the 3D world coordinate of the landing spot and corresponding 4 corners in the 2D image has been found, the homogeneous matrix obtained was calculated. The homogenous matrix was decomposed to get the rotation and translation vector between the world coordinate and the camera coordinate. The obtained position of the camera (which also represents the position of the MAV) has been tabulated as shown in Table 4. The ground truth values can also be seen in the table. Here the errors in x, y, z is within reasonable boundary when the yaw angle from which the picture was taken is not much different than in the landing pad. Since a valid constraint, that Y axis MAV should be pointing to Y axis of the landing pad during landing, can be added, this error can be eliminated. From this table it can be concluded that

Table 4. Comparison of final x, y & z positions obtained from the pose estimation with the ground truth.

	Ground truth (in meter)			Calculated result (in meter)			Error percent (Error/GT * 100%)		
S N	Height	X	Y	Height	X	Y	Z	X	Y
1	10.85	3.6	2.3	11.38	1.72	0.78	4.88	52.22	66.09
2	16.1	0.6	0.85	14.05	4.19	4.48	12.73	598.33	427.06
3	6.65	3.65	0.05	6.72	3.78	0.06	1.05	3.56	20.00
4	10.85	3.6	2.3	9.99	4.7	3.26	7.93	30.56	41.74
5	10.85	3.6	2.3	10.28	4.3	2.16	5.25	19.44	6.09

- Rows 1, 2 show high percentage of error due to high Yaw Difference between the landing spot and MAV. This can be eliminated if the landing is done with Y axis of MAV pointing in the Y axis of the world coordinate.
- Up to 8% of error in height estimation, and up to 30% and 41% error in X, Y estimations are obtained in normal conditions without any correction.
- It can be seen that when the image is corrected by pitch angle of the MAV, the error in X, Y and Z position reduces from 30%, 40% and 8% to 19%, 6% and 5% respectively. The row 4 depicts the result after normal calculation and row 5 depicts result of the same calculation after image correction by pitch angle.

After the pitch angle correction is done, it can be seen that the average error across many images in x, y, z position of the MAV is 12%, 13% and 4% respectively, which should be practically acceptable for calculating the position of the MAV using low cost approach described here. Hence, the correction of the captured image, by the pitch angle of the camera or the MAV, is recommended before the pose estimation calculation is done, for better approximation.

6 Conclusion

There are challenges for landing of Micro Unmanned aerial vehicle (MAV) and robust recognition of the landing pad and precise position control is necessary. This work proposes a new visual based approach for MAV for estimating the approximate x, y, z positions of the MAV from the landing spot using recorded camera images, thereby assisting in the landing of MAV. You Only Look Once (YOLO) v3 has been used for object detection of the landing spot in the image, which indicates sub section in the image where the landing spot can be found. Then Harris corner detector has been applied around the subsection, in order to get the four corners of the landing spot in the image. Then after some pre- processing, the pose estimation of the MAV from the planar landing spot has been done by decomposing the homogeneous matrix obtained from 4 points correspondence. The experiments were carried with Raspberry Pi and the estimation shows up to 4% of error in height and 12.5% error in X, Y position. The present work doesn't analyze the performance in adverse lighting condition. The techniques for mitigating the effect of low light and very bright light while taking images from the low-cost camera can be studied in future. Also, the pose estimation can be improved using stereo camera instead of the single camera.

Acknowledgement. This work has been supported by the University Grants Commission, Nepal under a Collaborative Research Grant (UGC Award No. CRG-74/75-Engg-01) for the research project "Establishment of a Disaster Telecommunications Research and Educational Facility Advancing a Scientifically Sound Disaster Telecommunication Infrastructure and Processes in Nepal".

References

1. Yang, S., Scherer, S.A., Zell, A.: An onboard monocular vision system for autonomous takeoff, hovering and landing of a micro aerial vehicle. J. Intell. Robot. Syst. **69**(1–4), 499–515 (2013)
2. Yang, S., Scherer, S.A., Schauwecker, K., Zell, A.: Autonomous landing of MAVs on an arbitrarily textured landing site using onboard monocular vision. J. Intell. Robot. Syst. **74** (12), 27–43 (2014)
3. Villa, D.K., Brandao, A.S., Sarcinelli-Filho, M.: A survey on load transportation using multirotor UAVs. J. Intell. Robot. Sys. **98**, 267–296 (2019)
4. Schipperijn, J., Kerr, J., Duncan, S., Madsen, T., Klinker, C.D., Troelsen, J.: Dynamic accuracy of GPS receivers for use in health research: a novel method to assess GPS accuracy in real-world settings. Front. Pub. Health **2**, 21 (2014)
5. Uijlings, J.R., Van De Sande, K.E., Gevers, T., Smeulders, A.W.: Selective search for object recognition. Int. J. Comput. Vis. **104**(2), 154–171 (2013)
6. Dalal, N., Triggs, B.: Histograms of oriented gradients for human detection. In: 2005 IEEE Computer Society Conference on Computer Vision and Pattern Recognition (2005)
7. Lowe, D.G.: Object recognition from local scale-invariant features. In: Proceedings of the Seventh IEEE International Conference on Computer Vision (1999)
8. Girshick, R.: Fast R-CNN. In: Proceedings of IEEE International Conference on Computer Vision (2015)

9. Redmon, J., Divvala, S., Girshick, R., Farhadi, A.: You only look once: unified, real-time object detection. In: Proceedings of the IEEE Conference on Computer Vision and Pattern Recognition (2016)
10. Liu, W., et al.: SSD: single shot multibox detector. In: Leibe, B., Matas, J., Sebe, N., Welling, M. (eds.) ECCV 2016. LNCS, vol. 9905, pp. 21–37. Springer, Cham (2016). https://doi.org/10.1007/978-3-319-46448-0_2
11. Joshi, B., Ohmi, K., Nose, K.: Comparative study of camera calibration models for 3D particle tracking velocimetry. Int. J. Innov. Comput. Inf. Control 9(5), 1971–1986 (2013)

A New Robotic Manipulator Calibration Method of Identification Kinematic and Compliance Errors

Phu-Nguyen Le[1] and Hee-Jung Kang[2](✉)

[1] Graduate School of Electrical Engineering, University of Ulsan, Ulsan 680-749, South Korea
[2] School of Electrical Engineering, University of Ulsan, Ulsan 680-749, South Korea
hjkang@ulsan.ac.kr

Abstract. In this work, a new robotic calibration method is proposed for reducing the positional errors of the robot manipulator. First, geometric errors of a robot are identified by using a conventional kinematic calibration model of the robot. Then, a radial basis function is constructed for compensating the compliance errors based on the effective torques for further increasing the positional precision of the robot. The enhanced positional accuracy of the robot manipulator in experimental studies that are carried on a YS100 robot illustrates the advantages of the suggested algorithm than the other techniques.

Keywords: Robot accuracy · Radial basis function · Robot calibration

1 Introduction

The robot manipulators are widely used in the industry. Although the robots are high repeatability, they are well-known by their low accuracy [1, 2]. The errors of the robot end-effector mostly come from geometric errors and non-geometric errors. The geometric errors are the results of misalignments, incorrect in manufacturing, and assembly robot. The non-geometric errors may come from many non-geometric sources, such as joint and link compliance, temperature variation, gear transmission, etc. Among the non-geometric errors, the compliance errors are dominant. These errors are caused by the flexibility of joints and links under the link self-gravity and external payload.

Geometric calibration methods are widely examined and become mature. The most famous kinematic calibration method, the D-H model is suggested by Denavit-Hartenberg [3–5]. This method is widely used in kinematic calibration by many researchers recently [6–8]. Moreover, the other geometric calibration methods are CPC model [9, 10], POE model [11, 12] and the zero-reference position method [13, 14]. However, these calibration methods do not consider the non-geometric errors. On the other hand, some studies used another approach to investigate joint compliance errors [15, 16]. However, these methods neglected effect of the geometric errors.

Some works have been proposed to deal with kinematic and compliance calibration. For instance, a calibration method to calibrate the geometric errors and

D.-S. Huang and P. Premaratne (Eds.): ICIC 2020, LNAI 12465, pp. 16–27, 2020.
https://doi.org/10.1007/978-3-030-60796-8_2

compensate the joint by radial basis function (RBF) [17] is proposed by Jang et al. However, the work [17] focused on calibrating the geometric parameters and compensating the compliance errors by compensating the joint ("joint level" calibration [1]). The work also needs to divide the robot working space into many subspaces and required many measurements and consuming a lot of time. Meggiolaro et al. proposed a method to approximate the compliance errors by a polynomial function of joint parameters and wrench using torque sensors [18]. Zhou and Hee-Jung proposed a method to simultaneously calibrate the geometric and joint stiffness parameters of the robot [19]. However, this method linearized the relationship between effective torques and joint compliance errors. Recently, some studies have been performed on joint stiffness calibration [20–22] with the need of the torques sensors.

This study proposed a new calibration algorithm for robotic manipulators. The method includes the kinematic calibration and non-geometric compensation with a RBF compensator that compensates for compliance errors based on the effective torques. It is assumed that the gravity compensation torques are nonlinearity related to the compliance errors. These relationships can be constructed by a RBF. The advantages of the suggested method are easy for implementing, removing the need for torque sensors, high ability to enhance the precision of the manipulator. These advantages are firmly confirmed by the experimental studies in contrasting with 2 other methods such as the conventional kinematic calibration and the method for simultaneously calibrate the geometric and joint stiffness parameters of the robot.

Following the introduction. Section 2 presents the kinematic model of the YS 100 robot. In Sect. 3, the geometrical and the gravity compensator using a Radial basis function that is based on the effective torques are presented. Sections 4 is devoted to the experimental calibration result of the proposed method in contrasting with other methods. Section 5 summarizes the abilities and advantages of the proposed method.

2 Kinematic Model of the YS100 Robot

YS100 is a 6 DOF serial robot [19]. The kinematic structure of it is briefly described in Fig. 1 and Table 1.

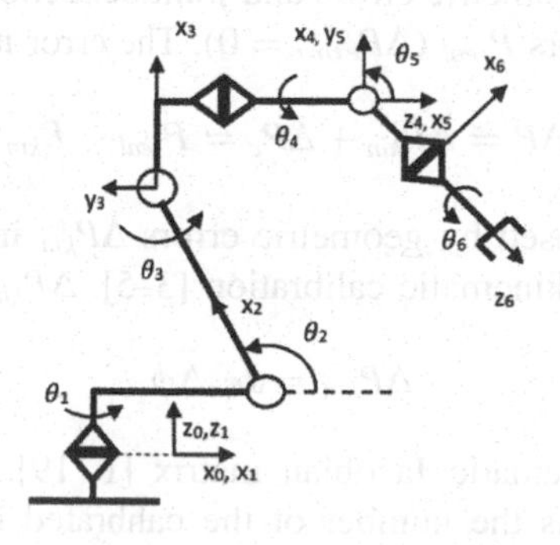

Fig. 1. Kinematic structure of the YS 100 robot.

The transformation that relates base frame {0} to tool frame{T}:

$$ {}^{0}_{E}T = {}^{0}_{1}T(\theta_1){}^{1}_{2}T(\theta_2){}^{2}_{3}T(\theta_3){}^{3}_{4}T(\theta_4){}^{4}_{5}T(\theta_5){}^{5}_{6}T(\theta_6){}^{6}_{E}T \quad (1) $$

The end-effector transformation:

$$ ({}^{6}_{T})T = Tr_X(a_6)Tr_Y(b_6)Tr_Z(d_T) \quad (2) $$

Table 1. Nominal D-H parameters of the Hyundai robot YS100.

i	α_{i-1}(deg)	a_{i-1}(m)	β_{i-1}(deg)	b_{i-1}(m)	d_i(deg)	θ_i(deg)
1	0	0	0	0	0.48	θ_1
2	90	0.32	–	–	0	θ_2
3	0	0.87	0	–	0	θ_3
4	90	0.2	–	–	1.03	θ_4
5	−90	0	–	–	0	θ_5
6	90	0	–	–	0.185	θ_6
T	–	0.2	–	0.05	0.5	–

3 Identification Kinematic Parameters and Compliance Compensation Based on the Effective Torques Using a Radial Basis Function

Assuming that the robot's end-effector position P_{real} is calculated by the following equation:

$$ P_{real} = P_{kin} + \Delta P_{kin} + \Delta P_c + \Delta P_{extra} \quad (3) $$

where P_{kin} is the position of the end effector calculated by the kinematic parameter,ΔP_{kin} is the position error caused by the geometric error, ΔP_c is the position error due to the joint compliance, and ΔP_{extra} is the positional residual error that is not modeled. Assuming that geometric errors and joint deflection errors are the main parts in causing the position errors P_{real} ($\Delta P_{extra} = 0$). The error model can be expressed as:

$$ \Delta P = \Delta P_{kin} + \Delta P_c = P_{real} - P_{kin} \quad (4) $$

The position errors caused by geometric errors ΔP_{kin} in the Eq. 4 could be identified by the conventional kinematic calibration [3–5]. ΔP_{kin} can be expressed as

$$ \Delta P_{kin} = J_{kin}\Delta\phi \quad (5) $$

where $J_{kin}(3 \times n)$ is a kinematic Jacobian matrix [1, 19]. $\Delta\phi$ is a $n \times 1$ kinematic parameter error vector. n is the number of the calibrated kinematic parameters. The total number of kinematic parameters is equal to 32. However, the 6 DOF revolute robot has several dependencies between some parameters. These dependency parameters are $\{\Delta\theta_1, \Delta\theta_0\}$, $\{\Delta d_1, \Delta d_0\}$, $\{\Delta d_3, \Delta d_2\}$, $\{\Delta z_T, \Delta d_6\}$, $\{(\Delta x_T, \Delta y_T), \Delta\theta_6\}$. In each pair, the parameter errors cannot be identified together. Therefore, the dependency parameters that are chosen to calibrate are $\{\Delta\theta_1$, Δd_1, Δd_3, Δx_T, Δy_T, $\Delta z_T\}$ while the other error parameter in each pair is set to the nominal parameter value. So, the number of calibrated kinematic is reduced to 27.

The Eq. 5 can be solved by the least-square method to overcome the effect of noise and uncertainty:

$$\Delta\phi = [(\mathrm{J^T J})^{-1}\mathrm{J^T}]\Delta P \tag{6}$$

The positional error ΔP is calculated by

$$\Delta P = P_m - P_{kin} \tag{7}$$

where $\mathrm{P_m}$ is the measured position vector and $\mathrm{P_{kin}}$ is the computed position vector by the recent kinematic parameters. The Eq. (6) is employed repetitive until the geometric parameters converge. Through the kinematic calibration process, the P_{kin} converges to the P^c_{kin} value. The position errors of the robot end-effector after kinematic calibration process are calculated by:

$$\Delta P_{res} = P_m - P^c_{kin} \tag{8}$$

Assuming that the position errors due to joint deflection errors are the main parts in causing these residual position errors ($\Delta P_{res} = \Delta P_c$). The joint deflections under link self-gravity and external payload are also assumed to be dominant in causing compliance errors. Therefore, the joint deflection errors can be calculated from the related effective torque of joints.

It should be noted that previous literatures [15, 19] constructed the compliance errors by linearizing the relationship of the effective torques and the joint compliances. However, there are some residual errors that could not be neglected caused by the nonlinear relation between joint torques and joint deflections. For further enhanced the robot precision, the relationship of the effective torque and the residual errors is constructed by a RBF in this paper. The RBF has 6 inputs that represent the total effective torque in 6 robot joints, 40 nodes in the hidden layer, and 3 nodes in the output layer that represent three elements of the position error vector.

The total effective torques in the robot $\mathrm{j_{th}}$ joint under related gravity forces are given as:

$$\tau_i = \sum_{j=i}^{N+1} \tau_{i,j} = \sum_{j=i}^{N+1} J^T_{\theta_{i,j}} F_j \tag{9}$$

where N = 6 is the number of DOF of the robot and $\mathrm{F_{N+1}}$ is the gravity force due to payload Here, the gravity force accompanying to $\mathrm{j_{th}}$ link is calculated by

$$F_j = [0 \quad 0 \quad -M_j g] \tag{10}$$

where M_j is the mass of the j_{th} link and g is the gravity coefficient. The transpose of the Jacobian matrix is used as a force transformation to find the effective joint torques $\tau_{i,j}$ in the i_{th} joint due to the gravity force in the j_{th} link. The Jacobian matrix is defined as

$$J_{\theta_{i,j}} = z_i \times l_{i,j} \tag{11}$$

where $l_{i,j}$ is the 3 × 1 vector between the origin of the i_{th} frame and the mass center of the j_{th} link.

The total effective torques are set to be the input of the RBF. Figure 2 shows the structure of the RBF. The output of the hidden node i in the RBF layer is calculated as follow:

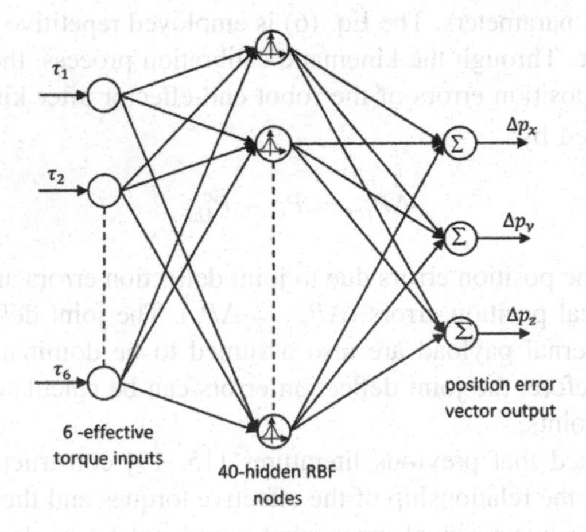

Fig. 2. Structure of the RBF.

$$o_j = e^{-n^2} \tag{12}$$

where n is a transfer function that describes the vector distance between the weight vector w_i and the input vector p, multiplied by the bias b_i.

$$n = \|\mathrm{w}_i - p\| b_i \tag{13}$$

The output layer is a linear function with 3 nodes in the output layer that represent three elements of the position error vector.

The output of the RBF is used to compensate for the compliance error(which is assumed to be the residual error $\Delta P_{res} = \Delta P_c$). Therefore, the residual error after compensated by the RBF is calculated by:

$$e = \Delta P_{res} - P_{nn} \tag{14}$$

In this work, the weights and bias of the RBF are trained by the MATLAB toolbox that creates a two-layer network. The hidden layer is the RBF layer (Eq. 12, 13). The output layer is a linear layer. At the beginning, there are no neurons in the hidden layer. The learning process is carried out following the steps below:

- Run the network and find the input vector with the greatest error.
- A RBF neuron is added with weights equal to that vector.
- The linear layer weights are redesigned to minimize error.
- Repeat until convergence.

In order to keep the RBF layer from increasing too much, the number of nodes in this layer are limited at 40 nodes. Overall, the suggested method could be described in the following flowchart (Fig. 3).

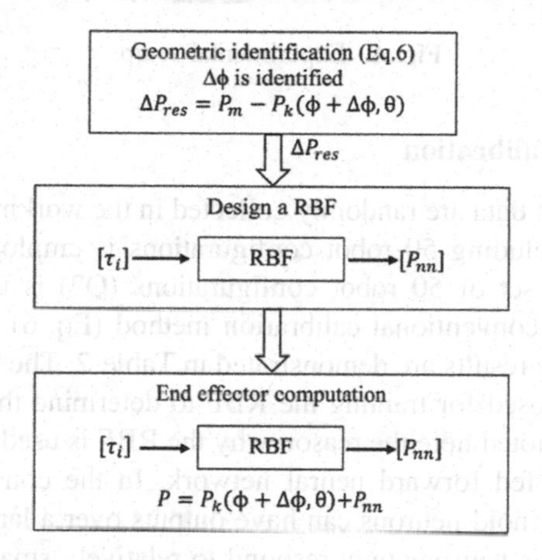

Fig. 3. Flowchart of the proposed method.

4 Experiment and Results

The experimental system is shown in Fig. 4. The 6 DOF robot manipulator (YS100). In this work, the mass of link j_{th} M_j (Eq. 10) is provided by the robot's manufacturer. The external payload weight is 110 kg. Therefore, the weight matrix is descried as follow:

$$M = \begin{bmatrix} 196.7 & 79.25 & 170.27 & 10.58 & 22.33 & 2.0 & 110 \end{bmatrix} \tag{15}$$

An API laser tracker (accuracy of 0.01 mm/m, repeatability of ± 0.006 mm/m) and an accompanying laser reflector are used to perform the calibration process. The proposed method (RBF-TCM) is used to calibrate the YS100 robot to show the advantage of the method in comparing with 2 others methods including the kinematic calibration method (KM) [3–5], the simultaneous identification of joint compliance and kinematic parameters methods (SKCM) [19] in the experimental study.

Fig. 4. Experimental setup.

4.1 Experimental Calibration

The robot configuration data are randomly collected in the working space and classified into 2 sets. Set Q1 including 50 robot configurations is employed in the calibration process and the other set of 50 robot configurations (Q2) is used in the validation process. By using the conventional calibration method (Eq. 6), 27 geometric parameters are identified. The results are demonstrated in Table 2. The residual errors and the computed torques are used for training the RBF to determine the weights and bias of the RBF. It should be noted here the reason why the RBF is used in this working rather than the conventional fed forward neural network. In the conventional feedforward neural network, the sigmoid neurons can have outputs over a large region of the input space, while radial basis neurons only respond to relatively small regions of the input space [23]. Therefore, the RBF could be said to be more stable in responding to noises and uncertainties inputs. However, the drawback of this method is that the larger the input space the more radial basis neurons are required [24]. The experimental calibration processes are carried out by 3 different calibration methods such as conventional kinematic calibration, SKCM, and RBF-TCM. The results of these calibration methods are shown in Fig. 5 and Table 3.

Table 2. D-H parameters of the Hyundai robot YS100.

i	α_{i-1}(deg)	a_{i-1}(m)	β_{i-1}(deg)	b_{i-1}(m)	d_i(deg)	θ_i(deg)
1	−0.1646	−0.016	0.4748	0.0493	0.4851	−0.3316
2	90.0578	0.3199	–	–	0	1.1517
3	0.0004	0.8704	0.0681	–	-0.0036	−1.6686
4	89.9919	0.2001	–	–	1.0272	−1.2079
5	−90.121	0.0003	–	–	−0.0017	−0.0017
6	89.9656	−0.0035	–	–	0.185	−1.8226
T	–	−0.28	–	0.0469	0.4219	–

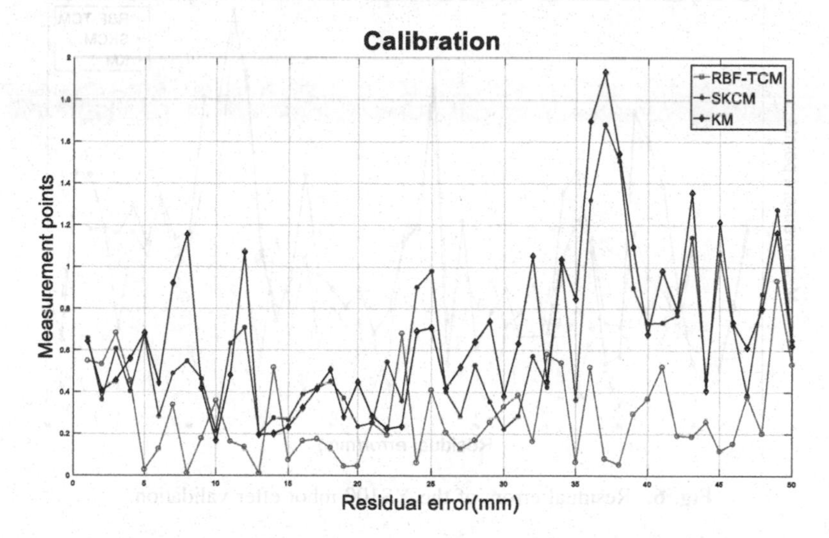

Fig. 5. Residual errors of the YS100 robot after calibration.

Table 3. The absolute position accuracy of the YS100 robot (Calibration).

	Mean (mm)	Maximum (mm)	Std. (mm)
Nominal robot model	13.5527	30.5911	6.0528
KM	0.6894	1.9318	0.4015
SKCM	0.6065	1.6811	0.3488
Proposed method	0.2785	0.9332	0.2095

The calibration results show that the precision of the robot after calibrated by the proposed method is dramatically reduced. By employing the RBF-TCM method, the position errors are lower than the results by other methods. In comparing to the conventional kinematic calibration method, the proposed method reduces the mean of position errors from 0.6894 mm to 0.2785 mm (precise increasing by 59.6%). It also increases the accuracy by 54.08% in comparison to the results generated by the SKCM method (from 0. 6065 mm to 0.2785 mm). The suggested algorithm also generates the lowest maximum position error (0.9332 mm), and the lowest standard deviation (0.2095 mm).

4.2 Experimental Validation Results

The proposed method should be validated by another robot configuration to demonstrate the ability of it over the working space. The robot configuration set Q2 that is totally different from Q1 is hired for the validation process with 3 different methods.

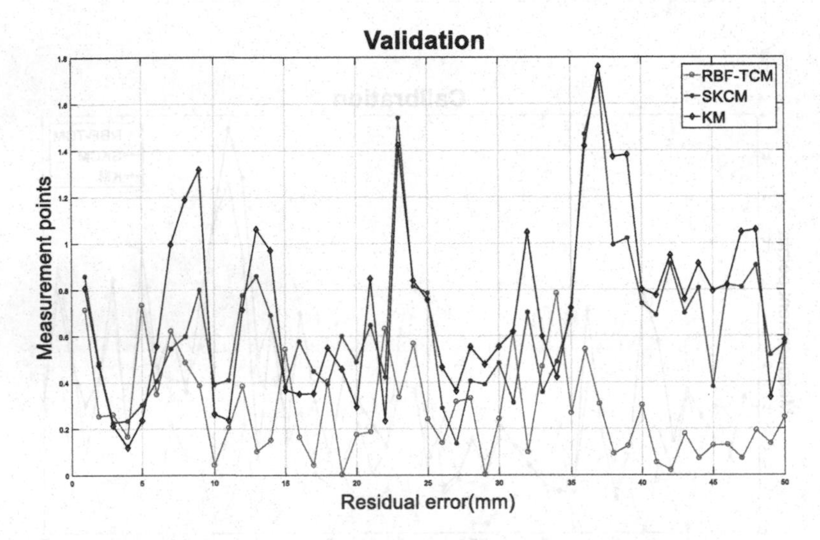

Fig. 6. Residual errors of the YS100 robot after validation.

Table 4. The absolute position accuracy of the YS100 robot (Validation).

	Mean (mm)	Maximum (mm)	Std. (mm)
Nominal robot model	14.1106	32.3303	5.9835
KM	0.7245	1.7584	0.3814
SKCM	0.6398	1.7031	0.3214
Proposed method	0.2802	0.7846	0.2084

By employing the method, the position errors are lower than the results by other methods in the validation process (Table 4 and Fig. 6). In comparing to the conventional kinematic calibration method, the proposed method reduces the mean of position errors from 0.7245 mm to 0.2802 mm (precise increasing by 61.33%). It also increases the accuracy by 56.21% in comparison to the results generated by the SKCM method (0.6398 mm to 0.2802 mm). The suggested algorithm also generates the lowest maximum position error (0.7846 mm), and the lowest standard deviation (0.2084 mm).

4.3 Discussion and Future Studying

In previous literatures [15, 19], the relationship of joint deflections of robot and the effective torques are linearized:

$$\Delta\theta_c = \tau C \tag{16}$$

where $\Delta\theta_c$ is the $N \times 1$ joint deflection vector, τ is the diagonal effective torque matrix, C is the $N \times 1$ joint compliance vector. Then, the Cartesian position errors due to the joint compliances can be modeled as:

$$\Delta P_c = J_e \Delta\theta_c = J_e \tau C \tag{17}$$

where J_e is vector of joint compliance parameters that is computed by the following published work [25]. $J_e\tau$ is the transformation matrix relating the joint compliance parameters and the deflections of robot end-effector. It should be noted here that the effective torques in the i_{th} joint is not only due to the gravity force related to the i_{th} link but also due to the gravity forces related to both the link after and the external load. In this work, the relationship of the effective torque τ and the positional errors due to the compliance errors ΔP_c could be constructed by a RBF for higher increasing the precision of the robot.

The work will be expanded in the future by implementing the optimizing method to select the calibration poses for better calibration results.

5 Conclusion

In this work, a new robotic calibration method is proposed for reducing the positional errors of the robot manipulator. First, geometric errors of a robot are identified by using a conventional kinematic calibration model of the robot. Then, a radial basis function is constructed for compensating the compliance errors based on the effective torques for further increasing the positional precision of the robot. By using a RBF, the relationship of the effective torque and the compliance errors is constructed for higher increasing the precision of the robot. The advantages of the suggested method are easy for implementing, removing the need for torque sensors, high ability to enhance the precision of the manipulator. These advantages are firmly confirmed by the experimental studies on a YS100 robot in contrasting with 2 other methods such as the conventional kinematic calibration and the method for simultaneously calibrate the geometric and joint stiffness parameters of the robot.

Acknowledgment. This research was supported by 2020 Research Fund of University of Ulsan, Ulsan, Korea.

References

1. Mooring, B.W., Roth, Z.S., Driels, M.R.: Fundamentals of Manipulator Calibration. Wiley, New York (1991)
2. Whitney, D.E., Lozinski, C.A., Rourke, J.M.: Industrial robot forward calibration method and results. J. Dyn. Syst. Meas. Control **108**, 1–8 (1986)
3. John, J.C., et al.: Introduction to Robotics: Mechanics and Control. Addison-Wesley, Read (1989)
4. Hayati, S., Mirmirani, M.: Improving the absolute positioning accuracy of robot manipulators. J. Robot. Syst. **2**, 397–413 (1985)
5. Hayati, S., Tso, K., Roston, G.: Robot geometry calibration. In: Proceedings of 1988 IEEE International Conference on Robotics and Automation, pp. 947–951 (1988)
6. Klug, C., Schmalstieg, D., Gloor, T., Arth, C.: A complete workflow for automatic forward kinematics model extraction of robotic total stations using the Denavit-Hartenberg convention. J. Intell. Robot. Syst. **95**, 311–329 (2019)
7. Faria, C., Vilaça, J.L., Monteiro, S., Erlhagen, W., Bicho, E.: Automatic Denavit-Hartenberg parameter identification for serial manipulators. In: IECON 2019-45th Annual Conference of the IEEE Industrial Electronics Society, pp. 610–617 (2019)
8. Morar, C.A., Hăgan, M., Doroftei, I., Marinca, Ş.: Analog matrix multiplier dedicated to the Denavit-Hartenberg algorithm. In: 2019 International Symposium on Signals, Circuits and Systems (ISSCS), pp. 1–4 (2019)
9. Zhuang, H., Roth, Z.S., Hamano, F.: A complete and parametrically continuous kinematic model for robot manipulators. In: Proceedings of the IEEE International Conference on Robotics and Automation, pp. 92–97 (1990)
10. Zhuang, H., Wang, L.K., Roth, Z.S.: Error-model-based robot calibration using a modified CPC model. Robot. Comput. Integr. Manuf. **10**, 287–299 (1993)
11. Okamura, K., Park, F.C.: Kinematic calibration using the product of exponentials formula. Robotica **14**, 415–421 (1996)
12. Chen, G., Kong, L., Li, Q., Wang, H., Lin, Z.: Complete, minimal and continuous error models for the kinematic calibration of parallel manipulators based on POE formula. Mech. Mach. Theory **121**, 844–856 (2018)
13. Gupta, K.C.: Kinematic analysis of manipulators using the zero reference position description. Int. J. Rob. Res. **5**, 5–13 (1986)
14. Cheng, L.-P., Kazerounian, K.: Study and enumeration of singular configurations for the kinematic model of human arm. In: Proceedings of the IEEE 26th Annual Northeast Bioengineering Conference (Cat. No. 00CH37114), pp. 3–4 (2000)
15. Dumas, C., Caro, S., Garnier, S., Furet, B.: Joint stiffness identification of six-revolute industrial serial robots. Robot. Comput. Integr. Manuf. **27**, 881–888 (2011)
16. Slavković, N.R., Milutinović, D.S., Kokotović, B.M., Glavonjić, M.M., Živanović, S.T., Ehmann, K.F.: Cartesian compliance identification and analysis of an articulated machining robot. FME Trans. **41**, 83–95 (2013)
17. Jang, J.H., Kim, S.H., Kwak, Y.K.: Calibration of geometric and non-geometric errors of an industrial robot. Robotica **19**, 311–321 (2001)
18. Meggiolaro, M.A., Dubowsky, S., Mavroidis, C.: Geometric and elastic error calibration of a high accuracy patient positioning system. Mech. Mach. Theory **40**, 415–427 (2005)
19. Zhou, J., Nguyen, H.-N., Kang, H.-J.: Simultaneous identification of joint compliance and kinematic parameters of industrial robots. Int. J. Precis. Eng. Manuf. **15**, 2257–2264 (2014)

20. Kamali, K., Joubair, A., Bonev, I.A., Bigras, P.: Elasto-geometrical calibration of an industrial robot under multidirectional external loads using a laser tracker. In: 2016 IEEE International Conference on Robotics and Automation (ICRA), pp. 4320–4327 (2016)
21. Müller, R., Scholer, M., Blum, A., Kanso, A.: Identification of the dynamic parameters of a robotic tool based on integrated torque sensors. In: 2019 23rd International Conference on Mechatronics Technology (ICMT), pp. 1–6 (2019)
22. Besset, P., Olabi, A., Gibaru, O.: Advanced calibration applied to a collaborative robot. In: 2016 IEEE International Power Electronics and Motion Control Conference (PEMC), pp. 662–667 (2016)
23. Xia, C., Liu, Y., Lei, B., Xiang, X.: Research on a generalized regression neural network model of thermocouple and it's spread scope. In: 2008 Fourth International Conference on Natural Computation, pp. 109–113 (2008)
24. Corino, V.D.A., Matteucci, M., Cravello, L., Ferrari, E., Ferrari, A.A., Mainardi, L.T.: Long-term heart rate variability as a predictor of patient age. Comput. Methods Programs Biomed. **82**, 248–257 (2006)
25. Nakamura, Y., Ghodoussi, M.: Dynamics computation of closed-link robot mechanisms with nonredundant and redundant actuators. Int. Conf. Robot. Autom. **5**, 294–302 (1989)

Person-Following Shopping Support Robot Using Kinect Depth Camera Based on 3D Skeleton Tracking

Md Matiqul Islam[1,2(✉)], Antony Lam[3], Hisato Fukuda[1], Yoshinori Kobayashi[1], and Yoshinori Kuno[1]

[1] Graduate School of Science and Engineering, Saitama University, Saitama, Japan
matiqul@cv.ics.saitama-u.ac.jp
[2] University of Rajshahi, Rajshahi-6205, Bangladesh
[3] Mercari, Inc., Roppongi Hills Mori Tower 18F, 6-10-1 Roppongi Minato-Ku, Tokyo 106-6118, Japan

Abstract. The lack of caregivers in an aging society is a major social problem. Without assistance, many of the elderly and disabled are unable to perform daily tasks. One important daily activity is shopping in supermarkets. Pushing a shopping cart and moving it from shelf to shelf is tiring and laborious, especially for customers with certain disabilities or the elderly. To alleviate this problem, we develop a person following shopping support robot using a Kinect camera that can recognize customer shopping actions or activities. Our robot can follow within a certain distance behind the customer. Whenever our robot detects the customer performing a "hand in shelf" action in front of a shelf it positions itself beside the customer with a shopping basket so that the customer can easily put his or her product in the basket. Afterwards, the robot again follows the customer from shelf to shelf until he or she is done with shopping. We conduct our experiments in a real supermarket to evaluate its effectiveness.

Keywords: Kinect camera · Supermarket · Person following · Elderly

1 Introduction

With the advancement of robotics technologies researchers have started to explore the application of service robots to our daily life. Assisted shopping is one application that can benefit many. Thus, many researchers have developed robotic shopping trolleys such as the one presented in Y. Kobayashi et al. [1]. In particular, [1] showed the benefits of using such robotic shopping systems in supporting the elderly. Only person following robot is not enough to support the elderly properly. If such robotic shopping trolleys could intelligently follow the customer by recognizing their behavior, it would be even more helpful. For this reason, we focus on recognizing the customers shopping behavior. By recognizing the customer's behavior, the robot could perform convenient tasks such as following the customer or moving to let the customer place goods in a basket. In our previous paper [2, 3] we developed an intelligent shopping support robot that can recognize customer shopping behavior using our developed GRU (Gated

D.-S. Huang and P. Premaratne (Eds.): ICIC 2020, LNAI 12465, pp. 28–37, 2020.
https://doi.org/10.1007/978-3-030-60796-8_3

Recurrent Unit) neural network. To develop these robots, we used an OpenPose [4] neural network model to detect a person's skeleton for person tracking and the LiDAR sensor to measure the distance from robot to person.

To operate our robot in a practical environment we must ensure three requirements: speed, accuracy and cost. The OpenPose based model we previously used does not fulfill these requirements. For this reason, in this paper, we replace the OpenPose model with a Kinect v2 depth camera [5] to get the following advantages:

1. The processing speed of the OpenPose model-based skeleton tracking is not fast enough(5frames/s) whereas Kinect based skeleton tracking is fast (30 frames/s). So, for realtime applications the Kinect based system is better.
2. The accuracy of shopping behavior recognition using the OpenPose model- based 2D skeleton data is not as high as the Kinect 3D skeleton data-based model. For the OpenPose model, we get 82% accuracy whereas using Kinect 3D skeleton based model we get 95% accuracy.
3. In our previous model, we used the LiDAR sensor to measure the distance between the robot and customer whereas the Kinect itself can measure the distance. So, it reduces the need for extra processing and cost.

Figure 1 shows our proposed robot. In our proposed system, we used a Kinect camera to find the location of the customer in a given shopping area and measure the distance from the robot to the customer so that it can easily follow the customer. We recognize the customer's shopping behavior using 3D skeleton data and according to the recognized "hand in shelf action", our robot takes on an appropriate position to the customer to put the product in the shopping basket.

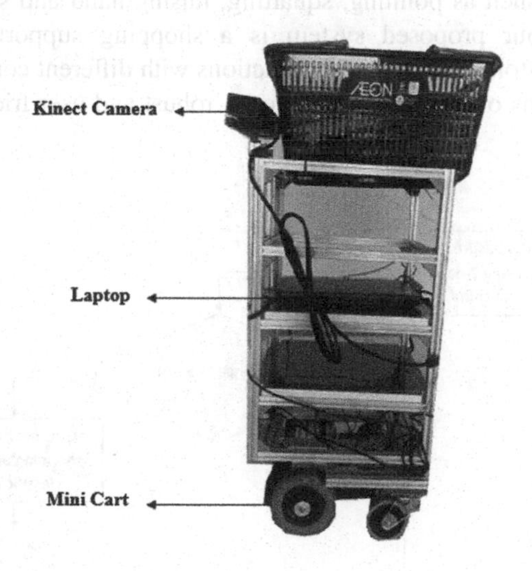

Fig. 1. Our person following a mini cart robot.

2 Related Work

Many researchers use stereo cameras to track people from moving platforms [6] and achieve person following through appearance models and stereo vision using a mobile robot [7]. This is a well-known method for person following robots. To follow a person, the robot must continuously receive two types of information, such as position and distance data.

The person following robot called "ApriAttenda" and "Nurse Following Robot" was created by T. Sonoura et al. and B. Ilias et al. in [8, 9]. The main task of these robots was to find and assign a person and continuously follow that person everywhere. Using a laser range finder (LRF), the OSAKA Institute of Technology has created a mobile robot named ASAHI, with semi-autonomous navigation using simple and robust person following behavior [10]. To follow a person T. Germa et al. have developed a mobile robot, named Rackham [11]. This robot uses one digital camera, one ELO touch screen, a pair of loudspeakers and an RFID system.

S. Nishimura et al. [12] developed an autonomous robotic shopping cart. This shopping cart can follow customers autonomously and transport goods. Kokhtsuka et al. [13] provide a conventional shopping cart with a laser range sensor to measure the distance from and the position of its user and develop a system to prevent collisions. Their robotic shopping cart also follows users to transport goods.

Hu et al. [14] proposed an action recognition system to detect the interaction between the customer and the merchandise on the shelf. The recognition of the shopping actions of the retail customers was also developed by using a stereo camera from the top view [15]. Lao et al. [16] used one surveillance camera to recognize customer's actions, such as pointing, squatting, raising hand and so on.

In this paper, our proposed system is a shopping support robot that reacts depending on the customer's hand in shelf actions with different conditions. Compared with the other systems ours is simple, low cost, robust and user friendly especially for the elderly.

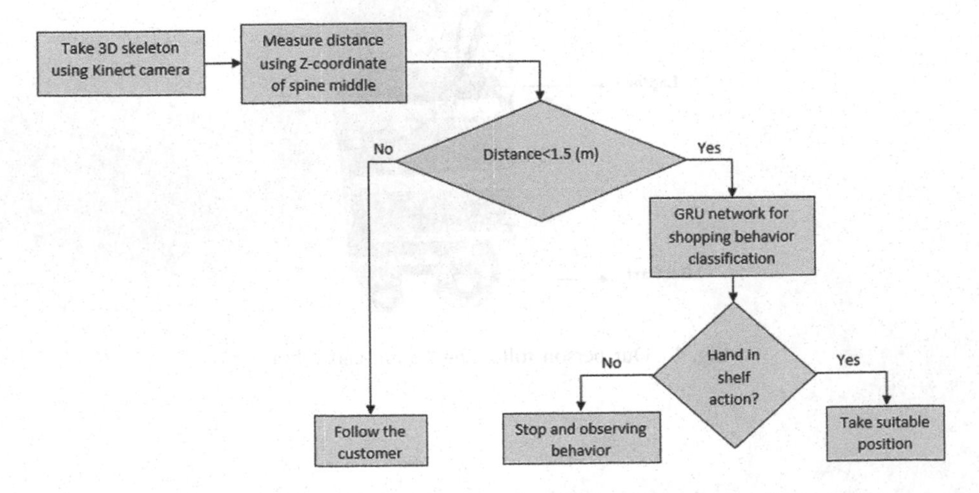

Fig. 2. Block diagram of our proposed shopping support robot.

3 Design Approach

We developed our system using the Xbox One Kinect V2 camera. Using this camera first we track the customer's 3D skeleton and recognize the customer's different shopping actions in front of the shelf using the GRU network. At the same time, we calculate the robot to customer distance using the Z-coordinate value of the middle of the spine. If the distance is less than 1.5 m our robot performs action recognition otherwise it just follows the customer. If our robot recognizes a "hand in shelf" action by the customer, it takes the proper position and helps the customer. Figure 2 shows the block diagram of our shopping support robot.

3.1 Person's Skeleton Tracking

In our system, we use the Kinect Version 2 camera to detect the 3D skeleton of a person. The skeleton has 25 joints as shown in Fig. 3.

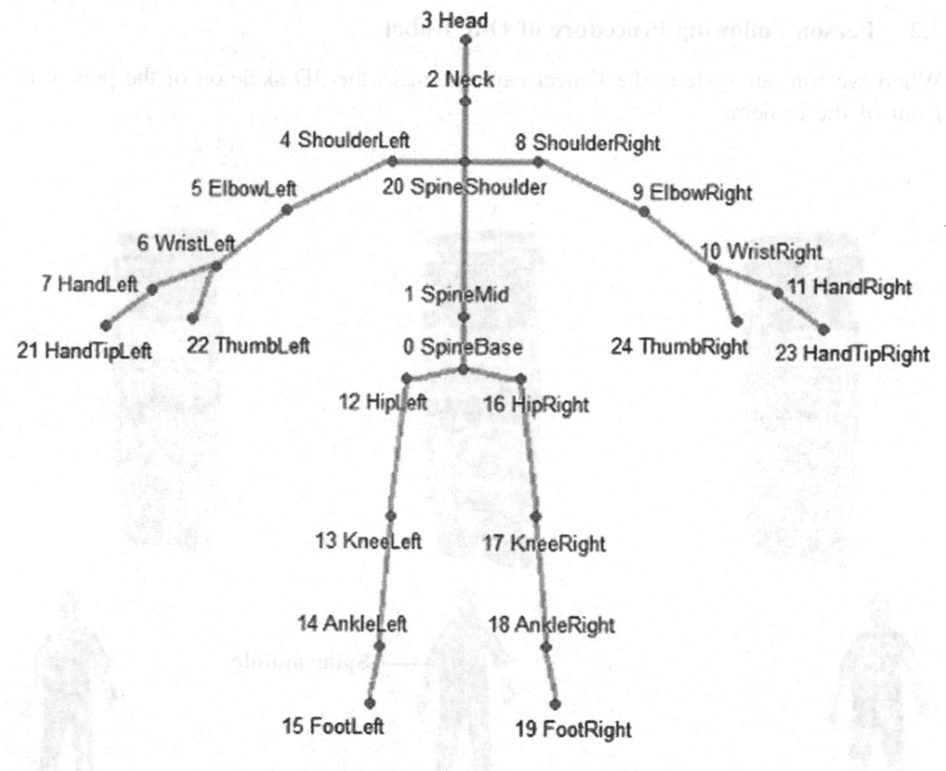

Fig. 3. Kinect V2 25-joint skeleton.

To find the 3D points of the joints in space the Kinect's camera coordinates use the Kinect's infrared sensor. The coordinate system is defined as according to Fig. 4.

The origin ($x = 0, y = 0, z = 0$) is located at the center of the IR sensor on Kinect. Positive X values tend towards to the sensor's left [from the sensor's POV].

Y moves in an upward direction (note that this direction is based on the sensor's tilt).
Z moves out in the direction the sensor is facing
1 unit = 1 m.

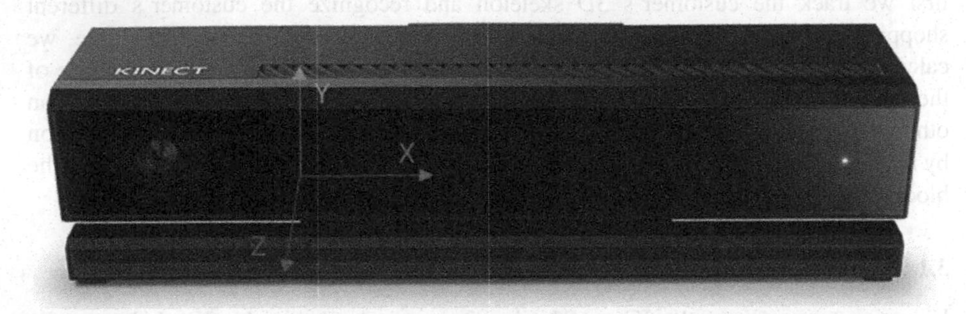

Fig. 4. Camera space coordinates from the Kinect SDK

3.2 Person Following Procedure of Our Robot

When we run our system, the Kinect camera tracks the 3D skeleton of the person in front of the camera.

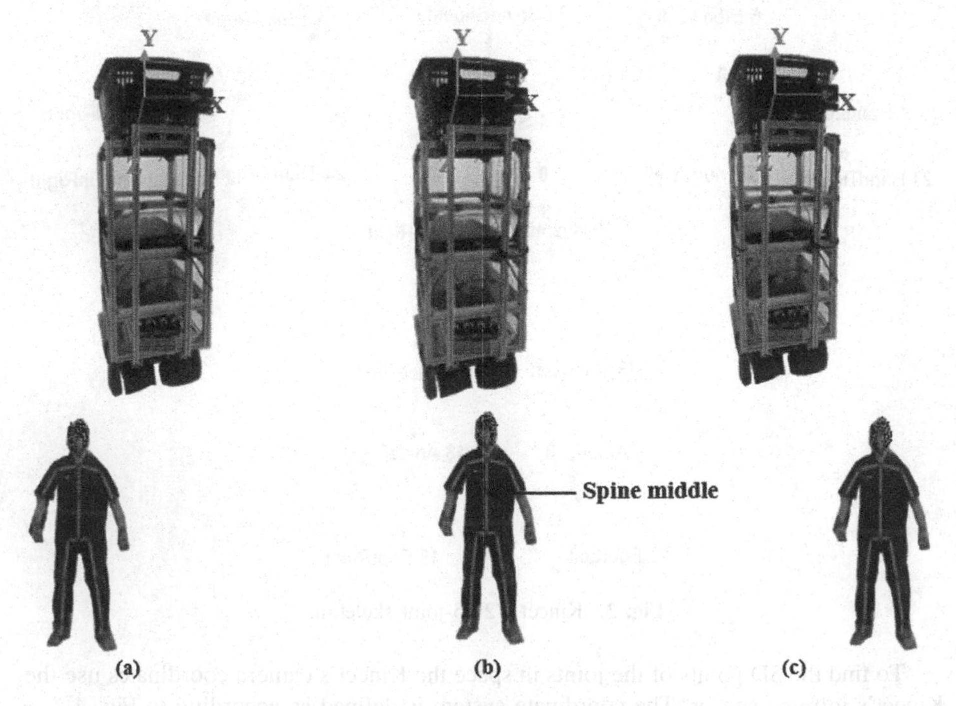

Fig. 5. A person following procedure.

To follow the customer, we take two values from detected skeleton joint points, sm. X and sm.Z, where sm.X is the "mid-spine" value of the X coordinate that represents the position of the person and sm.Z is the mid-spine value of the Z coordinate and represents the distance the robot to customer. Depending on this value, the appropriate command issued to our robot is given in Table 1.

We set the threshold value for the mid-spine value of X in the range (−0.179835 to 0.218678). If the tracked person's mid-spine value of X is in this range, we assume that the person's position is in the middle of the robot as shown in Fig. 5(b). In this the robot follows the person measuring the Z coordinate value of the mid-spine. If the mid-spine value of $X > 0.218678$, the robot assumes that the person is on the right side of the robot as shown in Fig. 5(c) and the robot then rotates right until it reaches the threshold value. When it reaches the threshold value, it again follows the person measuring the Z value of the mid-spine. If the mid-spine value of $X < -0.179835$, the robot assumes that the person is on the left side of the robot as shown in Fig. 5(a) and the robot then rotates left until it reaches the threshold value. When it reaches the threshold value, it again follows the person measuring the Z value of the mid-spine. In this way, our person following robot works using a Kinect camera.

Table 1. Different command conditions for our robot.

Command	Position
Forward	$0.218678 < sm.X < -0.179835$
Stop	$sm.Z < 1.5$ (m)
Left	$sm.X < -0.179835$
Right	$sm.X > 0.218678$

3.3 Shopping Behavior Action Recognition

Dataset Construction

We make a dataset to train our GRU network. The details of our GRU network are shown in our previous paper [1]. We take 3D joints of the skeleton to make the dataset for different shopping behaviors. We take 4 camera views to take the 3D skeleton joints data. We take 114,464 joints data for training and 33,984 joints data for the testing set.

Training the GRU Network

The details of the training specification are shown in Table 2. Figure 6 shows the plot of the model accuracy and loss over 50000 iterations. The results of the GRU network-based shopping behavior recognition OpenPose 2D skeleton data compared to the Kinect 3D skeleton data is shown in Table 3. Figure 7 shows the confusion matrix of different shopping behaviors.

Table 2. Training specification for our proposed GRU network.

Training parameters	Value
Batch size	512
Epochs	50000
Timesteps	32
No. of hidden layer	34
Learning rate	0.000220
Optimizer	Adam
Momentum	0.9

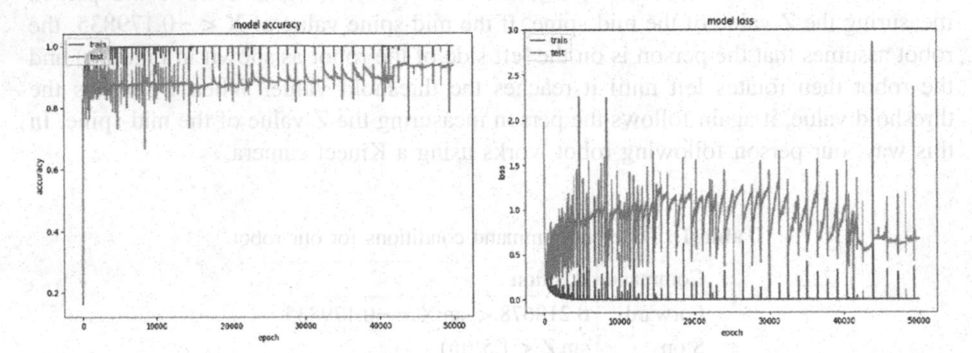

Fig. 6. The model accuracy and loss over 50000 iterations

Table 3. GRU network-based behavior recognition comparison using OpenPose 2D skeleton data and Kinect 3D skeleton data.

Shopping behavior	Precision		Recall		F1 score	
	Kinect 3D data	OpenPose 2D data	Kinect 3D data	OpenPose 2D data	Kinect 3D data	OpenPose 2D data
Reach to shelf	1.00	0.86	0.93	0.92	0.96	0.89
Retractfrom shelf	0.80	0.44	1.00	0.69	0.89	0.54
Hand in shelf	1.00	0.79	0.84	0.83	0.91	0.81
Inspect product	1.00	0.92	1.00	0.73	1.00	0.81
Inspect shelf	0.93	0.93	1.00	0.90	0.96	0.91
Avg/Total	**0.95**	**0.79**	**0.94**	**0.82**	**0.94**	**0.79**

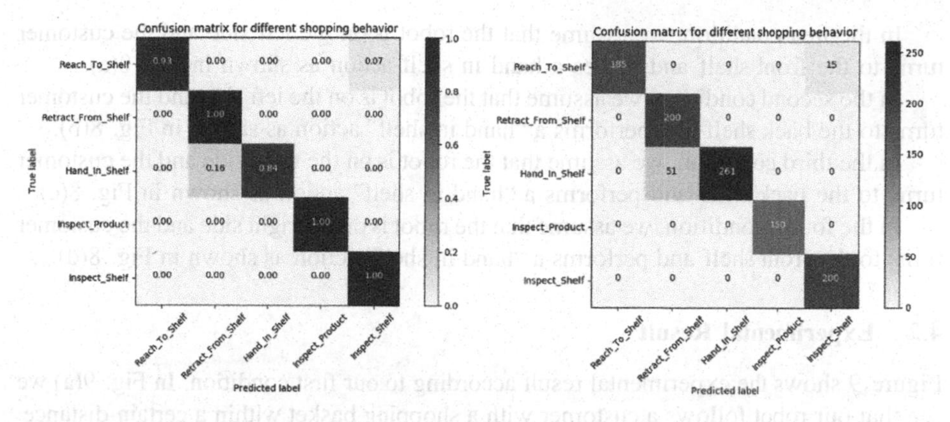

Fig. 7. The confusion matrix of different shopping behaviors.

4 Experiments

We experimented at an actual supermarket. The area was arranged with a two-sided shelf with different items. A person moved between the shelves and our robot followed that person. When the person performed a "hand in shelf" action, our robot took a suitable position so that the person could easily put his items in the basket.

4.1 Experimental Conditions

We conducted our experiment in four conditions. We define the different conditions below:

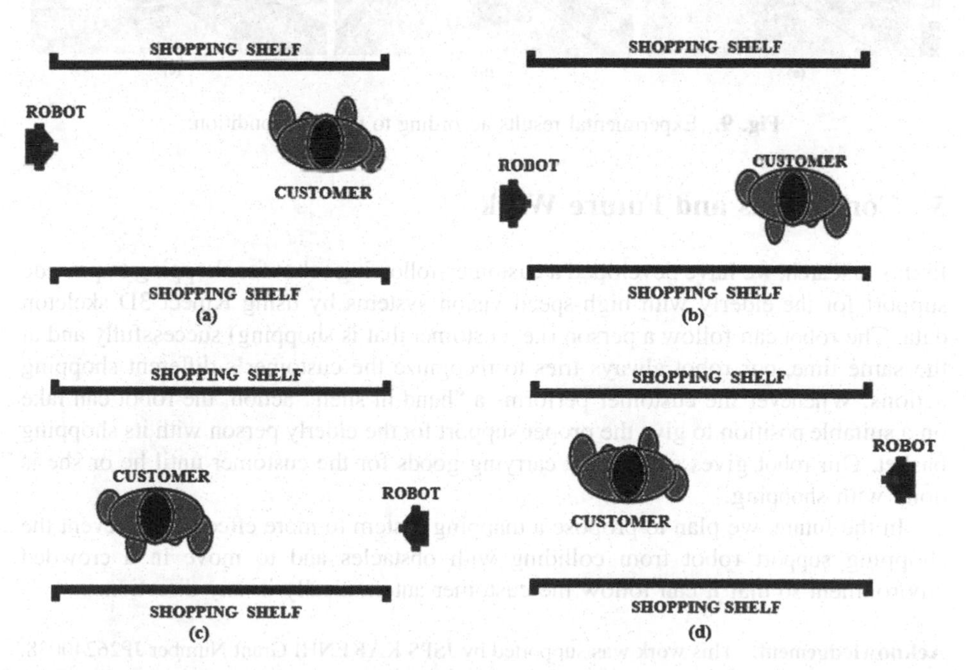

Fig. 8. (a) Customer performs a "hand in shelf" action on the front shelf. (b) Customer performs a "hand in shelf" action on the back shelf. (c) Customer performs a "hand in shelf" action on the back shelf. (d) Customer performs a "hand in shelf" action on the front shelf.

In the first condition, we assume that the robot is on the left side and the customer turns to the front shelf and makes a hand in shelf action as shown in Fig. 8(a).

In the second condition, we assume that the robot is on the left side and the customer turns to the back shelf and performs a "hand in shelf" action as shown in Fig. 8(b).

In the third condition, we assume that the robot is on the right side and the customer turns to the back shelf and performs a "hand in shelf" action as shown in Fig. 8(c).

In the fourth condition, we assume that the robot is on the right side and the customer turns to the front shelf and performs a "hand in shelf" action as shown in Fig. 8(d).

4.2 Experimental Result

Figure 9 shows the experimental result according to our first condition. In Fig. 9(a) we see that our robot follows a customer with a shopping basket within a certain distance. When the customer performs a "hand in shelf" action on the front shelf, we see that the robot is on the left side of the customer as shown in Fig. 9(b). Then our robot moves closer to the customer and changes its orientation according to the customer and the customer easily puts his product in the basket.

Fig. 9. Experimental results according to the first condition.

5 Conclusions and Future Work

In this research, we have developed a customer-following robot for shopping to provide support for the elderly with high-speed vision systems by using Kinect 3D skeleton data. The robot can follow a person (i.e. customer that is shopping) successfully and at the same time, our robot always tries to recognize the customer's different shopping actions. Whenever the customer performs a "hand in shelf" action, the robot can take on a suitable position to give the proper support for the elderly person with its shopping basket. Our robot gives support by carrying goods for the customer until he or she is done with shopping.

In the future, we plan to propose a mapping system to more effectively prevent the shopping support robot from colliding with obstacles and to move in a crowded environment so that it can follow the customer automatically in any direction.

Acknowledgement. This work was supported by JSPS KAKENHI Grant Number JP26240038.

References

1. Kobayashi, Y., Yamazaki, S., Takahashi, H., Fukuda, H., Kuno, Y.: Robotic shopping trolley for supporting the elderly. In: Lightner, N.J. (ed.) AHFE 2018. AISC, vol. 779, pp. 344–353. Springer, Cham (2019). https://doi.org/10.1007/978-3-319-94373-2_38
2. Islam, M.M., Lam, A., Fukuda, H., Kobayashi, Y., Kuno, Y.: An intelligent shopping support robot: understanding shopping behavior from 2D skeleton data using GRU network. ROBOMECH J. **6**(1), 1–10 (2019). https://doi.org/10.1186/s40648-019-0150-1
3. Islam, Md.M., Lam, A., Fukuda, H., Kobayashi, Y., Kuno, Y.: A person-following shopping support robot based on human pose skeleton data and LiDAR sensor. In: Huang, D.-S., Huang, Z.-K., Hussain, A. (eds.) ICIC 2019. LNCS (LNAI), vol. 11645, pp. 9–19. Springer, Cham (2019). https://doi.org/10.1007/978-3-030-26766-7_2
4. Cao, Z., Hidalgo, G., Simon, T., Wei, S.E., Sheikh, Y.: OpenPose: realtime multi-person 2D pose estimation using part affinity fields. arXiv preprint arXiv:1812.08008 (2018)
5. Microsoft Kinect SDK. http://www.microsoft.com/en-us/kinectforwindows/
6. Beymer, D., Konolige, K.: Tracking people from a mobile platform. In: Siciliano, B., Dario, P. (eds.) Experimental Robotics VIII Springer Tracts in Advanced Robotics, vol. 5. Springer, Heidelberg (2003). https://doi.org/10.1007/3-540-36268-1_20
7. Calisi, D., Iocchi, L., Leone, R.: Person following through appearance models and stereo vision using a mobile robot. In: VISAPP (Workshop on on Robot Vision), pp. 46–56, March 2007
8. Takafumi, S., Takashi, Y., Manabu, N., Hideichi, N., Seiji, T., Nobuto, M.: Person following robot with vision-based and sensor fusion tracking algorithm. In: Zhihui, X. (ed.) Computer Vision, p. 538 (2008)
9. Ilias, B., Nagarajan, R., Murugappan, M., Helmy, K., Awang Omar, A.S., Abdul Rahman, M.A.: Hospital nurse following robot: hardware development and sensor integration. Int. J. Med. Eng. Inform. **6**(1), 1–13 (2014)
10. Hiroi, Y., Matsunaka, S., Ito, A.: A mobile robot system with semi-autonomous navigation using simple and robust person following behavior. J. Man Mach. Technol. (JMMT) **1**(1), 44–62 (2012)
11. Germa, T., Lerasle, F., Ouadah, N., Cadenat, V., Devy, M.: Vision and RFID-based person tracking in crowds from a mobile robot. In: IEEE/RSJ International Conference on Intelligent Robots and Systems, pp. 5591–5596. IEEE (2009)
12. Nishimura S., Takemura H., Mizoguchi H.: Development of attachable modules for robotizing daily items-person following shopping cart robot. In: IEEE International Conference on Robotics and Biomimetics (ROBIO), pp 1506–1511. IEEE, New York (2007)
13. Kohtsuka, T., Onozato, T., Tamura, H., Katayama, S., Kambayashi, Y.: Design of a control system for robot shopping carts. In: König, A., Dengel, A., Hinkelmann, K., Kise, K., Howlett, R.J., Jain, L.C. (eds.) KES 2011. LNCS (LNAI), vol. 6881, pp. 280–288. Springer, Heidelberg (2011). https://doi.org/10.1007/978-3-642-23851-2_29
14. Hu, Y., Cao, L., Lv, F., Yan, S., Gong, Y., Huang, T.S.: Action detection in complex scenes with spatial and temporal ambiguities. In: IEEE 12th International Conference on Computer Vision, pp. 128–135. IEEE (2009)
15. Haritaoglu, I., Beymer, D., Flickner, M.: Ghost 3D: detecting body posture and parts using stereo. In: Proceedings of the Workshop on Motion and Video Computing, pp. 175–180. IEEE (2002)
16. Lao, W., Han, J., De With, P.H.: Automatic video-based human motion analyzer for consumer surveillance system. Consum. Electron. IEEE Trans. **55**(2), 591–598 (2009)

Intelligent Computing in Computer Vision

Real-Time Object Detection Based on Convolutional Block Attention Module

Ming-Yang Ban(✉), Wei-Dong Tian, and Zhong-Qiu Zhao

College of Computer and Information, Hefei University of Technology, Hefei, China
18326952576@163.com

Abstract. Object detection is one of the most challenging problems in the field of computer vision, the practicality of object detection requires accuracy and real-time. YOLOv3 is a good real-time object detection algorithm, but with insufficient recall rate and insufficient positioning accuracy. The Attention Mechanism in deep learning is similar to the attention mechanism of human vision, which is to focus attention on important points in many information, select key information, and ignore other unimportant information. In this paper, we integrate Convolutional Block Attention Module (CBAM) in YOLOv3 in order to improves the detection accuracy and keep real-time. Compared to a conventional YOLOv3, we experimentally show the effectiveness and accuracy of the proposed method on the PASCAL VOC and MS-COCO datasets.

Keywords: Object detection · Real-time · Attention mechanism

1 Introduction

Object detection is one of the fundamental problems of computer vision, forms the basis of many other computer vision tasks, such as image captioning [1], instance segmentation [2], object tracking [3], etc. Given one image, the purpose of object detection tries to find objects of certain target classes with precise localization and assign a corresponding class label to each object instance.

In recent years, object detection algorithm pays attention to high detection accuracy and real-time detection speed. Deep-learning based object detection algorithms, which can be classified into two categories: two-stage and one-stage detectors. Two-stage detectors, such as Fast R-CNN [4], Faster R-CNN [5], and FPN [6], conduct a first stage of region proposal generation, followed by a second stage of object classification and bounding box regression. These methods generally show a high accuracy but have a disadvantage of a slow detection speed and lower efficiency. One-stage detectors, such as SSD [7] and YOLO [8], conduct object classification and bounding box regression concurrently without a region proposal stage. These methods generally have a fast detection speed and high efficiency but a low accuracy.

The real-time object detection model represented by SSD and YOLO, has played a pretty detection effect in the industrial field and practical application scenarios. However, two algorithms consider the object detection process as a regression problem and cannot distinguish the foreground from the background well, which is prone to false

D.-S. Huang and P. Premaratne (Eds.): ICIC 2020, LNAI 12465, pp. 41–50, 2020.
https://doi.org/10.1007/978-3-030-60796-8_4

detection and missed detection. At present, the improvement of the object detection algorithm mainly includes: using a basic neural network capable of extracting richer features, fusing features of multiple scales for detection, or other methods for improving the network structure. Shen et al. [9] proposed a Stem Block structure that can improve the detection accuracy, and refer to the dense connection of DenseNet [10] on the basis of SSD. Fu et al. [11] Proposed a deeper ResNet-101 [16] network for feature extraction based on the SSD detection framework, and used a deconvolution layer to introduce additional large amounts of semantic information, which improved the ability of the SSD to detect small objects. In addition, Woo et al. [17] found that both channel and spatial relationships between convolution operations are modeled and weighted at the same time to screen out the required features better.

In YOLOv3 [13], each region in the entire feature map is treated equally, and each region's contribution to the final detection is considered to be the same. The extracted convolution features do not weight the different positions in the convolution kernel. As we can see in Fig. 1(b), some objects are picked. But one person is missed detection. Especially, A book is mistaken for a handbag. The surroundings of the object to be detected in the picture have complex and rich contextual information often in actual life scenarios, weighting the features of the target area can locate on the features to be detected better.

In this paper, we introduce a model, which is used for capturing the contextual information features and combining it with visual features for a better performance on object detection. We take YOLOv3 as the main network of our model, then we add an attention module for capturing key features. And In summary, the main contributions of this paper are summarized as follows:

(1) We propose a object detection model based on attention and test the improvement of YOLOv3 detection effect by several different attention mechanisms.
(2) We Improved the ability of YOLOv3 to extract and screen key features without affecting real-time performance.
(3) We comprehensively evaluate our method on PASCAL VOC and MS-COCO datasets and get better performance than the baseline method (YOLOv3).

2 Related Work

2.1 Attention Mechanism

Recently, the research on deep learning is growing esoteric, and many breakthroughs have been made in various fields. Neural networks based on attention have become a hot topic in recent object detection research. The Attention Mechanism in deep learning is similar to the attention mechanism of human vision, which is to focus attention on important points in many information, select key information, and ignore other unimportant information. The results are usually displayed in the form of probability maps or probability feature vectors. In principle, it is mainly divided into spatial attention models and channel attention models. Space and channel mixed attention model. Not all regions in the image contribute equally to the task. Only focus on task-

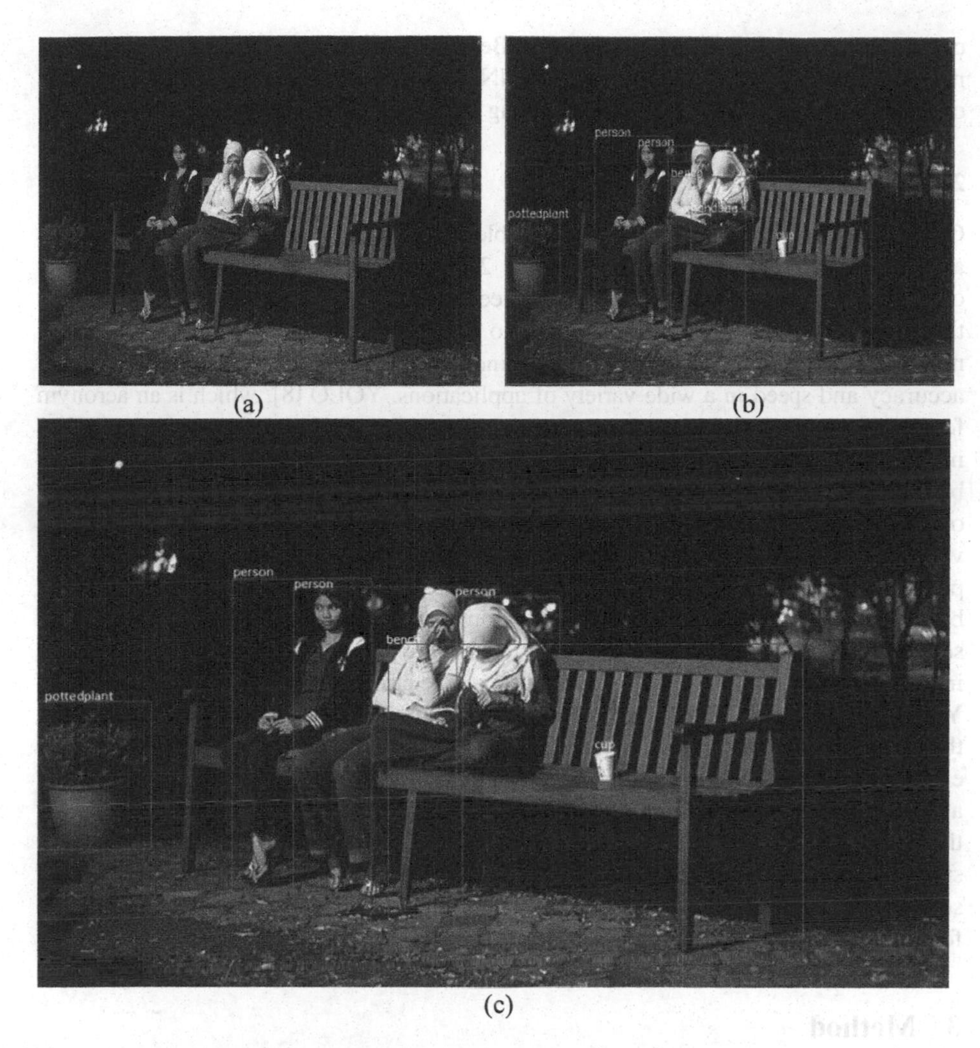

Fig. 1. The description of our motivation. (a) The image from the COCO validation set. (b) The output from YOLOv3. (c) The results of our method.

related regions, such as the main object of a picture. As each channel of a feature map is considered as a feature detector [18], channel attention focuses on 'what' is meaningful given an input image. Zhou et al. [19] suggest to use it to learn the extent of the target object effectively and Hu et al. [20] adopt it in their attention module to compute spatial statistics. Different from the channel attention, the spatial attention focuses on 'where' is an informative part, which is complementary to the channel attention. Given an input image, two attention modules, channel and spatial, compute complementary attention, focusing on 'what' and 'where' respectively. Considering this, two modules can be placed in a parallel or sequential manner. Woo et al. [17] proposed the convolutional block attention module (CBAM), which show consistent improvements in classification and detection performances. Because CBAM is a lightweight and general module, it can be integrated into any CNN architectures seamlessly with negligible overheads and is end-to-end trainable along with base CNNs.

2.2 YOLOv3

Object detection is an old fundamental problem in image processing, for which various approaches have been applied. But since 2012, deep learning techniques markedly outperformed classical ones. While many deep learning algorithms have been tested for this purpose in the literature, we chose to focus on one recent cutting-edge neural network architectures, namely YOLOv3, since it is proved to be successful in terms of accuracy and speed in a wide variety of applications. YOLO [8], which is an acronym for You Only Look Once, does not extract region proposals, but processes the complete input image only once using a fully convolutional neural network that predicts the bounding boxes and their corresponding class probabilities, based on the global context of the image. The first version was published in 2016. Later on in 2017, a second version YOLOv2 [12] was proposed, which introduced batch normalization, a retuning phase for the classifier network, and dimension clusters as anchor boxes for predicting bounding boxes. Finally, in 2018, YOLOv3 improved the detection further by adopting several new features, Contrary to Faster R-CNN's approach, each ground-truth object in YOLOv3 is assigned only one bounding box prior. These successive variants of YOLO were developed with the objective of obtaining a maximum mAP while keeping the fastest execution which makes it suitable for real-time applications. Special emphasis has been put on execution time so that YOLOv3 is equivalent to state-of-the-art detection algorithms like SSD in terms of accuracy but with the advantage of being three times faster [13]. The unified architecture is much faster compared to industry standards. The base YOLO model processes images in real-time at 45 frames per second. Since YOLO has faster inference times, this algorithm is possibly a good suit for real-time object detection.

3 Method

3.1 Network Architecture

In this section, the architecture of the object detection method proposed in this paper is shown in Fig. 2. We only integrate attention module within a ResBlock, through the screening of the transfer features, the information retained during the fusion of the residuals is more conducive to the reduction of training loss, and is conducive to the accuracy of positioning and classification. Our model makes prediction, like YOLOv3, at three scales. the results from by downsampling the size of the input image by 32, 16 and 8 separately.

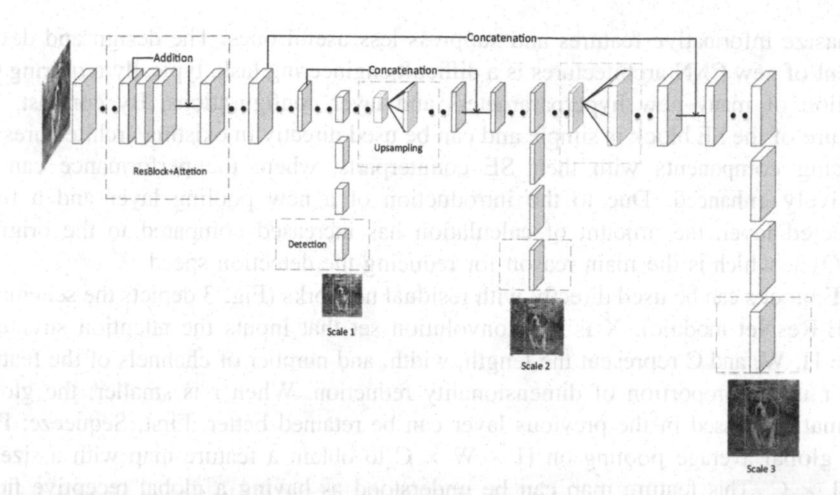

Fig. 2. The framework of our method.

3.2 Channel-Wise Attention

Hu et al. [20] proposed a mechanism that allows the network to perform feature recalibration, through which it can learn to use global information to selectively

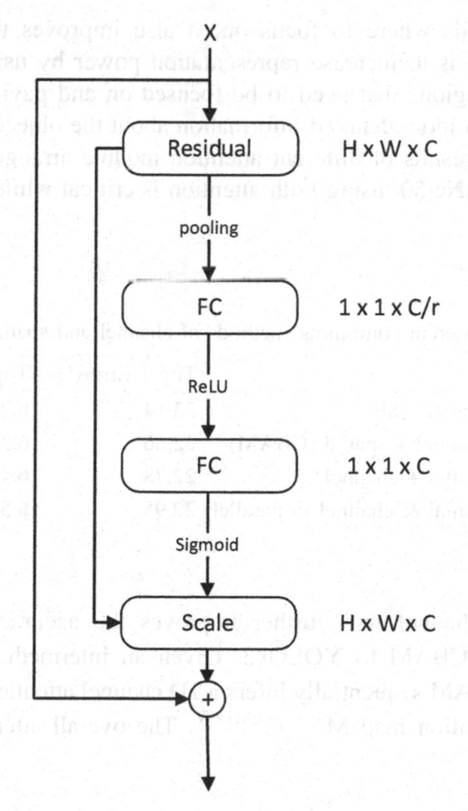

Fig. 3. The schema of the SE-ResNet module.

emphasize informative features and suppress less useful ones. The design and development of new CNN architectures is a difficult engineering task, typically requiring the selection of many new hyperparameters and layer configurations. By contrast, the structure of the SE block is simple and can be used directly in existing architectures by replacing components with their SE counterparts, where the performance can be effectively enhanced. Due to the introduction of a new pooling layer and a fully connected layer, the amount of calculation has increased compared to the original YOLOv3, which is the main reason for reducing the detection speed.

SE blocks can be used directly with residual networks (Fig. 3 depicts the schema of an SE-ResNet module). X is the convolution set that inputs the attention structure, where H, W, and C represent the length, width, and number of channels of the feature map, r is the proportion of dimensionality reduction. When r is smaller, the global information passed in the previous layer can be retained better. First, Sequeeze: Perform global average pooling on $H \times W \times C$ to obtain a feature map with a size of $1 \times 1 \times C$. This feature map can be understood as having a global receptive field. Then, Excitation: Use a fully connected neural network to perform a non-linear transformation on the results after Sequeeze. Finally, Feature recalibration: Use the results obtained by Excitation as weights and multiply the input features.

3.3 Convolutional Block Attention

Attention not only tells where to focus on, it also improves the representation of interests. Our purpose is to increase representation power by using attention mechanism: obtaining the regions that need to be focused on and paying more attention to these regions to obtain more detailed information about the objects. Table 1 show that Woo et al. [17] Test results of different attention module arrangement order on classification network ResNet50, using both attention is critical while the best-combining strategy.

Table 1. Different combining methods of channel and spatial attention.

Description	Top-1 Error(%)	Top-5 Error(%)
ResNet50 + channel (SE)	23.14	6.70
ResNet50 + channel + spatial (CBAM)	22.66	6.31
ResNet50 + spatial + channel	22.78	6.42
ResNet50 + spatial & channel in parallel	22.95	6.59

(i.e. sequential, channel-first) further improves the accuracy. We refer to this sequence and added CBAM to YOLOv3. Given an intermediate feature map $X \in \mathbb{R}^{C \times H \times W}$ as input, CBAM sequentially infers a 1D channel attention map $M_c \in \mathbb{R}^{C \times 1 \times 1}$ and a 2D spatial attention map $M_s \in \mathbb{R}^{1 \times H \times W}$. The overall attention process can be summarized as:

$$X' = M_c(X) \otimes X \tag{1}$$

$$X'' = M_s(X') \otimes X' \tag{2}$$

where $\otimes$ denotes element-wise multiplication. During multiplication, the attention values are broadcasted (copied) accordingly: channel attention values are broadcasted along the spatial dimension, and vice versa. X'' is the final refined output.

One can seamlessly integrate CBAM in any CNN architectures and jointly train the combined CBAM-enhanced networks. Figure 4 shows the exact position of module when integrated within a ResBlock. We apply CBAM on the convolution outputs in each block.

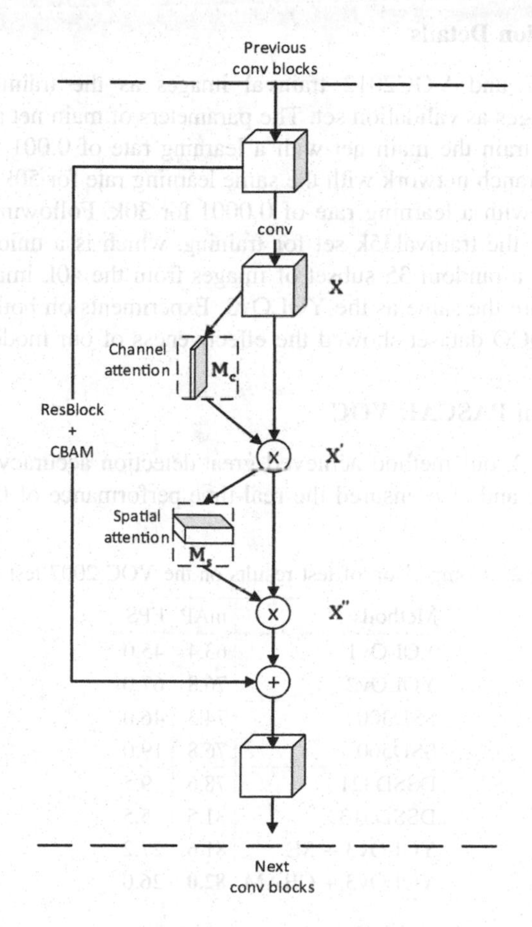

Fig. 4. CBAM integrated with a ResBlock in ResNet [21].

4 Experiments

4.1 Datasets

Our model is comprehensively evaluated on two well-known datasets, PASCAL VOC [14] and MS-COCO [15]. PASCAL VOC is a dataset which involves 20 categories. In object detection task, the performance is evaluated by mean average precision (mAP). We trained our model on VOC2007 and VOC2012. Besides, PASCAL VOC is consist of trainval images and test images, each of which contain about 5k images. PASCAL VOC needs to calculate the corresponding accuracy and recall rate when IoU = 0.5, and finally get the corresponding mAP. Different from PASCAL VOC, MS-COCO is a lager and more challenging dataset which focus on computer vision tasks. MS-COCO involves 80 categories. MS-COCO AP is averaged over multiple IoU thresholds between 0.5 (coarse localization) and 0.95 (perfect localization).

4.2 Implementation Details

We take VOC2007 and VOC2012 trainval images as the training set and take VOC20007 test images as validation set. The parameters of main net are pre-trained on ImagNet. First, we train the main net with a learning rate of 0.001 for 80k iteration. Then we train the branch network with the same learning rate for 50k. At last, we train the whole network with a learning rate of 0.0001 for 30k. Following the protocol in MS-COCO, we use the trainval35k set for training, which is a union of 80k images from train split and a random 35 subset of images from the 40k image val split. The parameter settings are the same as the YOLOv3. Experiments on both PASCAL VOC dataset and MS-COCO dataset showed the effectiveness of our model.

4.3 Evaluation on PASCAL VOC

As shown in Table 2, our method achieved great detection accuracy in the one-stage detection algorithm, and also ensured the real-time performance of the model.

Table 2. Comparison of test results on the VOC 2007 test set.

Methods	mAP	FPS
YOLOv1	63.4	45.0
YOLOv2	76.8	67.0
SSD300	74.3	46.0
SSD500	76.8	19.0
DSSD321	78.6	9.5
DSSD513	81.5	5.5
YOLOv3 + SE	81.6	27.2
YOLOv3 + CBAM	**82.0**	26.0

4.4 Evaluation on MS-COCO

According to Table 3, our method achieves 35.7% on test-dev score when the scale of input images is 608 × 608, which is 2.7% higher than the AP of YOLOv3. And our method maintains a high Frames Per Second (FPS) as well.

Table 3. Comparative results on MS-COCO test-dev set.

Methods	Avg. Precision, IoU:			FPS
	0.5:0.95	0.5	0.75	
YOLOv3	33.0	57.9	34.4	19.8
YOLOv3 + SE	34.5	58.3	34.8	16.2
YOLOv3 + CBAM	**35.7**	**58.6**	**35.6**	15.3

5 Conclusion

In this paper, we proposed a real-time object detection model, which is based on attention. In view of the disadvantages of low recall rate and insufficient positioning accuracy in YOLOv3, we tried three different attention mechanisms for the detection of YOLOv3. Especially the spatial attention and channel attention fusion module (CBAM) achieved the highest improvement effect on MS-COCO test dataset. Real-time and accurate object detection is suitable for real life, and improving these algorithms has the value of practical application.

Acknowledgement. This research was supported by the National Natural Science Foundation of China (Nos. 61672203, 61976079 & U1836102) and Anhui Natural Science Funds for Distinguished Young Scholar (No. 170808J08).

References

1. Wu, Q., Shen, C., Wang, P., Dick, A., van den Hengel, A.: Image captioning and visual question answering based on attributes and external knowledge. IEEE Trans. Pattern Anal. Mach. Intell. **40**(6), 1367–1381 (2018)
2. He, K., Gkioxari, G., Dollár, P., Girshick, R.: Mask R-CNN. In: ICCV, pp. 2980–2988. IEEE (2017)
3. Kang, K., et al.: T-CNN: tubelets with convolutional neural networks for object detection from videos. IEEE Trans. Circ. Syst. Video Technol. **28**(10), 2896–2907 (2018)
4. Girshick, R.: Fast R-CNN. In: International Conference on Computer Vision, pp. 1440–1448 (2015)
5. Ren, S., He, K., Girshick, R., Sun, J.: Faster R-CNN: towards real-time object detection with region proposal networks. In: Advances in Neural Information Processing Systems, pp. 91–99 (2015)
6. Lin, T.Y., Dollár, P., Girshick, R., et al.: Feature pyramid networks for object detection. In: Proceedings of the IEEE Conference on Computer Vision and Pattern Recognition, pp. 2117–2125 (2017)

7. Liu, W., et al.: SSD: Single Shot MultiBox Detector. In: Leibe, B., Matas, J., Sebe, N., Welling, M. (eds.) ECCV 2016. LNCS, vol. 9905, pp. 21–37. Springer, Cham (2016). https://doi.org/10.1007/978-3-319-46448-0_2
8. Redmon, J., Divvala, S., Girshick, R., et al.: You only look once: unified, real-time object detection. In: Proceedings of the IEEE Conference on Computer Vision and Pattern Recognition, pp. 779–788 (2016)
9. Shen, Z., Liu, Z., Li, J., et al.: DSOD: learning deeply supervised object detectors from scratch. In: IEEE International Conference on Computer Vision, pp. 1919–1927 (2017)
10. Huang, G., Liu, Z., et al.: Densely connected convolutional networks. In: IEEE Conference on Computer Vision and Pattern Recognition, pp. 2261–2269 (2017)
11. Fu, C.Y., Liu, W., Ranga, A., et al.: DSSD: deconvolutional single shot detector. arXiv preprint arXiv:1701.06659 (2017)
12. Redmon, J., Farhadi, A.: YOLO9000: better, faster, stronger. In: Computer Vision and Pattern Recognition, pp. 6517–6525. IEEE (2017)
13. Redmon, J., Farhadi, A.: YOLOv3: An incremental improvement. arXiv preprint arXiv: 1804.02767 (2018)
14. Zhu, F., Li, H., Ouyang, W., Yu, N., Wang, X.: Learning spatial regularization with image-level supervisions for multi-label image classification. In: Proceedings of the IEEE Conference on Computer Vision and Pattern Recognition, pp. 5513–5522 (2017)
15. Lin, T., Maire, M., Belongie, S., Hays, J., Perona, P., Ramanan, D., et al.: Microsoft COCO: common objects in context. In: ECCV, pp. 740–755 (2014)
16. He, K., Zhang, X., Ren, S., et al.: Deep residual learning for image recognition. In: Computer Vision and Pattern Recognition, pp. 770–778 (2016)
17. Woo, S., Park, J., Lee, J.Y., et al.: CBAM: convolutional block attention module. In: European Conference on Computer Vision, pp. 3–19 (2018)
18. Zeiler, M.D., Fergus, R.: Visualizing and understanding convolutional networks. In: Proceedings of European Conference on Computer Vision (ECCV) (2014)
19. Zhou, B., Khosla, A., Lapedriza, A., Oliva, A., Torralba, A.: Learning deep features for discriminative localization. In: Computer Vision and Pattern Recognition (CVPR) (2016)
20. Hu, J., Shen, L., Sun, G.: Squeeze-and-excitation networks. arXiv preprint arXiv:1709.01507 (2017)
21. He, K., Zhang, X., Ren, S., Sun, J.: Deep residual learning for image recognition. In: Proceedings of Computer Vision and Pattern Recognition (CVPR) (2016)

Image Super-Resolution Network Based on Prior Information Fusion

Cheng Ding, Wei-Dong Tian(✉), and Zhong-Qiu Zhao

College of Computer and Information, Hefei University of Technology, Hefei, China
2512673687@qq.com

Abstract. Research in the field of image super-resolution in recent years has shown that convolutional neural networks are conducive to improving the quality of image restoration. In the deep network, simply increasing the number of network layers cannot effectively improve the quality of image restoration, but increases the difficulty of training. Therefore, in this paper, we propose a new model. By using multi-layer convolution, the image segmentation map based on image texture is modulated into the network, and the attention mechanism is used to adjust the feature output of each layer. The output of each layer is used as hierarchical feature for global feature fusion. Finally, the attention mechanism is used to fuse the hierarchical features to improve the quality of image restoration.

Keywords: Image-resolution · Convolutional neural network · Prior information · Attention mechanism

1 Introduction

With the development of information technology, the field of artificial intelligence has been widely developed, and a lot of research has focused on computer vision [1, 2]. Single image super-resolution is a technology for obtaining a high-resolution image from one low-resolution image. This technology aims to provide image with better visual effect and dig out more image details. In the process of image acquisition, due to various restrictions on imaging conditions and methods, the imaging system cannot acquire all the information in the original image scene. How to improve the spatial resolution of images has always been a hot issue in the field of image processing. Image super-resolution is considered to be an effective method to solve this problem. Since there are multiple high-resolution solutions for a given low-resolution image, this type of problem is ill-posed. To solve this problem, researchers have proposed traditional interpolation-based algorithms [3, 4], reconstruction-based algorithms [5], and learning-based algorithms [6, 7]. These methods have not achieved good results.

In recent years, convolutional neural networks have proven their effectiveness in image super-resolution. By defining a convolutional neural network, the gradient descent algorithm is used to learn the correlation and correspondence between the two pairs of high-resolution and low-resolution images. The SRCNN [8] proposed by Dong et al., for the first time, applies a convolutional neural network to super-resolution

D.-S. Huang and P. Premaratne (Eds.): ICIC 2020, LNAI 12465, pp. 51–61, 2020.
https://doi.org/10.1007/978-3-030-60796-8_5

tasks, and it is better than the traditional methods in the past. Subsequently, many researchers began to gradually deepen the network to improve the quality of image restoration. With the introduction of the GAN [9] network, Perceptual loss that is more conducive to measuring image quality from a feature point of view is introduced, which eases blurring and oversmoothing artifacts and makes the image closer to reality.

However, all the above methods improve the quality of image restoration by increasing the depth of the network, which means simply increasing the computational complexity in exchange for the improvement of image quality. By analyzing some classic network structures, we find the following problems:

(1) **Unused image prior information.** Most convolutional neural network models continuously deepen the network and adjust the network structure to obtain better image restoration quality, but the prior information such as texture, brightness, and color contained in the image itself also affects image restoration. Without inputting stronger prior information, the existing methods cannot further dig out more useful information.
(2) **Equally process every detail of the image.** An image often contains multiple contents. The texture and other features contained in these contents have different characteristics. Treating the details of each image equally is not conducive to improving the restoration quality of the image.
(3) **Simple concatenation of the feature output of each layer.** Most classic neural network models treat the feature of each layer equally, and fuse deep and shallow features through simple concatenation. However, for different textures, suitable features need to be filtered to improve the quality of image restoration.

In order to solve the mentioned problems, we propose a new model based on prior information fusion, the main contributions of this paper are summarized as follows:

(1) We propose a new model based on prior information fusion. The segmentation map containing different texture feature information of the image is input into the network to guide the learning process of the network. The experimental results show the effectiveness of the proposed method.
(2) We apply different processing methods to different textures of the image by fusing prior information and applying multi-scale convolution kernels.
(3) We use the attention mechanism to assign weights to the features of each layer, and select feature layer that is more effective for the reconstruction part, instead of simply concatenating.

2 Related Work

2.1 Single-Image Super-Resolution

The current image super-resolution methods can be divided into three categories: interpolation-based, reconstruction-based, and learning-based. Traditional single-image super-resolution methods include interpolation-based algorithms such as bicubic interpolation and Lanczos resampling algorithms; reconstruction-based algorithms such as iterative back projection, maximum posterior probability, convex set projection; and learning-based Algorithms, such as neighborhood and local linear embedding methods, sparse coding and sparse coding networks. These methods either use the internal similarity of a single image or learn the mapping function of external low-resolution and high-resolution sample pairs. Although they focus on learning and optimizing the dictionary, the remaining steps of the above methods are rarely optimized or unified in a unified optimization framework consider.

Since the convolutional neural network was proposed, it has been used in most researches in the field of image processing [10–12]. SRCNN published by Dong et al. applies deep learning networks for the first time in the field of image super-resolution and is better than the traditional methods. Convolutional neural networks usually get better with deeper networks, but deeper networks are often difficult to train. VDSR [13] and DRCN [14] introduced the residual learning, and DRRN [15] solved the problems of difficult training and difficult convergence by introducing recursive blocks, and achieved significant accuracy improvements. However, all the above methods are to interpolate the image first, inevitably losing the details and increasing the amount of calculation. Extracting features from the original LR and improving the spatial resolution at the end of the network have become the main choices for deep convolutional network structures. FSRCNN [16] uses a deconvolution operation, and ESPCN [17] uses a sub-pixel convolution operation. After extracting features from low resolution images and then amplifying them, the speed is improved. For deeper networks, SRRESNT provides the solution. GAN networks have also been introduced in the direction of image super-resolution in recent years. SRGAN [18] network has brought Perceptual loss which is more conducive to measure image quality from features. EnhanceNet [19] introduced a GAN-based network model that combines automatic texture synthesis and perceptual loss. Although SRGAN and EnhanceNet alleviate blurring and oversmoothing artifacts to a certain extent, the results they get still feel unreal. By removing unnecessary modules in the traditional residual network, EDSR and MDSR [20] are obtained, and the effect is significantly improved.

3 Proposed Method

The complete structure of the network model proposed in this paper is shown in Fig. 1. We input the unprocessed low-resolution image I_{lr} directly into the network, and first extract the image features through a convolution layer with a kernel size of 3 × 3. The obtained image features and image prior information are simultaneously input into the prior information fusion layer, and the output of this layer contains the prior information of the image. Subsequently, MFEB (Multiple-scale Feature Extraction Block) with multiple kernel sizes will extract multi-scale features in the feature map. The output is divided into two paths, the first path goes deeper into the network, and the second path outputs hierarchical features, which will be used for the final hierarchical feature attention layer. Here we use 16 such MFEB layers. Hierarchical feature attention layer will be used as the last step in the feature extraction part. The attention mechanism will be applied to the hierarchical feature fusion instead of concatenation, allowing the model to focus more on information that is beneficial to improving the quality of recovery and reducing some redundant information. In the reconstruction part, we used a deconvolution layer to complete the upsampling process, and finally got I_{sr}.

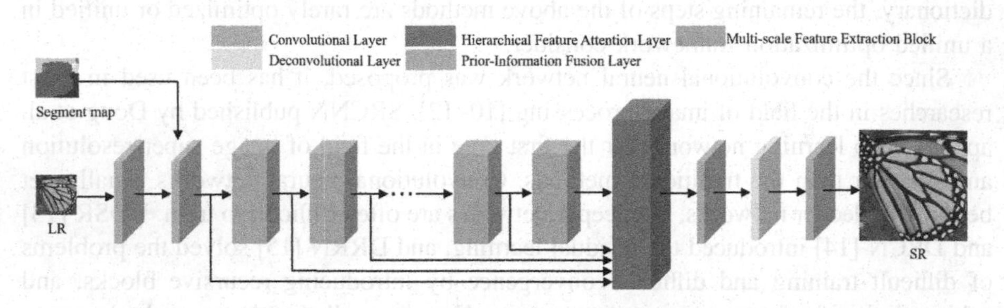

Fig. 1. The complete structure of the proposed network.

In recent years, there have been many researches on Loss function, including L_1, L_2 norm, perceptual loss, etc., and they have been introduced into the field of image super-resolution.

Compared with the L_1 loss, the L_2 loss assumes that the data conforms to a Gaussian distribution, but in fact the data usually has multiple peaks. The L_2 loss will make the model satisfy multiple peaks at the same time and produce an intermediate value, resulting in blurred images. At the same time, considering the efficiency of the network, we choose the L_1 loss function. Given N image pairs $\{I_{lr}^i, I_{hr}^i\}$, the L_1 loss function shows as following:

$$L(\theta) = \frac{1}{N}\sum_{i=1}^{N} \left\| I_{SR}^i - I_{HR}^i \right\|_1 \quad (1)$$

where θ denotes parameters of the network, and optimizes this parameter to find the final model. N represents the number of image pairs in the training data, and i represents the i-th optimization iteration. I_{SR}^i and I_{HR}^i represent the image generated by the network and ground truth respectively.

3.1 Priori-Information Fusion Layer

In this section we will describe the prior information fusion layer in detail. The main role of this layer is to fuse the image segmentation information obtained in advance into the image. The complete structure diagram is shown in Fig. 2.

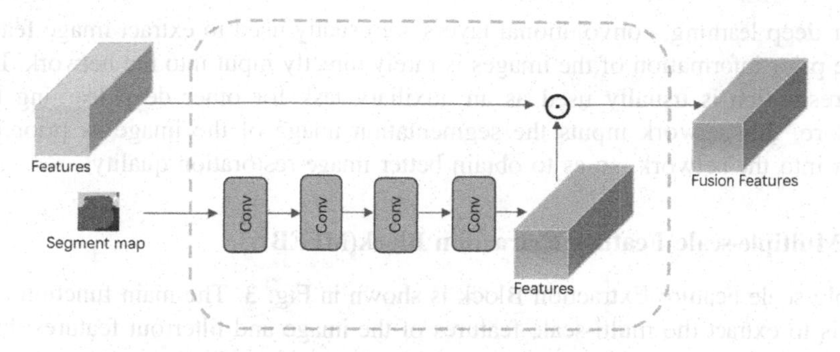

Fig. 2. Priori-information fusion layer

First, in order to obtain a priori information, we use the K-means clustering method to segment the image according to a certain texture distribution to obtain a texture-based segmentation map.

Then, we use 4 layers of convolutional layers to process the obtained segmentation map, obtain the prior information result with the same number of channels as the image features, and multiply it with the features.

In this way, we obtain the features that incorporate the prior information. And input it to the next layer for feature extraction. In this layer, all we use is a convolution with a kernel size of 1×1, setting the stride to 1, to ensure that the size of the obtained feature map remains the same.

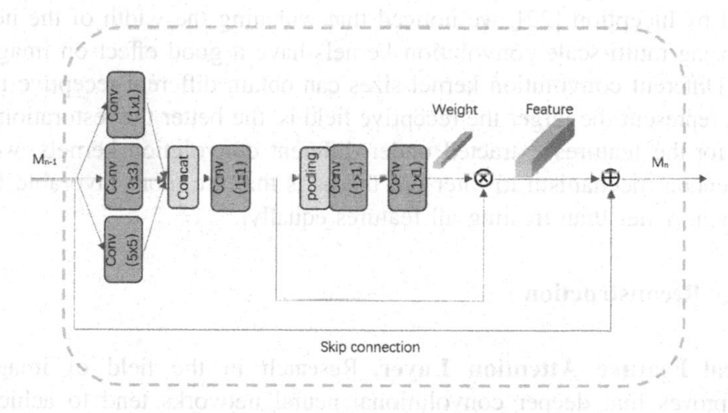

Fig. 3. Multiple-scale feature extraction block.

Traditional image super-resolution reconstruction methods usually use different prior images of natural images to estimate different original images. The prior information of common natural images has the characteristics of local smoothness, non-local self-similarity, and sparseness of natural images.

For deep learning, convolutional layers are usually used to extract image features, but the prior information of the images is rarely directly input into the network. Image super-resolution is usually used as an auxiliary task for other deep learning tasks. Therefore, this network inputs the segmentation image of the image as prior information into the network, so as to obtain better image restoration quality.

3.2 Multiple-scale Feature Extraction Block(MFEB)

Multiple-scale Feature Extraction Block is shown in Fig. 3. The main function of this block is to extract the multi-scale features of the image and filter out features that are beneficial to image reconstruction through the attention mechanism.

M_{n-1} denotes the output of the previous layer of MFEB. First use 3 convolutions with different kernel sizes: 1×1, 3×3, 5×5, and then concatenate the obtained features. Since the dimension of the features is three times the original, we use 1×1 convolution to reduce the features dimensions.

Then, we use the attention mechanism for feature selection. Inspired by RCAN [21], we first use a mean pooling feature to obtain a feature map with a size of 1×1 and a number of channels C, and then use two 1×1 convolution layers to reduce and increase the number of channels of the feature, thereby weight coefficient of size $1 \times 1 \times C$ is obtained.

Then multiply the obtained weight coefficient with the feature map. In this way, we have a feature map with weight coefficients, which can filter out features that are more conducive to image reconstruction under different scales of convolution kernels, and more conducive to image texture restoration.

Inspired by Inception [22], we noticed that widening the width of the neural network and using multi-scale convolution kernels have a good effect on image feature extraction. Different convolution kernel sizes can obtain different receptive fields, but they do not represent the larger the receptive field is, the better the restoration effect is. Therefore, for the features extracted under different convolution kernels, we use the channel attention mechanism to filter out the parts that are more favorable for image reconstruction, rather than treating all features equally.

3.3 Image Reconstruction

Hierarchical Feature Attention Layer. Research in the field of image super-resolution proves that deeper convolutional neural networks tend to achieve better experimental results, because deeper networks can often extract higher-level abstract features of the image, thereby recovering images with more texture.

However, these models only focus on deep features, ignoring the importance of shallow features for image restoration. Therefore, we use hierarchical feature attention layers to fuse the output of each layer of the MFEB and assign corresponding weights to screen out features that are more conducive to improving the quality of image restoration. The method adopted by the hierarchical feature attention layer is the same as the channel attention in MFEB.

Deconvolution Layer. In SRCNN, bicubic interpolation is used to first upsample the image to the required image size and perform feature extraction on larger images, but this will lead to a significant increase in computational complexity and more redundant information. The deconvolution operation in FSRCNN puts the upsampling part into the feature extraction and then directly extracts the features on the low-resolution image, reducing most of the calculation, so this model uses deconvolution for the upsampling operation.

4 Experiments

In this section, we will evaluate the performance of the model on several benchmark datasets and compare it with state-of-art methods. First, we will introduce the data set used, then we will detail the parameter settings and experimental details, and finally show the experimental results on the data set.

4.1 Datasets

Set5, set14, BSDS100, Urban100 and Manga109 is a common data set commonly used in the field of image super-resolution, and the image contains animals, people, plants, buildings and other types. We briefly introduce the content of each data set and other information, as shown in Table 1.

Table 1. Datasets used in our experiments.

Dataset	Amount	Contents
Set5 [23]	5	Human, bird, butterfly, etc.
Set14 [24]	14	Humans, animals, insects, etc.
BSDS100	100	Animal, building, food, etc.
Urban100 [25]	100	City, urban, etc.
Manga109 [26]	109	Manga volume

4.2 Parameters Setting

For the dataset, we augment the training dataset with operations such as scaling, rotation, and inversion. The training rate is set to 0.0001, and our model is optimized using stochastic gradient descent. In the final model, we used 16 MFEB blocks, and the output of each block was 64 feature maps. Padding was applied to the convolution to ensure that the size of the output feature map was consistent. The experiments were trained and tested using the Pytorch framework.

4.3 Evaluation on Datasets

To prove the effectiveness of our method, we tested the above dataset and compared it with the state-of-art method. The experimental results are shown in Table 2, 3, 4. PSNR is the peak signal-to-noise ratio and SSIM is the structural similarity.

From the table, we can see that the PSNR/SSIM of our experimental results on the 3 magnification scales have a certain improvement compared to the existing classic models LapSRN, VDSR, FSRCNN, etc., which verifies the effectiveness of our method.

From the results in the table, we can see that our proposed model has a relatively large improvement on the Urban100 dataset compared to the results of several other network experiments. The Urban100 dataset contains a large number of regular building structures. Through the prior image segmentation operation, it is possible to effectively segment the parts with uniform texture regularity. Therefore, the model has a good effect on image restoration with regular textures. However, the processing effect on other images including a large number of different texture blocks such as portraits and scenes needs to be improved (Fig. 4).

Fig. 4. The image of the experimental result with the scale of 2.

Table 2. Experimental results (x2).

Algorithm	Scale	Set5 PSNR/SSIM	Set14 PSNR/SSIM	BSDS100 PSNR/SSIM	Urban100 PSNR/SSIM	Manga100 PSNR/SSIM
ours	x2	**37.78/ 0.9589**	**33.51/ 0.9178**	**31.85/ 0.8902**	**31.59/ 0.9299**	**38.38/ 0.9762**
LapSRN	x2	**37.49/ 0.9562**	**33.01/ 0.9108**	**31.80/ 0.8899**	**30.41/ 0.9102**	**37.22/ 0.9759**
VDSR	x2	**37.52/ 0.9587**	**33.05/ 0.9121**	**31.90/ 0.8955**	**30.75/ 0.9143**	**37.19/ 0.9737**
FSRCNN	x2	**36.99/ 0.9558**	**32.59/ 0.9079**	**31.52/ 0.8909**	**29.79/ 0.9007**	**36.62/ 0.9694**
ESPCN	x2	*36.98/ 0.9545*	*32.42/ 0.9077*	*31.49/ 0.8925*	*29.78/ 0.9032*	*36.59 /0.9687*
SRCNN	x2	*36.65/ 0.9539*	*32.35/ 0.9059*	*31.39/ 0.8869*	*29.51/ 0.8951*	*35.67/ 0.9661*
Bicubic	x2	*33.65/ 0.9282*	*30.29/ 0.8669*	*29.57/ 0.8431*	*26.85/ 0.8429*	*30.79 /0.9333*

Table 3. Experimental results (x3).

Algorithm	Scale	Set5 PSNR/SSIM	Set14 PSNR/SSIM	BSDS100 PSNR/SSIM	Urban100 PSNR/SSIM	Manga100 PSNR/SSIM
ours	x3	**34.01/ 0.9251**	**30.01/ 0.8387**	**28.93/ 0.8081**	**27.31/ 0.8425**	**32.67/ 0.9412**
LapSRN	x3	**33.82/ 0.9219**	**29.85/ 0.8316**	**28.81/ 0.7979**	**27.05/ 0.8285**	**32.11/ 0.9301**
VDSR	x3	**33.65/ 0.9201**	**29.75/ 0.8301**	**28.83/ 0.7969**	**27.10/ 0.8289**	**32.00/ 0.9286**
FSRCNN	x3	**33.18/ 0.9140**	**29.39/ 0.8245**	**28.55/ 0.7933**	**26.41/ 0.8160**	**30.79 /0.9210**
ESPCN	x3	*33.01/ 0.9129*	*29.43/ 0.8255*	*28.22/ 0.7927*	*26.39/ 0.8159*	*30.78/ 0.9159*
SRCNN	x3	*32.41/ 0.9055*	*29.11/ 0.8199*	*28.29/ 0.7829*	*26.23/ 0.8021*	*30.49/ 0.9100*
Bicubic	x3	*30.39/ 0.8649*	*27.51/ 0.7715*	*27.11/ 0.7381*	*24.39/ 0.7310*	*26.95/ 0.8551*

Table 4. Experimental results (x4).

Algorithm	Scale	Set5 PSNR/SSIM	Set14 PSNR/SSIM	BSDS100 PSNR/SSIM	Urban100 PSNR/SSIM	Manga100 PSNR/SSIM
ours	x4	**31.79/ 0.8837**	**28.48/ 0.7757**	**27.41/ 0.7251**	**26.37/ 0.7761**	**29.76/ 0.9001**
LapSRN	x4	**31.42/ 0.8799**	**28.18/ 0.7629**	**27.29/ 0.7149**	**25.19/ 0.7559**	**29.01/ 0.8829**
VDSR	x4	**31.31/ 0.8789**	**28.09/ 0.7625**	**27.19/ 0.7165**	**25.11/ 0.7541**	**28.79/ 0.8799**
FSRCNN	x4	**30.61/ 0.8599**	**27.65/ 0.7469**	**26.89/ 0.7011**	**24.59/ 0.7279**	**27.79/ 0.8511**
ESPCN	x4	*30.59/ 0.8643*	*27.72/ 0.7559*	*26.91/ 0.7109*	*24.59/ 0.7359*	*27.59/ 0.8549*
SRCNN	x4	*30.41/ 0.8611*	*27.49/ 0.7489*	*26.99/ 0.7089*	*24.49/ 0.7221*	*27.59/ 0.8501*
Bicubic	x4	*28.41/ 0.8022*	*26.01/ 0.6959*	*25.97/ 0.6591*	*23.11/ 0.6587*	*24.88/ 0.7815*

5 Conclusion

In this paper, we propose an image super-resolution network that incorporates prior information, incorporates image segmentation maps obtained by clustering methods into the network, and uses the attention mechanism to improve the quality of image restoration. Experiments show that our method has good performance.

Acknowledgement. This research was supported by the National Natural Science Foundation of China (Nos. 61672203, 61976079 & U1836102) and Anhui Natural Science Funds for Distinguished Young Scholar (No. 170808J08).

References

1. Zhao, Z.Q., Xu, S.T., Liu, D., et al.: A review of image set classification. Neurocomputing **335**, 251–260 (2018)
2. Zhao, Z.Q., Glotin, H.: Diversifying image retrieval by affinity propagation clustering on visual manifolds. IEEE Multimedia **16**(99), 1 (2009)
3. Baghaie, A., Yu, Z.: Structure tensor based image interpolation method. AEU-Int. J. Electron. Commun. **69**(2), 515–522 (2015)
4. Chu, J., Liu, J., Qiao, J., Wang, X., Li, Y.: Gradient-based adaptive interpolation in super-resolution image restoration. In: 2008 9th International Conference on Signal Processing, vol. 415, pp. 1027–1030. IEEE (2008)
5. Elad, M., Hel-Or, Y.: A fast super-resolution reconstruction algorithm for pure translational motion and common space-invariant blur. IEEE Trans. Image Process. **10**(8), 1187–1193 (2001)
6. Yang, J., Wright, J., Huang, T.S., Ma, Y.: Image super-resolution via sparse representation. IEEE Trans. Image Process. **19**(11), 2861–2873 (2010)

7. Sahraee-Ardakan, M., Joneidi, M.: Joint dictionary learning for example based image super-resolution, arXiv preprint arXiv:1701.03420
8. Dong, C., Loy, C.C., He, K., Tang, X.: Image super-resolution using deep convolutional networks. IEEE Trans. Pattern Anal. Mach. Intell. **38**(2), 295–307 (2015)
9. Goodfellow, I.,et al.: Generative adversarial nets. In: Advances in Neural Information Processing Systems, vol. 445, pp. 2672–2680 (2014)
10. Zhao, Z., Wu, X., Lu, C., Glotin, H., Gao, J.: Optimizing widths with PSO for center selection of Gaussian radial basis function networks. Sci. China Inf. Sci. **57**(5), 1–17 (2013). https://doi.org/10.1007/s11432-013-4850-5
11. Zhao, Z.Q., Gao, J., Glotin, H., et al.: A matrix modular neural network based on task decomposition with subspace division by adaptive affinity propagation clustering. Appl. Math. Modell. **34**(12), 3884–3895 (2010)
12. Glotin, H., Zhao, Z.Q., Ayache, S.: Efficient image concept indexing by harmonic & arithmetic profiles entropy. In: IEEE International Conference on Image Processing. IEEE (2010)
13. Kim, J., Kwon Lee, J., Mu Lee, K.: Accurate image super-resolution using very deep convolutional networks. In: Proceedings of the IEEE Conference on Computer Vision and Pattern Recognition, pp. 1646–1654 (2016)
14. Kim, J., Kwon Lee, J., Mu Lee, K.: Deeply-recursive convolutional network for image super-resolution. In: Proceedings of the IEEE Conference on Computer Vision and Pattern Recognition, pp. 1637–1645 (2016)
15. Tai, Y., Yang, J., Liu, X.: Image super-resolution via deep recursive residual network. In: Proceedings of the IEEE Conference on Computer Vision and pattern recognition, vol. 465, pp. 3147–3155 (2017)
16. Dong, C., Loy, C.C., Tang, X.: Accelerating the super-resolution convolutional neural network. In: Leibe, B., Matas, J., Sebe, N., Welling, M. (eds.) ECCV 2016. LNCS, vol. 9906, pp. 391–407. Springer, Cham (2016). https://doi.org/10.1007/978-3-319-46475-6_25
17. Shi, W., et al.: Real-time single image and video super-resolution using an efficient sub-pixel convolutional neural network. In: Proceedings of the IEEE Conference on Computer Vision and Pattern Recognition, pp. 1874–1883 (2016)
18. Ledig, C., et al.: Photo-realistic single image super-resolution using a generative adversarial network. In: Proceedings of the IEEE Conference on Computer Vision and Pattern Recognition, pp. 4681–4690 (2017)
19. Sajjadi, M.S., Scholkopf, B., Hirsch, M.: EnhanceNet: single image super resolution through automated texture synthesis. In: Proceedings of the IEEE International Conference on Computer Vision, pp. 4491–4500 (2017)
20. Lim, B., Son, S., Kim, H., Nah, S., Mu Lee, K.: Enhanced deep residual networks for single image super-resolution. In: Proceedings of the IEEE Conference on Computer Vision and Pattern Recognition Workshops, vol. 485, pp. 136–144 (2017)
21. Zhang, Y., Li, K., Li, K., Wang, L., Zhong, B., Fu, Y.: Image super-resolution using very deep residual channel attention networks. In: Proceedings of the European Conference on Computer Vision (ECCV), pp. 286–301 (2018)
22. Szegedy, C., Liu, W., Jia, Y., et al.: Going deeper with convolutions (2014)
23. Bevilacqua, M., Roumy, A., Guillemot, C., Alberi-Morel, M.L.: Low-complexity single-image super-resolution based on nonnegative neighbor embedding. In: BMVC (2012)
24. Zeyde, R., Elad, M., Protter, M.: On single image scaleup using sparse-representations. In: International Conference on Curves and Surfaces (2010)
25. Huang, J.-B., Singh, A., Ahuja, N.: Single image super resolution from transformed self-exemplars. In: CVPR (2015)
26. Fujimoto, A., Ogawa, T., Yamamoto, K., Matsui, Y., Yamasaki, T., Aizawa, K.: Manga109 dataset and creation of metadata. In: MANPU (2016)

TFPGAN: Tiny Face Detection with Prior Information and GAN

Dian Liu(✉), Zhong-Qiu Zhao, and Wei-Dong Tian

College of Computer and Information, Hefei University of Technology, Hefei, China
381475649@qq.com

Abstract. This paper addresses two challenging tasks: detecting small faces in unconstrained conditions and improving the quality of very low-resolution facial images. Tiny faces are so fuzzy that the facial patterns are not clear or even ambiguous resulting in greatly reduced detection. In this paper, we proposed an algorithm to directly generate a clear high-resolution face from a blurry small one by adopting a generative adversarial network (GAN). Besides, we also designed a prior information estimation network which extracts the facial image features, and estimates landmark heatmaps respectively. By combining these two networks, we propose the end-to-end system that addresses both tasks simultaneously, i.e. both improves face resolution and detects the tiny faces. Extensive experiments on the challenging dataset WIDER FACE demonstrate the effectiveness of our proposed method in restoring a clear high-resolution face from a blurry small one, and show that the detection performance outperforms other state-of-the-art methods.

Keywords: Face detection · Super-resolution · GAN

1 Introduction

The purpose of this paper is to realize the two tasks of face detection and face super-resolution on very low-resolution facial images. This is important in many applications, like criminal investigation law and order, comprehensive management and other related fields. Because the human face belongs to the inherent characteristics of human beings, and has the advantages of uniqueness, stability, and convenience, it can quickly identify individual information. Therefore, the researches on face detection and face super resolution have been increased in recent years and have become a hot issue. Facial images can provide an important basis for the identity of the target, so it is an indispensable part of video intelligence analysis. However, in the actual monitoring video, due to the distance of human distance monitoring and the noise in the environment, this results in lower resolution of the face area, resulting in reduced image sharpness and reduced amount of critical information. resulting in a decrease in the recognition accuracy of the existing model. Therefore, the key to the problem is whether or not to develop an efficient super-resolution reconstruction technique for the face with a lower resolution, thereby improving the definition of the face and improving the recognition accuracy of the tiny face.

D.-S. Huang and P. Premaratne (Eds.): ICIC 2020, LNAI 12465, pp. 62–73, 2020.
https://doi.org/10.1007/978-3-030-60796-8_6

In recent years, deep networks have made great progress in the field of facial feature point detection in the cascade regression framework: Sun et al. [1] proposed cascaded deep convolutional neural network (DCNN) to gradually predict facial feature points; from coarse to The thin end-to-end recursive convolution system (MDM) [2] is similar to DCNN, but each stage takes the hidden layer feature of the previous stage as input. It is divided into several parts to alleviate the changes of facial features and return the coordinates of different parts separately. Heat map regression model generates probabilistic heat map for each feature point respectively, which is excellent in face feature point detection. Newell [3] used the heat map regression model, and designing a stacked hourglass network, extracting features from multiple scales to estimate the key points of the human body pose. The stacked hourglass network can repeatedly obtain the information contained in the images at different scales, which is more suitable for human faces. Feature point detection. Yang [4] used the standardized face supervised transformation and stacked hourglass network to obtain the predicted heat map, and achieved good results, which proved the superiority of the stacked hourglass network in facial feature detection. Wu et al. [5] used a face boundary heat map instead of a face feature point heat map to express the face geometry, proving the boundary importance of the information.

However, these studies only partially solve the facial features of different types and different facial features scale problems, and excessive stack hourglass combination network will affect the speed of detection. To deal with the nuisances in face detection, we propose a unified end-to-end convolutional neural network for better face detection and face super-resolution based on the facial prior estimation network and generative adversarial network (GAN) framework. In summary, the main contributions of this paper can be summarized as follows:

1. We propose the end-to-end system that addresses face super-resolution and face detection simultaneously, via integrating a sub-network for facial landmark localization through heatmap regression into a GAN-based super-resolution network.
2. The two-segment stacked hourglass network is used to extract the feature lines of different parts of the face to solve the problem of feature line extraction caused by image occlusion and low image resolution.
3. We demonstrate the effectiveness of our proposed method in restoring a clear high-resolution face from a blurry small face, and show that the detection performance outperforms other state-of-the-art approaches on the WIDER FACE dataset, especially on the most challenging Hard subset.

2 Related Work

2.1 Face Detection

The purpose of face detection task is to find out the corresponding positions of all faces in the image, and the output of the algorithm is the coordinates of the rectangular frame outside the face in the image, which may also include information such as attitude estimation such as tilt angle. From 1960s to now, we can divide the research of face detection algorithm into three stages. In the first stage, the early face detection algorithm uses the technology based on template matching, and the representative achievement is the method proposed by Henry a Rowley and others [6, 7]. They use neural network model for face detection. The second stage is the emergence of AdaBoost framework. In 2001, Viola and Jones designed a face detection algorithm [8], which uses simple Haar-like features and cascaded AdaBoost classifiers to construct detectors. It analyzes the differences between detection and classification problems, and uses asymmetric methods to improve weak classifiers.

The third stage begins with the advent of deep learning, at present, the most widely used face detection network in deep learning is the RCNN series network. RCNN [9] was proposed by Ross in 2014. In 2015, Microsoft Research Institute proposed r-cnn based on the spatial pyramid layer of SPP net [10] Fast RCNN [11], an improved version of the algorithm, has designed a pool layer structure of ROI pooling. The task of face detection also appeared in cascade CNN [12] and MTCNN [13] are the representative cascaded neural networks. These cascaded networks are designed with multiple stages to generate multi-scale image pyramid from the input image, and alternately complete multiple tasks such as face detection and face alignment. Literature [14] on CVPR 2017 trains different detectors according to the faces in different scales in order to keep the efficiency and train the detectors in a multi task way. According to this, J. Li et al. Proposed the perception generative adversarial networks [15] to use the adversary network (large-scale feature discriminator) to judge the similarity between large-scale features generated by small-scale targets and large-scale targets, so as to improve the accuracy of small target reconstruction and the recall rate of detection. In reference [16], GAN network is directly used in small face detection to reconstruct high-resolution large-scale face from fuzzy small-scale face and further improve the performance of the detector.

2.2 Facial Prior Knowledge

Many face SR methods use face prior knowledge to better super-resolve LR faces. Early techniques assumed that the face was in a controlled environment with little change Baker and Canard [17]. It is proposed to first understand the spatial distribution of the image gradient with respect to the frontal face image Wang et al. [18]. The mapping between the LR and HR faces is achieved by the eigen transformation. Kolouri et al. [19] studied the nonlinear Lagrangian model for HR facial images and enhanced the degraded images by finding the model parameters that best fit the given LR data Yang et al. [20]. A priori facial is combined by using a mapping between specific facial components. However, the matching between components is based on the results of landmark detection. When the down sampling factor is large, these results are difficult to estimate. Recently, deep convolutional neural networks have been successfully applied to face SR tasks. The literature such as [27–31] introduce the methods based on subspace learning and manifold learning, which are widely used in the visual task of human face Zhu et al. [21]. In a cascaded framework of tasks, super parsing very LR and unaligned faces. In their framework, face illusion and dense correspondence field estimates are also optimized. In addition, different from the above-mentioned method of performing face SR in a step-by-step manner, FSRNet [22] makes full use of face-characteristic heat maps and analytical diagrams in an end-to-end training manner.

3 Proposed Method

3.1 Network Architecture

The network structure of our TFPGAN model is shown in the Fig. 1. The model can be divided into three parts: the first part can be called a rough super-resolution reconstruction network, whose purpose is to slightly recover the low-resolution small-size face, because it is difficult to extract useful face information from too blurred images, so this paper decided to first sample these images through deconvolution to ensure the recovered image which can have more characteristic information. In the first stage, the generated results will be used for two purposes: one is to directly keep it as a part of the subsequent input image; the other is to serve as the input of the face prior information estimation network. Because the stacked hourglass network [3] can repeatedly obtain the information contained in the image at different scales, this paper uses the hourglass model to build a network to extract the face prior information, The feature map is represented in the form of Heatmap as the auxiliary task and prior information of the final face detector. And the last stage is Generative Adversarial model, we try to reconstruct the high-resolution large-scale face from the fuzzy small-scale face through GAN. Here we combine the heat image information with the original image concatenation as the input of the generator. In addition to the ability of judging Sr (generated image) and HR (real sample), another branch is designed to distinguish face image from background image, and a new loss function is proposed to train the two networks together to realize end-to-end.

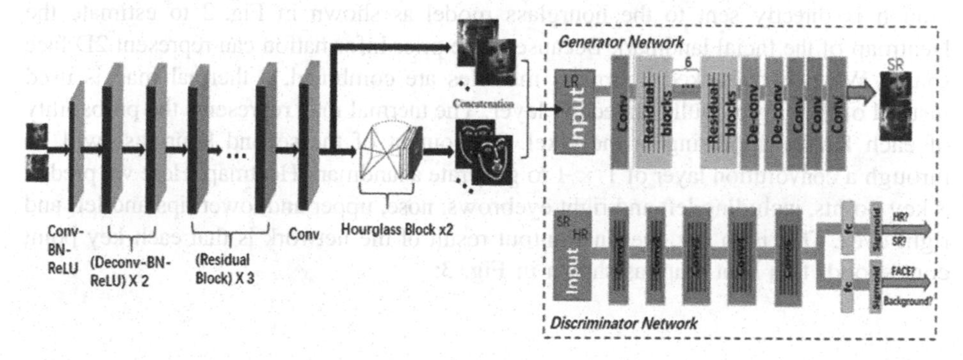

Fig. 1. The structure of our TFPGAN model.

3.2 Face Information Estimation Network

The perfect application of hourglass model [3] in human pose estimation enlightens us. When dealing with small-scale and low-resolution images, existing detectors such as MTCNN [13] often miss detection seriously. Many faces in the images will be mistakenly considered as background by the detectors and ignored. Considering this situation, this paper decided to use the network model to distinguish face region information and background region information. The main research is based on the particularity of face image. Any world object has different distribution in its shape and texture, and it is obvious that shape prior information is easier to represent than texture prior information, so this paper chooses shape information to model. Face image has many key points, such as eyes and nose, so it has geometry structure that other background pictures don't have. The most intuitionistic embodiment of these geometry structure is Heatmap. So, this paper uses the hourglass model mentioned above to build a network to extract face information, and uses the form of Heatmap to represent the feature map as the final face detector Supporting tasks and prior information. Considering that only the facial feature lines are preserved in the face image thermal image, and the key features extracted are not as many as the human body, so the facial point thermal image model only needs to use a few stacked hourglass networks to carry out point thermal image regression, this paper finds that the best effect is to use two hourglass models for the face information task, as shown in Fig. 2.

In the above figure, the small-scale low-resolution training samples will be sent to a sampling network. First, through a convolution layer with a convolution kernel size of 3×3 and a step size of 1, the size of the feature map will be kept unchanged while extracting features. Then, two 3×3 de-convolution layers are used to sample the size of the feature map 4 times to 64×64, and then three residual blocks are used to continue to extract the image depth Layer features, the final 3×3 convolution layer outputs the final image. It can be seen that the output image has a higher resolution than the original LR image. The generated course image will be input into two branches, one branch is used as the LR image of the final super-resolution network, and the other branch is directly sent to the hourglass model as shown in Fig. 2 to estimate the Heatmap of the facial landmark Because these prior information can represent 2D face shape. When two stacked hourglass modules are combined, a thermal map is used instead of the original full connection layer. The thermal map represents the probability of each feature appearing in the pixel. The output of the second hourglass will go through a convolution layer of 1×1 to generate a landmark Heatmap. Here we predict 8 key points, including left and right eyebrows, nose, upper and lower lips and left and right eyes. That is to say, the final output result of the network is that each key point corresponds to a heat map, as shown in Fig. 3:

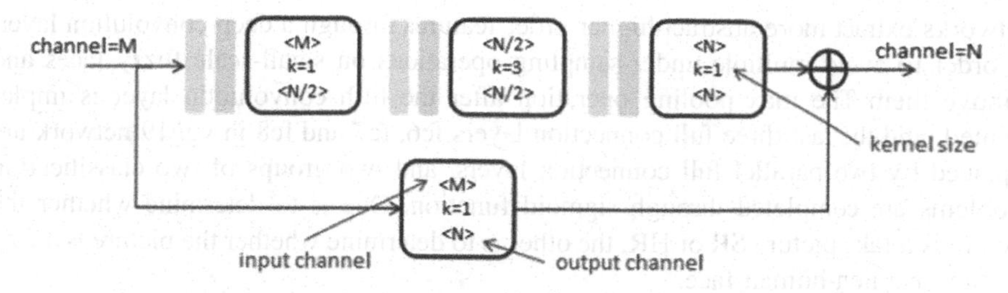

Fig. 2. The structure of Hourglass Module

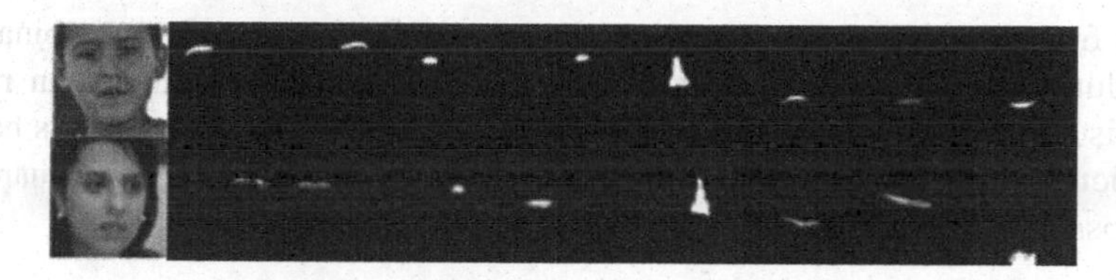

Fig. 3. Example of the facial landmark heatmap

3.3 TFPGAN

MTCNN and other model detectors are easy to treat the correct face image as the background image when processing the small-scale low-resolution image, resulting in a large number of missed detection situations. Now we use the small face samples and other background samples that are not detected by baseline through the above network model, so that the face image has the corresponding thermal image information, while the background image has no such prior information, Then we use these prior knowledge and super-resolution reconstruction network to restore these images to high-resolution images, and distinguish face and background thoroughly, so as to build a more accurate face detection model. In the previous face information estimation network, we have sampled the small size face image and background image, which makes the size of these training samples become larger, but the fine-grained information of these images is still insufficient, and the image pixels are still fuzzy. In addition to the ability of judging SR (generated image) and HR (real sample), we also have another distinguishing branch to distinguish face image and background image, and propose a new loss function to make the two networks Network joint training.

As shown in the figure, TFPGAN consists of the above structure. The LR image is input into the generator, after a 3×3 convolution layer, the number of channels of the feature map is set to 64. As mentioned before, the convolution core size is 3×3 and the step size is 1. The convolution layer design can keep the size of the feature map unchanged, which is inspired by the very successful resnet structure in the super-resolution task. Therefore, six residual blocks are used to extract features, and then use two previously mentioned de-conv (fractionally structured revolution) to sample the feature map 4 times of its size, and the last 3×3 rollup layers is used to generate the final high-definition image. The function of discriminant network D is similar to a classification network. The structure is modified according to vgg19. Vgg series networks extract more abstract higher-order features through a deep convolution layer. In order to avoid multiple under-sampling operations on small-scale fuzzy faces and remove them The max pooling operation after the fifth convolution layer is implemented, and the last three full connection layers fc6, fc7 and fc8 in vgg19 network are replaced by two parallel full connection layers, and two groups of two classification problems are completed through sigmoid function. One is to determine whether the picture is a fake picture SR or HR, the other is to determine whether the picture is a face picture or a non-human face.

3.4 Loss Function

Different from the original GAN network, the input of the generator is a combination of low-resolution image and some auxiliary heat map information, rather than random noise. It is known that the pixel wise loss function is used to calculate the loss between the predicted image and the target image pixel, so the pixel wise mean square error (MSE) loss is used here, and the loss formula is as follows:

$$L_{\text{MSE}} = \frac{1}{\text{n}} \sum\nolimits_{i=1}^{n} \left\| G_{\theta_G}(I_{LR}^i) - I_{HR}^i \right\|^2 \quad (1)$$

In which, I_{LR}^i and I_{HR}^i represent low-resolution image and high-resolution image with ground truth respectively, $D \equiv \left\{ (I_{LR}^i, I_{HR}^i) \right\}_i^n$ represents a large dataset of LR/HR image pairs. In order to obtain more realistic results, we introduce the adversarial loss in SRGAN [23] into objective loss, which is defined as follows:

$$L_{\text{adv}} = \frac{1}{\text{n}} \sum\nolimits_{i=1}^{n} \log(1 - D_\theta(G_{\theta_G}(I_{LR}^i))) \quad (2)$$

Here, the adversarial loss encourages the network to generate sharper high-frequency details for trying to fool the discriminator network. In Eq. 2, the $D_\theta(G_{\theta_G}(I_{LR}^i))$ is the probability that the reconstruction image $G_{\theta_G}(I_{LR}^i)$ is a natural super-resolution image.

In order to make the reconstructed images by the generator network easier to classify, we also introduce the classification loss [2] to the objective loss. Let $\{I_{LR}^i,\ i = 1, 2, \ldots, \text{n}\}$ and $\{I_{HR}^i,\ i = 1, 2, \ldots, \text{n}\}$ denote the small blurry images and the high-resolution real natural images respectively, and $\{y_i,\ i = 1, 2, \ldots, \text{n}\}$ represents the corresponding labels, where $y_n = 1$ or $y_n = 0$ indicates the image is the face or background respectively. The formulation of classification loss is like Eq. 3:

$$L_{\text{clc}} = \frac{1}{\text{n}} \sum\nolimits_{i=1}^{n} \log(y_i - D_{\theta_D}(G_{\theta_G}(I_{LR}^i)) + \log(y_i - D_{\theta_D}(I_{HR}^i))) \quad (3)$$

The classification loss plays two roles, where the first is to distinguish whether the high-resolution images, including both the generated and the natural real high-resolution images, are faces or non-faces in the discriminator network. The other role is to promote the generator network to reconstruct sharper images. Then we incorporate the adversarial loss Eq. 2 and classification loss Eq. 3 into the pixel-wise MSE loss

Eq. 1. The TFPGAN network can be trained by the objective function Eq. 4, where α and β are trade-off weights.

$$\min_{\theta_G}\max_{\theta_D} = \frac{1}{n}\sum\nolimits_{i=1}^{n} \alpha(\log(1 - D_{\theta_D}(G_{\theta_G}(I_{LR}^i))) + \log D_{\theta_D}(I_{HR}^i)) + L_{MSE} + \beta L_{clc} \quad (4)$$

4 Experiments

4.1 Datasets

WIDER FACE [24] data set is a benchmark data set of face detection, including 32203 images and 393703 human faces. These faces have a large range of changes in scale, posture and occlusion. The image selected by wire face mainly comes from the open dataset wire. The producers are from the Chinese University of Hong Kong. They chose 61 event categories of wider. For each category, they randomly selected 40%/10%/50% as the training, verification and test set. Participants can submit the prediction files and the wider face will give the evaluation results.

The WIDER FACE dataset is divided into three subsets, Easy, Medium, and Hard, based on the heights of the ground truth faces. The Easy/Medium/Hard subsets contain faces with heights larger than 50/30/10 pixels respectively. Compared to the Medium subset, the Hard one contains many faces with a height between 10–30 pixels. As expected, it is quite challenging to achieve good detection performance on the Hard subset.

4.2 Implementation Details

The weight of the neural network is initialized using a Gaussian distribution. Set the mean to 0 and the standard deviation to 0.01. The deviation is set to 0. To avoid undesirable local optima, we first train an MSE-based SR network to initialize the generator network. For the discriminator network, we employ the VGG19 model pretrained on ImageNet as our backbone network and we replace all the fc layers with two parallel fc layers. The fc layers are initialized by a zero-mean Gaussian distribution with standard deviation 0.1, and all biases are initialized with 0. The samples are labeled with face/background as follow-up training, and then the network is estimated by face prior information. The samples with face tags will generate corresponding face heat map through hourglass model, and be connected with the original image. In addition, the high-resolution and low-resolution image pairs in the training stage come from the low-resolution image obtained from the high-resolution image through bicubic interpolation and 4 times down sampling.

4.3 Evaluation on Datasets

In order to verify the performance of our proposed model, we compare our SRFGAN method with state-of-the-art methods on two public face detection benchmarks (i.e. WIDER FACE [24] and FDDB [25]).

4.3.1 Evaluation on WIDER FACE

We compare our method with the state-of-the-art face detectors [12–14, 16, 26, 27]. Table 1 shows the performance on WIDER FACE validation set. From Table 3, we see that our method achieves the highest performance (i.e. 87.39%) on the Hard subset. Compared to these CNN-based methods, the boosting of our performance mainly comes from three contributions: (1) our hourglass network can repeatedly obtain the information contained in the image at different scales, and we demonstrate that facial prior knowledge is significant for face super resolution, even without any advanced processing steps. (2) the Generative Adversarial Network learns finer details and reconstructs clearer images. Based on the clear super- resolution images, it is easier for the discriminator to classify faces or non-faces than depending on the low-resolution blurry images; (3) the classification loss Eq. 3 promotes the generator to learn a clearer face contour for easier classification.

Table 1. Results (The average precision (AP)) on the WIDER FACE dataset

Methods	Easy	Medium	Hard
Our Model	93.66	93.43	87.39
FaceGAN [16]	**94.40**	93.30	87.30
SSH [26]	**93.10**	92.10	84.51
HR [14]	**92.51**	91.10	80.60
MSCNN [12]	**69.16**	66.31	42.54
MTCNN [13]	*84.88*	82.50	59.82
CMS-RCNN	*89.94*	87.44	62.98
SFD [27]	*93.66*	92.47	85.86
Faster-RCNN	*92.15*	89.48	68.44
Faceness − WIDER	*71.89*	60.37	31.34

4.3.2 Evaluation on FDDB

We follow the standard metrics (i.e. precision at specific false positive rates) of the FDDB [25] and use this metric to compare with other methods. There are many unlabeled faces in FDDB, making precision not accurate at small false positive rates. Hence, we report the precision rate at 500 false positives. As shown in Fig. 4, we would like to note that the performance of SFD [27] is achieved after manually adding 238 unlabeled faces on the test set. However, we test our model on the original labeled test set. Under such an unfair condition, our method still gets the comparable performance, which further proves the effectiveness of our method. In Fig. 5, we show some detection results generated by our proposed method. It can be found that our face detector successfully finds almost all the faces, even though some faces are very small and blurred.

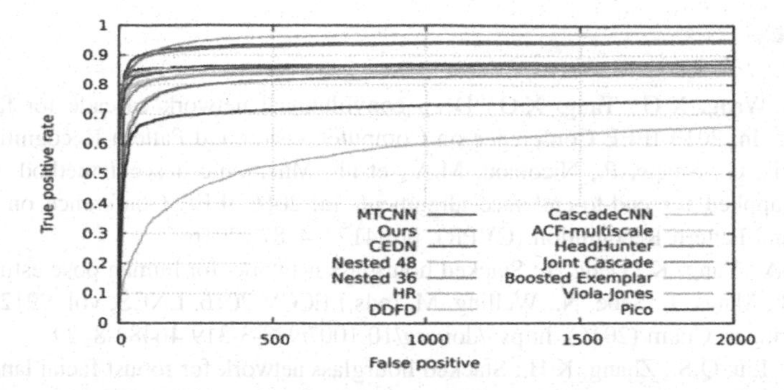

Fig. 4. On the FDDB dataset, we compare our method against many state-of-the-art methods. The precision rate with 500 false positives is reported.

Fig. 5. Examples detected by our detector.

5 Conclusion

In this paper, we proposed an algorithm to directly generate a clear high-resolution face from a blurry small one by adopting a TFPGAN model. Besides, we also designed a prior information estimation network which extracts the facial image features, and estimates landmark heatmaps respectively. In addition, we have introduced an additional classification branch in the discriminator network that can distinguish between fake/real and facial/background. extensive experiments on WIDER FACE and FDDB have shown that our approach has achieved significant improvements in both the hard, the easy and medium subset. We validated the proposed methods on two popular benchmarks. The promising performance over the state-of-the-art in terms of accuracy, model size and detection speed demonstrates the potential of our approach towards the real deployment on mobile devices.

Acknowledgment. This research was supported by the National Natural Science Foundation of China (Nos. 61672203 & 61976079) and Anhui Natural Science Funds for Distinguished Young Scholar (No. 170808J08).

References

1. Sun, Y., Wang, X.G., Tang, X.O.: Deep convolutional network cascade for facial point detection. In: 2013 IEEE Conference on Computer Vision and Pattern Recognition (2013)
2. Trigeorgis, G., Snape, P., Nicolaou, M.A., et al.: Mnemonic descent method a recurrent process applied for end-to-end face alignment. In: 2016 IEEE Conference on Computer Vision and Pattern Recognition (CVPR), pp. 4177–4187 (2016)
3. Newell, A., Yang, K., Deng, J.: Stacked hourglass networks for human pose estimation. In: Leibe, B., Matas, J., Sebe, N., Welling, M. (eds.) ECCV 2016. LNCS, vol. 9912, pp. 483–499. Springer, Cham (2016). https://doi.org/10.1007/978-3-319-46484-8_29
4. Yang, J., Liu, Q.S., Zhang, K.H.: Stacked hourglass network for robust facial landmark. In: 2017 IEEE Conference on Computer Vision and Pattern Recognition Workshops (CVPRW), pp. 2025–2033 (2017)
5. Wu, W.Y., Qian, C., Yang, S.: Look at boundary: a boundary-aware face alignment algorithm. In: IEEE CVF Conference on Computer Vision and Pattern Recognition
6. Rowley, H.A., Baluja, S., Kanade, T.: Neural network-based face detection. IEEE Trans. Pattern Anal. Mach. Intell. **20**, 23–38 (1998)
7. Rowley, H.A., Baluja, S., Kanade, T.: Rotation invariant neural network-based face detection. In: Computer Vision and Pattern Recognition (1998)
8. Viola, P., Jones, M.: Rapid object detection using a boosted cascade of simple features. In: Proceedings IEEE Conference on Computer Vision and Pattern Recognition (2001)
9. Girshick, R., Donahue, J., Darrell, T., Malik, J.: Rich feature hierarchies for accurate object detection and semantic segmentation. In: Proceedings of the IEEE Conference on Computer Vision and Pattern Recognition (CVPR) (2014)
10. He, K., Zhang, X., Ren, S., Sun, J.: Spatial pyramid pooling in deep convolutional networks for visual recognition. IEEE Trans. Pattern Anal. Mach. Intell. **37**(9), 1904–1916 (2015)
11. Girshick, R.: Fast R-CNN. In: ICCV (2015)
12. Qin, H., Yan, J., Li, X., Hu, X.: Joint training of cascaded cnn for face detection. In: CVPR (2016)
13. Zhang, K., Zhang, Z., Li, Z., Qiao, Y.: Joint face detection and alignment using multitask cascaded convolutional networks. IEEE Sig. Process. Lett. **23**(10), 1499–1503 (2016)
14. Hu, P., Ramanan, D.: Finding tiny faces. In: IEEE Conference on Computer Vision and Pattern Recognition (CVPR) (2017)
15. Li, J., Liang, X.D., Wei, Y.C., Xu, T.F., Feng, J.S., Yan, S.C.: Perceptual generative adversarial networks for small object detection. In: CVPR (2017)
16. Bai, Y., Zhang, Y.: Finding tiny faces in the wild with generative adversarial network. In: Proceedings of the IEEE Conference on Computer Vision and Pattern Recognition (CVPR) (2018)
17. Baker, S., Kanade, T.: Hallucinating faces. In: FG (2000)
18. Wang, X., Tang, X.: Hallucinating face by eigentransformation. IEEE TSMC Part C **35**(3), 425–434 (2005)
19. Kolouri, S., Rohde, G.K.: Transport-based single frame super resolution of very low resolution face images. In: CVPR (2015)
20. Yang, C.-Y., Liu, S., Yang, M.-H.: Structured face hallucination. In: CVPR (2013)
21. Zhu, S., Liu, S., Loy, C.C., Tang, X.: Deep cascaded bi-network for face hallucination. In: Leibe, B., Matas, J., Sebe, N., Welling, M. (eds.) ECCV 2016. LNCS, vol. 9909, pp. 614–630. Springer, Cham (2016). https://doi.org/10.1007/978-3-319-46454-1_37
22. Chen, Y., Tai, Y.: FSRNet: end-to-end learning face super-resolution with facial priors. In: CVPR (2018)

23. Ledig, C., et al.: Photo-realistic single image super-resolution using a generative adversarial network. arXiv preprint arXiv:1609.04802 (2016)
24. Yang, S., Luo, P., Loy, C.C., Tang, X.: WIDER FACE: a face detection benchmark. arXiv preprint arXiv:1511.06523
25. Jain, V., Learned-Miller, E.: FDDB: a benchmark for face detection in unconstrained settings. Technical report (2010)
26. Najibi, M., Samangouei, P., Chellappa, R., Davis, L.S.: SSH: single stage headless face detector. CoRR, abs/1708.03979 (2017)
27. Zhang, S., Zhu, X., Lei, Z., Shi, H., Wang, X., Li, S.Z.: Single shot scale-invariant face detector. CoRR, abs/1708.05237 (2017)
28. Zhao, Z.Q., Glotin, H.: Diversifying image retrieve by affinity propagation clustering on visual manifolds. IEEE Multimed. **16**, 1 (2009)
29. Zhao, Z.Q., Xu, S.T., Liu, D., Tian, W.D.: A review of image set classification. Neurocomputing **335**, 251–260 (2019)
30. Zhao, Z., Wu, X., Lu, C., Glotin, H., Gao, J.: Optimizing widths with PSO for center selection of Gaussian radial basis function networks. Sci. Chin. Inf. Sci. **57**(5), 1–17 (2013). https://doi.org/10.1007/s11432-013-4850-5
31. Zhao, Z.Q., Gao, J., Glotin, H., Wu, X.: A matrix modular neural network based on task decomposition with subspace division by adaptive affinity propagation clustering. Appl. Math. Model. **34**(12), 3884–3895 (2010)

Regenerating Image Caption with High-Level Semantics

Wei-Dong Tian(✉), Nan-Xun Wang, Yue-Lin Sun,
and Zhong-Qiu Zhao

School of Computer Science and Information Engineering,
Hefei University of Technology, Hefei 230601, Anhui, China
wdtian@hfut.edu.cn

Abstract. Automatically describing an image with a sentence is a challenging task in the crossing area of computer vision and natural language processing. Most existing models generate image captions by an encoder-decoder process based on convolutional neural network (CNN) and recurrent neural network (RNN). However, such a process employs low level pixel-level feature vectors to generate sentences, which may lead to rough captions. Therefore, in this paper, we introduce high-level semantics to generate better captions, and we propose a two-stage image captioning model: (1) generate initial captions and extract high-level semantic information about images; (2) refine initial captions with the semantic information. Empirical tests show that our model achieves better performance than different baselines.

Keywords: Image caption · High-level semantic · Convolutional neural network · Recurrent neural network · Encoder-decoder

1 Introduction

With the continuous development of the computer vision, the task of automatically generating image captions has gradually attracted people's attention. This is a particularly challenging task which requires the computer to understand image content and to generate a fine sentence as caption. Also, the caption should not only correctly describe the attributes of objects and relationships between objects in the image, but also ensure the correctness of the grammar and semantics.

Before neural network was widely applied, the generation of image captions mainly based on image retrieval and language template. Image retrieval-based methods is to query the image which is similar to the image visually, and take the caption of the queried image as the final output [1–3]. Language template-based methods firstly obtain the objects of the image, and then fill the language template to generate a description [4–6].

The current image captioning methods are mainly based on the encoder-decoder model [7, 9]. Similar to the machine translation task in natural language processing, images are taken as source sentences and translated into sentences. In this model, CNN is used to encode the image into feature vector, and RNN is used to decode the feature vector into caption.

D.-S. Huang and P. Premaratne (Eds.): ICIC 2020, LNAI 12465, pp. 74–86, 2020.
https://doi.org/10.1007/978-3-030-60796-8_7

Despite the variation in specific method, encoder-decoder models based on CNN + LSTM commonly take the global image feature vector as the initial value of the language model and generate captions which can make the language model to "feels" reasonable within a simple pass forward process. But for some complex image scene, this kind of model is still weak in visual grounding abilities (i.e., cannot associate high-level semantics to pixels in the image). They often tend to "look" at more different regions than humans and copy captions from training data [10]. Furthermore, there is a certain semantic gap between image and language, which makes it particularly tough to generate more appropriate descriptions. As shown in Fig. 1, although the corresponding relationship between image and caption can be learned relying on the simple pass forward process, the gap cannot be avoided and may lead to generating rough captions with semantic errors or omissions. To solve these problems, in this paper, we propose a two-stage model to generate the image caption. Firstly, the two-stage model uses a dual CNN structure to obtain the corresponding initial caption and high-level semantics. Then, in the second stage, the model uses an attention mechanism to refine the initial caption with the high-level semantics of the image.

Fig. 1. Illustration of high level semantic defects.

2 Related Work

Early image captioning approaches were based on machine learning to obtain corresponding image descriptions. For example, Ordonez et al. [3] compute a global representation of the image content and retrieved the relevant description in the dataset as the caption for the image based on the representation. Devlin et al. [1] proposed a K-nearest neighbor approach to obtain the K images closest to the image, and selected the best one from the description set of K images as the image caption. These methods based on image retrieval ensure the grammatical correctness of the captions but cannot guarantee their semantic consistency with the images. Farhadi et al. [4] and Kulkarni et al. [5] use some operators of image processing to extract image features and obtained objects, actions and attributes through SVM, then used CRF and pre-defined templates [6] to generate the description. Although these methods guarantee semantic correctness, they rely heavily on templates rules.

In recent years, with the rapid development of deep learning in computer vision [11–16], the method about image captioning based on neural networks has been proposed. Vinyals et al. [7] and Karpathy et al. [17] proposed the neural image captioner, which applies CNN to process images and represents the input image with a single feature vector, and uses an LSTM conditioned on this vector to generate words one by one. It achieves good results in syntactic correctness, semantic accuracy and generalization, thus this method lays the foundation of the encoder-decoder model based on CNN+LSTM to generate image caption. To generate better image captions, Xu et al. [9] further introduced an attention-based model that can learn where to look while generating corresponding words.

Despite the improvement over the model architectures, all these approaches adopt a one-pass forward process while decoding to generate captions. Although these models can generate a corresponding description for the image through the powerful learning ability of the neural network, they still cannot avoid the semantic gap between the image and the natural language. Besides, captions generated by decoding image feature vectors only once cannot be further polished, which may lead to incorrect semantics or lack of integrity. Therefore, we propose the second-pass decoder to generate and refine captions.

3 Model

We propose a neural network model for image captioning, there are four modules in our model. Figure 2 illustrates the overall framework for image captioning.

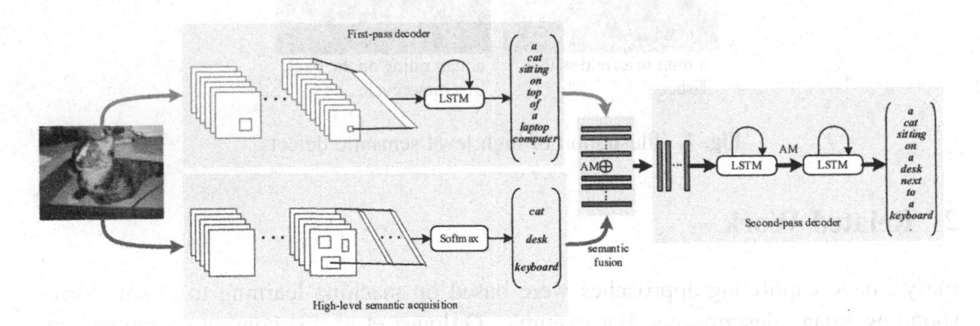

Fig. 2. The overall framework of the proposed model.

3.1 The First-Pass Decoder Module

Given an image *I*, the image is encoded into a set of feature vectors by the CNN and used as the initial input to the LSTM, and the LSTM is used to predict the next word until the entire initial caption is generated:

$$y'_{-1} = CNN(I) \tag{1}$$

$$h'_t = LSTM(h'_{t-1}, y'_{t-1}) \tag{2}$$

$$y'_t \sim p'_t = softmax(h'_t) \tag{3}$$

where I is an image, and the image I is only input once, at $t = -1$ [7], to inform the LSTM about the image contents. h'_t represents the hidden state of the LSTM at time step t, p'_t represents the probability of predicting the word at time step t, and y'_t is the ground-truth word encoded as word embedding at time step t. y'_{t-1} is the LSTM input vector and h'_t is the LSTM output vector.

In addition, in order to demonstrate the effectiveness of our model, we introduced the baseline method as the module to generate the initial caption. The specific method is shown in the training details.

3.2 The High-Level Semantics Acquisition Module

Given an image I, we use Faster R-CNN [18] to detect the image and obtain the corresponding high-level semantics, that is, specific objects and corresponding attributes. Faster R-CNN mainly consists of two stages to detect image objects. The first stage, described as a Region Proposal Network (RPN), generates object proposals. A small network is slid over features at an intermediate level of a CNN. Anchor boxes of multiple scales and aspect ratios are generated by K anchors at each sliding position. The anchor boxes are judged by categories or background and corrected by bounding box regression. Using non-maximum suppression (NMS) with an Intersection-over-Union (IoU) threshold, the top-N box proposals are selected as input to the second stage.

In the second stage, region of interest (RoI) pooling is used to extract feature maps for each box proposal. These feature maps are then fully connected and classified by softmax. Through this module, we can get the objects and attributes about the image. In Fig. 3, we provide some examples of the module output.

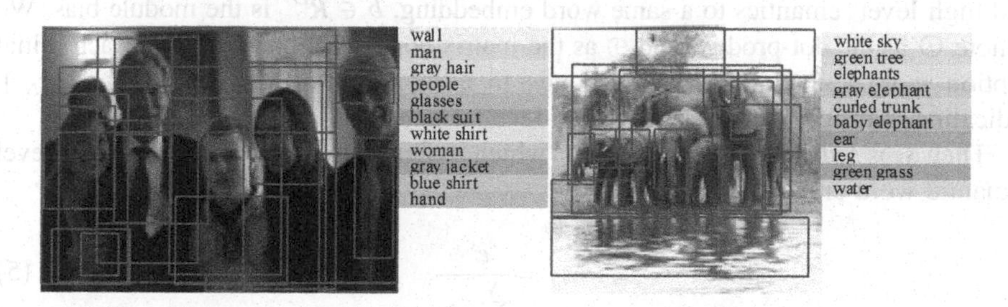

Fig. 3. Illustration of high level semantic information of images.

3.3 The Semantic Information Fusion Module

To generate a more perfect caption, we incorporate the high-level semantics of the image into the initial caption. We use the attention mechanism to calculate the correlation between the j-th word c_j in the initial caption and the high-level semantics, to get the rich attention vector a_j of the word c_j. Let $\boldsymbol{c} = [c_1, c_2, \ldots, c_L]$ be a initial caption with L words, where $c_j \in \boldsymbol{R}^{Dc}$, j is the position index in the initial caption, and $\boldsymbol{o} = [o_1, o_2, \ldots, o_N]$ represents the specific high-level semantic words, where $o_i \in \boldsymbol{R}^{Do}$, i is the i-th word and N is the number of high-level semantic words. The semantic information fusion module takes the word embedding $\boldsymbol{c}$ and $\boldsymbol{o}$ as input, which is shown in Fig. 4.

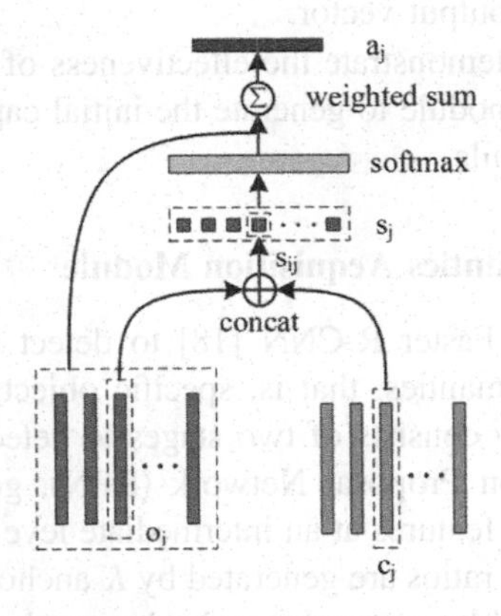

Fig. 4. The semantic information fusion module.

For each initial caption word embedding c_j and high-level semantic word embedding o_i, we calculate a score $s_{i,j}$ as follows:

$$s_{i,j} = V^T\{\tanh[(W_c \odot c_j) \oplus (W_o \odot o_i)] + b\} \tag{4}$$

where $W_c \in \boldsymbol{R}^{Dc}$, $W_o \in \boldsymbol{R}^{Do}$, $V \in \boldsymbol{R}^{Do}$ are learned parameter that map initial caption and high-level semantics to a same word embedding, $b \in \boldsymbol{R}^{DV}$ is the module bias. We denote $\odot$ as the dot-product and $\oplus$ as the matrix concatenation. Therefore each initial caption word embedding c_j corresponds to a score vector $\boldsymbol{s}_j = [s_{1,j}, s_{2,j}, \ldots, s_{N,j}]$, indicating matches with the image high-level semantics.

Then $\boldsymbol{s}_j$ is fed into a softmax layer, which assigns a weight $w_{i,j}$ for the high-level semantic word embedding o_i,

$$w_{i,j} = \frac{e^{s_{i,j}}}{\sum_{i=1}^{N} e^{s_{i,j}}} \tag{5}$$

We use the weighted sum operation to calculate the rich attention vector as follows,

$$a_j = \sum_{i=1}^{N} w_{i,j} o_i \tag{6}$$

the output of the attention mechanism is the attention features $\boldsymbol{a} = [a_1, a_2, \ldots, a_L]$, where $a_j \in \boldsymbol{R}^{Dc}$, corresponding to each word in the initial caption. Finally, each word $c_k \in \boldsymbol{c}$ of the initial caption is combined with the attention feature $a_k \in \boldsymbol{a}$ which is calculated from the referred information,

$$f_k = g(c_k, a_k) \tag{7}$$

where $g(.)$ is a function that returns a fusion vector f_k given the vectors c_k and the vectors a_k, and $\boldsymbol{f} = [f_1, f_2, \ldots, f_L]$, $f_j \in \boldsymbol{R}^{Dc}$.

3.4 The Second-Pass Decoder Module

We encode the vectors $\boldsymbol{f} = [f_1, f_2, \ldots, f_L]$ into a sequence of hidden states $[h_1{:}h_L]$, with Bi-directional Long-Short Term Memory (BiLSTM) net,

$$h_t = BiLSTM(h_{t-1}, f_{t-1}) \tag{8}$$

where h_t represents the hidden state of the BiLSTM at time step t. A BiLSTM consists of forward and backward LSTM's. The forward LSTM $\vec{h}$ reads the input vectors as it is ordered (from f_1 to f_L) and calculates a sequence of forward hidden states $\left[\vec{h}_1, \vec{h}_2, \ldots, \vec{h}_L\right]$. The backword LSTM $\overleftarrow{h}$ reads the input vectors in the reverse order (from f_1 to f_L), resulting in a sequence of backward hidden states $[\overleftarrow{h}_1, \overleftarrow{h}_2, \ldots, \overleftarrow{h}_L]$, i.e., $h_t = \left[\vec{h}_t; \overleftarrow{h}_t\right]$, which is shown in Fig. 5.

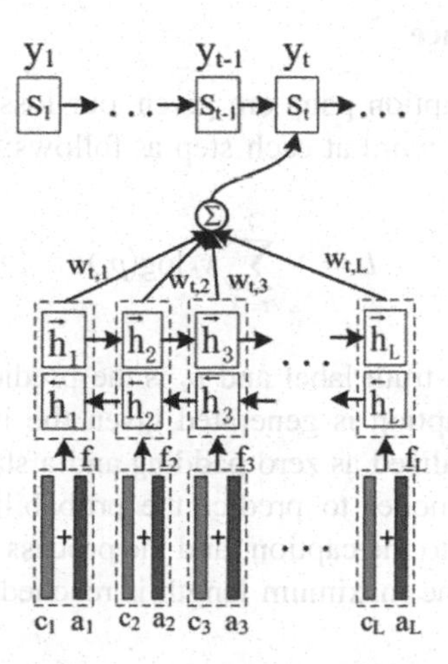

Fig. 5. The graphical illustration of the second-pass decoder module trying to generate the t-th target word y_t given a fusion vector $\boldsymbol{f} = [f_1, f_2, \ldots, f_L]$

Use Bahdanau Attention [19] to calculate hidden states $\boldsymbol{h}$ which is to get the text vector $z = [z_1, z_2, ..., z_T]$, and then decode the text vector z through the LSTM network to generate the sequence $\boldsymbol{y} = [y_1, y_2, .., y_T]$ of final caption,

$$z_t = \sum_{l=1}^{L} \alpha_{t,l} h_l \tag{9}$$

where the context vector z_t is computed as a weighted sum of these hidden state h_t, and the weight $\alpha_{t,l}$ of each hidden state h_l is computed by

$$e_{t,l} = \varphi(s_{t-1}, h_l) \tag{10}$$

$$\alpha_{t,l} = \frac{\exp(e_{t,l})}{\sum_{l=1}^{L} \exp(e_{t,l})} \tag{11}$$

where $\varphi(.)$ is an alignment model which scores how well the inputs around position l and the output at position t match. The score is based on the LSTM hidden state s_{t-1} and the l-th hidden state h_l of the input vector $\boldsymbol{f}$. We parametrize the alignment model as a feedforward neural network which is jointly trained with all the modules.

After each prediction, s_t is updated by

$$s_t = LSTM(s_{t-1}, y_{t-1}, z_t) \tag{12}$$

and the probability of the next word y_t can be defined as:

$$p(y_t | y1, y2, \ldots y_{t-1}, f) = softmax(s_t) \tag{13}$$

3.5 Training and Inference

During training, as image-caption pairs are given, our loss is the sum of the negative log-likelihood of the correct word at each step as follows:

$$L = -\sum_{t=1}^{T} y_t \log(p_t) \tag{14}$$

where y_t denotes the ground-truth label and p_t is the prediction probability.

During inference, the caption is generated given the image using a feed-forward process. The caption is initialized as zero padding and a start-token <S>, and is fed as the input sentence to the model to predict the probability of the next word. The predicted word is appended to the caption, and the process is repeated until the ending token <S> is predicted, or the maximum length is reached.

4 Experiments

4.1 Dataset and Experimental Setup

MSCOCO. MSCOCO is the most popular dataset for image captioning, comprising 82,783 training and 40,504 validation images. Each image is annotated with 5 reference captions. As in [9, 17], we randomly divide the images into 3 datasets, consisting of 5,000 validation and 5,000 testing images, and 113,287 training images.

Flickr30k. Flickr30k is another well-recognized dataset for image captioning, which contains 31,783 images. Each image in the dataset have 5 reference captions. Similar to the MSCOCO dataset, we randomly divide all the images into 3 datasets, consisting of 1,000 validation and 1,000 testing images, and the remaining 29,783 images as training images.

Baseline. To demonstrate the effectiveness of our model, we also present a baseline result. We have re-implemented the popular models of image caption using the same dataset partitioned on the MSCOCO dataset and the Flickr30k dataset: soft-ATT [9], Google NICv2 [20]. These trained models are used as the first-pass decoder module in our proposed model. The module generates captions of the test set and calculates corresponding values of metrics as the baseline result.

Implementation Details. We follow the preprocessing procedure in [17] for the captions, removing the punctuation and converting all characters to lower case. For the training set of the MSCOCO dataset, we discard words that occur fewer than 4 times in the ground truth, we get a vocabulary which contains 11,349 words. The same processing is performed on the Flickr30k dataset, and the resulting vocabulary contains 8,352 words.

We use the baseline model as the first-pass decoder module, which generates image captions for the image training, verification and test sets, respectively. These captions correspond to five different the ground truth and the generated three parts are used as the training, verification and test sets of the second-pass decoder module.

We use the trained Faster R-CNN as the high-level semantics acquisition module, use ResNet-101 to extract image feature vectors, and finally classify the specific categories and attributes of objects in the image.

In addition, we use different fusion methods $g(.)$ in the semantic information fusion module, where $g(.)$ includes addition, dot-product and MLB [21] to fuse feature vectors. The dimension of the feature vector f in this module is equal to the dimension of the word embedding, $D_c = D_o = D_V = 512$. The size of the hidden layer in the second-pass decoder module is 512, the size of the text vector z is 512, and $\lambda = 1e - 3$. We apply Adam optimizer with batch size 32 to train our model. The attention weights are initialized by the random normal initializer with stddev = 0.02. The model was implemented in TensorFlow.

Metrics. We compare the quality of the generated captions on the test set of MSCOCO and the test set of Flickr30k, in terms of CIDEr [22], BLEU-1,2,3,4 [23], ROUGE-L [24], METEOR [25] and SPICE [26]. Besides, the SPICE is an F-score of the matching tuples in predicted and reference scene graphs. It can be divided into meaningful subcategories. In our paper, we report the SPICE score as well as the subclass scores of objects, relations and attributes. For all metrics, higher values indicate better performance.

4.2 Experiment Results

We compare the quality of the generated captions on the test set of MSCOCO and the test set of Flickr30k. We report conventional evaluation metrics in Table 1 and the F-score for SPICE subclass properties in Table 2. Table 1 shows the conventional evaluation results on MSCOCO and Flickr30k. As far as the overall evaluation score is concerned, our model has been improved based on the different baselines, which indicates (CIDEr, BLEU-4, BLEU-1, ROUGE-L, METEOR, SPICE) that we can further improve the effect of image captions by combining the high-level semantics. Specifically, on Flickr 30K, the CIDEr result is 20.7% based on soft-ATT. In the semantic information fusion module, we use vector addition fusion (+) to raise it to 30.0%, dot-product fusion(*) to 29.9%, and MLB fusion (MLB) to 32.8%. The CIDEr score based on NICv2 was 36.5% increased to 38.8%, 38.4% and 39.9% under the fusion of addition (+), dot-product (*) and MLB, respectively, and the MLB fusion performed better, which further illustrates that finer-grained feature fusion helps to generate better captions.

Table 1. Performance of our proposed model and baseline models on MSCOCO and Flickr30k datasets. The (+), (*) and (MLB) are the feature vector fusion method of addition, dot-product and MLB, respectively. The bold is the highest value.

Dateset	Models	C	B-4	B-1	R-L	M	S
MSCOCO	soft-attend	63.4	20.3	62.8	46.2	20.1	13.0
	ours (+)	83.2	26.5	69.1	50.4	22.1	13.9
	outs (*)	82.7	26.2	69.7	50.3	22.2	14.4
	our (MLB)	**85.6**	**27.0**	**70.8**	**51.0**	**22.7**	**14.7**
	NICv2	94.8	**30.4**	70.9	53.0	25.1	17.6
	ours (+)	96.6	29.8	72.8	53.2	**25.8**	18.6
	ours (*)	95.0	29.1	71.9	53.1	25.6	18.7
	ours (MLB)	**97.7**	30.2	**73.9**	**53.2**	25.7	**18.9**
Flickr30k	soft-attend	20.7	13.6	58.0	38.3	14.5	7.3
	ours (+)	30.0	18.7	61.8	41.4	17.0	11.7
	outs (*)	29.9	19.8	63.4	41.3	17.0	11.2
	our (MLB)	**32.8**	**20.0**	**64.4**	**41.7**	**17.4**	**11.8**
	NICv2	36.5	18.2	62.1	42.1	17.7	10.9
	ours (+)	38.8	21.1	66.5	43.4	18.8	12.8
	ours (*)	38.4	20.4	65.4	42.8	18.6	12.5
	ours (MLB)	**39.9**	**22.0**	**66.7**	**43.7**	**19.0**	**13.2**

Table 2. Performance of our proposed model and the baseline models on SPICE measurement, for the two datasets. The (+), (*) and (MLB) are the feature vector fusion method of addition, dot-product and MLB, respectively. The bold is the highest value.

Dateset	Models	Ob	Re	At	Si	Co	Ca
MSCOCO	soft-attend	25.6	3.0	4.6	1.3	5.5	1.8
	ours (+)	26.7	2.6	6.8	2.4	12.9	0.8
	outs (*)	27.2	3.0	7.1	2.0	12.4	2.0
	our (MLB)	**27.6**	**3.0**	**7.6**	**2.7**	**14.6**	**2.4**
	NICv2	32.6	4.4	8.3	2.5	10.2	1.4
	ours (+)	34.7	**5.0**	8.7	**4.6**	15.2	2.3
	ours (*)	**34.8**	4.8	9.0	4.2	14.8	3.5
	ours (MLB)	34.7	4.9	**9.8**	3.2	**16.1**	**3.8**
Flickr30k	soft-attend	15.9	1.5	2.2	2.7	6.2	0.3
	ours (+)	22.7	**3.4**	6.1	4.3	14.8	0.6
	outs (*)	21.6	3.2	5.9	5.2	14.1	**1.5**
	our (MLB)	**23.0**	3.3	**6.5**	**5.6**	**16.0**	1.0
	NICv2	22.7	2.6	4.0	5.2	5.5	2.9
	ours (+)	24.8	3.6	6.8	**9.6**	14.3	1.2
	ours (*)	24.5	3.3	6.3	6.0	13.3	**3.6**
	ours (MLB)	**25.7**	**3.9**	**7.0**	5.1	**15.8**	2.3

Table 2 shows the SPICE scores and their various subclasses of F-score (Object, Relation, Attribute, Size, Color, Cardinality) on MSCOCO and Flickr30k. From the overall results, our proposed model has also been consistently improved on the baselines for SPICE subclass properties, especially the F-score of the object, attribute and color has a significant improvement, which shows that we can modify image captions by combining high-level semantics (object, attribute and color words).

Moreover, Fig. 6 presents several representative examples of captions produced by NICv2 and our model. For the first example, our model correct the wrong attribute in the caption generated by NICv2 model, such as "reading a book" to "looking at a cell phone", e.g., and our model appear to yield more accurate captions for the image.

In the fourth example, the caption generated by our model further improve the content of the image, such as "next to a tree". These examples offer further qualitative evidence that shows the utility and effectiveness of our model to improve the quality of image captions.

a man sitting on a bench reading a book .
a man sitting on a bench looking at a cell phone .

a stack of luggage sitting on top of a wooden floor .
a blue suitcase with a pair of shoes on a wooden floor .

a group of people sitting around a dinner table .
a group of people sitting at a table with plates of food .

a red truck parked in a tall lot .
a truck parked in a parking lot next to a tree .

a white toilet sitting in a bathroom next to a sink .
a white toilet in a bathroom with a toilet paper holder .

a man standing on a beach holding a surfboard .
a man in a wetsuit holding a surfboard .

a brown teddy bear sitting on top of a chair .
a white teddy bear sitting on a chair next to a book .

a parking clock tower with a sky background .
a clock tower with a flag on top of it .

Fig. 6. Examples of generated captions for image examples from MSCOCO. Red and blue text correspond to captions from NICv2 and our model. (Color figure online)

5 Conclusion

In this paper, we propose a regeneration model for image captioning. The proposed model decomposes the decoding process of image captioning into first-pass decoding and second-pass decoding. Firstly, the global feature vector of the image is decoded into the initial caption, and then the fusion feature vectors, which is obtained by combining the initial caption with the high-level semantics of the image, are decoded to generate the final caption.

Acknowledgement. This research was supported by the National Natural Science Foundation of China (Nos. 61672203, 61976079 & U1836102) and Anhui Natural Science Funds for Distinguished Young Scholar (No. 170808J08).

References

1. Devlin, J., Gupta, S., Girshick, R., Mitchell, M., Zitnick, C.L.: Exploring nearest neighbor approaches for image captioning. arXiv preprint arXiv:1505.04467 (2015)
2. Hodosh, M., Young, P., Hockenmaier, J.: Framing image description as a ranking task: data, models and evaluation metrics. J. Artif. Intell. Res. **47**, 853–899 (2013)
3. Ordonez, V., Kulkarni, G., Berg, T.L.: Im2text: describing images using 1 million captioned photographs. In: Shawe-Taylor, J., Zemel, R.S., Bartlett, P.L., Pereira, F., Weinberger, K.Q. (eds.) Advances in Neural Information Processing Systems, vol. 24, pp. 1143–1151. Curran Associates, Inc. (2011)

4. Farhadi, A., et al.: Every picture tells a story: generating sentences from images. In: Daniilidis, K., Maragos, P., Paragios, N. (eds.) ECCV 2010. LNCS, vol. 6314, pp. 15–29. Springer, Heidelberg (2010). https://doi.org/10.1007/978-3-642-15561-1_2
5. Kulkarni, G., et al.: Babytalk: understanding and generating simple image descriptions. IEEE Trans. Pattern Anal. Mach. Intell. **35**(12), 2891–2903 (2013)
6. Yang, Y., Teo, C.L., Daume III, H., Aloimonos, Y.: Corpus-guided sentence generation of natural images. In: Proceedings of the Conference on Empirical Methods in Natural Language Processing, pp. 444–454. Association for Computational Linguistics (2011)
7. Jia, X., Gavves, E., Fernando, B., Tuytelaars, T.: Guiding the long-short term memory model for image caption generation. In: Proceedings of the IEEE International Conference on Computer Vision, pp. 2407–2415 (2015)
8. Vinyals, O., Toshev, A., Bengio, S., Erhan, D.: Show and tell: a neural image caption generator. In: Proceedings of the IEEE Conference on Computer Vision and Pattern Recognition, pp. 3156–3164 (2015)
9. Xu, K., et al.: Show, attend and tell: Neural image caption generation with visual attention. In: International Conference on Machine Learning, pp. 2048–2057 (2015)
10. Das, A., Agrawal, H., Zitnick, L., Parikh, D., Batra, D.: Human attention in visual question answering: do humans and deep networks look at the same regions? Comput. Vis. Image Underst. **163**, 90–100 (2017)
11. Glotin, H., Zhao, Z.Q., Ayache, S.: Efficient image concept indexing by harmonic & arithmetic profiles entropy. In: 2009 16th IEEE International Conference on Image Processing (ICIP), pp. 277–280 (2009)
12. Zhao, Z.Q., Ming Cheung, Y., Hu, H., Wu, X.: Corrupted and occluded face recognition via cooperative sparse representation. Pattern Recogn. **56**, 77–87 (2016)
13. Zhao, Z.Q., Gao, J., Glotin, H., Wu, X.: A matrix modular neural network based on task decomposition with subspace division by adaptive affinity propagation clustering. Appl. Math. Model. **34**, 3884–3895 (2010)
14. Zhao, Z.Q., Glotin, H.: Diversifying image retrieval with affinity-propagation clustering on visual manifolds. IEEE MultiMed. **16**, 34–43 (2009)
15. Zhao, Z.Q., Wu, X., Lu, C., Glotin, H., Gao, J.: Optimizing widths with PSO for center selection of gaussian radial basis function networks. Sci. Chin. Inf. Sci. **57**, 1–17 (2013)
16. Zhao, Z.Q., Tao Xu, S., Liu, D., Tian, W., Jiang, Z.D.: A review of image set classification. Neurocomputing **335**, 251–260 (2019)
17. Karpathy, A., Fei-Fei, L.: Deep visual-semantic alignments for generating image descriptions. In: Proceedings of the IEEE Conference on Computer Vision and Pattern Recognition, pp. 3128–3137 (2015)
18. Ren, S., He, K., Girshick, R., Sun, J.: Faster R-CNN: towards real-time object detection with region proposal networks. In: Advances in Neural Information Processing Systems, pp. 91–99 (2015)
19. Bahdanau, D., Cho, K., Bengio, Y.: Neural machine translation by jointly learning to align and translate. arXiv preprint arXiv:1409.0473 (2014)
20. Vinyals, O., Toshev, A., Bengio, S., Erhan, D.: Show and tell: lessons learned from the 2015 MSCOCO image captioning challenge. IEEE Trans. Pattern Anal. Mach. Intell. **39**(4), 652–663 (2017)
21. Kim, J.H., On, K.W., Lim, W., Kim, J., Ha, J.W., Zhang, B.T.: Hadamard product for low rank bilinear pooling. arXiv preprint arXiv:1610.04325 (2016)
22. Vedantam, R., Lawrence Zitnick, C., Parikh, D.: Cider: consensus-based image description evaluation. In: Proceedings of the IEEE Conference on Computer Vision and Pattern Recognition, pp. 4566–4575 (2015)

23. Papineni, K., Roukos, S., Ward, T., Zhu, W.J.: Bleu: a method for automatic evaluation of machine translation. In: Proceedings of the 40th Annual Meeting on Association for Computational Linguistics, pp. 311–318. Association for Computational Linguistics (2002)
24. Lin, C.Y.: Rouge: a package for automatic evaluation of summaries. Text Summarization Branches Out (2004)
25. Denkowski, M., Lavie, A.: Meteor universal: language specific translation evaluation for any target language. In: Proceedings of the Ninth Workshop on Statistical Machine Translation, pp. 376–380 (2014)
26. Anderson, P., Fernando, B., Johnson, M., Gould, S.: SPICE: semantic propositional image caption evaluation. In: Leibe, B., Matas, J., Sebe, N., Welling, M. (eds.) ECCV 2016. LNCS, vol. 9909, pp. 382–398. Springer, Cham (2016). https://doi.org/10.1007/978-3-319-46454-1_24

Aggregated Deep Saliency Prediction by Self-attention Network

Ge Cao, Qing Tang, and Kang-hyun Jo(✉)

University of Ulsan, Ulsan 44610, Republic of Korea
acejo@ulsan.ac.kr

Abstract. The data-driven method has recently obtained great success on saliency prediction thanks to convolutional neural networks. In this paper, a novel end-to-end deep saliency prediction method named VGG-SSM is proposed. This model identifies three key components: feature extraction, self-attention module, and multi-level integration. An encoder-decoder architecture is used to extract the feature as a baseline. The multi-level integration constructs a symmetric expanding path that enables precise localization. Global information of deep layers is refined by a self-attention module which carefully coordinated with fine details in distant portions of a feature map. Each component surely has its contribution, and its efficiency is validated in the experiments. Additionally, In order to capture several quality factors, the loss function is given by a linear combination of some saliency evaluation metrics. Through comparison with other works, VGG-SSM gains a competitive performance on the public benchmarks, SALICON 2017 version. The PyTorch implementation is available at https://github.com/caoge5844/Saliency.

Keywords: Saliency prediction · Self-attention · Multi-level integration

1 Introduction

Capturing the salient area in a scene is an instinctive ability of human beings. For visual saliency, it describes the spatial location which attracts the observer most. When observing a graph without any special tasks, as an elusive process, humans can't pay attention to every portion with the same intensity. Many works show that computational saliency can be found usages in a wide range of applications like object recognition [1], tracking regions of interest [2], and image retargeting [3] and so on.

With the advent of the deep neural network, saliency prediction also achieved great success thanks to generous data-driven methods and large annotated datasets [4]. Generally, computational saliency models predict the probability distribution of the location of eye attention over the images. Visual saliency data are traditionally colected by eye-trackers [5], more recently with mouse clicks [4]. No matter which kind of method is used to collect the saliency data, where human observers look in the images is regarded as the ground truth to estimate the accuracy of the predicted saliency maps. Through the computation of the proposed model, the predictions use various evaluation metrics to evaluate how best of a saliency model. The work by [6] broadly classified the various metrics as location-based or distribution-based. Though a large variety of metrics to evaluate saliency prediction maps exist, the main difference between them

D.-S. Huang and P. Premaratne (Eds.): ICIC 2020, LNAI 12465, pp. 87–97, 2020.
https://doi.org/10.1007/978-3-030-60796-8_8

concerns the ground-truth representation. In this paper, seven different evaluation metrics are used to analyze and evaluate the proposed model.

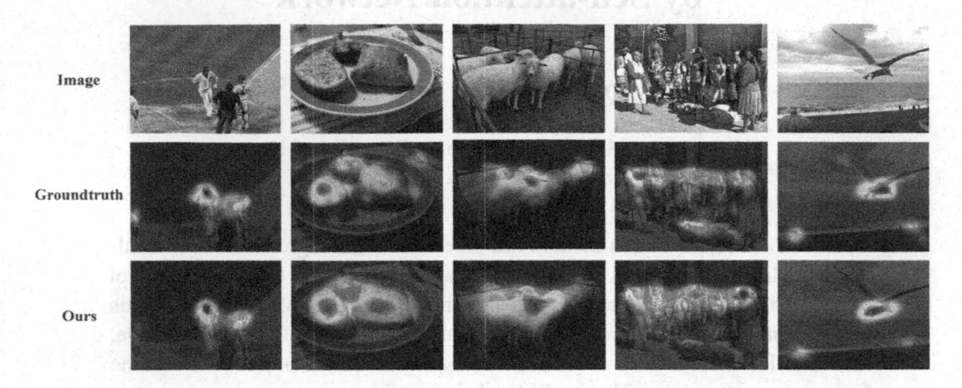

Fig. 1. Example results of the proposed method on images from SALICON dataset.

A novel end-to-end saliency prediction architecture is proposed to predict the saliency maps in this paper. Three key components in this architecture are identified respectively. First is the encoder-decoder architecture which directly extracts feature information. The second component is the self-attention module. The proposed model incorporates a Self-attention module that focuses on global, long-range dependencies to refine the details at every location. Each pixel in the feature maps can carefully coordinate with distant portions in the feature map, not limit to convolutional computation. In the third aspect, multi-level integration is constructed to reuse input feature maps for more local semantic information. Except for structural modify, the combination loss function outperform other loss function used single metric. The paper makes the following contributions:

1. This paper proposes a novel end-to-end saliency prediction method called VGG-SSM. The whole architecture is divided into separate components and analysis their efficiency respectively.
2. Self-attention module is incorporated with encoder-decoder based architecture to enhance global saliency information. The multi-level integration also improves the ability in local feature extraction.
3. The loss function used is formulated by some existing saliency metrics. The combined loss function makes multiple competing metrics be satisfied in concert.

Figure 1 shows examples of saliency maps predicted by the proposed method, which called the Saliency Self-attention Model (SSM), compared with ground truth saliency maps obtained from eye fixation. The proposed method is validated on publicly available datasets: SALICON. Experiments and evaluations results show that the proposed method improves the predictions.

The remaining content is organized as follows. Section 2 summarized the related work. The details of each component in the whole architecture and the loss functions used are introduced in Sect. 3. Section 4 provides the experiments details and results. Finally, Sect. 5 concludes the paper.

2 Related Work

Previous work on saliency prediction focused on low-level features. Far-reaching work by Itti [7] construct the first model to predict the saliency on images, which relied on color, intensity, orientation maps, and integrated them to get a global saliency map. After this seminal work, generous complementary methods about combining the low-level features were put forward. Judd [5] collected eye-tracking data to learn a model of saliency-based on low, middle, high-level features. Borji [8] combined low-level feature of previous best bottom-up models with top-down cognitive visual features and learn a direct mapping from those features to human eye fixations.

Same to other related fields of computer vision, deep learning solution achieved a far superior performance once it was proposed on saliency detection. And with the continuous progress of deep learning techniques, especially the success of Convolutional architectures, the performance of saliency detection is still steadily improving. *Ensemble of Deep Networks* (eDN) model by Vig et al. [9], one of the first proposals using a data-driving approach and richly-parameterized model, successfully predict image saliency map and outperform the previous work. After this proposal, many works based on convolutional neural networks emerged. Cornia et al. [10] explored combining CNN with recurrent architectures that focus on the most salient regions of the input image to iteratively refine the predicted saliency map. Pan et al. [11] introduced the Generative Adversarial Network into saliency detection. Their work used the generator to predict saliency maps that resemble the ground truth, and the discriminator to judge the authenticity of the saliency map. Recently, Reddy et al. [12] identified input features, multi-level integration, readout architecture, and loss function and proposed neater, minimal, more interpretable architecture, and achieved state-of-the-art performance on the SALICON [4], the largest eye-fixation dataset. This dataset contributed the availability of sufficient data and designed a mouse-contingent multi-resolutional paradigm to enable large-scale data collection.

This paper proposes a network architecture combining with attention mechanisms, which captures global dependencies. In particular, self-attention [13], also called intra-attention, applies in the natural language process, calculates the response at a position in a sequence by attending to all positions within the same sequence. Zhang et al. [14] introduced the self-attention module for image generation tasks. The proposed architecture also combines the self-attention module to efficiently find global and large-range dependencies within saliency maps.

3 Proposed Architecture

In this section, we introduce the proposed architectures, called SSM (Saliency Self-attention Model).

In general, the whole architecture adopts the convolutional encoder-decoder architecture. Section 3-A shows the detail of the network. The main innovation is the self-attention module, which is described in Sect. 3-B. Section 3-C shows the details of multi-level integration. The combination of evaluation metrics is used to evaluate the proposed network, and it is indicated in Sect. 3-D. Figure 2 shows the architecture of the proposal.

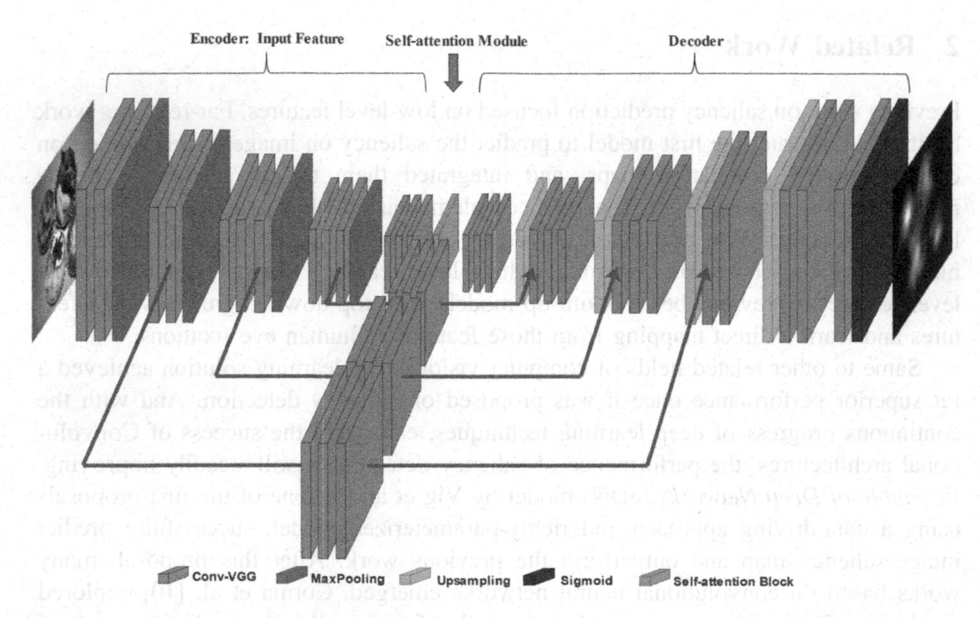

Fig. 2. The overview of the proposed Saliency Self-attention Model. After computing multi-scale feature maps on the inputs image through the encoder, a self-attention module based on attention mechanism is used to improve the global feature. Through the decoder, the model output the saliency prediction maps.

3.1 Overall Structure

The overall structure of the proposed network is introduced in this part. For saliency prediction, the fully convolutional framework achieves a great performance. As illustrated in Fig. 2, the whole network could be divided into three parts. The first is the feature maps extraction part, which can encode the input image and generate multi-scale feature maps. The second i the self-attention module we show in the next part. The third is the decoder, which upsamples the feature map to the same size with input image. The input size is initially resized to 256 $\times$256 and the initial channel is 3. In the encoder part, the network is identical in architecture to VGG16 [15] except the final max-pooling layer and three fully connected layers. Through the 13 convolutional layers and 4 max-pooling layers, the last layer of encoder have a small feature map with 16 $\times$16. And then the feature maps are fed into the self-attention module. For the decoder part, its layers' order is reversed with the encoder, with the max-pooling layers replaced by upsampling to successively restore feature maps' size. At the final of the network is a 1 $\times$1 convolutional layer with sigmoid non-linearity which ultimately produces the predicted saliency maps. There also have three U-Net like architecture that concatenates the same scale feature maps in encoder and decoder. Except for the weights of the encoder which are initialized with VGG-16 models pre-trained on ImageNet [16], other components' weights are randomly initialized. Hence VGG-SSM is used as the name for the proposed model.

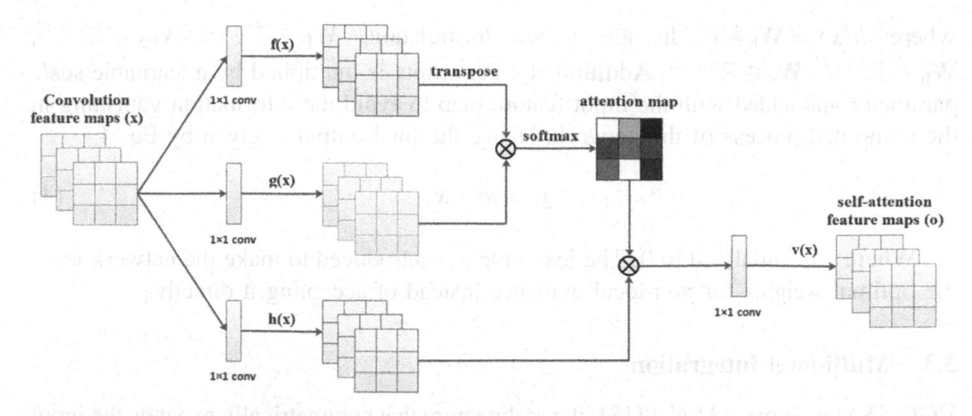

Fig. 3. The proposed self-attention module for VGG-SSM. The ⊗ denotes matrix multiplication.

3.2 Self-attention Module

Most saliency prediction models are built using CNN (Convolutional Neural Network) or RNN (Recurrent Neural Network). Unlike convolutional and recurrent operations, which both focus on building blocks that process local feature at a time, a non-local model [17] is adapted to combine self-attention with the previous part's network. Non-local means computing a weighted mean of all pixels in an image or a feature map. It allows distant pixels to contribute to the filtered response at a location based on patch appearance similarity. The self-attention module makes pixels in the feature map connect with all other pixels, no matter how distant. The approaches of the self-attention module are shown in Fig. 3. The input feature maps $\boldsymbol{x} \in \mathbb{R}^{H\times W\times C}$ from the last layers of the encoder is firstly transformed into two feature spaces with 1×1 convolution.

$$f(\boldsymbol{x}) = \mathbf{W_f} * \boldsymbol{x},\ g(\boldsymbol{x}) = \mathbf{W_g} * \boldsymbol{x} \tag{1}$$

where $*$ denotes convolutional opration, $\mathbf{W_f}$ and $\mathbf{W_g}$ are the 1×1 convolution kernels with C_1 channels. So $f(\boldsymbol{x})$ and $g(\boldsymbol{x})$ could be represented as $f(\boldsymbol{x}), g(\boldsymbol{x}) \in \mathbb{R}^{H\times W\times C_1}$. Then the attention map could be computed as Eq. 2.

$$\beta = \exp(\mathbf{s})/(\sum_{i=1}^{N} \exp(\mathbf{s})) \tag{2}$$

where $\boldsymbol{s} = f(\boldsymbol{x})^T g(\boldsymbol{x})$, in which $f(\boldsymbol{x})$ and $g(\boldsymbol{x})$ have been reshaped to $\{H \times W \times C_1\}$, $N = H \times W$. So after computing the softmax operation, the shape of β and $\boldsymbol{s}$ is the same with $\{H \times W, H \times W, C_1\}$. For memory efficiency, the method reduce the channel to $C_1 = C/k$ when computing 1×1 convolution, and choose $k = 8$ (i.e., $C_1 = C/8$) following [14] as the default value.

$$\boldsymbol{o} = \beta \otimes h(\boldsymbol{x})^T * \mathbf{W_v} \tag{3}$$

where $h(x) = \mathbf{W_h} * x$. In the above formulation, $\mathbf{W_f} \in \mathbb{R}^{C_1 \times C}$, $\mathbf{W_g} \in \mathbb{R}^{C_1 \times C}$, $\mathbf{W_h} \in \mathbb{R}^{C_1 \times C}$, $\mathbf{W_v} \in \mathbb{R}^{C_1 \times C}$. Additionally, the output is multiplied by a learnable scale parameter and added with the input feature map to avoid the information-vanishing in the computed process of the network. Hence the final output is given by Eq. 4.

$$y = \gamma o + x \tag{4}$$

Where γ is initialized to 0. The learnable γ is introduced to make the network learn the optimal weights for non-local evidence instead of accepting it directly.

3.3 Multi-level Integration

VGG-SSM employs a U-Net [18] like architecture that symmetrically expands the input feature maps after the first upsampling layer decoder. Feature maps in encoder and decoder with the same scale are concatenated to avoid information-vanishing. As shown in Fig. 2, there are three integrations in the whole architecture. Every step of expansion is composed of an upsampling of the feature map and concatenation with the same scale feature map from the encoder. Additionally, three 3 ×3 convolutional layers followed by ReLU are used to gradually extract deeper features at the original scale. The channels and scales are the same as the parameters of the convolutional layer before max-pooling.

3.4 Loss Function

The loss function evaluates the performance of the predicted saliency map compare with the ground truth. This paper uses a linear combination of three different saliency evaluation metrics: Kullback-Leibler Divergence (KLdiv), Pearson Cross-Correlation (CC), and Similarity (SIM). The new loss function is defined as follows:

$$L\left(\widehat{I},I\right) = \alpha \mathrm{KLdiv}\left(\widehat{I},I\right) + \beta CC\left(\widehat{I},I\right) + \gamma SIM\left(\widehat{I},I\right) \tag{5}$$

where $\widehat{I}$ and I are predicted saliency maps and the ground truth.

KLdiv is an information-theoretic measure of the difference between two probability distributions:

$$\mathrm{KLdiv}\left(\widehat{I},I\right) = \sum_i I log\left(\epsilon + \frac{I}{\widehat{I} + \epsilon}\right) \tag{6}$$

where i indexes the ith pixel and ϵ is a regularization constant. So KLdiv is computed on pixel-level.

CC is a statistical method used generally in the sciences for measuring how corrected or dependent two variables are.

$$CC(\hat{I},I) = \sigma(\hat{I},I)/(\sigma(\hat{I}) \times \sigma(I)) \tag{7}$$

where $\sigma\left(\widehat{I},I\right)$ denotes the covariance of $\widehat{I}$ and I.

SIM, also referred to as histogram intersection, measures the similarity between two distributions. SIM is computed as the sum of the minimum values at each pixel, after normalizing the input maps. Given a saliency map $\widehat{\boldsymbol{I}}$ and its ground truth $\boldsymbol{I}$:

$$SIM\left(\widehat{\boldsymbol{I}},\boldsymbol{I}\right) = \sum\nolimits_i min\left(\widehat{\boldsymbol{I}},\boldsymbol{I}\right), where \sum\nolimits_i \widehat{\boldsymbol{I}} = \sum\nolimits_i \boldsymbol{I} = 1 \quad (8)$$

iterating over discrete pixel location *i*.

The results of experiments using the proposed combined loss function are shown in Sect. 4-C.

4 Experiments and Results

The experiments' details and comparison results are shown in this section. Section 3-A shows the detail of the training process and other implementation details. Section 3-B describes the contributions of each component. The comparison between different loss functions is shown in Sect. 3-C. Finally, Sect. 3-D compares the proposed method with other state of the art. Here describe each part in detail.

4.1 Experimental Setup

Datasets: For training the proposed model and verify the results, we use the largest available dataset, SALICON [5] for saliency prediction. The dataset consists of 10,000 images for training, 5,000 images for validating, and 5,000 images for testing, taken from Microsoft COCO dataset [19]. We train the proposed model on SALICON datasets with 10,000 training images and use 5,000 images for validating. The ground truth maps are recorded by eye-tracker. It also provides the eye fixation simulated by mouse-click, but this part of the data is not used in the proposed method. The ground truth maps of test dataset are not available publicly, so the prediction only could be tested on the newest release, SALICON 2017, from the LSUN challenge.

Loss parameters: The parameters in the proposed loss function, α, β, γ are set to 10, -1 and -1 to balance the contribution of each components of loss function individually. Differently from the KLdiv loss which value should be minimized, the *CC* and the *SIM* loss is maximized to obtain the higher performance in saliency prediction. The values of the balancing weights are chosen by the target of obtaining good results on all evaluation metrics and by the numerical variation range single metrics have at convergence.

Evaluation metrics: This paper uses seven different evaluation metrics [6] adopted by SALICON to evaluate the proposed model. Among them, KLdiv, *CC* and *SIM* have been demonstrated in Sect. 3-D. AUC is the area under the ROC curve, the most widely used metric for evaluating saliency maps. The shuffled AUC metric (sAUC) samples negatives from other images, instead of uniformly at random. The Normalized Scanpath Saliency (NSS) is introduced to the saliency community as a simple correspondence measure between saliency maps and ground truth. Information Gain (IG) measures saliency model performance beyond systematic bias as an information theoretic metric.

Implementation Details: The training process resizes the input images into 256×256 resolution and trains VGG-SSM 30 epochs with the learning rate starting from 1e−4 and reducing after 3 epochs. The ADAM optimization algorithm is employed to train the whole network with the default batch size is set to 24. All the training and testing are conducted on one NVIDIA GeForce GTX 1080 Ti GPU with 11 GB memory.

4.2 Contribution of Each Component

The contributions of the self-attention module and the multi-level integration on SALICON test sets are described in this part. And the proposed combined loss function is used in the evaluation. To this end, this paper constructs three different components: the plain encoder-decoder architecture can be regarded as a baseline (This paper use VGGM to represent it), the self-attention module, and the multi-level integration. Table 1 illustrates the results of VGGM, VGGM plus self-attention module (Here use VGGSAM to represent), and the final version of the proposed model with all its components. As Table 1 shown, the results show that the overall architecture obtains the best grades on every evaluation metric and each component gives a great contribution to the final performance. It's obvious that the overall architecture makes a constant improvement on all metrics. For instance, the baseline achieved a result of 0.279 in terms of KLdiv, while it achieves a relative improvement of 5.0% with a self-attention module, and the result is improved by 1.5% when adding multi-level integration.

Table 1. Performance comparison of different version on test set of SALICON-2017.

Model	KLdiv ↓	CC ↑	AUC ↑	NSS ↑	SIM ↑	IG ↑	sAUC ↑
VGGM	0.279	0.854	0.858	1.839	0.745	0.750	0.727
VGGSAM	0.265	0.869	0.860	1.891	0.759	0.795	0.732
VGGSSM	**0.261**	**0.875**	**0.861**	**1.909**	**0.764**	**0.802**	**0.733**

4.3 Comparison Between Different Loss Functions

In this part, this paper verifies the effects of using different combinations of the loss function on SALICON validation set.

In Table 2, we compare the proposed loss function with its components individually as loss functions (KLdiv, CC, SIM). The results on SALICON validation set show the superiority of the proposed loss function. Although each single metric gain the best performance on its own evaluation term, the other evaluation terms obtain unsatisfactory results. Apparently, the combined loss function proposed to obtain an excellent trade-off among all the evaluation terms.

Table 2. Comparison between proposed loss function and its components using individually as loss function on Validation set of SALICON-2017.

Loss Function	KLdiv ↓	CC ↑	SIM ↑
KLdiv	**0.249**	0.872	0.764
CC	1.145	**0.881**	0.760
SIM	1.133	0.878	**0.773**
KLdiv+CC+SIM	0.251	0.876	0.769

Table 3 illustrates the result by adding CC and SIM to the KLdiv loss. Though we obtain better results when adding CC loss to KLdiv loss on CC evaluation metric, it brings reductions in other evaluation metrics. Higher performance can be achieved by adding CC and SIM terms to the loss. KLdiv+CC+SIM loss get all the results to value bold, which represent the best result upon different loss function.

Table 3. Comparison results between various loss functions on validation set of SALICON-2017.

Loss Function	KLdiv ↓	CC ↑	SIM ↑
KLdiv	0.249	0.872	0.764
KLdiv+CC	**0.247**	0.875	0.767
KLdiv+CC+SIM	0.251	**0.876**	**0.769**

4.4 Comparison with State-of-the-Art

The proposed models are compared with state of the art on SALICON test sets quantitatively. Table 4 shows the results in terms of KLdiv, CC, AUC, NSS, SIM, IG, and sAUC. VGG-SSM achieves great performance on two different metrics and outperforms other works by a large margin on KLdiv and IG. The proposed model also obtains competitive performance on other metrics.

Table 4. Performance comparison with state-of-the-art on test set of SALICON-2017.

Model	KLdiv ↓	CC ↑	AUC ↑	NSS ↑	SIM ↑	IG ↑	sAUC ↑
VGG-SSM (Ours)	**0.261**	0.875	0.861	1.909	0.764	**0.802**	0.733
EMLNET [20]	0.520	0.886	**0.866**	**2.050**	0.780	0.736	**0.746**
SAM-Resnet [10]	0.610	0.899	0.865	1.990	0.793	0.538	0.741
MSI-Net [21]	0.307	0.889	0.865	1.931	0.784	0.793	0.736
GazeNet [22]	0.376	0.879	0.864	1.899	0.773	0.720	0.736
ryanDINet [23]	0.777	**0.906**	0.864	1.979	**0.800**	0.347	0.742
Jinganu [23]	0.389	0.879	0.862	1.902	0.773	0.718	0.733
Lvjincheng [23]	0.376	0.856	0.855	1.829	0.705	0.613	0.726
Charleshuhy [23]	0.288	0.856	0.863	1.845	0.768	0.770	0.732

5 Conclusions

In this paper, a saliency self-attention Model VGG-SSM upon encoder-decoder architectures is proposed to predict saliency maps on natural images. This paper identifies three important components and does experiments to demonstrate the contribution of each part. The main novelty is the proposal of the self-attention module and its efficiency has been proved. Additionally, this paper compares the results of kinds of loss functions and validates the efficiency of combination loss function through an extensive evaluation. VGG-SSM achieves competitive results on SALICON test set. A similar method could be significant for other tasks that involve image refinement. Furthermore, the proposed model can be combined with a more recurrent network for potential further improvements.

References

1. Schauerte, B., Richarz, J., Fink, G.A.: Saliency-based identification and recognition of pointed-at objects. In: 2010 IEEE/RSJ International Conference on Intelligent Robots and Systems, pp. 4638–4643 (2010)
2. Frintrop, S., Kessel, M.: Most salient region tracking. In: 2009 IEEE International Conference on Robotics and Automation, pp. 1869–1874 (2009)
3. Takagi, S., Raskar, R., Gleicher, M.: Automatic image retargeting, vol. 154, no. 01, pp. 59–68 (2005)
4. Jiang, M., Huang, S., Duan, J., Zhao, Q.: SALICON: saliency in context, no. 06 (2015)
5. Judd, T., Ehinger, K., Durand, F., Torralba, A.: Learning to predict where humans look. In: 2009 IEEE 12th International Conference on Computer Vision, pp. 2106–2113 (2009)
6. Bylinskii, Z., Judd, T., Oliva, A., Torralba, A., Durand, F.: What do different evaluation metrics tell us about saliency models? arXiv e-print, arXiv:1604.03605 (2016)
7. Itti, L., Koch, C., Niebur, E.: A model of saliency-based visual attention for rapid scene analysis. IEEE Trans. Pattern Anal. Mach. Intell. **20**(11), 1254–1259 (1998)
8. Borji, A.: Boosting bottom-up and top-down visual features for saliency estimation. In: 2012 IEEE Conference on Computer Vision and Pattern Recognition, pp. 438–445 (2012)
9. Vig, E., Dorr, M., Cox, D.: Large-scale optimization of hierarchical features for saliency prediction in natural images. In: 2014 IEEE Conference on Computer Vision and Pattern Recognition, pp. 2798–2805 (2014)
10. Cornia, M., Baraldi, L., Serra, G., Cucchiara, R.: Predicting human eye fixations via an LSTM-based saliency attentive model. IEEE Trans. Image Process. **27**(10), 5142–5154 (2018)
11. Pan, J., et al.: SalGAN: Visual Saliency Prediction with Generative Adversarial Networks. arXiv e-prints, arXiv:1701.01081 (2017)
12. Reddy, N., Jain, S., Yarlagadda, P., Gandhi, V.: Tidying Deep Saliency Pre diction Architectures. arXiv e-prints, arXiv:2003.04942 (2020)
13. Parikh, A.P., Täckström O., Das, D., Uszkoreit, J.: A Decomposable Attention Model for Natural Language Inference. arXiv e-prints, arXiv:1606.01933 (2016)
14. Zhang, H., Goodfellow, I., Metaxas, D., Odena, A.: Self-Attention Generative Adversarial Networks. arXiv e-prints, arXiv:1805.08318 (2018)
15. Simonyan, K., Zisserman, A.: Very Deep Convolutional Networks for Large-Scale Image Recognition. arXiv e-prints, arXiv:1409.1556 (2014)

16. Russakovsky, O., et al.: ImageNet large scale visual recognition challenge. Int. J. Comput. Vis. **115**(3), 211–252 (2015). https://doi.org/10.1007/s11263-015-0816-y
17. Wang, X., Girshick, R., Gupta A., He, K.: Non-local Neural Networks. arXiv e-prints, arXiv:1711.07971 (2017)
18. Ronneberger, O., Fischer, P., Brox, T.: U-Net: convolutional networks for biomedical image segmentation. arXiv e-prints, arXiv:1505.04597 (2015)
19. Lin, T.-Y., et al.: Microsoft COCO: Common Objects in Context. *arXiv e-print*, arXiv:1405.0312 (2014)
20. Jia, S., Bruce, N.D.B.: EML-NET: An Expandable Multi-layer NETwork for Saliency Prediction. arXiv e-prints, arXiv:1805.01047 (2018)
21. Kroner, A., Senden, M., Driessens, K., Goebel, R.: Contextual Encoder-Decoder Network for Visual Saliency Prediction. arXiv e-prints, arXiv:1902.06634 (2019)
22. Che, Z., Borji, A., Zhai, G., Min, X., Guo, G., Le Callet, P.: How is gaze influenced by image transformations? dataset and model. IEEE Trans. Image Process. **29**, 2287–2300 (2020)
23. LSUN 2017. https://competitions.codalab.org/competitions/17136#results

Identification of Diseases and Pests in Tomato Plants Through Artificial Vision

Ernesto García Amaro(✉), Jair Cervantes Canales,
Josué Espejel Cabrera, José Sergio Ruiz Castilla,
and Farid García Lamont

Universidad Autónoma del Estado de México (UAEMEX), Jardín Zumpango s/n, Fraccionamiento El Tejocote, Texcoco, Estado de México, Mexico
ernestogarciaamaro@gmail.com
https://www.uaemex.mx

Abstract. The extraction of characteristics, currently, plays an important role, likewise, it is considered a complex task, allowing to obtain essential descriptors of the processed images, differentiating particular characteristics between different classes, even when they share similarity with each other, guaranteeing the delivery of information not redundant to classification algorithms. In this research, a system for the recogntion of diseases and pests in tomato plant leaves has been implemented. For this reason, a methodology represented in three modules has been developed: segmentation, feature extraction and classification; as a first instance, the images are entered into the system, which were obtained from the Plantvillage free environment dataset; subsequently, two segmentation techniques, Otsu and PCA, have been used, testing the effectiveness of each one; likewise, feature extraction has been applied to the dataset, obtaining texture descriptors with the Haralick and LBP algorithm, and chromatic descriptors through the Hu moments, Fourier descriptors, discrete cosine transform DCT and Gabor characteristics; finally, classification algorithms such as: SVM, Backpropagation, Naive Bayes, KNN and Random Forests, were tested with the characteristics obtained from the previous stages, in addition, showing the performance of each one of them.

Keywords: Tomato diseases · Artificial vision · Feature extraction

1 Introduction

Currently, México has an important role in exporting a great diversification of open-air crops; with a production percentage of 97.7%, and protected agriculture (greenhouses) of 2.3%; in addition, of the total tomato production, 60.8% [1] are obtained from protected environments; Chiapas is the main producer of coffee in protected environments, Guanajuato of broccoli, Mexico City of christmas eve and Sinaloa of tomato; therefore, in the country the number of greenhouses has increased; achieving an increase in production per plant and fruit quality; these results have been get with the implementation of new automated methods for the care of greenhouses, such as: controlling temperature, humidity and lighting; impacting on the care of planting,

D.-S. Huang and P. Premaratne (Eds.): ICIC 2020, LNAI 12465, pp. 98–109, 2020.
https://doi.org/10.1007/978-3-030-60796-8_9

nutrition, growth and harvest of it. However; producers have reported economic declines, due to diseases y pests that have attacked tomato plants. Some of the most common diseases in tomato plants, the following are considered: root rot, bacterial cancer of the tomato, freckle and bacterial spot, leaf mold, gray mold, early blight, late blight and dusty ashes [2], presented by variations in humidity, drought, temperature, residues of previous crops, wind, insects, overcast and negligence of greenhouse operators; diagnosing the disease, through the root, stem, leaf or fruit. After the identification of any anomaly in the plant, the producer goes to experts to accurately diagnose the disease, which is considered a late detection and with a certain degree of progress; likewise, the recommended dose of some pesticide or fungicide is applied to control and/or eliminate the disease, generating additional expenses. One of the main causes of loss of tomato production is the inaccurate identification of pests and diseases; for this reason, artificial vision algorithms have identified early and accurately: leaf mold, late blight, early blight, bacterial spot, septoria leaf spot, target spot, tomato mosaic virus, tomato yellow leaf curl virus, spider mites two-spotted and a completely healthy class, in tomato plant leaves, avoiding the excessive application of chemical products to combat diseases and pests, reducing the impact on plants and humans, in addition, contributing to the decrease in production loss and reducing financial losses.

2 State of the Art

Today, computer science, has been dedicated to solving problems in the environment in which we live, likewise, digital image processing and machine learning, among others, are considered areas that have stood out and have become fundamental techniques for this purpose. In this section, a study of investigations focused on the agricultural area is proposed, likewise, the works focus on the proposal of methods to solve different topics, such as: classification and recognition of leaves and identification of diseases and pests in leaves of different plants; solved with computer vision techniques. Plants, in their gender diversity, are currently of great importance, since they have a primary role for all living beings, and their development in all their environment; therefore, researchers who are in charge of the study and classification of plants, make the detection, through ocular methods, considering an inaccurate procedure; however, in literature, there are works focused on leaves recognition, achieving its mission with its own proposals with deep learning techniques, specifically, convolutional neural networks CNN, comparing performance with existing ones [3, 4]; on the other hand, the identification and classification of plants, through leaves that share similarities with each other, is a complex task, in previous works, this problem has been solved with the implementation of extraction techniques and selection of characteristics, considering color, shape and texture, classifying with machine learning algorithms, obtaining favorable results [5–8].

In addition, numerous works have been carried out; where, different methods are proposed to detect and classify diseases in leaves of different plants, through computer vision techniques [9]; likewise, in previous works, researchers have contributed in the field of segmentation, in color images and grayscale, considered an area with great opportunities, since it is still rigorously studied, both in controlled and non-controlled environments controlled; therefore, in-depth reviews of work related to color image segmentation have been developed [10], being a topic with a lot of impact, since it influences the performance of the classification algorithms; in addition, other works have segmented with the implementation of modified fully-convolutional networks FCNs through the leaves [11].

As previously mentioned, crops are affected by the unwanted arrival of pests [12] and diseases [13], both in protected environments and in the open air, likewise, this directly impacts production, reducing the financial balances of producers; therefore, computer science has made a great contribution trying to solve this problem, however, the resolution has not reached the top. In previous investigations, they have developed disease detection in different plants, using GWT feature extraction techniques and classifying with support vector machines SVM [14], on the other hand, improvements to CNN models have been proposed, based on CNN VGG, recognizing diseases, through the leaves [15] and the trunk of the plant [16].

Also, with the wave of deep learning implementation, it has taken a lot of strength and they have sought to solve multidisciplinary problems; however, and without lagging behind, CNN networks have been evaluated for the detection of diseases and pests in tomato plants [17]; as well as, both deep learning and machine learning techniques have been merged for the same purpose [18].

On the other hand, in the literature, the systems of detection and identification of diseases in tomato plants, through the leaves, deep learning, have turned to see with great momentum, since CNN networks have been implemented and evaluated, monitoring the performance of each proposed architecture [13, 19, 20]; without leaving behind, the development of robotic systems in conjunction with methods of computer vision [21]. The development of this research, has stood out for the low computational cost compared to [13, 20], since for training and testing, they use additional hardware or completely dedicated computing; likewise, for this proposed system, a portable computer equipment with mid-level characteristics has been used, considering taking the application to a real and/or mobile environment.

3 Methodology

In this section, a modular system was proposed that allows the precise identification of diseases and pests in tomato plant leaves, based on the implementation of characteristic extraction techniques and artificial intelligence algorithms, contributing to the reduction of financial losses and the excessive application of chemical products in crops, reducing their consumption in humans and plants. The adopted method is represented by three modules, segmentation, feature extraction and classification. For this work, a portable computer equipment has been used, with the following characteristics: MacBook Pro, Intel Core i5 2.6 GHz, 8 GB of memory (Fig. 1).

3.1 Segmentation

The experimentation applied in this section was executing the segmentation algorithm adaptive border Otsu [22, 23] and a phase of principal component analysis PCA [21]. By successfully segmenting an image, the system uses only the region of interest,

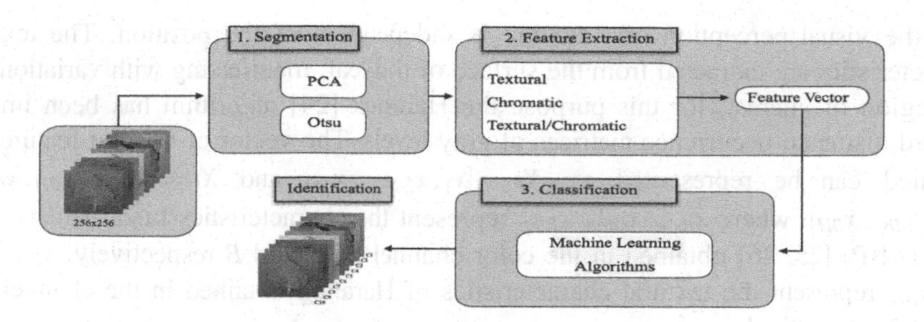

Fig. 1. Methodology used.

determining its edges and calculating properties by extracting features of textural, chromatic and textural/chromatic. Likewise, the segmentation results with both techniques are very similar, in some cases identical, so it was not necessary to use any technique such as: Probabilistic Rand Index (PRI), Variation of Information (VoI) and Boundary Displacement Error (BDE). In the Fig. 2, is shows the execution of segmentation methods on images of tomato plant leaves, in row one, there are images in RGB format; in row two, images segmented with the PCA algorithm and finally in row three, segmented images with Otsu.

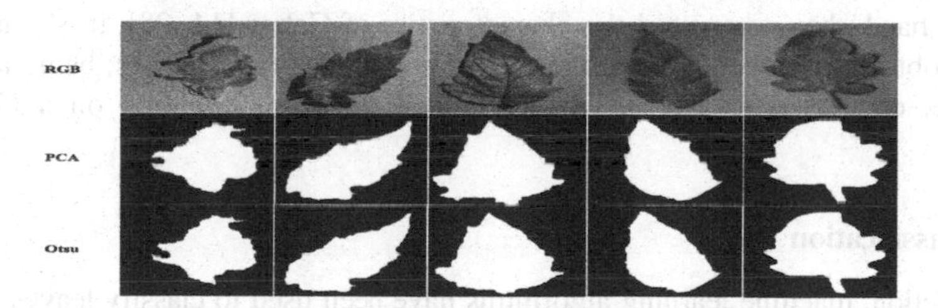

Fig. 2. Segmentation with PCA and Otsu.

3.2 Feature Extraction

The extraction of characteristics is a delicate process and is considered a cornerstone for machine learning algorithms, the correct implementation of extraction methods, define the descriptors gathered for the process of recognition of diseases and pests in tomato plant leaves. Furthermore, the characteristics obtained are invariant to scaling, rotation and translation, allowing the classifier to recognize objects despite having different size, position and orientation. Developing a comparative analysis with two methods of extraction of characteristics; textural features, chromatic features and the combination of both, textural/chromatic, measuring the performance of the system with machine learning classifiers on the Plantvillage dataset.

Textural Features. These structures give rise to a property that can be roughness, harshness, granulation, fineness, smoothness, among others. The texture is invariant to displacements, because it repeats a pattern along a surface, therefore, it is explained why the visual perception of a texture is independent of the position. The texture characteristics are extracted from the surface of the leaf, manifesting with variations in the region of interest, for this purpose the Haralick [24] algorithm has been implemented, using co-occurrence matrices of gray levels. The vector of textural features X_t obtained can be represented as: $Xt = [x_1, x_2, \ldots, x_{85}]$ and $Xt = [x_{Rlbp}, x_{RH}, x_{Glbp}, x_{GH}, x_{Blbp}, x_{BH}]$; where $x_{Rlbp}, x_{Glbp}, x_{Blbp}$ represent the characteristics Local Binary Patterns (LBP) [25, 26] obtained in the color channel R, G and B respectively, x_{RH}, x_{GH} and x_{BH} represent the textural characteristics of Haralick obtained in the channels R, G and B respectively.

Chromatic Features. Color characteristics provide a lot of information, and can be extracted starting from a specific color space, basically, are obtained starting from three primary channels, such as RGB, hue saturation value HSV and grayscale, among others, locating descriptors through different algorithms, considering: Hu moments, Fourier descriptors, discrete cosine transform DCT y characteristics of Gabor. The Hu moments [27], integrate information of the color variable of the region of interest; likewise, other characteristics were obtained with the Fourier descriptors, calculated using: $d_u = |F(u)|$; where $F(u)$ is calculated for $u = 1, \ldots, N$, where N is the number of descriptors to calculate. The discrete cosine transform DCT, use base transformations and cosine functions of different wavelengths. A particularity about DCT in relation to the discrete Fourier transform DFT, is the limitation to the use of real coefficients. On the other hand, they were used the characteristics of Gabor [14, 28], it is considered another robust technique, used for the extraction of features in images; being a hybrid technique, composed of the nucleus of the Fourier transformation on a Gaussian function.

3.3 Classification

In this section, machine learning algorithms have been used to classify leaves images; identifying ten different classes, including eight diseases, a plague and a completely healthy class; likewise, the performance of each of them is measured. The classifier, support vector machines (SVM), ANN Backpropagation, Naive Bayes, K-Nearest Neighbours (KNN) and Random Forests, were tested with different feature extraction techniques.

Support Vector Machines (SVM). SVM is one of the most widely used classification methods in recent years. The main characteristic that identifies it, is the use of kernels when working in non-linear sets, the absence of local minima, the solution depends on a small subset of data and the discriminatory power of the model constructed by optimizing the separability margin between the ten classes. When this is not possible, a function called Kernels is used, which transforms the input space to a highly dimensional space, where the sets can be linearly separated after the transformation. However, the choice of a function is restricted to those that satisfy the Mercer conditions [29].

ANN Backpropagation. Humans, to solve problems of daily life, take prior knowledge, acquired from the experience of some specific area, likewise, artificial neural networks, collect information on solved problems to build models or systems that can make decisions automatically. The multiple connections between neurons, form an adaptive system, the weights of which are updated using a particular learning algorithm. One of the most used ANN algorithms and the one that was implemented in this work was backpropagation (BP); which in general, performs the learning and classification process in four points; initialization of weights, forward spread, backward spread and the updating of weights. For further analysis of the BP algorithm, refer to [30].

Naive Bayes. A Bayesian classifier uses a probabilistic approach to assign the class to an example. Be C the class of an object, that belongs to a set of m class $(C_1, C_2, \ldots, C_m)$ and X_k is object with k characteristics $X_k = [x_1, x_2, \ldots, x_k]$, for this case, the set of characteristics defines a specific disease. For further analysis of the algorithm, refer to [31].

K-Nearest Neighbours (KNN). KNN, classifies a new point in the dataset, based on Euclidean distance, finding the k closest distances to the object to classify, later, the class of the closest point in the dataset is assigned by majority vote [32].

Random Forests. Random Forests is an algorithm composed of decision tree classifiers, each tree depends on the values of a random vector con with sampling independently and with the same distribution for all trees in the forests. Generalization error for forests converges to a limit, as the number of trees in the forest increases. When a model is generalized and fails, depends on the strength of individual trees in the forest and the correlation between them. By randomly selecting features to divide each node, error rates occur that compare favorably with the Adaboost algorithm but are more robust with respect to noise. For further analysis of the algorithm, refer to [33].

4 Experimental Results

In this section, the description of the metrics, the used dataset is presented, and the analysis of the results obtained, product of the experimentation developed for this proposed method. The selection of parameters is a very essential step, since a good selection of parameters has a considerable effect on the performance of the classifier. In all the classifiers used, the optimal parameters were obtained through cross validation. In the experiments carried out, cross validation with $k = 10$ was used to validate results, that is, 10 tests were performed with 90% and 10% of the data for training and testing respectively.

4.1 Metrics

In the experimental results presented in this research the evaluation metrics used for classification were the following.

Accuracy: represents the portion of instances that are correctly classified, out of the total number of cases, $Acc = \frac{TP+TN}{TP+TN+FP+FN}$; Precision: for each classifier used, performance was evaluated with this metric, getting the correct values of the classifier between the total of the dataset, $\Pr ecision = \frac{TP}{TP+FP}$; Recall: other metric used, is Recall, where, represents the number of positive predictions divided by the number of positive class values in the test data. Recall can be thought of as a measure of a classifiers completeness, $Recall = \frac{TP}{TP+FN}$; F-Measure: can be interpreted as a weighted average of the precision and recall, where an its best value is 1 and worst score is 0, $F - Measure = \frac{2*precision*recall}{precision + recall}$; true positive rate (TP Rate), $TP\ Rate = \frac{TP}{TP+FN}$; false positive rate (FP Rate), $FP\ Rate = \frac{FP}{FP+TN}$; MCC: is Matthews Correlation Coefficient, is used in machine learning as a measure of the quality of binary classifications, $MCC = \frac{(TP*TN)-(FP*TN)}{\sqrt{(TP+FP)(TP+FN)(TN+FP)(TN+FN)}}$; ROC Area: is Receiver Operating Characteristic, the ROC curve is defined by sensitivity, which is the true positive rate and 1-specificity, which is the false positive rate; and PRC Area: is Precision Recall Curve, is a plot of precision of positive predictive value against the Recall.

4.2 Dataset

In the process of the development of a disease or pest, the symptoms and the sign are factors that appear on the surface of the leaf, likewise, between these two stages there are some very similar visual characteristics; therefore, it is a complex task for machine learning algorithms to discriminate between classes. In the experimentation developed in this work, a free environment dataset has been used, Plantvillage [13, 18–20]; composed of ten different classes, eight diseases, such as: C1 = tomato mosaic virus with 373 images, C2 = leaf mold with 952, C3 = early blight with 1000, C4 = target spot with 1404, C7 = septoria leaf spot with 1771, C8 = late blight with 1908, C9 = bacterial spot with 2127 and C10 = tomato yellow leaf curl virus with 5357; and a plague, C6 = spider mites two-spotted with 1676 images, in addition to the C5 = completely healthy class with 1591 images; adding a total of 18159 processed images, the images are in a RGB color space with dimensions of 256x256 pixels, see Fig. 3. In the Fig. 3, some classes visually share color and texture characteristics; for example: classes C1, C2, C3, C6 and C10 have color characteristics in common; classes C3, C7 and C8 share color and texture characteristics; finally classes C7 and C9 have small brown spots; however, despite the similarities, the algorithms used in this work have managed to discriminate each class. The Plantvillage dataset has demonstrated its effectiveness; even when it's out of balance; this has been achieved under the implementation of performance metrics, defined in Sect. 4.1.

4.3 Experimental Results

In this section, the results of the different experiments carried out are shown, displaying the models build times, see Table 1; detailed accuracy by class, see Table 2 and Table 5; percent correctly classified instances, see Table 3 and Table 4, and comparison of results with two segmentation methods, see graph of the Fig. 4.

Build Times of the Models. The Table 1, shows the performance of the construction times of the models, considering the Naive Bayes, Backpropagation, KNN, Random Forests and SVM algorithms. The classifiers performed this calculation with different

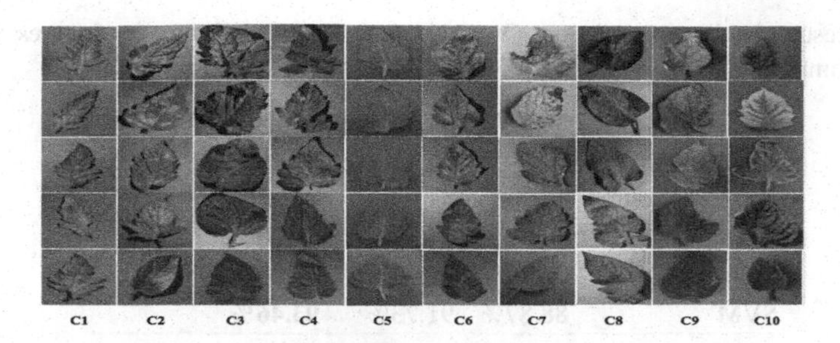

Fig. 3. Dataset Plantvillage.

Table 1. Build times of the models.

	Otsu			PCA		
Classifier	T	C	T/C	T	C	T/C
Naive Bayes	0.32	0.92	1.13	0.3	0.97	1.14
Backpropagation	507.2	7403.48	12817.64	499.51	7718.28	12417.69
KNN	0.01	0.01	0.01	0.15	0.02	0.01
Random Forests	21.59	34.63	32.07	22.46	34.3	32.41
SVM	1754.61	914.96	570.3	1982.99	528.07	459.66

Table 2. Results by class, with segmentation Otsu, textural/chromatic features and classification SVM.

Class	TP Rate	FP Rate	Precision	Recall	F-Measure	MCC	ROC Area	PRC Area
C1	0.923	0.002	0.901	0.923	0.912	0.910	0.997	0.897
C2	0.914	0.006	0.884	0.914	0.899	0.893	0.986	0.855
C3	0.854	0.010	0.831	0.854	0.842	0.833	0.969	0.758
C4	0.899	0.010	0.881	0.899	0.890	0.881	0.983	0.837
C5	0.981	0.002	0.981	0.981	0.981	0.979	0.998	0.981
C6	0.929	0.007	0.929	0.929	0.929	0.922	0.991	0.898
C7	0.890	0.009	0.903	0.890	0.896	0.887	0.983	0.861
C8	0.879	0.012	0.893	0.879	0.886	0.872	0.975	0.837
C9	0.950	0.006	0.954	0.950	0.952	0.945	0.990	0.928
C10	0.977	0.008	0.982	0.977	0.979	0.970	0.994	0.977

techniques of processing digital of images, applied to the dataset, such as: segmentation Otsu and PCA; extraction of characteristics, considering, T = Textural, C = Chromatic and the combination of both T/C = Textural/Chromatic. The results show that the shortest processing time was obtained by the KNN algorithm, while Backpropagation was the most costing. For these results, the unit of measurement is expressed in seconds.

Table 3. Results with segmentation Otsu, textural features, chromatic features and both textural/chromatic.

Classifier	Textural	Chromatic	Textural/Chromatic
Naive Bayes	37.86%	38.26%	40.37%
Backpropagation	81.44%	87.69%	82.21%
KNN	72.39%	74.59%	79.66%
Random Forests	76.17%	79.67%	81.95%
SVM	88.87%	91.73%	**93.46%**

Table 4. Results with segmentation PCA, textural features, chromatic features and both textural/chromatic.

Classifier	Textural	Chromatic	Textural/Chromatic
Naive Bayes	39.12%	44.76%	46.13%
Backpropagation	81.81%	88.27%	81.46%
KNN	73.57%	76.10%	80.23%
Random Forests	77.27%	80.77%	82.77%
SVM	89.81%	92.71%	**93.86%**

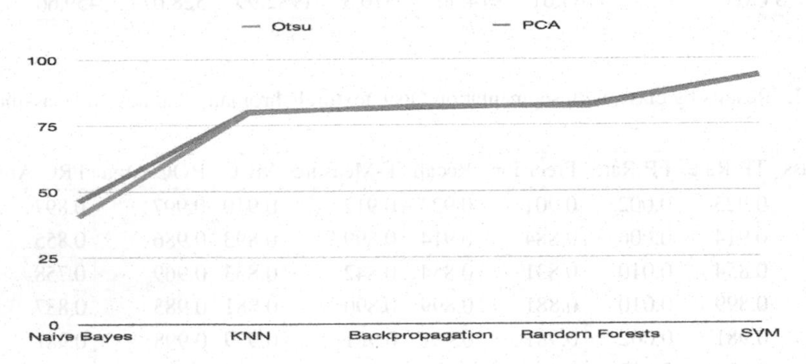

Fig. 4. Graph of results with two segmentation methods.

Results with the Otsu Segmentation Method. In this part of the article, are shows the results obtained from executing the experimentation with the method Otsu. The Table 2, contains the detail of percentages of accuracy by class, these results, are product of the tests of the highest percentage obtained, including techniques of feature extraction of textural/chromatic, classifying with the SVM algorithm. The metrics displayed, are described in Sect. 4.1.

In the following results, Otsu was applied, in addition, SVM, Backpropagation, Naive Bayes, KNN and Random Forests; were tested with textural features, chromatic and the combination of both textural/chromatic. The best performing algorithm was SVM, obtaining a percent correctly classified instances for textural features, 88.87%, for chromatic features, 91.73% and for textural/chromatic features, 93.46%, see Table 3.

Table 5. Results by class, with segmentation PCA, textural/chromatic features and classification SVM.

Class	TP Rate	FP Rate	Precision	Recall	F-Measure	MCC	ROC Area	PRC Area
C1	0.910	0.002	0.877	0.910	0.893	0.891	0.997	0.874
C2	0.936	0.005	0.914	0.936	0.925	0.921	0.991	0.903
C3	0.853	0.012	0.805	0.853	0.828	0.818	0.967	0.738
C4	0.903	0.009	0.890	0.903	0.897	0.888	0.985	0.853
C5	0.982	0.002	0.984	0.982	0.983	0.982	0.999	0.984
C6	0.918	0.008	0.916	0.918	0.917	0.909	0.990	0.887
C7	0.916	0.006	0.933	0.916	0.924	0.917	0.986	0.895
C8	0.881	0.010	0.909	0.881	0.894	0.882	0.977	0.850
C9	0.950	0.006	0.955	0.950	0.952	0.945	0.991	0.930
C10	0.981	0.007	0.984	0.981	0.982	0.974	0.995	0.980

Results with the PCA Segmentation Method. In this part of the article, are shows the results obtained from executing the experimentation with the method PCA. The Table 5, contains the detail of percentages of accuracy by class, these results, are product of the tests of the highest percentage obtained, including techniques of feature extraction of textural/chromatic, classifying with the SVM algorithm. The metrics displayed, are described in Sect. 4.1.

In the following results, PCA was applied, in addition, SVM, Backpropagation, Naive Bayes, KNN and Random Forests; were tested with textural features, chromatic and the combination of both textural/chromatic. The best performing algorithm was SVM, obtaining a percent correctly classified instances for textural features, 89.81%, for chromatic features, 92.71% and for textural/chromatic features, 93.86%, see Table 4.

In the Fig. 4, the best results of this research are observed, considering the algorithms, the percent correctly classified instances and the features textural/chromatic. The orange line belongs to the tests with segmentation Otsu, and the green line to the tests with segmentation PCA. The highest percent was obtained with segmentation PCA, except, in the backpropagation algorithm.

5 Conclusion

Derived of analysis, the applied digital image processing techniques, the integration of classification algorithms and the experimentation carried out; it was shown, that by combining the segmentation PCA method, the conjunction of textural/chromatic feature extraction, and the SVM classification process, the system has achieved a performance of 93.86% respectively. The main contribution of the method developed in this research, is the identification of diseases and pests in tomato plant leaves early and accurately, reducing the financial losses and the excessive application of chemical products, minimizing the affectation to plants and human beings; likewise, the proposed system can be implemented in a real and/or mobile environment, since the computational cost is low compared with other works, and can be executed in a portable computer equipment, without requiring additional hardware.

References

1. INEGI: Encuesta nacional agropecuaria 2017. Report ENA. 2017, Instituto Nacional de Estadística y Geografía, México (2017). http://www.beta.inegi.org.mx/proyectos/encagro/ena/2017/
2. CESAVEG: Campaña manejo fitosanitario del jitomate. Comité Estatal de Sanidad Vegetal de Guanajuato, A.C., Irapuato Guanajuato, 2016 edn. (2016)
3. Jiao, Z., Zhang, L., Yuan, C.-A., Qin, X., Shang, L.: Plant leaf recognition based on conditional generative adversarial nets. In: Huang, D.-S., Bevilacqua, V., Premaratne, P. (eds.) ICIC 2019. LNCS, vol. 11643, pp. 312–319. Springer, Cham (2019). https://doi.org/10.1007/978-3-030-26763-6_30
4. Zheng, Y., Yuan, C.-A., Shang, L., Huang, Z.-K.: Leaf recognition based on capsule network. In: Huang, D.-S., Bevilacqua, V., Premaratne, P. (eds.) ICIC 2019. LNCS, vol. 11643, pp. 320–325. Springer, Cham (2019). https://doi.org/10.1007/978-3-030-26763-6_31
5. Ayala Niño, D., Ruiz Castilla, J.S., Arévalo Zenteno, M.D., D. Jalili, L.: Complex Leaves Classification with Features Extractor. In: Huang, D.-S., Jo, K.-H., Huang, Z.-K. (eds.) ICIC 2019. LNCS, vol. 11644, pp. 758–769. Springer, Cham (2019). https://doi.org/10.1007/978-3-030-26969-2_72
6. Cervantes, J., Garcia Lamont, F., Rodriguez Mazahua, L., Zarco Hidalgo, A., Ruiz Castilla, J.S.: Complex identification of plants from leaves. In: Huang, D.-S., Gromiha, M.M., Han, K., Hussain, A. (eds.) ICIC 2018. LNCS (LNAI), vol. 10956, pp. 376–387. Springer, Cham (2018). https://doi.org/10.1007/978-3-319-95957-3_41
7. Jalili, L.D., Morales, A., Cervantes, J., Ruiz-Castilla, J.S.: Improving the performance of leaves identification by features selection with genetic algorithms. In: Figueroa-García, J.C., López-Santana, E.R., Ferro-Escobar, R. (eds.) WEA 2016. CCIS, vol. 657, pp. 103–114. Springer, Cham (2016). https://doi.org/10.1007/978-3-319-50880-1_10
8. Cervantes, J., Taltempa, J., García-Lamont, F., Castilla, J.S.R., Rendon, A.Y., Jalili, L.D.: Análisis comparativo de las técnicas utilizadas en un sistema de reconocimiento de hojas de planta. Revista Iberoamericana de Automática e Informática Industrial RIAI **14**(1), 104–114 (2017). https://doi.org/10.1016/j.riai.2016.09.005
9. Dhingra, G., Kumar, V., Joshi, H.D.: Study of digital image processing techniques for leaf disease detection and classification. Multimed. Tools Appl. **77**(15), 19951–20000 (2017). https://doi.org/10.1007/s11042-017-5445-8
10. Garcia-Lamont, F., Cervantes, J., López, A., Rodriguez, L.: Segmentation of images by color features: a survey. Neurocomputing **292**, 1–27 (2018). https://doi.org/10.1016/j.neucom.2018.01.091
11. Wang, X.-f., Wang, Z., Zhang, S.-w.: Segmenting crop disease leaf image by modified fully-convolutional networks. In: Huang, D.-S., Bevilacqua, V., Premaratne, P. (eds.) ICIC 2019. LNCS, vol. 11643, pp. 646–652. Springer, Cham (2019). https://doi.org/10.1007/978-3-030-26763-6_62
12. Gutierrez, A., Ansuategi, A., Susperregi, L., Tubío, C., Rankić, I., Lenza, L.: A benchmarking of learning strategies for pest detection and identification on tomato plants for autonomous scouting robots using internal databases. J. Sens. **2019**, 1–15 (2019). https://doi.org/10.1155/2019/5219471
13. Zhang, K., Wu, Q., Liu, A., Meng, X.: Can deep learning identify tomato leaf disease? Adv. Multimed. **2018**, 1–10 (2018). https://doi.org/10.1155/2018/6710865
14. Prasad, S., Kumar, P., Hazra, R., Kumar, A.: Plant leaf disease detection using gabor wavelet transform. In: Panigrahi, B.K., Das, S., Suganthan, P.N., Nanda, P.K. (eds.) SEMCCO 2012. LNCS, vol. 7677, pp. 372–379. Springer, Heidelberg (2012). https://doi.org/10.1007/978-3-642-35380-2_44

15. Fang, T., Chen, P., Zhang, J., Wang, B.: Identification of apple leaf diseases based on convolutional neural network. In: Huang, D.-S., Bevilacqua, V., Premaratne, P. (eds.) ICIC 2019. LNCS, vol. 11643, pp. 553–564. Springer, Cham (2019). https://doi.org/10.1007/978-3-030-26763-6_53
16. Hang, J., Zhang, D., Chen, P., Zhang, J., Wang, B.: Identification of apple tree trunk diseases based on improved convolutional neural network with fused loss functions. In: Huang, D.-S., Bevilacqua, V., Premaratne, P. (eds.) ICIC 2019. LNCS, vol. 11643, pp. 274–283. Springer, Cham (2019). https://doi.org/10.1007/978-3-030-26763-6_26
17. Fuentes, A., Yoon, S., Kim, S., Park, D.: A robust deep-learning-based detector for real-time tomato plant diseases and pests recognition. Sensors **17**(9), 2022 (2017). https://doi.org/10.3390/s17092022
18. Shijie, J., Peiyi, J., Siping, H., sLiu Haibo: Automatic detection of tomato diseases and pests based on leaf images. In: 2017 Chinese Automation Congress (CAC). IEEE, October 2017. https://doi.org/10.1109/cac.2017.8243388
19. Suryawati, E., Sustika, R., Yuwana, R.S., Subekti, A., Pardede, H.F.: Deep structured convolutional neural network for tomato diseases detection. In: 2018 International Conference on Advanced Computer Science and Information Systems (ICACSIS). IEEE, October 2018. https://doi.org/10.1109/icacsis.2018.8618169
20. Durmus, H., Gunes, E.O., Kirci, M.: Disease detection on the leaves of the tomato plants by using deep learning. In: 2017 6th International Conference on Agro-Geoinformatics. IEEE, August 2017. https://doi.org/10.1109/Agro-Geoinformatics.2017.8047016
21. Schor, N., Bechar, A., Ignat, T., Dombrovsky, A., Elad, Y., Berman, S.: Robotic disease detection in greenhouses: combined detection of powdery mildew and tomato spotted wilt virus. IEEE Robot. Autom. Lett. **1**(1), 354–360 (2016). https://doi.org/10.1109/lra.2016.2518214
22. Gonzalez, R.C., Woods, R.E., Eddins, S.L.: Digital Image Processing using MATLAB. Pearson Education India, London (2004)
23. Sonka, M., Hlavac, V., Boyle, R.: Image Processing. Analysis and Machine Vision. Springer, Heidelberg (1993). https://doi.org/10.1007/978-1-4899-3216-7
24. Haralick, R.M., Shanmugam, K., Dinstein, I.: Textural features for image classification. IEEE Trans. Syst. Man Cybern. SMC **3**(6), 610–621 (1973). https://doi.org/10.1109/tsmc.1973.4309314
25. He, D.C., Wang, L.: Texture unit, texture spectrum and texture analysis. In: 12th Canadian Symposium on Remote Sensing Geoscience and Remote Sensing Symposium. IEEE. https://doi.org/10.1109/igarss.1989.575836
26. Wang, L., He, D.C.: Texture classification using texture spectrum. Pattern Recogn. **23**(8), 905–910 (1990). https://doi.org/10.1016/0031-3203(90)90135-8
27. Hu, M.K.: Visual pattern recognition by moment invariants. IEEE Trans. Inf. Theor. **8**(2), 179–187 (1962). https://doi.org/10.1109/tit.1962.1057692
28. Gabor, D.: Theory of communication. J. Inst. Electr. Eng. **93**, 429–457 (1946)
29. Vapnik, V.: An overview of statistical learning theory. IEEE Trans. Neural Netw. **10**(5), 988–999 (1999). https://doi.org/10.1109/72.788640
30. Rumelhart, D.E., Hinton, G.E., Williams, R.J.: Learning representations by back-propagating errors. Nature **323**(6088), 533–536 (1986). https://doi.org/10.1038/323533a0
31. John, G.H., Langley, P.: Estimating continuous distributions in bayesian classifiers (1995)
32. Aha, D.W., Kibler, D., Albert, M.K.: Instance-based learning algorithms. Mach. Learn. **6**(1), 37–66 (1991). https://doi.org/10.1007/bf00153759
33. Breiman, L.: Random forests. Mach. Learn. **45**(1), 5–32 (2001)

Depth Guided Attention for Person Re-identification

Md Kamal Uddin[1,2](✉), Antony Lam[3], Hisato Fukuda[1], Yoshinori Kobayashi[1], and Yoshinori Kuno[1]

[1] Graduate School of Science and Engineering, Saitama University, Saitama, Japan
{kamal, fukuda, kuno}@cv.ics.saitama-u.ac.jp, yosinori@hci.ics.saitama-u.ac.jp
[2] Noakhali Science and Technology University, Noakhali, Bangladesh
[3] Mercari, Inc., Roppongi Hills Mori Tower 18F, 6-10-1 Roppongi Minato-ku, Tokyo 106-6118, Japan
antonylam@cv.ics.saitama-u.ac.jp

Abstract. Person re-identification is an important video-surveillance task for recognizing people from different non-overlapping camera views. Recently it has gained significant attention upon the introduction of different sensors (i.e. depth cameras) that provide the additional information irrespective of the visual features. Despite recent advances with deep learning models, state-of-the-art re-identification approaches fail to leverage the sensor-based additional information for robust feature representations. Most of these state-of-the-art approaches rely on complex dedicated attention-based architectures for feature fusion and thus become unsuitable for real-time deployment. In this paper, a new deep learning method is proposed for depth guided re-identification. The proposed method takes into account the depth-based additional information in the form of an attention mechanism, unlike state-of-the-art methods of complex architectures. Experimental evaluations on a depth-based benchmark dataset suggest the superiority of our proposed approach over the considered baseline as well as with the state-of-the-art.

Keywords: Re-identification · Depth guided attention · Triplet loss

1 Introduction

In recent years, Person re-identification (Re-id) has gained great attention in both the computer vision community and industry because of its practical applications, such as in forensic search, multi-camera tracking and public security event detection. Person re-identification is still a challenging task in computer vision due to the variation of person pose, misalignment, different illumination conditions and diverse cluttered backgrounds. Figure 1 shows a typical person re-identification system, where the task is to match the unknown probe with a set of known gallery images captured over non-overlapping cameras. It can be clearly observed from Fig. 1 that background clutter here in the scene works as the source of noisy information. And the trained model could be suffering from over-fitting as noisy information could propagate to it as salient features.

D.-S. Huang and P. Premaratne (Eds.): ICIC 2020, LNAI 12465, pp. 110–120, 2020.
https://doi.org/10.1007/978-3-030-60796-8_10

State-of-the-art approaches in Re-id deal with this problem by relying on different attention-based mechanisms [1–4]. All state-of-the-art attention-based Re-id approaches can be placed into two categories: whole body attention and part-based attention. In the former case, methods focused on whole body attention, fully focused on the foreground while part-based methods focus more on local body parts. In all of these cases, methods rely on complex dedicated architectures which hinder the processes to deploy them in real world applications due to their large and over-parametrized models. Moreover, these methods are mainly based on RGB input that do not leverage additional information from other sources such as depth images.

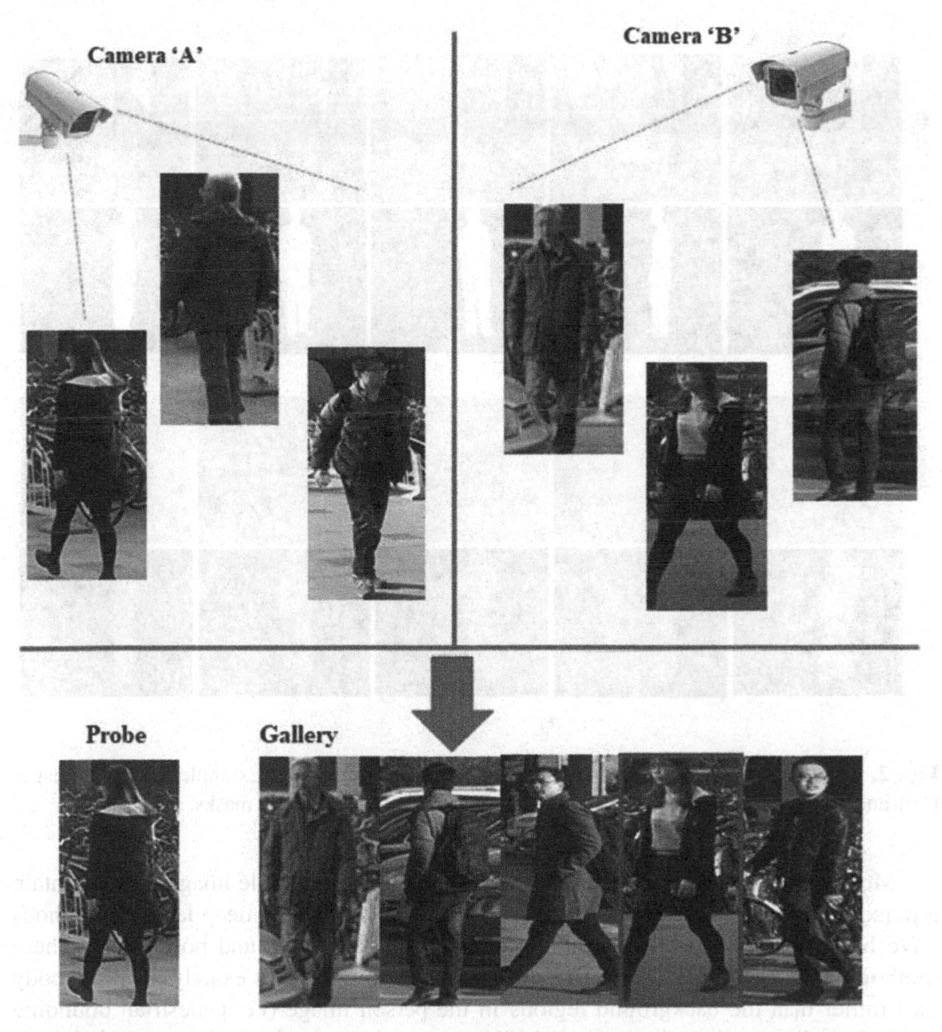

Fig. 1. Illustration of challenges for a typical re-identification system. Sample images are taken from [23].

In our work, we emphasize how to extract discriminative and robust features using a depth sensor-based camera (e.g. Microsoft Kinect) when an individual appears on different cameras with diverse cluttered backgrounds. Specifically, whenever videos are recorded with a Kinect camera (i.e. RGB-D sensor) for each person, the Kinect SDK provides RGB frames, depth frames, the person's segmentation mask and skeleton data [5] with low computational effort.

In this paper, we introduce depth guided binary segmentation masks to construct masked-RGB images (i.e. foreground images), where masked-RGB images retain the whole-body part of a person with different viewpoint variations and pose (see Fig. 2). In this work, we also focus on long-term person re-identification for RGB-D sensors with different pose variations of a person, which is suited to our proposed approach.

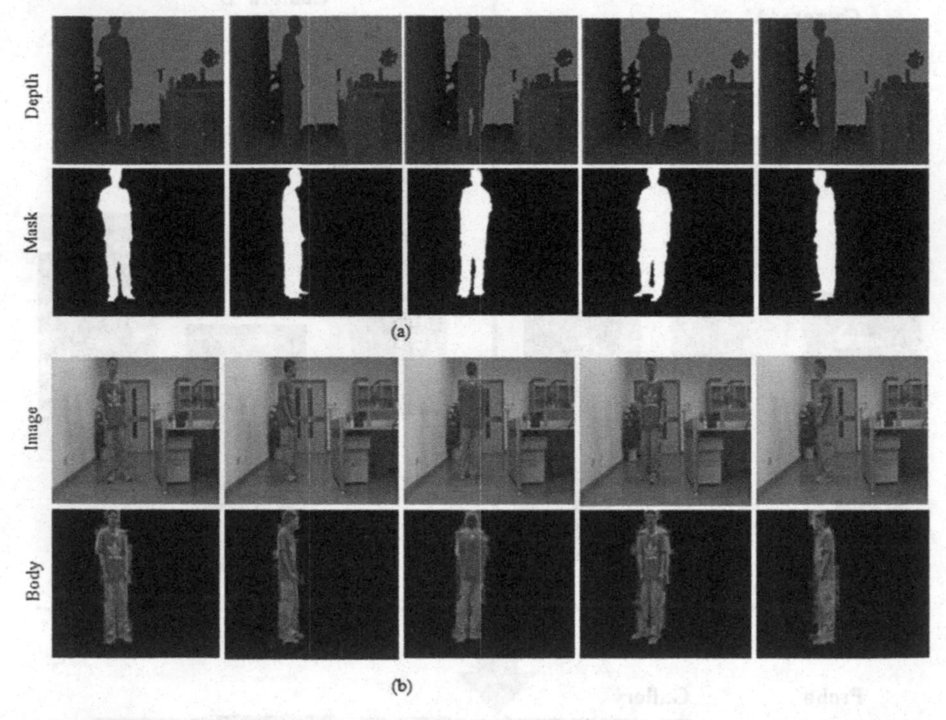

Fig. 2. (a) Illustration of depths and their corresponding masks. (b) Examples of RGB images [19] and their corresponding body regions extracted directly with the masks.

Most previous methods directly learn features from the whole image which contain a person's body with a cluttered background. Recently, several deep learning methods have been proposed to learn features from the body parts [6] and pose [7, 8]. These methods have been proved effective through extracting features exactly from the body part rather than the background regions in the person image (i.e. pedestrian bounding box). It indicates that eliminating the background clutter in each person image is helpful for improving the performance of person re-identification.

This paper also proposes a new deep learning Re-id framework that takes into account the additional information from the depth domain, thanks to the depth camera. Unlike past methods, the proposed approach exploits the advantage of using the depth image to generate a person's segmentation mask that helps us to develop deep learning methods which focus only on the foreground.

We evaluated the proposed method on the publicly available RGB-D dataset RobotPKU RGBD-ID. Experimental results show the effectiveness of our proposed method. The contributions of this paper can be summarized as follows:

1. We introduce a depth guided (DG) attention-based person re-identification framework. The key component of this framework is the depth-guided foreground extraction that helps the model to dynamically select the more relevant convolutional filters of the backbone CNN architecture.
2. Extensive experiments show the effectiveness of the proposed method in a depth-based benchmark re-identification dataset.

2 Related Work

In this section, we first review some related works in person re-identification, especially for whole body attention and part-based attention which are the most related to our work.

There are some state-of-the-art methods in the Re-id task [1–4] to handle the background clutter problem in RGB images using whole body attention and part-based attention mechanisms. The first key ingredient of these approaches is human body mask generation which is very costly in computation. These methods obtain human body masks using different deep learning based image segmentation models such as FCN [9], Mask R-CNN [10], JPPNet [11], and Dense Pose [12]. In [1], the authors generate binary segmentation masks corresponding to the body and background regions with an FCN [9] based segmentation model which is trained on labeled human segmentation datasets such as [13, 14]. The authors also designed a contrastive attention model which is guided by these binary masks and finally generated whole body-aware and background-aware attention maps. *Chen et al.* [2] proposes a mask-guided two stream CNN model for person Re-id, which explicitly makes use of one stream from the foreground and another one from the original image. To separate the foreground person from the background, authors apply an off-the-shelf instance segmentation method, FCIS [15] on the whole image, and then designate the person to the correct mask via majority vote. In [3], the authors propose a human semantic parsing model that learns to segment the human body into multiple semantic regions and then use them to exploit local cues for person re-identification.

Recently, *Cai et al.* [4] proposes a multi-scale body-part mask guided attention network to improve Re-id performance. The authors creatively use the masks of different parts of the body to guide attention learning in the training phase. All the above state-of-the-art approaches heavily depend on very complex dedicated attention-based architectures which involve large computational costs. For this reason, it is difficult to deploy for them in real time scenarios.

In contrast to the above works, we propose a new deep learning Re-id framework that takes into account the additional information from the depth domain and introduce a depth guided attention mechanism for person re-identification with less computational effort.

3 Proposed Method

In this section, we present our proposed depth guided attention-based person Re-id in detail. First, we describe the overall framework of our method, then we present our triplet-based convolutional neural networks (CNNs) structure.

3.1 The Overall Framework

Our proposed pipeline is illustrated in Fig. 3. Our Re-id framework consists of two states: depth guided body segmentation and triplet loss for re-identification.

In the first stage, we extract the foreground part of each image with the help of depth guided person segmentation masks. Once the foreground has been separated, then we feed the extracted body part T into the CNN model for feature mapping. For a given mask I_m and corresponding RGB image I_{rgb}, we separate the foreground after performing following operation,

$$T = I_m \otimes I_{rgb} \quad (1)$$

where $\otimes$ represents the element-wise product.

In the second stage, we describe the whole training procedure for Re-id with CNN blocks. All the CNN blocks share parameters (i.e. weights and biases). During training, three CNNs take triplet examples (i.e. three foreground images), which is denoted as $T_i = \left(T_i^a, T_i^p, T_i^n\right)$ and forming the *i-th* triplet, where superscript 'a' indicates the anchor image, 'p' indicates positive image and 'n' indicates negative image. 'a' and 'p' come from the same person while 'n' is from a different person. Foreground images are fed into the CNN model and maps the triplets T_i from the raw image space into a learned feature space $F_i = \left(F_i^a, F_i^p, F_i^n\right)$. For details, when a sample image is fed into the CNN model, it maps to the deep feature space $F = \varphi(x)$, where $\varphi(\cdot)$ represents the mapping function of the whole CNN model and x is the input representation of the corresponding image T.

3.2 Triplet Loss

The CNN model is trained by a triplet loss function. In particular, the triplet loss has been shown to be effective in state-of-the-art person Re-id systems [16, 17]. The triplet loss function aims to reduce the distance of feature vectors (i.e. F_i^a and F_i^p) taken from the same person (i.e. a and p) and enlarge the distance between different persons (i.e. a and n). It is defined as

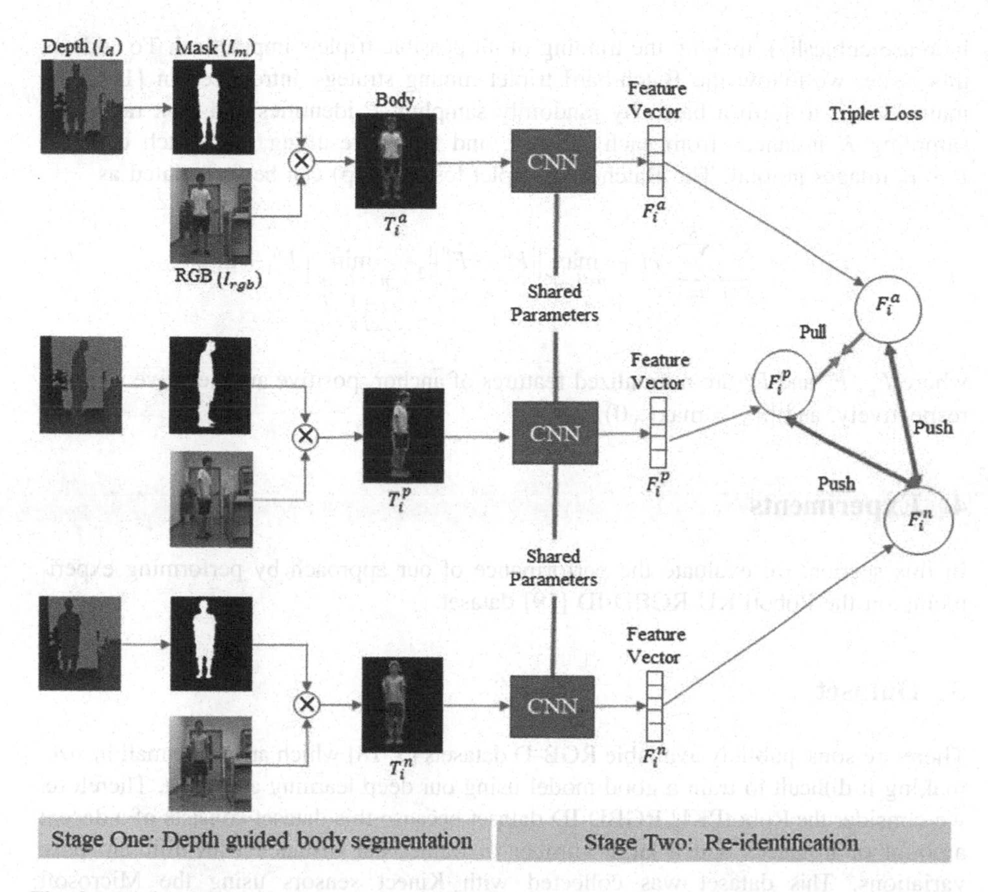

Fig. 3. Triplet training framework for re-identification. It is composed of two stages: 1) Depth guided body segmentation and 2) Body segmented images are fed into three CNN models with shared parameters, where the triplet loss aims to pull the instances of the same person closer and at the same time, push the instances of different persons farther from each other in the learned feature space.

$$L_{trp} = \max\left\{0, \|F_i^a - F_i^p\|_2^2 - \|F_i^a - F_i^n\|_2^2 + m\right\} \tag{2}$$

where $\|\cdot\|_2^2$ is the squared Euclidean distance and m is a predefined margin which regularizes the distance. In our work, we train our model with margin $m = 0.3$. We use the Euclidean distance in our all experiments because the authors in [16] notice that using the squared Euclidean distance makes the optimization more prone to collapsing, whereas using an actual (non-squared) Euclidean distance is more stable.

Triplet generation is crucial to the final performance of the system. When the CNN is trained with the triplet inputs for a large-scale dataset then there can be an enormous possible number of combinations of triplet inputs (because triplet combinations increase cubically), making the training of all possible triplets impractical. To address this issue, we follow the Batch-hard triplet mining strategy introduced in [16]. The main idea is to form a batch by randomly sampling P identities and then randomly sampling K instances from each identity, and thus a resulting mini-batch contains $P \times K$ images in total. The Batch-hard triplet loss (BHtrp) can be formulated as

$$L_{BHtrp} = \sum_{i=1}^{P} \sum_{a=1}^{K} [m + \max_{p=1...K} \left\| F_i^a - F_i^p \right\|_2 - \min_{\substack{n=1...K \\ j=1...P \\ j \neq i}} \left\| F_i^a - F_j^n \right\|_2]_+ \tag{3}$$

where F_i^a, F_i^p and F_i^n are normalized features of anchor, positive and negative samples respectively, and $[.]_+ = \max(.,0)$.

4 Experiments

In this section, we evaluate the performance of our approach by performing experiments on the RobotPKU RGBD-ID [19] dataset.

5 Dataset

There are some publicly available RGB-D datasets [5, 18] which are very small in size, making it difficult to train a good model using our deep learning approach. Therefore, we consider the RobotPKU RGBD-ID dataset because this dataset consists of a decent amount of instances and a large number of frames per instance with different pose variations. This dataset was collected with Kinect sensors using the Microsoft Kinect SDK. There are 180 video sequences of 90 people, and for each person still and walking sequences were collected in two separate indoor locations.

Data Pre-processing. Depth sensor-based cameras can capture depth images of a person within a particular range. In situations where depth sensors cannot capture depth frames properly, our system cannot extract the foreground part of the RGB image (see Fig. 4). Therefore, in our experiment, we consider only those RGB frames that have proper depth images of a person which can generate proper masks. After pre-processing, we obtain about 7,109 frames for training and 6,958 frames for testing, which come from 46 and 44 different identities respectively. We note that this is not a serious limitation as our system still covers a wide range of real world use cases.

5.1 Evaluation Protocol

We use cumulative matching characteristic (CMC) for quantitative evaluation, which is common practice in the Re-id literature. For our experimental dataset, we randomly

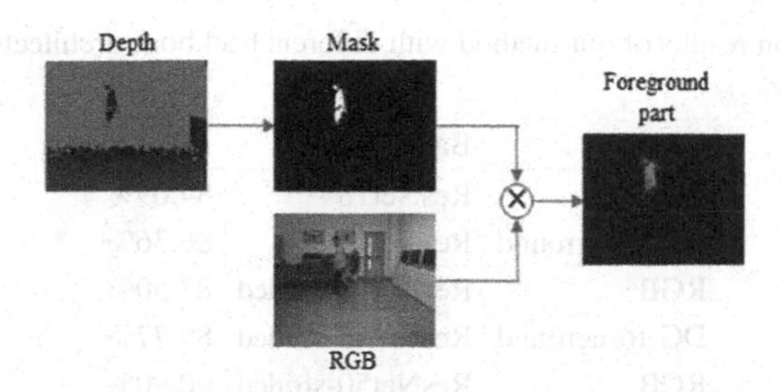

Fig. 4. Illustration of the limitation of depth sensor to capture the depth frame of a distant person and their corresponding person segmentation mask.

select about half of the people for training, and the remaining half for testing. In the testing phase, for each query image, we first compute the distance between the query image and all the gallery images using the Euclidean distance with the features extracted by the trained network, and then return the top n images which have the smallest distance to the query image in the gallery set. If the returned list contains an image featuring the same person as that in the query image at the *k-th* position, then this query is considered as rank k. In all our experiments, rank 1 result is reported.

5.2 Implementation Details

In our experiments, we use ResNet-18 [20] as well as ResNet50 [20] as the backbone CNN model. We use ResNet18 because it takes less memory and is computationally efficient, and the parameters are pre-trained on the ImageNet dataset. Following the state-of-the-art methods, we also did our experiments using ResNet50. We train our model with stochastic gradient descent with a momentum of 0.9, weight decay of 5×10^{-4}, and initial learning rate of 0.01. The batch size is set to $32 \times 4 = 128$, with 32 different persons and 4 instances per person in each mini-batch. In our implementation, we follow the common practice of using random horizontal flips during training [22]. We resize all the images to 256×128. Our framework is implemented on the Pytorch [21] platform.

5.3 Experimental Evaluation

In this section, we report our experimental results on the RobotPKU RGBD-ID dataset. To demonstrate the effectiveness of our method using the additional information available from the depth domain, first we evaluate our proposed approach with different backbone architectures (such as ResNet50 and ResNet18) and variants of the original backbones. Second, we compare our approach with the available state-of-the-art methods for the given dataset.

Evaluation with Different Backbone. The goal of this experimental evaluation to check the effectiveness of our proposed method for different backbone architectures. As we already mentioned, we choose ResNet50 and ResNet18 as our backbone architectures. We also try different variants of those backbone architectures. To do so, we adopt the stride version of ResNet50 and ReNet18 by changing the stride of the last

Table 1. Comparison results of our method with different backbone architectures on RobotPKU dataset.

Method	Backbone	Rank-1
RGB	ResNet18	84.09%
DG foreground	ResNet18	86.36%
RGB	ResNet18-strided	87.50%
DG foreground	ResNet18-strided	89.77%
RGB	ResNet50-strided	90.90%
DG foreground	ResNet50-strided	92.04%

convolutional layer from 2 to 1, which basically increases the resolution of the final activation layers. We report our results in Table 1.

Table 1 reports the rank-1 accuracy rate of the methods on the experimental dataset. We can make the following observations from these reported results:

ResNet50-strided indeed outperforms the original ResNet18 and ResNet18-strided for both scenarios in all the measures, which confirms our claims that increasing resolution on the final activation does affect the re-identification accuracy. The rank-1 performance improvement of the ResNet18-strided version over the original ResNet18 is 3.41% on both RGB and depth guided (DG) foreground images. From the above results, we can also see that our depth guided approach outperforms RGB for all the backbone CNN architectures.

Our proposed depth-guided foreground approach consistently works well for both versions of the considered backbone CNNs. The margin of improvement of our proposed approach considering original backbone architectures are relatively higher than their strided version. This implies that the finer details introduced by the proposed architecture on backbone architectures further improves the re-identification accuracy.

Comparison with Representative State-of-the-Art Methods. The aim of these experiments is to analyze and compare the effectiveness of our proposed depth-guided foreground method to relevant state-of-the-art methods. Table 2 reports the comparative performances of our methods with the state-of-the-art methods. Though some state-of-the-art methods [24, 25] performed experiments with this dataset, but all of these are cross-modality matching (i.e. RGB-Depth matching). The performance of the cross-modality matching is very low, around 20%, that's why we do not include the results in this report.

Our proposed approach considerably outperforms the state-of-the-art in all the measures. Among the alternatives, SILTP [19] performs worse while using handcrafted features which are mostly biased by the color or textures. The margin of improvement over the high performing state-of-the-art FFM (feature funnel model) is 14.1%. In FFM, the authors use both appearance and skeleton information provided by RGB-D sensors. The performance of the state-of-the-art methods varies significantly depending on their backbone architectures. We demonstrate the results of our method using different backbones and its variants in the previous section. Nevertheless, our proposed

Table 2. Comparison with other methods on RobotPKU dataset.

Method	Rank-1
HSV [19]	69.79%
SILTP [19]	46.71%
Concatenation [19]	72.95%
Score-level [19]	74.95%
FFM [19]	77.94%
RGB + ResNet18-strided **(Ours)**	87.50%
DG foreground + ResNet18-strided **(Ours)**	89.77%
RGB + ResNet50-strided **(Ours)**	90.90%
DG foreground + ResNet50-strided **(Ours)**	92.04%

approach consistently outperforms the state-of-the-art methods irrespective to their backbone architectures.

Our proposed approach does not rely on complex dedicated architectures for extracting foreground as it does in most of the state-of-the-art works. Thus, our proposed approach is computationally efficient and provides better recognition accuracy using depth data, which can be useful to deploy in real-time applications.

6 Conclusions

In this paper, we have presented a depth guided attention-based re-identification system. The key component of this framework is the depth-guided foreground extraction that helps the model to dynamically select the more relevant convolutional filters of the backbone CNN architecture, for enhanced feature representation and inference. Our proposed framework requires minimal modification to the backbone architecture to train the backbone network. Experimental results with a particular implementation of the framework (Resnet50 and Resnet18 with triplet loss) on the benchmark dataset indicate that the proposed framework can outperform related state-of-the-art methods. Moreover, our proposed architecture is general and can be applied with a multitude of different feature extractors and loss functions.

References

1. Song, C., Huang, Y., Ouyang, W., Wang, L.: Mask-guided contrastive attention model for person re-identification. In: CVPR (2018)
2. Chen, D., Zhang, S., Ouyang, W., Yang, J., Tai, Y.: Person search via a mask-guided two-stream CNN model. In: ECCV (2018)
3. Kalayeh, M.M., Basaran, E., Gökmen, M., Kamasak, M.E., Shah, M.: Human semantic parsing for person re-identification. In: CVPR (2018)
4. Cai, H., Wang, Z., Cheng, J.: Multi-scale body-part mask guided attention for person re-identification. In: CVPR (2019)

5. Munaro, M., Fossati, A., Basso, A., Menegatti, E., Van Gool, L.: One-shot person re-identification with a consumer depth camera. In: Gong, S., Cristani, M., Yan, S., Loy, C.C. (eds.) Person Re-Identification. ACVPR, pp. 161–181. Springer, London (2014). https://doi.org/10.1007/978-1-4471-6296-4_8
6. Li, D., Chen, X., Zhang, Z., Huang, K.: Learning deep context-aware features over body and latent parts for person re-identification. In: CVPR (2017)
7. Kumar, V., Namboodiri, A., Paluri, M., Jawahar, C.V.: Pose-aware person recognition. In: CVPR (2017)
8. Su, C., Li, J., Zhang, S., Xing, J., Gao, W., Tian, Q.: Pose-driven deep convolutional model for person re-identification. In: ICCV (2017)
9. Long, J., Shelhamer, E., Darrell, T.: Fully convolutional networks for semantic segmentation. In: CVPR (2015)
10. He, K., Gkioxari, G., Dollár, P., Girshick, R.: Mask R-CNN. In: ICCV (2017)
11. Liang, X., Gong, K., Shen, X., Lin, L.: Look into person: joint body parsing & pose estimation network and a new benchmark. IEEE Trans. Pattern Anal. Mach. Intell. **41**(4), 871–885 (2018)
12. Alp Guler, R., Trigeorgis, G., Antonakos, E., Snape, P., Zafeiriou, S., Kokkinos, I.: Densereg: fully convolutional dense shape regression in-the-wild. In: CVPR (2017)
13. Song, C., Huang, Y., Wang, Z., Wang, L.: 1000 fps human segmentation with deep convolutional neural networks. In: ACPR (2015)
14. Wu, Z., Huang, Y., Yu, Y., Wang, L., Tan, T.: Early hierarchical contexts learned by convolutional networks for image segmentation. In: ICPR (2014)
15. Li, Y., Qi, H., Dai, J., Ji, X., Wei, Y.: Fully convolutional instance-aware semantic segmentation. In: CVPR (2017)
16. Hermans, A., Beyer, L., Leibe, B.: In defense of the triplet loss for person re-identification. arXiv preprint arXiv:1703.07737
17. Almazan, J., Gajic, B., Murray, N., Larlus, D.: Re-id done right: towards good practices for person re-identification. arXiv preprint arXiv:1801.05339
18. Munaro, M., Basso, A., Fossati, A., Van Gool, L., Menegatti, E.: 3D reconstruction of freely moving persons for re-identification with a depth sensor. In: IEEE International Conference on Robotics and Automation (ICRA), pp. 4512–4519 (2014)
19. Liu, H., Hu, L., Ma, L.: Online RGB-D person re-identification based on metric model update. CAAI Trans. Intell. Technol. **2**(1), 48–55 (2017)
20. He, K., Zhang, X., Ren, S., Sun, J.: Deep residual learning for image recognition. In: CVPR (2016)
21. Paszke, A., Gross, S., Chintala, S., Chanan, G.: Pytorch: tensors and dynamic neural networks in python with strong GPU acceleration (2017). https://pytorch.org/
22. Ahmed, E., Jones, M., Marks, T.K.: An improved deep learning architecture for person re-identification. In: CVPR (2015)
23. Wei, L., Zhang, S., Gao, W., Tian, Q.: Person transfer GAN to bridge domain gap for person re-identification. In: CVPR (2018)
24. Hafner, F.M., Bhuiyan, A., Kooij, J.F., Granger, E.: RGB-depth cross-modal person re-identification. In: AVSS (2019)
25. Hafner, F., Bhuiyan, A., Kooij, J.F., Granger, E.: A cross-modal distillation network for person re-identification in RGB-depth. arXiv preprint arXiv:1810.11641

Improved Vision Based Pose Estimation for Industrial Robots via Sparse Regression

Diyar Khalis Bilal[1,2], Mustafa Unel[1,2](✉), and Lutfi Taner Tunc[1,2]

[1] Faculty of Engineering and Natural Sciences, Sabanci University, Istanbul, Turkey
{diyarbilal,munel,ttunc}@sabanciuniv.edu

[2] Integrated Manufacturing Technologies Research and Application Center, Sabanci University, Istanbul, Turkey

Abstract. In this work a monocular machine vision based pose estimation system is developed for industrial robots and the accuracy of the estimated pose is improved via sparse regression. The proposed sparse regression based method is used improve the accuracy obtained from the Levenberg-Marquardt (LM) based pose estimation algorithm during the trajectory tracking of an industrial robot's end effector. The proposed method utilizes a set of basis functions to sparsely identify the nonlinear relationship between the estimated pose and the true pose provided by a laser tracker. Moreover, a camera target was designed and fitted with fiducial markers, and to prevent ambiguities in pose estimation, the markers are placed in such a way to guarantee the detection of at least two distinct non parallel markers from a single camera within ±90° in all directions of the camera's view. The effectiveness of the proposed method is validated by an experimental study performed using a KUKA KR240 R2900 ultra robot while following sixteen distinct trajectories based on ISO 9238. The obtained results show that the proposed method provides parsimonious models which improve the pose estimation accuracy and precision of the vision based system during trajectory tracking of industrial robots' end effector.

Keywords: Machine vision · Pose estimation · Industrial robots · Trajectory tracking · Sparse regression

1 Introduction

In the near future industrial robots are projected to replace CNC machines for machining processes due to their flexibility, lower prices and large working space. The required accuracy for robotic machining is around ±0.20 mm based on aerospace specifications, but in reality, only accuracies around 1 mm are obtained [1]. Therefore, the robot's relatively low accuracy hinders them from being used in high precision applications.

Some works in literature proposed implementation of static calibration or usage of secondary high accuracy encoders installed at each joint for increasing the accuracy of industrial robots [2, 3]. However, disturbances acting on the robots during processes are not taken into account in static calibration methods, and installation of secondary

D.-S. Huang and P. Premaratne (Eds.): ICIC 2020, LNAI 12465, pp. 121–132, 2020.
https://doi.org/10.1007/978-3-030-60796-8_11

encoders is very expensive and not feasible for all robots. Thus, real time path tracking and correction based on visual servoing is a feasible alternative to achieve the desired accuracies in manufacturing processes [4]. Many works in literature utilize highly accurate sensors such as laser trackers or photogrammetry sensors in the feedback loop of visual servoing [5, 6]. However, these sensors are very expensive and sometimes more than the industrial robot. Hence, relatively cheaper alternatives based on monocular camera systems were proposed by many works in literature. Nissler et al. [7] proposed utilization of AprilTag markers attached to the end effector of a robot. In their work they used optimization techniques to reduce positioning tracking errors to less than 10 mm. However, they used only planar markers thus faced rank deficiency problems in pose estimation and their work was not evaluated during trajectory tracking. Moreover, two data fusion methods based on multi sensor optimal information algorithms (MOIFA) and Kalman filter (KF) were proposed by Liu et al. [8]. These methods were used for fusing orientation data acquired from a digital inclinometer and position data obtained from a photogrammetry system during positioning of a KP 5 Arc Kuka robot's end effector at seventy six points in a one meter cube space. However, they did not report orientation errors and did not evaluate their approach for trajectory tracking. In general, these works in literature assume the dynamics or kinematics of the industrial robots are known in the proposed eye in hand approaches. As for the KF type methods, they assume a linear dynamic process model along with the process and measurement noise to be known as well. Some works in literature utilized extended Kalman filter (EKF) [9], and adaptive Kalman filter (AKF) [10] to overcome these shortcomings in the estimation of an industrial robot's pose. However, an accurate dynamic process model required for EKF is hard to obtain, and in the proposed AKF based methods measurement noise and time varying effects due to the robot's trajectories are not considered, which in turn degrades their effectiveness. In these cases, data driven modeling techniques that can take into account all kinds of sensor errors, sensor noise and uncertainties have been found to be more effective [11–14].

In this work, an eye to hand camera based pose estimation system is developed for industrial robots through which a target object trackable with a monocular camera with $\pm 90°$ in all directions is designed. The designed camera target (CT) is fitted with fiducial markers where their placement guarantees the detection of at least two non-planar markers from a single frame, thus preventing ambiguities in pose estimation.

Moreover, a data driven modeling method based on sparse regression is proposed for improving the pose estimated by the Levenberg Marquardt (LM) based algorithm [15], where the ground truth is obtained from a laser tracker. Using the proposed method, one can train all the camera based systems using a single laser tracker in a factory where several industrial robots are required to perform the same task.

The rest of the manuscript is structured as follows: In Sect. 2, a method for improving vision based pose estimation based on sparse regression is presented. The effectiveness of the proposed approach is validated by an experimental study in Sect. 3 where design and detection of the camera target for pose estimation are also described, followed by the conclusion in Sect. 4.

2 Improved Vision Based Pose Estimation Using Sparse Regression

This work proposes to improve the pose estimation accuracy of vision based systems through a data driven approach based on sparse regression. Using this method existing camera based systems can be made to provide better accuracies when trained using the ground truth pose $(T_X, T_Y, T_Z, \alpha, \beta, \gamma)$ such as the one provided by a laser tracker. In order to formulate this problem under a sparse regression framework, the inputs and ground truth of the system needs to be determined properly. The ground truth in pose estimation problem can obtained through the highly accurate laser tracker systems. As for inputs, the estimated pose $\left(\widehat{T}_X, \widehat{T}_Y, \widehat{T}_Z, \widehat{\alpha}, \widehat{\beta}, \widehat{\gamma}\right)$ provided by the vision system can be obtained through standard pose estimation algorithms in literature such as the Levenberg Marquardt (LM) based algorithm [15].

As for the proposed method based on sparse regression, this work builds upon the work presented by Brunton et al. in which they formulated sparse identification of nonlinear dynamics (SINDy) [16] for discovering governing dynamical equations from data. They leverage the fact that only a few terms are usually required to define dynamics of a physical system. Thus, the equations become sparse in a high dimensional nonlinear function space. Their work is formulated for dynamic systems where large data is collected for determining a function in state space which defines the equations of motion. In their formulation, they collect a time-history of the state $X(t)$ and its derivative from which candidate nonlinear functions are generated. These functions can be constants, higher order polynomials, sinusoidal functions,..., etc. Afterwards, they formulate the problem as sparse regression and propose a method based on sequential thresholded least-squares algorithm [16] to solve it. This method is a faster and robust alternative to the least absolute shrinkage and selection operator (LASSO) [17] which is an $\ell 1$-regularized regression that promotes sparsity. Using their proposed method, the sparse vectors of coefficients defining the dynamics can be determined, showing which nonlinearities are active in the physical system. This results in parsimonious models that balance accuracy with model complexity to avoid overfitting.

However, in this work the sparse regression problem is formulated for sparse identification of nonlinear statics (SINS). In particular, the relationship between the pose estimated by the vision system and the pose provided by the laser tracker is assumed to be represented by the following static nonlinear model:

$$Y = \Psi(X)\Phi \tag{1}$$

where

$$X = \begin{bmatrix} x_1(t_1) & \cdots & x_6(t_1) \\ \vdots & \ddots & \vdots \\ x_1(t_m) & \cdots & x_6(t_m) \end{bmatrix} \text{ and } Y = \begin{bmatrix} y_1(t_1) & \cdots & y_6(t_1) \\ \vdots & \ddots & \vdots \\ y_1(t_m) & \cdots & y_6(t_m) \end{bmatrix} \quad (2)$$

$$\Psi(\mathrm{X}) = \begin{bmatrix} 1 & X & X^{P2} \end{bmatrix} \quad (3)$$

$$X^{P2} = \begin{bmatrix} x_1^2(t_1) & x_1(t_1)x_2(t_1) & \cdots & x_2^2(t_1) & x_2(t_1)x_3(t_1) & \cdots & x_6^2(t_1) \\ \vdots & \vdots & \ddots & \vdots & \vdots & \ddots & \vdots \\ x_1^2(t_m) & x_1(t_m)x_2(t_m) & \cdots & x_2^2(t_m) & x_2(t_m)x_3(t_m) & \cdots & x_6^2(t_m) \end{bmatrix} \quad (4)$$

where x_1 to x_6 are the $\widehat{T}_X, \widehat{T}_Y, \widehat{T}_Z, \widehat{\alpha}, \widehat{\beta}$ and $\widehat{\gamma}$ estimated by the LM based pose estimation algorithm, y_1 to y_6 are the ground truth $T_X, T_Y, T_Z, \alpha, \beta$, and γ measured by the laser tracker, Φ contains the sparse vectors of coefficients, X^{P_2} denotes the quadratic nonlinearities in the variable X, and $\Psi(X)$ is the library consisting of candidate nonlinear functions of the columns of X.

Each column of the augmented library $\Psi(X)$ represents a candidate function for defining the relationship between the estimated and the ground truth pose. There is total freedom in choosing these functions and in this work the augmented library was constructed using up to 2nd order polynomials (X^{P_2}) with cross terms and thus the resulting size of the sparse regression problem using m samples is as follows:

$$Y_{mx6} = \Psi(X_{mx6})_{mx28}\Phi_{28x6} \quad (5)$$

The sequential thresholded least-squares based algorithm proposed by Brunton et al. [16] starts with finding a least squares solution for Φ and then setting all of its coefficients smaller than a threshold value (λ) to zero. After determining the indices of the remaining nonzero coefficients, another least squares solution for Φ onto the remaining indices is obtained. This procedure is performed repeatedly for the new coefficients using the same λ until the nonzero coefficients converge. This algorithm is computationally efficient and rapidly converges to a sparse solution in a small number of iterations. Moreover, only a single parameter λ is required to determine the degree of sparsity in Φ. The overall flowchart of the proposed method is shown in Fig. 1.

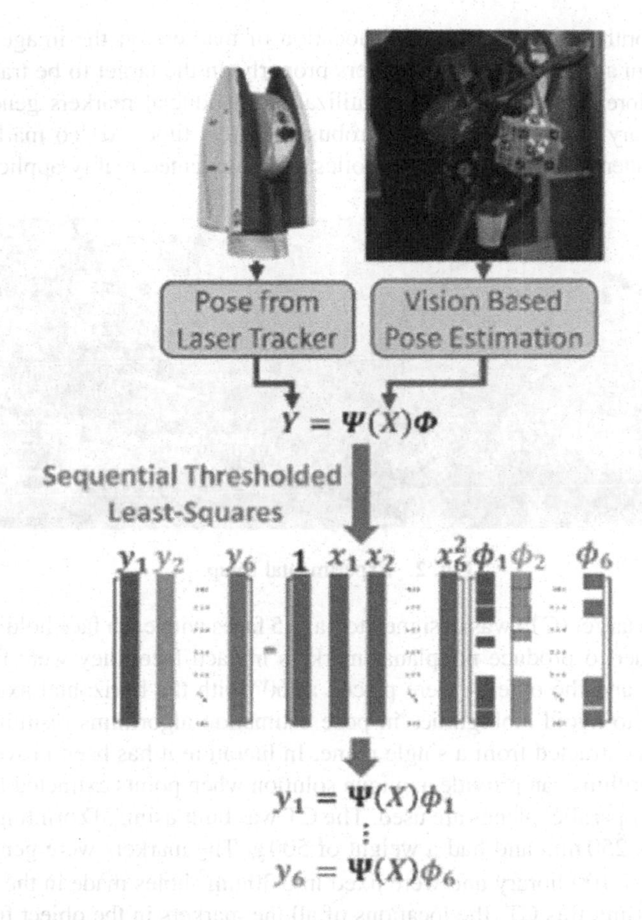

Fig. 1. The proposed sparse identification of nonlinear statics (SINS) for improving vision based pose estimation.

3 Experimental Results

In this section the design of the camera target for pose estimation, detection of the camera target and improved pose estimation results using the proposed method will be presented.

3.1 Design of the Camera Target for Pose Estimation

In this work the pose of a KUKA KR240 R2900 ultra robot's end effector was tracked in real time using a vision based pose estimation system utilizing a Basler acA2040–120 um camera and was compared with the measurements obtained from a Leica AT960 laser tracker as shown in Fig. 2. The laser tracker works in tandem with the T-MAC probe which is rigidly attached to the end effector and the system has an accuracy of ±10 μm. A target object fitted with markers was designed and fixed to the end effector of the robot so as to estimate its pose from the camera. Since vision based pose estimation algorithms require the exact location of markers on the image plane, it is crucial to design and distribute the markers properly on the target to be tracked by the camera. Therefore, this work proposes utilization of fiducial markers generated from the ArUco library that can be detected robustly in real time. ArUco markers are 2D barcode like patterns usually used in robotics and augmented reality applications [18].

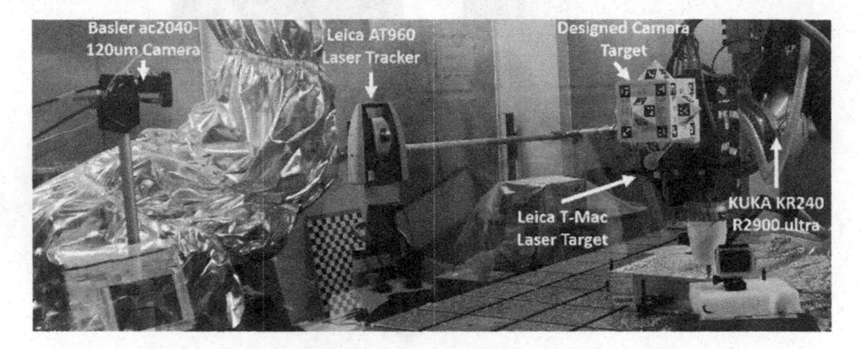

Fig. 2. Experimental setup.

The camera target (CT) was designed to have 5 faces with each face holding 8 ArUco markers. In order to produce nonplanar markers in each face, they were fitted with 4 planar markers and the other 4 were placed at 60° with the horizontal axis. This was designed so as to avoid ambiguities in pose estimation algorithms resulting from the usage of points extracted from a single plane. In literature it has been proven that pose estimation algorithms can provide a unique solution when points extracted from at least two distinct non-parallel planes are used. The CT was built using 3D printing with a size of $250 \times 234 \times 250$ mm and had a weight of 500 g. The markers were generated from ArUco's $4 \times 4 \times 100$ library and were fixed into 30mm^2 holes made in the constructed target object. Using this CT, the locations of all the markers in the object frame can be obtained from the CAD model and used in the vision based pose estimation algorithms.

3.2 Detection of the Camera Target

In the experiments, the vision based pose estimation and synchronization of data with the laser tracker was performed in LabVIEW [19] software. The images were acquired from the Basler ac2040–120 um camera at 375 Hz with a resolution of 640×480 pixels. These images were then fed into the python [20] node inside LabVIEW where the ArUco marker detection and Levenberg Marquardt based pose estimation algorithms were both operated at 1000 Hz. Moreover, the proposed method can work at 6000 Hz for a single frame as well. Therefore, the total processing time[1] for each image is 0.00216 s or about 463 Hz. The estimated pose of the camera target (CT) as well as the detected markers are shown in Fig. 3. These results clearly show that the designed CT allows the detection of multiple nonplanar markers with a viewing angle of ±90° from all sides, hence rank deficiency problem is prevented in the pose estimation algorithm.

[1] Tested on a workstation with Intel Xeon E5-1650 CPU @ 3.5 GHz and 16 GB RAM.

Fig. 3. (a)–(d) Samples showing marker detection (detected corners are in red) and estimated pose (red, green, blue coordinate axes) of the target object with respect to the camera frame. (Color figure online)

3.3 Pose Estimation Results

In order to evaluate the accuracy and precision of the camera based system, a trajectory tracking experiment based on ISO 9238 standard was conducted using a KUKA KR240 R2900 robot. The accuracy and repeatability of industrial robots are typically evaluated using the ISO 9238 standard during which the robot is tasked with following a set of trajectories multiple times while changing or not changing the orientation of the robot's end effector. To evaluate the effectiveness of the proposed SINS algorithm and the constructed vision based system, the robot's end effector was set to follow 16 distinct trajectories based on the ISO 9238 standard while changing its orientation continuously. As per the ISO 9238 guidelines, each of these trajectories contained 5 specific points at which the robot was stopped for 5 s and the experiment took 105.9 min to complete.

First the LM based pose estimation algorithm was implemented for the trajectory tracking of the KUKA KR240 R2900 robot's end effector. Then, the proposed sparse identification of nonlinear statics (SINS) method was used to improve the pose estimated by the LM based algorithm. In order to evaluate the robustness of the proposed method, the training phase was performed three times using 30%, 50%, and 70% of the data and was validated on the remaining 70%, 50%, and 30% of the data based the time series cross validation [21] approach. The training was performed for 10 iterations using a threshold value (λ) of 0.001 for the each of the three aforementioned cases and the obtained results are tabulated in Table 1, 2 and 3 for the trajectory tracking based on ISO 9238. The errors given in these tables which are denoted as E_X, E_Y, E_Z, E_{Roll}, E_{Pitch}, and E_{Yaw} are the absolute errors between the ground truth pose provided by the laser tracker and the estimated pose by the LM based algorithm and improved with SINS. These tracking errors are given in *mm* for translation (E_X, E_Y, E_Z) and in degrees (°) for orientation (E_{Roll}, E_{Pitch}, E_{Yaw}).

Table 1. Pose tracking errors during trajectory tracking based on ISO 9238, trained with 30% of the dataset and validated on the rest.

Training size	30% of the dataset					
Error for the validation set (70% of the dataset)	E_X (mm)	E_Y (mm)	E_Z (mm)	E_{Roll} (°)	E_{Pitch} (°)	E_{Yaw} (°)
LM	9.84 (9.86)	7.30 (6.61)	16.44 (14.07)	0.93 (0.33)	1.02 (0.89)	1.15 (0.72)
LM with SINS	8.01 (8.98)	6.19 (5.76)	11.62 (9.80)	0.20 (0.18)	0.85 (0.78)	0.56 (0.46)

The () below the errors contain their standard deviation

Table 2. Pose tracking errors during trajectory tracking based on ISO 9238, trained with 50% of the dataset and validated on the rest.

Training size	50% of the dataset					
Error for the validation set (50% of the dataset)	E_X (mm)	E_Y (mm)	E_Z (mm)	E_{Roll} (°)	E_{Pitch} (°)	E_{Yaw} (°)
LM	9.85 (9.87)	7.35 (6.62)	16.23 (13.60)	0.92 (0.32)	1.01 (0.88)	1.14 (0.71)
LM with SINS	7.85 (8.70)	6.04 (5.72)	10.32 (9.20)	0.19 (0.17)	0.82 (0.74)	0.53 (0.46)

The () below the errors contain their standard deviation

Table 3. Pose tracking errors during trajectory tracking based on ISO 9238, trained with 70% of the dataset and validated on the rest.

Training size	70% of the dataset					
Error for the validation set (30% of the dataset)	E_X (mm)	E_Y (mm)	E_Z (mm)	E_{Roll} (°)	E_{Pitch} (°)	E_{Yaw} (°)
LM	10.11 (10.20)	7.39 (6.78)	15.794 (13.69)	0.91 (0.33)	1.04 (0.87)	1.10 (0.67)
LM with SINS	7.98 (8.98)	6.01 (5.84)	9.66 (8.67)	0.19 (0.17)	0.81 (0.73)	0.51 (0.46)

The () below the errors contain their standard deviation

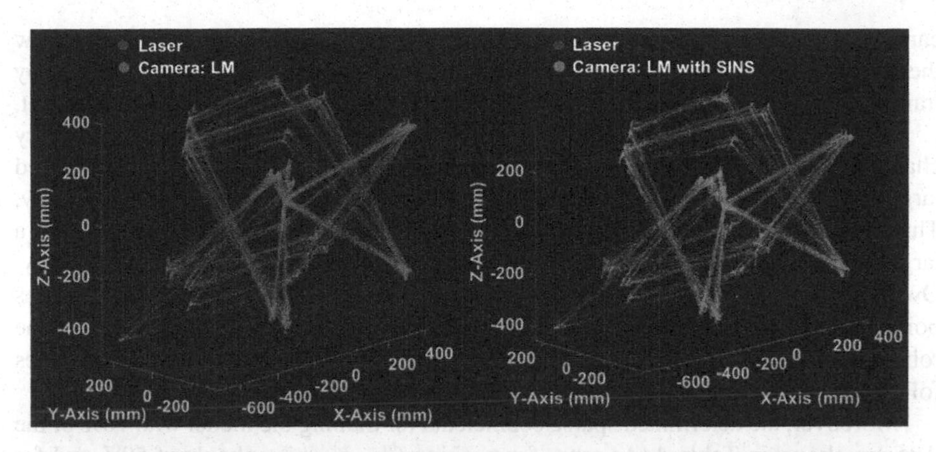

Fig. 4. Position tracking results based on ISO 9238. (Color figure online)

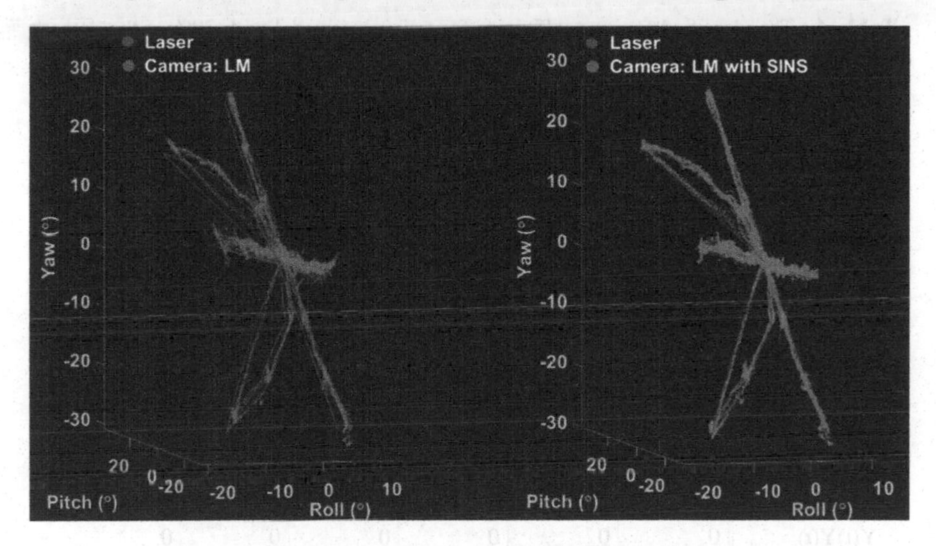

Fig. 5. Orientation tracking results based on ISO 9238. (Color figure online)

As seen from the errors in these tables, the proposed method is able to reduce the position tracking errors at least by 1.23, 1.18, and 1.42 times and up to 1.26, 1.23, and 1.64 times for X, Y, and Z axes, respectively when compared with the pure LM based algorithm using 30% and 70% of the data for training the models. This is in addition to reducing the standard deviation of the position errors by up to 1.14, 1.16, and 1.58 times for X, Y, and Z axes, respectively. Furthermore, the orientation tracking errors were reduced by at least 4.65, 1.20, and 2.05 times and up to 4.79, 1.28, and 2.16 times for Roll, Pitch and Yaw axes, respectively. Moreover, the standard deviation of orientation errors were reduced by up to 1.94, 1.19, and 1.46 times for the Roll, Pitch and Yaw axes, respectively. From these results, it is seen that the proposed method is able to improve the position and orientation tracking accuracies even when 30% of the data is used for training the proposed method, thus proving its robustness.

Figure 4 and Fig. 5 show the position and orientation trajectories of the laser target as tracked by the laser tracker in blue. The gray trajectories are the ones estimated by the camera system using LM based pose estimation algorithm and the red trajectories show the improved pose by the proposed SINS method. These images were obtained by training the proposed method with 70% of the data and evaluating it on the whole dataset.

It should be noted that the conducted experiment based on ISO 9238 is very challenging for vision based pose estimation due to the distance between the tracked target and the camera increasing a lot, thus decreasing the estimated pose's accuracy. This is particularly the case in the conducted experiment due to the robot covering a large working space of 1140 × 610 × 945 mm along the X, Y, and Z axes, respectively. Owing to this and the fact that the camera had to be placed 1 m away from the closes point of the work space due to viewing angle restrictions, the distance between the robot's end effector and the camera changed from 1 m to 3 m during the 16 trajectories followed by the robot, thus making the position errors relatively high.

Moreover, the determined sparse coefficients for training the model with 70% of the data are shown in Table 4. As seen, for position (ϕ_1, ϕ_2, ϕ_3) only about 50% and for

Table 4. The identified sparse coefficients for training a model with 70% of the data.

	ϕ_1	ϕ_2	ϕ_3	ϕ_4	ϕ_5	ϕ_6
1	−0.54955	5.483865	−2.34268	−0.80253	0.169695	−0.76172
X(t)	0.984231	0.01329	0.006688	0	0	0
Y(t)	−0.00315	0.994628	−0.00959	0	−0.00201	0
Z(t)	0.001783	−0.00849	0.934572	0	0	0
Roll(t)	2.207604	−1.73696	1.395375	0.889916	−0.15587	−0.17946
Pitch(t)	0.008375	−0.18872	0.4609	−0.01473	0.980488	−0.008
Yaw(t)	0.519546	−0.77316	0.382094	−0.01947	−0.06671	0.892436
X(t)X(t)	0	0	0	0	0	0
X(t)Y(t)	0	0	0	0	0	0
X(t)Z(t)	0	0	0	0	0	0
X(t)Roll(t)	0	−0.00318	0	0	0	0
X(t)Pitch(t)	0	0	0	0	0	0
X(t)Yaw(t)	0	−0.00111	0	0	0	0
Y(t)Y(t)	0	0	0	0	0	0
Y(t)Z(t)	0	0	0	0	0	0
Y(t)Roll(t)	−0.00285	0	−0.00246	0	0	0
Y(t)Pitch(t)	0	0	0	0	0	0
Y(t)Yaw(t)	0	0	0	0	0	0
Z(t)Z(t)	0	0	0	0	0	0
Z(t)Roll(t)	0	0	0	0	0	0
Z(t)Pitch(t)	0	0	0	0	0	0
Z(t)Yaw(t)	0	0	0	0	0	0
Roll(t)Roll(t)	0.129671	−0.33664	0.133981	−0.0037	−0.00789	−0.02765
Roll(t)Pitch(t)	−0.11072	0.008094	−0.12339	−0.00193	0.018478	0.00901
Roll(t)Yaw(t)	0.085	−0.23532	0.099387	0	−0.00359	−0.02075
Pitch(t)Pitch(t)	−0.00346	−0.00202	0.004847	0	0	0
Pitch(t)Yaw(t)	−0.01809	−0.07036	0	0.006763	0.005202	−0.0072
Yaw(t)Yaw(t)	0.006045	−0.03945	0.021693	0	0	−0.00299

orientation (ϕ_4, ϕ_5, ϕ_6) only around 30% of the coefficients are active. This makes the model sparse in the space of possible functions thus determining only the fewest terms to accurately represent the data. Furthermore, such a method is very intuitive in that one can clearly see the coefficients defining the nonlinear relationship and thus provides more insight into the structure of the problem at hand. Besides, training such a model in MATLAB [22] took only 0.35, 0.68, and 0.87 s for 30%, 50%, and 70% of the data containing 63551 samples.

4 Conclusion

In this work a monocular machine vision based system was developed for estimating the pose of industrial robots' end effector in real time. A camera target guaranteeing the detectability of at least two non-parallel markers within ±90° in all directions of the camera's view was designed and fitted with fiducial markers. Moreover, sparse identification of nonlinear statics (SINS) based on sparse regression was proposed to determine a model with the least number of active coefficients relating the pose estimated by Levenberg-Marquardt (LM) to ground truth pose provided by a laser tracker. Thus, providing a parsimonious model to increase the accuracy and precision of the vision based pose estimation.

The proposed method was validated by tracking an industrial robot's end effector for 16 distinct trajectories based on ISO 9238. The trajectories were followed by a KUKA KR240 R2900 ultra robot and the ground truth data was provided by the Leica AT960 laser tracker. As seen from the experimental results, the proposed method was able to reduce the position tracking errors by up to 1.26, 1.23, and 1.64 times for X, Y, and Z axes, respectively when compared with the pure LM based algorithm. This is in addition to reducing the orientation tracking errors by up to 4.79, 1.28, and 2.16 times for Roll, Pitch and Yaw axes, respectively. Moreover, by using the proposed method the standard deviation of the position errors were reduced by up to 1.14, 1.16, and 1.58 times for X, Y, and Z axes, respectively. All the while reducing the standard deviation of the orientation errors by up to 1.94, 1.19, and 1.46 times for the Roll, Pitch and Yaw axes, respectively. Therefore, the proposed method is able to increase the accuracy and precision of the standard LM based pose estimation algorithm during trajectory tracking of industrial robots' end effector.

The determined sparse coefficients for training the model showed that only about 50% of the coefficients were active for position improvement, whereas for orientation, only around 30% of the coefficients were active. Thus, only the most important terms accurately representing the data were determined using the proposed method. This resulted in obtaining simple and robust models very fast, where one can clearly see the coefficients defining the nonlinear static system.

Acknowledgement. This work was funded by TUBITAK with grant number 217M078.

References

1. Klimchik, A., Ambiehl, A., Garnier, S., Furet, B., Pashkevich, A.: Efficiency evaluation of robots in machining applications using industrial performance measure. Robot. Comput.-Integr. Manuf. **48**, 12–29 (2017)
2. Devlieg, R.: Expanding the use of robotics in airframe assembly via accurate robot technology. SAE Int. J. Aerosp. **3**(1846), 198–203 (2010)
3. Keshmiri, M., Xie, W.F.: Image-based visual servoing using an optimized trajectory planning technique. IEEE/ASME Trans. Mechatron. **22**(1), 359–370 (2016)
4. Hashimoto, K.: A review on vision-based control of robot manipulators. Adv. Robot. Int.: J. Robot. Soc. Japan **17**(10), 969–991 (2003)
5. Shu, T., Gharaaty, S., Xie, W., Joubair, A., Bonev, I.A.: Dynamic path tracking of industrial robots with high accuracy using photogrammetry sensor. IEEE/ASME Trans. Mechatron. **23** (3), 1159–1170 (2018)
6. Comet project. https://comet-project.eu/results.asp. Accessed 07 Aug 2020
7. Nissler, C., Stefan, B., Marton, Z.C., Beckmann, L., Thomasy, U.: Evaluation and improvement of global pose estimation with multiple apriltags for industrial manipulators. In: 2016 IEEE 21st International Conference on Emerging Technologies and Factory Automation (ETFA), pp. 1–8. IEEE (2016)
8. Liu, B., Zhang, F., Qu, X.: A method for improving the pose accuracy of a robot manipulator based on multi-sensor combined measurement and data fusion. Sensors **15**(4), 7933–7952 (2015)
9. Janabi-Sharifi, F., Marey, M.: A Kalman-filter-based method for pose estimation in visual servoing. IEEE Trans. Robot. **26**(5), 939–947 (2010)
10. D'Errico, G.E.: A la kalman filtering for metrology tool with application to coordinate measuring machines. IEEE Trans. Ind. Electron. **59**(11), 4377–4382 (2011)
11. Alcan, G.: Data driven nonlinear dynamic models for predicting heavy-duty diesel engine torque and combustion emissions. Ph.D. thesis, Sabanci University (2019)
12. Mumcuoglu, M.E., et al.: Driving behavior classification using long short term memory networks. In: 2019 AEIT International Conference of Electrical and Electronic Technologies for Automotive (AEIT AUTOMOTIVE), pp. 1–6. IEEE (2019)
13. Alcan, G., et al.: Estimating soot emission in diesel engines using gated recurrent unit networks. IFAC-PapersOnLine **52**(5), 544–549 (2019)
14. Aran, V., Unel, M.: Gaussian process regression feedforward controller for diesel engine airpath. Int. J. Automot. Technol. **19**(4), 635–642 (2018)
15. Darcis, M., Swinkels, W., Guzel, A.E., Claesen, L.: PoseLab: a levenberg-marquardt based prototyping environment for camera pose estimation. In: 2018 11th International Congress on Image and Signal Processing, BioMedical Engineering and Informatics (CISP-BMEI), pp. 1–6. IEEE (2018)
16. Brunton, S.L., Proctor, J.L., Kutz, J.N.: Discovering governing equations from data by sparse identification of nonlinear dynamical systems. Proc. Nat. Acad. Sci. **113**(15), 3932–3937 (2016)
17. James, G., Witten, D., Hastie, T., Tibshirani, R.: An introduction to statistical learning, vol. 112. Springer, New York (2013)
18. Romero-Ramirez, F.J., Muñoz-Salinas, R., Medina-Carnicer, R.: Speeded up detection of squared fiducial markers. Image Vis. Comput. **76**, 38–47 (2018)
19. LabVIEW. https://www.ni.com/en-tr/shop/labview.html. Accessed 07 Aug 2020
20. Python. https://www.python.org/. Accessed 07 Aug 2020
21. Hyndman, R.J., Athanasopoulos, G.: Forecasting: Principles and Practice. OTexts, Melbourne (2018)
22. Matlab. https://www.mathworks.com/products/matlab.html. Accessed 07 Aug 2020

LiDAR-Camera-Based Deep Dense Fusion for Robust 3D Object Detection

Lihua Wen and Kang-Hyun Jo(✉)

School of Electrical Engineering, University of Ulsan, Ulsan, South Korea
wenlihua@islab.ulsan.ac.kr, acejo@ulsan.ac.kr

Abstract. For the camera-LiDAR-based three-dimensional (3D) object detection, image features have rich texture descriptions and LiDAR features possess objects' 3D information. To fully fuse view-specific feature maps, this paper aims to explore the two-directional fusion of arbitrary size camera feature maps and LiDAR feature maps in the early feature extraction stage. Towards this target, a deep dense fusion 3D object detection framework is proposed for autonomous driving. This is a two stage end-to-end learnable architecture, which takes 2D images and raw LiDAR point clouds as inputs and fully fuses view-specific features to achieve high-precision oriented 3D detection. To fuse the arbitrary-size features from different views, a multi-view resizes layer (MVRL) is born. Massive experiments evaluated on the KITTI benchmark suite show that the proposed approach outperforms most state-of-the-art multi-sensor-based methods on all three classes on moderate difficulty (3D/BEV): Car (75.60%/88.65%), Pedestrian (64.36%/66.98%), Cyclist (57.53%/57.30%). Specifically, the DDF3D greatly improves the detection accuracy of hard difficulty in 2D detection with an 88.19% accuracy for the car class.

Keywords: Two directional fusion · 3D object detection · Autonomous driving

1 Introduction

This paper focuses on 3D object detection, which is a fundamental and key computer vision problem impacting most intelligent robotics perception systems including autonomous vehicles and drones. To achieve robust and accurate scene understanding, autonomous vehicles are usually equipped with various sensors (e.g. camera, Radar, LiDAR) with different functions, and multiple sensing modalities can be fused to exploit their complementary properties. However, developing a reliable and accurate perception system for autonomous driving based on multiple sensors is still a very challenging task.

Recently, 2D object detection with the power of deep learning has drawn much attention. LiDAR-based 3D object detection also becomes popular with deep learning. Point clouds generated by LiDAR to capture surrounding objects and return accurate depth and reflection intensity information to reconstruct the objects. Since the sparse and unordered attributes of point clouds, representative works either convert raw point clouds into bird-eye-view (BEV) pseudo images [1–4], 2D front view images [2], or structured voxels grid representations [5–7]. Some references [8–10] directly deal with

D.-S. Huang and P. Premaratne (Eds.): ICIC 2020, LNAI 12465, pp. 133–144, 2020.
https://doi.org/10.1007/978-3-030-60796-8_12

raw point clouds by multi-layer perceptron (MLP) to estimate the 3D object and localization. However, due to the sparsity of point clouds, these LiDAR-based approaches suffer information loss severely in long-range regions and when dealing with small objects.

On the other hand, 2D RGB images provide dense texture descriptions and also enough information for small objects based on high resolution, but it is still hard to get precise 3D localization information due to the loss of depth information caused by perspective projection, particularly when using monocular camera [11–13]. Even if using stereo images [14], the accuracy of the estimated depth cannot be guaranteed, especially under poor weather, dark and unseen scenes. Therefore, some approaches [15–19] have attempted to take the mutual advantage of point clouds and 2D images. However, they either utilize *Early Fusion*, *Late Fusion*, or *Middle Fusion* is shown in Fig. 1 to shallowly fuse two kinds of features from 2D images and point clouds. Their approaches make the result inaccurate and not stable.

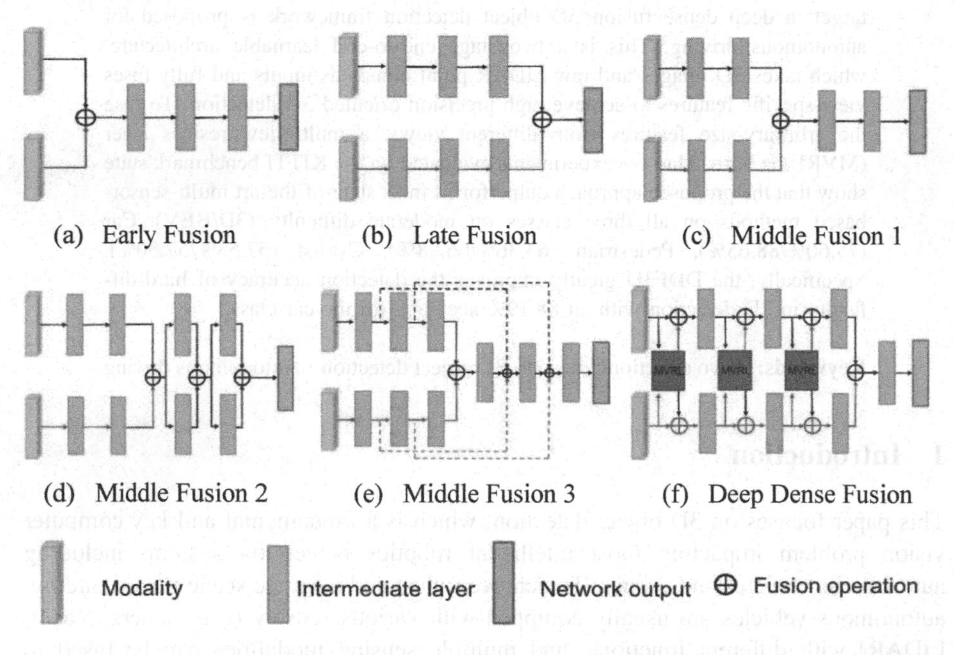

Fig. 1. A comparison of existed fusion methods and the deep dense fusion (proposed). Compared with methods (a–e), the deep dense fusion moves forward to the feature extraction phase and becomes denser. The proposed fusion method fully integrates each other's characteristics.

MV3D [2] and AVOD [6] fuse region-based multi-modal features at the region proposal network (RPN) and detection stage, the *local* fusion method causes the loss of semantic and makes its results inaccurate. Conversely, ContFusion [20] proposed a *global* fusion method to fuse BEV features and image features from different feature levels, it verifies the superiority of the full fusion of 2D image and point clouds.

However, ContFusion [20] is only unidirectional fusion. Based on logical experience, a bidirectional fusion will be even more superior than the unidirectional fusion. The challenge lies in the fact that the image feature is dense at discrete state, while LiDAR points are continuous and sparse. Thus, fusing them in both directions is non-trivial.

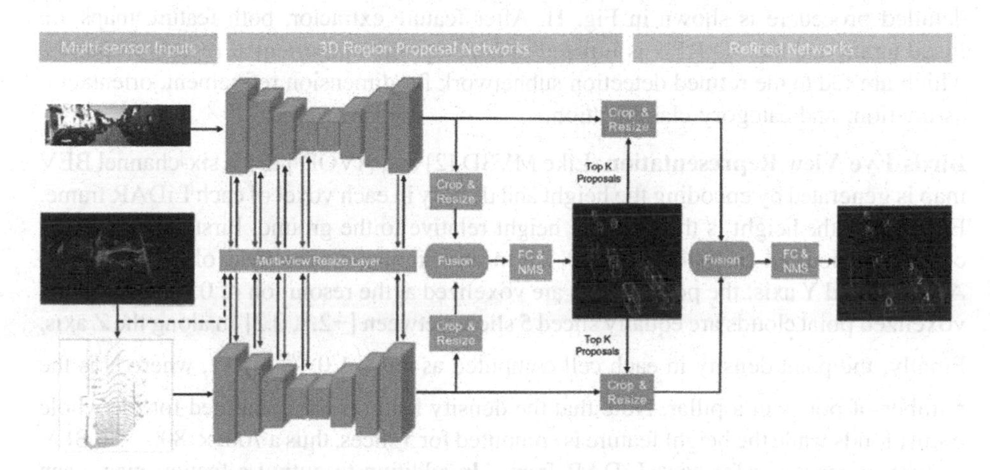

Fig. 2. The architecture of deep dense fusion 3D object detection network.

This paper proposes a two-stage multi-sensor 3D detector, called DDF3D, which fuses image feature and BEV feature at different levels of resolution. The DDF3D is an end-to-end learnable architecture consisting of a 3D region proposal subnet (RPN) and a refined detector subnet in the order illustrated in Fig. 2. First, the raw point clouds are partitioned into six-channel pseudo images and 2D images are cropped based on the central region. Second, two identical fully convolutional networks are used to extract view-specific features and fuse them by the MVRL simultaneously. Third, 3D anchors are generated from BEV, and anchor-dependent features from different views are fused to produce 3D non-oriented region proposals. Finally, the proposal-dependent features are fused again and fed to the refined subnetwork to regress dimension, orientation, and classify category.

Here, the contributions in this paper have summarized in 3 points as follows:

1. A highly efficient multi-view resizes layer (MVRL) designed to resize the features from BEV and camera simultaneously, which makes real-time fusion of multiple view-specific feature maps possible.
2. Based on the MVRL, a deep dense fusion method is proposed to fully fuse view-specific feature maps at different levels of resolution synchronously. The fusion method allows different feature maps to be fully fused during feature extraction, which greatly improves the detection accuracy of small size object.
3. The proposed architecture achieves a higher and robust 3D detection and localization accuracy for car, bicycle, and pedestrian class. Especially the proposed architecture greatly improves the accuracy of small classes on both 2D and 3D.

2 The Proposed Architecture

The main innovation of proposed DDF3D, depicted in Fig. 2, is to fully fuse view-specific features simultaneously based on the MVRL, and the fused features are fed into the next convolutional layers at BEV stream and camera stream respectively, the detailed procedure is shown in Fig. 1f. After feature extractor, both feature maps are fused again and the 3D RPN is utilized to generate 3D non-oriented region proposals, which are fed to the refined detection subnetwork for dimension refinement, orientation estimation, and category classification.

Birds Eye View Representation Like MV3D [2] and AVOD [15], a six-channel BEV map is generated by encoding the height and density in each voxel of each LiDAR frame. Especially, the height is the absolute height relative to the ground. First, the raw point clouds are located in $[-40, 40] \times [0, 70]$ m and limited to the field of camera view. Along X and Y axis, the point clouds are voxelized at the resolution of 0.1 m. Then, the voxelized point clouds are equally sliced 5 slices between $[-2.3, 0.2]$ m along the Z axis. Finally, the point density in each cell computed as $min\left(1.0, \frac{\log(N+1)}{\log 64}\right)$, where N is the number of points in a pillar. Note that the density features are computed for the whole point clouds while the height feature is computed for 5 slices, thus a $700 \times 800 \times 6$ BEV feature is generated for each LiDAR frame. In addition to output a feature map, each LiDAR frame also outputs the voxelized points to construct the MVRL.

2.1 The Feature Extractor and Multi-view Resize Layer

This section will introduce the feature extractor and MVRL. The MVRL is used to resize the view-specific features at a different resolution, then view-specific features are concatenated with the features resized from different views. Finally, the fused features are fed into the next convolutional layers.

The Multi-view Resize Layer. To fuse feature maps from different perspectives is not easy since the feature maps from different views are of different sizes. Also, fusion efficiency is a challenge. So, the multi-view resize layer is designed to bridge multiple intermediate layers on both sides to resize multi-sensor features at multiple scales with highly efficient. The inputs of MVRL contains three parts: the source BEV indices $I_{bev/ori}$ and LiDAR points P_{ori} obtained during a density feature generation, the camera feature f_{cam}, and the BEV feature f_{bev}. The workflow of the MVRL shown in Fig. 3. The MVRL consists of data preparation and bidirectional fusion. In data preparation, the voxelized points P_{ori} are projected onto the camera plane, the process is formulated as Eq. 1 and Eq. 2, and the points P_{cam} in original image size 360×1200 are kept. The points $P_{cam/fusion}$ in image size $H_i \times W_i$ are used to obtained image indices $I_{camcam/fusion}$ based on bilinear interpolation. A new BEV index $I_{bev/fusion}$ are obtained based on BEV indices $I_{bev/ori}$ and BEV size $H_b \times W_b$. Then, a sparse tensor T_s with $H_b \times W_b$ shape is generated by image indices $I_{cam/fusion}$ and BEV indices $I_{bev/fusion}$. Finally, a feature multiplies the sparse tensor to generate the feature which can be fused by a camera feature map or an image feature map formulated as Eq. 3 and Eq. 4.

$$(\mathrm{u}, \mathrm{v})^{\mathrm{T}} = \mathrm{M} \cdot (\mathrm{x}, \mathrm{y}, \mathrm{z})^{\mathrm{T}}, \tag{1}$$

$$\mathrm{M} = \mathrm{P}_{\mathrm{rect}} \cdot \begin{pmatrix} \mathrm{R}_{\mathrm{velo}}^{\mathrm{cam}} & \mathrm{t}_{\mathrm{velo}}^{\mathrm{cam}} \\ 0 & 1 \end{pmatrix}, \tag{2}$$

$$\mathrm{f}_{\mathrm{b2c}} = \mathrm{S}\left(\mathrm{Matmul}\left(\mathrm{T}_{\mathrm{s}}^{-1}, \mathrm{f}_{\mathrm{bev}}\right)\right), \tag{3}$$

$$\mathrm{f}_{\mathrm{c2b}} = \mathrm{R}\left(\mathrm{Matmul}\left(\mathrm{T}_{\mathrm{s}}, \mathrm{G}\left(\mathrm{f}_{\mathrm{cam}}, \mathrm{I}_{\mathrm{cam/fusion}}\right)\right)\right), \tag{4}$$

where (x, y, z) is a LiDAR point coordinate and (u, v) is image coordinate, $\boldsymbol{P_{rect}}$ is a project matrix, $\boldsymbol{R_{velo}^{cam}}$ is the rotation matrix from LiDAR to the camera, $\boldsymbol{t_{velo}^{cam}}$ is a translation vector, **M** is the homogeneous transformation matrix from LiDAR to the camera, **S** and **G** represent scatter operation and gather operation, respectively, **Matmul** means multiplication, **R** means reshape operation, $\mathrm{f}_{\mathrm{b2c}}$ is the feature transferred from BEV to the camera, Conversely, $\mathrm{f}_{\mathrm{c2b}}$ is the feature transferred from the camera to BEV.

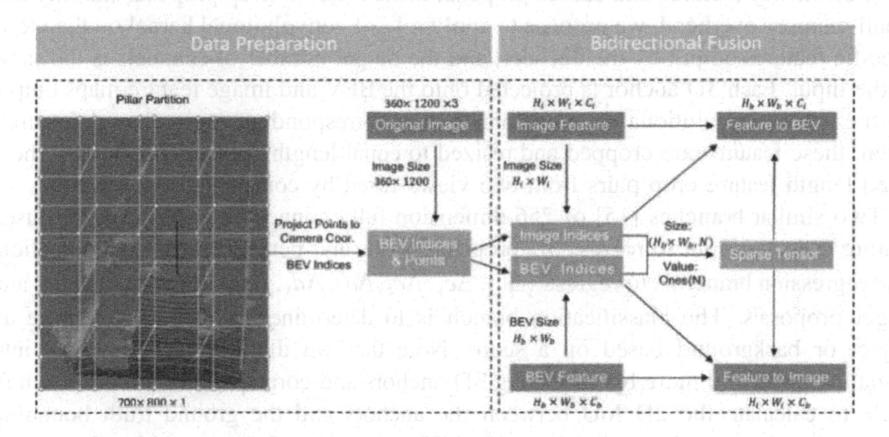

Fig. 3. Multi-view resize layer: it includes data preparation and bidirectional fusion.

The Feature Extractor. The backbone network follows a two-stream architecture [22] to process multi-sensor data. Specifically, it uses two identical CNNs to extract features from both of 2D image and BEV representation in this paper. Each CNNs includes two parts: an encoder and a decoder. VGG-16 [23] is modified and simplified as the encoder. The convolutional layers from conv-1 to conv-4 are kept, and the channel number is reduced by half. In the feature extraction stage, the MVRL is used to resize two-side features. A little of information is retained for small classes such as cyclists and pedestrians in the output feature map. Therefore, inspired by FCNs [24] and Feature Pyramid Network (FPN) [25], a decoder is designed to up-sample the features back to the original input size. To fully fuse the view-specific features, The MVRL is used again to resize features after decoding. The final feature map has powerful semantics with a high resolution, and are fed into the 3D RPN and the refined subnetwork.

2.1.1 3D Region Proposal Network

3D Anchor Generation and Fusion. Unlike MV3D [2], this paper directly generates 3D plane-based anchors like AVOD [15] and MMF [22]. The 3D anchors are parameterized by the centroid (c_x, c_y, c_z) and axis aligned dimensions (d_x, d_y, d_z). The (c_x, c_y) pairs are sampled at intervals of 0.5 m in the BEV, while c_z is a fixed value that is determined according to the height of the sensor related to the ground plane. Since this paper does not regress the orientation at the 3D proposal stage, the (d_x, d_y, d_z) are transformed from (*w, l, h*) of the prior 3D boxes based on rotations. Furthermore, the (*w, l, h*) are determined by clustering the training samples for each class. For the car case, each location has two sizes of anchors. While each location only has one size of anchor for pedestrians and cyclists.

3D Proposal Generation. AVOD [15] reduces the channel number of BEV and image feature maps to 1, and aims to process anchors with a small memory overhead. The truncated features are used to generate region proposals. However, the rough way loses most of the key features and causes proposal instability. To keep proposal stability and small memory overhead, we propose to apply a 1×1 convolutional kernel on the view-specific features output by the decoder, and the output number of channels is the same as the input. Each 3D anchor is projected onto the BEV and image feature maps output by the 1×1 convolutional layer to obtain two corresponding region-based features. Then, these features are cropped and resized to equal-length vectors, e.g. 3×3. These fixed-length feature crop pairs from two views fused by concatenation operation.

Two similar branches [15] of 256-dimension fully connected layers take the fused feature crops as input to regress 3D proposal boxes and perform binary classification. The regression branch is to regress $(\Delta c_x, \Delta c_y, \Delta c_z, \Delta d_x, \Delta d_y, \Delta d_z)$ between anchors and target proposals. The classification branch is to determine an anchor to capture an object or background based on a score. Note that we divide all 3D anchors into negative, positive, ignore by projecting 3D anchors and corresponding ground-truth to BEV to calculate the 2D IoU between the anchors and the ground truth bounding boxes. For the car class, anchors with IoU less than 0.3 are considered negative anchors, while ones with IoU greater than 0.5 are considered positive anchors. Others are ignored. For the pedestrian and cyclist classes, the object anchor IoU threshold is reduced to 0.45. The ignored anchors do not contribute to the training objective [21].

The loss function in 3D proposal stage is defined as follows:

$$Loss = \lambda L_{cls} + \gamma L_{box}, \tag{5}$$

where L_{cls} is the focal loss for object classification and L_{box} is the smooth *l*1 loss for 3D proposal box regression, $\lambda = 1.0$, $\gamma = 5.0$ are the weights to balance different tasks.

Followed by two task-specific branches, 2D non-maximum suppression (NMS) at an IoU threshold of 0.8 in BEV is used to remove redundant 3D proposals and the top 1,024 3D proposals are kept during the training stage. At inference time, 300 3D proposals are kept for the car class, and 1,024 3D proposals are used for cyclist and pedestrian class.

2.2 The Refined Network

The refined network aims to further optimize the detection based on the top K non-oriented region proposals and the features output by the two identical CNN to improve the final 3D object detection performance. First, the top K non-oriented region proposals are projected onto BEV and image feature maps output by feature extractors to obtain two corresponding region-based features. The region-based features are cropped and resized to 7×7 equal-length shapes. Then, the paired fixed-length crops from two views are fused with element-wise mean method. The fused features are fed into a three parallel fully connected layers for outputting bounding box regression, orientation estimation, and category classification, simultaneously. MV3D [2] proposes an 8-corner encoding, however, it does not take into account the physical constraints of a 3D bounding box. Like AVOD [15], a plane-based 3D bounding box is represented by a 10-dimensional vector to remove redundancy and keep the physical constraints. Ground truth boxes and 3D anchors are defined by $(x_1 \cdots x_4, y_1 \cdots y_4, h_1, h_2, \theta)$. The corresponding regression residuals between 3D anchors and ground truth are defined as follows:

$$\Delta \mathrm{x} = \frac{\mathrm{x_c^g} - \mathrm{x_c^a}}{\mathrm{d^a}}, \Delta \mathrm{y} = \frac{\mathrm{y_c^g} - \mathrm{y_c^a}}{\mathrm{d^a}}, \Delta \mathrm{h} = \log\left(\frac{\mathrm{h^g}}{\mathrm{h^a}}\right), \tag{6}$$

$$\Delta\theta = \sin(\theta^{\mathrm{g}} - \theta^{\mathrm{a}}), \tag{7}$$

where $d^a = \sqrt{(x_1 - x_2)^2 + (y_4 - y_1)^2}$ is the diagonal of the base of the anchor box.

The localization loss function and orientation loss function [7] as follows:

$$\mathrm{L_{box}} = \sum\nolimits_{\mathrm{b} \in (\mathrm{x_1 \cdots x_4, y_1 \cdots y_4, h_1, h_2}, \theta)} \mathrm{Smooth_{L1}}(\Delta \mathrm{b}), \tag{8}$$

$$\mathrm{L_{dir}} = \sum \mathrm{Smooth_{L1}}(\Delta\theta). \tag{9}$$

For the object classification loss, the focal loss is used:

$$\mathrm{L_{cls}} = -\alpha_{\mathrm{a}}(1 - \mathrm{p^a})^{\gamma}\log(\mathrm{p^a}), \tag{10}$$

where $\mathrm{p^a}$ is the class probability of an anchor, we set $\alpha = 0{:}25$ and $\gamma = 2$, the total loss for the refined network is, therefore,

$$\mathrm{Loss} = \frac{1}{\mathrm{N_{pos}}}(\beta_1 \mathrm{L_{box}} + \beta_2 \mathrm{L_{cls}} + \beta_3 \mathrm{L_{dir}}), \tag{11}$$

Where $\mathrm{N_{pos}}$ is the number of positive anchors and $\beta_1 = 5.0$, $\beta_2 = 1.0$, $\beta_3 = 1.0$.

In refined network, 3D proposals are only considered in the evaluation of the regression loss if they have at least a 0.65 2D IoU in bird's eye view with the ground-truth boxes for the car class (0.55 for pedestrian/cyclist classes). NMS is used at a threshold of 0.01 to choose out the best detections.

3 Experiments and Results

3.1 Implementation Details

Due to the 2D RGB camera images are with different size, the images are center-cropped into a uniform size of 1200 × 360. Each point clouds are voxelized as a 700 × 800 × 6 BEV pseudo image. For data augmentation, it flips images, voxelized pseudo images, and ground-truth labels horizontally at the same time with a probability of 0.5 during the training. The DDF3D model is implemented by TensorFlow on one NVIDIA 1080 Ti GPU with a batch size of 1. Adam is the optimizer. The DDF3D model is trained for a total of 120K iterations with the initial learning rate of 0.0001, and decayed by 0.1 at 60K iterations and 90K iterations. The whole training process takes only 14 h, and the DDF3D model is evaluated from 80K iterations to 120K iterations every 5K iterations.

3.2 Quantitative Results

Table 1. Comparison of the 3D Object and BEV performance of DDF3D with state-of-the-art 3D object detectors.

Class	Method	Time	3D AP (%)			BEV AP (%)		
			E	M	H	E	M	H
Car	MV3D [2]	0.36	71.29	62.68	56.56	86.55	78.10	76.67
	AVOD [15]	0.10	84.41	74.44	68.65	–	–	–
	F-PointNet [17]	0.17	83.76	70.92	63.65	88.16	84.02	76.44
	ContFusion [20]	0.06	84.58	72.33	67.50	**93.84**	86.10	**82.00**
	MCF3D [16]	0.16	84.11	75.19	**74.23**	88.82	86.11	79.31
	Proposed (Ours)	0.12	**84.65**	**75.60**	68.64	89.81	**88.65**	79.88
Ped.	MV3D [2]	0.36	–	–	–	–	–	–
	AVOD [15]	0.10	–	58.80	–	–	–	–
	F-PointNet [17]	0.17	70.00	61.32	53.59	**72.38**	66.39	59.57
	ContFusion [20]	0.06	–	–	–	–	–	–
	MCF3D [16]	0.16	68.54	**64.93**	59.47	68.56	64.98	59.55
	Proposed (Ours)	0.12	**70.04**	64.36	**59.55**	70.05	**66.98**	**59.66**
Cyc.	MV3D [2]	0.36	–	–	–	–	–	–
	AVOD [15]	0.10	–	49.70	–	–	–	–
	F-PointNet [17]	0.17	77.15	56.49	**53.37**	**81.82**	**60.03**	**56.32**
	ContFusion [20]	0.06	–	–	–	–	–	–
	MCF3D [16]	0.16	78.18	51.06	50.43	78.18	51.09	50.45
	Proposed (Ours)	0.12	**79.19**	**57.53**	50.99	79.19	57.30	50.99

To showcase the superiority of the deep dense fusion method, this paper compares its approach with the existing state-of-the-art fusion methods (MV3D [2], AVOD [15], F-pointNet [17], ContFusion [20], MCF3D [16]) based on the RGB images and point clouds as inputs only. Table 1 shows the comparing results on the 3D and BEV performance measured by the AP. According to KITTI's metric, the DDF3D increases 0.41% in 3D performance and 2.54% in BEV performance in the "Moderate" difficulty on the car class, respectively. For pedestrian/cyclist classes, DDF3D achieves 2.00% growth in BEV performance on the "Moderate" difficulty for pedestrian class and 1.04% growth in 3D performance on the "Moderate" difficulty for cyclist class. In the easy difficulty of 3D performance, DDF3D surpasses the second-best 1.50% for the pedestrian class and 1.01% for the cyclist, respectively. However, F-pointNet [17] is slightly better than DDF3D in BEV performance for cyclist. F-pointNet [17] utilizes the ImageNet weights to fine-tune its 2D detector, whereas DDF3D model is trained from scratch. Some 2D detection results in RGB images, 3D detection results are illustrated in Fig. 4.

Fig. 4. Visualizations of DDF3D results on RGB images, point clouds.

3.3 Ablation Study

To analyze the effects of optimal deep dense fusion, an ablation is conducted on KITTI's validation subset with massive experiments. Table 2 shows the effect of varying different combinations of the deep dense fusion method on the performance measured by the AP. As shown in Fig. 2, Each encoder has 4 convolution blocks in order: Conv1, Conv2, Conv3, Conv4. Each decoder also has 4 deconvolution blocks in order: Deconv1, Deconv2, Deconv3, Deconv4. To ensure the DDF3D high efficiency, the combinations of deep dense fusion are only designed shown in Table 2.

Table 2. Ablation study for the combinations of the deep dense method on KITTI's validation subset. All results are in moderate difficulty in the car class.

Deep dense fusion					2D	3D	BEV
Deconv4	Conv4	Conv3	Conv2	Conv1			
					86.93	72.43	85.78
√					88.11	72.73	86.06
√	√				88.28	74.02	86.25
√	√	√			**88.96**	**75.60**	**88.65**
√	√	√	√		88.57	73.55	86.08
√	√	√	√	√	87.44	72.18	86.49

To explore the effects of fusion method in different directions, two more sets of experiments are conducted based on the best combinations in Table 2. The first set of the experiment only projects features from BEV to the camera view. In contrast, the second set of the experiment only projects feature from camera view to BEV. Table 3 demonstrates that two-way fusion method is better than one-way fusion. The effect of different fusion methods on it is very limited for 2D and BEV performance, but they have a significant impact on the accuracy of 3D detection.

Table 3. Ablation study for the fusion method in different directions. **B2C** means the one-way fusion from BEV to the camera view. **C2B** means the one-way fusion from the camera view to BEV. **Both** mean bidirectional fusions.

Method	2D (%)			3D (%)			BEV (%)		
	E	M	H	E	M	H	E	M	H
B2C	90.00	87.97	86.33	83.12	74.18	68.08	89.30	85.17	78.84
C2B	89.67	87.07	86.15	82.08	72.56	66.96	88.51	84.98	78.77
Both	**90.33**	**88.96**	**88.19**	**84.65**	**75.60**	**68.64**	**89.81**	**88.65**	**79.88**

Besides, the DDF3D model converges faster and the experimental values keep steadily after 80K iterations. Based on the attribute, the model can be checked good or not good within 12 h. Figure 5 shows the evaluation results are extracted every 5K iterations from 80K iterations to 120K iterations on the validation subset.

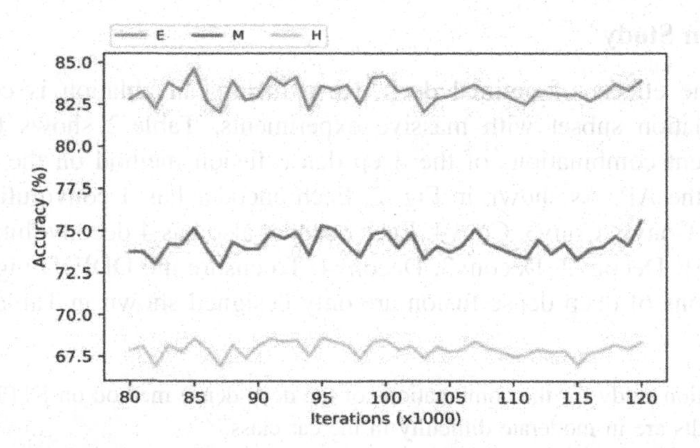

Fig. 5. 3D detection accuracy of DDF3D for car class from 80K iterations to 120K iterations. The *light coral color*, *medium aquamarine color*, and *Navajo white color* denote the *Easy*, *Moderate*, *Hard* difficulty respectively. (Color figure online)

4 Conclusion

This work proposed DDF3D, a full fusion 3D detection architecture. The proposed architecture takes full consideration of the mutual advantages of RGB images and point clouds in the feature extraction phase. The deep dense fusion is two-directional fusion at the same time. A high-resolution feature extractor with the full fusion features, the proposed architecture greatly improves 3D detection accuracy, specifically for small objects. Massive experiments on the KITTI object detection dataset, DDF3D outperforms the state-of-the-art existing method in among of 2D, 3D, and BEV.

References

1. Li, B., Zhang, T., Xia, T.: Vehicle detection from 3D lidar using fully convolutional network. In: Robotics: Science and Systems XII (2016)
2. Chen, X., Ma, H., Wan, J., Li, B., Xia, T.: Multi-view 3D object detection network for autonomous driving. In: 2017 IEEE Conference on Computer Vision and Pattern Recognition (CVPR), pp. 6526–6534 (2017)
3. Caltagirone, L., Scheidegger, S., Svensson, L., Wahde, M.: Fast LIDAR-based road detection using fully convolutional neural networks. In: 2017 IEEE Intelligent Vehicles Symposium (IV), Los Angeles, CA, pp. 1019–1024 (2017)
4. Yang, B., Luo, W., Urtasun, R.: PIXOR: real-time 3D object detection from point clouds. In: 2018 IEEE/CVF Conference on Computer Vision and Pattern Recognition, Salt Lake City, UT, pp. 7652–7660 (2018)
5. Zhou, Y., Tuzel, O.: VoxelNet: end-to-end learning for point cloud based 3D object detection. In: 2018 IEEE/CVF Conference on Computer Vision and Pattern Recognition, pp. 4490–4499 (2018)
6. Wen, L., Jo, K.-H.: Fully convolutional neural networks for 3D vehicle detection based on point clouds. In: Huang, D.-S., Jo, K.-H., Huang, Z.-K. (eds.) ICIC 2019. LNCS, vol. 11644, pp. 592–601. Springer, Cham (2019). https://doi.org/10.1007/978-3-030-26969-2_56
7. Yan, Y., Mao, Y., Li, B.: Second: sparsely embedded convolutional detection. Sensors **18**, 3337 (2018)
8. Charles, R.Q., Su, H., Kaichun, M., Guibas, L.J.: Pointnet: deep learning on point sets for 3d classification and segmentation. In: 2017 IEEE Conference on Computer Vision and Pattern Recognition (CVPR), pp. 77–85 (2017)
9. Qi, C.R., Yi, L., Su, H., Guibas, L.J.: Pointnet++: deep hierarchical feature learning on point sets in a metric space. In: Advances in Neural Information Processing Systems, vol. 30, pp. 5099–5108 (2017)
10. Lang, A.H., Vora, S., Caesar, H., Zhou, L., Yang, J., Beijbom, O.: PointPillars: fast encoders for object detection from point clouds. In: 2019 IEEE/CVF Conference on Computer Vision and Pattern Recognition (CVPR), Long Beach, CA, USA, pp. 12689–12697 (2019)
11. Chen, X., Kundu, K., Zhang, Z., Ma, H., Fidler, S., Urtasun, R.: Monocular 3D object detection for autonomous driving. In: IEEE CVPR (2016)
12. Xu, B., Chen, Z.: Multi-level fusion based 3D object detection from monocular images. In: 2018 IEEE/CVF Conference on Computer Vision and Pattern Recognition, Salt Lake City, UT, pp. 2345–2353 (2018)

13. He, T., Soatto, S.: Mono3D++: Monocular 3D vehicle detection with two-scale 3D hypotheses and task priors. In: Proceedings of the AAAI Conference on Artificial Intelligence, vol. 33, pp. 8409–8416, July 2019
14. Li, P., Chen, X., Shen, S.: Stereo R-CNN based 3d object detection for autonomous driving. In: 2019 IEEE/CVF Conference on Computer Vision and Pattern Recognition (CVPR), pp. 7636–7644 (2019)
15. Ku, J., Mozifian, M., Lee, J., Harakeh, A., Waslander, S.L.: Joint 3D proposal generation and object detection from view aggregation. In: 2018 IEEE/RSJ International Conference on Intelligent Robots and Systems (IROS), pp. 1–8 (2018)
16. Wang, J., Zhu, M., Sun, D., Wang, B., Gao, W., Wei, H.: MCF3D: multi-stage complementary fusion for multi-sensor 3D object detection. IEEE Access **7**, 90801–90814 (2019)
17. Qi, C.R., Liu, W., Wu, C., Su, H., Guibas, L.J.: Frustum pointnets for 3D object detection from RGB-D data. In: 2018 IEEE/CVF Conference on Computer Vision and Pattern Recognition, pp. 918–927 (2018)
18. Xu, D., Anguelov, D., Jain, A.: PointFusion: deep sensor fusion for 3D bounding box estimation. In: 2018 IEEE/CVF Conference on Computer Vision and Pattern Recognition, Salt Lake City, UT, pp. 244–253 (2018)
19. Du, X., Ang, M.H., Karaman, S., Rus, D.: A general pipeline for 3D detection of vehicles. In: 2018 IEEE International Conference on Robotics and Automation (ICRA), pp. 3194–3200 (2018)
20. Liang, M., Yang, B., Wang, S., Urtasun, R.: Deep continuous fusion for multi-sensor 3D object detection. In: Ferrari, V., Hebert, M., Sminchisescu, C., Weiss, Y. (eds.) ECCV 2018. LNCS, vol. 11220, pp. 663–678. Springer, Cham (2018). https://doi.org/10.1007/978-3-030-01270-0_39
21. Girshick, R., Fast R-CNN. In: 2015 IEEE International Conference on Computer Vision (ICCV), pp. 1440–1448 (2015)
22. Liang, M., Yang, B., Chen, Y., Hu, R., Urtasun, R.: Multi-task multi-sensor fusion for 3D object detection. In: The IEEE Conference on Computer Vision and Pattern Recognition (CVPR), June 2019
23. Simonyan, K., Zisserman, A.: Very deep convolutional networks for large-scale image recognition. In: International Conference on Learning Representations (2015)
24. Long, J., Shelhamer, E., Darrell, T.: Fully convolutional networks for semantic segmentation. In: 2015 IEEE Conference on Computer Vision and Pattern Recognition (CVPR), Boston, MA, pp. 3431–3440 (2015)
25. Lin, T., Dollar, P., Girshick, R., He, K., Hariharan, B., Belongie, S.: Feature pyramid networks for object detection. In: 2017 IEEE Conference on Computer Vision and Pattern Recognition (CVPR), pp. 936–944 (2017)

PON: Proposal Optimization Network for Temporal Action Proposal Generation

Xiaoxiao Peng[1,2,3], Jixiang Du[1,2,3(✉)], and Hongbo Zhang[3]

[1] Department of Computer Science and Technology, Huaqiao University, Quanzhou, China
jxdu@hqu.edu.cn

[2] Fujian Key Laboratory of Big Data Intelligence and Security, Huaqiao University, Quanzhou, China

[3] Xiamen Key Laboratory of Computer Vision and Pattern Recognition, Huaqiao University, Quanzhou, China

Abstract. Temporal action localization is a challenging task in video understanding. Although great progress has been made in temporal action localization, the most advanced methods still have the problem of sharp performance degradation when an action proposal generated. Most methods use sliding windows method or simply group frames according to frame-level scores. These methods are not enough to provide accurate action boundary and maintain reasonable temporal structure. In order to solve these problems, we propose a novel proposal optimization network to generate start score, end score, action score and regression score, and then remove the redundancy by NMS algorithm. In the proposed method, we introduce a metric loss function to maintain the temporal structure of action proposal in the training process. To verify the effectiveness of the proposed method, we have made comparative experiments on ActivityNet-1.3 dataset respectively, and the proposed method has surpassed some of the state-of-the-art methods on the dataset.

Keywords: Temporal action localization · Action proposal · Proposal Optimization Network

1 Introduction

In recent years, with the rapid development of digital audio equipment, more and more video data appear in people's lives. Video contains more complex information than image, and video analysis has become the focus of people's research. Temporal action proposal [1–7] aims to capture video temporal segments that are likely to contain an action in untrimmed video. This task is the key technology of temporal action localization [8–13] and even video action analysis, such as action recognition [14–16], video caption [17], spatial-temporal action detection [18].

To achieve high quality action proposal, according to literature [6], action proposal generation should (1) generate temporal proposals with flexible action duration and precise action boundaries to cover all ground truth action instances exactly and exhaustively; (2) generate credible confidence scores so that proposals can be retrieved

D.-S. Huang and P. Premaratne (Eds.): ICIC 2020, LNAI 12465, pp. 145–157, 2020.
https://doi.org/10.1007/978-3-030-60796-8_13

properly. One method [1, 4] is often used is sliding window which generates segment proposals by uniform sampling or manually-predefined over the video frame sequence, then a binary classifier is used to evaluate the confidence score of each video segments. Although such methods can get proposals with various temporal spans, the generated proposals naturally have imprecise boundary. The second thread work [3, 11], tackle the action proposals via evaluating the action score at the frame level. these methods intensively evaluate the confidence score of each video frame, then the consecutive frames with similar score are combined by a specific algorithm (watershed algorithm [11]) to form a candidate action proposal. Compared with the previous method based on segment level, frame-level based methods have more accurate action boundaries than previous method. However, this method is prone to generate unreliable confidence score for long video segments, resulting in a lower recall rate.

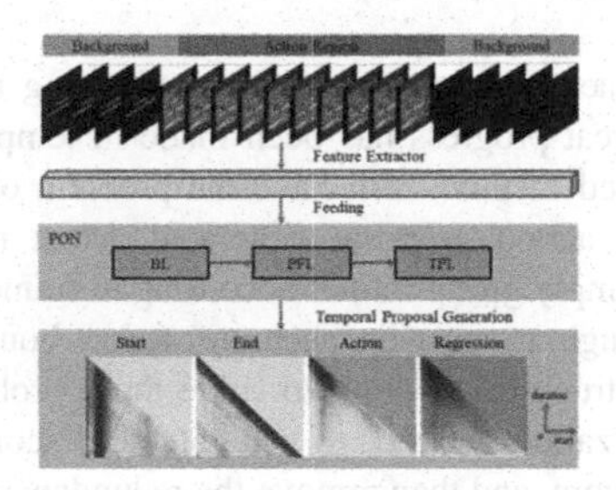

Fig. 1. Proposal Optimization Network mainly consist of three layers: Base Layer (BL), Proposal Feature Layer (PFL), Temporal Predict Layer (TPL). PON densely evaluates all action proposals by generating simultaneously four score maps.

Complementary temporal action proposal [3] designs a three stages network, which includes proposals initialization, complementary proposal generation, proposal boundary adjustment and ranking. Boundary sensitive network [6] adopts a "local to global" scheme for action proposal. In practice, adjacent detected boundary points will be paired to form valid proposals. Both of these two methods are three stage models. However, the modules in different stages are trained independently, and it lacks overall optimization of the model. The second drawback is the lack of temporal position information model for generating action proposals, which conveys the temporal ordering information of the video sequence. That is, the start point must match the end point and there must be an order between them.

In order to address the mentioned problems, we proposed a proposal optimization network (PON) for generating efficient temporal action proposals. This network is an end-to-end unified optimization network framework, rather than staged training. we construct a two-dimensional temporal matrix to represent all possible temporal action proposals in video, and output four different confidence scores at each position of the matrix to evaluate the quality of each proposal. Figure 1 illustrates an overview of our architecture for temporal action proposal generation.

1. We proposed a unified and optimized network for end-to-end training PON, which represents the temporal action proposal as a two-dimensional matrix, and trains four different types of confidence scores on the two-dimensional matrix to evaluate all proposals.
2. In order to keep the temporal structure information in training, we introduce a novel IoU loss function based on metric learning for joint optimization.

2 Related Work

Temporal Action Localization. The task of temporal action localization is to locate the exact time stamp of each action instance in an untrimmed video and recognize the category of the action instance. Similar to object detection, temporal action localization method can be divided into two categories. The first is the single shot temporal action localization [8, 9], the advantages of this method are simple and very fast, the disadvantage is that its accuracy is not satisfactory. The second is a two-stage approach [10–12], temporal action localization can be divided into two stages: temporal action proposal generation and action classification. Although in recent years, with the development of deep learning technology, the action recognition model has achieved excellent performance, temporal action localization positioning has been unsatisfactory. Recently, many works pay attention to improving the performance of temporal action localization model, however, many researchers regard the proposal generation algorithm as the bottleneck of temporal action localization. The main drawback of these two stage approaches is the indirect optimization strategy, which may result in a sub-optimal solution.

Temporal Action Proposal Generation. As aforementioned, the key to improve the performance of temporal action localization lies in the quality of temporal action proposal. Different from the two-dimensional image object proposals, the label of temporal action proposal is fuzzy and has certain degree of fault tolerance. For temporal proposal generation task, most previous works [1–4] adopt top-down fashion to generate proposals with predefined duration and interval, such as sliding windows, where the main drawback is the lack of boundary precision and duration flexibility. What's more, there are also works that generate action proposals in a bottom-up way. TAG [11] generates frame level action score via a binary classifier, then product proposals using watershed algorithm, however, lack of reliable confidence score for retrieval. BSN [6] generates action proposals via a three-stage network. It is a simple way to generate action proposals, which lacks effective modeling of temporal and localization information. In this work, we proposed a Proposal Optimization Network (PON) to model temporal position information, and an effective loss function is introduced to maintain this temporal structure in training process.

3 Proposal Approach

3.1 Problem Formulation

We can denote an untrimmed video V as frame set $V = \{v_t\}_{t=1}^{T}$. Where v_t is the $t-th$ frame in V, and T is the total number of frames in this video. Annotation of video V is composed by a set of action instances $\Omega_g = \{(s_n, e_n, c_n)\}_{n=1}^{N_g}$ Where N_g is the total number of action instances in video V, the three letters in the formula represent the start time, end time and action category of the action instance respectively. Start and end time labels are accurate to frame level, and $c_n \in [1, k]$, k is the total number of action categories. Different from action localization, in temporal action proposal generation task, it is not need to use action category label. During prediction, the target is to accurately generate the proposals set $\Omega_p = \{(s_n, e_n)\}_{n=1}^{N_p}$ for the test video.

3.2 Video Feature Extraction

Recent models of temporal action proposal generation are all based on the visual features of the raw video. In the construction of visual features, 3D Convolutional Networks (C3D) [14] and Two-Stream Network (TSN) [15] are generally selected. Two-Stream Network is used to extract the visual features of the video since it achieves great action recognition precision. It is widely used in many video Understanding models [12, 13]. According to the above method, given an untrimmed video V with length of T, we can extract visual feature sequence $F = \{f_t\}_{n=1}^{t_\alpha}$, where $t_\alpha = T/\alpha$, α is the regular sampling interval, which aims to reduce computation cost. $f_t \in R^C$ is the visual feature of frame t from network. C is the visual feature dimension. In order to solve the problem of different length of untrimmed video, we divide into one or more different snippets with $t\alpha$ length by feature scale.

3.3 Proposal Optimization Network

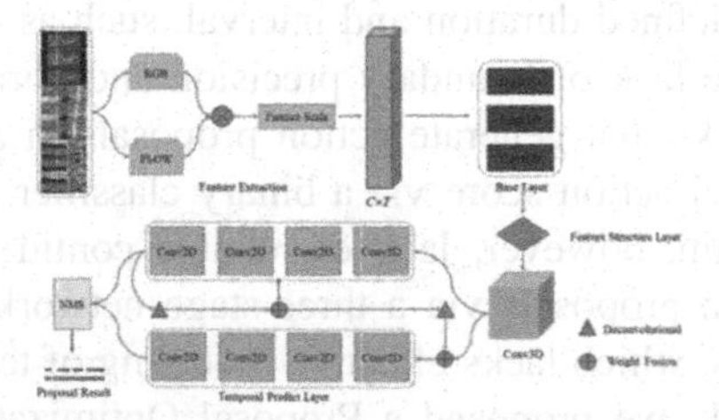

Fig. 2. An overview of our Proposal Optimization Network (PON) architecture. The proposed method uses two-stream features offline via feature extraction network. Then feed the features into the base layer. And then construct all the action proposal features in the video via feature structure layer. Finally, predict four score map to determine action proposals.

In this section, based on the above analysis, we propose a Proposal Optimization Network (PON) which can effectively capture the context information of long untrimmed video to generate high-quality temporal action proposals. Figure 2 illustrates an overview of our architecture for temporal action proposal generation. It consists of three main layers: base layer, feature structure layer and temporal predict layer. These three layers will be described in detail in the following subsections.

Base Layer. In this part, we introduce the functions of the base layer. The basic element of the base layer is one-dimension convolutional. One dimensional convolutional is suitable for processing the temporal correlation feature of speech and video. Multi-Stage Temporal Convolutional Network (MS-TCN) [19] practice one-dimensional convolutional module to efficiently predict frame level action probability. It is proved that one-dimensional convolutional has great potential in the field of temporal. The Base layer consists of three one-dimensional convolutions. The convolution kernel size is 3 and the stride is 1. The output channels of first two convolutions are 256, and the last one is 128. The purpose is to simplify the input control feature sequence and expand the temporal receptive field. In order to facilitate the subsequent matrix processing, we will set up an observation window with length of l_d to truncate the untrimmed video with length of l_u. We can define such a window as $w = \{t_{w,s}, t_{w,e}, \Omega_w, \Omega_e\}$, where $t_{w,s}$, $t_{w,e}$ are the start time and end time of the observation windows w, Ω_{w}, F_w are annotations and feature representation within the window separately. The input of the base layer is the scaling feature $F^{D \times T}$ of Two stream network and the output is after the feature dimension is reduced to 128, which is the basic feature of the subsequent construction of the temporal action proposal matrix.

Feature Structure Layer. For an arbitrary input feature f_i, the PFG layer is able to product a proposal features whose shape is $f_p \in R^{C \times N \times D \times T}$. Specially, as shown in Fig. 3, we multiply the video feature features output by the base layer and a mask constructed in the temporal dimension to obtain the sampling features of each proposal. In the past, the others work used to sample the features of the proposal serially. However, these may ignore the context relationship between some proposals. With the built-in vectorization mechanism in the python library, we can simplify the operation between matrices and calculate the features of the candidate proposals in a video simultaneously. In Particular, since our features are scaled to a fixed scale, some temporal points may not be in the range during the sampling process. As shown in Eq. 1 and Eq. 2, we use Eq. 1 to fuzzy sample the so-called positions that are not integral points, and then we can get the temporal action proposals feature by multiplying the video features with all mask matrix vectorization points. The whole computing process is parallel, so the temporal action proposals have abundant temporal context information.

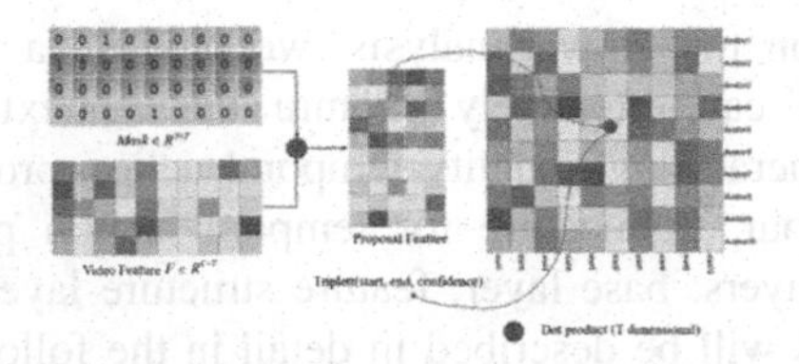

Fig. 3. Illustration of the feature structure layer. A mask matrix and visual feature are constructed to perform dot product in temporal dimension to obtain the feature representation of each action proposals. After a series of dot product operations to get all the action proposal features of video.

$$w_t = \begin{cases} 1 - \text{mod}(t_n) & \text{if } t_n = lower(t_n) \\ \text{mod}(t_n) & \text{if } t_n = upper(t_n) \\ 0 & \text{if } t_n = others \end{cases} \tag{1}$$

$$F = \sum_{i=1}^{D} \sum_{j=1}^{T} \sum_{t=1}^{T} f^{C \times T} \cdot [w^{N \times T}]^T \tag{2}$$

Temporal Predict Layer. The temporal prediction layer is the core part of the network. The goal is to generate four different types of confidence maps, which are start confidence, end confidence, action confidence and regression confidence. The start and end confidence scores determine the boundary of temporal action proposals. It is the evaluation score at the frame level. The action confidence score and the regression confidence score ensure the integrity of the action, which is the evaluation score at the proposal level. We design a novel network structure for generating the four confidence maps proposed above.

Table 1. Convolution module configs

Convolution	Kernel	Stride	Activation	Output
conv2D	(1, 1)	(0, 0)	RELU	(128, D, T)
conv2D	(3, 3)	(1, 1)	RELU	(128, D, T)
conv2D	(3, 3)	(1, 1)	RELU	(128, D, T)
conv2D	(1, 1)	(0, 0)	RELU	(2, D, T)
conv3D	(32, 1, 1)	(32, 0, 0)	RELU	(512, 1, D, T)

Table 1 is the specific convolution configs, proposal feature constructed by the input PFG layer of the network $F \in R^{C \times N \times D \times T}$, where N is set to 32, and then this feature is subjected to 3D convolution. This operation is to reduce the number of samples N to 1, while increasing the number of hidden layer units from 128 to 512. Then the proposal features are reduced from $R^{C \times 1 \times D \times T}$ to $R^{C \times D \times T}$ via a pytorch compression operation. After the compression operation, the 2D convolution module is used to generate the starting confidence score map $M_s \in R^{D\times T}$ and the ending confidence map $M_e \in R^{D\times T}$. RELU activation followed by all previous convolutions, and the final output convolution layer is activated by sigmoid. And M_s, M_e are trained using binary classification in Boundary Sensitive Network [6]. On the generation of action score map and regression score map, inspired by U-Net [21], we use the method of inverse convolution up sampling at the output of the start and end confidence map. They are fused with the features of the upper layers to obtain the fused features. This feature is fed into a network the same as the previous start and end confidence map generation. The output is action score map M_a and regression score map M_r. These two maps share the same label map, but are monitored using different loss functions. M_a is trained using binary classification function and M_r is trained using MSE loss function.

3.4 Training of PON

In Proposal Optimization Network, to jointly learn action confidence map, regression confidence, start confidence map and end confidence map. A unified multitask loss is further proposed. The training details will be covered in detail in this section.

Training Data Construction. Given an untrimmed video V, we can extract the feature F of length t_α via a two-stream framework. To reduce the computational overhead, we use a window of length l_w to truncate the feature sequence, with an overlap rate of 60%. These windows contain at least one action instance are kept for training, thus, a training set can be represented as $\Omega = \sum_{v=1}^{N_v} \{\{w_n\}_{n=1}^{N_w}\}_v$, where $N_w \geq 1, N_v$ is the total of video in the training set.

Label Construction. Given the annotation $\Omega_g = \{\Omega_i = (t_{si}, t_{ei})\}_{i=1}^{N_g}$ of a video, we compose start label $g^s \in R^{D\times T}$ for auxiliary PON start confidence map classification loss. Similarly, we compose end label $g^e \in R^{D\times T}$ for auxiliary PON end confidence map classification loss. On the g^s/g^e, we define r^s/r^e as $r^s = [t_s - \frac{2}{5}\cdot d, t_s + \frac{2}{5}\cdot d]$, where d is the length of proposal. We define the ground-truth region corresponding to the above region as $r_{t_n} = [t_n - l_f, t_n + l_f]$ and l_f, so we calculate overlap ratio IOR as g_t^s/g_t^e. Thus, we can get 2D labels g^s and g^e. For a proposal $\rho_{i,j} = (t_s = t_i, t_e = t_j + t_i)$ in confidence map, where t_i is start time, t_j is duration of an action proposal. We calculate the temporal Intersection-over-Union (IoU) on the confidence to all the ground-truth in this video, and use the maximum value as the label of the action integrity confidence score. There can generate $g^a \in R^{D\times T}$, $g^r \in R^{D\times T}$.

Loss Function. The proposal optimization network is a multitask network. Hence, its loss function consists of the following parts:

$$Loss = L_{start} + L_{act} + L_{end} + \lambda \cdot L_{reg} + \eta \cdot L_{met} \tag{3}$$

Where L_{start}/L_{end} is the loss function to supervise the starting/end confidence map. L_{act} is the loss function defined for action score map generation in action-level, and L_{reg} is the loss function defined for regression confidence score map generation used to evaluate the integrity of action. L_{met} is define as the IoU loss function that supervises two types of confidence map. We set the constant λ to 10 and the constant η to 2 via several experiments. On the boundary-type confidence score, we use the binary logistic function [6] since the binary classification loss is more suitable for the boundary type confidence score generation. The L_{start} and L_{end} loss function are shown below:

$$\frac{1}{l_w}\sum_{i=1}^{l_w}\left(\alpha^{+}\cdot b_i\cdot\log(p_i)+\alpha^{-}\cdot(1-b_i)\cdot\log(1-p_i)\right) \tag{4}$$

$$l^{+}=\sum_{i=1}^{l_g} g_i,\ \alpha^{+}=\frac{l_w}{l^{+}},\ \alpha^{-}=\frac{l_w}{l^{-}}$$

$$b_i=\begin{cases}1 & if\ (g_i-\theta_{IoP})>0\\ 0 & if\ (g_i-\theta_{IoP})=0\\ -1 & if\ (g_i-\theta_{IoP})<0\end{cases} \tag{5}$$

As shown in Eq. 4, where l_w is the length of window, p_i is the predicted probability value. The specific equivalent conversion formula of b_i is shown in Eq. 5, g_i is the ground-truth, and b_i is a binary function for $(g_i-\theta_{IoP})$ to convert matching score. For θ_{IoP}, we uniformly set it to 0.5 in the calculation. In order to alleviate the problem of positive and negative samples in training. In the implementation process, we can transform the dimensional matrix into a long vector, and then calculate the loss function. For the L_{act} loss function, it is similar to binary classification loss function, the difference is that it shares the label map with L_{reg}. With introduce the L_{reg} loss function in detail:

$$L_{reg}=\frac{1}{L^2}\sum_{i=1}^{L}\sum_{j=1}^{L} mse(p_{ij}-g_{ij}) \tag{6}$$

As shown in Eq. 6, we use MSE loss to regress the integrity score of proposal level since it has better robustness to noise points than smooth-L1 loss function.

In order to maintain the time structure of movements in training, we introduce the matric loss function L_{met} and propose an Algorithm 1 to calculate it.

Algorithm 1: Measurement loss calculation algorithm

Step1: input $p^s_{i,j}, p^e_{i,j} p^a_{i,j} p^r_{i,j}$.

Step2: perform produce operation, get the score of $p^b_{i,i}, p^c_{i,j}$.

Step3: select the top 20 points in $p^b_{i,j}$ **and** $p^c_{i,j}$ **respectively, map 20 action proposal**

$\{\mathrm{pro}^b_i\}^{20}_{i=1}$ *and* $\{\mathrm{pro}^c_i\}^{20}_{i=1}$.

Step4: calculate the IoU loss of these 20 pairs of action proposals:

$$L_{met} = \sum_{p=1}^{20} \frac{pro^b_p \cap pro^c_p}{pro^b_p \cup pro^c_p}$$

Inference of PON. In the inference process of Proposal Optimization Network, we generate two of confidence scores: boundary type and integrity type score. Boundary type score includes start score and end score. Integrity type score includes action score and regression score. We take the multiplication of these four score as the evaluation score of the action proposals:

$$p_f = p^s_{i,j} \cdot p^e_{i,j} \cdot p^a_{i,j} \cdot p^r_{i,j} \tag{7}$$

In the right side of Eq. 7 is the confidence score each category corresponding to the position in the confidence map. After getting the confidence score of each action proposal, we need remove redundant proposals to achieve higher recall with fewer proposals, where Non-maximum suppression (NMS) algorithm is widely adopted for this purpose. In PON, we apply Soft-NMS algorithm to remove redundant action proposals. The final action proposal can be represented $\Gamma_p = \{\Phi_n = \{t_s, t_e, p_f\}\}$. In order to reduce the computation, we take the top 100 scores as the input of the Soft-NMS algorithm.

4 Experiments

4.1 Dataset and Implementation

ActivityNet. It contains 19994 videos labeled in 200 classes. It is divided into train validation and test with a ratio 0.5, 0.25, 0.25. Compared with Thumos14, most of the videos in ActivityNet-1.3 contain activity instances of a single class instead of sparsely distributed.

Implementation Details. As the mentioned in Sect. 3, we use two-stream network to encode the visual feature, which used ResNet [20] as spatial stream and BNInception

[20] as the temporal stream. For ActivityNet, we use the submission scheme proposed in [20], which have been proved to be effective in extracting the features of long and untrimmed video on ActivityNet Challenge 2017. During the training, we adopt the method of transfer learning set. The interval of snippets is set to 16 on ActivityNet-1.3 and 6 on Thumos14.

Evaluation Metric. We use the average recall (AR) and average number of proposal (AN) per video curve as the evaluation metric. A proposal is a true positive if the temporal intersection over union(tIoU) between the proposal and ground-truth segment is greater than the given threshold (e.g., tIoU > 0.5). AR is defined as the mean of all recall values using tIoU between 0.5 and 0.9(inclusive) with a step size of 0.05. AN is defined as the total number of proposals divided by the number of videos in the testing subset.

4.2 Experimental Results on ActivityNet

In order to verify the effectiveness of our model, we use the official evaluation indicators to evaluate the performance of action proposal generation. We have made comparison experiments on the validation set and test set of ActivityNet dataset respectively. We further compare our PON model with other state-of-the-art models on the validation set of ActivityNet-1.3. From Table 2, in the validation set, our model achieves excellent performance. The AR@100 is improved from 74.54 to 75.78 and AUC is improved from 66.43 to 67.32. On the test set, we submit the result file to the official server of ActivityNet challenge. Due to the limitation of server related index calculation, we only get AUC results, our approach is 1.32% higher than BSN.

Table 2. Comparison results between our approach and other state-of-the-art temporal action generation approaches on ActivityNet-1.3 dataset.

Method	AR@100	AUC (val)	AUC (test)
TCN [8]	–	59.58	61.56
ANET [21]	–	63.12	–
MSRA [32]	–	63.13	64.18
SSAD [9]	73.01	64.40	64.80
CTAP [4]	73.17	65.72	–
BSN [6]	74.16	66.17	66.26
MGG [7]	74.54	66.43	66.47
OURS	**75.78**	**67.32**	**67.59**

In the process of inference, time is mainly consumed in two parts: one is the generation of confidence map, the other is post-processing. Compared with BSN, the inference of BSN consists of four parts: Temporal evaluation, Proposal Generation, Proposal evaluation and post-processing. The time consumed by PGM is the main part, because this process needs to traverse all the temporal points to form temporal action proposals and reconstruct proposal feature. Our method does not need redundant PGM module and TEM module, which greatly reduces the inference time.

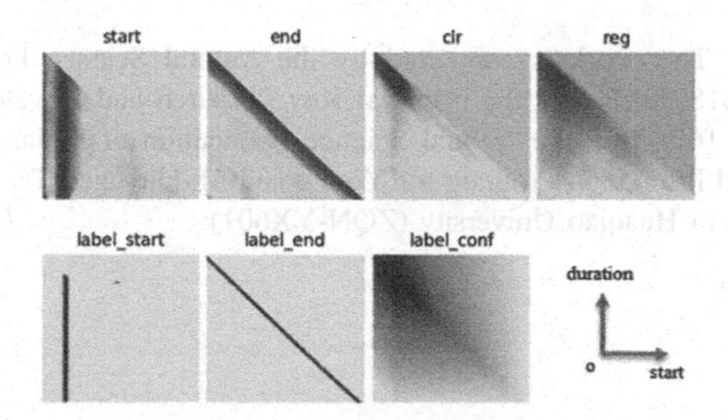

Fig. 4. This is visualization of prediction result of PON in single action instance video.

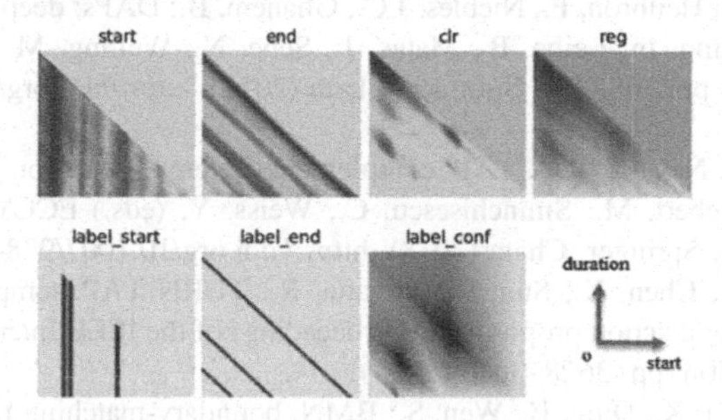

Fig. 5. This is the visualization of prediction results of PON on videos with multiple action instances.

We visualized the prediction results on two representative samples. As shown in Fig. 5 and Fig. 6, they are the results of simple samples and difficult samples. Our prediction effect is very great on the simple sample, since the movements on these samples are relatively single, reasoning call well distinguish them. However, in the difficult samples, the performance is not satisfactory, the essential reason is that the action boundary in the video is fuzzy, and the adjacent two action boundaries are easy to affect each other. Which is also one of the difficulties in the traditional detection field.

5 Conclusion

In this work, we introduce a novel proposal optimization network for temporal action proposal generation. In the proposed algorithm, a two-level fusion network architecture is built to generate four confidence scores. To train the proposed network, we propose a novel metric loss function to correct temporal association and ensure the temporal structure of action proposal. Our Proposal Optimization Network has achieved state-of-the-art performance compared with other competitive methods on ActivityNet-1.3. However, in the generation of boundary scores, the dependence and mutual restriction between scores are expected to be further studied in the further.

Acknowledgement. This work is supported by the Natural Science Foundation of China (No. 61673186 and 61871196) and the National Key Research and Development Program of China (No. 2019YFC1604700), the Natural Science Foundation of Fujian Province of China (No. 2019J01082) and Promotion Program for Young and Middle-aged Teacher in Science and Technology Research of Huaqiao University (ZQN-YX601).

References

1. Buch, S., Escorcia, V., Shen, C., Ghanem, B., Carlos Niebles, J.: SST: single-stream temporal action proposals. In: Proceedings of the IEEE Conference on Computer Vision and Pattern Recognition, pp. 2911–2920 (2017)
2. Escorcia, V., Caba Heilbron, F., Niebles, J.C., Ghanem, B.: DAPs: deep action proposals for action understanding. In: Leibe, B., Matas, J., Sebe, N., Welling, M. (eds.) ECCV 2016. LNCS, vol. 9907, pp. 768–784. Springer, Cham (2016). https://doi.org/10.1007/978-3-319-46487-9_47
3. Gao, J., Chen, K., Nevatia, R.: CTAP: complementary temporal action proposal generation. In: Ferrari, V., Hebert, M., Sminchisescu, C., Weiss, Y. (eds.) ECCV 2018. LNCS, vol. 11206, pp. 70–85. Springer, Cham (2018). https://doi.org/10.1007/978-3-030-01216-8_5
4. Gao, J., Yang, Z., Chen, K., Sun, C., Nevatia, R.: TURN TAP: temporal unit regression network for temporal action proposals. In: Proceedings of the IEEE International Conference on Computer Vision, pp. 3628–3636 (2017)
5. Lin, T., Liu, X., Li, X., Ding, E., Wen, S.: BMN: boundary-matching network for temporal action proposal generation. In: Proceedings of the IEEE International Conference on Computer Vision, pp. 3889–3898 (2019)
6. Lin, T., Zhao, X., Su, H., Wang, C., Yang, M.: BSN: boundary sensitive network for temporal action proposal generation. In: Ferrari, V., Hebert, M., Sminchisescu, C., Weiss, Y. (eds.) ECCV 2018. LNCS, vol. 11208, pp. 3–21. Springer, Cham (2018). https://doi.org/10.1007/978-3-030-01225-0_1
7. Liu, Y., Ma, L., Zhang, Y., Liu, W., Chang, S.F.: Multi-granularity generator for temporal action proposal. In: Proceedings of the IEEE Conference on Computer Vision and Pattern Recognition, pp. 3604–3613 (2019)
8. Dai, X., Singh, B., Zhang, G., Davis, L.S., Qiu Chen, Y.: Temporal context network for activity localization in videos. In: Proceedings of the IEEE International Conference on Computer Vision. pp. 5793–5802 (2017)
9. Lin, T., Zhao, X., Shou, Z.: Single shot temporal action detection. In: Proceedings of the 25th ACM international conference on Multimedia, pp. 988–996 (2017)
10. Shou, Z., Wang, D., Chang, S.F.: Temporal action localization in untrimmed videos via multi-stage CNNs (2016)
11. Xiong, Y., Zhao, Y., Wang, L., Lin, D., Tang, X.: A pursuit of temporal accuracy in general activity detection. arXiv preprint arXiv:1703.02716 (2017)
12. Xu, H., Das, A., Saenko, K.: R-C3d: Region convolutional 3D network for temporal activity detection. In: Proceedings of the IEEE International Conference on computer Vision, pp. 5783–5792 (2017)

13. Yang, X., Yang, X., Liu, M.Y., Xiao, F., Davis, L.S., Kautz, J.: Step: spatiotemporal progressive learning for video action detection. In: Proceedings of the IEEE Conference on Computer Vision and Pattern Recognition, pp. 264–272 (2019)
14. Ji, S., Xu, W., Yang, M., et al.: 3D convolutional neural networks for human action recognition. IEEE Trans. Pattern Anal. Mach. Intell. **35**(1), 221–231 (2013)
15. Wang, L., et al.: Temporal segment networks: towards good practices for deep action recognition. In: Leibe, B., Matas, J., Sebe, N., Welling, M. (eds.) ECCV 2016. LNCS, vol. 9912, pp. 20–36. Springer, Cham (2016). https://doi.org/10.1007/978-3-319-46484-8_2
16. Carreira, J., Zisserman A.: Quo vadis, action recognition? A new model and the kinetics dataset. In: Proceedings of the IEEE Conference on Computer Vision and Pattern Recognition, pp. 6299–6308 (2017)
17. Venugopalan, S., Rohrbach, M., Donahue, J., Mooney, R., Darrell, T., Saenko, K.: Sequence to sequence-video to text. In: Proceedings of the IEEE International Conference on Computer Vision, pp. 4534–4542 (2015)
18. Yang, X., Yang, X., Liu, M.-Y., Xiao, F., Davis, L.S., Kautz, J.: Step: spatio-temporal progressive learning for video action detection. In: Proceedings of the IEEE Conference on Computer Vision and Pattern Recognition, pp. 264–272 (2019)
19. Farha, Y.A., Gall, J.: MS-TCN: multi-stage temporal convolutional network for action segmentation. In: Proceedings of the IEEE Conference on Computer Vision and Pattern Recognition, pp. 3575–3584 (2019)
20. Caba Heilbron, F., Escorcia, V., Ghanem, B., Carlos Niebles, J.: Activitynet: a large-scale video benchmark for human activity understanding. In: Proceedings of the IEEE Conference on Computer Vision and Pattern Recognition, pp. 961–970 (2015)
21. Yao, T., et al.: Msr asia msm at activitynet challenge 2017: trimmed action recognition, temporal action proposals and densecaptioning events in videos. In: CVPR ActivityNet Challenge Workshop (2017)

Intelligent Computing in Communication Networks

A Second-Order Adaptive Agent Network Model for Social Dynamics in a Classroom Setting

Kasper Nicholas, Eric Zonneveld, and Jan Treur(✉)

Social AI Group, Vrije Universiteit Amsterdam, Amsterdam, The Netherlands
kasperanicholas@gmail.com, eric.zonneveld0@gmail.com,
j.treur@vu.nl

Abstract. Alcohol consumption among young adolescents is problematic as health implications and behavioural changes are common consequences. Another problematic factor among young adolescents is the amount of delinquencies committed. In this paper an adaptive social agent network model using friendship relationships as predictor for alcohol consumption and amount of delinquencies committed is explored. The proposed agent network model was empirically validated using classroom data on young adolescents gathered by Knecht in Dutch schools. The agent network model is second-order adaptive and applies a bonding by homophily adaptation principle with adaptive adaptation speed describing clustering in friends networks based on the two aforementioned factors respectively.

Keywords: Social dynamics · Agent network model · Second-order adaptive

1 Introduction

Drinking alcohol on a daily basis is a widely accepted practice and is often seemingly inseparable from social interactions in many modern day societies. However, health consequences due to alcohol consumption are a common occurrence, and can be especially problematic at a young age [1, 5–7, 10, 12, 14, 15, 19]. Both online and offline interaction among young people have been shown to contribute to increased alcohol consumption [10, 12]. As alcohol is often considered a social cohesive, it is logical to study the relation between social networks and alcohol consumption in order to assess any potential causality. Especially during early adolescence do social interactions among peers become increasingly important influential factors. A phenomenon closely related to this is bonding-by-homophily, which expresses the increased occurrence of interactions between similar persons compared to dissimilar persons [2, 13, 16]. An extensive study done by Knecht et al. [10] in 2010 aims to shed light on the effects of this bonding principle and potential social influences among adolescent friends in Dutch schools. They observed that, despite the varying friendship dynamics and individual drinking behaviour, the social network dynamics are a recurring element for the prediction of alcohol consumption among young adolescents. In a similar way they analysed the relation to delinquencies committed by young adolescents.

D.-S. Huang and P. Premaratne (Eds.): ICIC 2020, LNAI 12465, pp. 161–173, 2020.
https://doi.org/10.1007/978-3-030-60796-8_14

The conceptual idea of social agent networks is a useful basis to formalise and analyse the above processes computationally. Computational formalisation can be a good basis for the development of tools to explore the complex and phenomena as described above and support societal decision making. One of the proposed computational methods to study such social dynamics in agent networks is the Network-Oriented Modelling approach presented in [18]. This generic AI modelling approach is based on (adaptive) causal relations incorporating a continuous time dimension to model agent network dynamics and adaptivity; it is briefly introduced in Sect. 2. A number of studies have shown that this computational modeling method can be used to model a variety of social networks; e.g., [3, 4, 8, 11, 18]. Together with the scientific domain literature indicated above, this modeling perspective provides an adequate multidisciplinary research background for the work reported here.

As a main contribution, in this paper a second-order adaptive social agent network model addressing the above processes is described (in Sect. 3) and explored by simulation experiments (in Sect. 4). The social agent network model incorporates the social contagion principle, the first-order adaptive bonding by homophily principle, and a second-order principle for adaptive speed of first-order adaptation. The agent network model was verified by means of mathematical analysis of equilibria (see Sect. 5). Validation was done by the analysis of a data set on alcohol consumption and delinquencies among Dutch high school students [10]; this is discussed in Sect. 6. In order to assess the formation of friendships based on both factors, subsequent parameter tuning was performed by means of Simulated Annealing [9]. The second-order adaptivity and the validation using these empirical data distinguish this work from existing work. Finally, Sect. 7 is a discussion.

2 The Adaptive Modelling Approach for Agent Networks

Network-Oriented Modelling [18] based on temporal-causal networks uses nodes and connections between these nodes as a basic representation. The former are interpreted as states, or states variables, whereas the latter model causal relationships between such states, and have weights as labels. Both states and (in adaptive networks) connections are allowed to vary over time, and thus give rise to dynamics within the network and adaptivity of the network. The design of a network model on a conceptual level is specified as a labeled graph or in a conceptual role matrix specification format [18]. Table 1 summarizes the main concepts. Firstly, *states* and *connections* between them representing causal impacts of the states upon each other. Secondly, the notion of a *connection weight* expresses the strength of impact of a connection. *Combination functions* are used to aggregate the combined influence of states on a given state, and *speed factors* represent the rate of change of a given state with respect to time.

Table 1. An overview of the important concepts in conceptual temporal-causal networks.

Concepts	Notation	Explanation
States and connections	X, Y, $X \rightarrow Y$	Denotes the nodes and edges in the conceptual representation of a network
Connection weights	$\boldsymbol{\omega}_{X,Y}$	A connection between states X and Y has a corresponding *connection weight*. In most cases: $\boldsymbol{\omega}_{X,Y} \in [-1, 1]$
Aggregating multiple impacts on a state Y	$\mathbf{c}_Y(..)$	Each state has a *combination function* which is responsible for combining causal impacts of all states from which Y gets incoming connections
Timing of the effect of causal impact	$\boldsymbol{\eta}_Y$	The *speed factor* determines how fast a state changes by any aggregated causal impact. In most cases $\boldsymbol{\eta}_Y \in [0, 1]$

A large variety of combination functions can be used for different states in temporal-causal networks, providing sufficient flexibility for the aggregation of causal impact of states upon one another. The choice of a combination function largely depends on the application at hand, and can further be varied between states in the same system. Combination functions that are often used are briefly elaborated upon in Table 3. The numerical representations based on the above defined conceptual framework is presented in Table 2.

Table 2. Overview of the numerical representations for temporal-causal network models.

Concept	Representation	Explanation
State values over time t	$Y(t)$	At each time point t any state Y has a real number value, usually in [0, 1]
Single causal impact	$\mathbf{impact}_{X,Y}(t) = \boldsymbol{\omega}_{X,Y}\, X(t)$	For every time point t state X with connection to state Y impacts Y, through connection weight $\boldsymbol{\omega}_{X,Y}$
Aggregating multiple impacts	$\mathbf{aggimpact}_Y(t)$ $= \mathbf{c}_Y(\mathbf{impact}_{X_1,Y}(t), \ldots, \mathbf{impact}_{X_k,Y}(t))$ $= \mathbf{c}_Y(\boldsymbol{\omega}_{X_1,Y}X_1(t), \ldots, \boldsymbol{\omega}_{X_k,Y}X_k(t))$	The aggregated causal impact of multiple states X_i on Y at t, is determined using a combination function $\mathbf{c}_Y(V_1, \ldots, V_k)$ applied to the single causal impacts
Timing of the causal effect	$Y(t+\Delta t) = Y(t)$ $+ \boldsymbol{\eta}_Y[\mathbf{aggimpact}_Y(t) - Y(t)]\, \Delta t$ $= Y(t) + \boldsymbol{\eta}_Y[\mathbf{c}_Y(\boldsymbol{\omega}_{X_1,Y}X_1(t),$ $\ldots, \boldsymbol{\omega}_{X_k,Y}X_k(t)) - Y(t)]\, \Delta t$	The speed factor $\boldsymbol{\eta}_Y$ determines how fast a state changes upon aggregated impact of the states X_i from which state Y has incoming connections

The obtained *difference equation* or its equivalent *differential equation* in the last row of Table 2 is important for both simulation and mathematical analysis of a network model. When a (first-order) *adaptive* temporal-causal network model is considered, network characteristics such as the connection weights, speed factors and combination functions can explicitly be represented themselves as network states, called reification states [18], and thus also evolve over time according to a difference equation of the type presented above for them. Such (first-order) reification states can be depicted as a separate level in the network picture, called first-order reification level. As this process of network reification can be repeated to obtain higher-order adaptation, in this way multiple levels can be distinguished, as illustrated for second-order adaptation by the example network model in Fig. 1.

Several examples of combination functions in adaptive temporal-causal networks are used in literature such as [3, 8, 11]; see Table 3 for some of them. The first is the *identity* **id(.)** for a single state impacting another state. The second is the *scaled sum* **ssum(..)** with a scaling factor λ. The third is the *advanced logistic sum* $\mathbf{alogistic}_{\sigma,\tau}$**(..)** with parameters **σ** for the steepness and **τ** for the threshold. These and other combination functions are further explained in [18], Chap. 2. A fourth combination function discussed is the *simple linear homophily function* **slhomo,(..)**, where is a homophily modulation factor and is a homophily tipping point. For an in depth derivation of this function Sect. 3. In Table 3 a mathematical representation of each of the four combinations functions is presented. In the $\mathbf{slhomo}_{\alpha,\tau}(V_1, V_2, W)$ combination function, the variable W stands for the value of the connection weight reification state used in the adaptive network.

Table 3. An overview of some combination functions. The latter two are used in the adaptive social agent network model presented here.

Combination function	Description	Formula $c_Y(..)$ =
$\mathbf{id}(V)$	Identity	V
$\mathbf{ssum}_\lambda(V_1, \ldots, V_k)$	Scaled sum	$\frac{V_1 + \ldots + V_k}{\lambda}$ with $\lambda > 0$
$\mathbf{alogistic}_{\sigma,\tau}(V_1, \ldots, V_k)$	Advanced logistic sum	$\left[\frac{1}{1+e^{-\sigma(V_1+\ldots+V_k-\tau)}} - \frac{1}{1+e^{\sigma\tau}}\right](1+e^{-\sigma\tau})$
$\mathbf{slhomo}_{\alpha,\tau}(V_1, V_2, W)$	Standard linear homophily	$W + \alpha\,(\tau - \lvert V_1 - V_2 \rvert)(1 - W)W$

3 The Adaptive Agent Network for Bonding by Homophily

In this section the second-order adaptive social agent network model for bonding by homophily is introduced. In [18] it is shown how the design of adaptive agent network models can be addressed in a principled manner. First, note that at the base level (non-adaptive) dynamics of the states in the form of social contagion is modeled. By this, the state values mutually affect each other through the connections with their weights. In this way there is a causal pathway from connection weights to state values. Next, a first form of adaptivity (first-order adaptation) addresses the dynamics of the connection weights between two persons A and B. In particular, bonding by homophily describes how these connections are affected by the activation levels of the states A and B. To make this more precise, the effect of the state activation levels on the connection weights must be determined. This is where the homophily principle is detailed further:

- Activation values close to one another exert an upward pressure on connection weight $\omega_{A,B}$; activation levels for A and B distant from each other exert downward pressure on $\omega_{A,B}$

In this way, a causal pathway occurs from state values to connection weights. Therefore there is a circular causal relation between state values and connection weights. In other words, it becomes hard to distinguish between the causes and the consequences, as is discussed more in depth in [2, 15–17].

To incorporate the described effect of bonding by homophily on the connection weights in a numerical manner, for any agents A and B in the base level network, representations in terms of first-order network reification states $\mathbf{W}_{B,A}$ and their network characteristics are required. The change in $\mathbf{W}_{B,A}$ will depend on a yet to be chosen combination function $\mathbf{c}_{\mathbf{W}_{B,A}}(V_1, V_2, W)$; the general difference equation becomes

$$\mathbf{W}_{B,A}(t+\Delta t) = \mathbf{W}_{B,A}(t) + \mathbf{\eta}_{\mathbf{W}_{B,A}}\left[\mathbf{c}_{\mathbf{W}_{B,A}}(\mathrm{A}(t),\ \mathrm{B}(t), \mathbf{W}_{B,A}(t)) - \mathbf{W}_{B,A}(t)\right]\Delta t$$

and the *differential equation* becomes:

$$\mathrm{d}\mathbf{W}_{B,A}(t)/\mathrm{dt} = \mathbf{\eta}_{\mathbf{W}_{B,A}}\left[\mathbf{c}_{\mathbf{W}_{B,A}}(\mathrm{A}(t),\ \mathrm{B}(t), \mathbf{W}_{B,A}(t)) - \mathbf{W}_{B,A}(t)\right]$$

As shown in [18], Chap. 13, a simple linear homophily combination function with connection weight reification state $\mathbf{W}_{B,A}$ for the connection from agent A to agent B can be obtained by the following combination function (also shown in Table 3):

$$\mathbf{c}_{\mathbf{W}}(V_1, V_2, W) = \mathbf{slhomo}_{\alpha,\tau}(V_1, V_2, W) = W + \alpha(\tau - |V_1 - V_2|)(1 - W)W$$

The parameters α and τ can be chosen as required for the model at hand. The term $W(1 - W)$ ensures adequate bounding within [0, 1].

On top of the first-order adaptive social agent network model described above, a second-order adaptation level is built. This is used to make the adaptation speed of the first-order adaptation, adaptive itself. For this, second-order reification states $\mathbf{H}_{\mathbf{W}_{B,A}}$ are introduced, indicating in a dynamic manner the speed of change for the connection weight $\mathbf{W}_{B,A}$ from agent B for agent A. For the states $\mathbf{H}_{\mathbf{W}_{B,A}}$ the combination function $\mathbf{alogistic}_{\sigma,\tau}(V_1, \ldots, V_k)$ was used.

So, within the adaptive social agent network model, each of the agents is modeled by a *three-level agent model* for a second-order adaptive social agent, consisting of a number of states and their connections:

- *base state* Y for the agent
- *first-order reification states* $\mathbf{W}_{X_1,Y}, \ldots, \mathbf{W}_{X_k,Y}$ for the weights of all of Y's adaptive incoming connections
- *second-order reification states* $\mathbf{H}_{\mathbf{W}_{X_1,Y}}, \ldots, \mathbf{H}_{\mathbf{W}_{X_k,Y}}$ for the adaptive learning rates of each of the incoming connections for Y

For a more detailed overview of the connectivity within this adaptive agent model, see Table 4; for a simple example for only one first-order reification state and one second-order reification state, see the pink oval in Fig. 1.

Table 4. Overview of the different types of states, their roles, and their connectivity (as used in the simulations)

State		Role	Connectivity in the network
Name	Number		
A,.., I	X_1, …, X_9	Base states for the different agents	• All mutually connected (72 black arrows in the base plane) • For each Y of them 8 incoming connections from the first-order reification states $\mathbf{W}_{X_i,Y}$ (8 blue downward arrows) • For each Y of them 8 outgoing connections to $\mathbf{W}_{X_i,Y}$ and 8 outgoing connections to $\mathbf{H}_{\mathbf{W}_{X_i,Y}}$ (16 blue upward arrows)
$\mathbf{W}_{X,Y}$	X_{10}, …, X_{81}	Connection weight reification states for the base connections from X to Y	• An outgoing connection to Y to provide the adaptive weight for the connection from X to Y (red downward arrow) • An outgoing connection to $\mathbf{H}_{\mathbf{W}_{X,Y}}$ (blue upward arrow) • Three incoming connections: from X and Y (blue upward arrow) and from itself
$\mathbf{HW}_{X,Y}$	X_{82}, …, X_{153}	Speed factor reification states for states $\mathbf{W}_{X,Y}$	• An outgoing connection to $\mathbf{W}_{X,Y}$ to provide the adaptive speed factor for $\mathbf{W}_{X,Y}$ (red downward arrow) • Four incoming connections from X and Y, and from $\mathbf{W}_{X,Y}$ (blue upward arrows) and from itself

Note that the three-layered social adaptive agent model for agent Y as a whole only has incoming connections from X_1, …, X_k; internally these inputs are processed in parallel at each of the three levels. Note that the red downward arrows define special effects according to the role played by the reification state. For example, in the difference equation from Table 2, for reification state $\mathbf{W}_{B_i,A}$ (playing the role of connection weight) its value is used as connection weight $\boldsymbol{\omega}_{B_i,A}$, and for reification state $\mathbf{H}_{\mathbf{W}_{B_i,A}}$ (playing the role of speed factor) its value is used as speed factor $\boldsymbol{\eta}_{\mathbf{W}_{B_i,A}}$. This is explained in more technical detail in [18], Chaps. 9 and 10.

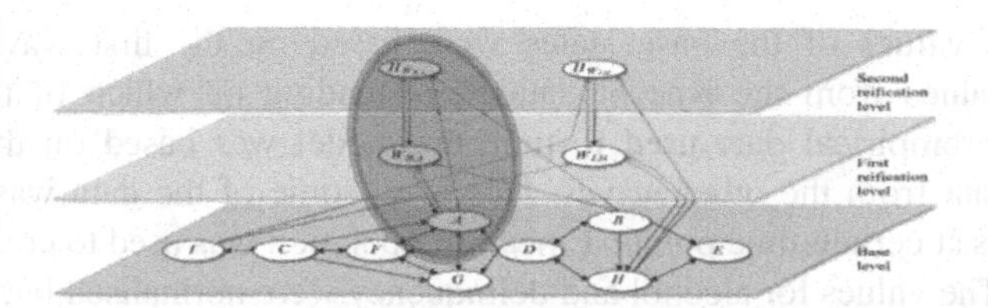

Fig. 1. A conceptual representation of the designed second-order adaptive social agent network model. Two example states are given for the first and second reification level to illustrate the effect of each state. The actual agent network model used in the simulations contains on each reification level not 2 but 9*8 = 72 reification states for the weights of all connections between the base level states and their speed factors. The pink oval depicts what together forms a three-layered model for one second-order adaptive social agent. (Color figure online)

4 Simulation Scenarios for the Agent Network Model

The designed agent network model has been compared to empirical data from one classroom from the Knecht data set [10]. To make the validation of the model more feasible, the group was split on sex and only the friendship, alcohol and delinquency data related to the male students were kept. In Table 5 all empirical data used are shown.

Table 5. The data on alcohol consumption and delinquency at four points in time with preprocessing as explained in the paper.

Alcohol	Wave 1	Wave 2	Wave 3	Wave 4	Delinquency	Wave 1	Wave 2	Wave 3	Wave 4
A	0.1	0.1	0.1	0.1	A	0.3	0.3	0.5	0.3
B	0.1	0.1	0.1	0.1	B	0.1	0.1	0.3	0.1
C	0.3	0.3	0.5	0.7	C	0.1	0.3	0.5	0.3
D	0.1	0.1	0.5	0.9	D	0.1	0.3	0.3	0.1
E	0.1	0.1	0.1	0.1	E	0.1	0.1	0.2	0.3
F	0.1	0.1	0.1	0.1	F	0.3	0.1	0.3	0.1
G	0.1	0.1	0.5	0.9	G	0.7	0.1	0.5	0.7
H	0.1	0.1	0.1	0.1	H	0.1	0.3	0.1	0.1
I	0.1	0.1	0.3	0.7	I	0.1	0.3	0.5	0.3

The initial values for the model were taken from the first wave of the remaining data. Connections in the base network mean that student X sees student Y as a friend. Since the network is directed, friendships are not always reciprocated. The network initially consists of 9 nodes and 23 edges, giving an average in-degree of 2.556. The nodes with the highest in-degree are A, F and G, which have an in-degree of 4. Two strongly connected components can be identified with B, D, E and H making up the first component and A, C, F, G, I the second component. As the friendship connections should be able to change over time each student was to some extent connected to every other student. The initial values of the connection weights that initially exist were set at 0.9, and the initial values of the other connection weights are set to the low value 0.1.

The initial values of the base states were based on the first wave alcohol or delinquency values from the Knecht dataset, depending on which of the two were simulated. The empirical data used to tune the model was based on the alcohol or delinquency data from the other waves. However, some of the data was missing for certain students at certain time points. Linear interpolation was used to create values for those entries. The values for alcohol and delinquency were normalised between 0.1 and 0.9 for the model. The initial values of the 72 (first-order) **W** states are either 0.1 or 0.9, depending on if initially a recognized friendship exists between the nodes in the network. The initial values of all 72 (second-order) **H** states were 0.1.

5 Empirical Validation of the Adaptive Agent Network Model

Running the model described in the previous section with hand-set values for the network characteristics until $t = 20$ with $\Delta t = 0.1$ for alcohol and delinquency values separately provided results as shown in Fig. 2 with a Root Mean Square Error (RMSE) between the empirical data and results of 0.7017 and 0.6866 for alcohol and delinquency respectively. Clustering was observed, but the results do not correspond to the empirical data, which is why a high RMSE occurs.

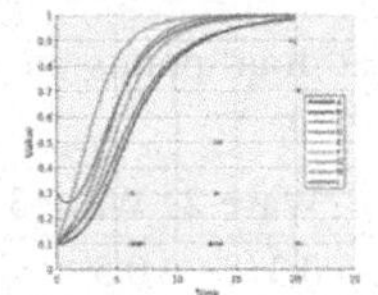

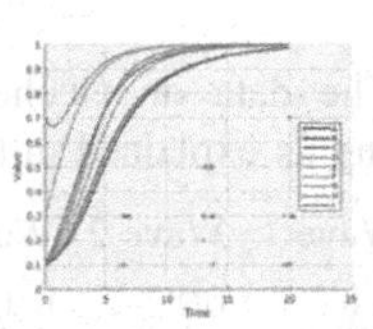

Fig. 2. The 9 state values of the network are presented for both alcohol (on the left) and delinquencies (on the right) with respect to time.

In order to reduce the RMSE, the model was tuned more systematically to get values for the network characteristics that correspond more to the real world. In the first experiment, the speed factors of the base states were tuned using Simulated Annealing with roughly 5000 iterations, while keeping all other parameters the same [9]. This give the results for alcohol and for delinquency shown in Fig. 3. The RMSE and resulting parameter values are shown in Table 6. The RMSE for these models is significantly lower than for the simple models. Note that the predicting alcohol usage has a slightly higher RMSE than the delinquency variant. Also observe that the speed factors are low for students that do not change their alcohol consumption or the amount of delinquencies they commited. Clusters are being formed in both simulation scenarios.

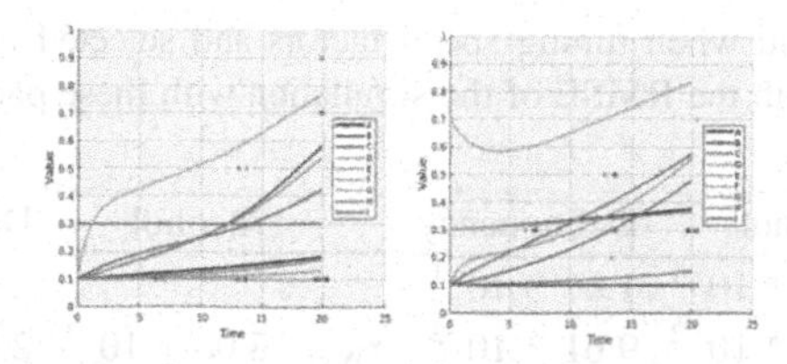

Fig. 3. The 9 state values of the network are presented for both alcohol (on the left) and delinquencies (on the right) with respect to time. The tuning of the speed values is done in both instances.

Table 6. Parameters found when tuning speed factors for alcohol and delinquency, along with the RMSE of the simulation with these parameters compared to the empirical data.

	RMSE	η_A	η_B	η_X	η_Δ	η_E	η_Φ	η_Γ	η_H	η_I
Alcohol	$1.85 * 10^{-1}$	$7.02 * 10^{-3}$	$9.90 * 10^{-1}$	$1.78 * 10^{-3}$	$2.19 * 10^{-1}$	$6.02 * 10^{-3}$	$7.42 * 10^{-5}$	$9.95 * 10^{-1}$	$9.57 * 10^{-3}$	$2.24 * 10^{-1}$
Delinquency	$1.83 * 10^{-1}$	$8.09 * 10^{-3}$	$2.49 * 10^{-3}$	$1.36 * 10^{-1}$	$1.26 * 10^{-2}$	$9.74 * 10^{-1}$	$7.02 * 10^{-3}$	$3.60 * 10^{-1}$	$6.99 * 10^{-3}$	$1.43 * 10^{-1}$

For a next experiment, again the speed factors for the base states were tuned but additionally the tipping points $\boldsymbol{\tau}$ for a few selected **W** states were tuned. These **W** states were chosen in the following manner. The nodes with the highest in-degrees were identified, being A, F and G. The students these nodes represent are thus liked by a lot of classmates. Then the outgoing edges from these nodes present in Fig. 1 were used to optimize the influence they have on the network. Again simulated annealing for roughly 5000 iterations was used to find the optimal values.

This results in the RMSE and parameters shown in Table 7 and the simulations with these parameters as shown in Fig. 4. The RMSE for both simulations is lower than in the previous experiment. The alcohol usage based model has a slightly lower RMSE than the delinquency variant. In both cases 3 clusters are formed.

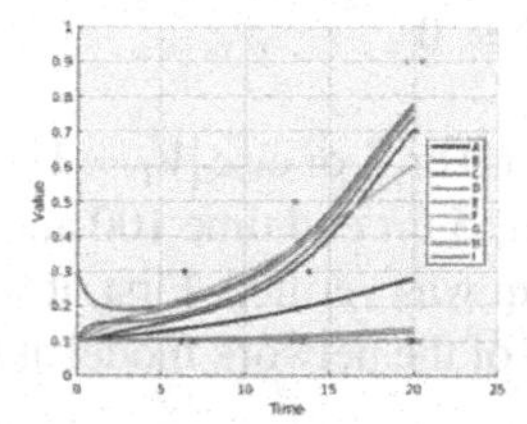

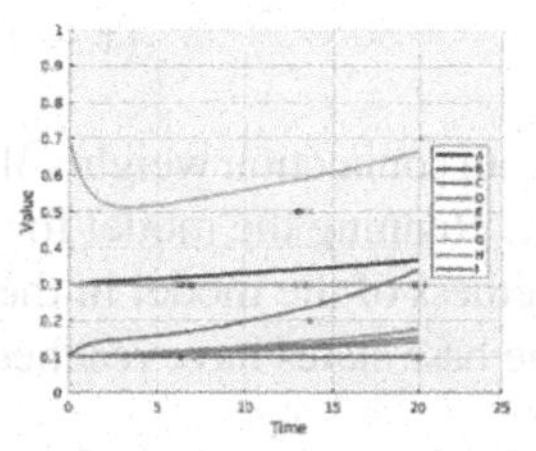

Fig. 4. The 9 state values of the network are presented for both alcohol (on the left) and delinquencies (on the right) with respect to time. The tuning of both the speed values and several tipping points was done in both instances.

Table 7. Parameters found when tuning speed factors and selected tipping points for alcohol and delinquency, along with the RMSE of the simulation with these parameters compared to the empirical data.

	Alcohol	Delinquency		Alcohol	Delinquency
RMSE	$1.57 * 10^{-1}$	$1.67 * 10^{-1}$			
η_A	$1.93 * 10^{-2}$	$9.61 * 10^{-3}$	$\tau_{W_{A,F}}$	$3.08 * 10^{-1}$	$2.53 * *10^{-1}$
η_B	$6.41 * 10^{-1}$	$2.54 * 10^{-5}$	$\tau_{W_{A,G}}$	$2.86 * 10^{-1}$	$9.78 * 10^{-1}$
η_C	$9.70 * 10^{-1}$	$9.30 * 10^{-3}$	$\tau_{W_{F,A}}$	$9.98 * 10^{-1}$	$9.65 * 10^{-3}$
η_D	$9.86 * 10^{-1}$	$1.84 * 10^{-2}$	$\tau_{W_{F,C}}$	$6.73 * 10^{-1}$	$3.54 * 10^{-1}$
η_E	$2.26 * 10^{-4}$	$2.77 * 10^{-2}$	$\tau_{W_{F,G}}$	$4.09 * 10^{-2}$	$7.52 * 10^{-1}$
η_F	$3.36 * 10^{-3}$	$3.31 * 10^{-5}$	$\tau_{W_{G,A}}$	$4.47 * 10^{-2}$	$2.73 * 10^{-2}$
η_G	$9.15 * 10^{-2}$	$9.82 * 10^{-1}$	$\tau_{W_{G,F}}$	$3.66 * 10^{-1}$	$1.07 * 10^{-2}$
η_H	$3.37 * 10^{-3}$	$9.31 * 10^{-3}$			
η_I	$9.51*10^{-1}$	$9.82*10^{-1}$			

6 Mathematical Verification of the Model

To verify that the model is mathematically correct, in [18], Chap. 12 it is observed that a state has a stationary point (i.e., d$Y(t)$/dt = 0) if and only if

$$\eta_Y = 0 \quad \text{or} \quad \mathbf{aggimpact}_Y(t) = Y(t)$$

The model is in equilibrium at t if *all* states have a stationary point at t. Furthermore, from [18], Chap. 3, Sect. 3.6.1 it is concluded that (assuming speed $\mathbf{H} > 0$) for the standard linear homophily function obtain the following equilibrium equation is obtained:

$$W = \mathbf{slhomo}_{\alpha,\tau}(V_1, V_2, W) = W + \alpha(\tau - |V_1 - V_2|)(1 - W)W$$

which for $\alpha > 0$ is equivalent to:

$$(\tau - |V_1 - V_2|)\,(1 - W)W = 0$$

Thus, all connection weights W should equal to either 0 or 1, or else $|V_1 - V_2|$ must be equal to τ. Running the model for the simulation in Sect. 5 for end time 100 and $\Delta t = 1$, the correctness of the model in the emerging equilibrium was verified. First, it was analysed if the base states have reached an equilibrium state of the network model at $t = 100$.

Table 8. Equilibrium analysis for all 9 base states, two first-order reification states and two second-order reification states in the social agent network model.

Base state X_i	A	B	C	D	E	F	G	H	I
$X_i(t)$	1.0000	1.0000	1.0000	1.0000	1.0000	1.0000	1.0000	1.0000	1.0000
$\mathbf{aggimpact}_{X_i}$	1.0000	1.0000	1.0000	1.0000	1.0000	1.0000	1.0000	1.0000	1.0000
deviation	$3 * 10^{-10}$	$6 * 10^{-10}$	$4 * 10^{-10}$	$4 * 10^{-10}$	$4 * 10^{-10}$	$3 * 10^{-10}$	$3 * 10^{-10}$	$3 * 10^{-10}$	$5 * 10^{-10}$
2nd order state X_i	$\mathbf{H}_{\mathbf{W}_{A,B}}$	$\mathbf{H}_{\mathbf{W}_{I,H}}$	1st order state X_i					$\mathbf{W}_{A,B}$	$\mathbf{W}_{I,H}$
$X_i(t)$	0.9886	0.9886	$X_i(t)$					0.9993	0.9993
$\mathbf{aggimpact}_{X_i}$	0.9886	0.9886	**value for equilibrium equation**					1	1
deviation	$-2.3 * 10^{-7}$	$-2.1 * 10^{-7}$	**deviation**					$7.2 * 10^{-4}$	$6.7 * 10^{-4}$

As seen in Table 8 all base states have reached a stationary point, since **aggimpact**$_Y(t) = Y(t)$. Next, it was analysed if the homophily function in the model is mathematically correct by evaluating the W states or the τ parameter value of the homophily function of these states. The complete analysis of all 144 W and H states is required for a complete mathematical verification of the proposed model. Including the verification is however beyond the scope of here presented work. Instead, two W and H states were analysed to show mathematical validity of sample states. However, the similarity between the states in both reification levels respectively indicates that the calculated validity is likely to hold for all states. The chosen first-order reification states are $\mathbf{W}_{A,B}$ and $\mathbf{W}_{I,H}$. State $\mathbf{W}_{A,B}$ has a value of 0.9993 at equilibrium, giving an error from 1 of $1 - \mathbf{W}_{A,B} = 7.2289*10^{-4}$. State $\mathbf{W}_{I,H}$ has a value of 0.9993 at equilibrium, giving an error from 1 of $1 - \mathbf{W}_{I,H} = 6.7210*10^{-4}$. The results for the corresponding **H** states are shown in Table 8 lower part. These results are further in accordance with the aforementioned theory.

7 Discussion

In this paper the dynamic and adaptive relation between friendships and alcohol consumption and the committing of delinquencies were analysed computationally by means of a second-order adaptive social agent network model. Also other network models have been previously proposed as a means to simulate social networks and the homophily principle (e.g. [2, 4, 11, 18]); however, these models are only first-order adaptive whereas in the current paper a second-order adaptive agent network model was used. In [18], also a second-order adaptive agent model was presented. However, that was a single cognitive agent model and the focus was on metaplasticity as known in Cognitive Neuroscience. In contrast, the current paper addresses a multi-agent case of a social agent network model. In [18], Sect. 6 also an adaptive social network model was modeled and simulated which is second-order adaptive. However, the second-order adaptation there has focus on adaptive tipping points and not on adaptive learning speed as in the current paper.

To provide a good fit to empirical data, initially the speed factors of the base states were tuned by Simulated Annealing for both alcohol and delinquency data. Thereafter, also multiple tipping points of the simple linear homophily combination function wre tuned. These tipping points were chosen by picking the three nodes with the highest in-degree. The model that predicted alcohol usage was slightly more accurate than the delinquency variant with RMSE values of $1.6*10^{-1}$ and $1.7*10^{-1}$ respectively. The difference in RMSE was, however, relatively insignificant. This shows that both alcohol consumption and the committing of delinquencies have been adequately illustrated by the proposed model. In both cases clustering is observed, but more for alcohol consumption compared to the committing of delinquencies, as is shown in Fig. 4. This is likely due the similarity in initial values for alcohol consumption, whereas a larger spread is observed in the initial values of the data on delinquencies. A subsequent mathematical verification further indicates the mathematical validity of the model.

A number of aspects could be further improved upon in future work. The first of which is the inclusion of more parameters in the tuning process. Secondly, the inclusion of data on a complete class or multiple classes could further improve the accuracy of clustering behaviour in the model compared to the fitted data. Additionally, a multicriteria homophily can be used to combine alcohol and delinquency data in one model, alongside other demographic information that is present in de dataset. A final improvement to the model can be to focus on the inclusion of different adaptivity mechanisms. This could for instance be implemented by creating extra adaptive states for other combination function parameters.

References

1. Ali, M.M., Dwyer, D.S.: Social network effects in alcohol consumption among adolescents. Addict. Behav. **35**(4), 337–342 (2010)
2. Aral, S., Muchnik, L., Sundararajan, A.: Distinguishing influence-based contagion from homophily-driven diffusion in dynamic networks. PNAS **106**(51), 21544–21549 (2009)
3. Blankendaal, R., Parinussa, S., Treur, J.: A temporal-causal modelling approach to integrated contagion and network change in social networks. In: Proceedings of the 22nd European Conference on Artificial Intelligence, ECAI 2016, pp. 1388–1396. IOS Press (2016)
4. van den Beukel, S., Goos, S.H., Treur, J.: An adaptive temporal-causal network model for social networks based on the homophily and more-becomes-more principle. Neurocomputing **338**, 361–371 (2019)
5. Christiansen, B.A., Smith, G.T., Roehling, P.V., Goldman, M.S.: Using alcohol expectancies to predict adolescent drinking behavior after one year. J. Consult. Clin. Psychol. **57**(1), 93–99 (1989)
6. Henneberger, A.K. Mushonga, D.R., Preston, A.M.: Peer influence and adolescent substance use: a systematic review of dynamic social network research. Adolesc. Res. Rev. (2020). https://doi-org.vu-nl.idm.oclc.org/10.1007/s40894-019-00130-0
7. Huang, G.C., et al.: Peer influences: the impact of online and offline friendship networks on adolescent smoking and alcohol use. J. Adolesc. Health **54**(5), 508–514 (2014)
8. Kappert, C., Rus, R., Treur, J.: On the emergence of segregation in society: network-oriented analysis of the effect of evolving friendships. In: Nguyen, N.T., Pimenidis, E., Khan, Z., Trawiński, B. (eds.) ICCCI 2018. LNCS (LNAI), vol. 11055, pp. 178–191. Springer, Cham (2018). https://doi.org/10.1007/978-3-319-98443-8_17
9. Kirkpatrick, S., Gelatt, C.D., Vecchi, M.P.: Optimization by simulated annealing. Science **220**(4598), 671–680 (1983)
10. Knecht, A.B., Burk, W.J., Weesie, J., Steglich, C.: Friendship and alcohol use in early adolescence: a multilevel social network approach. J. Res. Adolesc. **21**(2), 475–487 (2011)
11. Kozyreva, O., Pechina, A., Treur, J.: Network-oriented modeling of multi-criteria homophily and opinion dynamics in social media. In: Staab, S., Koltsova, O., Ignatov, D.I. (eds.) SocInfo 2018. LNCS, vol. 11185, pp. 322–335. Springer, Cham (2018). https://doi.org/10.1007/978-3-030-01129-1_20
12. McCreanor, T., Lyons, A., Griffin, C., Goodwin, I., Barnes, H.M., Hutton, F.: Youth drinking cultures, social networking and alcohol marketing: Implications for public health. Crit. Public Health **23**(1), 110–120 (2013)
13. McPherson, M., Smith-Lovin, V., Cook, J.M.: Birds of a feather: homophily in social networks. Ann. Rev. Sociol. **27**(1), 415–444 (2001)

14. Montgomery, S.C., Donnelly, M., Bhatnagar, P., Carlin, A., Kee, F., Hunter, R.F.: Peer social network processes and adolescent health behaviors: a systematic review. Prev. Med. **130**, 105900 (2020)
15. Mundt, M.P., Mercken, L., Zakletskaia, L.: Peer selection and influence effects on adolescent alcohol use: a stochastic actor-based model. BMC Pediatr. **12**(1), 115 (2012)
16. Shalizi, C.R., Thomas, A.C.: Homophily and contagion are generically confounded in observational social network studies. Sociol. Methods Res. **40**(2), 211–239 (2011)
17. Steglich, Ch., Snijders, T.A.B., Pearson, M.: Dynamic networks and behavior: separating selection from influence. Sociol. Methodol. **40**(1), 329–393 (2010)
18. Treur, J.: Network-Oriented Modeling for Adaptive Networks: Designing Higher-Order Adaptive Biological, Mental and Social Network Models. Springer, Cham (2020)
19. Zhang, J., Centola, D.: Social networks and health: new developments in diffusion, online and offline. Ann. Rev. Sociol. **45**(1), 91–109 (2019)

Intelligent Control and Automation

A Fast Terminal Sliding Mode Control Strategy for Trajectory Tracking Control of Robotic Manipulators

Anh Tuan Vo, Hee-Jun Kang(✉), and Thanh Nguyen Truong

School of Electrical Engineering, University of Ulsan, Ulsan 44610, South Korea
hjkang@ulsan.ac.kr

Abstract. This paper proposes a fast terminal sliding mode control strategy for trajectory tracking control of robotic manipulators. Firstly, to degrade the chattering behavior and speed up the fast response of the conventional Terminal Sliding Mode Control, a novel robust, reaching control law with two variable power components is introduced. With this proposal, the state error variables of the system quickly converge on the sliding surface whether their initial value is far or near to the sliding surface. Secondly, a Fast Terminal Sliding Mode surface is designed to guarantee that the system states arrive at the equilibrium and stabilize along the sliding surface with rapid convergence speed. The result is a new control strategy is formed based on the suggested reaching control law and the above sliding surface. Thanks to this hybrid method, the control performance expectations are guaranteed such as faster convergence, robustness with exterior disturbance and dynamic uncertainties, high tracking accuracy, and finite-time convergence. Moreover, the asymptotic stability and finite-time convergence of the control system are fully confirmed by Lyapunov theory. Finally, computer simulation examples are performed to verify the effectiveness of the suggested control strategy.

Keywords: Terminal Sliding Mode Control · Fast Terminal Sliding Mode Control · Finite-Time control · Robotic manipulators

1 Introduction

Robotics are increasingly substituting for humans in the areas of social life, in various industrial fields, production, exploring the ocean and space, or implementing other complex works [1, 2]. To carry out smoothly and reliably the above tasks, to improve productivity, product quality, the mechanical systems of robotic systems must have a more advanced design. Consequently, this leads to an increase in the complexity of the kinematic structure and mathematical model when there is an additional occurrence of system uncertainties.

Sliding Mode Control (SMC) is an effective control technique because of its simple design and robust properties against the effects of uncertain components and external disturbances [3–5]. However, SMC has three main drawbacks (1) the chattering obstacle in the conventional SMC generates oscillation in the control signal leading to vibration in the production system, undesired heat and even generating instability;

D.-S. Huang and P. Premaratne (Eds.): ICIC 2020, LNAI 12465, pp. 177–189, 2020.
https://doi.org/10.1007/978-3-030-60796-8_15

(2) the conventional SMC does not offer convergence in finite-time; (3) the SMC theory is based on the asymptotic stability and underpinned by the Lipschitz condition in the Ordinary Differential Equations. However, the very nature of the asymptotical stability indicates that, in the evolution of system dynamics, the closer to the equilibrium, the slower the state convergence. This means that the system state would never approach equilibrium. While this may not be a problem in real applications, that means, if much higher accuracy is demanded, greater control torque would be needed which may not be feasible if control devices are limited. Terminal Sliding Mode Control (TSMC) is introduced to solve the convergence obstacle in finite time, improve transient performance [6, 7]. However, in several situations, TSMC does not offer the desired performance with initial state variables far from the equilibrium point. And, it has not solved the chattering and slow convergence as well as creating a new problem that is singularity phenomenon.

To solve the concerns of SMC and TSMC in a synchronized method, Fast Terminal Sliding Mode Control (FTSMC) [8, 9] or Nonsingular FTSMC (NFTSMC) has been developed successfully for robot systems or several nonlinear systems [10–13, 19]. It thoroughly solves the problems, such as singularity, finite-time convergence, and slow convergence rate. Unfortunately, the chattering phenomena have not yet been tackled as the controllers based FTSMC or NFTSMC still use a robust reaching control law to compensate for uncertain terms.

For chattering defect, researchers have focused a lot of effort to develop methods that eliminate chattering. The first technique has been introduced as the boundary layer [14]. However, control errors may be increased because of the influences of the boundary. The second method is known as High-Order Sliding Mode (HOSMC) [15, 16]. Because the level of the chattering is generated corresponding to the magnitude of the sliding gain, value is selected to be greater than the bound value of uncertainty and disturbance. Method to degrade the chattering is to degrade the impact of the uncertainty or disturbance by adding a continuous compensation component. Other methods can be noted, such as Super-Twisting Algorithm (STA) [17, 18], or Full-Order Sliding Mode Control (FOTSMC) [11, 20]. These methods use an integral of the control input to reject the chattering. However, the selection way of sliding gain is the same as SMC.

Based on the mentioned analysis, our paper proposes an advanced FTSMC with the following contributions for robot manipulators: 1) the proposed controller has a simple design, powerful properties and high application for the robot arms; 2) offers finite-time convergence and faster transient performance without singularity problem in controlling; 3) inherits the advantages of FTSMC in the aspects of robustness against system uncertainties and exterior disturbances as well as its fast convergence; 4) a new reaching control was proposed and evidence of finite-time convergence was sufficiently demonstrated by Lyapunov theory; 5) the precision of the designed system was further improved in the trajectory tracking control; 6) the proposed controller shows the smoother control torque commands with lesser chattering.

This paper is outlined as follows. The problem formulation is stated in Sect. 2. Section 3 gives an overview of the proposed control strategy. Computer simulation examples are performed to evaluate the influence of the designed controller for 2-DOF robot manipulator in Sect. 4 and the performance of the designed controller is also discussed along with the performance of different control algorithms, including SMC and FTSMC. Section 5 presents conclusions.

2 Problem Formulation

2.1 A Dynamic Model of Robotic Manipulators

According to [21], the robot dynamic model is described as:

$$M(q)\ddot{q} + C(q,\dot{q})\dot{q} + G(q) + \Delta(q,\dot{q}) = \tau \tag{1}$$

where $q, \dot{q}$, and $\ddot{q} \in R^n$ correspond to the position vector, velocity vector, and acceleration vector at each joint of the robot. $M(q) \in R^{n\times n}$ is inertia matrix, $C(q,\dot{q}) \in R^{n\times 1}$ is the matrix from the centrifugal force and Coriolis, $G(q) \in R^{n\times 1}$ represents the gravity force matrix, and $\tau \in R^{n\times 1}$ is the designed control input of actuators, $\Delta(q,\dot{q})$ is the vector of the lumped system uncertainties and exterior disturbances and this vector is described as:

$$\Delta(q,\dot{q}) = \Delta M(q)\ddot{q} + \Delta C(q,\dot{q})\dot{q} + \Delta G(q) + F_r(\dot{q}) + \tau_d \tag{2}$$

where $F_r(\dot{q}) \in R^{n\times 1}$ represents friction force matrix, $\tau_d \in R^{n\times 1}$ represents an exterior disturbance matrix. $\Delta M(q), \Delta C(q,\dot{q})$, and $\Delta G(q)$ are dynamic uncertain components.

The robot model in Eq. (1) can be transformed into a class of second-order nonlinear system as follows:

$$\begin{cases} \dot{x}_1 = x_2 \\ \dot{x}_2 = \Pi(x) + b(x)u + \Delta(x) \end{cases} \tag{3}$$

where $x = [x_1, x_2]^T$, $x_1 = q$, $x_2 = \dot{q}$, $u = \tau$, $\Pi(x) = M^{-1}(q)[-C(q,\dot{q})\dot{q} - G(q)]$ is the nominal robot model of the robot without exterior disturbances and dynamic uncertainties, $\Delta(x)$ represents the lumped system uncertainties, and $b(x) = M^{-1}(q)$.

The target of this paper here is that the proposed controller has a simple design, powerful properties and high application for the robot arms; the trajectory position variables will quickly correctly track the desired trajectory, with the control performance expectations such as faster convergence and high tracking accuracy, under a robust control input without chattering phenomena.

The following assumption is fundamental for the control design approach.

$$|\Delta(x)| \le \bar{\Delta} \tag{4}$$

where $\bar{\Delta}$ is assigned as a positive constant.

3 Design Procedure of Tracking Control Strategy

The design approach in this paper is like the conventional SMC or TSMC. Therefore, a novel control strategy is developed for robot system (1) in this section, which is performed by two following major steps.

3.1 Design of Fast Terminal Sliding Mode Surface

In the first step, the following fast terminal sliding surface is prior designed to ensure that the tracking error variables arrive at the equilibrium and stabilize along the sliding surface with rapid convergence speed (refer to [8]).

$$s = \dot{e} + \alpha e + \beta e^{[\gamma]} \tag{5}$$

where $e_i = x_i - x_d, (i = 1, 2, \cdots, n)$ is the tracking positional error, x_d is described as the prescribed trajectory value (let $x_d = q_d$), α, β, and γ are positive constants.

When $s = 0$, it results in $\dot{e} = -\alpha e - \beta |e|^{\gamma}\text{sgn}(e)$, which will reach $e = 0$ in finite-time for $s = 0$ by properly selecting γ, i.e.

$$t_s = \alpha^{-1}(1-\gamma)^{-1}\left(\ln\left(\alpha|e(0)|^{1-\gamma} + \beta - \ln\beta\right)\right) \tag{6}$$

3.2 Design of the Proposed Control Strategy

In the second step, the proposed controller is synthesized according to the following procedure.

The time derivative of the sliding surface is calculated along with system (3) as follows:

$$\dot{s} = \Pi(x) + b(x)u + \Delta(x) - \ddot{x}_d + \alpha\dot{e} + \beta\gamma e^{[\gamma-1]}\dot{e} \tag{7}$$

To achieve the expectations for control performance, the control input system is designed for robot manipulator as:

$$u = -b^{-1}(x)\left(u_{eq} - u_r\right) \tag{8}$$

where u_{eq} is designed as

$$u_{eq} = \Pi(x) - \ddot{x}_d + \alpha\dot{e} + \beta\gamma e^{[\gamma-1]}\dot{e} \tag{9}$$

and a novel reaching control law is proposed as:

$$\dot{s} = -\kappa_1|s|^{\mu_1}\text{sgn}(s) - \kappa_2|s|^{\mu_2}\text{sgn}(s) \tag{10}$$

with

$$\mu_1(s) = k_0 + k_1 \tanh(s^r) - k_2 \tanh(\lambda s^2) \tag{11}$$

$$\mu_2(s) = \begin{cases} k & \text{if } |s| \geq 1 \\ 1 & \text{if} |s| < 1 \end{cases} \tag{12}$$

where $\kappa_1, \kappa_2, \lambda > 0$, $0 < k_2 < k_0 < 1$, $k_1 > 1$, $\mu = k_0 + k_1 - k_2 > 1$, and r is a positive even number. Therefore, u_r can be obtained since $\dot{s} = 0$.

$$u_r = -\kappa_1 |s|^{\mu_1} \operatorname{sgn}(s) - \kappa_2 |s|^{\mu_2} \operatorname{sgn}(s) \tag{13}$$

The novel reaching control law with two variable power components has strong adaptive capability. $\mu_1(s)$ is a constructed nonlinear function, and $\mu_2(s)$ is a piecewise function. By designing suitable parameters r and λ, Eq. (10) is equivalent to the following expression:

$$\begin{cases} \dot{s} = -\kappa_1 |s|^{\mu} \operatorname{sgn}(s) - \kappa_2 |s|^{\mu} \operatorname{sgn}(s), & |s| \geq 1 \\ \dot{s} = -\kappa_1 |s|^{k_0 - k_2} \operatorname{sgn}(s) - \kappa_2 s, & 0 < |s| < 1 \\ \dot{s} = -\kappa_1 |s|^{k_0} \operatorname{sgn}(s) - \kappa_2 s, & \text{near} |s| = 0 \end{cases} \tag{14}$$

3.3 Stability Analysis of the Proposed Reaching Control Law

Theorem 1. *The sliding mode variables s and $\dot{s}$ described in (10) can converge to the equilibrium in finite time.*

Proof. To confirm the correctness of Theorem 1, the following Lyapunov function is considered:

$$V_1 = \frac{1}{2} s^2 \tag{15}$$

Using Eq. (10), the time derivative of the Lyapunov function is derived as follows:

$$\dot{V}_1 = s\dot{s} = -\kappa_1 |s|^{\mu_1 + 1} \operatorname{sgn}(s) - \kappa_2 |s|^{\mu_2 + 1} \operatorname{sgn}(s) < 0 \tag{16}$$

When condition, including $V_1 > 0$ and $\dot{V}_1 < 0$ are satisfied, the sliding surface is accessible. Therefore, the state variables of the system reach the sliding surface in a finite time and it is considered in the following phases:

When $|s(0)| > 1$, the reaching phase includes two stages: $s(0) \to |s| = 1$ and $|s| = 1 \to s = 0$. The convergence time in the two stages are computed.

Stage 1: $s(0) \to |s| = 1$, $\mu_1(s) = \mu$, and $\mu_2(s) = \mu$. In this phase, both terms of (10) play the role, then the convergence time is computed by:

$$\int_{0}^{t_1} dt = \int_{1}^{s(0)} \frac{1}{(\kappa_1+\kappa_2)s^{\mu}} d(|s|) \tag{17}$$

then,

$$t_1 = \frac{1-|s(0)|^{1-\mu}}{(\kappa_1+\kappa_2)(\mu-1)} \tag{18}$$

Stage 2: $|s| = 1 \rightarrow s = 0$, the convergence time is approximately computed by:

$$\int_{0}^{t_2} dt \approx \int_{0}^{1} \frac{1}{\kappa_1 |s|^{k_0-k_2} + \kappa_2 |s|} d(|s|) \tag{19}$$

then,

$$t_2 \approx \frac{1}{\kappa_1(1-k_0+k_2)} \ln\left(1+\frac{\kappa_1}{\kappa_2}\right) \tag{20}$$

Therefore, the convergence time is obtained as:

$$t_r = t_1 + t_2 \approx \frac{1-|s(0)|^{1-\mu}}{(\kappa_1+\kappa_2)(\mu-1)} + \frac{1}{\kappa_1(1-k_0+k_2)} \ln\left(1+\frac{\kappa_1}{\kappa_2}\right) \tag{21}$$

To accommodate the effects of the lumped system uncertainties with the faster convergence, the reaching control law (13) is modified as:

$$u_r = -\kappa_1 |s|^{\mu_1} \operatorname{sgn}(s) - \kappa_2 |s|^{\mu_2} \operatorname{sgn}(s) - (\bar{\Delta}+\rho)\operatorname{sgn}(s) \tag{22}$$

in which ρ is a positive constant, the term of $(\bar{\Delta}+\rho)\operatorname{sgn}(s)$ is applied to compensate the effects of the lumped system uncertainties. Therefore, the proposed control system yields

$$u = -b^{-1}(x)\begin{pmatrix} \Pi(x) - \ddot{x}_d + \alpha\dot{e} + \beta\gamma e^{[\gamma-1]}\dot{e} + \kappa_1|s|^{\mu_1}\operatorname{sgn}(s) \\ + \kappa_2|s|^{\mu_2}\operatorname{sgn}(s) + (\bar{\Delta}+\rho)\operatorname{sgn}(s) \end{pmatrix} \tag{23}$$

Theorem 2. *For the uncertain dynamic system in state space as Eq. (3) if the control input signal is constructed (23) and a suitable finite time FTSM surface is selected as in Eq. (5). Consequently, the state variables of the system (3) will quickly approach the sliding function in finite-time* t_r *and then stabilize around zero within finite-time* $t = t_r + t_s$.

3.4 Stability Analysis of the Proposed Control Strategy

Proof. Applying the control input (23) to the sliding surface (7), we can gain

$$\dot{s} = \Delta(x) - \left(\kappa_1|s|^{\mu_1} + \kappa_2|s|^{\mu_2} + \left(\bar{\Delta} + \rho\right)\right)\mathrm{sgn}(s) \tag{24}$$

To confirm the correctness of Theorem 2, the following Lyapunov function is considered:

$$V_2 = \frac{1}{2}s^2 \tag{25}$$

With Eq. (24), the time derivative of the Lyapunov function V_2 is derived as follows:

$$\dot{V}_2 = s\dot{s} = -\left(\kappa_1|s|^{\mu_1} + \kappa_2|s|^{\mu_2} + \bar{\Delta} + \rho - |\Delta(x)|\right)|s| < 0 \tag{26}$$

It is seen that $\dot{V}_2$ is a negative definite. Consequently, the system will be guaranteed, and the tracking error variables will converge to zero in finite-time. Therefore, Theorem 2 is confirmed.

Block diagram of the designed control strategy is shown in Fig. 1.

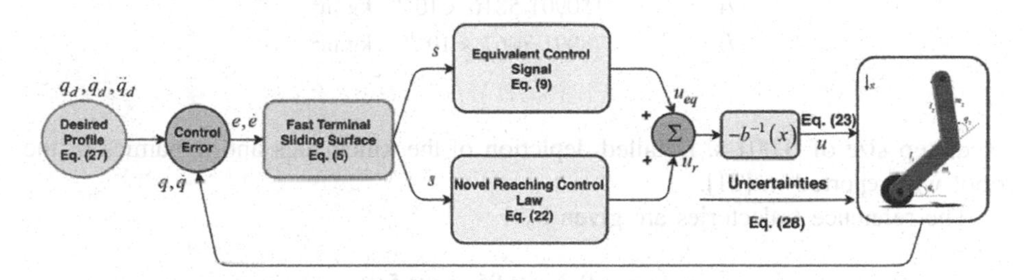

Fig. 1. Block diagram of the proposed control strategy.

4 Computer Simulation Results and Discussion

To verify the improved performance of the designed strategy in overcoming the outstanding issues of SMC and FTSMC as well as to demonstrate the effectiveness and applicability of the proposed control method, a two-link manipulator as shown in Fig. 2 is employed and its essential parameters are reported in Table 1. The simulations are performed in the MATLAB/Simulink environment using ODE5 dormand-prince with a

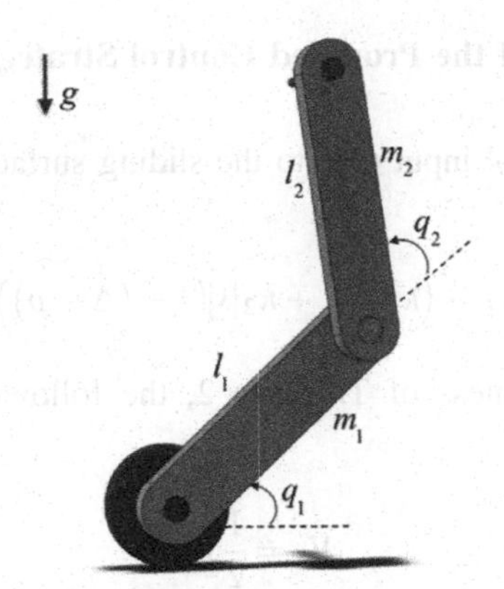

Fig. 2. Configuration of the two-link robotic system.

Table 1. The essential parameters of the robotic system.

Parameters	Value	Unit
m_1	11.940596	kg
m_2	7.400618	kg
l_1	0.3	m
l_2	0.3	m
l_{c1}	0.193304	m
l_{c2}	0.193304	m
I_1	$150901.5816 \times 10^{-6}$	kg.m^2
I_2	$78091.7067 \times 10^{-6}$	kg.m^2

fixed-step size of 0.001 s. Detailed depiction of the kinematics and dynamics of the robot was reported in [21].

The reference trajectories are given by:

$$\begin{cases} x = 0.3 + 0.05\cos(0.5t) \\ y = 0.15 + 0.05\sin(0.5t) \end{cases} \tag{27}$$

To test the robustness against the effects of uncertain components, the system uncertainties and exterior disturbances are assumed as:

$$F_r(\dot{q}) + \tau_d = \begin{bmatrix} -0.5\sin(\dot{q}_1) - 0.5\dot{q}_1 \\ -0.1\sin(\dot{q}_2) - 0.5\dot{q}_2 \end{bmatrix} + \begin{bmatrix} 3\sin(t) \\ \sin(t) \end{bmatrix} \tag{28}$$

The control strategies, including SMC [4] and FTSMC [8], have the corresponding control torque as follows:

$$u_{SMC} = -b^{-1}(x)\big(\Pi(x) - \ddot{x}_d + \alpha\dot{e} + (\bar{\Delta} + \rho)\mathrm{sgn}(s)\big) \tag{29}$$

$$u_{FTSMC} = -b^{-1}(x)\Big(\Pi(x) - \ddot{x}_d + \alpha\dot{e} + \beta\gamma e^{[\gamma-1]}\dot{e} + (\bar{\Delta} + \rho)\mathrm{sgn}(s)\Big) \tag{30}$$

Control parameters of different control systems, including SMC, FTSMC and the proposed controller are shown in Table 2.

Table 2. Control parameters of different control systems.

Control strategy	Control parameters	Value of control parameter
SMC	$\alpha, \bar{\Delta}, \rho$	$5, 7, 0.001$
FTSMC	$\alpha, \bar{\Delta}, \rho, \beta, \gamma$	$15, 7, 0.001, 5, 1.3$
Proposed strategy	$\alpha, \bar{\Delta}, \rho, \beta, \gamma$	$15, 0.1, 0.001, 5, 1.3$
	$\kappa_1, \kappa_2, k_0, k_1, k_2, r, \lambda$	$20, 50, 0.3, 0.5, 0.1, 4, 1.5$

Figure 3 exhibits the prescribed path and actual path of end-effector under three different control strategies, including SMC, FTSMC, and the proposed control strategy. The end-effector of the robotic system is controlled to follow a circular path. As seen in Fig. 3, it is seen that all three controllers seem to provide similar good trajectory tracking performance. Figure 4 and Fig. 5 exhibit the tracking errors of the end effector

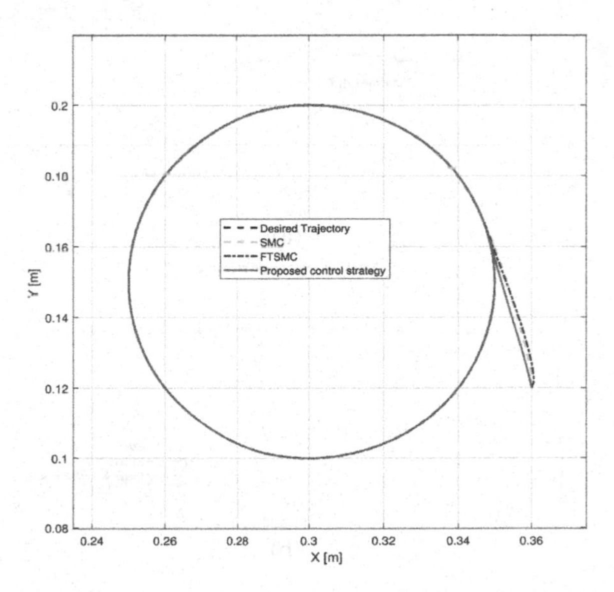

Fig. 3. The prescribed path and actual path of the end-effector under the three different control methods.

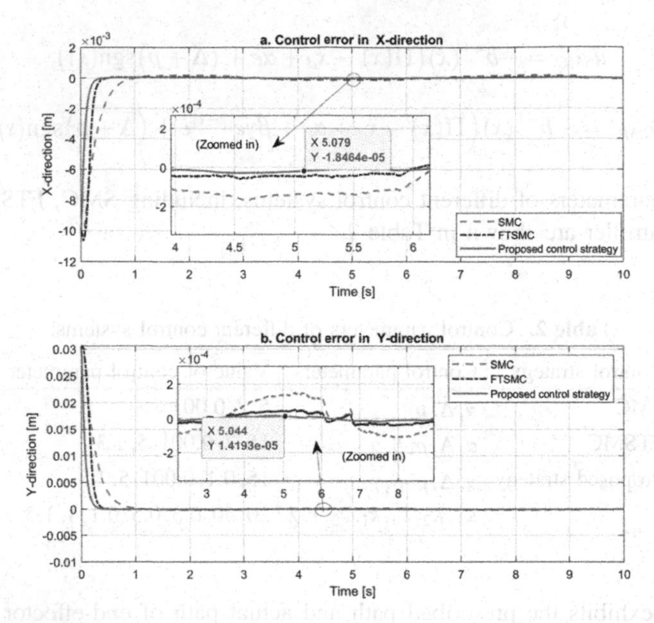

Fig. 4. Control errors in Cartesian coordinates: a) control error in X-direction, b) control error in Y-direction.

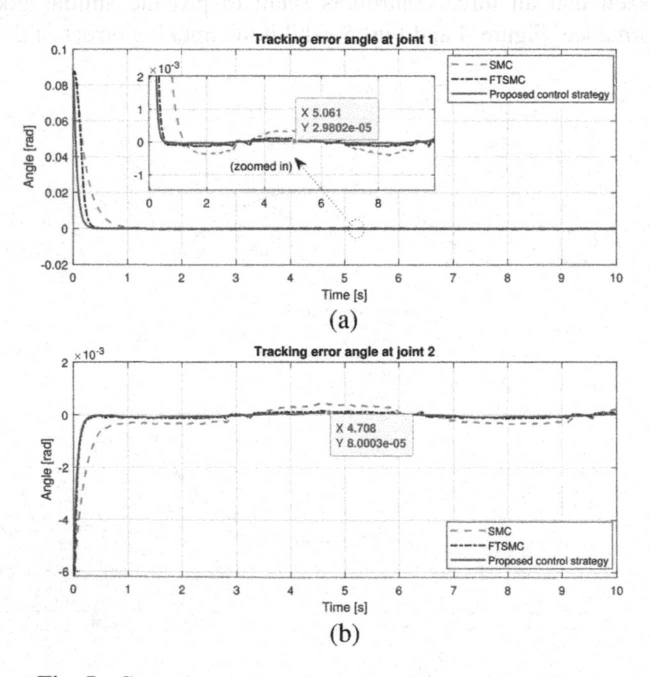

Fig. 5. Control errors at Joints: a) at Joint 1, b) at Joint 2.

in the X- axis, Y- axis, and control errors at joints, respectively. From Fig. 4 and Fig. 5, SMC provides the worst tracking performance among the three control methodologies, the tracking errors produced by FTSMC are smaller than the tracking errors offered by SMC. Specifically, the proposed control strategy offers the smallest tracking errors compared with SMC and FTSMC.

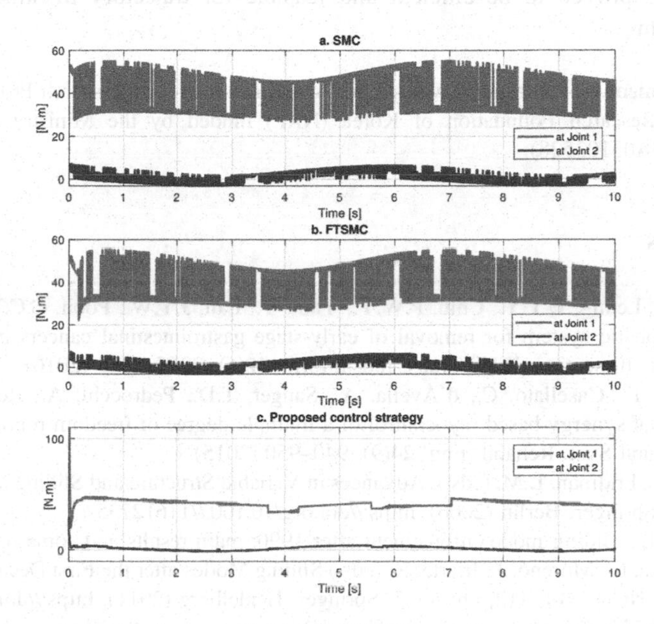

Fig. 6. Control input actions: a) SMC, b) FTSMC, and c) proposed control strategy.

The control signals for three control manners, including SMC, FTSMC and the proposed control strategy are illustrated in Fig. 6. In Fig. 6a and 6b, SMC and FTSMC show a discontinuous control signal with serious chattering. On the contrary, the proposed control strategy seems to show a continuous control signal with impressive small chattering behavior, as illustrated in Fig. 6c.

From the simulation results, it can be concluded that the proposed strategy shows the best performance among the four in terms of tracking positional accuracy, small steady state error, fast response speed, and weak chattering behavior.

5 Conclusion

From theoretical evidence, simulation results, and comparison with SMC and FTSMC, the proposed control strategy The proposed controller shows a significant improvement in control performance compared to the two compared methods: (1) offers finite-time convergence and faster transient performance without singularity problem in controlling; (2) inherits the advantages of FTSMC in the aspects of robustness against system uncertainties and exterior disturbances as well as its fast convergence; (3) a new reaching control was proposed and evidence of finite-time convergence was sufficiently demonstrated by Lyapunov theory; (4) the precision of the designed system was further improved in the trajectory tracking control; (5) the proposed controller shows the smoother control torque commands with lesser chattering; (6) The proposed controller has a simple design suitable for applying to an actual system. To sum up, the designed controller has proven to be efficient and feasible for trajectory tracking control of robotic systems.

Acknowledgement. This research was supported by Basic Science Research Program through the National Research Foundation of Korea (NRF) funded by the Ministry of Education (2019R1D1A3A03103528).

References

1. Lau, K.C., Leung, E.Y.Y., Chiu, P.W.Y., Yam, Y., Lau, J.Y.W., Poon, C.C.Y.: A flexible surgical robotic system for removal of early-stage gastrointestinal cancers by endoscopic submucosal dissection. IEEE Trans. Ind. Inform. **12**(6), 2365–2374 (2016)
2. Lunardini, F., Casellato, C., d'Avella, A., Sanger, T.D., Pedrocchi, A.: Robustness and reliability of synergy-based myocontrol of a multiple degree of freedom robotic arm. IEEE Trans. Neural Syst. Rehabil. Eng. **24**(9), 940–950 (2015)
3. Colet, E.F., Fridman, L.M. (eds.): Advances in Variable Structure and Sliding Mode Control, vol. 334. Springer, Berlin (2006). https://doi.org/10.1007/11612735
4. Fridman, L.: Sliding mode enforcement after 1990: main results and some open problems. In: Fridman, L., Moreno, J., Iriarte, R. (eds.) Sliding Modes after the First Decade of the 21st Century. LNCIS, vol. 412, pp. 3–57. Springer, Heidelberg (2011). https://doi.org/10.1007/978-3-642-22164-4_1
5. Truong, T.N., Kang, H.J., Le, T.D.: Adaptive neural sliding mode control for 3-DOF planar parallel manipulators. In: Proceedings of the 2019 3rd International Symposium on Computer Science and Intelligent Control, pp. 1–6, 2019 September
6. Yu, X., Zhihong, M.: Fast terminal sliding-mode control design for nonlinear dynamical systems. IEEE Trans. Circuits Syst. I: Fundam. Theory Appl. **49**(2), 261–264 (2002)
7. Wu, Y., Yu, X., Man, Z.: Terminal sliding mode control design for uncertain dynamic systems. Syst. Control Lett. **34**(5), 281–287 (1998)
8. Yang, L., Yang, J.: Nonsingular fast terminal sliding mode control for nonlinear dynamical systems. Int. J. Robust Nonlinear Control **21**(16), 1865–1879 (2011)
9. Yu, S., Yu, X., Shirinzadeh, B., Man, Z.: Continuous finite-time control for robotic manipulators with terminal sliding mode. Automatica **41**(11), 1957–1964 (2005)
10. Vo, A.T., Kang, H.J.: Neural integral non-singular fast terminal synchronous sliding mode control for uncertain 3-DOF parallel robotic manipulators. IEEE Access **8**, 65383–65394 (2020)
11. Vo, A.T., Kang, H.J.: Adaptive neural integral full-order terminal sliding mode control for an uncertain nonlinear system. IEEE Access **7**, 42238–42246 (2019)
12. Vo, A.T., Kang, H.J.: An adaptive neural non-singular fast-terminal sliding-mode control for industrial robotic manipulators. Appl. Sci. **8**(12), 2562 (2018)
13. Li, H., Dou, L., Su, Z.: Adaptive nonsingular fast terminal sliding mode control for electromechanical actuator. Int. J. Syst. Sci. **44**(3), 401–415 (2013)

14. Suryawanshi, P.V., Shendge, P.D., Phadke, S.B.: A boundary layer sliding mode control design for chatter reduction using uncertainty and disturbance estimator. Int. J. Dyn. Control **4**(4), 456–465 (2015). https://doi.org/10.1007/s40435-015-0150-9
15. Goel, A., Swarup, A.: Chattering free trajectory tracking control of a robotic manipulator using high order sliding mode. In: Bhatia, S., Mishra, K., Tiwari, S., Singh, V. (eds.) Advances in Computer and Computational Sciences. AISC, vol. 553, pp. 753–761. Springer, Singapore (2017). https://doi.org/10.1007/978-981-10-3770-2_71
16. Zhang, Y., Li, R., Xue, T., Liu, Z., Yao, Z.: An analysis of the stability and chattering reduction of high-order sliding mode tracking control for a hypersonic vehicle. Inf. Sci. **348**, 25–48 (2016)
17. Seeber, R., Horn, M., Fridman, L.: A novel method to estimate the reaching time of the super-twisting algorithm. IEEE Trans. Autom. Control **63**(12), 4301–4308 (2018)
18. Anh Tuan, V., Kang, H.J.: A new finite time control solution for robotic manipulators based on nonsingular fast terminal sliding variables and the adaptive super-twisting scheme. J. Comput. Nonlinear Dyn. **14**(3), 10 (2019)
19. Vo, A.T., Kang, H.J.: A novel fault-tolerant control method for robot manipulators based on non-singular fast terminal sliding mode control and disturbance observer. IEEE Access **8**, 109388–109400 (2020)
20. Vo, A.T., Kang, H.J.: A chattering-free, adaptive, robust tracking control scheme for nonlinear systems with uncertain dynamics. IEEE Access **7**, 10457–10466 (2019)
21. Craig, J.J.: Introduction to Robotics: Mechanics and Control. Pearson Education India, New Delhi (2009)

An Active Disturbance Rejection Control Method for Robot Manipulators

Thanh Nguyen Truong, Hee-Jun Kang(✉), and Anh Tuan Vo

School of Electrical Engineering, University of Ulsan, Ulsan 44610, South Korea
hjkang@ulsan.ac.kr

Abstract. This paper proposed an active disturbance control method for tracking control of robot manipulators. Firstly, all of the system uncertainties and external disturbances are considered as an extended variable and a disturbance observer is used to exactly approximate this total uncertainty. Therefore, accurate information is provided for the control loop and chattering behavior in the control input is significantly reduced. Next, to improve the response speed and tracking accuracy, a sliding mode control is synthesized by combining the non-singular fast terminal sliding mode control and the designed observer. The proposed is reconstructed using backstepping control to obtain the asymptotic stability for the whole control system based on Lyapunov theory. Finally, the examples are simulated to demonstrate the effectiveness of the proposed control method.

Keywords: Backstepping control · Sliding Mode Control · Non-singular fast terminal sliding mode control · Disturbance observer · Robotic manipulators

1 Introduction

Robots play an important role in nowadays. However, robots have a certain complexity in geometric structure and dynamics as well as there are always exist uncertain components and external noise affecting the robot system. Therefore, controlling robots to achieve high performance is really a challenge for researchers until now. There are lots of suggested controllers for the robot such as PID controllers [1, 2], Computed Torque Control (CTC) [3] adaptive control [4] optimal control [5], Sliding Mode Control (SMC) [6, 7], Backstepping control [8], and so on. These controllers are marked with a simple design, easy to apply to real systems. However, these controllers have drawbacks when they only provide a reasonable control performance. In cases where high precision is required, or under the presence of uncertainty and disturbance components, the above controllers do not provide the desired performance. Among these controllers, SMC can be said to be more applications than the rest of other controllers. SMC possesses characteristics such as robustness against uncertainties, simple design, and stable operation that suit the requirements of the actual robot system. Unfortunately, SMC also has some disadvantages that need to be overcome, including chattering, finite time convergence, and slow convergence rate in the presence of a large number of uncertain components.

D.-S. Huang and P. Premaratne (Eds.): ICIC 2020, LNAI 12465, pp. 190–201, 2020.
https://doi.org/10.1007/978-3-030-60796-8_16

To speed up convergence in a finite time, Non-singular Fast Terminal Sliding Mode Control (NFTSMC) has been proposed [9–11]. With NFTSMC, most of the disadvantages of traditional SMC have been solved except for chattering. NFTSMC can provide the desired control performance such as high tracking accuracy, fast finite-time convergence, and robustness against system uncertainties.

With the remaining drawback is chattering, there are many methods reported in control theory such as High-Order Sliding Mode Control (HOSMC) [12, 13], super-twisting method [14, 15], Full-Order Sliding Mode Control (FOSMC) [16, 17], boundary layer [18], disturbance observer [19–21], and so on.

From the aforementioned assessments, the motivation of this paper is to develop an active disturbance rejection control algorithm for robot arms. The controller must achieve the following control objectives

1. The selected disturbance observer exactly approximates this total uncertainty (including dynamical uncertainties and external disturbances) to feed for closed-loop controller.
2. The controller inherits the strengths of SMC, backstepping control, and NFTSMC as simplicity, efficiency with uncertain components, global asymptotic stability, fast response speed, and high tracking accuracy.
3. The controller is designed to overcome the chattering disadvantages of SMC and NFTSMC-based control methods.
4. The stability of the proposed controller has been completely verified by Lyapunov theory and computer simulation results.

The rest of this article has the following presentation. The problem statements facilitated for the proposed control law are presented in Sect. 2. Section 3 explains the design approach of the suggested controller to obtain the desired performance and to overcome the limitations of the conventional SMC. Then, the designed control algorithm is applied for a two-link robot system in Sect. 4. Next, reviews and discussions are discussed to investigate positional control errors, convergence time rate, fast response, and chattering decrease. Finally, conclusions of this paper are given in Sect. 5.

2 Problem Statement

Consider the dynamic of n-degree-of-freedom (DOF) robotic manipulators are described as:

$$M(\theta)\ddot{\theta} + C(\theta,\dot{\theta})\dot{\theta} + G(\theta) + F(\dot{\theta}) + \tau_d = \tau \tag{1}$$

where $\theta = [\,\theta_1 \quad \theta_2 \quad \ldots \quad \theta_n\,]^T \in \Re^n$, $\dot{\theta} = \left[\,\dot{\theta}_1 \quad \dot{\theta}_2 \quad \cdots \quad \dot{\theta}_n\,\right]^T \in \Re^n$, and $\ddot{\theta} = \left[\,\ddot{\theta}_1 \quad \ddot{\theta}_2 \ldots \quad \ddot{\theta}_n\,\right]^T \in \Re^n$ represents the joint angle position, the joint angle velocity and the joint angle acceleration, respectively. $M(\theta) = \hat{M}(\theta) + \Delta M(\theta) \in \Re^{n\times n}$ is a real inertia matrix, $C(\theta,\dot{\theta}) = \hat{C}(\theta,\dot{\theta}) + \Delta C(\theta,\dot{\theta}) \in \Re^{n\times n}$ consists of real Coriolis and real centrifugal force, and $G(\theta) = \hat{G}(\theta) + \Delta G(\theta) \in \Re^{n\times 1}$ is a real gravity matrix. $F(\dot{\theta}) \in \Re^{n\times 1}$ and $\tau_d \in \Re^{n\times 1}$ are friction and external disturbance matrices, respectively. $\tau \in \Re^{n\times 1}$ is control input vector. $\hat{M}(\theta) \in \Re^{n\times n}$, $\hat{C}(\theta,\dot{\theta}) \in \Re^{n\times n}$ and $\hat{G}(\theta) \in \Re^{n\times 1}$ are estimated matrices of M, C, and G, respectively. $\Delta M(\theta) \in \Re^{n\times n}$, $\Delta C(\theta) \in \Re^{n\times n}$, and $\Delta G(\theta) \in \Re^{n\times n}$ are the estimated error matrices of the dynamic model.

The real dynamic model of n-DOF robotic manipulator can be represented as:

$$\hat{M}(\theta)\ddot{\theta} + \hat{C}(\theta,\dot{\theta})\dot{\theta} + \hat{G}(\theta) + \Omega = \tau \tag{2}$$

where $\Omega = \Delta M(\theta)\ddot{\theta} + \Delta C(\theta,\dot{\theta})\dot{\theta} + \Delta G(\theta) + F(\dot{\theta}) + \tau_d \in \Re^{n\times 1}$ is the lumped system uncertainties and external disturbance.

From the Eq. (2) we can rearrange as:

$$\ddot{\theta} = \hat{M}^{-1}(\theta)\tau + \Psi(\theta,\dot{\theta}) + D \tag{3}$$

where $\Psi(\theta,\dot{\theta}) = \hat{M}^{-1}(\theta)\Big(-C(\theta,\dot{\theta})\dot{\theta} - G(\theta)\Big) \in \Re^{n\times 1}$, $D = -\hat{M}^{-1}(\theta)\Omega \in \Re^{n\times 1}$.

The dynamic Eq. (3) can be transferred into second-order state space model as follow:

$$\begin{cases} \dot{x}_1 = x_2 \\ \dot{x}_2 = \Psi(x_1,x_2) + \hat{M}^{-1}(x_1)u + D \end{cases} \tag{4}$$

where $x_1 = \theta \in \Re^{n\times 1}$ and $x_2 = \dot{\theta} \in \Re^{n\times 1}$ are state vectors of system, $u = \tau \in \Re^{n\times 1}$ is the control input vector.

The main objective of this paper is to design a control input to provide a high control performance in the presence of uncertainties and external disturbance.

3 Design Procedure of Tracking Control Algorithm

3.1 Design and Analysis of Disturbance Observer

From the system (4), the lumped system uncertainties and external disturbances D are estimated by the following disturbance observer:

$$\begin{cases} \dot{\hat{D}} = -\lambda_1(\hat{\omega} - \dot{x}_1) \\ \dot{\hat{\omega}} = \Psi(x_1,x_2) + \hat{M}^{-1}(x_1)u + \hat{D} - \lambda_2(\hat{\omega} - \dot{x}_1) \end{cases} \tag{5}$$

where $\lambda_1 > 0$ and $\lambda_2 > 0$ are the observer gains. $\hat{\omega}$ and $\hat{D}$ are the estimated values of x_2 and D, respectively.

Consider a Lyapunov function candidate as

$$V_1 = 0.5\tilde{\omega}^T\tilde{\omega} + 0.5\lambda_1^{-1}\tilde{D}^T\tilde{D} \tag{6}$$

where $\tilde{\omega} = x_2 - \hat{\omega} \in \Re^{n\times 1}$ and $\tilde{D} = D - \hat{D} \in \Re^{n\times 1}$ are the approximation errors of velocity x_2 and lumped system uncertainties and external disturbances D, respectively.

Assumption 1. *We assume that the estimation error of lumped system uncertainties and external disturbances is bounded by*

$$\|\tilde{D}\| \leq \Sigma \tag{7}$$

where Σ is a positive constant.

Differentiating V_1 with respect to time yields

$$\dot{V}_1 = \tilde{\omega}^T\dot{\tilde{\omega}} + \lambda_1^{-1}\tilde{D}^T\dot{\tilde{D}} = \tilde{\omega}^T\left(\dot{x}_2 - \dot{\hat{\omega}}\right) + \lambda_1^{-1}\tilde{D}^T\left(\dot{D} - \dot{\hat{D}}\right) \tag{8}$$

Substituting Eqs. (4) and (5) into Eq. (8), we have

$$\begin{aligned}\dot{V}_1 &= \tilde{\omega}^T\left(\Psi(x_1, x_2) + \hat{M}^{-1}(x_1)u + D - \left(\Psi(x_1, x_2) + \hat{M}^{-1}(x_1)u + \hat{D} - \lambda_2(\hat{\omega} - \dot{x}_1)\right)\right)\\ &\quad + \lambda_1^{-1}\tilde{D}^T\left(\dot{D} + \lambda_1(\hat{\omega} - \dot{x}_1)\right)\\ &= \tilde{\omega}^T\left(D - \hat{D} + \lambda_2(\hat{\omega} - \dot{x}_1)\right) + \lambda_1^{-1}\tilde{D}^T\left(\dot{D} + \lambda_1(\hat{\omega} - \dot{x}_1)\right)\\ &= -\lambda_2\tilde{\omega}^T\tilde{\omega} + \frac{1}{\lambda_1}\tilde{D}^T\dot{D} = -\lambda_2\sum_{i=1}^{n}\tilde{\omega}_i^2 + \frac{1}{\lambda_1}\sum_{i=1}^{n}\tilde{D}_i\dot{D}_i\end{aligned} \tag{9}$$

We assume that $\|D\|$ is bounded ($\|D\| \leqslant \Theta$). When the value of λ_1 is large leading to $\lambda_1^{-1}\sum_{i=1}^{n}\dot{D}_i \approx 0$. Similarly, when the value of λ_2 is large, we can get:

$$\dot{V}_1 - \frac{1}{\lambda_1}\sum_{i=1}^{n}\tilde{D}_i\dot{D}_l - \lambda_2\sum_{i=1}^{n}\tilde{\omega}_i^2 \leq 0 \tag{10}$$

From Eq. (10), we can see that the lumped system uncertainties and external disturbances are exactly estimated by the designed disturbance observer.

3.2 Design of the Proposed Control Method

Let $x_d \in \Re^{n\times 1}$ be the desired state vector. For the dynamic system (4), the position control error (x_e) and velocity control error (x_{de}) are defined as

$$\begin{cases} x_e = x_1 - x_d \in \Re^{n\times 1} \\ x_{de} = x_2 - \dot{x}_d \in \Re^{n\times 1} \end{cases} \tag{11}$$

The design procedure of the proposed controller is developed according to the approach of Backstepping control as follow.

Step1: To realize the position error to zero ($x_e \rightarrow 0$).

The Lyapunov function is chosen as

$$V_2 = 0.5\sum\nolimits_{i=1}^{n} x_{ei}^2 \tag{12}$$

The time derivative of Lyapunov function (12) is

$$\dot{V}_2 = \sum\nolimits_{i=1}^{n} x_{ei}x_{dei} = \sum\nolimits_{i=1}^{n} x_{ei}(x_{2i} - \dot{x}_{di}) \tag{13}$$

To realize $\dot{V}_2 \leq 0$, we let $x_{2i} = s_i + \dot{x}_{di} - \alpha_1|x_{ei}|^{\beta_1}\text{sgn}(x_{ei}) - \alpha_2|x_{ei}|^{\beta_2}\text{sgn}(x_{ei})$, $i = 1, 2, \ldots, n$, that is

$$s_i = x_{dei} + \alpha_1|x_{ei}|^{\beta_1}\text{sgn}(x_{ei}) + \alpha_2|x_{ei}|^{\beta_2}\text{sgn}(x_{ei}), i = 1, 2, \ldots, n \tag{14}$$

where $s = [\, s_1 \quad s_2 \quad \ldots \quad s_n \,]^T \in \Re^{nx1}$ is the non-singular fast terminal sliding variable, with α_1, α_2, β_1, and β_2 are positive constants. Therefore, we have

$$\begin{aligned}\dot{V}_2 &= \sum\nolimits_{i=1}^{n} x_{ei}\left(s_i + \dot{x}_{di} - \alpha_1|x_{ei}|^{\beta_1}\text{sgn}(x_{ei}) - \alpha_2|x_{ei}|^{\beta_2}\text{sgn}(x_{ei}) - \dot{x}_{di}\right) \\ &= -\alpha_1\sum\nolimits_{i=1}^{n}|x_{ei}|^{\beta_1+1} - \alpha_2\sum\nolimits_{i=1}^{n}|x_{ei}|^{\beta_2+1} + \sum\nolimits_{i=1}^{n} x_{ei}s_i\end{aligned} \tag{15}$$

If $s_i = 0$, $i = 1, 2, \ldots, n$, then $\dot{V}_2 \leq 0$. Therefore, the next step is required.

Step2: To design the control input u which control x_e, s, $\tilde{\omega}$ and $\tilde{D}$ to zero.

The Lyapunov function is selected as

$$V_3 = V_1 + V_2 + 0.5s^T s \tag{16}$$

The time derivation of sliding mode function is expressed as

$$\dot{s}_i = \dot{x}_{dei} + \beta_1\alpha_1|x_{ei}|^{\beta_1-1}x_{dei} + \beta_2\alpha_2|x_{ei}|^{\beta_2-1}x_{dei},\ i = 1, 2, 3, \ldots, n \tag{17}$$

To simplify, let $Z_i = \dot{x}_{dei} + \beta_1\alpha_1|x_{ei}|^{\beta_1-1}x_{dei} + \beta_2\alpha_2|x_{ei}|^{\beta_2-1}x_{dei}$, $i = 1, 2, 3, \ldots, n$. Equation (17) in matrix form as

$$\dot{s} = \dot{x}_{de} + Z \tag{18}$$

where $Z = [\, Z_1 \quad Z_2 \quad \ldots \quad Z_n \,]^T \in \Re^{n\times 1}$.

Substituting Eq. (4) into Eq. (18), we have

$$\dot{s} = \Psi(x_1, x_2) + \hat{M}^{-1}(x_1)u + D - \ddot{x}_d + Z \tag{19}$$

The time derivative of Lyapunov function (16) obtains

$$\dot{V}_3 = \dot{V}_1 + \dot{V}_2 + s^T\dot{s} \tag{20}$$

Substituting Eqs. (15) and (19) into Eq. (20), we gain

$$\begin{aligned}\dot{V}_3 = & \dot{V}_1 - \alpha_1 \sum_{i=1}^{n} |x_{ei}|^{\beta_1+1} - \alpha_2 \sum_{i=1}^{n} |x_{ei}|^{\beta_2+1} \\ & + \sum_{i=1}^{n} x_{ei} s_i + s^T(\Psi(x_1,x_2) + M^{-1}(x_1)u + D - \ddot{x}_d + Z)\end{aligned} \tag{21}$$

Based on Eq. (21), the proposed controller is designed as

$$u = \hat{M}(x_1)\left(\ddot{x}_d - \Psi(x_1,x_2) - \hat{D} - Z - x_e - \kappa s - (\Sigma + \eta)\text{sgn}(s)\right) \tag{22}$$

where κ is a positive value, η is a small positive value, the term $\hat{D}$ is designed as (5).

The block diagram of proposed controller is shown in Fig. 1.

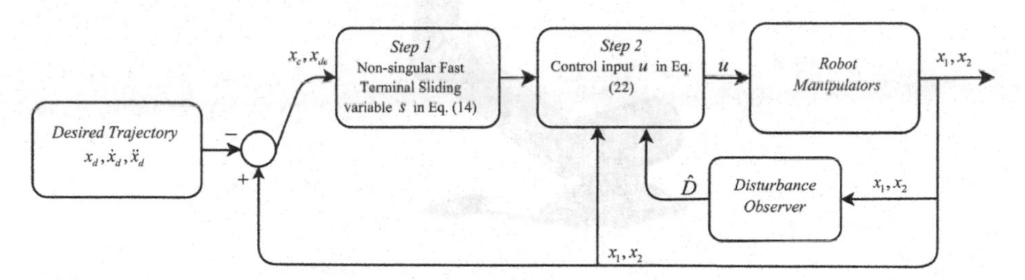

Fig. 1. Block diagram of the proposed controller.

Substituting the designed controller (22) into (21) yields

$$\begin{aligned}\dot{V}_3 &= \dot{V}_1 - \alpha_1 \sum_{i=1}^{n} |x_{ei}|^{\beta_1+1} - \alpha_2 \sum_{i=1}^{n} |x_{ei}|^{\beta_2+1} + s^T\left(-\kappa s + \tilde{D} - (\Sigma + \eta)\text{sgn}(s)\right) \\ &\le \dot{V}_1 - \alpha_1 \sum_{i=1}^{n} |x_{ei}|^{\beta_1+1} - \alpha_2 \sum_{i=1}^{n} |x_{ei}|^{\beta_2+1} - \kappa \sum_{i=1}^{n} s_i^2 - \eta \sum_{i=1}^{n} |s_i| \le 0\end{aligned} \tag{23}$$

The first term $\dot{V}_1 \le 0$ has been proved by Eq. (10). We see that $\dot{V}_3$ is a negative semidefinite. Therefore, the proposed controller is asymptotically stable in the presence of system uncertainties and external disturbance.

4 Numerical Simulation Results and Discussion

To verify the performance of the proposed controller, we employ it for a 2-DOF robotic manipulator. Simulations were performed on the MATLAB-SIMULNIK environment. The mechanical model of the 2-DOF robotic manipulator is designed on SOLID-WORKS software and embedded into the SIMULINK environment through the Simscape Multibody Link tool. In this way, the mechanical model of the 2-DOF robotic manipulator is completely identical to the actual model. The 3D SOLIDWORK model of 2-DOF robotic manipulator is shown in Fig. 2 and the robot parameters are given in Table 1. In the MATLAB/SIMULINK environment, the configuration of the model is set under a fixed-step (ODE5 dormand-prince) with a fundamental sample time of 0.001 s. To examine advanced capabilities and superior effect of the proposed system, the proposed system applies to the above robotic manipulator and its control performance is compared to CTC, SMC and NFTSMC.

Fig. 2. 3D SOLIDWORK model of the 2-DOF robotic manipulator.

The CTC has the following control torque

$$u = \hat{M}(x_1)\left(\ddot{x}_d - \Psi(x_1, x_2) - K_p x_e - K_v x_{de}\right) \tag{24}$$

where K_p and K_v are positive constants.

The SMC has the following control torque

$$u = \hat{M}(x_1)(\ddot{x}_d - \Psi(x_1, x_2) - c x_{de} - (\Theta + \eta)\mathrm{sgn}(s)) \tag{25}$$

where $s = x_{de} + c x_e$ with c is a positive constant. η is a small positive constant.

The NFTSMC has the following control torque

$$u = \hat{M}(x_1)(\ddot{x}_d - \Psi(x_1, x_2) - Z - (\Theta + \eta)\mathrm{sgn}(s)) \tag{26}$$

where s and Z are designed as (14), (18), respectively. η is a small positive constant.

Remark 1. *The sliding value* Θ *in Eqs. (25) and (26) is a positive constant and it should be selected bigger than the upper bound value of lumped system uncertainties and external disturbances D.*

Table 1. The design parameters of the robot system.

Parameters	Value	Unit
m_1	11.940596	kg
m_2	7.400618	kg
l_1	0.3	m
l_2	0.3	m
l_{c1}	0.193304	m
l_{c2}	0.153922	m
I_1	$150901.5816 \times 10^{-6}$	$kg.m^2$
I_2	$78091.7067 \times 10^{-6}$	$kg.m^2$

The desired trajectory of a robot's end-effector is designed to track the following circle:

$$\begin{cases} x_d = 0.3 + 0.05\cos(t) \\ y_d = 0.15 + 0.05\sin(t) \end{cases} \tag{27}$$

The effects of the considered friction and external disturbance are given by

$$F(\dot{\theta}) + \tau_d = \begin{bmatrix} -0.5\sin(\theta_1) - 0.5\dot{\theta}_1 \\ -0.1\sin(\theta_2) - 0.5\dot{\theta}_2 \end{bmatrix} + \begin{bmatrix} 3\sin(t) \\ \sin(t) \end{bmatrix} \tag{28}$$

The selection of control parameters for the controllers ensures good results and they are reported in Table 2.

Table 2. The control parameters of four different controllers.

Control System	Control Parameters	Value
CTC	K_p, K_v	600, 100
SMC	c, Θ, η	5, 7, 0.01
NFTSMC	α_1, α_2, β_1, β_2, Θ, η	5, 3, 0.9, 1.4, 7, 0.01
Proposed Controller	α_1, α_2, β_1, β_2, κ, Σ, η, λ_1, λ_2	5, 3, 0.9, 1.4, 40, 0.3, 0.01, 20000, 1500

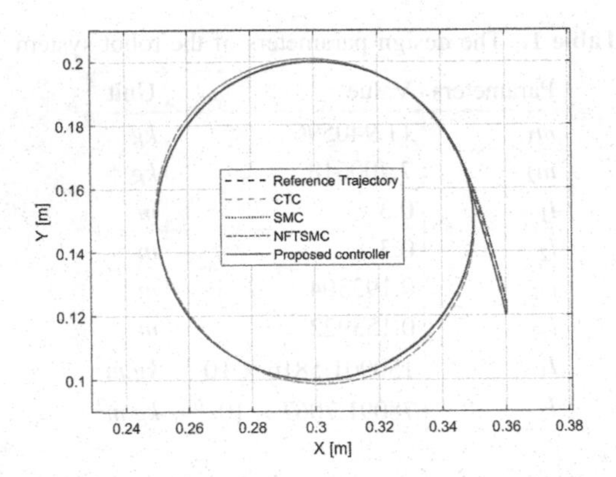

Fig. 3. The position desired trajectory and actual position trajectory of end-effector under four controllers.

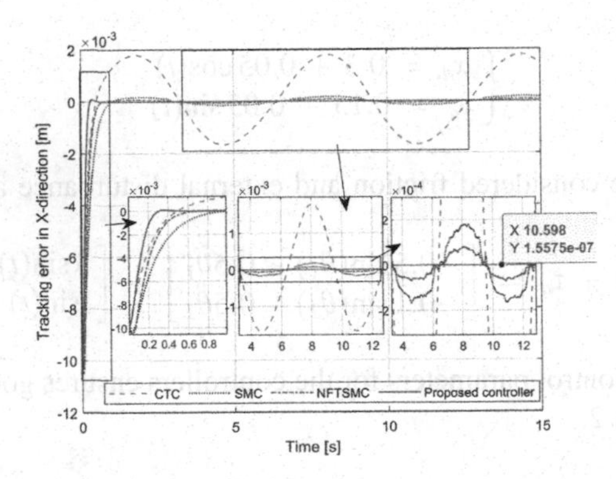

Fig. 4. The tracking error of end-effector in X-direction of four controllers.

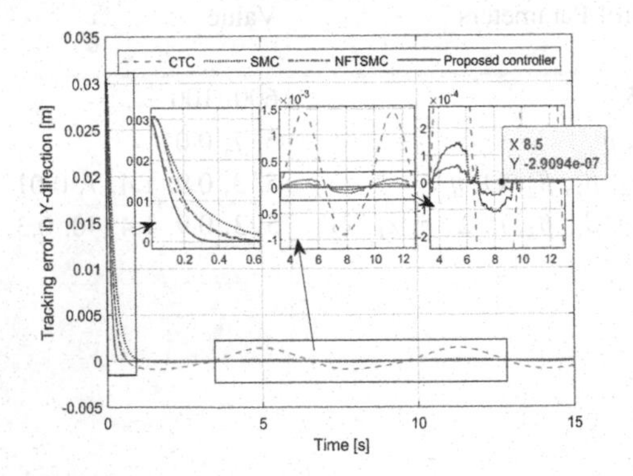

Fig. 5. The tracking error of end-effector in Y-direction of four controllers.

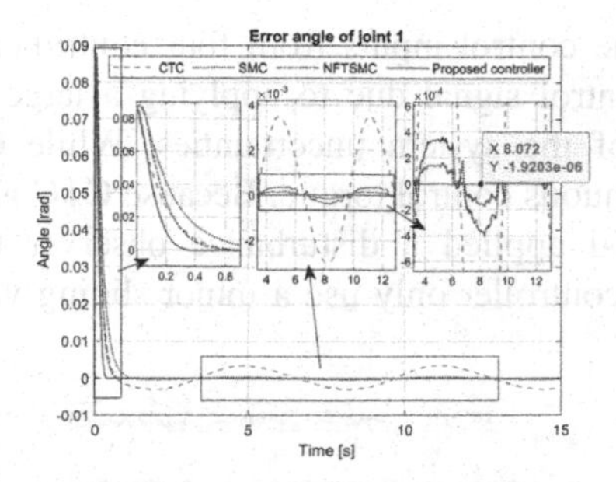

Fig. 6. The tracking error of Joint 1 under four controllers.

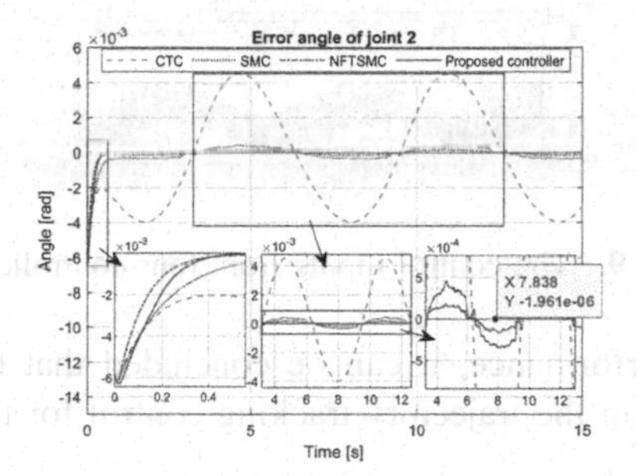

Fig. 7. The tracking error of Joint 2 under four controllers.

Figures (3), (4), (5) and (6), (7) exhibit the result of the trajectory tracking performance from four difference controllers. CTC provides the worst tracking performance among 4 controllers. NFTSMC has better performance than CTC and SMC as well as it has faster speed of stabilization and convergence. Especially, the proposed controller demonstrates outstanding capabilities when it provides the best control performance and fastest convergence speed.

Figure (8) shows the assumed disturbance and the result of the estimated disturbance. It is clearly that the uncertain components and disturbances affecting the robotic system were accurately estimated. Therefore, we can yield an exact robot model to improve control performance and reduce chattering.

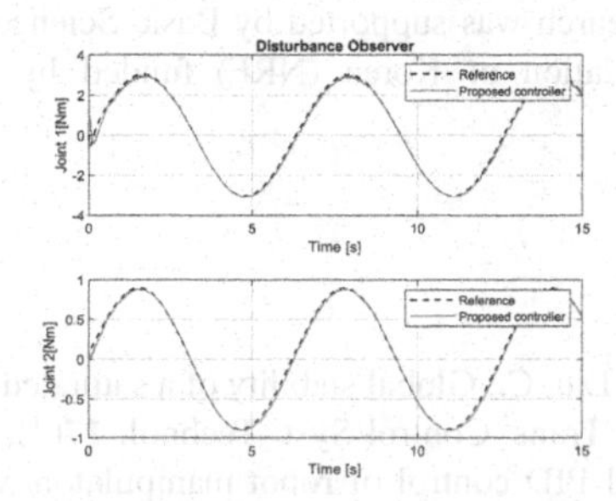

Fig. 8. Estimated disturbance and assumed disturbance.

Figure (9) displays the control inputs from four controllers. SMC and NFTSMC show a discontinuous control signal due to applying a large sliding value for compensation of the effect of the system uncertainties. While CTC and the proposed controller provide a continuous control torque. Because CTC is a continuous controller and the proposed method applied a disturbance observer to estimate the system uncertainties lead to our controller only use a minor sliding value.

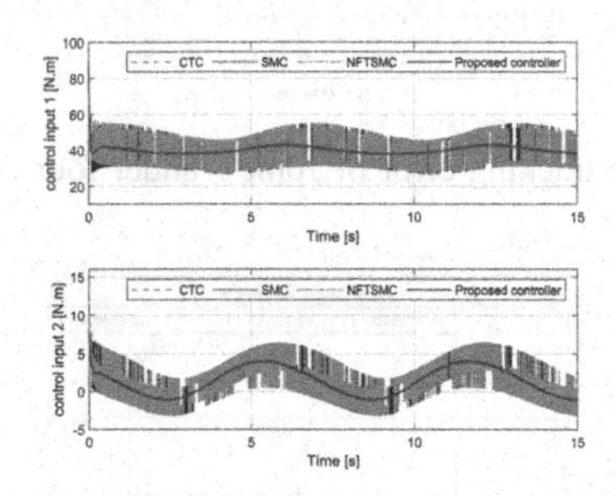

Fig. 9. The control inputs from four controllers.

From the control performance, it can be concluded that the proposed controller provides high efficiency in the trajectory tracking control for the robot arms.

5 Conclusion

This paper proposed an active disturbance control method for tracking control of robot manipulators. The proposed control method obtained main contributions as follows: 1) the selected disturbance observer exactly approximates this total uncertainty (including dynamical uncertainties and external disturbances) to feed for closed-loop controller; 2) the controller inherits the strengths of SMC, backstepping control, and NFTSMC as simplicity, efficiency with uncertain components, global asymptotic stability, fast response speed, and high tracking accuracy; 3) the controller is designed to overcome the chattering disadvantages of SMC and NFTSMC-based control methods; 4) the stability of the proposed controller has been completely verified by Lyapunov theory and computer simulation results.

Acknowledgement. This research was supported by Basic Science Research Program through the National Research Foundation of Korea (NRF) funded by the Ministry of Education (2019R1D1A3A03103528).

References

1. Sun, D., Hu, S., Shao, X., Liu, C.: Global stability of a saturated nonlinear PID controller for robot manipulators. IEEE Trans. Control Syst. Technol. **17**(4), 892–899 (2009)
2. Yu, W., Rosen, J.: Neural PID control of robot manipulators with application to an upper limb exoskeleton. IEEE Trans. Cybern. **43**(2), 673–684 (2013)

3. Song, Z., Yi, J., Zhao, D., Li, X.: A computed torque controller for uncertain robotic manipulator systems: fuzzy approach. Fuzzy Sets Syst. **154**(2), 208–226 (2005)
4. Craig, J.J., Hsu, P., Sastry, S.S.: Adaptive control of mechanical manipulators. Int. J. Robot. Res. **6**(2), 16–28 (1987)
5. Lin, F., Brandt, R.D.: An optimal control approach to robust control of robot manipulators. IEEE Trans. Robot. Autom. **14**(1), 69–77 (1998)
6. Islam, S., Liu, X.P.: Robust sliding mode control for robot manipulators. IEEE Trans. Ind. Electron. **58**(6), 2444–2453 (2010)
7. Truong, T.N., Kang, H.J., Le, T.D.: Adaptive neural sliding mode control for 3-DOF planar parallel manipulators. In: Proceedings of the 2019 3rd International Symposium on Computer Science and Intelligent Control, pp. 1–6 (2019)
8. Van, M., Mavrovouniotis, M., Ge, S.S.: An adaptive backstepping nonsingular fast terminal sliding mode control for robust fault tolerant control of robot manipulators. IEEE Trans. Syst. Man Cybern.: Syst. **49**(7), 1448–1458 (2018)
9. Yang, L., Yang, J.: Nonsingular fast terminal sliding-mode control for nonlinear dynamical systems. Int. J. Robust Nonlinear Control **21**(16), 1865–1879 (2011)
10. Jin, M., Lee, J., Chang, P.H., Choi, C.: Practical nonsingular terminal sliding-mode control of robot manipulators for high-accuracy tracking control. IEEE Trans. Ind. Electron. **56**(9), 3593–3601 (2009)
11. Vo, A.T., Kang, H.J.: An adaptive terminal sliding mode control for robot manipulators with non-singular terminal sliding surface variables. IEEE Access **7**, 8701–8712 (2018)
12. Laghrouche, S., Plestan, F., Glumineau, A.: Higher order sliding mode control based on integral sliding mode. Automatica **43**(3), 531–537 (2007)
13. Zhang, Y., Li, R., Xue, T., Liu, Z., Yao, Z.: An analysis of the stability and chattering reduction of high-order sliding mode tracking control for a hypersonic vehicle. Inf. Sci. **348**, 25–48 (2016)
14. Anh Tuan, V., Kang, H.J.: A new finite time control solution for robotic manipulators based on nonsingular fast terminal sliding variables and the adaptive super-twisting scheme. J. Comput. Nonlinear Dyn. **14**(3) (2019)
15. Chalanga, A., Kamal, S., Fridman, L.M., Bandyopadhyay, B., Moreno, J.A.: Implementation of super-twisting control: super-twisting and higher order sliding-mode observer-based approaches. IEEE Trans. Ind. Electron. **63**(6), 3677–3685 (2016)
16. Doan, Q.V., Le, T.D., Vo, A.T.: Synchronization full-order terminal sliding mode control for an uncertain 3-DOF planar parallel robotic manipulator. Appl. Sci. **9**(9), 1756 (2019)
17. Feng, Y., Zhou, M., Zheng, X., Han, F., Yu, X.: Full-order terminal sliding-mode control of MIMO systems with unmatched uncertainties. J. Franklin Inst. **355**(2), 653–674 (2018)
18. Chen, M.S., Hwang, Y.R., Tomizuka, M.: A state-dependent boundary layer design for sliding mode control. IEEE Trans. Autom. Control **47**(10), 1677–1681 (2002)
19. Li, S., Yang, J., Chen, W.H., Chen, X.: Disturbance Observer-Based Control: Methods and Applications. CRC Press, Boca Raton (2014)
20. Liu, J., Wang, X.: Advanced sliding mode control for mechanical systems, pp. 206–210. Springer, Beijing (2012). https://doi.org/10.1007/978-3-642-20907-9
21. Zhang, J., Liu, X., Xia, Y., Zuo, Z., Wang, Y.: Disturbance observer-based integral sliding-mode control for systems with mismatched disturbances. IEEE Trans. Ind. Electron. **63**(11), 7040–7048 (2016)

A Fault Tolerant Control for Robotic Manipulators Using Adaptive Non-singular Fast Terminal Sliding Mode Control Based on Neural Third Order Sliding Mode Observer

Van-Cuong Nguyen[1] and Hee-Jun Kang[2(✉)]

[1] Graduate School of Electrical Engineering, University of Ulsan, Ulsan 44610, South Korea
[2] School of Electrical Engineering, University of Ulsan, Ulsan 44610, South Korea
hjkang@ulsan.ac.kr

Abstract. This paper proposes a fault tolerant control technique for uncertain faulty robotic manipulators when only position measurement is available. First, a neural third-order sliding mode observer is utilized to approximate the system velocities, the lumped uncertainties and faults, in which the radial basis function neural network is employed to approximate the observer gains. Then, the obtained information is applied to design a non-singular fast terminal sliding mode control to deal with the effect of the lumped uncertainties and faults. In addition, an adaptive law is used to approximate the sliding gain in switching control law. The controller-observer method can provide superior features such as high tracking precision, less chattering phenomenon, finite-time convergence, and robustness against the lumped uncertainties and faults without the requirement of its prior knowledge. The stability and finite-time convergence of the proposed technique are proved in theory by using the Lyapunov function. To verify the usefulness of the proposed strategy, computer simulations for a 2-link serial robotic manipulator are performed.

Keywords: Fault tolerant control · Neural Third-Order Sliding Mode Observer · Non-singular fast terminal sliding mode control · Radial basis function neural network

1 Introduction

Robot manipulators are very popular and have many applications in industry. However, it is very difficult to get the robot's exact dynamics in realization because of the dynamic uncertainties such as payload changes, frictions, and external disturbances. In some special cases, faults happen when the robot is operating. They are the big challenges in both theoretical and practical control. To deal with this problem, sliding mode control (SMC) is one of the most popular controllers that have been widely used because of its standout properties such as easy design procedure, robustness against the effect of uncertainties and faults [1–3]. Besides the great advantages, some problems

D.-S. Huang and P. Premaratne (Eds.): ICIC 2020, LNAI 12465, pp. 202–212, 2020.
https://doi.org/10.1007/978-3-030-60796-8_17

still exist in conventional SMC such as chattering phenomenon, velocity and upper bound of the lumped uncertainties and faults requirement, and the finite-time convergence cannot be guaranteed.

To archive the finite-time convergence, the terminal SMC (TSMC) has been proposed [4–6]. Although the TSMC provides higher precision and finite-time convergence; unfortunately, it includes two main drawbacks that are slower convergence time compared with the SMC and singularity problem. In order to solve each of the two drawbacks, the fast TSMC (FTSMC) [7–9] and the nonsingular TSMC (NTSMC) [10–12] have been developed, separately. To eliminate them simultaneously, the nonsingular fast TSMC (NFTSMC) that provides superior properties such as singularity elimination, high tracking accuracy, finite-time convergence, and robustness against the lumped uncertainties and faults has been proposed [13–16].

In order to decrease the chattering phenomenon and eliminate the velocity measurement requirement, the third-order sliding mode observer (TOSMO) has been performed to approximate the system velocities and the lumped uncertainties and faults with high accuracy and less chattering phenomenon [17, 18]. The obtained estimation information is applied to compensate for the effect of the lumped uncertainties and faults; therefore, the switching control element now plays the role to deal with the effects of the estimation error instead; consequently, the chattering is reduced.

Although the prior knowledge of the lumped uncertainties and faults is not needed to design the controller, it is remained in the observer design process. To deal with this problem, the radial basis function neural network (RBFN), which is well-known with abilities to approximate parameter with high accuracy, fast learning ability, and simple structure [19, 20], is used to approximate the observer gains of TOSMO.

In this paper, a neural TOSMO is performed to estimate system velocities and the lumped uncertainties and faults. The obtained estimation signal is applied to design an NFTSMC to deal with its effects. In addition, an adaptive law is utilized to approximate the switching gain to completely remove the requirement of the precision of the observer. The proposed controller-observer technique can provide high tracking precision, less chattering phenomenon.

The construction of this paper is as follows. After the introduction part, the dynamic equation of a n-link serial robotic manipulator is introduced in Sect. 2. Then, the neural TOSMO is designed for robotic manipulators in Sect. 3. Next, an adaptive NFTSM controller is proposed in Sect. 4. In Sect. 5, the computer simulations of the proposed controller-observer strategy are performed for a 2-link serial robotic manipulator. Finally, some conclusions are given in Sect. 6.

2 Problem Statement

Consider a serial n-link uncertain faulty robotic manipulator with dynamic equation as following

$$\ddot{\theta} = M^{-1}(\theta)\left[\tau - C\left(\theta,\dot{\theta}\right)\dot{\theta} - G(\theta) - F\left(\theta,\dot{\theta}\right) - \tau_d\right] + \Phi(t) \qquad (1)$$

where $\theta, \dot{\theta}, \ddot{\theta} \in \mathbb{R}^n$ represent robot joints' position, velocity, and acceleration, respectively. $M(\theta) \in \mathbb{R}^{n \times n}$ denotes the inertia matrix, $C\left(\theta, \dot{\theta}\right) \in \mathbb{R}^n$ denotes the Coriolis and centripetal forces, and $G(\theta) \in \mathbb{R}^n$ denotes the gravitational force term. $\tau \in \mathbb{R}^n$ denotes the control input signal. $F\left(\theta, \dot{\theta}\right) \in \mathbb{R}^n, \tau_d \in \mathbb{R}^n$, and $\Phi(t) \in \mathbb{R}^n$ denote the friction vector, disturbance vector, and unknown faults, respectively.

The robotic system (1) can be rewritten as

$$\ddot{\theta} = \mathrm{H}\left(\theta, \dot{\theta}\right) + M^{-1}(\theta)\tau + \Delta\left(\theta, \dot{\theta}, t\right) \tag{2}$$

where $\Delta\left(\theta, \dot{\theta}, t\right) = M^{-1}(\theta)\left[-F\left(\theta, \dot{\theta}\right) - \tau_d\right] + \Phi(t)$ represent the lumped uncertainties and faults and $\mathrm{H}\left(\theta, \dot{\theta}\right) = M^{-1}(\theta)\left[-C\left(\theta, \dot{\theta}\right)\dot{\theta} - G(\theta)\right]$.

To simply in the designing process, the system (2) is rewritten in state space as

$$\begin{aligned} \dot{x}_1 &= x_2 \\ \dot{x}_2 &= \mathrm{H}(x) + M^{-1}(x_1)u(t) + \Delta(x, t) \end{aligned} \tag{3}$$

where $x_1 = \theta$, $x_2 = \dot{\theta}$, $x = \begin{bmatrix} x_1^T & x_2^T \end{bmatrix}^T$, $u(t) = \tau$.

The objective of this paper is to design a fault tolerant controller that has ability to deal with effect of lumped uncertainties and faults without the requirement of its prior knowledge. In addition, only position measurement is available. The controller strategy is designed based on the following assumptions.

Assumption 1: The lumped uncertainties and faults are bounded as

$$|\Delta(x, t)| \leq \bar{\Delta} \tag{4}$$

Assumption 2: The time derivative of the lumped uncertainties and faults are bounded as

$$\left|\dot{\Delta}(x, t)\right| \leqslant \bar{\Lambda} \tag{5}$$

where $\bar{\Delta}$ and $\bar{\Lambda}$ are unknown positive constants.

3 Design of Neural Third-Order Sliding Mode Observer

The neural TOSMO is designed for the robotic system (3) as [20]

$$\begin{aligned} \dot{\hat{x}}_1 &= \hat{\alpha}_1|x_1 - \hat{x}_1|^{2/3} sign(x_1 - \hat{x}_1) + \hat{x}_2 \\ \dot{\hat{x}}_2 &= \mathrm{H}(\hat{x}) + M^{-1}(\hat{x}_1)u(t) + \hat{\alpha}_2|x_1 - \hat{x}_1|^{1/3} sign(x_1 - \hat{x}_1) + \hat{z} \\ \dot{\hat{z}} &= \hat{\alpha}_3 sign(x_1 - \hat{x}_1) \end{aligned} \tag{6}$$

where $\hat{x}$ is the estimation of the system states, x, and $\hat{\alpha}_i$ is the estimation of observer gains.

$$\hat{\alpha}_i = W_i^T \Xi_i(E),\ i = 1, 2, 3 \tag{7}$$

where $E = \left[e_1^T\, e_2^T\right]^T$ denotes the input of the neural network, in which $e_1 = x_1 - \hat{x}_1$ and $e_2 = \dot{\hat{x}}_1 - \hat{x}_2$.

The RBFN is utilized as activation function

$$\Xi_i(E) = \exp\left(\frac{\|E - c_{ij}\|^2}{\sigma_{ij}^2}\right) \tag{8}$$

where σ_{ij} is the spread factor, c_{ij} is the center vector, and $j = 1, 2, \ldots, \mathrm{N}$ represents the number of nodes in the hidden layer.

The neural network weigh is updated by the following law

$$\dot{\hat{W}}_i = \lambda_i \Xi_i(E)\|E\| \tag{9}$$

where λ_i represent the learning rate.

We can get the estimation errors by subtracting (6) from (3)

$$\begin{aligned} \dot{\tilde{x}}_1 &= -\hat{\alpha}_1|\tilde{x}_1|^{2/3}\ sign(\tilde{x}_1) + \tilde{x}_2 \\ \dot{\tilde{x}}_2 &= -\hat{\alpha}_2|\tilde{x}_1|^{1/3}\ sign(\tilde{x}_1) + \Delta(x, t) - \hat{z} \\ \dot{\hat{z}} &= \hat{\alpha}_3 sign(\tilde{x}_1) \end{aligned} \tag{10}$$

where $\tilde{x} = x - \hat{x}$, defining $\tilde{x}_3 = -\hat{z} + \Delta(x, t)$, the estimation errors (10) can be rewritten as

$$\begin{aligned} \dot{\tilde{x}}_1 &= -\hat{\alpha}_1|\tilde{x}_1|^{2/3} sign(\tilde{x}_1) + \tilde{x}_2 \\ \dot{\tilde{x}}_2 &= -\hat{\alpha}_2|\tilde{x}_1|^{1/3} sign(\tilde{x}_1) + \tilde{x}_3 \\ \dot{\tilde{x}}_3 &= -\hat{\alpha}_3 sign(\tilde{x}_1) + \dot{\Delta}(x, t) \end{aligned} \tag{11}$$

After the transition time, the estimated states will converge to the actual states ($\hat{x}_1 = x_1,\ \hat{x}_2 = x_2$), the third term of the estimation errors (11) becomes

$$\dot{\tilde{x}}_3 = -\hat{\alpha}_3 sign(\tilde{x}_1) + \dot{\Delta}(x, t) \equiv 0 \tag{12}$$

The lumped uncertainties and faults can be rebuilt as

$$\hat{\Delta}(\hat{x},t) = \int \hat{\alpha}_3 sign(\tilde{x}_1) \tag{13}$$

We can see that Eq. (13) includes an integral element; therefore, the estimation of the uncertainties and faults can be directly rebuilt and the chattering of the estimated signal is partially eliminated. Further, the use of RBF neural network to approximate the observer gains help eliminate the need of prior knowledge of the lumped uncertainties and faults.

4 Design of Adaptive Non-singular Fast Terminal Sliding Mode Controller

The position tracking and velocity errors are respectively defined as

$$\begin{aligned} e &= x_1 - x_d \\ \dot{e} &= x_2 - \dot{x}_d \end{aligned} \tag{14}$$

where x_d, and $\dot{x}_d$ represent the expected trajectories and velocities.

A non-singular fast terminal sliding surface is selected as in [21]

$$s = \dot{e} + \int_0^t \left(\kappa_2 |\dot{e}|^{\beta_2} sign(\dot{e}) + \kappa_1 |e|^{\beta_1} sign(e)\right) dt \tag{15}$$

where constants κ_1, κ_2 are the sliding gains that are selected such that the polynomial $\kappa_2 p + \kappa_1$ is Hurwitz and β_1, β_2 can be selected as

$$\begin{aligned} \beta_1 &= (1-\varepsilon, 1),\ \varepsilon \in (0,1) \\ \beta_2 &= \frac{2\beta_1}{1+\beta_1} \end{aligned} \tag{16}$$

An adaptive NFTSM control law is proposed as following

$$u = -M(x_1)\left(u_{eq} + u_c + u_{asw}\right) \tag{17}$$

$$u_{eq} = \mathrm{H}(x) + \kappa_2 |\dot{e}|^{\beta_2} sign(\dot{e}) + \kappa_1 |e|^{\beta_1} sign(e) - \ddot{x}_d \tag{18}$$

$$u_c = \hat{\Delta}(\hat{x},t) = \int \hat{\alpha}_3 sign(\tilde{x}_1) \tag{19}$$

$$u_{asw} = \left(\hat{K} + \mu\right) sign(s) \tag{20}$$

where μ is chosen as a small positive constant and $\hat{K}$ is the estimation of the desired switching gain, K^*, and is obtained by an adaptive law as follows

$$\dot{\hat{K}} = \begin{cases} k|s|, & \text{if } |s| \geq \eta \\ 0, & \text{else} \end{cases} \tag{21}$$

where k is an arbitrary positive constant and η is a sufficiently small constant.

Theorem 1: Consider the uncertain faulty robotic manipulator systems in (3), if the control input signal is designed as (17–20), then the system is stable and the tracking error converges to zero in finite-time.

Proof:

The derivative of the sliding surface is taken as following

$$\begin{aligned} \dot{s} &= \ddot{e} + \kappa_2|\dot{e}|^{\beta_2}\, sign(\dot{e}) + \kappa_1|e|^{\beta_1}\, sign(e) \\ &= \dot{x}_2 - \ddot{x}_d + \kappa_2|\dot{e}|^{\beta_2}\, sign(\dot{e}) + \kappa_1|e|^{\beta_1}\, sign(e) \\ &= -\ddot{x}_d + \mathrm{H}(x) + M^{-1}(x_1)u(t) + \Delta(x,t) + \kappa_2\left|\hat{\dot{e}}\right|^{\alpha_2} sign\left(\hat{\dot{e}}\right) + \kappa_1|e|^{\alpha_1}\, sign(e) \end{aligned} \tag{22}$$

Substituting the control law (17–20) into (22) yields

$$\dot{s} = -(\hat{K} + \mu)sign(s) + d(\tilde{x}, t) \tag{23}$$

where $d(\tilde{x}, t) = \Delta(x, t) - \hat{\Delta}(\hat{x}, t)$ is the estimation error, $|d(\tilde{x}, t)| \leq K^*$.

A Lyapunov function is selected as follows

$$V = \frac{1}{2}s^T s + \frac{1}{2k}\tilde{K}^T\tilde{K} \tag{24}$$

where $\tilde{K} = \hat{K} - K^*$ denotes the estimation error of switching gain.

Taking its derivative and substituting the derivative of the sliding surface from (23) yields

$$\dot{V} = s^T\dot{s} + \frac{1}{k}\tilde{K}^T\dot{\tilde{K}} \tag{25}$$

The time derivative of $\tilde{K}$ is taken as

$$\dot{\tilde{K}} = \dot{\hat{K}} - \dot{K}^* = \dot{\hat{K}} \tag{26}$$

Substituting the result in (23) and (26) into (25), yields

$$\dot{V} = s^T\left(-(\hat{K} + \mu)sign(s) + d(\tilde{x}, t)\right) + \frac{1}{k}\left(\hat{K} - K^*\right)\dot{\hat{K}} \tag{27}$$

Substituting the adaptive law (21) into (27), we can get

$$\begin{aligned}\dot{V} &= s^T\left(-(\hat{K}+\mu)sign(s)+d(\tilde{x},t)\right)+\frac{1}{k}(\hat{K}-K^*)k|s| \\ &= -(\hat{K}+\mu)|s|+d(\tilde{x},t)s+(\hat{K}-K^*)|s| \leq -\mu|s|<0,\ \forall s\neq 0\end{aligned} \tag{28}$$

Therefore, the Theorem 1 is proved.

In this paper, we assume that the tachometers are absent in the robotic systems. To keep the system works normally, we apply the obtained estimation of velocities from the neural TOSMO, which is introduced in Sect. 3; therefore, the adaptive NFTSM control law (17–20) become

$$u_{eq} = \mathrm{H}(x)+\kappa_2|\dot{e}|^{\beta_2}\,sign(\dot{e}) + \kappa_1|e|^{\beta_1}\,sign(e) - \ddot{x}_d \tag{29}$$

$$u_c = \hat{\Delta} = \int \dot{\alpha}_3 sign(\tilde{x}_1) \tag{30}$$

$$u_{asw} = (\hat{K}+\mu)sign(\hat{s}) \tag{31}$$

where $\hat{s} = \hat{e} + \int_0^t\left(\kappa_2|\hat{e}|^{\alpha_2}\,sign(\hat{e}) + \kappa_1|e|^{\alpha_1}\,sign(e)\right)dt$ with $\hat{e} = \hat{x}_2 - \dot{x}_d$.

5 Simulation Results

In this part, the usefulness of the controller-observer technique is demonstrated by performing simulation for a 2-link serial robotic manipulator with the dynamic model as follows

$$\ddot{\theta} = M^{-1}(\theta)\left[\tau - C\left(\theta,\dot{\theta}\right)\dot{\theta} - G(\theta) - F\left(\theta,\dot{\theta}\right) - \tau_d\right] + \Phi(t)$$

Inertia term

$$M(\theta) = \begin{bmatrix} m_1l_{c1}^2+m_2(l_1^2+l_{c2}^2+2l_1l_{c2}cos(\theta))+I_1+I_2 & m_1l_{c2}^2+m_2l_{c2}l_1cos(\theta)+I_2 \\ m_1l_{c2}^2+m_2l_{c2}l_1cos(\theta)+I_2 & m_2l_{c2}^2+I_2 \end{bmatrix}$$

Coriolis and centripetal term

$$C(\theta,\dot{\theta}) = \begin{bmatrix} -2m_2l_1l_{c2}sin(\theta)\dot{\theta}_1\dot{\theta}_2 - m_2l_1l_{c2}sin(\theta_2)\dot{\theta}_2^2 \\ m_2l_1l_{c2}sin(\theta_2)\dot{\theta}_1^2 \end{bmatrix}$$

Gravitational term

$$G(\theta) = \begin{bmatrix} m_1gl_{c1}\cos(\theta_1)+m_2g(l_1\cos(\theta_1)+l_{c2}\cos(\theta_1+\theta_2)) \\ m_2l_{c2}g\cos(\theta_1+\theta_2) \end{bmatrix}$$

The parameter values of robot are given as $m_1 = 1.5(kg), m_2 = 1.3(kg), l_1 = 1(m)$, $l_2 = 0.8(m), l_{c1} = 0.5(m), l_{c2} = 0.4(m), I_1 = 1(kgNm^2)$, and $I_2 = 0.8(kgNm^2)$.

The friction, disturbance, and fault are assumed as

$$F\left(\theta, \dot{\theta}\right) = \begin{bmatrix} 1.9\cos(2\dot{q}_1) \\ 0.53\sin(\dot{q}_2 + \pi/3) \end{bmatrix} \quad \tau_d = \begin{bmatrix} 1.2\sin(3q_1 + \pi/2) - \cos(t) \\ -1.14\cos(2q_2) + 0.5\sin(t) \end{bmatrix}$$

$$\Phi(t) = \varphi\left(t - T_f\right)\Psi(t) = \varphi\left(t - T_f\right)\begin{bmatrix} -12.5\sin(\pi t/7) \\ 13.7\cos(\pi t/5 + \pi/2) \end{bmatrix}$$

where $\varphi\left(t - T_f\right) = diag\left\{\varphi_1\left(t - T_f\right), \varphi_2\left(t - T_f\right), \ldots, \varphi_n\left(t - T_f\right)\right\}$ denotes the time profile of fault and T_f is occurrence time. With $\varphi_i\left(t - T_f\right) = \begin{cases} 0 & if\ t \le T_f \\ 1 - e^{-\varsigma_i(t - T_f)} & if\ t \ge T_f \end{cases}$, $\varsigma_i > 0$ is the evolution rate of fault.

In order to validate the usefulness of the proposed controller, a comparison with an adaptive SMC based on neural TOSMO is performed, which is designed based on the conventional sliding function as

$$\hat{s} = \hat{\dot{e}} + ce \tag{32}$$

where $e = x_1 - x_d$ and $\hat{\dot{e}} = \hat{x}_2 - \dot{x}_d$.

The control law is considered as

$$u = -M(x_1)\left(u_{eq} + u_c + u_{asw}\right) \tag{33}$$

$$u_{eq} = \mathrm{H}(x) + c\hat{\dot{e}} - \ddot{x}_d \tag{34}$$

$$u_c = \hat{\Delta} = \int \alpha_3 sign(\tilde{x}_1) \tag{35}$$

$$u_{asw} = \left(\hat{K} + \mu\right)sign(\hat{s}) \tag{36}$$

In the simulation, the parameters of the NFTSMC and the conventional SMC are chosen as $\kappa_1 = diag(15, 15)$, $\kappa_2 = diag(10, 10)$, $\beta_1 = 1/2$, $\beta_2 = 2/3$, $c = diag(3, 3)$. The parameters of the adaptive law are chosen as $k = 0.5$, $\eta = 0.05$. To estimate the observer gains, three RBFNs are employed, in which each hidden layer includes 20 neurons. The parameters of the neural network are chosen as $\sigma_i = 20$, $\lambda_i = 2.5$, $i = 1, 2, 3$.

For the simulation results, the obtained estimation of the neural TOSMO are compared with those of the TOSMO. The estimation of system velocity at each joint is shown in Fig. 1. The results shown that, the neural TOSMO can estimate system velocity with high accuracy. Although the requirement of the prior knowledge of the lumped uncertainties and faults is eliminated, the neural TOSMO can provide a little higher estimation accuracy compared with the results of the TOSMO. In term of the lumped uncertainties and faults, the estimation result is presented in Fig. 2. As we can see that the estimations accuracy of the two observers are approximately the same. Nevertheless, since it must take time for the estimation process of the RBFNs, the convergence time of the neural TOSMO is a little slower.

To show the effectiveness of the proposed adaptive NFTSMC control method, a comparison with an adaptive SMC, which is design based on the conventional sliding function, are performed. The position tracking error of the two control methods are presented in Fig. 3. Thanks to the superior control properties of the NFTSMC, the proposed control provides higher tracking accuracy and faster dynamic response compared with the adaptive SMC. The control input torque is shown in the Fig. 4. As we can see in the results, both two control methods provide control input torque with less chattering phenomenon. This result is the consequence of compensation of the lumped uncertainties and faults, thus the switching variable in the switching control law is now very small, which leads to the chattering phenomenon reducing.

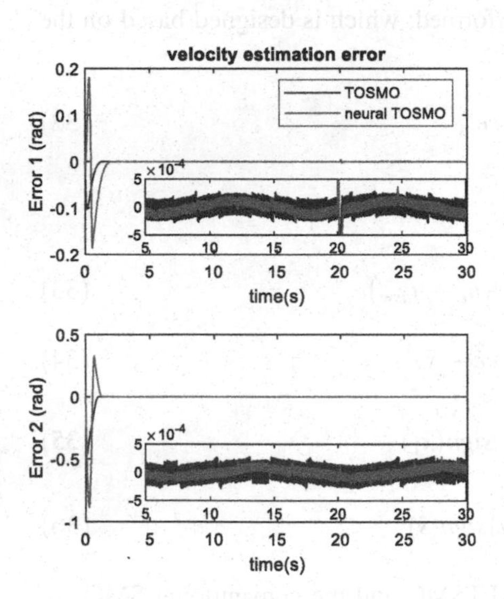

Fig. 1. Velocity estimation error at each joint.

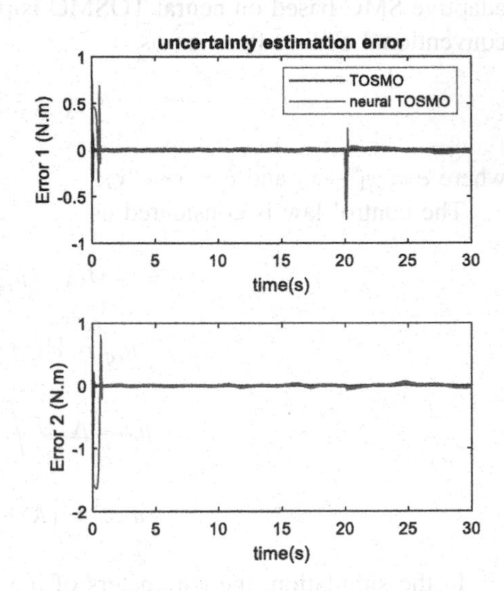

Fig. 2. Lumped uncertainties and faults estimation error at each joint.

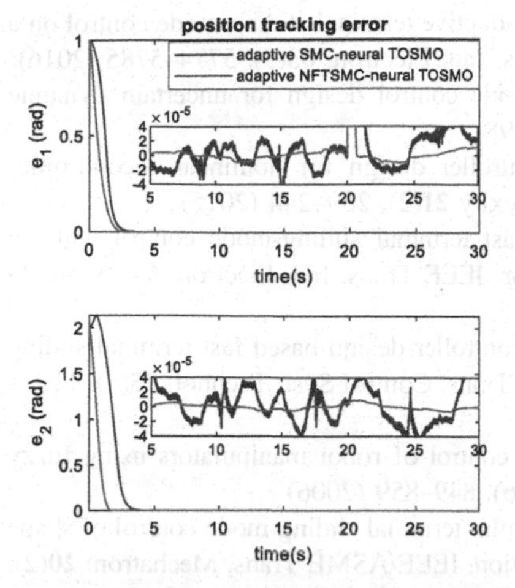

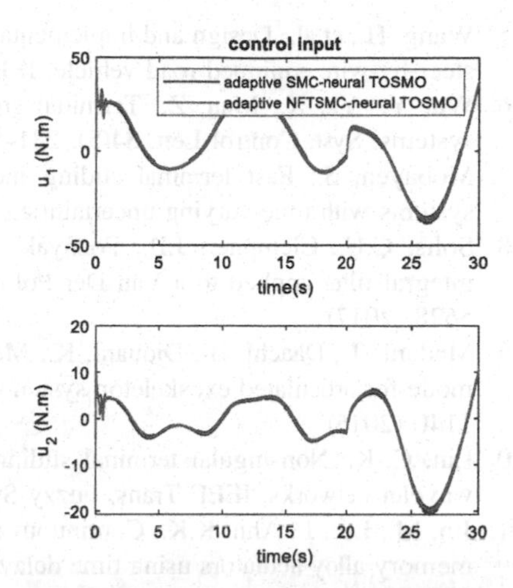

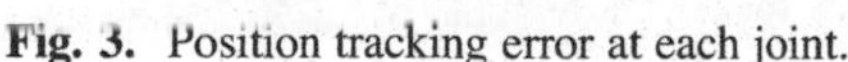

Fig. 3. Position tracking error at each joint.

Fig. 4. Control input signal at each joint.

6 Conclusions

This paper proposes a fault tolerant control approach using an adaptive NFTSMC based on a neural TOSMO for uncertainty faulty robotic manipulators with only position measurement. The neural TOSMO offers high precision of estimation and less chattering, in which the RBFN is performed to eliminate the requirement of the prior knowledge of the lumped uncertainties and the faults in designing of the observer. The adaptive NFTSMC improves the tracking error accuracy and reduces the chattering phenomenon. The system stability and finite-time convergence is guaranteed in theory of Lyapunov function. The computer simulation on a 2-link robot confirm the usefulness of the proposed controller-observer approach.

Acknowledgement. This work was supported by the Basic Science Research Program through the National Research Foundation of Korea (NRF) funded by the Ministry of Education under Grant 2019R1D1A3A03103528.

References

1. Utkin, V.I.: Sliding Modes in Control and Optimization. Springer, Heidelberg (2013)
2. Islam, S., Liu, X.P.: Robust sliding mode control for robot manipulators. IEEE Trans. Ind. Electron. **58**(6), 2444–2453 (2010)
3. Vo, A.T., Kang, H.-J., Nguyen, V.-C.: An output feedback tracking control based on neural sliding mode and high order sliding mode observer. In: 2017 10th International Conference on Human System Interactions (HSI), pp. 161–165 (2017)
4. Zhihong, M., Paplinski, A.P., Wu, H.R.: A robust MIMO terminal sliding mode control scheme for rigid robotic manipulators. IEEE Trans. Autom. Control **39**(12), 2464–2469 (1994)

5. Wang, H., et al.: Design and implementation of adaptive terminal sliding-mode control on a steer-by-wire equipped road vehicle. IEEE Trans. Ind. Electron. **63**(9), 5774–5785 (2016)
6. Wu, Y., Yu, X., Man, Z.: Terminal sliding mode control design for uncertain dynamic systems. Syst. Control Lett. **34**(5), 281–287 (1998)
7. Mobayen, S.: Fast terminal sliding mode controller design for nonlinear second-order systems with time-varying uncertainties. Complexity **21**(2), 239–244 (2015)
8. Solis, C.U., Clempner, J.B., Poznyak, A.S.: Fast terminal sliding-mode control with an integral filter applied to a Van Der Pol oscillator. IEEE Trans. Ind. Electron. **64**(7), 5622–5628 (2017)
9. Madani, T., Daachi, B., Djouani, K.: Modular-controller-design-based fast terminal sliding mode for articulated exoskeleton systems. IEEE Trans. Control Syst. Technol. **25**(3), 1133–1140 (2016)
10. Lin, C.-K.: Nonsingular terminal sliding mode control of robot manipulators using fuzzy wavelet networks. IEEE Trans. Fuzzy Syst. **14**(6), 849–859 (2006)
11. Jin, M., Lee, J., Ahn, K.K.: Continuous nonsingular terminal sliding-mode control of shape memory alloy actuators using time delay estimation. IEEE/ASME Trans. Mechatron. **20**(2), 899–909 (2014)
12. Eshghi, S., Varatharajoo, R.: Nonsingular terminal sliding mode control technique for attitude tracking problem of a small satellite with combined energy and attitude control system (CEACS). Aerospace Sci. Technol. **76**, 14–26 (2018)
13. Yang, L., Yang, J.: Nonsingular fast terminal sliding-mode control for nonlinear dynamical systems. Int. J. Robust Nonlinear Control **21**(16), 1865–1879 (2011)
14. Van, M.: An enhanced robust fault tolerant control based on an adaptive fuzzy PID-nonsingular fast terminal sliding mode control for uncertain nonlinear systems. IEEE/ASME Trans. Mechatron. **23**(3), 1362–1371 (2018)
15. Anh Tuan, V., Kang, H.-J.: A new finite time control solution for robotic manipulators based on nonsingular fast terminal sliding variables and the adaptive super-twisting scheme. J. Comput. Nonlinear Dyn. **14** (3), (2019)
16. Van, M., Mavrovouniotis, M., Ge, S.S.: An adaptive backstepping nonsingular fast terminal sliding mode control for robust fault tolerant control of robot manipulators. IEEE Trans. Syst. Man Cybern.: Syst. **49**(7), 1448–1458 (2018)
17. Van, M., Kang, H.-J., Suh, Y.-S., Shin, K.-S.: Output feedback tracking control of uncertain robot manipulators via higher-order sliding-mode observer and fuzzy compensator. J. Mech. Sci. Technol. **27**(8), 2487–2496 (2013). https://doi.org/10.1007/s12206-013-0636-3
18. Chalanga, A., Kamal, S., Fridman, L.M., Bandyopadhyay, B., Moreno, J.A.: Implementation of super-twisting control: super-twisting and higher order sliding-mode observer-based approaches. IEEE Trans. Ind. Electron. **63**(6), 3677–3685 (2016)
19. Hoang, D.-T., Kang, H.-J.: Fuzzy neural sliding mode control for robot manipulator. In: Huang, D.-S., Han, K., Hussain, A. (eds.) ICIC 2016. LNCS (LNAI), vol. 9773, pp. 541–550. Springer, Cham (2016). https://doi.org/10.1007/978-3-319-42297-8_50
20. Nguyen, V.-C., Vo, A.-T., Kang, H.-J.: Continuous PID sliding mode control based on neural third order sliding mode observer for robotic manipulators. In: Huang, D.-S., Huang, Z.-K., Hussain, A. (eds.) ICIC 2019. LNCS (LNAI), vol. 11645, pp. 167–178. Springer, Cham (2019). https://doi.org/10.1007/978-3-030-26766-7_16
21. Nguyen, V.-C., Vo, A.-T., Kang, H.-J.: A non-singular fast terminal sliding mode control based on third-order sliding mode observer for a class of second-order uncertain nonlinear systems and its application to robot manipulators. IEEE Access **8**, 78109–78120 (2020)

Fuzzy PID Controller for Adaptive Current Sharing of Energy Storage System in DC Microgrid

Duy-Long Nguyen[1] and Hong-Hee Lee[2(✉)]

[1] Graduate School of Electrical Engineering, University of Ulsan, Ulsan, South Korea

[2] School of Electrical Engineering, University of Ulsan, Ulsan, South Korea
hhlee@mail.ulsan.ac.kr

Abstract. In DC microgrid, conventional droop control is widely used to perform current sharing of distributed energy storage system. Although, this method has distributed and reliable characteristic, it cannot achieve accurate current sharing due to mismatched line resistances. Moreover, thermal effect causes these line resistances to change over long-term operation, which makes unequal current sharing more seriously. To overcome this problem, adaptive virtual resistance is applied in order to achieve accurate current sharing among energy storage system in DC microgrid. The virtual resistance is regulated by means of a Fuzzy PID controller in this paper. Although Fuzzy controller is widely used in literature, it has not been realized for achieving accurate current sharing in DC MG. Thanks to Fuzzy PID controller, the dynamic response becomes faster and the stability of the microgrid system are improved in comparison with the conventional PID controller. The proposed method is validated through the simulation using Matlab and Simulink.

Keywords: DC microgrid · Power sharing · Droop control · Fuzzy logic control

1 Introduction

In order to provide optimal and reliable operation for power systems, the concept of microgrids (MGs) has been introduced as an aggregated entity to integrate and utilize distributed generators (DGs) based on renewable energy sources (RES) such as solar PV, wind turbine, hydrogen power [1]. MGs can be distinguished as alternative current (AC) and direct current (DC) MGs [2]. In comparison with AC MG, DC MG can achieve higher efficiency by reducing number of ac/dc or dc/ac conversion stages due to direct interface with many types of RES and ESS [3]. Furthermore, in DC power system, there is no harmonic problem, reactive power sharing, or synchronization, which leads to simpler controller compared with AC MG [4]. Consequently, DC MG becomes more attractive in these days.

Figure 1 shows the typical configuration of a DC MG where all units including DGs, battery, and loads are connected to a common DC bus. To mitigate the power fluctuation caused by RES, battery unit (BU) system is employed. Due to distributed

D.-S. Huang and P. Premaratne (Eds.): ICIC 2020, LNAI 12465, pp. 213–223, 2020.
https://doi.org/10.1007/978-3-030-60796-8_18

connection with DC MG, these BUs may suffer from unequal current sharing which leads to overuse of a certain BU and reduce the life time of BU. This problem is more serious when line resistance is mismatch and changed in long time operation. To solve this problem, many methods have been proposed [5–7].

Belonging to decentralized method, droop control scheme is generally used to achieve cooperative control for various sources in MG [8]. In this method, the current sharing is determined by droop gain which is calculated from rated capacity of source. Although this method is simple and reliable, it is hard to achieve accurate current sharing due to missing information from other units [9].

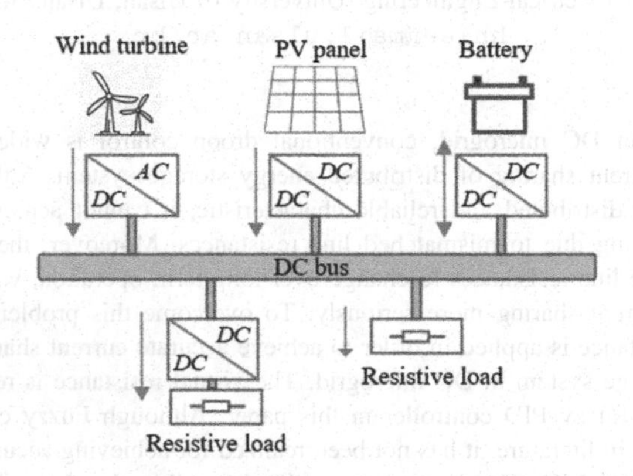

Fig. 1. Typical configuration of DC microgrid

To solve these problems, centralized control scheme with hierarchical structure is proposed [10]. In this approach, based on local information of DG such as output voltage, output current, a central controller calculates compensated value and transmits them to local controllers to achieve desired power management such as accurate current sharing, power balancing or operation mode change [11]. However, this method experiences serious problem such as single point of failure in which the central controller is broken and all system may be corrupt [12]. This disadvantage can be solved with the aid of distributed control scheme [13]. In this approach, local controllers exchange information with each other controllers through low bandwidth communication network to operate coordinately. Even though any communication link failure occurs, the system can maintain full functionality [11]. Therefore, distributed control is more reliable and robust in comparison with centralized control.

In this paper, based on the distributed control scheme, we propose an enhanced fuzzy proportional-integral-derivative (Fuzzy PID) controller to regulate virtual resistance adaptively in order to achieve accurate current sharing among energy storage system (ESS) in DC MG. Thanks to Fuzzy PID controller, virtual resistance is adjusted adaptively to compensate variation of line resistance, that leads to attain accurate current sharing of battery system in MG. Compared to the conventional PID controller, the proposed Fuzzy PID controller has better dynamic performance such as faster transient response, smaller overshoot, and guarantees system stability in spite of the load change. Although Fuzzy controller is widely used in literature, it has not been performed for achieving accurate current sharing in DC MG. The effectiveness of the proposed Fuzzy PID controller is verified by simulation in Matlab and Simulink.

2 Droop Control and Distributed Control Scheme

2.1 Droop Control Scheme

For simple analysis, DC MG with two BUs is considered as shown in Fig. 2. The droop controlled battery sources can be modeled as a DC voltage source V_{nom} with a virtual resistance R_v. Besides, there are line resistance R_{line1}, R_{line2} and load resistance R_{load}.

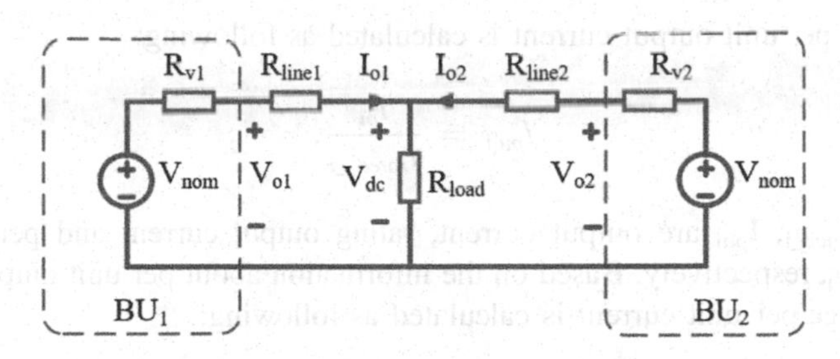

Fig. 2. Droop control scheme for DC MG.

From Fig. 2, we have

$$I_{o1} = \frac{V_{nom}R_{e2}}{R_{e1}R_{e2} + R_{e1}R_{load} + R_{e2}R_{load}}, \tag{1}$$

$$I_{o2} = \frac{V_{nom}R_{e1}}{R_{e1}R_{e2} + R_{e1}R_{load} + R_{e2}R_{load}}, \tag{2}$$

where I_{o1}, I_{o2} are output current of BU_1, BU_2, respectively, R_{e1} and R_{e2} are equivalent resistance with $R_{e1} = R_{v1} + R_{line1}$, $R_{e2} = R_{v2} + R_{line2}$.

From (1) and (2), to achieve accurate current sharing $I_{o1} = I_{o2}$, R_{e1} and R_{e2}should be equal ($R_{e1} = R_{e2}$):

$$R_{v1} + R_{line1} = R_{v2} + R_{line2} \tag{3}$$

When line resistances R_{line1} and R_{line2} are not constant due to thermal effect, conventional droop with constant R_{v1} and R_{v2} cannot achieve accurate current sharing.

2.2 Distributed Control Scheme

Based on the conventional droop control, the distributed droop control for accurate current sharing is shown in Fig. 3. To achieve accurate current sharing, droop gain is change adaptively to ensure the condition (3) regardless of line resistance change.

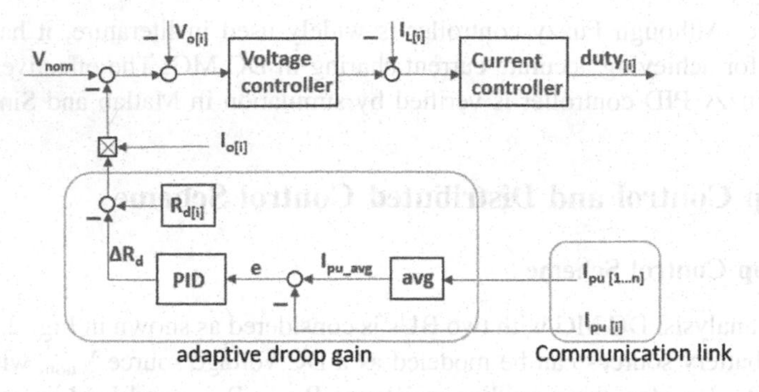

Fig. 3. Distributed control scheme for DC MG.

In Fig. 3, per unit output current is calculated as following:

$$I_{pu[i]} = \frac{I_{oi}}{I_{rated_i}}, \tag{4}$$

where, I_{oi}, I_{rated_i}, I_{pui} are output current, rating output current and per unit output current of BU_i, respectively. Based on the information about per unit output current of all BU, average per unit current is calculated as following:

$$I_{pu_avg} = \frac{\sum I_{pu[1\ldots n]}}{n} \tag{5}$$

Average per unit current I_{pu_avg} and per unit output current $I_{pu[i]}$ are used to feed PID controller to change adaptively droop gain $R_{d[i]}$ through $\Delta R_{d[i]}$. This is illustrated in Fig. 4. If output current is smaller than average current, output of PID is positive and droop gain decreases, and output current I_1 will increase to track average current. Reversely, if output current is larger than average current, output of PID becomes negative, droop gain increases, output current decreases, and average value is finally obtained.

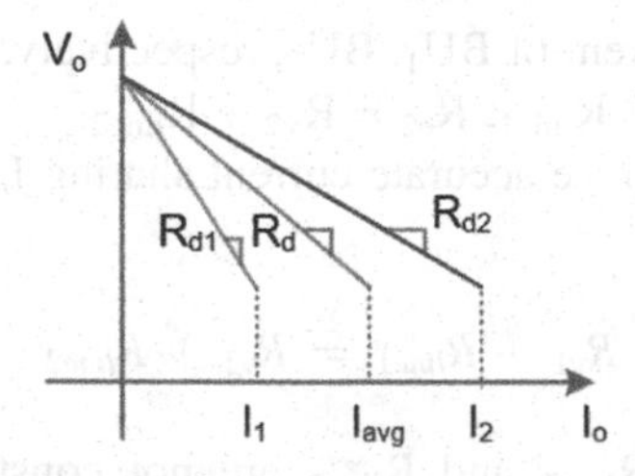

Fig. 4. Adaptive droop gain to track average value

3 Proposed Fuzzy PID Controller for Accurate Current Sharing in DC Microgrid

Figure 5 shows control system to explain a design process of Fuzzy PID controller, where r, y, e, u are reference value, output value, error, and control signal, respectively.

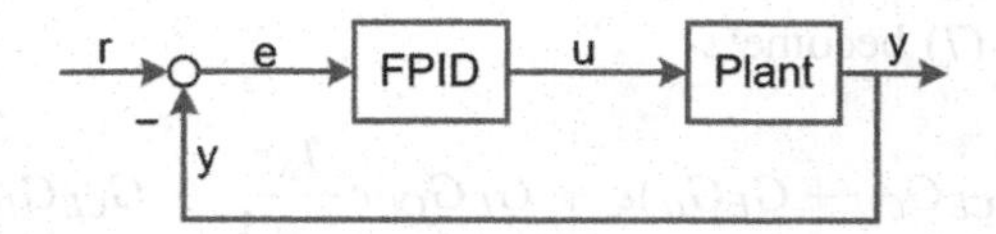

Fig. 5. Feedback control system with Fuzzy PID controller.

Structure of Fuzzy PID controller is shown in Fig. 6 [14, 15].

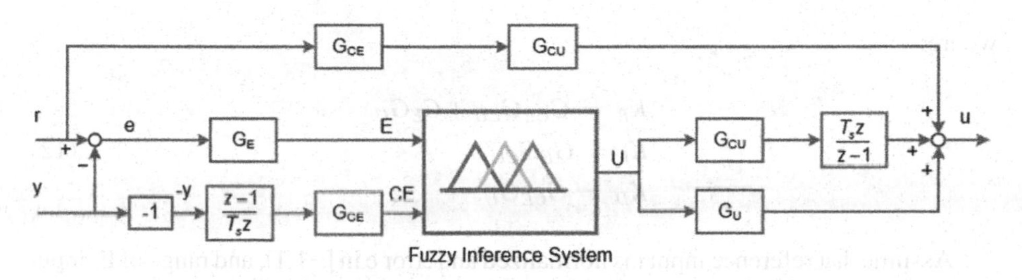

Fig. 6. The overall structure of Fuzzy PID controller.

In Fig. 6, G_{CE}, G_{CU}, G_E, G_U are gain factors. In order to design these gains, we assume that output of Fuzzy Inference System U depends on E and CE linearly:

$$U = E + CE \tag{6}$$

From (6) and Fig. 6, control input u can be expressed as

$$u = (G_{CE}G_{CU} + G_E G_U)e + G_E G_{CU} e \frac{T_S z}{z-1} - G_{CE} G_U y \frac{z-1}{T_S z} \tag{7}$$

Notice that:

$$\begin{aligned} -\frac{y(k) - y(k-1)}{T_S} &= \frac{(r - y(k)) - (r - y(k-1))}{T_S} \\ &= \frac{e(k) - e(k-1)}{T_S} \end{aligned} \tag{8}$$

We have

$$-y\frac{z-1}{T_S z} = e\frac{z-1}{T_S z} \tag{9}$$

Substituting (9), (7) become:

$$u = (G_{CE}G_{CU} + G_E G_U)e + G_E G_{CU} e \frac{T_S z}{z-1} + G_{CE} G_U e \frac{z-1}{T_S z} \tag{10}$$

Equation (10) is the form of PID controller:

$$u = K_P e + K_I e \frac{T_S z}{z-1} + K_D e \frac{z-1}{T_S z} \tag{11}$$

where:

$$\begin{aligned} K_P &= G_{CE}G_{CU} + G_E G_U \\ K_I &= G_E G_{CU} \\ K_D &= G_{CE} G_U \end{aligned} \tag{12}$$

Assume that reference input r is normalized and error e in [−1,1], and range of E (input of Fuzzy Inference System) is [−α,α] where α is a predetermined constant. Then, the factor G_E which scales e to E is α. From (12) and G_E, we can calculate the values of G_{CU}, G_U, and G_{CE}. However, if the gains in (12) are applied to the current sharing controller, the performance of Fuzzy PID controller is exactly same as conventional PID controller. To improve performance of Fuzzy PID controller, we have to change linear relation (6) to nonlinear relation by modifying membership function input and change rule [15].

Finally, proposed Fuzzy PID controller can be applied to achieve accurate current sharing in DC MG as in Fig. 7.

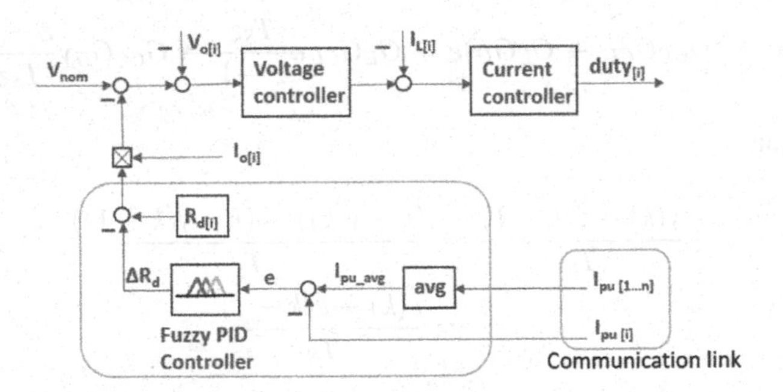

Fig. 7. Fuzzy PID controller for accurate current sharing.

4 Design Example

To verify the effectiveness of the proposed Fuzzy PID controller, DC MG with 3 BUs is evaluated by Matlab and Simulink. Each BU consists of a boost converter with the parameters given as: V_{in} = 100 V; V_{nom} = 200 V; $L_1 = L_2 = L_3$ = 1 mH; $C_{in} = C_{out}$ = 1000 uF; R_{line1} = 0.1 Ω; R_{line2} = 0.2 Ω; R_{line3} = 0.3 Ω; $I_{rated1} = I_{rated2} = I_{rated3}$ = 10A;fsw = 20 kHz; T_{sample} = 50 μs.

Because input current is normalized to per unit, the error e is within the range [−1, 1]. The input range of Fuzzy Inference System is chosen in [−20, 20], G_E becomes 20. After tuning PID controller for accurate current sharing, PID gains are selected as: $K_P = 10$, $Ki = 50$, $K_D = 0$. From (12), we obtain $G_U = 0$, $G_{CU} = 2.5$, $G_{CE} = 4$.

To satisfy linear Eq. (6), membership functions for Fuzzy input E and CE are chosen as triangle form with Negative (N), Zero (Z), Positive (P) as in Fig. 8.

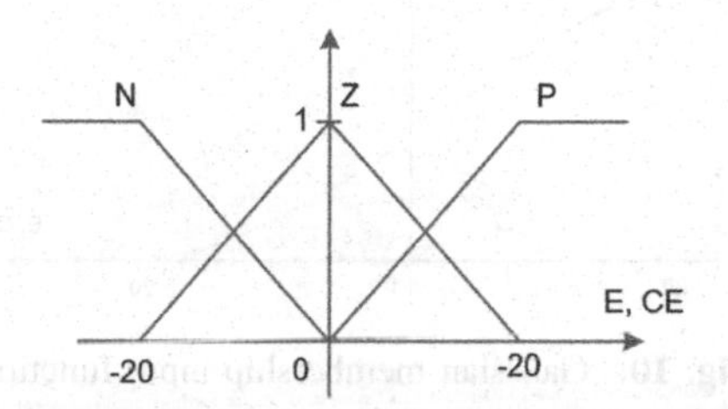

Fig. 8. Triangle membership input function

Membership functions for Fuzzy output U are chosen as following: Large Negative (LN) is −40, Medium Negative (MN) is −20, Zero (Z) is 0, Medium Positive (MP) is 20, Large Positive (LP) is 40.

The Fuzzy control rule are defined in Table 1.

Table 1. Fuzzy control rule for linear input output mapping

E	CE		
	N	Z	P
N	LN	MN	Z
Z	MN	Z	MP
P	Z	MP	LP

By these predefined membership function and Fuzzy control rule, linear input output mapping are received as shown in Fig. 9.

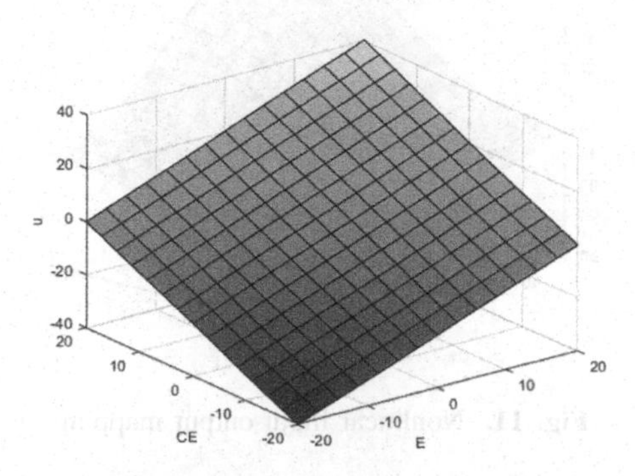

Fig. 9. Linear input output mapping.

To improve the performance of Fuzzy PID controller, the relationship between input and output is modified to nonlinear characteristic by modifying input membership function and Fuzzy control rule [15]. Membership function is chosen as Gaussian function as shown in Fig. 10.

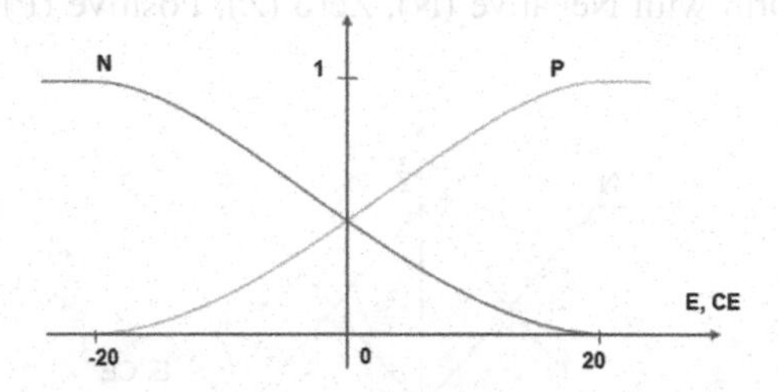

Fig. 10. Gaussian membership input function

Output membership functions are chosen as following: Negative (N) is −40, Zero (Z) is 0, and Positive (P) is 40. Then, Fuzzy control rule in Table 1 is changed to the rules in Table 2.

Table 2. Fuzzy Control rule for nonlinear input output mapping

E	CE	
	N	P
N	N	Z
P	Z	P

From these modified membership function and Fuzzy control rule, nonlinear input output relation is plotted as shown in Fig. 11.

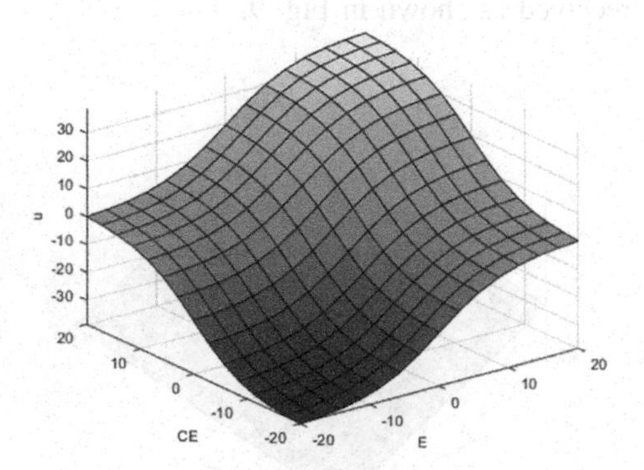

Fig. 11. Nonlinear input output mapping

In Fig. 11, when error or change of error is large, gains become higher than in case of linear input output mapping, and then nonlinear Fuzzy PID controller achieve faster performance comparing with linear Fuzzy PID controller. Since performance of linear Fuzzy PID controller is exactly the same as conventional PID controller, nonlinear Fuzzy PID controller is better performance comparing with conventional PID controller

5 Simulation Result

Figure 12 shows the dynamic performance of the system with the conventional PID and Fuzzy PID controllers. From 0 to 0.2 s, when the current sharing controller is not active, the per unit current of each BU is different due to the different line resistance. At 0.2 s, accurate current sharing scheme becomes active, and the Fuzzy PID shows better performance with faster response and shorter settling time comparing to the conventional PID controller.

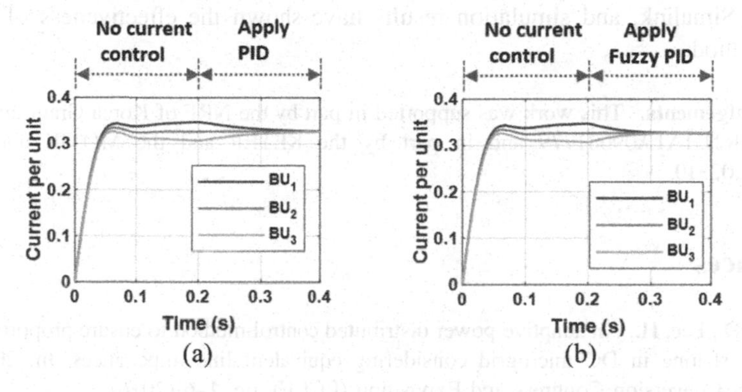

Fig. 12. Dynamic respone of system with (a) conventional PID controller (b) Fuzzy PID controller.

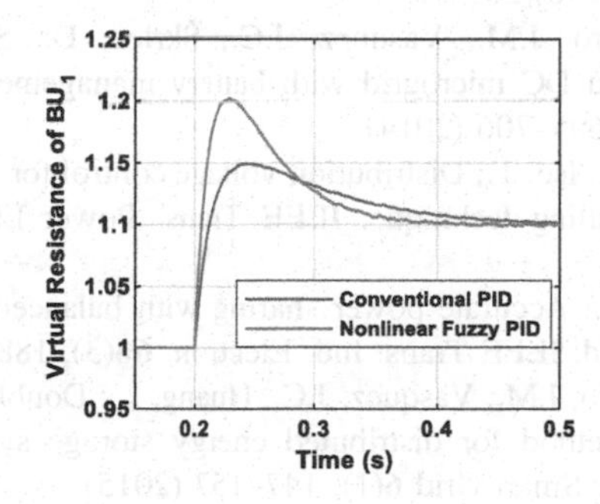

Fig. 13. Virtual resistance of BU1 with conventional PID and Nonlinear Fuzzy PID controller

Figure 13 shows the virtual resistance of BU_1 with conventional PID and Fuzzy PID controller. Before 0.2 s, current sharing controller is not active, the virtual resistance is 1 Ω. After 0.2 s, the current sharing controller is active and then the virtual resistance is regultated to share current equally. At the first time, when the error is large, Fuzzy PID controller has higher gain than traditional PID controller, reversely when error is small the Fuzzy PID has lower gain than PID controller. This explains that Fuzzy PID controller has faster and therefore better performance than conventional PID controller.

6 Conclusion

This paper introduced accurate current sharing method based on the fuzzy logic controller when battery unit system is used in DC microgrids. Thanks to Fuzzy PID controller, virtual resistance is regulated adaptively and currents among battery unit system are shared equally. The Fuzzy PID controller within on linear characteristic brings enhanced performance in comparison with the conventional PID controller. The validity of the proposed Fuzzy PID controller is evaluated through the simulation with Matlab& Simulink, and simulation results have shown the effectiveness of the proposed method.

Acknowledgements. This work was supported in part by the NRF of Korea Grant under Grant NRF-2018R1D1A1A09081779 and in part by the KETEP and the MOTIE under Grant 20194030202310.

References

1. Dam, D., Lee, H.: An adaptive power distributed control method to ensure proportional load power sharing in DC microgrid considering equivalent line impedances. In: 2016 IEEE Energy Conversion Congress and Exposition (ECCE), pp. 1–6 (2016)
2. Wang, P., Jin, C., Zhu, D., Tang, Y., Loh, P.C., Choo, F.H.: Distributed control for autonomous operation of a three-port AC/DC/DS hybrid microgrid. IEEE Trans. Ind. Electron. **62**(2), 1279–1290 (2015)
3. Dragičević, T., Guerrero, J.M., Vasquez, J.C., Škrlec, D.: Supervisory control of an adaptive-droop regulated DC microgrid with battery management capability. IEEE Trans. Power Electron. **29**(2), 695–706 (2014)
4. Kakigano, H., Miura, Y., Ise, T.: Distribution voltage control for DC microgrids using fuzzy control and gain-scheduling technique. IEEE Trans. Power Electron. **28**(5), 2246–2258 (2013)
5. Hoang, K.D., Lee, H.H.: Accurate power sharing with balanced battery state of charge in distributed DC microgrid. IEEE Trans. Ind. Electron. **66**(3), 1883–1893 (2019)
6. Lu, X., Sun, K., Guerrero, J.M., Vasquez, J.C., Huang, L.: Double-quadrant state-of-charge-based droop control method for distributed energy storage systems in autonomous DC microgrids. IEEE Trans. Smart Grid **6**(1), 147–157 (2015)
7. Lu, X., Sun, K., Guerrero, J.M., Vasquez, J.C., Huang, L.: State-of-charge balance using adaptive droop control for distributed energy storage systems in DC microgrid applications. IEEE Trans. Ind. Electron. **61**(6), 2804–2815 (2014)

8. Gu, Y., Xiang, X., Li, W., He, X.: Mode-adaptive decentralized control for renewable DC microgrid with enhanced reliability and flexibility. IEEE Trans. Power Electron. **29**(9), 5072–5080 (2014)
9. Nguyen, D.L., Lee, H.H.: Fuzzy PID controller for accurate power sharing in DC microgrid. In: Lecture Notes in Computer Science (including subseries Lecture Notes in Artificial Intelligence and Lecture Notes in Bioinformatics) (2019)
10. Jin, C., Wang, P., Xiao, J., Tang, Y., Choo, F.H.: Implementation of hierarchical control in DC microgrids. IEEE Trans. Ind. Electron. **61**(8), 4032–4042 (2014)
11. Dragičević, T., Lu, X., Vasquez, J.C., Guerrero, J.M.: DC microgrids—part I: a review of control strategies and stabilization techniques. IEEE Trans. Power Electron. **31**(7), 4876–4891 (2016)
12. Dam, D., Lee, H.: A power distributed control method for proportional load power sharing and bus voltage restoration in a DC microgrid. IEEE Trans. Ind. Appl. **54**(4), 3616–3625 (2018)
13. Anand, S., Fernandes, B.G., Guerrero, J.: Distributed control to ensure proportional load sharing and improve voltage regulation in low-voltage DC microgrids. IEEE Trans. Power Electron. **28**(4), 1900–1913 (2013)
14. Xu, J.-X., Hang, C.-C., Liu, C.: Parallel structure and tuning of a fuzzy PID controller. Automatica **36**(5), 673–684 (2000)
15. Jantzen, J.: Tuning Of Fuzzy PID Controllers (1998)

Deep Learning Based Fingerprints Reduction Approach for Visible Light-Based Indoor Positioning System

Huy Q. Tran and Cheolkeun Ha(✉)

Robotics and Mechatronics Lab, Ulsan University, Ulsan 44610, Republic of Korea
cheolkeun@gmail.com

Abstract. Received signal strength and fingerprints based indoor positioning algorithm has been commonly used in recent studies. The actual implementation of this method is, however, quite time-consuming and may not be possible in large spaces, mainly because a large number of fingerprints should be collected to maintain high positioning accuracy. In this work, we first propose the deep learning-based fingerprints reduction approach to reduce the data collection workload in the offline mode while ensuring low positioning error. After estimating the extra fingerprints using a deep learning model, these new fingerprints combine with the initially collected fingerprints to create the whole training dataset for the real estimation process. In the online mode, the final estimated location is determined using the combination of trilateration and k-nearest neighbors. The experiment results showed that mean positioning errors of 1.21 cm, 6.86 cm, and 7.51 cm are achieved in the center area, the edge area, and the corner area, respectively.

Keywords: Fingerprints · Visible light positioning · Deep learning · Trilateration · k-nearest neighbors

1 Introduction

Wireless indoor positioning system has recently become one of the popular research areas. The transmission signals now are more diverse, such as Wireless Local Area Network (WLAN), Radio Frequency Identification (RFID), Zigbee, Bluetooth, and visible light. Visible light-based indoor positioning has grown rapidly thanks to the utilization of available infrastructure as the demand for LED lights is increasing. In addition to the advantage of availability, low cost and high positioning accuracy are also the premise to attract the attention of researchers [1].

The implementation of indoor positioning systems (IPS), using any kind of signal, encounters some technical difficulties, especially for ensuring high positioning accuracy and stability. Indoor visible light positioning (VLP) is no exception. The development of various VLP-based positioning algorithms is, therefore, one of the most common approaches to improve the performance of IPS. For wireless IPS, commonly used positioning techniques include received signal strength (RSS) (i.e., in conjunction with trilateration, fingerprint, proximity), time of arrival/time difference of arrival

D.-S. Huang and P. Premaratne (Eds.): ICIC 2020, LNAI 12465, pp. 224–234, 2020.
https://doi.org/10.1007/978-3-030-60796-8_19

(i.e., in conjunction with trilateration, multilateration), angle of arrival [1, 2]. In reality, each method has its pros and cons. This depends on the applicable application environment and the accuracy requirements of each specific system. Among the mentioned methods, RSS based fingerprints technique is considered as the most popular technique because of its simplicity and acceptable positioning accuracy for indoor positioning applications [3].

In this work, by applying deep learning (DL) technique to estimate the extra fingerprints from the initial 49 fingerprint points, we greatly reduce the time-consuming measurement of sample points, which could be seen as one of the most serious limitations of fingerprint scheme in the offline mode. Additionally, we only employ DL to decrease the number of collected fingerprints without using it for estimating the current position of considering mobile objects in the online mode. This process helps the proposed system become more practical for real-time applications. After estimating the extra fingerprints, we combine the initially measured fingerprints with the later estimated fingerprints to create a full dataset for the next estimation in the online mode.

Not only first propose the DL technique to improve the performance of the data collection process in the offline mode, but we also combine the traditional trilateration method with k-nearest neighbors (kNN) to determine the estimated coordinate of the followed object. In the online mode, the first position was computed thanks to the trilateration method. Though the positioning accuracy is quite low, this method helps to find the estimated position in a very short time. After achieving the first estimated location, we identify points within the specific range based on a certain threshold. Then, we apply kNN to figure out the final estimated position. The use of the threshold helps to limit the number of fingerprints when calculating Euclidean distances with kNN algorithm, which can lead to a heavy computation work.

Based on the above two methods (i.e., DL in the offline mode, trilateration, and kNN in the online mode), we have reduced the workload in the online mode, while ensuring the real-time capability and the positioning accuracy in the online mode. The experiment results showed that the mean error in the central area, near the edge area, and near the corner area are 1.21 cm, 6.86 cm, and 7.51 cm, respectively.

The rest of this work is organized as follows: Sect. 2 presents the recent related works in the VLP field. The system model and the proposed method are thoroughly described in Sect. 3. In Sect. 4, experiment results and discussion are discussed.

2 Related Works

Artificial intelligence (AI) has been applied to several recent research on the positioning field, regardless of the type of signal, including WLAN, RFID, Zigbee, and Bluetooth [4–6]. In addition to applying for the above wireless signals, AI algorithms have been exploiting for VLP system [7–9]. In [7], the authors combined the fingerprint technique with extreme learning machine to simulate and experiment with their approach. Both the simulation and the experiment results showed that the system could achieve a positioning error of 2.11 cm and 3.65 cm, respectively. In previous works, fingerprints-based AI algorithms were developed for our VLP applications. In [8], we proposed a novel approach for the multipath reflection problem which produces a negative effect on the positioning accuracy out of the center area (i.e., the corners and edges). To enhance the positioning accuracy in these areas, we applied the kNN-RF model which create the positioning accuracy nearly five times better than the traditional kNN algorithm. In most of the cases, fingerprints-based AI algorithms were used to estimate the position of the mobile object thanks to the regression function of the supervised learning methods. In [9], some famous dual-function machine learning algorithms have been used for VLP. In this approach, we first apply the classification function to divide the entire floor into two specific areas (i.e., the center area and the edge area) based on the optical power distribution of the four light-emitting diode (LED) lights. Then, we utilize the regression function of each AI algorithm to determine the coordinate of the considering location. The simulation showed that, in the best case, the proposed approach achieved a 52.55% and 78.26% improvement in positioning accuracy and computation time, respectively.

RSS based fingerprints method is quite impractical, although this technique is preferred by many researchers because of its simplicity. The main problem is the data collection process. Especially when we need to collect a huge number of fingerprints within a very large space. As we all know, the more the number of fingerprints we gather, the more accurate the system achieves. Recently, researchers have tried to minimize the number of fingerprints, while maintaining positioning accuracy as high as possible. Based on this approach, Fakhrul Alam *et al.* [10] adopted a creative approach using a calibrated propagation model. In this way, the Lambertian order was regenerated relying on 12 offline measurements. The experiment results showed that a mean error of 2.7 cm could be achieved with the proposed technique and only 12 offline fingerprints. Meanwhile, a slight difference in the mean error could be seen when applying 187 offline datasets. The difference between the two cases in terms of the number of data is not trivial. In addition to obtaining nearly the same positioning accuracy, the proposed method helped reduce the tedious time to collect all the fingerprints and increase the practicality of the system. Similarly, Haiqi Zhang *et al.* [11] applied the Bayesian regularization based on the Levenberg-Marquardt algorithm to estimate the position of 100 unknown points in a space of 1.8 m × 1.8 m × 2.1 m. By using only 20 initial training points, the authors proved that the mean positioning accuracy is 3, 4 cm, 4.35 cm, and 4.58 cm for the diagonal set, the arbitrary set, and the even set, respectively. The above positive results encourage us to build an indoor visible light system thanks to RSS data and a relatively small number of fingerprints. Fingerprints reduction not only simplify the data collection process but also make the system more practical.

3 System Model and the Proposed Method

3.1 System Model

The LED system is always suspended from the ceiling or the wall. In addition to serving the lighting purpose, the light from LED is also modulated for signal transmission. The transmitted signal is absorbed by the photosensor or specific camera.

The photodetectors convert the optical signal into photocurrent and the output data of the photodetector is, therefore, illustrated as follows [12]:

$$y(t) = \mu_L(t) \otimes \mu_c(t) \otimes x(t) + N_G(t) \quad (1)$$

where $\mu_L(t)$, $\mu_c(t)$ is the impulse response of the LED and the channel, respectively; $x(t)$ is the input data; $N_G(t)$ is additive white Gaussian noise.

In this work, the proposed VLP system is described in detail in Fig. 1, where the transmitted signals are controlled by the Arduino Uno board and four drivers. The transmitted optical power is received by a photodiode. These signals are transmitted based on the time-division multiplexing technique [13], where each signal from each LED light is transmitted at every 1.33 ms intervals within an 11 ms cycle. The received data in the optical sensor are amplified and are displayed on the oscilloscope. Then all the received data are stored on the personal computer for later processing and estimation.

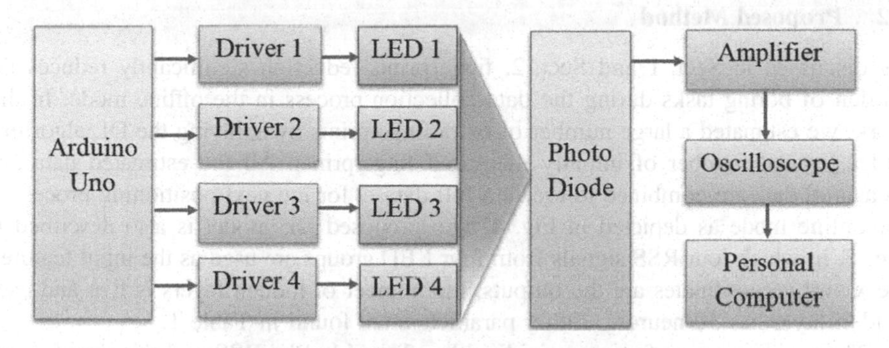

Fig. 1. The structure of the VLP system

Fig. 2. Real VLP system

To evaluate the performance of the proposed system, we designed an experimental space of 1.2 m × 1.2 m × 1.2 m, where four LED groups were hung on the aluminum frame as shown in Fig. 2. Other important components in the proposed system are the photosensor and the amplifier, which was depicted in Fig. 3. The specific parameters of the optical transmitter and receiver are described in detail in Table 1.

Fig. 3. Optical receiver

3.2 Proposed Method

As discussed in Sect. 1 and Sect. 2, fingerprints reduction significantly reduces the burden of boring tasks during the data collection process in the offline mode. In this work, we estimated a large number of extra fingerprints by applying the DL algorithm and a limited number of initially measured fingerprints. All the estimated data and measured data are combined to create a full dataset for the next positioning process in the online mode as depicted in Fig. 4. The proposed DL model is also described in Fig. 5, in which four RSS signals from four LED groups are used as the input features, the x and y coordinates are the outputs, the number of hidden layers is five and each hidden layer has 50 neurons. Other parameters are found in Table 1.

The positioning performance is directly affected by the RSS and the signal distribution of all the fingerprints. In Fig. 6 and Fig. 7, we depict the RSS distribution before and after applying the DL algorithm. The result in Fig. 7 showed that the optical power distribution after estimating by DL is perfectly balanced. This potential result is displayed in more detail in the positioning step in the online mode.

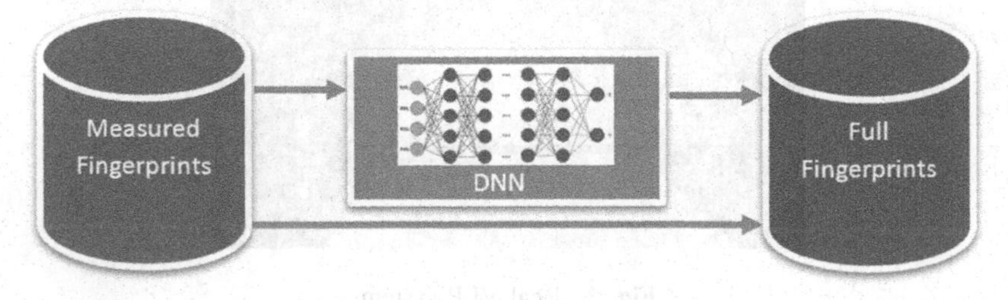

Fig. 4. Data collection in the offline mode

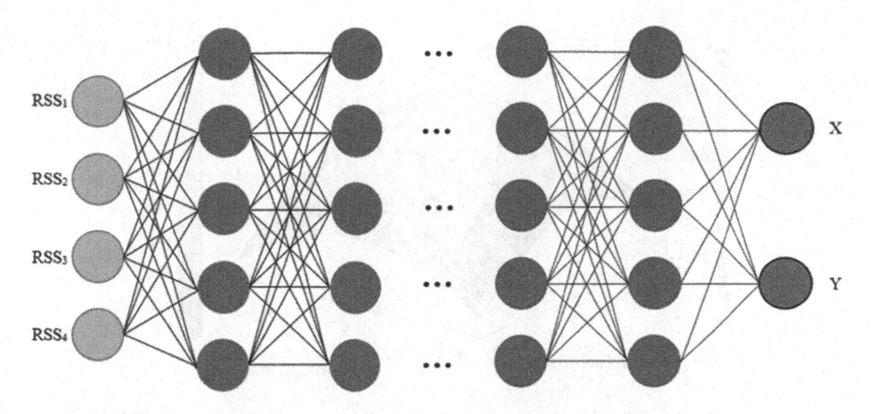

Fig. 5. The structure of the DNN model

Table 1. Parameters of the DL model

Parameters	Value
Training algorithm	Backpropagation
Activation function	Relu
Solver	Adam
Number of hidden layers	5
Hidden layer size	50
Alpha parameter	1e−5
Regularization	L2
Learning rate	1e−3
Percentage of samples for validation	10%
Percentage of samples for testing	20%

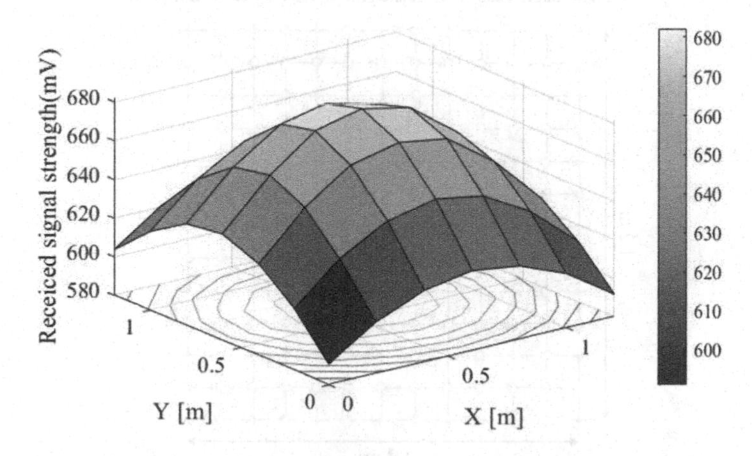

Fig. 6. Optical power distribution with real collection

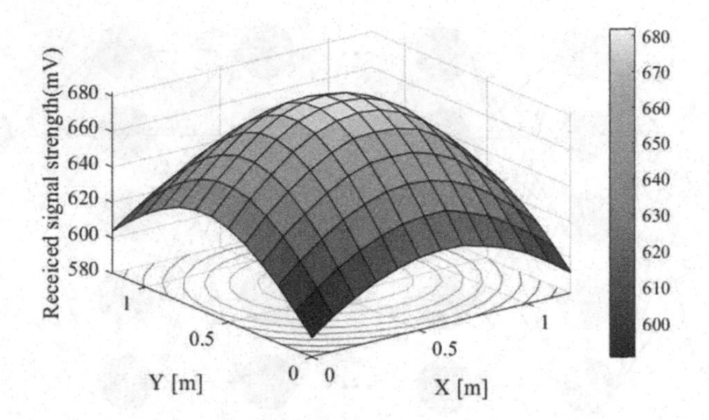

Fig. 7. Practical optical power distribution after estimating by DL

In the online mode, we first use the trilateration technique to determine the initial position of the considering point. Then, a threshold value is suggested thanks to the mean positioning error as shown in Fig. 8. Based on this threshold, we specify a fixed range of fingerprints that is used as the input data for the next step when we apply kNN to find the final estimated position. The reason for first determining the small range of input data instead of using the whole dataset is a time-consuming problem when applying kNN to the whole data. In this paper, a non-parametric technique, namely kNN, is used to estimate the position of the mobile objects. By finding the nearest neighbors which nearly have the same optical power intensity with the RSS of the considering position, the kNN algorithm is always one of the simplest approaches. The suitable number of nearest points depends on many factors, such as the type of signal, noises, experimental space, etc. In this scheme, we set this k value to 3 as the best value for the whole system.

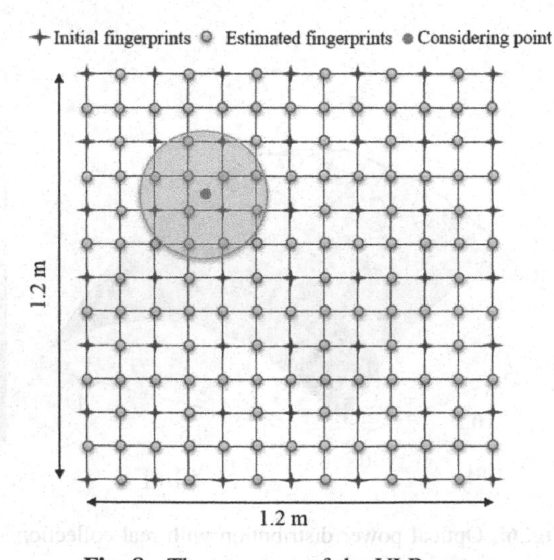

Fig. 8. The structure of the VLP system

The initial locations of the considering point are derived from (2):

$$\begin{cases} r_1^2 = (x - x_1)^2 + (y - y_1)^2 \\ r_2^2 = (x - x_2)^2 + (y - y_2)^2 \\ r_3^2 = (x - x_3)^2 + (y - y_3)^2 \end{cases} \quad (2)$$

where (x, y, z) represent the target coordinate, (x_i, y_i, z_i) represent the LED location, and (r_i, r_i, r_i) are the projection of the distance between each LED and the receiver on the floor.

After calculating the threshold value, we apply kNN to estimate the final location of the considering point as:

$$x_f = \frac{\sum_{i=1}^{k} x_i}{k}$$

$$y_f = \frac{\sum_{i=1}^{k} y_i}{k} \quad (3)$$

4 Experiment Results

To evaluate the effectiveness of the proposed algorithm to the positioning performance, we conducted experiments in three specific areas, including the central area, the edge area, and the corner area (see Fig. 9). In each area, we chose three random points to figure out its real locations. As shown in Fig. 10, the positioning accuracy is very satisfactory, in which the mean positioning errors near the center, the edge and corner

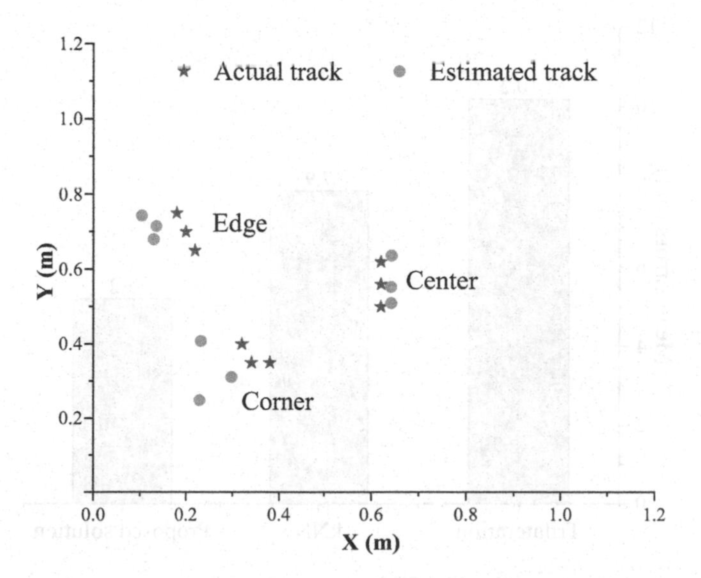

Fig. 9. Distribution of actual and estimated points

are 1.21 cm, 6.86 cm, and 7.51 cm, respectively. The difference in positioning accuracy between the central area and the remaining areas is quite reasonable because the central area is always where received the highest optical intensity from all the LEDs, and also where the impact of noise is lowest. To further demonstrate the prominence of the proposed approach, we provided a comparison between our solution and other popular methods including kNN and trilateration. As depicted in Fig. 11, our approach outperformed the others and achieved the best positioning error, 5.2 cm, though we used a limited number of fingerprints.

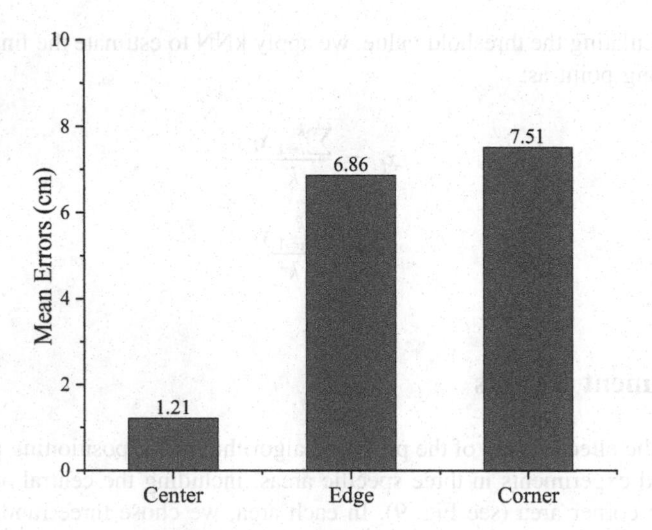

Fig. 10. Positioning errors in specific areas

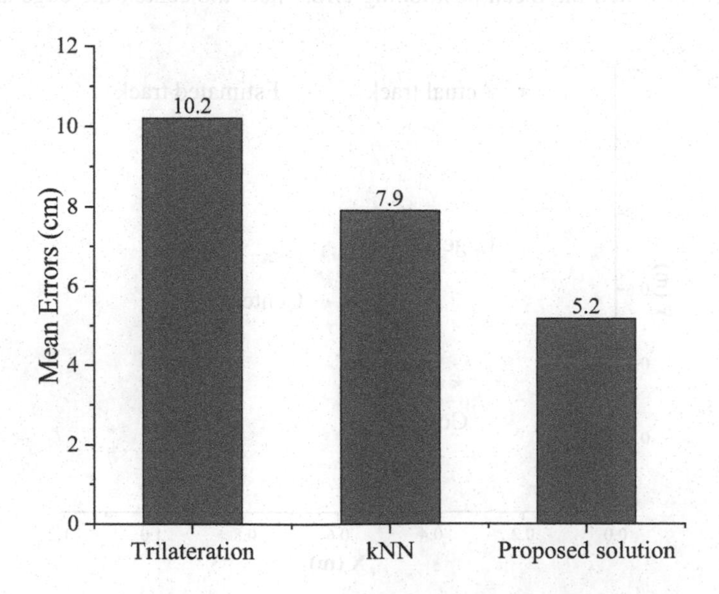

Fig. 11. Performance comparison

5 Conclusion

Collecting a large number of fingerprints is a tedious and impractical task in real applications with a large range of positioning. However, this method is simple to implement with relatively high accuracy. It is, therefore, still a potential and beloved approach. To overcome the mentioned problem, we first apply the DL algorithm to create extra fingerprints based on a limited number of previously collected points. The initially collected data and the DL-based estimated data are used for the process of determining the coordinates of the mobile object in the real phase thanks to the combination of trilateration and kNN. The estimated results proved that the proposed technique achieved positioning errors of 1.21 cm for the central area, 6.86 cm for the edge area, and 7.51 cm for the corner area.

This consequence set the stage for us to study more specialized algorithms to further reduce the number of fingerprints while maintaining or improving positioning accuracy at a higher level.

Acknowledgments. This work was supported by the KHNP (Korea Hydro & Nuclear Power Co., Ltd.) Research Fund Haeorum Alliance Nuclear Innovation Center of Ulsan University.

References

1. Liu, H., Darabi, H., Banerjee, P., Liu, J.: Survey of wireless indoor positioning techniques and systems. IEEE Trans. Syst. Man Cybern. Part C (Appl. Rev.) **37**(6), 1067–1080 (2007)
2. Zhuang, Y., et al.: A survey of positioning systems using visible LED lights. IEEE Commun. Surv. Tutor. **20**(3), 1963–1988 (2018). thirdquarter
3. Zhao, C., Zhang, H., Song, J.: Fingerprint and visible light communication based indoor positioning method. In: 2017 9th International Conference on Advanced Infocomm Technology (ICAIT), Chengdu, pp. 204–209 (2017)
4. Aikawa, S., Yamamoto, S., Morimoto, M.: WLAN finger print localization using deep learning. In: 2018 IEEE Asia-Pacific Conference on Antennas and Propagation (APCAP), Auckland, pp. 541–542 (2018)
5. Shen, L., Zhang, Q., Pang, J., Xu, H., Li, P.: PRDL: relative localization method of RFID tags via phase and RSSI based on deep learning. IEEE Access **7**, 20249–20261 (2019)
6. Ou, C., et al.: A ZigBee position technique for indoor localization based on proximity learning. In: 2017 IEEE International Conference on Mechatronics and Automation (ICMA), Takamatsu, pp. 875–880 (2017)
7. Chen, Y., Guan, W., Li, J., Song, H.: Indoor real-time 3-D visible light positioning system using fingerprinting and extreme learning machine. IEEE Access **8**, 13875–13886 (2020)
8. Tran, H.Q., Ha, C.: Fingerprint-based indoor positioning system using visible light communication—a novel method for multipath reflections. Electronics **8**, 63 (2019)
9. Tran, H.Q., Ha, C.: Improved visible light-based indoor positioning system using machine learning classification and regressio. Appl. Sci. **9**, 1048 (2019). https://doi.org/10.3390/app9061048
10. Alam, F., Chew, M.T., Wenge, T., Gupta, G.S.: An accurate visible light positioning system using regenerated fingerprint database based on calibrated propagation model. IEEE Trans. Instrum. Meas. **68**(8), 2714–2723 (2019)

11. Zhang, H., et al.: High-precision indoor visible light positioning using deep neural network based on the Bayesian regularization with sparse training point. IEEE Photonics J. **11**(3), 1–10 (2019). Art no. 7903310
12. Ghassemlooy, Z., Popoola, W., Rajbhandari, S.: Optical wireless communications, system and channel modeling with MATLAB. CRC Press, Boca Raton (2012). ISBN 9781439851883
13. Yasir, M., Ho, S., Vellambi, B.N.: Indoor positioning system using visible light and accelerometer. J. Lightwave Technol. **32**(19), 3306–3316 (2014). https://doi.org/10.1109/jlt.2014.2344772

Intelligent Data Analysis and Prediction

Anomaly Detection for Time Series Based on the Neural Networks Optimized by the Improved PSO Algorithm

Wenxiang Guo[1,2], Xiyu Liu[1,2(✉)], and Laisheng Xiang[2]

[1] Academy of Management Science, Shandong Normal University, Jinan, China
xyliu@sdnu.edu.cn
[2] College of Business, Shandong Normal University, Jinan, China

Abstract. Anomaly detection has been a popular research area for a long time due to its ubiquitous nature. Deep neural networks such as Long short-term memory (LSTM) and Convolutional Neural Networks (CNN) have been applied successfully to time series prediction, however, is not commonly used in time series anomaly detection and the performance of these algorithms depends heavily on their hyperparameter values. It is important to find an efficient method to get the optimal values. In this work, we combine LSTM with CNN, and propose a new framework (IPSO-CLSTM) to automatically optimize hyperparameters for time series anomaly detection tasks using improved particle swarm optimization (IPSO), which is capable of fast convergence when compared with others evolutionary approaches. Our experimental results show that IPSO-CLSTM can automatically find good hyperparameter values and achieve quality performance comparable to the state-of-the-art designs.

Keywords: Anomaly detection · Time series · Long short-term memory · Particle swarm optimization

1 Introduction

Anomaly detection refers to the problem of finding instances or patterns in data that deviate from normal behavior and is widely studied in many fields, such as fault detection in industrial systems. The reason that why anomaly detection important is because anomalies often indicate critical, and actionable information. For instance, anomaly detection can help realize predicted maintenance using sensor data from the manufacturing industry. However, anomaly detection is considered a hard problem [1]. In this paper, anomaly detection for time series is studied which presents its own unique challenges, this is mainly due to the issues inherent in time series analysis. In fact, time series anomaly detection is related to time series prediction, as anomalies are points or sequences which deviate from expected values.

A significant amount of work has been performed in the area of time series anomaly detection. Auto-regressive moving average, auto-regressive integrated moving average, vector auto-regression are some models for time series anomaly detection in statistic field. ARIMA model is very effective when there are clear trendor auto-correlation in

D.-S. Huang and P. Premaratne (Eds.): ICIC 2020, LNAI 12465, pp. 237–246, 2020.
https://doi.org/10.1007/978-3-030-60796-8_20

the data. However, the real situation is far more complex that is impacted by multiple factors including economic phenomena, media effects. One approach for anomaly detection is building a prediction model and use the difference between the predicted value and true value to calculate the anomaly score [2]. A variety of prediction models have been used. In [3] a simple window-based approach is used to calculate the median of recent values as the predicted value, and a threshold is to label the outliers. In [4] the authors build a one-step-ahead prediction model, data is considered as an anomaly point if it falls outsides a prediction interval computing using the standard deviation of the prediction errors.

Deep learning has become one of the most popular machine learning techniques. The ability of learning high-level representations related to the data is why deep learning popular. These representations are learned automatically from data with little or no need of manual feature engineering and domain expertise. For sequential and temporal data, LSTM has become the most widely used model for its ability of learning long-range patterns. However, performance of these algorithms depends heavily on their hyperparameter values and is difficult to choose, intelligent algorithms such as PSO is used here to automatically choose the parameters. In this paper, we propose a novel unsupervised anomaly detection algorithm (IPSO-CLSTM) which predicting the next timestamp using a window of time series(used as a context), this algorithm combines the LSTM and CNN as a feature extractor. The objective of IPSO-CLSTM is to robustly detect anomalies. The following are the main contributions of this paper.

1) The proposed IPSO-CLSTM is a novel unsupervised deep-learning based algorithm which is capable of detecting point anomalies, and discords in univariant as well as multivariate time series data.
2) This proposedcan search the vital parameters automatically via improved PSO which introduces a nonlinear decreasing assignment method to improve the inertia weight.
3) We evaluated the proposed anomaly detection algorithm on three well-known benchmarks, IPSO-CLSTM shows better performance compared to other anomaly detection algorithms.

The rest of the paper is organized as follows. Section 2 gives an overview of related works. Section 3 proposes the IPSO-CLSTM. Section 4 provides a detailed evaluation of the proposed algorithm on three benchmarks. Finally, Sect. 5 concludes the paper and sketches direction for possible future work.

2 Related Works

2.1 Long Short-Term Memory and Convolutional Neural Network

LSTM (long short-term memory) is a recurrent neural network architecture that has been adopted for time series forecasting and has become the mainstream structure of RNNs at present. LSTM addresses the problem of vanishing gradient by replacing the self-connected hidden units with memory blocks [5]. The architecture of an LSTM is shown in Fig. 1 and the mathematical form of LSTM is given below, the hidden state h_t given input x_t is computed as follows:

$$\begin{aligned} z_t &= \tanh(W^z x_t + R^z h_{t-1} + b^z) \quad (input) \\ i_t &= \sigma(W^i x_t + R^i h_{t-1} + b^i) \quad (input\ gate) \\ f_t &= \sigma(W^f x_t + R^f h_{t-1} + b^f) \quad (forget\ gate) \\ o_t &= \sigma(W^o x_t + R^o h_{t-1} + b^o) \quad (output\ gate) \\ s_t &= z_t \odot i_t + s_{t-1} \odot f_t \quad (cell\ gate) \\ h_t &= \tanh(s_t) \odot o_t \quad (output) \end{aligned} \tag{1}$$

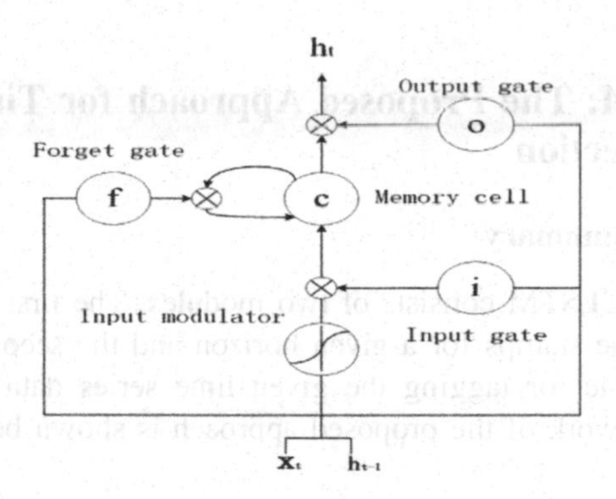

Fig. 1. The architecture of an LSTM.

CNN is a type of artificial neural network that has been widely used in computer vision (CV) and natural language processing (NLP). As the name indicates, the network employs a convolution operation. Normally, CNN consists of convolutional layers, pooling layers, and fully connected layers. The convolutional operation is normally denoted as asterisk:

$$s(t) = (x * w)(t) \tag{2}$$

The new function s can be described as a smoothed estimate or a weighted average of the function $x(\tau)$ at the time stamp t, where weighting is given by $w(-\tau)$ shifted by amount t. One dimensional convolutional is defined as:

$$s(t) = \sum_{\tau=-\infty}^{\infty} x(\tau)w(t-\tau) \tag{3}$$

2.2 Particle Swarm Optimization Algorithm

Particle swarm optimization (PSO) [6] was proposed by Kennedy and Eberhart in 1995 and is a kind of heuristic evolutionary algorithm based on swarm intelligence.

Each particle decides its path based on its previous best position (pbest) and global best position (gbest) among all the particles. The update equations for any particle i of the swarm in the i_{th} iteration are given below:

$$\begin{aligned} v_{ij}(t+1) = v_{ij}(t) + c_1 \times r_{1(ij)} \times (pbest_{ij}(t) - x_{ij}(t)) \\ + c_2 \times r_{2(ij)} \times (gbest_j(t) - x_{ij}(t)) \end{aligned} \tag{4}$$

$$x_{ij}(t+1) = x_{ij}(t) + v_{ij}(t+1) \tag{5}$$

where, $x_{ij}(t)$ and $v_{ij}(t)$ denotes the position and velocity of the particle in dimension, c_1 and c_2 are acceleration constants, $r_{1(ij)}$ and $r_{2(ij)}$ are the uniform random numbers in the range [0,1].

3 IPSO-CLSTM: The Proposed Approach for Time Series Anomaly Detection

3.1 Architecture Summary

The proposed IPSO-CLSTM consists of two modules. The first module, Time Series Predictor predicts time stamps for a given horizon and the second module, Anomaly Detector is responsible for tagging the given time series data points as normal or abnormal. The framework of the proposed approach is shown below (Fig. 2).

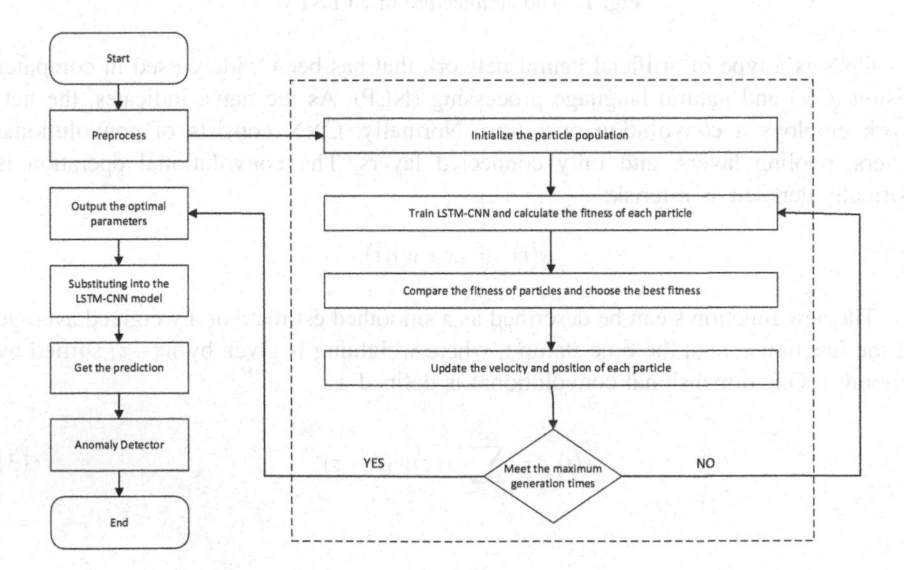

Fig. 2. Architecture summary of the proposed approach

3.2 Time Series Predictor:CLSTM Optimized by the Improved PSO

Long Short-Term Memory works well for time series prediction. However, LSTM lacks the ability to extract local contextual information. In order to improve the performance, and due to its parameter efficiency of CNN, 1D Conv is integrated into this paper. Furthermore, the performance while getting an optimal or near to optimal solution shows a significant difference due to different parameters value of CNN-LSTM, such as the important parameters look_back which means the time window, cell numbers of LSTM and 1D CNN. In this paper, there are single hidden layer LSTM, one convolution layer and a fully connected layer. PSO is a typical evolutionary algorithm, the inertia weight w is used to control the influence of the migration velocity on the current particle velocity, which is manifested as the performance of the PSO. Traditional inertia weight assignment is linear assignment, which as the number of iterations decreases linearly, the local search ability of the PSO will be worse. In this paper, the nonlinear decrement assignment method is adopted.

$$w = w_{\max} - (w_{\max} - w_{\min}) * \sqrt{\frac{i}{iter_max}} \tag{6}$$

where w_{max} and w_{min} represent the maximum and minimum inertia weight respectively, i is the current iteration number and iter_max is the maximum iteration number.

In the PSO optimization phrase, particles were initialized (look_back, node1, node2). Look_back represents the time window, node1 is the number of nodes contained in the hidden layer in the LSTM with a single hidden layer, nodel2 is the number of nodes included in the 1D Conv. In this article, the training data are input to the neural network for training, internal parameters of LSTM are trained using Adam. At each iteration, the system computes velocity and a new position for each particle. The root mean square error (RMSE) of the training data is selected as the individual fitness function, and the minimum fitness value is taken as the iterative target of the improved PSO. The improved PSO was used to find the best parameters to be optimized to determine the optimal prediction model.

3.3 Anomaly Detector

Once the former stage gives the prediction of the next timestamp x_{t+1}. This model performs the function of detecting the anomalies. The difference between the actual and predicted value is calculated. On new data, the log probability densities (PDs) of errors are calculated and used as anomaly scores, with lower values indicating a greater likelihood of the observation being an anomaly. Here a Threshold value is given to determine a time stamp anomaly or normal which is required in most of the anomaly detection algorithm.

4 Experiments

IPSO-CLSTM has been tested on 3 datasets described in Sect. 4.2. The optimal parameters of each dataset were found via the improved particle swarm optimization. The specific experimental environment is the Anaconda platform using Python, TensorFlow 1.12.0 and Keras 2.2.4.

4.1 Dataset Description

In this section, we chose three real-world datasets from different domains. These datasets have been used in previous works on anomaly detection.

a. *YAHOO webscope S5 datasets*

Yahoo in the United States provides a dataset consisting of 367 time series, each of which consists of almost 1,500 data points (https://research.yahoo.com/). The dataset contains four classes. Class A1 contains traffic data from actual web services [7], but classes A2, A3, and A4 contain synthetic anomaly data One example time series is shown in Fig. 3(a).

b. *Numenta Machine Temperature Dataset*

NAB is a benchmark for evaluating algorithms for anomaly detection in real-time applications. The machine temperature dataset is included in the NAB and contains temperature sensor readings of an internal component of a large industrial machine. There are four anomalies with known causes. The data is shown in Fig. 3(b), with anomalies indicated in red.

c. *NASA Space Shuttle Valve Dataset*

The time series in this data set is current measurements on a Marotta MPV- 41 series valve. These valves are used to control the flow of fuel on the space shuttle. In this data set, some subsequences are normal whereas a few are abnormal. Originally, each subsequence consists of 1, 000 data points. The data is shown in Fig. 3(c).

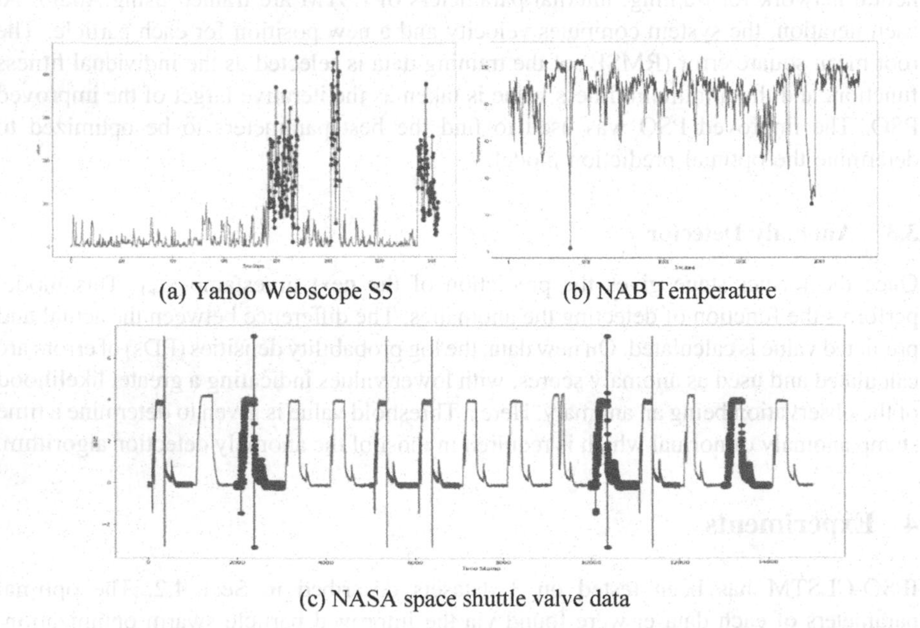

(a) Yahoo Webscope S5 (b) NAB Temperature

(c) NASA space shuttle valve data

Fig. 3. Example plots of three benchmark datasets used. (Color figure online)

4.2 Evaluation Metrics

In this paper, various models, including the proposed model are evaluated using precision, recall, and F-score. An error occurs if an abnormal instance is marked as a normal instance or vice versa. The former type of error is a false negative (FN) and the latter is false positive (FP). True positive (TP) and true negative (TN) defined similarly.

$$precision = \frac{TP}{TP + FP} \tag{7}$$

$$recall = \frac{TP}{TP + FN} \tag{8}$$

$$F - Score = 2 * \frac{precision * recall}{precision + recall} \tag{9}$$

4.3 Experiment Results

a. *Yahoo Webscope S5 datasets*

The performance of IPSO-CLSTM comparing to other detection algorithms is shown in Table 1 and the example results for the YahooWebscope S5 dataset are shown in Fig. 5. IPSO-CLSTM detects five out of five anomalies in the example time series. We can also find that the algorithm proposed possesses better performance comparing to other algorithms in the whole Yahoo benchmark. In the process of optimizing the LSTM-CNN by PSO, optimal parameter of the model is set as node1 = 111, node2 = 57, look_back = 29. The threshold of −12.5 was necessary to detect the anomalies. The change of the fitness value shown in Fig. 4 indicates the improved PSO has a faster convergence rate.

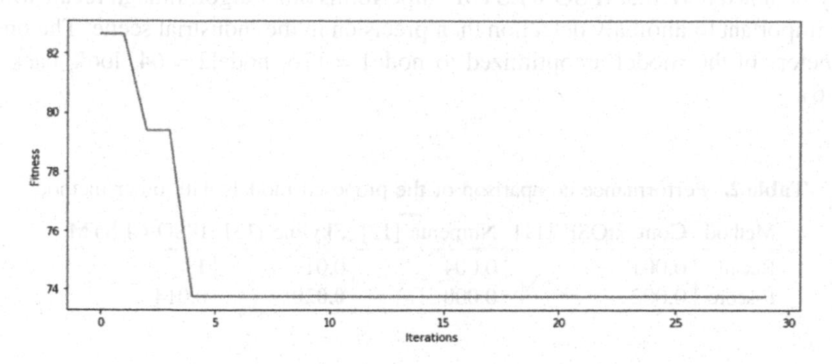

Fig. 4. The change of fitness value

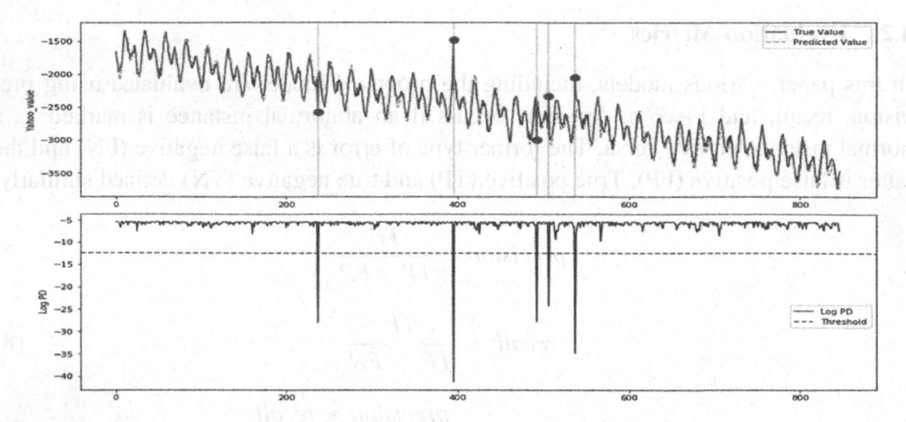

Fig. 5. Example result on Yahoo dataset. True anomalies are highlighted by red markers. Shaded in peach denotes detections made by the IPSO-CLSTM algorithm. (Color figure online)

Table 1. Performance comparison of the proposed models with other methods

Sub-benchmark	Yahoo EGADS [9]	Twitter Anomaly Detection [10]	LSTM	IPSO-CLSTM
A1	0.47	0.48	0.44	**0.60**
A2	0.58	0	**0.97**	0.90
A3	0.48	0.27	0.72	**0.87**
A4	0.29	0.33	0.59	**0.68**

b. *Numenta Machine Temperature Dataset*

Tables 2 shows the experimental results including the recall and F-score of a wide range of algorithms on the machine temperature dataset, the proposed method performed well. It may be noted here that IPSO-CLSTM outperforms other algorithms in recall which is more important to anomaly detection then precision in the industrial scene. The optimal parameters of the model areoptimized to node1 = 116, nodel2 = 64, look_back = 16 (Fig. 6).

Table 2. Performance comparison of the proposed models with other methods

Method	ContextOSE [11]	Numenta [12]	Skyline [13]	IPSO-CLSTM
Recall	0.001	0.004	0.01	**1**
F-score	0.002	0.008	**0.020**	0.014

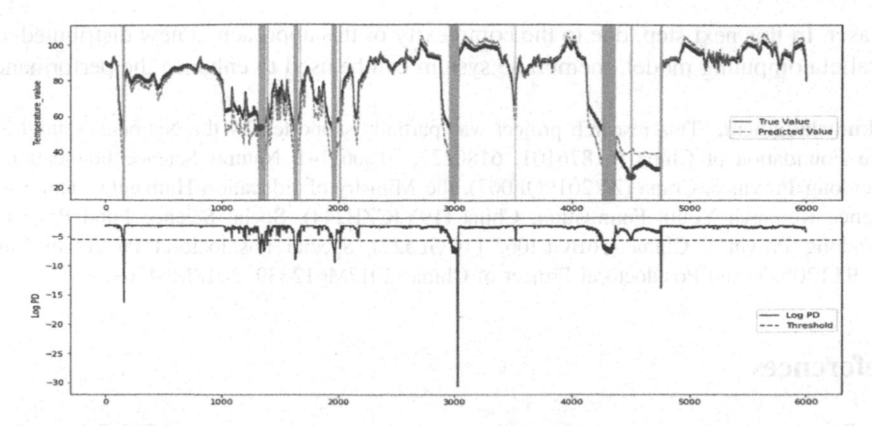

Fig. 6. Test set result on NAB dataset. Anomalies are highlighted by red markers. Shaded in peach denotes detections made by the IPSO-CLSTM algorithm. (Color figure online)

c. *NASA space shuttle valve data set*

In the previous experiments, we have shown that IPSO-CLSTM has the capability of detecting point anomalies in time series data. In this section, we show that IPSO-CLSTM is also applicable to time series discord detection. Time series discords are subsequences of a longer time series, which are different from the rest of the subsequences [8]. In this dataset, each subsequence consists of 1, 000 data points, experiment result below shows that IPSO-CLSTM is also applicable to detect time series discords (Fig. 7).

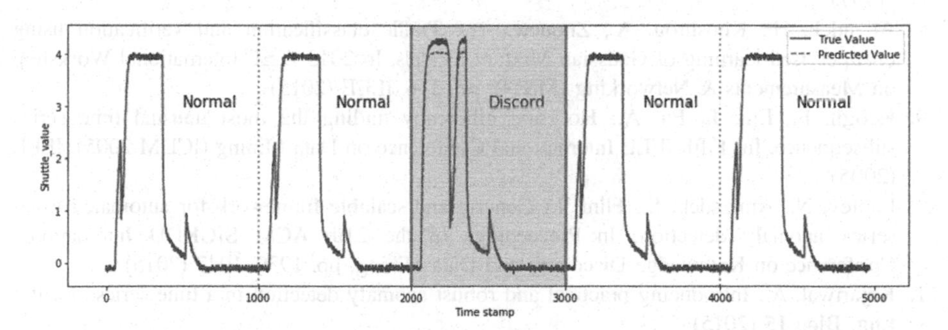

Fig. 7. Shaded in peach denotes detections made by the IPSO-CLSTM.

5 Conclusion

In this paper, we presented a deep-learning based anomaly detection approach for time series data. Not only the point anomaly but also the time series discord can be detected. In this algorithm, improved particle swarm optimization is used to automatically choose the optimal parameters. Experiment results show that the architecture possesses better performance than other anomaly detection algorithms in different benchmark dataset. In this next step, due to the complexity of this approach, a new distributed and parallel computing model, membrane system can be used to enhance the performance.

Acknowledgment. This research project was partially supported by the National Natural Science Foundation of China (61876101, 6180223, 61806114), Natural Science Foundation of Shandong Province, China (ZR2019QF007). the Ministry of Education Humanities and Social Science Research Youth Foundation, China (19YJCZH244), Social Science Fund Project of Shandong Province, China (16BGLJ06, 11CGLJ22), Special Postdoctoral Project of China (2019T120607) and Postdoctoral Project of China (2017M612339, 2018M642695).

References

1. Chandola, V., Banerjee, A., Kumar, V.: Anomaly detection: a survey. ACM Comput. Surv. (CSUR) **41**(3), 1–58 (2009)
2. Campos, G.O., et al.: On the evaluation of unsupervised outlier detection: measures, datasets, and an empirical study. Data Min. Knowl. Disc. **30**(4), 891–927 (2016). https://doi.org/10.1007/s10618-015-0444-8
3. Gupta, M., Gao, J., Aggarwal, C.C., Han, J.: Outlier detection for temporal data: a survey. IEEE Trans. Knowl. Data Eng. **26**(9), 2250–2267 (2013)
4. Basu, S., Meckesheimer, M.: Automatic outlier detection for time series: an application to sensor data. Knowl. Inf. Syst. **11**(2), 137–154 (2007)
5. Hill, D.J., Minsker, B.S.: Anomaly detection in streaming environmental sensor data: a data-driven modeling approach. Environ. Model Softw. **25**(9), 1014–1022 (2010)
6. Kennedy, J., Eberhart, R.: Particle swarm optimization. In: Proceedings of ICNN 1995-International Conference on Neural Networks, pp. 1942–1948. IEEE (1995)
7. Hochreiter, S., Schmidhuber, J.: Long short-term memory. Neural Comput. **9**(8), 1735–1780 (1997)
8. Alizadeh, H., Khoshrou, A., Zuquete., A.: Traffic classification and verification using unsupervised learning of Gaussian Mixture Models. In 2015 IEEE International Workshop on Measurements & Networking (M&N), pp. 1–6. IEEE (2015)
9. Keogh, E., Lin, J., Fu, A.: Hot sax: efficiently finding the most unusual time series subsequence. In: Fifth IEEE International Conference on Data Mining (ICDM 2005). IEEE (2005)
10. Laptev, N., Amizadeh, S., Flint, I.: Generic and scalable framework for automated time-series anomaly detection. In: Proceedings of the 21th ACM SIGKDD International Conference on Knowledge Discovery and Data Mining, pp. 1939–1947 (2015)
11. Kejariwal, A.: Introducing practical and robust anomaly detection in a time series. Twitter Eng. Blog **15** (2015)
12. Contextual Anomaly Detector (2015). https://github.com/smirmik/CAD
13. Lavin, A., Ahmad, S.: Evaluating Real-time anomaly detection algorithms–the numenta anomaly benchmark. In: 2015 IEEE 14th International Conference on Machine Learning and Applications (ICMLA), pp. 38–44. IEEE (2015)
14. Skyline (2013). https://github.com/etsy/skyline

An Integration Framework for Liver Cancer Subtype Classification and Survival Prediction Based on Multi-omics Data

Zhonglie Wang[1,2], Rui Yan[1,2], Jie Liu[3], Yudong Liu[2], Fei Ren[2](✉), Chunhou Zheng[1](✉), and Fa Zhang[2](✉)

[1] College of Computer Science and Technology, Anhui University, Hefei, China
zhengch99@126.com

[2] High Performance Computer Research Center, Institute of Computing Technology, Chinese Academy of Sciences, Beijing, China
{renfei,zhangfa}@ict.ac.cn

[3] Institutes of Physical Science and Information Technology, Anhui University, Hefei, China

Abstract. Accurate prediction is helpful to the treatment of liver cancer. In this paper, we propose a method based on a combination of deep learning and network fusion to predict the survival subtype of liver cancer, of which Univariate Cox-PH regression model was used twice. We integrated RNA sequencing, miRNA sequencing, DNA methylation data and clinical data of liver cancer from TCGA to infer two survival subtypes. We then also constructed an XGBoost supervised classification model to predict the survival subtype of the new sample. Experimental results show that our model gives two subgroups with significant survival differences and Concordance index. We also use two additional confirmation cohorts downloaded from the GEO database to verify our multi-omics model. We found highly expressed stemness marker genes *CD24*, *KRT19* and *EPCAM* and the tumor marker gene *BIRC5* in two survival subgroups. Our method has great clinical significance for the prediction of HCC prognosis.

Keywords: Deep learning · Network fusion · Survival subtype · Supervised classification model · Multi-omics data

1 Introduction

Liver cancer is the second leading cause of death worldwide and is one of the malignant tumors [1] that seriously threatens human health. Over the past 40 years, the incidence and mortality of liver cancer is the fastest growing of all cancers in the United States [2]. Hepatocellular carcinoma (HCC) is the most common type of liver cancer, and its occurrence has a great relationship with hepatitis C virus, hepatitis B virus and nonalcoholic teatohepatitis [3].

If we can set up the classification standard of HCC and carry out more accurate prognosis treatment and management for different patients, the survival of patients will be improved significantly. So far there has been a lot of research on liver cancer to

D.-S. Huang and P. Premaratne (Eds.): ICIC 2020, LNAI 12465, pp. 247–257, 2020.
https://doi.org/10.1007/978-3-030-60796-8_21

determine the molecular subtype of liver cancer [4, 5]. In 2017, Zhu et al. proposed an effective survival prediction method based on pathological images, named Whole Slide Histopathological Images Survival Analysis framework (WSISA), and applied it to the survival prediction of glioma and non-small cell lung cancer, the research results confirmed that this method could significantly improve the accuracy of prediction [6]. In 2018, Sun, D. et al. proposed a learning-based calculation method by combining pathological images and genomic biomarkers to predict the survival period based on deep learning, which exceeded the prediction accuracy of human experts using current clinical standards, and provided an innovative method for objective, accurate and comprehensive prediction of patient results [7]. In 2019, Dong R. et al. proposed a machine learning method based on methylation data of liver cancer to predict the survival of patients with liver cancer, the author use a three-category method to predict overall survival of patients with HCC [8]. Also, based on multi-omics data, various methods has been developed to identify molecular subtypes of liver cancer, for example, Chandhary et al. combined RNAseq data, miRNA data, methylation data, and clinical survival data of total 360 liver cancer patients to predict the prognosis of liver cancer [9]. To our best knowledge, few researcher considers the survival status of patients in study of molecular subtype. The survivals have great clinical significance for study of molecular subtype indeed, and large difference in survival often have a big impact on molecular subtype. Therefore, combining the survival data with the multi-omics data in the research of HCC will improve the prediction accuracy of molecular subtype.

According to the above description, in this paper, we proposed a framework based on deep learning and network fusion on the multi-omics data and survival data of HCC. First, we scored each feature on these multi-omics data based on the Univariate Cox-PH model [10] and then selected features with P value < 0.05. Secondly, we use a deep autoencoder to reduce the dimension of the multi-omics data of HCC and extract features, then used again the Univariate Cox-PH model to score the features extracted by autoencoder output and select the features with P value < 0.05. These selected features and extracted features previously were stacked up to prevent from losing some important information by autoencoder. Finally, the multi-omics data were fused through a similar network based on the radial basis function kernel to form a new data matrix. Experimental results showed that our model give two subgroups with significant survival differences and Concordance index, which is very helpful for the clinical prognosis and treatment of HCC.

2 Methods

2.1 Experimental Design

In terms of experimental design, our experiment is mainly divided into three steps. The first step is the data processing. The second step uses deep learning and network fusion to predict the survival of TCGA data. The third step is to build an XGBoost classifier based on the survival subtypes obtained in the second step, use the training data to train the classification model, and then use the confirmation cohorts processed on GEO to verify the model (Fig. 1).

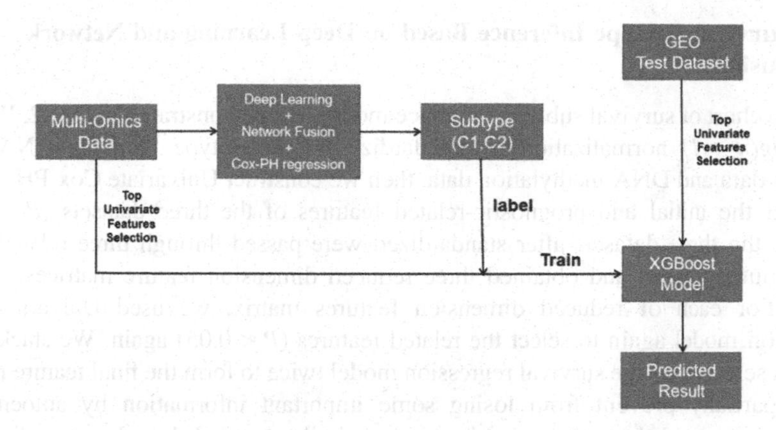

Fig. 1. General flow chart.

2.2 Multi-omics Data Process

In this paper, we first downloaded RNA data, miRNA data and methylation data and clinical data of 364 HCC patients from the TCGA database. And we also mined GSE14520 and GSE31384 from the GEO database as confirmation cohorts. For the RNA data and miRNA data of 364 HCC patients, we preprocessed them using *log (n + 1)* normalization to eliminate the effects caused by singular samples data. For methylation data, we mapped the CpG islands to within 1500 bp before the transcription start site of the gene, then their methylation values were averaged. For data processing, if more than 20% of patients have zero values, we delete the relate biological feature, if more than 20% of missing features in one sample, we deleted the corresponding sample [9]. Finally, we filled the missing values using the software *R*.

In addition, we also downloaded the GSE14520 (S = 222) and GSE31384 (S = 166) datasets and corresponding clinical data from the GEO database as confirmation cohorts. We use the same data processing method of TCGA datasets to process the above two Confirmation cohorts (as shown in Table 1).

Table 1. Statistics of training datasets from TCGA and Confirmation cohorts from GEO database.

Datasets	Data type	Samples	Features
Train	RNA-seq	364	21,617
	miRNA	364	415
	Methylation	364	27,304
Confirmation cohort	*GSE14520*	222	10,787
	GSE31384	166	169

2.3 Survival Subtype Inference Based on Deep Learning and Network Fusion

The flowchart of survival subtype inference model was demonstrated in Fig. 2. We first used *log(n + 1)* normalization to standardize the three type datasets, RNA data, miRNA data and DNA methylation data, then we construct Univariate Cox-PH models to select the initial and prognostic-related features of the three datasets ($P < 0.05$). Second, the three datasets after standardized were passed through three related regularized autoencoder, and obtained three reduced dimension feature matrices, respectively. For each of reduced dimension features matrix, we used Univariate Cox regression model again to select the related features ($P < 0.05$) again. We stacked the features selected by the survival regression model twice to form the final feature matrix, which partially prevent from losing some important information by autoencoder. Finally, we putted these three matrices into a similar network based on a radial basis function kernel to fuse the three sets of omics data to form a single matrix, and used the spectral clustering approach [11] to obtain the subtype cluster. In this way, we can obtain a label for the survival risk group of each patient.

To integrate the multi-omics data, we presented a modified similar network fusion (SNF) based on radial basis function kernel (RBF). In 2014, Wang et al. proposed a similar network fusion (SNF) method for integrating multi-omics data [12]. To handle examples when the relationship between class labels and feature is nonlinear, we modified the above-mentioned first step based on it, using a radial basis function kernel to construct a similarity weight matrix.

Before building a similar weight matrix, we first normalize the data:

$$P = \frac{p - E(p)}{\sqrt{Var(p)}} \tag{1}$$

Where p is biological feature, P is its corresponding biological feature after normalization, and $Var(p)$ and $E(p)$ represent the variance and empirical mean of the feature p.

Given M omics data, for the m omics data, we have n samples $\{x_1, x_2, \ldots, x_n\}$, where $p \times m$ represents the corresponding characteristics of the m omics data. $W(i, j)$ represents the similarity between the ith sample and the jth sample:

$$W(i,j) = \frac{1}{\sqrt{2\pi}\theta_{ij}} \exp\left(-\frac{\|x_i - x_j\|^2}{2\theta_{ij}^2}\right) \tag{2}$$

θ_{ij}^2 is defined as:

$$\theta_{ij}^2 = \frac{1}{3}\left(\frac{1}{k}\sum_{r\in\lambda_i} \|x_i - x_r\|^2 + \frac{1}{k}\sum_{r\in\lambda_j} \|x_j - x_r\|^2 + \|x_i - x_j\|^2\right) \tag{3}$$

This similarity measure is based on a radial basis function kernel [13]. θ_{ij}^2 is a normalization factor that passes the ith and jth samples to the sum of the squares of their nearest neighbors and the squared distance controls the density of the two samples [12, 14, 15]. λ_i represents its k nearest neighbors.

After constructed the similarity weights $W(i, j)$ between samples, we applied the SNF method to integrated the multi-omics data. The detail of SNF can be found in [12].

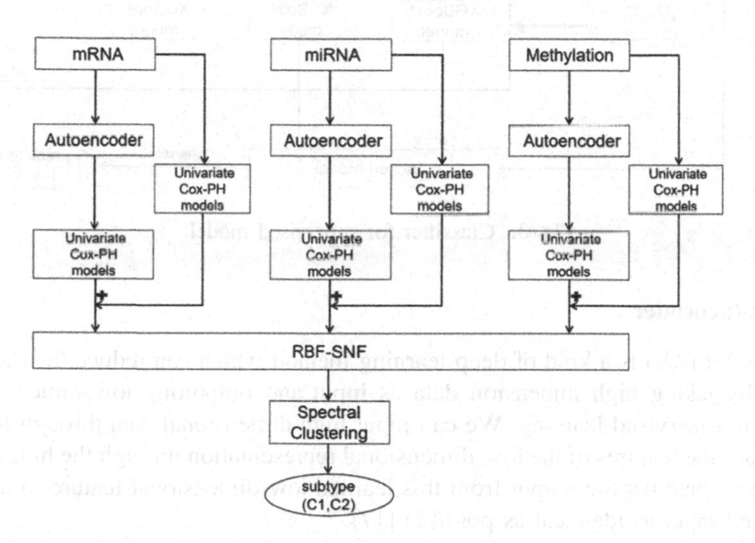

Fig. 2. Survival subtype inference model based on deep learning and network fusion.

2.4 Construction of XGBoost Classifier Based on Survival Subtype

For each multi-omics matrix generated above, we use Univariate Feature Selection approach [16] to select the top K features according to the clustering labels, and then stack the features selected from the three sets of data together. Finally, we trained four XGBoost classifiers, three classifiers for mRNA, miRNA, DNA methylation, respectively, the fourth classifier for stacked features of multi-omics data. We use the datasets GSE14520 and GSE31384 mined from the GEO database as the confirmation cohorts for the RNA-seq and miRNA trained classifiers, respectively. The model of the classifier is shown in Fig. 3. For the two confirmation cohorts, we first select the common features in the training set samples, and normalize the data using the method same as the multi-omics data normalization. From the Table 1, we know that the common feature of GSE14,520 and RNA-seq is 10,787, and the common feature of GSE31384 and miRNA is 169. In the study, we need to select the same K features based on clustering labels for both training sets and two cohorts. In this way, the two cohorts will input as the validation dataset to test the model and the classification result will be finally obtained. Here, we set the value of K (50–100), and find that when the value of K is set 50, the training model could obtain the best prediction results.

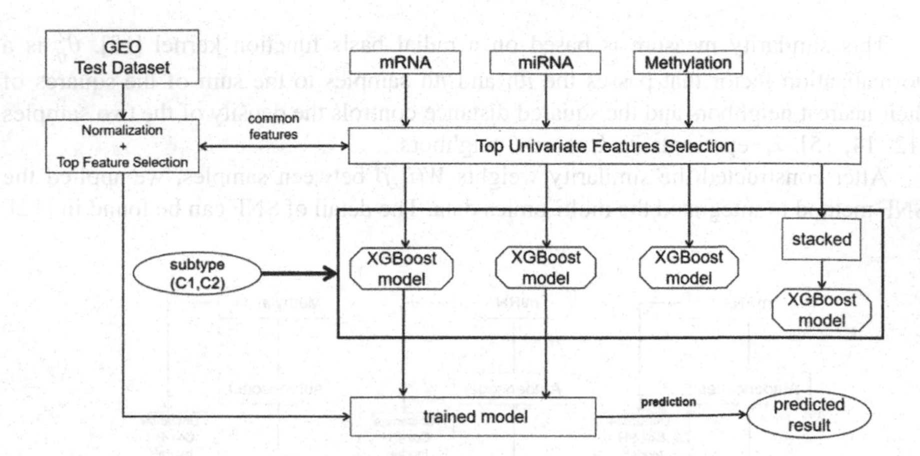

Fig. 3. Classifier for supervised model.

2.5 Autoencoder

Autoencoder (AE) is a kind of deep learning method which can reduce the dimension of data by taking high dimension data as input and outputting low dimension data through unsupervised learning. We can input high-dimensional data through the input layer, learn the features of the low-dimensional representation through the hidden layer, and then reconstruct the output from this learned low-dimensional feature to make the output and input as identical as possible [17].

Given *m* dimensional input data $X = (x_1, \ldots, x_m)$, the autoencoder mainly reconstructs the output by passing the input through continuous hidden layers. In terms of a layer *i*, for activation function, we use *Relu* between the input and the output layer, so we have:

$$y = f_i(x) = \mathrm{R}elu(W_i \cdot x + b_i) \quad (4)$$

x, y are two vectors, and their sizes are *n, m,* respectively, *x* is the value of a single feature of *X*. The size of W_i is $m \times n$, which is a weight matrix, and b_i is an offset of W_i of size *m*. For autoencoder with *k* layers, the output x' is defined as follows:

$$x' = f_1 \bullet f_2 \bullet f_3 \cdots f_{k-1} \bullet f_k(x) \quad (5)$$

Where $f_1 \bullet f_2(x) = f_1(f_2(x))$, $f_2 \bullet f_3 = f_2(f_3(x))$, ..., $f_{k-2} \bullet f_{k-1} = f_{k-2}(f_{k-1}(x))$ and so on, $f1 \bullet f2$ is the composed function of f_1 *with* f_2.

To train the autoencoder, in the last layer, we add the mean square error (MSE) as a function to measure the error between input and output of the autoencoder:

$$MSE(x, x') = \frac{1}{2}\sum_{i=1}^{k} \|x - x'\|^2 \quad (6)$$

To prevent overfitting, we add the *L2* regularization penalty coefficient β_w to the weight W_i, and the loss function becomes as follows:

$$\mathrm{L}(x, x') = MSE(x, x') + \sum_{i=1}^{k} \beta_w \|W_i\| \tag{7}$$

3 Results

3.1 Evaluation Metrics

We use Log-rank *P* value and Concordance index to evaluate the prediction results. We draw the corresponding Kaplan-Meier survival curves [18] for the two survival risk subtypes of the prediction results.

Log-Rank *P* Value. We calculate the Log-rank *P* value using the software *R survival* package to evaluate the difference between two risk groups.

Concordance Index. For C-index [19], we calculated the corresponding C-index for this model using R *survcomp* package, and used the C-index to evaluate the fitting ability and prediction performance of the network we proposed. The value of C-index is calculated as follows:

$$c = \frac{1}{u} \sum_{i \in \{1 \ldots U | \delta_i = 1\}} \sum_{b_j > b_i} O\left[Y_i \hat{0} > Y_j \hat{0}\right] \tag{8}$$

where u is the number of comparable pairs, *O[.]* is the indicator function, b is the actual value. The range of C-index is 0–1, 1 means the best effect.

3.2 Prediction of Two Different Survival Subtypes

We use the *Keras* library of python to implement the above mentioned autoencoder with two hidden layers, we set the number of nodes in the two hidden layers to 100 and 50, respectively. We obtain two subtypes with significant survival differences (Log-rank *P* value = 1.93e−09; as shown in Fig. 4A), and the model fitting index (C-index = 0.78). In order to verify the effectiveness of this classifier in predicting survival, we used two sets of data from GEO, which are GSE14520 (S = 222, Log-rank *P* value = 5.73e−06; as shown in Fig. 4B) and GES31384 (S = 166, Log-rank *P* value = 7.18e−04; as shown in Fig. 4C) to confirm this model. Finally, we also compared our results with those of other models [9] (as shown in Table 2). Whether it's Log-rank *P* value or C-index, our experimental results are obviously better than others' experimental results.

As can be seen from the Fig. 4, the red curve represents C1 subtype, the blue curve represents C2 subtype. Through the two curves in the Fig. 4, the survival state of C1 subtype is better than that of C2 subtype, and the survival curve is very significant, for

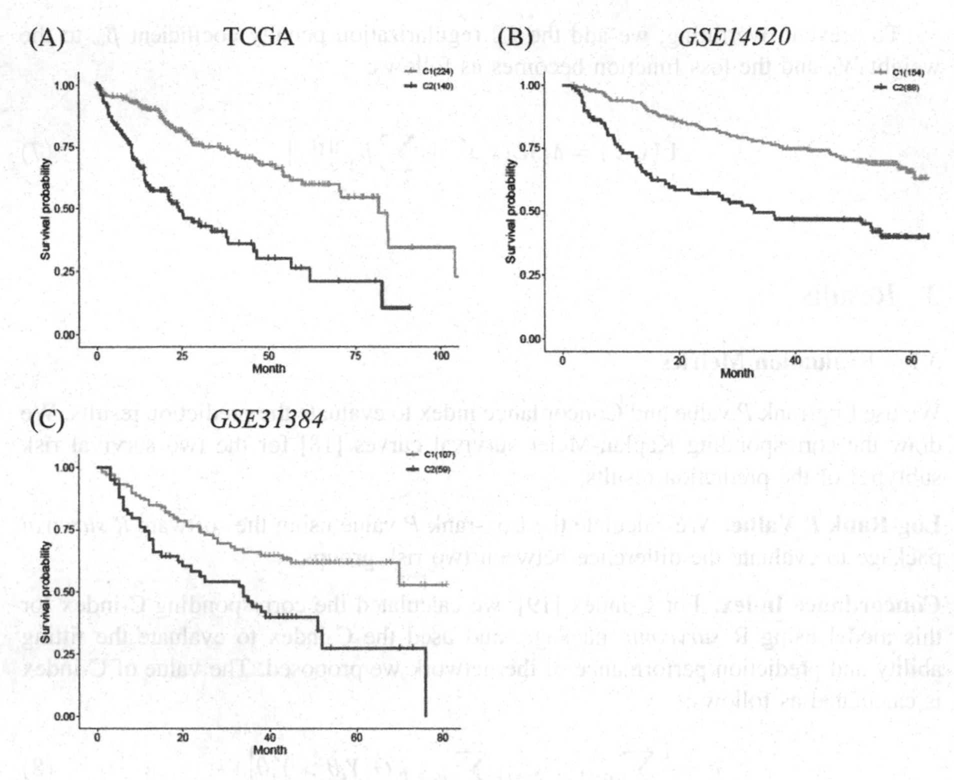

Fig. 4. Significant survival profiles for TCGA cohort and two confirmation cohorts (GSE14520 cohort and GSE31384 cohort). (Color figure online)

Table 2. Log-rank *P* value and C-index for train and two confirmation cohorts.

Omic data models	Data type	Log-rank *P* value	C-index
Method by Chaudhary et al. [9]	3-omics data	7.13e−06	0.68
	NCI cohort	1.05e−03	0.67
	Chinese cohort	8.49e−04	0.69
Our model	3-omics data	1.93e−09	0.78
	GSE14520	5.73e−06	0.74
	GSE31384	7.18e−04	0.70

the training cohort and two confirmation cohorts, there are significant differences between the two survival subtypes. For the survival curves of two survival subtypes, our results are better than others' results, so it can be seen that compared with other published models, the prediction effect of our model has significantly improved.

3.3 Function Analysis with HCC Survival Subgroups

For differential gene expression analysis, we can identify 1465 up-regulated genes and 930 down-regulated genes, including tumor marker genes BIRC5 ($P = 2.07\text{e}{-41}$) and stemness marker genes CD24 ($P = 2.83\text{e}{-11}$), KRT19 ($P = 2.82\text{e}{-26}$), and EPCAM ($P = 1.01\text{e}{-6}$). In addition, we also found 28 genes (*SLC2A2, AQP9, RGN, SULT2A1, CRYL1, SERPINC1, PAH, CDO1, PLG, APOC3, CYP27A1, PFKFB3, TM4SF1, ACSL5, RGS2, HN1, SERPINA10, CYB5A, EPHX2, SPHX2, RGS1, ADH1B, LECT2, TBX3, RNASE4, ALDOA, ADH6, SLC38A1*) are different between the two survival risk groups we identified and have a strong connection with liver cancer survival [20].

We also performed Kyoto Encyclopedia of Genes and Genomes (KEGG) pathway analysis on two subgroups (as shown in Fig. 5, A, B). *PI3K-Akt signaling pathway, cell cycle signaling pathway, p53 signaling pathway* and so on are enriched with cancer related way in aggressive subtype (C2). There were some related pathways in lower risk survival subtype (C1), such as, *Drug metabolism-cytochrome P450*, *metabolic pathways* and *fatty acid degradation* and so on. These pathways are of great significance for studying the prognosis of liver cancer.

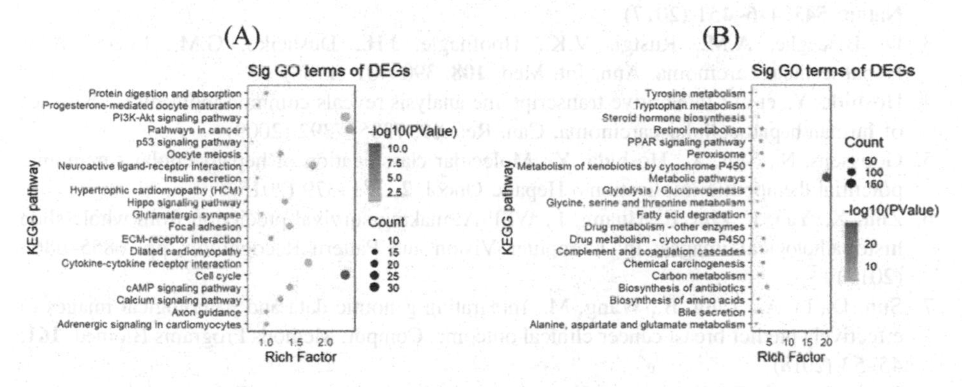

Fig. 5. KEGG pathway analysis of DEGs for the two subtypes. (A, B) KEGG pathway analysis for C1,C2.

4 Conclusion

In this paper, we proposed an integration framework based on deep learning and network fusion on the multi-omics data and survival data of liver cancer to predict the survival of HCC patients, and have identified two subtypes with significant survival differences (Log-rank P value = 1.93e−09, C-index = 0.78) at the molecular level. Then we train an XGBoost classifier, and then use the new data to verify the classification model, we get GES31384 (Log-rank P value = 7.18e−04, C-index = 0.70) and GSE14520 (Log-rank P value = 5.73e−06, C-index = 0.74). The results show that our experimental method has achieved good prediction and classification results. Our method is very helpful to the clinical treatment and prognosis of patients. Moreover, the C2 subtype is enriched with *PI3K-Akt signaling pathway, cell cycle signaling pathway, p53 signaling pathway* and so on. The stemness markers (*CD24, KRT19, EPCAM*) that associated with the C2 subtype. And we also found that there are 28 genes different between the two survival risk groups we identified and have a strong connection with liver cancer survival.

Combining the survival data with the multi-omics data in the research of HCC will improve the prediction accuracy of molecular subtype. Our experiment is very helpful for the clinical prognosis and treatment of HCC.

Acknowledgments. This work was supported by grants from the NSFC projects Grant (No. U1611263, U1611261 and 61932018) and the National Natural Science Foundation of China (Nos. U19A2064 and 61873001).

References

1. Torre, L.A., Bray, F.I., Siegel, R.L., Ferlay, J., Lortettieulent, J., Jemal, A.: Global cancer statistics, 2012. CA Cancer J. Clin. **65**, 87–108 (2012)
2. Abbosh, C., et al.: Phylogenetic ctDNA analysis depicts early-stage lung cancer evolution. Nature **545**, 446–451 (2017)
3. Di Bisceglie, A.M., Rustgi, V.K., Hoofnagle, J.H., Dusheiko, G.M., Lotze, M.T.: Hepatocellular carcinoma. Ann. Int. Med. **108**, 390–401 (1988)
4. Hoshida, Y., et al.: Integrative transcriptome analysis reveals common molecular subclasses of human hepatocellular carcinoma. Can. Res. **69**, 7385–7392 (2009)
5. Goossens, N., Sun, X., Hoshida, Y.: Molecular classification of hepatocellular carcinoma: potential therapeutic implications. Hepatic Oncol. **2**, 371–379 (2015)
6. Zhu, X., Yao, J., Zhu, F., Huang, J.: WSISA: making survival prediction from whole slide histopathological images. In: Computer Vision and Pattern Recognition, pp. 6855–6863 (2017)
7. Sun, D., Li, A., Tang, B., Wang, M.: Integrating genomic data and pathological images to effectively predict breast cancer clinical outcome. Comput. Methods Programs Biomed. **161**, 45–53 (2018)
8. Dong, R., et al.: Predicting overall survival of patients with hepatocellular carcinoma using a three-category method based on DNA methylation and machine learning. J. Cell Mol. Med. **23**, 3369–3374 (2019)
9. Chaudhary, K., Poirion, O., Lu, L., Garmire, L.X.: Deep learning-based multi-omics integration robustly predicts survival in liver cancer. Clin. Cancer Res. **24**, 1248–1259 (2017)
10. Cox, D.R.: Regression models and life-tables. J. Roy. Stat. Soc.: Ser. B (Methodol.) **34**, 187–202 (1972)
11. Ng, A.Y., Jordan, M.I., Weiss, Y.: On spectral clustering: analysis and an algorithm. In: Neural Information Processing Systems, pp. 849–856 (2001)
12. Wang, B., et al.: Similarity network fusion for aggregating data types on a genomic scale. Nat. Methods **11**, 333–337 (2014)
13. Buhmann, M.D.: Radial Basis Functions: Theory and Implementations. Cambridge University Press, Cambridge (2003)
14. Wang, B., Jiang, J., Wang, W., Zhou, Z., Tu, Z.: Unsupervised metric fusion by cross diffusion. In: Computer Vision and Pattern Recognition, pp. 2997–3004 (2012)

15. Yang, X., Bai, X., Latecki, L.J., Tu, Z.: Improving shape retrieval by learning graph transduction. In: Forsyth, D., Torr, P., Zisserman, A. (eds.) ECCV 2008. LNCS, vol. 5305, pp. 788–801. Springer, Heidelberg (2008). https://doi.org/10.1007/978-3-540-88693-8_58
16. Dash, M., Liu, H.: Feature selection for classification. In: Intelligent Data Analysis, pp. 131–156 (1997)
17. Hinton, G.E., Salakhutdinov, R.: Reducing the dimensionality of data with neural networks. Science **313**, 504–507 (2006)
18. Bland, J.M., Altman, D.G.: Survival probabilities (the Kaplan-Meier method). Br. Med. J. **317**, 1572–1580 (1998)
19. Steck, H., Krishnapuram, B., Dehingoberije, C., Lambin, P., Raykar, V.C.: On ranking in survival analysis: bounds on the concordance index. In: Neural Information Processing Systems, pp. 1209–1216 (2007)
20. Kim, S.M., et al.: Sixty-five gene-based risk score classifier predicts overall survival in hepatocellular carcinoma. Hepatology **55**, 1443–1452 (2012)

Short-Term Rainfall Forecasting with E-LSTM Recurrent Neural Networks Using Small Datasets

Cristian Rodriguez Rivero[1(✉)], Julián Pucheta[2], Daniel Patiño[3], Paula Otaño[4], Leonardo Franco[5], and Gustavo Juarez[6]

[1] University of Amsterdam, Science Park 904, 1098XH Amsterdam, The Netherlands
c.m.rodriguezrivero@uva.nl
[2] Universidad Nacional de Córdoba, Córdoba, Argentina
[3] INAUT-UNSJ, San Juan, Argentina
[4] Universidad Tecnológica Nacional – FRC, Córdoba, Argentina
[5] University of Malaga, Malaga, Spain
[6] Universidad Nacional de Tucumán, San Miguel de Tucumán, Argentina

Abstract. This paper proposes an ensemble of forecasting methods based on neural networks/recurrent neural networks (E-LSTM). The aim of the algorithm is to help organizing the planting cycle using short-term rainfall forecasts when the data are taken from a single observation point. The computational models are carried out for univariate rainfall time series by of multi-step prediction horizons in combination of nonlinear autoregressive models (NAR) modified by several approaches such energy associated to series, subsampling methods and their combinations, which are heuristically modified by Bayesian inference and statistical roughness in the learning process. The study analyses and compares the relative advantages and limitations of each algorithm against the aforementioned to forecast rainfall from 1 to 6 months ahead. Simulation results illustrate the effectiveness of the E-LSTM approach through different series classified by their statistical roughness in both, the learning process and the validation test using the SMAPE and RMSE metrics. Comparisons also are made by adding fractional Gaussian noise to highlight the performance and constraints of the ensemble approach.

Keywords: Rainfall · Time series forecasting · Ensemble · LSTM recurrent neural networks

1 Introduction

Precipitation Forecast is important for many decision makers who are sensitive to the occurrence [1]. The complexity of systematically predicting rainfall time series with a small amount of observations is an exciting challenge that misinterpretation of what the forecast represents can actually lead to poorer decision-making. One key scope to see is whether a model incorporating localized short or long rainfall data can provide reasonably simple but better method for farmers to improve the odds in their favor of

D.-S. Huang and P. Premaratne (Eds.): ICIC 2020, LNAI 12465, pp. 258–270, 2020.
https://doi.org/10.1007/978-3-030-60796-8_22

starting cultivation for the next six months while taking risks during the crop cycle [2]. The motivation of this paper is to reliably present state-of-the-art methods to estimate water availability useful for guiding the crop in the semi-arid regions of Argentina. The scope of this research is to contribute with the development of ensemble of ***neural networks-based robust algorithms to predict short-term rainfall series, mainly from single geographical points***, which is considered one of the most likely to receive these benefits, but not the only one. The aforementioned tools provide opportunities of changing the distribution of planting cycle based on spatial and temporal variability of water amount in a short-medium term that follows the closed-loop scheme as in [3] (Fig. 1).

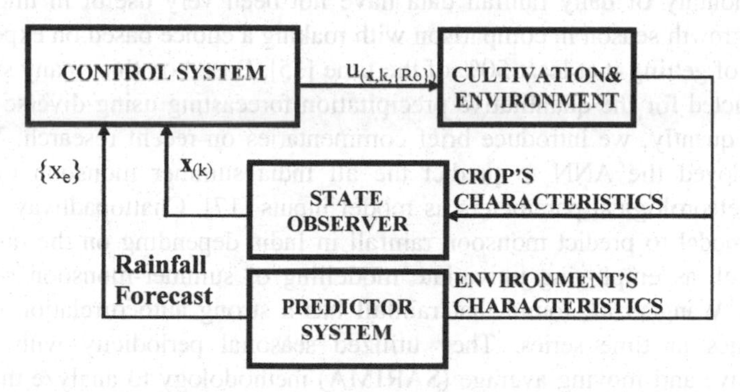

Fig. 1. Representation of the close-loop control system scheme based on guiding the crop.

This paper, while recognizing the difficult task of accurately predicting long-short rainfall series, primarily due to the seasonal pattern or characteristic that rainfall series present [4]. Thus, it is proposed to actually exploit the hypothesis of an *fBm*-based-behavior as an underlying generator process by the estimation of the Hurst parameter. This can be interpreted like the statistical roughness of the series in this research [5]. Having saying that, and in view of the highly nonlinear relationships governing the rainfall phenomenon, it is assumed that long-short-long term rainfall forecasting can be modeled as a trace of a stochastic process [6], showing that scaling models of rain can provide attractive parsimonious representations over a wide range of scales and are supported by theoretical arguments and empirical evidence which rainfall exhibits stochasticity and scale-invariant symmetry [7]. Some recent developments in stochastic rainfall analysis drive towards scaling issues by using wavelet transforms [8].

In order to give an idea of statistical roughness of rainfall series and determines its long-short stochastic dependence, the Hurst parameter is used for tuning the neural networks parameters as well. The limitation of the neural networks methods is that though these demonstrated preeminent performance of ANNs [9], these algorithms were not evaluated against statistical methods, using small set of heterogeneous time series. This ignored evidence within the forecasting field on how to design valid and reliable empirical evaluations [10].

However, due to rainfall forecasting involves a rather complex nonlinear data pattern; there are plenty of novel forecasting approaches using different computational intelligence and statistical methods to improve the horizon of the forecast [11]. Recently, due to the significant progress in the fields of pattern recognition methodology, ANNs and their hybrid combinations are the most employed to forecast rainfall [12], since the idea behind the proposed hypothesis is toward an automatic rainfall time series forecasting [13]. In fact, what is important to notice is that rainfall forecast is far from being satisfactory, even though nowadays it is possible to access big data sources. The crop yield instead, depends on a number of weather factors but is more commonly influenced by rainfall [14]. Farmers have historically had no management tools that can effectively assist them in predicting the monsoon due to climate change. Even faithfully recorded monthly or daily rainfall data have not been very useful in improving the upcoming growth season in comparison with making a choice based on experience and the chance of getting it at least 50% of the time [15]. Based on this, many studies have been conducted for the quantitative precipitation forecasting using diverse techniques [16]. Subsequently, we introduce brief commentaries on recent research. Venkatesan *et al.* employed the ANN to predict the all India summer monsoon rainfall with different meteorological parameters as model inputs [17]. Chattopadhyay constructed an ANNs model to predict monsoon rainfall in India depending on the rainfall series alone as well as employing univariate modelling of summer-monsoon rainfall time series [18]. Wang *et al.* found that rainfall has a strong autocorrelation of seasonal characteristics in time series. They utilized seasonal periodicity with a seasonal autoregressive and moving average (SARIMA) methodology to analyze the statistical data of precipitation [19]. Studies related to rainfall forecasts have been conducted in Argentina, particularly in the center and northwestern region using Fourier analysis and linear/nonlinear regression models based on time series of meteorological variables recorded at specifics points. In a recent survey, Rodriguez Rivero *et al.* [20, 21] and Pucheta *et al.* [22] reports several applications of neural networks based forecasting approaches facing the long-short-term stochastic dependences.

In this paper, an ensemble of methods called E-LSTM is grounded on Bayesian Approaches such as (RNN-BA) [23] a LSTM recurrent neural network with a full-connected layer to form a regression model for prediction in order capture the temporal relationship among time series data using the Bayes information of the weights updated following a heuristic approach to adjust and update the number of iteration of the RNNs using the roughness of the series measured by the Hurst parameter, plus the addition of Bayesian Enhanced (BEA) [20] and Bayesian Enhanced Modified (BEMA) [24] techniques, mostly used to predict time series data from well-known chaotic systems.

Such approaches have been utilized to investigate to what extent can an ensemble of artificial neural networks actually be accurate and precise in term of rainfall prediction horizon for farmer's decision-making. To study the performance and the limitation, we compare it against nine models, Bayesian Enhanced Ensemble Approach (BEEA) [25], Bayesian Enhanced Modified Combined Approach (BEMCA) [26], BEMA [24], BEMA [27], BEA [20], BEMA [28], Energy Associated to Series (EAS) [29], Sub Sampling Nonparametric Methods (SUB2-3) [30], Neural Network Modified (NNMod.) [31], where the BA and NNMod. were used as a baseline.

The experiments results indicate an improvement of the E-LSTM approach throughout rainfall time series, thus confirming the applicability using small datasets. The remainder is structured as follows: Sect. 2 provides the methodology implemented with a particular interest in the ensemble approach employed, and with Sect. 3 showing the experimental setup using datasets of monthly rainfall. Section 4 shows the prediction results with statistical evaluation of all models. Lastly, Sect. 5 presents some discussions and concluding remarks.

2 Methodology

2.1 Ensemble of LSTM RNNs and ANNs Modified Approaches

To take advantage of the characteristic that ensembles owns, we follow an extension of the BEEA approach [25], which makes use of two out of three techniques deemed in this research, such as the BEA [20] that consist of the Bayes assumption used to update a prior distribution into a posterior distribution by incorporating the information driven as likelihood function from fractional Brownian provided by neural networks weights from observed data in order to generate point forecasts into a predictive distribution for the future values. The second is the BEMA [24] that comprises of the BEA method modified by *Renyi* entropy. The *Renyi* entropic information is used to update the neural network parameters in order to combine this with a prior distribution inferred into a posterior distribution provided by neural networks weights from observed data in order to generate point and interval forecasts by combining all the information and sources of uncertainty into a predictive distribution for the future values. The latter is the LSTM RNN-BA [21] that heuristically adjusts the weights of the neural nets modified by Bayesian learning and statistical roughness (H parameter measured in the series). The proposed model combination is depicted in Fig. 2 as follow:

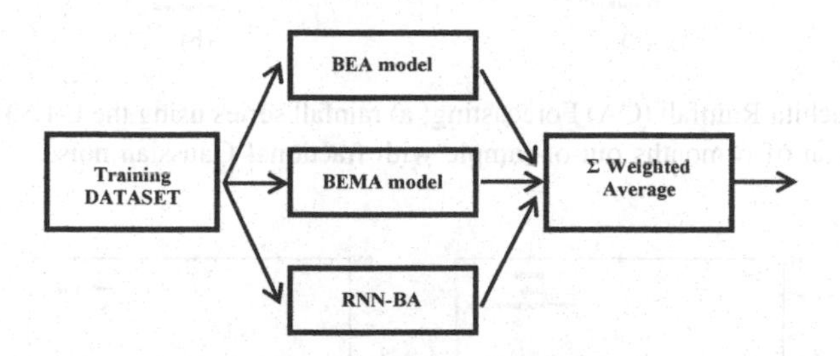

Fig. 2. Ensemble of Bayesian approches (E-LSTM).

The reason behind the models selected was due to the fact that their performance to capture the long-short term dependence as a counterpart amongst other approaches (Fig. 3).

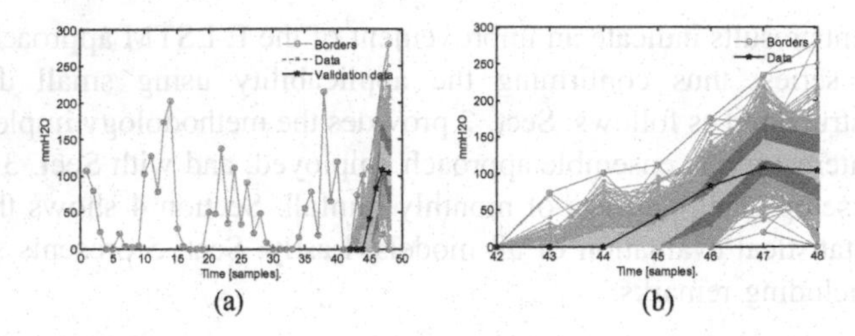

Fig. 3. Alta Gracia Rainfall (AG) Forecasting; a) rainfall series using the E-LSTM approach, b) prediction horizon of 6 months out-of-sample with fractional Gaussian noise.

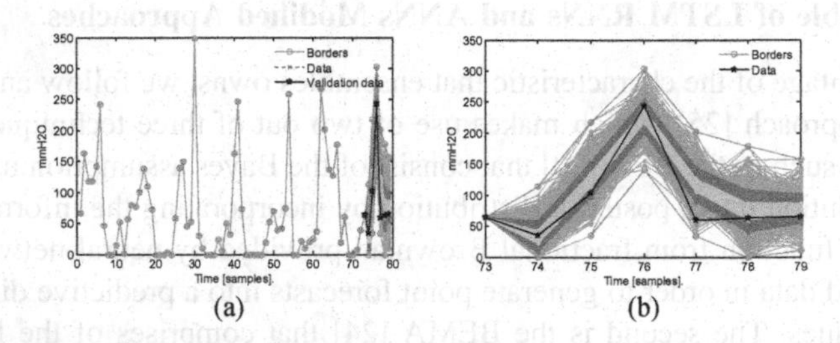

Fig. 4. Balnearia Rainfall (BAL) Forecasting; a) using the E-LSTM approach, b) prediction horizon of 6 months out-of-sample with fractional Gaussian noise.

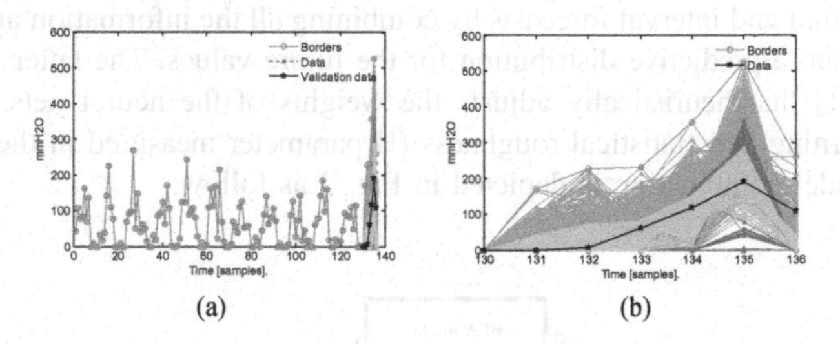

Fig. 5. Calamuchita Rainfall (CA) Forecasting; a) rainfall series using the E-LSTM approach, b) prediction horizon of 6 months out-of-sample with fractional Gaussian noise.

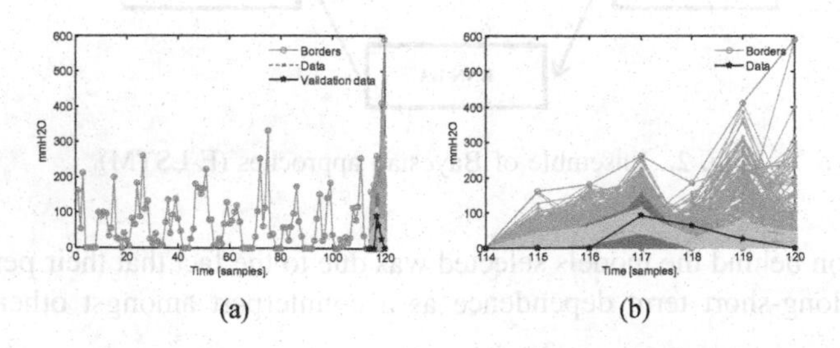

Fig. 6. Las Varillas Rainfall (LV) Forecasting; a) rainfall series using the E-LSTM approach, b) prediction horizon of 6 months out-of-sample with fractional Gaussian noise.

3 Experimental Setup

Four short monthly cumulative rainfall series from Province of Cordoba, Argentina were analyzed in this study. The stations are from Calamuchita (−31.914218, −64.310961), Alta Gracia (−31.679583, −64.430465), Balnearia (−31.016841, −62.643746) and Las Varillas (−31.869773, −62.749611) collected by farmers. The monthly rainfall series data were collected by agricultural producers between January 2000 and December 2014. Hereafter, the following acronyms are used for rainfall-series, Calamuchita (CA), Alta Gracia (AG), Balnearia (BAL) and Las Varillas (LV).

The monthly rainfall average was corroborated by area-weighted observations at weekly bulletin from National Institute of Agricultural Technology (INTA) measured uniformly upon the province. Each rainfall series is partitioned into three parts as training set, non-dependent cross-validation set and testing set [32].

Table 1. Length and statistical roughness measured by the Hurst parameter of rainfall series.

Series	N (Length)	H (Statistical roughness)
Alta Gracia (AG)	48	0.31
Calamuchita	136	0.007
Balnearia (BAL)	79	0.29
Las Varillas (LV)	120	0.064

The training set served the model training and the testing set was used to evaluate the performances of models. The non-dependent cross-validation set aimed to implement an early stopping approach for avoiding the overfitting. The same data partition format was adopted in four rainfall series: the first half of the entire data as training set and the first half of the remaining data as cross-validation set and the other half as testing set [33]. The length of each series is n as shown in Table 1.

3.1 Hosking Method for Rainfall Time Series

The Hosking method is an algorithm to compute Fractional Gaussian Noise (FGN) [34]. This method is based on the key fact that a particular sample can be completely computed given its past. In other words, the method generates x_{n+1} given x_o, ..., x_n recursively. From this result, we then just have to apply the algorithm recursively until we produce enough samples. The forecast performance is shown by predicting 6 future values from each time series simulated by a Monte Carlo of 500 trials with fractional Gaussian noise to specify the variance. The fractional noise was generated by the Hosking method with the H parameter estimated from the data time series. The mean and the variance of 500 trials of the forecasted horizon values are shown at each rainfall time series. Such outcomes for one (30%) and two (69%) sigma are shown in Fig. 4 (b), Fig. 5 (b). Figure 6 (b) and Fig. 7 (b).

In the experiments, we have split the data into two parts: the training set and the test set. In the training phase, each of the individual models is trained with parameters optimization given by each filter. This means that every model is constructed by its optimal values of their respective parameters. The next subsection shows the set-up of the BA-RNN used in the ensemble model.

3.2 Recurrent Neural Networks Setup (BA-RNN)

The startup of the RNN is as follow:

- $l_x = 25$ (length of the input)
- Gate Non-linearity = sigmoid;
- Input-Output non-linearity = tanh;
- Training epochs = 50;
- Iteration per epoch $i_t = l_x$;
- Learning rate w = 2e−3;
- Training percent = 0.80;
- Number of hidden neurons = 512;
- Dropout rate = 1.0;
- Weight decay = 1e−8;

3.3 Forecast Error Metrics

There are some commonly used accuracy measures whose scale depends on the scale of the data. These are useful when comparing different methods on the same set of data, but should not be used, for example, when comparing across data sets that have different scales. The most commonly used scale-dependent measures are based on the absolute error or squared [35].

The computational results with addition of fractional Gaussian noise in the time series are presented in Table 2 showing metrics of 6-out-of-sample forecast horizon. The performance of various models during calibration and validation were measured by the Symmetric Mean Absolute Percent Error (SMAPE) proposed in the most of metric evaluation [36], defined by

$$SMAPE_S = \frac{1}{n}\sum_{t=1}^{n}\frac{|X_t - F_t|}{(|X_t| + |F_t|)/2} \cdot 100 \quad (1)$$

where t is the observation time, n is the size of the test set, s is each time series, X_t and F_t are the actual and the forecasted time series values at time t respectively. The SMAPE of each series s calculates the symmetric absolute error in percent between the actual X_t and its corresponding forecast value F_t, across all observations t of the test set of size n for each time series s. The Root Mean Square Error (RMSE) is as follows,

$$RMSE_S = \sqrt{\frac{1}{n}\sum_{t=1}^{n}(X_t - F_t)^2} \quad (2)$$

where is calculated by dividing the difference between actual value X_t and forecasted value F_t (known as the forecasting error) by the actual value X_t, where i is the series, t is the forecast period and s is the forecasting method.

4 Results

Although the comparison was performed on LSTM-RNNs and ANNs-based filters, the experimental results confirm that E-LSTM method can effectively outperform in terms of SMAPE and RMSE indices show the experimental results from Alta Gracia, Calamuchita, Balnearia and La Sevillana.

Table 2 show the experimental results from Alta Gracia, Calamuchita, Balnearia and La Sevillana. The outcomes of E-LSTM are quite slightly better than the other methods proposed across the SMAPE and RMSE indexes (highlighted in red). Furthermore, we observed that there were relationships involving trade-offs between how short the series is and the roughness of the series (measured by Hurst Parameter). The shorter the series is, the worse the prediction is. Nevertheless, the performance was not considerably worse than cited by [27]. The assessment shows that even though the E-LSTM has a significance improvement measured by SMAPE and RMSE index across all series, it means that model averaging has a great impact on prediction values instead of using a single predictor model, such as BEEA when dealing with small dataset.

5 Discussion and Conclusion

This paper reports the results of an ensemble algorithm E-LSTM that outputs the average of three ANNs models, the BA-RNNs that heuristically adjusts the parameter of the recurrent neural network's weight taking into account Bayesian learning and the Hurst parameter associated to the rainfall series in the training process plus the BEA and BEMA aforesaid before. The study analyzed and compared the relative advantages and limitations of each time-series predictor filter technique. The election of BEMA and BEA to jointly build the ensemble forecasting was because of their good performance in capturing the short-term dependency of the series.

Although the comparison was performed on LSTM-RNNs and ANNs-based filters, the experimental results confirm that E-LSTM method can effectively outperform in terms of SMAPE and RMSE indices.

Table 2. Comparison of Forecasting Methods using Fractional Gaussian Noise of 6 Months out-of-sample.

Series	Method	Real Mean	Forecasted Mean	SMAPE	RMSE
AG	E-LSTM	47.78	47.69	0.15	2.35
	BEEA	47.78	47.95	0.25	3.45
	BEMCA	47.78	48.15	0.29	5.19
	BEMAmod	47.78	48.05	0.27	5.08
	BEMA	47.78	46.95	0.29	5.14
	BEA	47.78	47.20	0.31	6.01
	BA	47.78	67.21	0.94	17.76
	EAS	246.76	221.08	3.54	75.21
	SUB$_{2-3}$	47.78	32.73	1.01	18.59
	NNMod.	47.78	13.84	0.89	17.91
CA	E-LSTM	69.57	59.55	0.11	6.55
	BEEA	69.57	56.41	0.18	7.88
	BEMCA	69.57	52.12	0.22	8.10
	BEMAmod	69.57	54.77	0.23	8.23
	BEMA	69.57	55.23	0.23	8.45
	BEA	69.57	51.2	0.22	8.63
	BA	69.57	41.64	0.35	13.50
	EAS	525.56	424.39	0.53	23.80
	SUB$_{2-3-6}$	69.57	63.08	0.55	15.82
	NNMod.	69.57	52.90	0.23	8.12
BAL	E-LSTM	90.28	93.10	0.16	5.43
	BEEA	90.28	94.53	0.19	6.11
	BEMCA	90.28	99.10	0.21	6.45
	BEMAmod	90.28	98.02	0.23	7.01
	BEMA	90.28	98.91	0.26	7.08
	BEA	90.28	102.34	0.25	7.35
	BA	90.28	48.54	0.69	23.65
	EAS	332.14	338.20	0.80	25.59
	SUB$_{2-3}$	90.28	58.64	0.53	19.63
	NNMod.	90.28	79.94	0.96	27.81
LV	E-LSTM	25.85	22.73	0.19	5.38
	BEEA	25.85	21.45	0.21	6.11
	BEMCA	25.85	28.23	0.22	6.34
	BEMAmod	25.85	21.67	0.23	6.56
	BEMA	25.85	20.93	0.24	6.76
	BEA	25.85	30.56	0.24	7.28
	BA	25.85	20.93	0.32	9.93
	EAS	187.34	174.47	0.42	12.12
	SUB$_{2-3-6}$	25.85	35.63	0.95	26.27
	NNMod.	25.85	32.33	0.41	13.34

The outcomes of E-LSTM are quite slightly better than the other methods proposed across the SMAPE and RMSE indexes (highlighted in red). Furthermore, we observed that there were relationships involving trade-offs between how short the series is and the roughness of the series (measured by Hurst Parameter) shown in Fig. 7 by averaging the SMAPE and RMSE across all prediction horizons of each rainfall series in comparison.

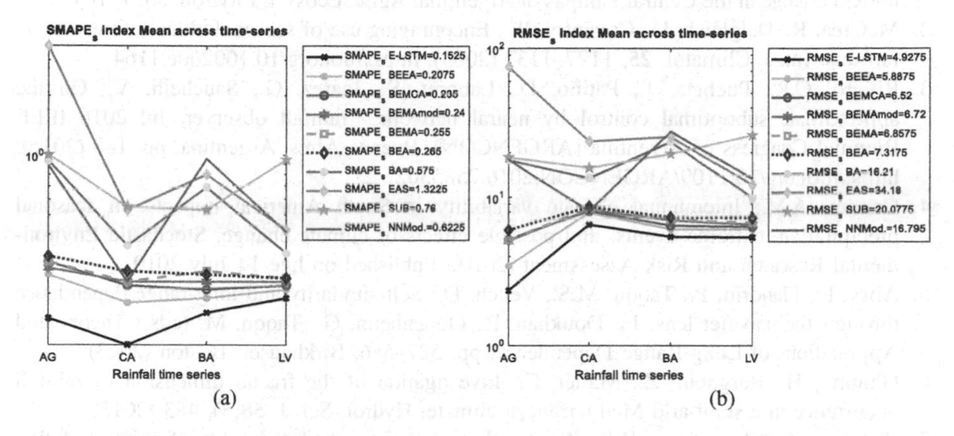

Fig. 7. a) SMAPE performance throughout forecast horizons and methods, b) RMSE performance throughout forecast horizons and methods.

The shorter the series is, the worse the prediction is. Nevertheless, the performance was not considerably worse than cited by [28]. The assessment shows that even though the E-LSTM has a significance improvement measured by SMAPE and RMSE index across all series, it means that model averaging has a great impact on prediction values instead of using a single predictor model, such as BEEA when dealing with small rainfall dataset. From the comparison one can intuitively deduct that the uncertainties in the data can be modeled assuming a priori function of the series nonlinearity as a trace of an *fBm,* that generates the underlying process. Despite of the nature of the time series, which not only include nonlinearity, nonstationary, noise, but and in particular, limit quantity of data. The discussion of how ensemble models can successfully approximate the quantitative dynamics of the time series data due to changes in the parameters associated with single methods dealing with latent variables remains open for study, mainly. This research acknowledges when applying neural networks to rainfall time series modeling, that important issues appears such as proper input variable selections, and balancing bias/variance trade-off of a model. In fact, when sufficient data is available for training and validation, the increase in the SMAPE index is shown particularly in Table 2. As expected, the short series has much worse performance than the longer.

Other important drawn conclusion is that empirical simulations have demonstrated that no technique alone is sufficient, but a combination of selected models and techniques yield superior results. As a way forward we will focus on the use of inverse entropy as a technique to modify heuristically the training of the ensemble learning algorithms.

Acknowledgement. The authors wish to thank FNWI at University of Amsterdam and Agronomist Engineers Ernesto Carreño, Nicolas Bernaldez Brunt, Cecilia Scalerandi and Monica Piccardi for providing crop yield information at Cordoba, Argentina.

References

1. Viglizzo, E., Roberto, Z., Filippin, M., Pordomingo, A.: Climate variability and agroecological change in the Central Pampas of Argentina. Agric. Ecosyst. Environ. **55**, 7–16 (1995)
2. McCrea, R., Dalgleish, L., Coventry, W.: Encouraging use of seasonal climate forecasts by farmers. Int. J. Climatol. **25**, 1127–1137 (2005). https://doi.org/10.1002/joc.1164
3. Rivero, C.R., Pucheta, J., Patiño, D., Laboret, S., Juárez, G., Sauchelli, V.: On the approximate suboptimal control by neural network - rainfall observer. In: 2016 IEEE Biennial Congress of Argentina (ARGENCON), Buenos Aires, Argentina, pp. 1–8 (2016). https://doi.org/10.1109/ARGENCON.2016.7585302
4. Grimm, A.M.: Interannual climate variability in South America: impacts on seasonal precipitation, extreme events, and possible effects of climate change. Stochastic Environmental Research and Risk Assessment (2010). Published on line 14 July 2010
5. Abry, P., Flandrin, P., Taqqu, M.S., Veitch, D.: Self-similarity and long-range dependence through the wavelet lens. In: Doukhan, P., Oppenheim, G., Taqqu, M. (eds.) Theory and Applications of Long-Range Dependence, pp. 527–556. Birkhäuser, Boston (2003)
6. Ghanmi, H., Bargaoui, Z., Mallet, C.: Investigation of the fractal dimension of rainfall occurrence in a semi-arid Mediterranean climate. Hydrol. Sci. J. **58**(3), 483 (2013)
7. Beecham, S., Chowdhury, R.K.: Temporal characteristics and variability of point rainfall: a statistical and wavelet analysis. Int. J. Climatol. **30**, 458–473 (2010)
8. Venugopal, V., Roux, S.G., Foufoula-Georgiou, E., Arnéodo, A.: Scaling behavior of high resolution temporal rainfall: new insights from a wavelet-based cumulant analysis. Phys. Lett. A **348**(3–6), 335 (2006)
9. Crone, S.F., Hibon, M., Nikolopoulos, K.: Advances in forecasting with neural networks? Empirical evidence from the NN3 competition on time series prediction. Int. J. Forecast. **27** (3), 635–660 (2011)
10. Fildes, R., et al.: Generalizing about univariate forecasting methods: further empirical evidence. Int. J. Forecast. **14**, 339–358 (1998)
11. Yen, M., Liu, D., Hsin, Y., et al.: Application of the deep learning for the prediction of rainfall in Southern Taiwan. Sci. Rep. **9**, 12774 (2019). https://doi.org/10.1038/s41598-019-49242-6
12. Wu, C.L., Chau, K.W.: Prediction of rainfall time series using modular soft computing methods. Eng. Appl. Artif. Intell. **26**, 997–1007 (2013)
13. Yan, W.: Toward automatic time-series forecasting using neural networks. IEEE Trans. Neural Netw. Learn. Syst. **23**(7), 1028–1039 (2012)
14. Grimm, A., Barros, V., Doyle, M.: Climate variability in Southern South America associated with El Niño and La Niña events. J. Clim. **13**, 35–58 (2000). ISSN 0894 8755
15. Nnaji, A.O.: Forecasting seasonal rainfall for agricultural decision-making in northern Nigeria. Agric. Forest Meteorol. **107**, 193–205 (2001)
16. Xu, L., Chen, N., Zhang, X., Chen, Z.: A data-driven multi-model ensemble for deterministic and probabilistic precipitation forecasting at seasonal scale. Clim. Dyn. **54**, 3355–3374 (2020). https://doi.org/10.1007/s00382-020-05173-x

17. Venkatesan, C., Raskar, S.D., Tambe, S.S., Kulkarni, B.D., Keshavamurty, R.N.: Prediction of all India summer monsoon rainfall using error-back-propagation neural networks. Meteorol. Atmos. Phys. **62**(3–4), 225–240 (1997)
18. Chattopadhyay, S., Chattopadhyay, G.: Univariate modelling of summer-monsoon rainfall time series: comparison between ARIMA and ARNN. C.R. Geosci. **342**, 100–107 (2010)
19. Wang, S., Feng, J., Liu, G.: Application of seasonal time series model in the precipitation forecast. Math. Comput. Model. **58**, 677–683 (2013)
20. Rivero, C.R., Patiño, D., Pucheta, J., Sauchelli, V.: A new approach for time series forecasting: bayesian enhanced by fractional brownian motion with application to rainfall series. Int. J. Adv. Comput. Sci. Appl. (IJACSA) **7**(3) (2016). https://doi.org/10.14569/IJACSA.2016
21. Rivero, C.R., et al.: Bayesian inference for training of long short term memory models in chaotic time series forecasting. In: Orjuela-Cañón, A., Figueroa-García, J., Arias-Londoño, J. (eds.) Applications of Computational Intelligence: ColCACI 2019, vol. 1096, pp. 197–208. Springer, Cham (2019). https://doi.org/10.1007/978-3-030-36211-9_16
22. Pucheta, J., Alasino, G., Salas, C., Herrera, M., Rivero, C.R.: Stochastic analysis for short- and long-term forecasting of latin american country risk indexes. In: Arabnia, H.R., Daimi, K., Stahlbock, R., Soviany, C., Heilig, L., Brüssau, K. (eds.) Principles of Data Science. TCSCI, pp. 249–272. Springer, Cham (2020). https://doi.org/10.1007/978-3-030-43981-1_12
23. Rivero, C.R., et al.: Time series forecasting using recurrent neural networks modified by bayesian inference in the learning process. In: 2019 IEEE Colombian Conference on Applications in Computational Intelligence (ColCACI), Barranquilla, Colombia, pp. 1–6 (2019). https://doi.org/10.1109/ColCACI.2019.8781984
24. Rivero, C.R., Pucheta, J.A., Laboret, S., Sauchelli, V., Patiño, D.: Short time series prediction: Bayesian enhanced modified approach with application to cumulative rainfall series. Int. J. Innov. Comput. Appl. **7**(3), 153–162 (2016). https://doi.org/10.1504/IJICA.2016.078730
25. Rivero, C.R., et al.: Bayesian enhanced ensemble approach (BEEA) for time series forecasting. In: 2018 IEEE Biennial Congress of Argentina (ARGENCON), San Miguel de Tucumán, Argentina, pp. 1–7 (2018). https://doi.org/10.1109/ARGENCON.2018.8646177
26. Rivero, C.R., Pucheta, J., Tupac, Y., Franco, L., Juárez, G., Otaño, P.: Time-series prediction with BEMCA approach: application to short rainfall series. In: 2017 IEEE Latin American Conference on Computational Intelligence (LA-CCI), Arequipa, Peru, pp. 1–6 (2017). https://doi.org/10.1109/LA-CCI.2017.8285721
27. Rivero, C.R., Pucheta, J., Baumgartner, J., Laboret, S., Sauchelli, V.: Short-series prediction with BEMA approach: application to short rainfall series. IEEE Lat. Am. Trans. **14**(8), 3892–3899 (2016). https://doi.org/10.1109/TLA.2016.7786377
28. Rivero, C.R., Pucheta, J., Herrera, M., Sauchelli, V., Laboret, S.: Time series forecasting using bayesian method: application to cumulative rainfall. IEEE Lat. Am. Trans. **11**(1), 359–364 (2013). https://doi.org/10.1109/TLA.2013.6502830
29. Rivero, C., Pucheta, J., Laboret, S., et al.: Energy associated tuning method for short-term series forecasting by complete and incomplete datasets. J. Artif. Intell. Soft Comput. Res. **7** (1), 5–16 (2016). https://doi.org/10.1515/jaisrc-2017-0001
30. Pucheta, J., Rivero, C.R., Herrera, M., Salas, C., Sauchelli, V.: Rainfall forecasting using sub sampling nonparametric methods. IEEE Lat. Am. Trans. **11**(1), 646–650 (2013). https://doi.org/10.1109/TLA.2013.6502878
31. Pucheta, J., Patiño, D., Kuchen, B.: A statistically dependent approach for the monthly rainfall forecastfrom one point observations. In: Li, D., Zhao, C. (eds.) CCTA 2008. IAICT, vol. 294, pp. 787–798. Springer, Boston (2009). https://doi.org/10.1007/978-1-4419-0211-5_1

32. Bergmeir, C., Hyndman, R., Koo, B.: A note on the validity of cross-validation for evaluating autoregressive time series prediction. Comput. Stat. Data Anal. **120**(C), 70–83 (2018)
33. Bergmeir, C., Benitez, J.M.: On the use of cross validation for time series predictor evaluation. Inf. Sci. **191**, 192–213 (2012)
34. Hosking, J.R.M.: Modeling persistence in hydrological time series using fractional differencing. Water Resour. Res. **20**(12), 1898–1908 (1984)
35. Makridakis, S., Spiliotis, E., Assimakopoulos, V.: Statistical and machine learning forecasting methods: concerns and ways forward. PLoS ONE **13**(3), 1–26 (2018)
36. Goodwin, P., Lawton, R.: On the asymmetry of the symmetric MAPE. Int. J. Forecast. **15**(4), 405 ± 408 (1999). http://dx.doi.org/10.1016/S0169-2070(99)00007-2

A Highly Efficient Biomolecular Network Representation Model for Predicting Drug-Disease Associations

Han-Jing Jiang[1,2,3], Zhu-Hong You[1(✉)], Lun Hu[1], Zhen-Hao Guo[1,2,3], Bo-Ya Ji[1,2,3], and Leon Wong[1,2,3]

[1] The Xinjiang Technical Institute of Physics and Chemistry, Chinese Academy of Sciences, Urumqi 830011, China
zhuhongyou@ms.xjb.ac.cn
[2] University of Chinese Academy of Sciences, Beijing 100049, China
[3] Xinjiang Laboratory of Minority Speech and Language Information Processing, Urumqi, China

Abstract. Identification of drug-disease association is crucial for drug development and reposition. However, discovering drugs which are associated with diseases from *in vitro* testing is costly and time-consuming. Accumulating evidence showed that computational approaches can complement biological and clinical experiments for this identification task. In this work, we propose a novel computational method Node2Bio for predicting drug-disease associations using a highly efficient biomolecular network representation model. Specifically, we first construct a large-scale biomolecular association network (BAN) by integrating the associations among drugs, diseases, proteins, miRNAs and lncRNAs. Then, the network embedding model node2vec is used to extract network behavior features of drug and disease nodes. Finally, the feature vectors are taken as inputs for the XGboost classifier to predict potential drug-disease associations. To evaluate the prediction performance of the proposed method, five-fold cross-validation tests are performed on a widely used SCMFDD-S dataset. The experimental results demonstrate that our method achieves competitive performance with a high AUC value of 0.8569, which suggests that our method is a useful tool for identification of drug-disease associations.

Keywords: Drug-disease associations · Drug reposition · Drug-disease association · Node2Bio · Biomolecular network

1 Introduction

Drug-disease association is almost involved in the entire process of drug repositioning, providing a theoretical basis for the discovery of new drug efficacy. Therefore, it is a prospective task to explore as many new drug-disease associations as possible. In recent years, several computational methods of drug-disease association based on drug target information, drug structure information, disease semantic information and other information sources have been proposed. For example, some methods use disease, drug and drug target to predict drug-disease associations (TL-HGBI). Drug - disease

D.-S. Huang and P. Premaratne (Eds.): ICIC 2020, LNAI 12465, pp. 271–279, 2020.
https://doi.org/10.1007/978-3-030-60796-8_23

association prediction based on drug target information is a popular method [1]. Drug targets are also considered to be one of the sources of information for predicting drug-disease interactions, but the computational conditions for these methods are that the drug can find the corresponding drug target information. In these methods, a three-layer heterogeneous network is typically constructed using drugs, diseases, and drug targets, and the network is constructed based on the distribution of similarity measures [2]. Combining multiple associated sources of information provides more insight into predictive drug-disease association than using only drug targets as sources of information [3]. Therefore, how to effectively integrate more information sources has attracted wide attention [4].

Inspired by graph representation learning, we re-examine some basic relational prediction problems from the perspective of graphs to find better solutions. Graph is a basic and commonly used data structure. Many scenes in the real world can be abstracted into a graph structure, such as social network, traffic network, etc. [5]. The biomolecule in the cell can also be viewed as a graph structure, with the association of different types of biomolecules forming the edges of the graph and the biomolecules serving as the nodes of the graph [6]. Using graph theory to develop reliable bio-association graph technology to solve bio-association prediction problem will have a subversive impact on current bioinformatics research [7]. There is no doubt that the seamless integration of graph with biomacromolecules will drive the development of the post-genomic era [8].

The prediction of nodes and edges is an important task in network analysis [9]. In the node classification task, the most likely node label in the prediction network is the first task [10]. For example, in the drug-target interaction network, the focus is on predicting the functional labeling of drugs [11]. Similarly, in a molecular association network, we want to predict whether a pair of nodes in the network should have an edge that connects them [12, 13]. Predicting nodes and edges can help us discover new interactions between drugs and diseases [14]. Node2vec is an algorithm framework for learning the continuous feature representation of nodes in a network [15]. It defines a flexible concept of node network domain and designs a biased random walk process to effectively explore different network domains [16].

Computational methods used to find new drugs and disease associations can solve the problem of high cost and low efficiency, so it has important practical significance [17]. Based on the similarity of biomolecular association network and graph structure, this paper proposes a biomolecular network representation learning model to predict drug-disease association [18]. The model is based on the biomolecular network representation method Node2Bio and XGboost classifier [19].

The biomolecular network consists of two parts: nodes (drugs, diseases, proteins, ncRNA (miRNA, lncRNA)) and edges (the relationship of nodes) [20]. Each node can be represented in two ways: attribute information of the node (such as the molecular fingerprint of the drug and the phenotype of the disease) and a vector of relationships with other nodes in the network embedding [21]. Finally, all node features are integrated to form feature descriptors and imported into the XGboost classifier to predict the association of each drug with all diseases [22]. It is worth noting that although the main purpose is to predict drug-disease association, our proposed molecular association network model and iterative update algorithm can be applied to other prediction problems as well [23].

2 Materials and Methods

2.1 Nine Kinds of Molecular Associations

To build a molecular association network, we need to download drugs, diseases, lncRNA, miRNA and protein information from different data sources. Then the feature vectors of drug, disease, lncRNA, miRNA and protein were calculated by different methods. All known interactions are derived from existing databases [24]. Drugs and diseases are downloaded from the CTD database and drugs SMILE is downloaded from DrugBank [25]. Zhang et al. collated 18,416 drug-disease associations from the CTD database and named this data set "SCMFDD-S" [26]. Drug-protein associations were collected from the DrugBank database for a total of 11,107 associations. The Protein-protein association is based on 19,237 associations in the STRING dataset [27]. The Protein-disease association was collected from the DisGeNET [28] database and a total of 25,087 associations were collected. A total of 690 lncRNA-protein associations were collected from the LncRNA2Target [29] database. A total of 1264 lncRNA-disease associations were collected from the LncRNADisease [30] database and the lncRNASNP2 [31] database. 4494 miRNA-protein associations were collected from miRTarBase [32]. The miRNA-disease association was collected from HMDD [33] for a total of 16,427. 8374 miRNA-lncRNA associations were downloaded from lncRNASNP2 [31].

2.2 Disease MeSH Descriptors and Directed Acyclic Graph

In this study, we used the MeSH disease descriptor downloaded from the National Library to calculate the semantic similarity of the disease. This representation is described by a directed acyclic graph (DAG), in which nodes in the DAG represent disease, and the ends of each edge are the parent and child nodes, respectively [34]. If the disease $p(j)$ is the parent of the disease $p(i)$, the disease $p(i)$ can be described as:

$$DAG_{p(i)} = \left(p(i), N_{p(i)}, E_{p(i)}\right) \tag{1}$$

where $N_{p(i)}$ represents the set of points for all diseases. $E_{p(i)}$ contains all the edges in $DAG_{p(i)}$.

In $DAG_{p(i)}$ of disease s, the contribution of any ancestral disease $p(i)$ to disease s is as the formula:

$$\begin{cases} D_{p(i)}(s) = 1 & if\ s = p(i) \\ D_{p(i)}(s) = \max\{\beta \cdot D_{p(i)}(\acute{s}) | \acute{s} \in children\ of\ s\} & if\ s \neq p(i) \end{cases} \tag{2}$$

In addition, disease $p(i)$ contributes 1 to its own semantic value. Therefore, the semantic value $DV(p(i))$ of the disease $p(i)$ is defined as follows:

$$DV(p(i)) = \sum\nolimits_{s \in N_{p(i)}} D_{p(i)}(s) \tag{3}$$

We hypothesized that the more DAG Shared between diseases, the higher the semantic similarity score. The DAG similarity value $SV_1(p(i), p(j))$ of the disease $p(i)$ and disease $p(j)$ is calculated as:

$$SV_1(p(i),p(j)) = \frac{\sum_{s \in N_{p(i)} \bigcap N_{p(j)}} (D_{p(i)}(s) + D_{p(j)}(s))}{DV(p(i)) + DV(p(j))} \quad (4)$$

2.3 Stacked Autoencoder

Stacked auto-encoder (SAE) is a multi-layer neural network and is a deep learning model that uses modular units to create deep neural networks [35]. The purpose of Auto-encoder is to make the value of the output as close as possible to the value of the input. Given a drug molecular fingerprint set x, autoencoder input x through an expression to determine the mapping of hidden:

$$Y = \sigma(W_1 x + b_1) \quad (5)$$

where σ denotes the logistic sigmoid. Y is the result of the hidden representation, and x is the reconstructed vector after mapping:

$$\acute{x} = \sigma(W_2 x + b_2) \quad (6)$$

The stack auto-encoder is a combination of multiple autoencoders. The principle is to use the output of the first layer of the autoencoder as the input of the next layer of the autoencoder, and so on, to obtain the output of the last layer of the auto-encoder. In this paper, a drug fingerprint obtains a descriptor representing a structural feature by a stacked autoencoder.

2.4 NcRNA and Protein Sequence

We chose to encode the sequence using a 64 ($4 \times 4 \times 4$) dimensional vector encoding ncRNA and analyzed it with K-mer, where k is taken as 3. The 3-mer mode is a sliding window containing 3 nucleotides to analyze each transcription. In the initial state, the number of occurrences of all patterns is set to 0. If the window matches exactly the string in the transcript, the count is incremented by 1 and the slide continues. Finally, divide the number of occurrences by the length of the sequence to get the normalized frequency.

The article by Shen et al. [36] proposes that protein sequences can be encoded into four classes based on the polar side chains of the amino acids. Each protein sequence is characterized by a 3-mer. The ncRNA uses the same normalized frequency calculation method.

2.5 Node Representation

In the molecular association network, many nodes and edges are involved in the prediction task. We chose node2vec to learn the continuous feature representation of nodes in the network [37]. Suppose just traversed go from edge (t, v) to node v. Assume that the transition probability of the next step edge (v, x) is π_{vx}. We set the unnormalized transition probability to $\pi_{vx} = \alpha_{pq}(t,x) \cdot \omega_{vx}$, where d_{tx} represents the shortest path distance between nodes t and x:

$$\alpha_{pq}(t,x) = \begin{cases} \frac{1}{p} & if\ d_{tx} = 0 \\ 1 & if\ d_{tx} = 1 \\ \frac{1}{q} & if\ d_{tx} = 2 \end{cases} \tag{7}$$

2.6 XGBoost

XGBoost algorithm has been widely applied in the field of bioinformatics. XGBoost is an integration of several weak classifiers, in this case the CART regression tree model. The objective function of XGBoost is defined as:

$$Obj = \sum\nolimits_{m=1}^{n} l(y_m, \hat{y}_m) + \sum\nolimits_{k=1}^{K} \Omega(f_k) \tag{8}$$

$$\Omega(f) = \Upsilon T + 0.5\lambda\|\omega\|^2 \tag{9}$$

Here l is a differentiable convex loss function that measures the difference between the prediction $\widehat{y_m}$ and the target y_m. The complexity of the Ω penalty model. The newly generated tree is to fit the residual error predicted last time. When t trees are generated, the prediction score is:

$$\hat{y}_m^{(t)} = \hat{y}_m^{(t-1)} + f_t(x_m) \tag{10}$$

The target function is updated to:

$$\mathcal{L}^{(t)} = \sum\nolimits_{m=1}^{n} l\left(y_m, \hat{y}_m^{t-1} + f_t(x_m)\right) + \Omega(f_t) \tag{11}$$

In general, a second order approximation can be used to quickly optimize the target. The approximate objective function is:

$$\mathcal{L}^{(t)} \simeq \sum\nolimits_{m=1}^{n} \left[l\left(y_m, \hat{y}^{t-1}\right) + g_m f_t(x_m) + \frac{1}{2} h_m f_t^2(x_m) \right] + \Omega(f_t) \tag{12}$$

where g_m is the first derivative and h_m is the second derivative.

$$g_m = \partial_{\hat{y}^{(t-1)}} l\left(y_m, \hat{y}^{t-1}\right) \tag{13}$$

$$h_m = \partial^2_{\hat{y}^{(t-1)}} l\left(y_m, \hat{y}^{t-1}\right) \tag{14}$$

Since the prediction score of the former $t - 1$ tree and the residual of y do not affect the optimization of the objective function, the objective function can be simplified as:

$$\tilde{\mathcal{L}}^{(t)} = \sum\nolimits_{m=1}^{n}\left[g_m f_t(x_m) + \frac{1}{2}h_m f_t^2(x_m)\right] + \Omega f(t) \tag{15}$$

3 Results and Discussion

3.1 Evaluation Criteria

In order to verify the predictive power of our model. Five-fold cross-validation was performed to verify. All samples were first randomly divided into nearly the same number of five subsets. Each time four subsets are used as a training set and the remaining subsets are used as test sets, the process is repeated five times so that each subset can be used as a test set. Finally, the average of the five groups was taken as the final result. Several evaluation criteria used in our study to estimate the predictive power of our model, including sensitivity (Sen.), specificity (Spec.), precision (Prec.) accuracy (Acc.) and Matthews correlation coefficient (MCC). The calculation method is as follows:

$$Sen. = \frac{TP}{TP + FN} \tag{16}$$

$$Spec. = \frac{TN}{FP + TN} \tag{17}$$

$$Prec. = \frac{TP}{TP + FP} \tag{18}$$

$$Acc. = \frac{TP + TN}{TP + TN + FP + FN} \tag{19}$$

$$MCC = \frac{TP \times TN - FP \times FN}{\sqrt{(TP + FP)(TP + FN)(TN + FP)(TN + FN)}} \tag{20}$$

For further evaluation, we also compute the receiver operating characteristic (ROC) curve, sum up the ROC curve in a numerical way, and calculate the area under the ROC curve (AUC). We compute the precision-recall (PR) curve and calculate the area under the PR curve (AUPR).

4 Results and Discussion

4.1 Five-Fold Cross-Validation on SCMFDD-S Dataset

We performed five-fold cross-validation on the SCMFDD-S data set to evaluate the performance of Node2Bio in predicting drug-disease association [38]. The process of cross-validation is to divide the data set into five equal parts, select a different set as the test set each time, and the remaining four sets as the training set, and repeat the experiment five times [39]. Node2Bio yielded an average accuracy of 77.42%, sensitivity of 75.25%, specificity of 79.59%, precision of 78.67%, Matthews correlation coefficient of 54.90% and AUC of 85.69% with standard deviations of 0.24%, 1.01%, 0.74%, 0.41%, 0.46% and 0.12% [40]. To evaluate the performance of Node2Bio, we compare it to some related methods of NTSIM-C. The comparison method uses the same data set for five-fold cross-validation. The experimental results represented by AUC are shown in Table 1. The results from experiments demonstrate that the performance of Node2Bio is significantly better than the related methods of NTSIM-C. Unlike the comparison method, Node2Bio combines nine molecular associations and integrates related information from a cellular perspective to achieve significant predictive effects.

Table 1. AUC comparison of Node2Bio-based method with different methods

Methods	AUC (%)
NTSIM-C-target	84.40
NTSIM-C-enzyme	84.50
NTSIM-C-pathway	85.00
NTSIM-C-substructure	84.70
NTSIM-C-drug-drug interaction	84.30
Node2Bio	85.69

5 Conclusion

In this study, we proposed a computational method for predicting drug-disease associations using a highly efficient biomolecular network representation model. The proposed method leverages multiple types of relational data that are biologically associated and constructs a heterogeneous network on which a graph embedding technique, node2vec, is applied for feature extraction. Using the embedding feature as inputs, we adopted the XGboost algorithm to do classification for drug-disease association. The experimental results are the proposed method to be effective, robust and superior to existing methodologies. It is anticipated that the model we trained can be applied to predict drug effects on different kinds of diseases on a large scale.

Funding. This work is supported by the Xinjiang Natural Science Foundation under Grant 2017D01A78.

Conflict of Interest. The authors declare that they have no conflict of interest.

References

1. Chen, Z.-H., et al.: Identification of self-interacting proteins by integrating random projection classifier and finite impulse response filter. BMC Genom. **20**(13), 1–10 (2019)
2. Zheng, K., Wang, L., You, Z.-H.: CGMDA: an approach to predict and validate MicroRNA-disease associations by utilizing chaos game representation and LightGBM. IEEE Access **7**, 133314–133323 (2019)
3. Wang, L., et al.: Identification of potential drug–targets by combining evolutionary information extracted from frequency profiles and molecular topological structures. Chem. Biol. Drug Des. (2019)
4. Jiang, H.-J., You, Z.-H., Zheng, K., Chen, Z.-H.: Predicting of drug-disease associations via sparse auto-encoder-based rotation forest. In: Huang, D.-S., Huang, Z.-K., Hussain, A. (eds.) ICIC 2019. LNCS (LNAI), vol. 11645, pp. 369–380. Springer, Cham (2019). https://doi.org/10.1007/978-3-030-26766-7_34
5. Zheng, K., et al.: DBMDA: a unified embedding for sequence-based miRNA similarity measure with applications to predict and validate miRNA-disease associations. Mol. Ther.-Nucleic Acids **19**, 602–611 (2020)
6. Guo, Z.-H., You, Z.-H., Yi, H.-C.: Integrative construction and analysis of molecular association network in human cells by fusing node attribute and behavior information. Mol. Ther.-Nucleic Acids **19**, 498–506 (2020)
7. Wang, M.-N., et al.: LDGRNMF: LncRNA-disease associations prediction based on graph regularized non-negative matrix factorization. Neurocomputing (2020)
8. Wang, M.-N., et al.: GNMFLMI: graph regularized nonnegative matrix factorization for predicting LncRNA-MiRNA interactions. IEEE Access **8**, 37578–37588 (2020)
9. Wong, L., et al.: LNRLMI: linear neighbour representation for predicting lncRNA-miRNA interactions. J. Cell Mol. Med. **24**(1), 79–87 (2020)
10. Hu, P., et al.: Learning multimodal networks from heterogeneous data for prediction of lncRNA-miRNA interactions. IEEE/ACM Trans. Computat. Biol. Bioinform. (2019)
11. Huang, Y.-A., et al.: ILNCSIM: improved lncRNA functional similarity calculation model. Oncotarget **7**(18), 25902 (2016)
12. You, Z.-H., et al.: Highly efficient framework for predicting interactions between proteins. IEEE Trans. Cybern. **47**(3), 731–743 (2016)
13. Huang, Y.-A., Chan, K.C., You, Z.-H.: Constructing prediction models from expression profiles for large scale lncRNA–miRNA interaction profiling. Bioinformatics **34**(5), 812–819 (2018)
14. Zheng, K., et al.: iCDA-CGR: Identification of circRNA-disease associations based on Chaos Game Representation. PLoS Comput. Biol. **16**(5), e1007872 (2020)
15. Guo, Z.-H., Yi, H.-C., You, Z.-H.: Construction and comprehensive analysis of a molecular association network via lncRNA–miRNA–disease–drug–protein graph. Cells **8**(8), 866 (2019)
16. Guo, Z.-H., et al.: A learning-based method for lncRNA-disease association identification combing similarity information and rotation forest. iScience **19**, 786–795 (2019)
17. Jiang, H.-J., You, Z.-H., Huang, Y.-A.: Predicting drug – disease associations via sigmoid kernel-based convolutional neural networks. J. Transl. Med. **17**(1), 1–11 (2019)
18. Jiang, H.-J., Huang, Y.-A., You, Z.-H.: SAEROF: an ensemble approach for large-scale drug-disease association prediction by incorporating rotation forest and sparse autoencoder deep neural network. Sci. Rep. **10**(1), 1–11 (2020)
19. Huang, Y.-A., et al.: Graph convolution for predicting associations between miRNA and drug resistance. Bioinformatics **36**(3), 851–858 (2020)

20. Wang, Y., et al.: A high efficient biological language model for predicting protein–protein interactions. Cells **8**(2), 122 (2019)
21. Wang, Y., et al.: Predicting protein interactions using a deep learning method-stacked sparse autoencoder combined with a probabilistic classification vector machine. Complexity **2018** (2018)
22. Wang, L., et al.: Combining high speed ELM learning with a deep convolutional neural network feature encoding for predicting protein-RNA interactions. IEEE/ACM Trans. Comput. Biol. Bioinform. (2018)
23. Huang, Y.-A., You, Z.-H., Chen, X.: A systematic prediction of drug-target interactions using molecular fingerprints and protein sequences. Curr. Protein Pept. Sci. **19**(5), 468–478 (2018)
24. Huang, Y.-A., et al.: Sequence-based prediction of protein-protein interactions using weighted sparse representation model combined with global encoding. BMC Bioinform. **17** (1), 184 (2016)
25. Wishart, D.S., et al.: DrugBank 5.0: a major update to the DrugBank database for 2018. Nucleic Acids Res. **46** (2018)
26. Zhang, W., et al.: Predicting drug-disease associations by using similarity constrained matrix factorization. BMC Bioinform. **19**(1), 233 (2018)
27. Szklarczyk, D., et al.: The STRING database in 2017: quality-controlled protein–protein association networks, made broadly accessible. Nucleic Acids Res. **45** (2017)
28. Pinero, J., et al.: DisGeNET: a comprehensive platform integrating information on human disease-associated genes and variants. Nucleic Acids Res. **45** (2017)
29. Jiang, Q., et al.: LncRNA2Target: a database for differentially expressed genes after lncRNA knockdown or overexpression. Nucleic Acids Res. **43**(Database issue), D193 (2015)
30. Geng, C., et al.: LncRNADisease: a database for long-non-coding RNA-associated diseases. Nucleic Acids Res. **41**(Database issue), D983–D986 (2013)
31. Miao, Y.R., et al.: lncRNASNP2: an updated database of functional SNPs and mutations in human and mouse lncRNAs. Nucleic Acids Res. **46**(Database issue), D276–D280 (2018)
32. Chou, C.H., et al.: miRTarBase update 2018: a resource for experimentally validated microRNA-target interactions. Nucleic Acids Res. **46**(Database issue) (2017)
33. Yang, L., et al.: HMDD v2.0: a database for experimentally supported human microRNA and disease associations. Nucleic Acids Res. **42**(Database issue), D1070 (2014)
34. Wang, D., et al.: Inferring the human microRNA functional similarity and functional network based on microRNA-associated diseases. Bioinformatics **26**(13), 1644–1650 (2010)
35. Jiang, H.-J., Huang, Y.-A., You, Z.-H.: Predicting drug-disease associations via using gaussian interaction profile and kernel-based autoencoder. Biomed. Res. Int. **2019**, 11 (2019)
36. Shen, J., et al.: Predicting protein-protein interactions based only on sequences information. Proc. Natl. Acad. Sci. U.S.A. **104**(11), 4337–4341 (2007)
37. Grover, A., Leskovec, J.: node2vec: scalable feature learning for networks. In: ACM SIGKDD International Conference on Knowledge Discovery & Data Mining (2016)
38. Guo, Z.-H., et al.: MeSHHeading2vec: a new method for representing MeSH headings as vectors based on graph embedding algorithm. Briefings Bioinform. (2020)
39. Huang, Y.-A., et al.: Prediction of microbe–disease association from the integration of neighbor and graph with collaborative recommendation model. J. Transl. Med. **15**(1), 1–11 (2017)
40. Guo, Z.-H., et al.: A learning based framework for diverse biomolecule relationship prediction in molecular association network. Commun. Biol. **3**(1), 1–9 (2020)

DAAT: A New Method to Train Convolutional Neural Network on Atrial Fibrillation Detection

Jian Zhang[1], Juan Liu[1,2](✉), Pei-Fang Li[1], and Jing Feng[1,2]

[1] School of Computer Science, Wuhan University, Wuhan 430072, Hubei, China
liujuan@whu.edu.cn
[2] College of Artificial Intelligence, Wuhan University, Wuhan 430072, Hubei, China

Abstract. Atrial fibrillation (AF) is a common disease in elderly people which is associated with high morbidity. Detecting AF with electrocardiogram (ECG) recordings benefits them for early diagnose and treatment. Lots of models based on convolutional neural network (CNN) have been proposed for such purpose. However, how to train such models so as to get better performance still remains challenging. In this paper, we put forward a dynamic attention assistant training (DAAT) process for CNN model training, which can not only improve the accuracy of verified strong ResNet on AF detection task, but also help hardly trained DenseNet to get a good performance under the precondition of a low proportion of positive AF samples, which usually occurs in real tasks. The training process works even when some attention layers have already been utilized within convolutional layers like SENet. (The source code can be downloaded from https://github.com/mszjaas/DAAT).

Keywords: Atrial fibrillation · CNN · ECG · Dynamic attention assistant training · Training method

1 Introduction

Atrial fibrillation (AF) is the most common cardiac arrhythmia affecting approximately 3% of the adult population, with great prevalence in elderly people [1]. This may come from a right atrial re-entrant activity caused by an AF-maintaining substrate that right heart produces when people are under diseases of lungs and pulmonary circulation, which induces right atrial fibrosis and conduction abnormalities [2]. AF can often go unnoticed and yet is a risk factor for stroke [3]. It is also associated with an increasing risk of dementia and cognitive impairment [4]. AF detection can help with early diagnose for patients to be aware and treated as soon as possible.

Main characteristics of AF lay on the absence of P-wave or irregular heart rate variability [5]. Currently, the holter is a commonly used technology to detect AF by monitoring electrocardiogram (ECG) signals. However, the traditional analysis of ECGs is mainly done by experienced cardiologists, which is laborious and inefficient. It is required to automatically analyze ECGs by using machine learning techniques.

D.-S. Huang and P. Premaratne (Eds.): ICIC 2020, LNAI 12465, pp. 280–290, 2020.
https://doi.org/10.1007/978-3-030-60796-8_24

Now that deep learning models such as convolutional neural networks (CNN) have achieved great success in many medical tasks [6], some researchers have proposed CNN based methods for predicting AF via ECGs and achieved good results. Most of the existing researches mainly focus on network architecture, model fusing and feature extracting while few efforts are devoted to the model training, especially when sufficient amount of data is not available.

Inspired by the work in [7], in this paper we propose a dynamic attention assistant training (DAAT) process for assisting training CNN models with small positive data. The key idea of DAAT is to introduce an assistant symmetric architecture during the training process of, so that the model can reach to a good state by gradually reducing the attention-like constraint. After the assistant architecture helps the model learn more from the training data to get better performance, it can then be removed from the network. (The source code can be downloaded from https://github.com/mszjaas/DAAT). In order to verify the utility of DAAT, three CNN models are chosen to train for the AF-detection task: ResNet [8], DenseNet [9] and SENet [10]. Both the static attention assistant training (SAAT) and the DAAT are performed to train the models, and the second method is proven more effective.

2 Related Work

2.1 Improving Accuracy of CNN Models on AF Detection Task

Feature extracting and model fusing are two common ways to improve the accuracy of CNN models on the AF detection task. Tran et al. introduces the feature of ECG sequence into multi-layer perceptron and the raw data into residual blocks followed by Long Short-Term Memory (LSTM) and fuses the output into fully connected layers [11]. Kharshid *et al.* extracts a 188-dimensions feature and feeds it into a residual convolutional neural network for AF classification [12]. Lai *et al.* use both R-R interval and F-wave frequency spectrum as features for training a CNN model [13]. Shen *et al.* combines multi-classifier extracted features and neural network extraction features for atrial fibrillation classification [14]. Zhu *et al.* pre-processes the ECG sequence and detects AF with a 94-layer SE-ResNet [15]. Some also try to propose training method for this task. Shi *et al.* pre-trains a multiple-input deep neural network by labeled samples and post-train it by continuously fine-tuning on AF data [16].

2.2 Training Methods on CNN Models

There are three main methods to train a CNN model to perform better. (1) Add some new limits to guide the training process. Zhang *et al.* proposes a hierarchical guidance and regularization learning framework to utilize multiple wavelet features for training CNN model [17]. Dong *et al.* use a weighted cross entropy loss to assist CNN model in learning more from positive rather than negative labels [18]. (2) Do data augmentation or pre-treatment. Liu *et al.* trains a CNN model by under-sampling the negative data and use weighted cross entropy loss and focal loss to guide training [19]. Zheng *et al.* Proposes a two-stage data augmentation method to improve accuracy of CNN on image classification [20]. (3) Transfer training with large size of unlabeled data. Huang *et al.* proposes a two-stage transfer learning strategy in which they firstly initialize their CNN model by learning texture from source data and then transfer it to train by target data [21].

3 Method

3.1 Twin Convolutional Model Architecture for Training

Similar to [7], a twin convolutional model (TCM) architecture is constructed as in Fig. 1. For a CNN model to be trained, a copy of it is constructed to make the TCM with two inputs. For every ECG sequence in training set, a reversed sequence is obtained by flipping it horizontally. The pair of original and its reversed sequence act as the inputs to the original and the copy CNNs of the TCM each followed by a global average pooling (GAP) layer, a full-join classification (FC) layer and a softmax layer.

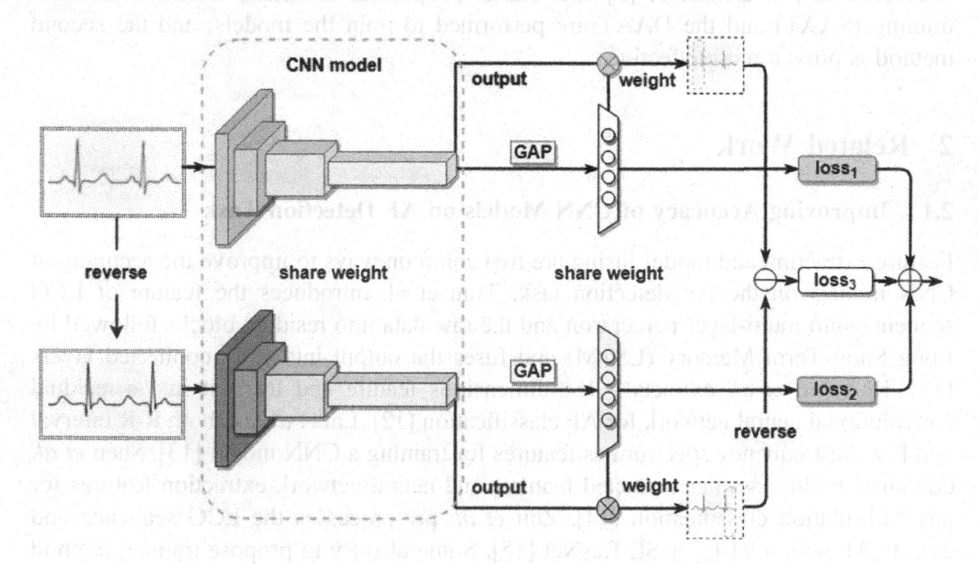

Fig. 1. Architecture of the twin-convolutional-model. $loss_1$, $loss_2$ and $loss_3$ refer to $loss_{orgin}$, $loss_{reverse}$ and $loss_{assit}$ in Eq. (3).

Let the output $X \in R^{N \times C \times T}$ of TCM, and the weight $W \in R^{L \times C}$ of FC, in which *N, C, T, L* refers to number of batch, channel, sequence and label respectively. *W* can be interpreted as the weight of each channel for each label. The class activation mapping *M* can be computed by *X* and *W* as Eq. (1):

$$M_l(t) = \sum_{c=1}^{C} W(l, c) X_l(t) \quad (1)$$

$M_l(t)$ corresponds to the heatmap of attention at time slot *t* for the classifier to point to a label [22]. Since two output *X*s come from symmetrical inputs, *M* can also be supposed symmetrical. So M is flipped horizontally again and calculate the variance of it to another M, and define the loss of attention as Eq. (2):

$$loss_{assist} = \frac{1}{LT}\sum_{l=1}^{L}\sum_{t=1}^{T}\left|M_l^1(t) - flip(M_l^2(t))\right|^2 \quad (2)$$

Then the total loss function can be defined as Eq. (3)

$$loss_{total} = loss_{label} + k * loss_{assist};\ loss_{label} = loss_{origin} + loss_{reverse} \quad (3)$$

where k is a weight parameter of $loss_{assist}$, which adjusts the influence of the attention loss item.

3.2 Dynamic Attention Assistant Training Process

Generally, the models are trained by fixing a predetermined value for the parameter k, which is call as SAAT process. In order to get a good model with high performance, an appropriate value of k is needed. Too large or too small k would imports too much or too less assistance to training model than an appropriate one.

Different from the usual one, this research proposes the dynamic attention assistant training (DAAT) process which is not just modifying the parameter k, but also balancing the assisting for training and the impacting on the strength of model. The process contains two stages: pre-training and refining. In the pre-training stage, k begins with an initial value k_0, and then dynamically decreases along with the epochs of the model training; in the filtering stage, the pre-trained model is refined by being trained other epochs without TCM architecture.

By the way, the parameters of the models will update as the average of corresponding new ones during each batch. Best parameters are remained at the end of each epoch. After training, the original trained CNN model can be used on the AF-detection task by simply removing the copy one.

4 Evaluation Experiment

In order to evaluate whether the proposed DAAT method can help to get a good AF-detection model from unbalanced data where the labeled AF data only accounts for a small proportion, three widely used CNN models are chosen and respectively installed in the TCM training architecture (Fig. 1). Two kinds of training strategies (SAAT and DAAT) are carried out and the comparing results are shown in this section.

4.1 Data Set

The experiment data set comes from physionet2017 database [23], containing 8528 ECG recordings of four types (Normal, AF, Other rhythm, Noisy). The number and proportion of the four types of data are shown in Table 1, from which we can see that the data is very unbalanced and the proportion of AF data is very small. Though this paper mainly focuses on the detection of AFs, a four-class classifier is used without doing any class combination. Furthermore, No data augmentation is performed on AF class in the training process. The original length of the ECG sequence is extended or clipped to 9,216 as a power of 2 to fit the CNN models in which extension is implemented by appending 0 s.

Table 1. Profile of ECG data

Type	Number	Proportion (%)
Normal	5,154	60.5
AF	771	9.0
Other rhythm	2,557	30.0
Noisy	46	0.5
Total	8,528	100

Each weight of constitutional kernel is initialized as in [24]. Adam optimizer is used with learning rate 0.0001 for model training [25]. Each training goal is to train the model to get higher overall accuracy. Overall classifying accuracy and its F1 score on AF class are investigated. F1 score is calculated as [23].

4.2 Experiment Settings

During SAAT process, different k values are chosen to observe how the weight of $loss_{assit}$ affects model training. Each model is trained for 300 epochs.

During DAAT process, the initial value of k is set referring to the best one in the SAAT process. (That doesn't means a SAAT process is necessary before DAAT. The initial value of k can also be set as any other values at different magnitude tentatively.) During the pre-training stage, k declines at an exponential rate of 10^{-1} each 50 epochs, and the model is totally pre-trained 200 epochs. Then in the refining stage, the pre-trained model in the TCM architecture will be retrieved and trained another 100 epochs with the original ECG sequence.

4.3 CNN Models Chosen for Training

Three CNN models are chosen to do the validation experiments: ResNet [8], DenseNet [9] and SENet [10]. ResNet has been used for the same task with data augmentation and integrating some other data; DenseNet is hardly trained and performed not very well in this task; SENet contains lots of attention process between convolutional layers. Since the assistant training method looks like working with an attention mechanism at the output of last convolutional layer, the method is also performed on SENet to verify it works when attention processes has been used within convolutional layers. For each kind of CNN, the TCM training architecture is constructed to train the model.

5 Result

5.1 Influence of the Attention on the Performance of the Model

The performances of models trained by SAAT process are shown in Fig. 2, illustrating that the changing tendencies of performance along with k are similar for three models: too large or too small k provides less assistance for training model than an appropriate one. As $-lgk$ decreases from large value (corresponding to the increasing of k), the total accuracy along with F1 score of AF rises, which means the $loss_{assit}$ contributes to total loss and SAAT helps the model to be trained better. However, the model performs worse when k continues increasing. This issue may arise out of the less significance of $loss_{label}$ and more dominance of $loss_{assit}$ in the total loss, in which case models may learn less from labels and show poorer performance. Accurately, three models achieve the highest accuracies when k is set to 10^{-3}, 10^{-5}, and 10^{-2} respectively.

The tendency mentioned above seems not so strong on ResNet as it is already strong enough for AF-detection task and SAAT doesn't help improve a lot. While as for DenseNet which can hardly be trained for AF detection (DenseNet always stays over-fitting during the 300 epochs of training by common methods, while no over-fitting happened when using DAAT or SAAT), k over 10^{-5} can assist to detect AF powerfully. SENet can also benefit obviously from SAAT in the same trend regardless of the asynchrony of largest accuracy and F1 score of AF.

Note that only label loss works when k is zero (not shown), and the result is simply like doubling number of ECG sequences. While models get lower accuracy, which comes to the conclusion that models achieve better performance not because of data augmentation but the help of $loss_{assist}$.

5.2 Models Trained by DAAT

Different from SAAT, DAAT process dynamically balances the significance of $loss_{assit}$ and $loss_{origin}$. In the pre training stage, weight of the assisting of $loss_{assit}$ decreases, which leads the models to learn more from labels. From Table 2 the classifying accuracy of three models trained by DAAT (even just the models pre-trained for 200 epochs) can be higher than the best one trained by SAAT. Moreover, both SAAT and DAAT make use of the assistant attention mechanism to train the models, thus the models can learn more and achieve better performance than the original one. Table 2 also shows that the attention assistant training method can promisingly improve the performance of DensNet which is hard to train by using traditional methods.

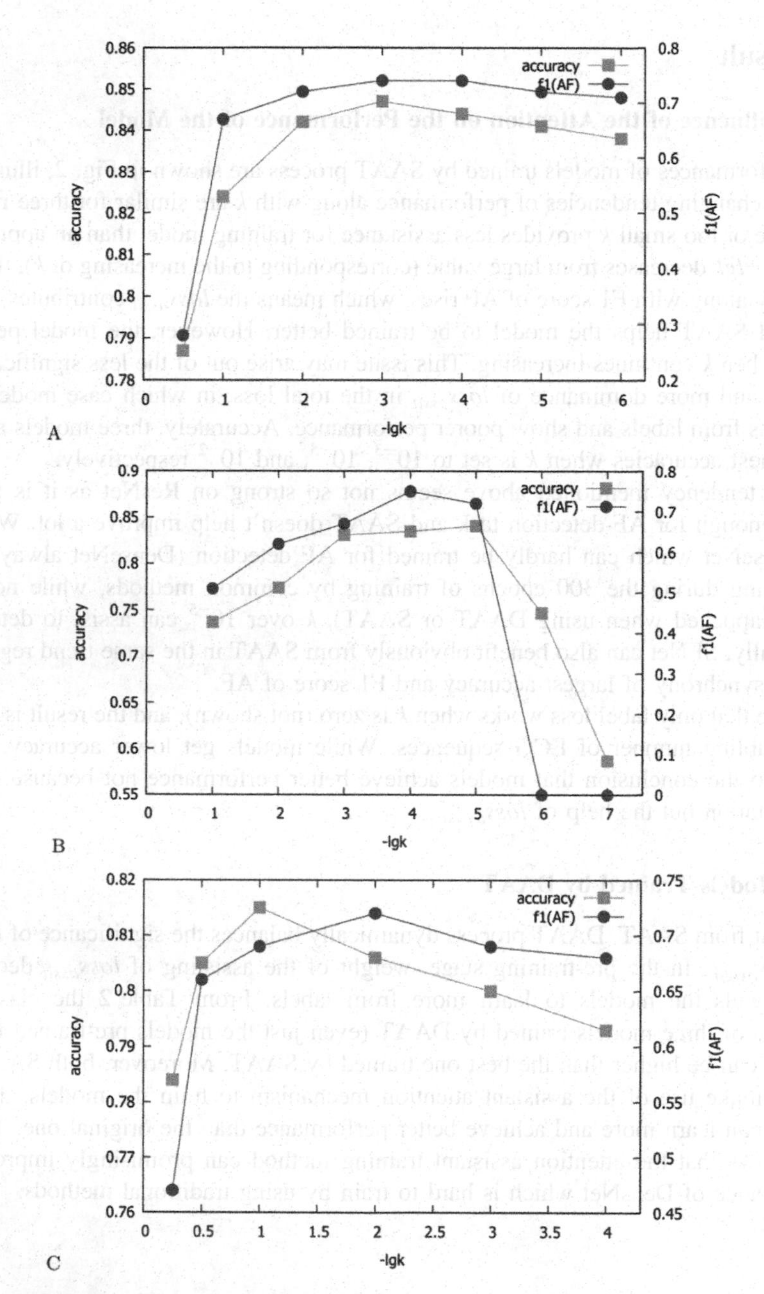

Fig. 2. Performance of the models trained by SAAT at different *–log(k)*. A. Resnet; B. DenseNet; C. SENet.

Training performance of these three models can still improve a little after 100 epochs of post-training in the common training way (see Table 2). Especially the accuracy of SENet increases 0.018 than before. These indicates that the training method can help train a CNN model to a good initial state in which it can be trained better by common training method after tearing away the assistant facilities.

As for k, no special values are tuned in DAAT and SAAT. DAAT performs a little better than SAAT with best value of k. It can be due to the gradual loosing restriction of symmetry but not a proper value of k, since k in DAAT declines lower than the best value of k in SAAT at the beginning of training and the model gets better accuracy and F1 score. $Loss_{assist}$ introduces symmetry into the training of model parameters, whereas ECG records are in fact not symmetrical, which leads to the slightly weaker performance of model trained by SAAT process with best value of k.

Table 2. Comparing results of models trained by different strategies

Type	Epoch	Accuracy	F1(AF)
ResNet	300	0.839	0.720
ResNet(SAAT:BEST)	300	0.847	0.740
ResNet(DAAT:pre-training)	200	0.853	0.770
ResNet(DAAT)	**300**	**0.860**	**0.770**
DenseNet	300	0.587	0.000
DenseNet(SAAT:BEST)	300	0.841	0.720
DenseNet(DAAT:pre-training)	200	0.845	0.770
DenseNet(DAAT)	**300**	**0.848**	**0.780**
SENet	300	0.766	0.620
SENet(SAAT:BEST)	300	0.815	0.690
SENet(DAAT:pre-training)	200	0.810	0.680
SENet(DAAT)	**300**	**0.824**	**0.690**

5.3 Heatmap of the Output of TCM

The channel-combined outputs of TCM M or the above mentioned class activation mapping (see Eq. 1) for both original and reversed inputs are reviewed manually as Guo *et al.* do [7]. In their result, hot regions giving high number in the outputs can overlap symmetrically, which can be regard as the attention a model used to judge and classify. While this can be hardly seen except for some short records with long padding zeros (not shown in paper). There is an example in Fig. 3. The brighter in heatmap means the higher value of M. As is supposed, the value of M in subplot b should overlap the flipped output of reversed input in subplot c and all of the longer RR interval in subplot a, which acts as a hint for AF event, should be brighter in subplots b and c as the interval in dashed red square do. This result may be due to two reasons. Firstly, sequence region with AF can hardly be tagged because doctors still need to see a long interval of ECG sequence and compare each interval to judge if AF happened in that record. That is to say the attention spreads on the whole length rather than on some specific regions. Secondly, ECG records are one-dimension data, of which the attention may perform in other ways rather than the channel-combined output. The improvement of model's training performance may owe to some more complex reasons related to symmetry we introduce into the models by attention assisting training process.

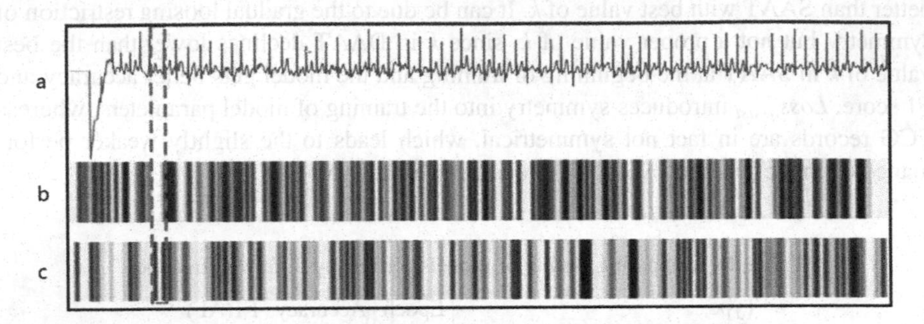

Fig. 3. Example of attention heatmap of label AF for DenseNet. a. A sample of input ECG sequence; b. Attention heatmap for model with origin input ECG sequence; c. Flipped attention heatmap for the model with flipped input ECG sequence.

6 Conclusion

In this paper, a DAAT process is proposed for training CNN models on AF detection task by which the trained model can be retrieved from the TCM architecture for using as a common model. It works on verified excellent ResNet and improves its accuracy without data augmentation. It also helps DenseNet to perform well which used to be a little weak on AF detection task with a small amount of AF positive labels. DAAT can also improve the performance of the CNN model which has used attention within convolutional layers such as SENet. This method is promising for training models on more tasks related to ECG classification especially when data is poor balanced. That can be an alternative strategy when some models can hardly be trained in a normal way.

Acknowledgement. The work was supported by the Major Projects of Technological Innovation in Hubei Province (2019AEA170), the Frontier Projects of Wuhan for Application Foundation (2019010701011381).

References

1. Kirchhof, P., et al.: 2016 ESC guidelines for the management of atrial fibrillation developed in collaboration with EACTS. Europace **18**(11), 1609–1678 (2016)
2. Hiram, R., et al.: Right atrial mechanisms of atrial fibrillation in a rat model of right heart disease. J. Am. Coll. Cardiol. **74**(10), 1332–1347 (2019)
3. Soto, J.T., Ashley, E.A.: Deepbeat: a multi-task deep learning approach to assess signal quality and arrhythmia detection in wearable devices. arXiv: Signal Processing (2020)
4. Saglietto, A., Scarsoglio, S., Ridolfi, L., Gaita, F., Anselmino, M.: Higher ventricular rate during atrial fibrillation relates to increased cerebral hypoperfusions and hypertensive events. Sci. Rep. **9**(1), 1–9 (2019)

5. Silvafilarder, M.D., Marzbanrad, F.: Combining template-based and feature-based classification to detect atrial fibrillation from a short single lead ECG recording. In: 2017 Computing in Cardiology (CinC), pp. 1–4. IEEE (2017)
6. Mane, D., Kulkarni, U.V.: A survey on supervised convolutional neural network and its major applications. In: Deep Learning and Neural Networks: Concepts, Methodologies, Tools, and Applications, pp. 1058–1071. IGI Global (2020)
7. Guo, H., Zheng, K., Fan, X., Yu, H., Wang, S.: Visual attention consistency under image transforms for multi-label image classification. In: Proceedings of the IEEE Conference on Computer Vision and Pattern Recognition, pp. 729–739 (2019)
8. Andreotti, F., Carr, O., Pimentel, M.A., Mahdi, A., De Vos, M.: Comparing feature-based classifiers and convolutional neural networks to detect arrhythmia from short segments of ECG. In: 2017 Computing in Cardiology (CinC), pp. 1–4. IEEE (2017)
9. Huang, G., Liu, Z., Der Maaten, L.V., Weinberger, K.Q.: Densely connected convolutional networks. In: Proceedings of the IEEE Conference on Computer Vision and Pattern Recognition, pp. 2261–2269 (2017)
10. Hu, J., Shen, L., Albanie, S., Sun, G., Wu, E.: Squeeze-and-excitation networks. IEEE Trans. Pattern Anal. Mach. Intell. 1 (2019)
11. Tran, L., Li, Y., Nocera, L., Shahabi, C., Xiong, L.: Multifusionnet: atrial fibrillation detection with deep neural networks. AMIA Summits Translational Sci. Proc. **2020**, 654 (2020)
12. Kharshid, A., Alhichri, H.S., Ouni, R., Bazi, Y.: Classification of short-time single-lead ECG recordings using deep residual CNN. In: 2019 2nd International Conference on New Trends in Computing Sciences (ICTCS), pp. 1–6. IEEE (2019)
13. Lai, D., Zhang, X., Zhang, Y., Heyat, M.B.B.: Convolutional neural network based detection of atrial fibrillation combing RR intervals and f-wave frequency spectrum. In: 2019 41st Annual International Conference of the IEEE Engineering in Medicine and Biology Society (EMBC), pp. 4897–4900. IEEE (2019)
14. Shen, M., Zhang, L., Luo, X., Xu, J.: Atrial fibrillation detection algorithm based on manual extraction features and automatic extraction features. In: IOP Conference Series: Earth and Environmental Science, vol. 428, p. 012050. IOP Publishing (2020)
15. Zhu, J., Zhang, Y., Zhao, Q.: Atrial fibrillation detection using different duration ECG signals with SE-ResNet. In: 2019 IEEE 21st International Workshop on Multi-media Signal Processing (MMSP), pp. 1–5. IEEE (2019)
16. Shi, H., Wang, H., Qin, C., Zhao, L., Liu, C.: An incremental learning system for atrial fibrillation detection based on transfer learning and active learning. Comput. Methods Programs Biomed. **187**, 105219 (2020)
17. Zhang, Z., Xu, C., Yang, J., Tai, Y., Chen, L.: Deep hierarchical guidance and regularization learning for end-to-end depth estimation. Pattern Recogn. **83**, 430–442 (2018)
18. Dong, Q., Zhu, X., Gong, S.: Single-label multi-class image classification by deep logistic regression. In: Proceedings of the AAAI Conference on Artificial Intelligence, vol. 33, pp. 3486–3493 (2019)
19. Liu, Y., You, X.: Specific action recognition method based on unbalanced dataset. In: 2019 IEEE 2nd International Conference on Information Communication and Signal Processing (ICICSP), pp. 454–458. IEEE (2019)
20. Zheng, Q., Yang, M., Tian, X., Jiang, N., Wang, D.: A full stage data augmentation method in deep convolutional neural network for natural image classification. Discret. Dyn. Nat. Soc. **2020**(2), 1–11 (2020)
21. Huang, S., Lee, F., Miao, R., Si, Q., Lu, C., Chen, Q.: A deep convolutional neural network architecture for interstitial lung disease pattern classification. Med. Biol. Eng. Compu. **58**(4), 725–737 (2020). https://doi.org/10.1007/s11517-019-02111-w

22. Zhou, B., Khosla, A., Lapedriza, A., Oliva, A., Torralba, A.: Learning deep features for discriminative localization. In: Proceedings of the IEEE Conference on Computer Vision and Pattern Recognition, pp. 2921–2929 (2016)
23. Clifford, G.D., et al.: AF classification from a short single lead ECG recording: the physionet/computing in cardiology challenge 2017. In: 2017 Computing in Cardiology (CinC), pp. 1–4. IEEE (2017)
24. He, K., Zhang, X., Ren, S., Sun, J.: Delving deep into rectifiers: surpassing human-level performance on imagenet classification. In: Proceedings of the IEEE International Conference on Computer Vision, pp. 1026–1034 (2015)
25. Kingma, D.P., Ba, J.: Adam: a method for stochastic optimization. arXiv preprint arXiv: 1412.6980 (2014)

Prediction of lncRNA-Disease Associations from Heterogeneous Information Network Based on DeepWalk Embedding Model

Xiao-Yu Song[1], Tong Liu[1], Ze-Yang Qiu[3], Zhu-Hong You[2(✉)], Yue Sun[1], Li-Ting Jin[1], Xiao-Bei Feng[1], and Lin Zhu[1]

[1] School of Electronic and Information Engineering, Lanzhou Jiaotong University, Lanzhou 730070, Gansu, China
sxy9998@126.com, lt837163867@163.com

[2] Xinjiang Technical Institute of Physics and Chemistry, Chinese Academy of Sciences, Urumqi 830011, China
zhuhongyou@ms.xjb.ac.cn

[3] School of Mechanical Engineering, Lanzhou Jiaotong University, Lanzhou 730070, Gansu, China

Abstract. Long non-coding RNA is a class of non-coding RNAs, with a length of more than 200 nucleotides. A large number of studies have shown that lncRNAs are involved in various life processes of the Human body and play an important role in the occurrence, development, and treatment of Human diseases. However, it is time-consuming and laborious to identify the associations between lncRNAs and diseases by traditional methods. In this paper, we propose a novel computational method to predict lncRNA-disease associations based on a heterogeneous information network. Specifically, the heterogeneous information network is constructed by integrating known associations among drugs, proteins, lncRNA, miRNA and diseases. After that, the network embedding method Online Learning of Social Representations (DeepWalk) is employed to learn vector representation of nodes in heterogeneous information network. Finally, we trained the random forest classifier to classify and predict the relationship between lncRNA and disease. As a result, the proposed method achieves average AUC of 0.8171 using five-fold cross-validation. The experimental results show that our method performs better than existing approaches, so it can be a useful tool for predicting disease-related lncRNA.

Keywords: lncRNA-disease associations · DeepWalk · Heterogeneous information network · Network embedding · Protein sequence

1 Introduction

LncRNA participates in a variety of life processes, not only participating in various processes of organisms, but also closely related to the occurrence and development of diseases [1–9]. Studying the relationship between lncRNA and diseases can deepen people's understanding of the human disease mechanism at the lncRNA level and further improve the medical level of human beings [10–25]. By analyzing the known lncRNA- disease association, it is very

D.-S. Huang and P. Premaratne (Eds.): ICIC 2020, LNAI 12465, pp. 291–300, 2020.
https://doi.org/10.1007/978-3-030-60796-8_25

meaningful to develop a new calculation method to predict the potential lncRNA-disease association and to provide the most likely lncRNA-disease association for experimental verification [26–39]. the current lncRNA-disease prediction methods can be divided into three categories [40]: The method based on machine learning, the method based on network model and the method based on non-known lncRNA-disease correlation data. Specifically, Chen *et al.* [41] uses a semi-supervised learning algorithm to predict the potential correlation between lncRNA-disease and proposes a LRLSLDA model. In this model, the known lncRNA-disease correlation data are used for the first time to calculate the Gaussian nuclear similarity of long non-coding RNA Gaussian nuclear similarity disease. Yang *et al.* [42] constructed the lncRNA-disease bipartite graph network in the work and applied the propagation algorithm to predict the network. This is the first method of prediction based on the network model. Liu *et al.* [43] predicted lncRNA-disease association by integrating human long non-coding RNA expression profile data, gene expression profile data and gene-disease correlation data. This is also the first way to predict without using standard lncRNA-disease data. As a rapidly developing discipline, the complex network has an important position in various fields. Using a complex network model to predict lncRNA-disease will be an important research direction in the future. Therefore, based on the heterogeneous information network, this paper predicts the association of lncRNA-disease through the network model based on sequence and network embedding.

2 Materials and Methods

2.1 Dataset

To ensure that heterogeneous information networks are more abundant, we download positive training dataset data from several different databases, after data processing, as shown in Table 1, at the same time, an equal number of unknown associations are randomly generated as negative samples, and then positive samples and negative samples are used as training sets. Finally, the heterogeneous information network includes 105962 pairs of known associated data.

Table 1. The details of nine kinds of associations in the heterogeneous information network.

Relationship type	Database	Number of associations
miRNA-lncRNA	lncRNASNP2 [44]	8374
miRNA-protein	miRTarBase [45]	4944
miRNA-disease	HMDD [46]	16427
lncRNA-disease	LncRNADisease [47], LncRNASNP2 [44], Lnc2Cancer [48]	1680
Drug-protein	DrugBank [49]	11107
Protein-disease	DisGeNET [50]	25087
Drug-disease	CTD [51]	18416
Protein-protein	STRING [52]	19237
lncRNA-protein	LncRNA2Target [53]	690
Total	N/A	105962

2.2 LncRNA and Protein Sequence

The attributes of molecular nodes can be represented by sequences, and the sequence information of lncRNA is downloaded from the NONCODE [54] database. In the process of coding the lncRNA and protein sequence, we get inspiration from the article of Shen *et al.* [55], according to the polarity of the side chain, we divide 20 amino acids into 4 categories. Therefore, a 64 ($4 \times 4 \times 4$) dimensional vectors is chosen to encode the lncRNA sequence, where each feature represents the normalized frequency of the corresponding 3-polymer (such as ACG, CAU, UUG) that appears in the RNA sequence. Therefore, in the process of sequence representation, we use 64 dimensional vector to represent the lncRNA coding sequence, in which each feature represents the normalized frequency of the corresponding 3-mer in the lncRNA sequence.

2.3 Disease MeSH Descriptors and Directed Acyclic Graph

Medical Subject Headings (MeSH) is a tool widely used in medical information retrieval. The purpose of this paper is to index periodical literature and books in the field of life science.

In the MeSH descriptor hierarchy, the highest categories are: Anatomy [A], Biology [B], Disease [C], Chemicals [D], etc. In this structure, any disease can be represented by a directed acyclic graph (DAG) [56] generated by MeSH, which can effectively describe its characteristics. For disease A, DAG is denoted as DAG (A) = (D (A), E (A)), where D (A) includes nodes representing the disease itself and its ancestors, and E (A) consists of corresponding direct edges from the parent node to the child node, representing the relationship between the two nodes.

In the same way, as described in reference [56], the contribution of each disease semantics term to disease *D* is numerically studied as follows:

$$\begin{cases} D1_D(D) = 1 \\ D1_D(t) = max\{\Delta * D1_D(t') | t' \subset children\ of\ t\}\ if\ t \neq D \end{cases} \tag{1}$$

Where $\Delta \in [0, 1]$ is the attenuation factor of the semantic contribution. The DAG (D), disease D for the disease is clear, and the contribution of disease D to itself is 1. The semantic score of disease D is defined by the following formula.

$$DV1(D) = \sum\nolimits_{t \in N_D} D1_D(t) \tag{2}$$

Assuming that the DAG of disease A and disease B share nodes, the semantic similarity between disease A and disease B defined as equal diseases is calculated:

$$S1(A,B) = \frac{\sum_{t \in (N_A \cap N_B)} (D1_A(t) + D1_B(t))}{DV1(A) + DV1(B)} \tag{3}$$

S1 is the semantic similarity matrix of disease.

2.4 Node Representation

In this paper, the heterogeneous information network we build is a complex network composed of multiple nodes and edges. Here, we use DeepWalk [57] to represent nodes in a heterogeneous information network as 64-dimensional vectors. DeepWalk is a new way to learn the representation of nodes in a network. We used the method of language modeling in the social network, so we can use the method of deep learning. It can represent not only the nodes but also the topological relations between the nodes, that is, the social relations of the social network.

When we input the network, the nodes with similar relationships represent the more similar nodes in the topological relationship, and the output is the low-dimensional vector of the nodes. Our goal is to map each node in the network into a low-dimensional vector. That is, a vector is used to represent each node in the network, and it is hoped that these vectors can express the relationship in the nodes in the network.

DeepWalk algorithm is a model in word2vec with the help of language, skip-gram to learn the vector representation of nodes. The nodes in the network are simulated as words in the language model, and the sequence of nodes (which can be obtained by random walk) is simulated as sentences in the language as the input of skip-gram.

First, the probability of the next node is estimated based on the nodes included in the random walk:

$$\Pr(v_i|(v_1, v_2, \ldots v_{i-1})) \tag{4}$$

Then, need a mapping function: a mapping function Φ: $v \in V \rightarrow R^{|V| \times d}$. This mapping function represents the hidden social representation between nodes.

$$P_r(v_i|(\Phi(v_1), \Phi(v_2), \ldots, \Phi(v_{i-1}))) \tag{5}$$

Finally optimized by the SkipGram module:

$$minimize \; - logP_r(\{v_{i-w}, \ldots, v_{i+w}\} \backslash v_i | \varphi(v_i)) \tag{6}$$

2.5 Random Forest Classifier

Random forest (RF) is a classifier consisting of an arboreal classifier collection, RF utilizes two efficient machine-learning techniques, bagging and random feature selection. Every tree is trained in bagging on a bootstrap sample of the training results, and predictions are made by majority tree vote. RF is yet another bagging growth. RF automatically chooses a subset of features to be separated at each node as a tree expands, instead of using all the features. Thanks to the excellent performance of the random forest classifier, it was selected for training to classify and predict the potential association between lncRNA and diseases.

2.6 Heterogeneous Information Network

This paper is based on a heterogeneous information network, is to predict the association relationship of lncRNA-disease through a network model based on sequence and network embedding. The flow chart of the model we built to predict potential lncRNA-disease associations is shown in Fig. 1. The network is composed of two parts, including nodes and edges. Nodes are composed of miRNA, lncRNA, disease, protein, and drug, and the edges consist of relationships between nodes. Determining the relationship between lncRNA and disease nodes in complex networks contributes to a comprehensive understanding of biological life activities [58–60], We collected nine kinds of molecular correlation data, such as drug-protein, protein-disease, lncRNA-protein, miRNA-lncRNA, and after sorting the data, five kinds of nodes are obtained: lncRNA, protein, drug, miRNA and disease, they construct heterogeneous information networks in the form of network nodes. The specific details are that we take the known associations obtained from the database as positive samples, and then randomly select an equal number of unknown associations as negative samples and positive and negative samples as training sets. Then the random forest with excellent performance is selected as the classifier for experimental training verification and testing. Our method is evaluated by five-fold cross-validation, and the effect is good. This method effectively combines the attribute characteristics and behavior information of nodes to obtain robust prediction performance. The heterogeneous information network constructed by five kinds of nodes provides a more comprehensive perspective for biology.

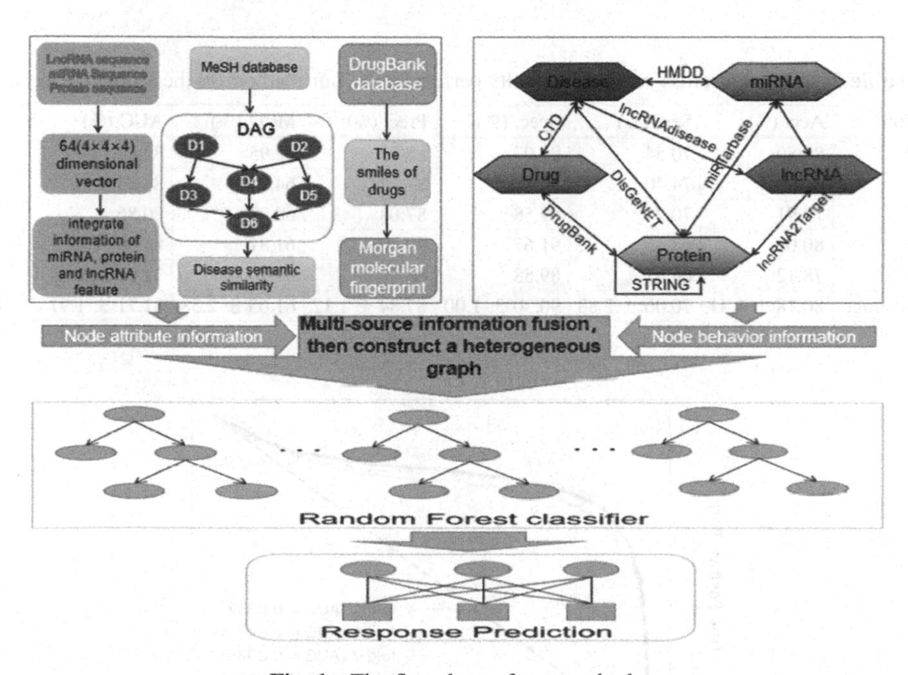

Fig. 1. The flowchart of our method.

3 Experimental Results

When using five-fold cross-validation, the whole data set is composed of the same amount of positive and negative samples. First, the data set is randomly divided into five equal subsets, and then the randomly selected subset is used as the test set. Then the remaining four subsets are integrated to form the training set, and the training set is used to construct the classifier. It is worth mentioning that each time only the current training set data is cross-verified, 80% of the total edge is embedded into the node, which can protect the test information. But this method may isolate some nodes in the network. It can also better simulate the real environment, provide support for researchers, and help manual experiments by exploring unknown areas.

The use of ACC, Sen, Spec, Prec, and MCC constitutes a broader set of evaluation criteria, and the evaluation of the results of the model is comprehensive and fair. The details are shown in Table 2, when we apply the proposed framework to the entire network to predict arbitrary associations, The average results of Acc, Sen, Spec, Prec, MCC, and AUC were 80.18%, 70.06%, 90.30%, 87.84%, 61.65%, and 81.71%, respectively. The details of the results of performing five-fold cross-validation using our method are shown in Fig. 2. Each point on the receiver operating characteristic curve reflects the sensitivity to the same signal stimulus and is a general standard for evaluating the model. The host operation property curve (AUC) is a graphical area surrounded by ROC. The transverse and longitudinal coordinates are false positive (FPR) and true positive (TPR), respectively. And then we draw the ROC and calculate the AUC, for visual evaluation and five-fold cross-validation of our model.

Table 2. Five-fold cross-validation results performed by our method on the whole datasets.

Fold	Acc. (%)	Sen. (%)	Spec. (%)	Prec. (%)	MCC (%)	AUC (%)
0	80.80	70.54	91.07	88.76	62.95	82.52
1	81.99	74.40	89.58	87.72	64.74	84.24
2	79.91	70.24	89.58	87.08	60.97	80.86
3	80.06	68.45	91.67	89.15	61.81	81.96
4	78.12	66.67	89.58	86.49	57.79	78.95
Average	80.18 ± 1.41	70.06 ± 2.88	90.30 ± 1.00	87.84 ± 1.12	61.65 ± 2.58	81.71 ± 1.97

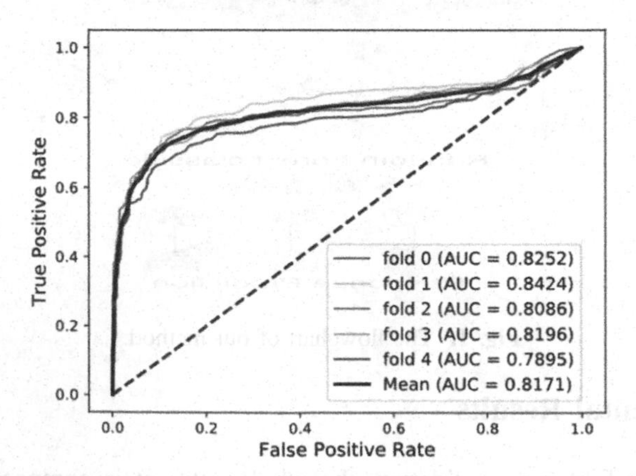

Fig. 2. The ROC curves of our method in lncRNA-disease association prediction under five-fold cross-validation.

4 Conclusion

Previous works show that lncRNAs play an important role in the occurrence and development of the disease. Identification of disease-related lncRNAs could help researchers to understand the mechanism of disease more deeply at the molecular level. In this paper, a novel computational method is proposed to predict lncRNA-disease associations based on a heterogeneous information network. It is worth noting that we use the attribute information and behavior information of nodes in the network to form feature vectors. The attribute information of lncRNA is a 64-dimensional feature vector encoded by the *k*-mer method, where each feature vector represents the normalized frequency of the corresponding 3-mer in the lncRNA sequence. For behavior information, the network embedding method DeepWalk can be used to globally represent the relationship between lncRNA and disease. At the same time, a random forest classifier in used for classification and prediction tasks, and the experimental results show that our method achieves better results than existing approaches.

Acknowledgements. This work is supported in part by the National Science Foundation of China, under Grants 61722212, in part by the Pioneer Hundred Talents Program of Chinese Academy of Sciences, The authors all anonymous reviewers for their constructive advices.

References

1. Taft, R.J., Pheasant, M., Mattick, J.S.: The relationship between non-protein-coding DNA and eukaryotic complexity. BioEssays **29**, 288–299 (2007)
2. Esteller, M.: Non-coding RNAs in human disease. Nat. Rev. Genet. **12**, 861–874 (2011)
3. Wang, M.-N., You, Z.-H., Wang, L., Li, L.-P., Zheng, K.: LDGRNMF: lncRNA-disease associations prediction based on graph regularized non-negative matrix factorization. Neurocomputing (2020)
4. Wang, M., You, Z., Li, L., Wong, L., Chen, Z., Gan, C.: GNMFLMI: graph regularized nonnegative matrix factorization for predicting LncRNA-MiRNA interactions. IEEE Access **8**, 37578–37588 (2020)
5. Zheng, K., You, Z.-H., Wang, L., Zhou, Y., Li, L.-P., Li, Z.-W.: DBMDA: a unified embedding for sequence-based mirna similarity measure with applications to predict and validate mirna-disease associations. Mol. Ther.-Nucleic Acids **19**, 602–611 (2020)
6. Zheng, K., You, Z.-H., Wang, L., Li, Y.-R., Wang, Y.-B., Jiang, H.-J.: MISSIM: improved miRNA-disease association prediction model based on chaos game representation and broad learning system. In: Huang, D.-S., Huang, Z.-K., Hussain, A. (eds.) ICIC 2019. LNCS (LNAI), vol. 11645, pp. 392–398. Springer, Cham (2019). https://doi.org/10.1007/978-3-030-26766-7_36
7. Chen, Z.-H., You, Z.-H., Guo, Z.-H., Yi, H.-C., Luo, G.-X., Wang, Y.-B.: Prediction of drug-target interactions from multi-molecular network based on deep walk embedding model. Front. Bioeng. Biotechnol. **8**, 338 (2020)
8. Chen, Z.-H., You, Z.-H., Li, L.-P., Wang, Y.-B., Qiu, Y., Hu, P.-W.: Identification of self-interacting proteins by integrating random projection classifier and finite impulse response filter. BMC Genom. **20**, 1–10 (2019)

9. Ji, B.-Y., You, Z.-H., Cheng, L., Zhou, J.-R., Alghazzawi, D., Li, L.-P.: Predicting miRNA-disease association from heterogeneous information network with GraRep embedding model. Sci. Rep. **10**, 1–12 (2020)
10. Chen, X., Yan, C.C., Zhang, X., You, Z.: Long non-coding RNAs and complex diseases: from experimental results to computational models. Brief. Bioinform. **18**, 558–576 (2016)
11. You, Z., et al.: PBMDA: a novel and effective path-based computational model for miRNA-disease association prediction. PLOS Comput. Biol. **13**, e1005455 (2017)
12. You, Z., Zhou, M., Luo, X., Li, S.: Highly efficient framework for predicting interactions between proteins. IEEE Trans. Syst. Man Cybern. **47**, 731–743 (2017)
13. Chen, X., Huang, Y., You, Z., Yan, G., Wang, X.: A novel approach based on KATZ measure to predict associations of human microbiota with non-infectious diseases. Bioinformatics **33**, 733–739 (2016)
14. Huang, Y., Chan, K.C.C., You, Z.: Constructing prediction models from expression profiles for large scale lncRNA-miRNA interaction profiling. Bioinformatics **34**, 812–819 (2018)
15. Li, S., You, Z., Guo, H., Luo, X., Zhao, Z.: Inverse-free extreme learning machine with optimal information updating. IEEE Trans. Syst. Man Cybern. **46**, 1229–1241 (2016)
16. Wang, L., You, Z., Li, Y., Zheng, K., Huang, Y.: GCNCDA: a new method for predicting circrna-disease associations based on graph convolutional network algorithm. bioRxiv 858837 (2019)
17. Ma, L., et al.: Multi-neighborhood learning for global alignment in biological networks. IEEE/ACM Trans. Comput. Biol. Bioinform. 1 (2020)
18. Wong, L., You, Z.H., Guo, Z.H., Yi, H.C., Cao, M.Y.: MIPDH: A Novel Computational Model for Predicting microRNA–mRNA Interactions by DeepWalk on a Heterogeneous Network (2020)
19. Wang, Y., You, Z., Li, L., Chen, Z.: A survey of current trends in computational predictions of protein-protein interactions. Front. Comput. Sci. **14**(4), 1–12 (2020). https://doi.org/10.1007/s11704-019-8232-z
20. Guo, Z.H., et al.: MeSHHeading2vec: a new method for representing MeSH headings as vectors based on graph embedding algorithm. Briefings Bioinform. (2020)
21. Jiang, H.J., Huang, Y.A., You, Z.H.: SAEROF: an ensemble approach for large-scale drug-disease association prediction by incorporating rotation forest and sparse autoencoder deep neural network. Entific Rep. **10**, 4972 (2020)
22. Guo, Z., You, Z., Yi, H.: Integrative construction and analysis of molecular association network in human cells by fusing node attribute and behavior information. Mol. Ther. Nucleic Acids **19**, 498–506 (2020)
23. Wang, Y., You, Z., Yang, S., Yi, H., Chen, Z., Zheng, K.: A deep learning-based method for drug-target interaction prediction based on long short-term memory neural network. BMC Med. Inform. Decis. Mak. **20**, 49 (2020)
24. Yi, H.C., You, Z.H., Guo, Z.H., Huang, D.S., Kcc, C.: Learning representation of molecules in association network for predicting intermolecular associations. IEEE/ACM Trans. Comput. Biol. Bioinform. 1 (2020)
25. Huang, Y., Hu, P., Chan, K.C.C., You, Z.: Graph convolution for predicting associations between miRNA and drug resistance. Bioinformatics **36**, 851–858 (2019)
26. Li, J., Shi, X., You, Z., Chen, Z., Fang, M.: Using weighted extreme learning machine combined with scale-invariant feature transform to predict protein-protein interactions from protein evolutionary information. In: International Conference on Intelligent Computing (2020)
27. Yi, H.C., You, Z.H., Cheng, L., Zhou, X., Wang, Y.B.: Learning distributed representations of RNA and protein sequences and its application for predicting lncRNA-protein interactions. Comput. Struct. Biotechnol. J. **18**, 20–26 (2019)

28. Wong, L., Huang, Y., You, Z., Chen, Z., Cao, M.: LNRLMI: linear neighbour representation for predicting lncRNA-iRNA interactions. J. Cell. Mol. Med. **24**, 79–87 (2019)
29. Hu, P., Huang, Y., Chan, K.C.C., You, Z.: Learning multimodal networks from heterogeneous data for prediction of lncRNA-miRNA interactions. IEEE/ACM Trans. Comput. Biol. Bioinform. 1 (2019)
30. Li, Z., Nie, R., You, Z., Cao, C., Li, J.: Using discriminative vector machine model with 2DPCA to predict interactions among proteins. BMC Bioinform. **20**, 694 (2019)
31. Jiang, H., You, Z., Huang, Y.: Predicting drug–disease associations via sigmoid kernel-based convolutional neural networks. J. Transl. Med. **17**, 1–11 (2019)
32. Guo, Z., You, Z., Wang, Y., Yi, H., Chen, Z.: A learning-based method for LncRNA-disease association identification combing similarity information and rotation forest. iScience **19**, 786–795 (2019)
33. Yi, H., et al.: ACP-DL: a deep learning long short-term memory model to predict anticancer peptides using high-efficiency feature representation. Mol. Ther. Nucleic Acids **17**, 1–9 (2019)
34. Wang, L., et al.: Identification of potential drug-targets by combining evolutionary information extracted from frequency profiles and molecular topological structures. Chem. Biol. Drug Des. (2019)
35. Li, J., ct al.: An efficient attribute-based encryption scheme with policy update and file update in cloud computing. IEEE Trans. Ind. Inf. **15**, 6500–6509 (2019)
36. Hu, L., Hu, P., Yuan, X., Luo, X., You, Z.: Incorporating the coevolving information of substrates in predicting HIV-1 protease cleavage sites. IEEE/ACM Trans. Comput. Biol. Bioinform. 1 (2019)
37. An, J., You, Z., Zhou, Y., Wang, D.: Sequence-based prediction of protein-protein interactions using gray wolf optimizer–based relevance vector machine. Evol. Bioinform. **15**, 117693431984452 (2019)
38. Wang, L., et al.: LMTRDA: using logistic model tree to predict MiRNA-disease associations by fusing multi-source information of sequences and similarities. PLOS Computat. Biol. **15**, e1006865 (2019)
39. Zhu, H., You, Z., Shi, W., Xu, S., Jiang, T., Zhuang, L.: Improved prediction of protein-protein interactions using descriptors derived from PSSM via gray level co-occurrence matrix. IEEE Access **7**, 49456–49465 (2019)
40. Chen, X., Xie, D., Zhao, Q., You, Z.H.: Long non-coding RNAs and complex diseases: from experimental results to computational models. Briefings Bioinform. **558** (2017)
41. Chen, X., Yan, G.-Y.: Novel human lncRNA–disease association inference based on lncRNA expression profiles. Bioinformatics (2013)
42. Yang, X., et al.: A network based method for analysis of lncRNA-disease associations and prediction of lncRNAs implicated in diseases. PLoS ONE **9**, e87797 (2014)
43. Liu, M., Chen, X., Chen, G., Cui, Q., Yan, G.: A computational framework to infer human disease-associated long noncoding RNAs. Plos One **9** (2014)
44. Miao, Y., Liu, W., Zhang, Q., Guo, A.: lncRNASNP2: an updated database of functional SNPs and mutations in human and mouse lncRNAs. Nucleic Acids Res. **46** (2018)
45. Chou, C.-H., et al.: miRTarBase update 2018: a resource for experimentally validated microRNA-target interactions. Nucleic Acids Res. (2017)
46. Huang, Z., Shi, J., Gao, Y., Cui, C., Zhang, S.: HMDD v3.0: a database for experimentally supported human microRNA-disease associations. Nucleic Acids Res. **47**, D1013–D1017 (2018)
47. Chen, G., et al.: LncRNADisease: a database for long-non-coding RNA-associated diseases. Nucleic Acids Res. **41**, D983–D986 (2013)

48. Ning, S., et al.: Lnc2Cancer: a manually curated database of experimentally supported lncRNAs associated with various human cancers. Nucleic Acids Res. **44**, 980–985 (2016)
49. Wishart, D.S., et al.: DrugBank 5.0: a major update to the DrugBank database for 2018. Nucleic Acids Res. **46**, D1074 (2018)
50. Janet, P., et al.: DisGeNET: a comprehensive platform integrating information on human disease-associated genes and variants. Nucleic Acids Res. D833–D839 (2017)
51. Davis, A.P., et al.: The comparative toxicogenomics database: update 2019. Nucleic Acids Res. **47** (2019)
52. Szklarczyk, D., et al.: The STRING database in 2017: quality-controlled protein–protein association networks, made broadly accessible. Nucleic Acids Res. **45** (2017)
53. Cheng, L., et al.: LncRNA2Target v2.0: a comprehensive database for target genes of lncRNAs in human and mouse. Nucleic Acids Res. **47**, D140–D144 (2019)
54. Fang, S.S., et al.: NONCODEV5: a comprehensive annotation database for long non-coding RNAs. Nucleic Acids Res. **46**(D1), D308–D314 (2017)
55. Shen, J., et al.: Predicting protein–protein interactions based only on sequences information. Proc. Natl. Acad. Sci. U.S.A. **104**, 4337–4341 (2007)
56. Wang, D., Wang, J., Lu, M., Song, F., Cui, Q.: Inferring the human microRNA functional similarity and functional network based on microRNA-associated diseases. Bioinformatics **26**, 1644–1650 (2010)
57. Perozzi, B., Alrfou, R., Skiena, S.: DeepWalk: online learning of social representations. In: Knowledge Discovery and Data Mining, pp. 701–710 (2014)
58. Guo, Z.H., Yi, H.C., You, Z.H.: Construction and comprehensive analysis of a molecular association network via lncRNA–miRNA –disease–drug–protein graph. Cells **8**, 866 (2019)
59. Hrdlickova, B., De Almeida, R.C., Borek, Z., Withoff, S.: Genetic variation in the non-coding genome: involvement of micro-RNAs and long non-coding RNAs in disease. BBA – Mol. Basis Dis. **1842**, 1910–1922 (2014)
60. Barabási, A.L., Oltvai, Z.N.: Network biology: understanding the cell's functional organization. Nat. Rev. Genet. **5**, 101 (2004)

Phishing Attacks and Websites Classification Using Machine Learning and Multiple Datasets (A Comparative Analysis)

Sohail Ahmed Khan[1], Wasiq Khan[2](✉), and Abir Hussain[2]

[1] The University of Sheffield, Sheffield S10 2TN, UK
[2] Liverpool John Moores University, Liverpool L3 5UG, UK
w.khan@ljmu.ac.uk

Abstract. Phishing attacks are the most common type of cyber-attacks used to obtain sensitive information and have been affecting individuals as well as organizations across the globe. Various techniques have been proposed to identify the phishing attacks specifically, deployment of machine intelligence in recent years. However, the algorithms and discriminating factors used in these techniques are very diverse in existing works. In this study, we present a comprehensive analysis of various machine learning algorithms to evaluate their performances over multiple datasets. We further investigate the most significant features within multiple datasets and compare the classification performance with the reduced dimensional datasets. The statistical results indicate that random forest and artificial neural network outperform other classification algorithms, achieving over 97% accuracy using the identified features.

Keywords: Phishing attacks · Cyber security · Phishing emails · Information security · Security and privacy · Phishing classification · Phishing websites detection

1 Introduction

Phishing in general, is a fraud in which a target (e.g., person) or multiple targets are contacted by email, telephone or text message, by a fraudster or cybercriminals [1]. These cybercriminals pose as a legitimate and reputable entity or a person and try to convince individuals to provide their sensitive data such as passwords, identity information, bank or credit card details etc. The provided information is then used to gain access to important accounts or services and can result in identity theft and financial loss. Phishing is popular among fraudsters due to its simplicity to trick users for clicking a malicious link that can break a computer's defence systems or can cause to bypass modern authentication systems.

Variety of attributes have been used to identify a phished webpage such as use of IP address in the URL, abnormal URL (special symbols in the URL) and many more [2]. However, a naive computer user can easily be tricked into considering a fake webpage as a legitimate webpage. Various techniques have been employed to deal with phishing attacks and distinguishing the phishing webpages automatically. For instance, blacklist-

D.-S. Huang and P. Premaratne (Eds.): ICIC 2020, LNAI 12465, pp. 301–313, 2020.
https://doi.org/10.1007/978-3-030-60796-8_26

based detection technique keeps a list of websites' URLs that are categorized as phishing sites. If a web-page requested by a user exists in the formed list, the connection to the queried website is blocked [2].

The webpage feature-based approach (i.e., visual features) [3] examines the abnormalities in webpages such as, the disagreement between a website's identity and its structural features. Likewise, machine learning (ML) based approaches rely on classification algorithms such as support vector machines (SVM) [4] and decision trees (DT) [5] to train a model that can later automatically classify the fraudulent websites at run-time without any human intervention. In phishing detection, ML algorithms try to make sense of the given training data by learning patterns that are present within the dataset. Current state-of-the-art ML algorithms take different features into account while making predictions such as, URL text features, domain name features, and web content features etc. We employ supervised ML algorithms in order to learn patterns in given datasets and classify phishing and legitimate websites accurately. Most of the existing studies focuses single classifiers and/or a single dataset however, it would be helpful to investigate different classifiers while using multiple dataset with variety of attributes. Likewise, investigation of the most significant features (i.e. attributes) within the multiple datasets might be of special interest.

This manuscript entails a comprehensive review of different ML algorithms for the phishing web-sites classification. Compared to existing research, we present a comparative study that performs the comprehensive analysis and comparison of different techniques for the classification of phishing websites. We used three different datasets to train, test and validate multiple classification algorithms including DT [4], SVM [5], random forest (RF)[6], na ̈ive Bayes (NB) [7], k-nearest neighbours (KNN) [8] and artificial neural networks (ANN) [9], to distinguish the phishing websites from legitimate websites. We further employ well-known Principal Component Analysis (PCA) [10] for dimensionality reduction and achieves approximately similar classification performance as compared to using full attributes within the dataset. In addition, we investigated the level of significance for all attributes within the three datasets using the PCA based component loadings. Rest of the manuscript is organized as follows. Section 2 addresses the existing work in this domain while Sect. 3 comprises the proposed methodology. Section 4 presents the discussion and comparison of results achieved followed by Sect. 5 which concludes the study and presents directions for future work.

2 Literature Review

Phishing websites detection is a crucial step towards countering online fraud. Recent technological advancements have been made with the use of ML and data science methods within the diverse application domains including aerospace [11], speech processing [12], healthcare technologies [13, 14], border security [15], object recognition [16], cybercrime detection [17], smart city [18] and so on. Likewise, there have been many technological developments in the domain of cyber security specifically to automatically detect the phishing attacks, but there is still room for a lot of improvements in this regard. Malicious attackers are coming up with new techniques, and phishing incidents are on a rise [19]. Several detection strategies have been devised in order to counter phishing attacks. For instance, in [20], authors used dataset [21] containing 30 different types of features with 11055 instances. Authors trained five different classification algorithms i.e. prism, NB, K*, RF, and ANN. They achieved best results 98.4% and 95.2% with RF, both with and without feature selection respectively. The study achieved some very good results however, as it uses single dataset only, the results are not reliable enough to be universal. Zhang et al. [22] employed ANN to detect phishing emails. The dataset they used was comprised of approximately 8762 emails, out of which 4560 were phishing emails and rest were non-phishing or legitimate. They trained a feedforward neural network using resilient propagation training and compared the performance with other ML models. They found that while maintaining highest recall, ANN had 95% accuracy [23], which makes ANNs excellent at distinguishing phishing emails whilst misclassifying a slight percentage of legitimate emails. The study focuses the use of ML to classify phishing emails and did not considers the phishing websites classification.

A study presented in [24] employed a rule-based classification technique for the detection of phishing websites. The authors used 4 different classification algorithms where the study indicated that using the feature reduction algorithm and classification-based association together produced the efficient performance. The study only relied on features which were occurring in high frequency that can be misleading sometimes, as higher frequency does not always guarantee higher importance. Karnik et al., [25] used the SVM in combination with cluster ensemble to classify phishing and malware website URLs. Training is performed through SVM using kernel functions (linear, radial, polynomial and sigmoid). With the proposed technique, the SVM model predicted correctly with 95% accuracy. The study only takes URL-based and textual features into consideration and does not consider any other features such as host based and content (iframes etc.) features. Likewise, other ML algorithms could be used for the comparative analysis to achieve more reliable findings.

A meta-heuristic based nonlinear regression algorithm for feature selection and phishing website detection is introduced in [26]. For classification, non-linear regression based on harmony search technique and SVM are deployed. Phishing dataset from UCI's machine learning repository [21] is used which contains 11055 instances and 30 features. This study did achieve some interesting results however, relied on single dataset only that contains 11055 instances only. In [27], Sahingoz et al., proposed a real time phishing detection system based on 7 different classification algorithms and Natural Language Processing (NLP) based features (i.e. word vectors, NLP based and Hybrid features). They found the RF algorithm based only on NLP features to be the best performer with an accuracy of 97.98%. A similar work is presented in [28] that proposes phishing website classification based on a hybrid model to overcome the problem posed by phishing websites. They used the dataset [21] from UCI's machine learning repository which comprises of 30 features and 11055 total instances. The system achieved 97.75% accuracy using DT (J48) and ensemble method. Similar to other works, only single dataset was employed to train and test the algorithms. No feature selection/dimensionality reduction mechanism was implemented. A heuristic based phishing detection technique is proposed in [29] which uses dataset of 3,000 phishing site URLs and 3,000 legitimate site URLs. Authors employed several ML algorithms including, KNN, RF, NB, SVM, and DT. Study indicated the RF to be the highest performer in all three performance measurements with an accuracy of 98.23%. However, this work also considers URL based features and does not consider other features such as contents or domain-based features. Also, the training and test sets are very small comprising only 6000 instances.

Aforementioned research studies demonstrate a considerable amount of work has already been done in phishing websites classification using different ML based techniques. Researchers have employed different techniques in order to predict phishing websites efficiently and with better accuracy. However, it would be helpful to analyse the ML algorithms' performances over the multiple datasets as well as over the reduced features from all dataset to investigate the impact of dimensionality reduction on the classification performances and the most significant features in these datasets. This work therefore employs PCA for the attribute analysis and dimensionality reduction on the three datasets and compares the classifiers' performances with results achieved using non-compressed feature sets.

3 Methodology

3.1 Datasets

We used three different datasets in this study to investigate the ML algorithms' performances as well as the feature importance within the three datasets. Dataset 1 [30] comprises of 48 different features obtained from 5000 different phishing webpages and 5000 different legitimate webpages. The webpages were downloaded during the time period between January to May 2015 and from May to June 2017. This dataset is labelled with binary labels e.g. 0 for legitimate, and 1 for phishing. Dataset 2 is obtained from University of California, Irvine's Machine Learning Repository [21]. This dataset contains 30 different features which uniquely identify phishing and legitimate websites. The target variable is binary, –1 for phishing and 1 for legitimate. The dataset is populated from different sources, some are PhishTank archive, google search engine, and MillerSmiles archive. This dataset contains mostly the same features as dataset 1 with some additional features. In total, it contains 11055 distinct website entries out of which 6157 are legitimate websites and 4898 are phishing websites. The dataset features are normalized and given values from –1 to 1, where –1 represents phishing, 0 represents suspicious and 1 means legitimate. Dataset 3 [31] is obtained from University of California, Irvine's Machine Learning Repository [9] and contains different features related to legitimate and phishing websites. This dataset contains data from 1353 different websites collected from different sources containing 702 phishing URLs, 548 legitimate URLs, 103 suspicious URLs records. This is a multi-class dataset, i.e. three different class labels where –1 means phishing, 0 means suspicious, 1 means legitimate. Suspicious represents a webpage can be either phishing or legitimate.

3.2 Experimental Design

The experiments are designed by utilizing different ML and data analytics libraries including Scikit Learn [32], KERAS [33], Numpy [34] and Pandas [35]. Five ML algorithms namely SVM [4], DT[5], RF[6], NB [7], KNN [8] and ANN [9] were employed along with the PCA [10] based feature importance measure as well as reduced dimensions. For the baseline experimental setup, recursive classification trials are conducted to compare the classifiers' performances for the model tuning and configurations such as kernel (e.g. radial, polynomial), cost, gamma, ntree, number of neurons in each layer, batch size, and time stamp. Standard 10-fold cross-validation train/test trials were run by partitioning the entire dataset into training and testing proportions of 70% and 30%, respectively. It was ensured that the test data contains fair distribution for all classes. Following the baseline experimental results, the classifiers' parameters were set imperially to get the optimal performance. Following experiments are designed with a consistent classifiers' configurations:

- Train and test the five ML algorithms over the individual datasets (i.e. dataset 1, dataset 2 and dataset 3) using 10-fold CV to compare the performances.
- Train and test the five ML algorithms over the PCA based dimension reduced datasets (PCs covering 90% of variance distribution) using 10-fold CV to compare the performances.

Additional experiments are conducted to investigate the attribute/feature importance in each individual dataset that represent the most distinguishing attributes to classify the phishing websites. The models' performances are assessed using various gold standards including accuracy, specificity, precision, recall and F1-score. Algorithm 1 summarizes the experimental steps carried out to conduct the above-mentioned experiments (A, B).

Algorithm 1: Experimental setup in proposed study
- Let S be a set of attributes for the phishing website dataset where $S=\{RandomString, RandomString, NumUnderscore ...\}$ - Let C_1, C_2, C_3 be the set of the website's classes as *Legitimate*, *Phishing*, *Suspicious* where: $Phishing = \{c_1: c_1 \in S$ & website type = -1$\}$ $Legitimate = \{c_2: c_2 \in S$ & flood severity = 1$\}$ $Suspicious = \{c_3: c_3 \in S$ & flood severity = 0$\}$ - Let S be the set of PCA components where: $S = \{s: \forall s \in S, \exists th \Rightarrow PC(s) > overall\ threshold\ variance\}$ $Training = \{t \in S\}$ where Training is 70% of S $Test = \{ts \in S \,\&\, ts \notin Training\}$ where Test is 30% of S $Validation = \{vs \in S \,\&\, vs \notin Training \,\&\, ts \notin test\}$ where *Validation* is 20% of *ts* For every selected ML algorithm determined $E[Accuracy_{C1,C2,C3}] = \{S: S \Rightarrow ML(Training, Validation, Test)\}$

3.3 Feature Importance and Dimensionality Reduction

One of the well-known dimensionality reduction technique is PCA [10] that have successfully been deployed in various application domains [16]. Major aim of the PCA is to transform a large dataset containing large number of features/variables to a lower dimension which still holds most of the information contained in the original high dimensional dataset. The interesting property of PCA is the attribute loadings that can also be used for the identification of attribute importance within the original dataset. We utilized PCA for the dimensionality reduction as well as calculation of feature importance score to investigate the most distinguishing features within all three datasets we used in this study.

Figure 1 represents the distributions for first two PCs with respect to target class, original attributes and corresponding impacts of the target classes within the dataset 3. These plots also indicate the non-linearity of the problem specifically in terms of first two PCs covering the highest variances within the overall principal components. However, the plots help to understand the corresponding influence of the variables within the datasets on the classification of phishing and legitimate websites. For instance, in Fig. 1, *'web-traffic'* has a clear impact on class '1' while *'ssl-final-state'* influences the '–1' class. The first two PCs cover approximately 53% of the overall PCs variance.

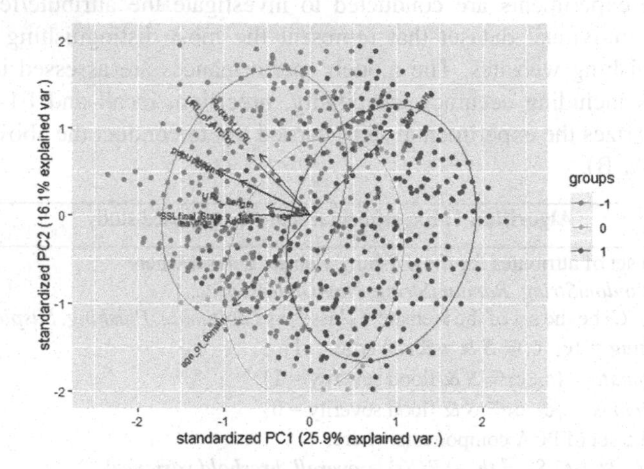

Fig. 1. First two PCA components' distributions in Dataset 3 w.r.t target classes

The correlation coefficient between the dataset attributes is represented by the principal components' loadings (i.e. obtained through PCA). The component rotations provide the maximized sum of variances of the squared loadings. The absolute sum of component rotations gives the degree of importance for the attributes in dataset.

Figure 2 demonstrates the attribute/feature importance in dataset 2 which is calculated through the PCs loadings. The result indicates a clear variation in the importance measure of variables that might be helpful to eliminate the unnecessary features from the dataset. For instance, *'having-sub-domain'* and *'age-of-domain'* are indicated

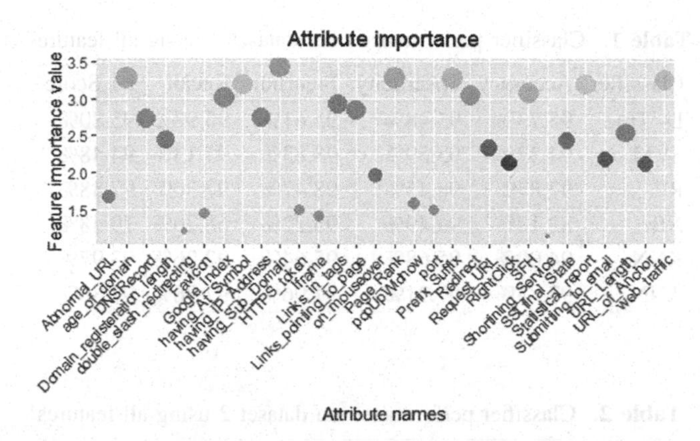

Fig. 2. Measure of feature importance within the Dataset 2 using PCA based attribute loadings. High value on y-axis represents higher importance and vice versa. Circle size increases with increasing attribute importance

the top-ranked variables compared *to 'double-slash-redirecting'* and *'shortening-service'* which are indicated the least important variables within the dataset 2.

4 Results and Discussion

Results are achieved following the experimental design for two experiments (A and B) and the variables rankings in three datasets. Table 1, Table 2 and Table 3 presents the statistical results achieved by different ML algorithms over the three datasets without dimensionality reductions.

4.1 Classification Performance Using Original Datasets and ML Algorithms

Table 1 presents the classifiers outcomes for dataset 1. The highest accuracy is achieved by RF and ANN indicating 97.87% and 97.83% respectively while NB indicated the lowest accuracy (82.17%). It can be observed that the NB classifier has very high specificity (96.49%) however, very low recall (68.2%) which indicate poor compromise between the sensitivity and specificity from NB, therefore affecting the overall accuracy of the classifier.

Table 2 indicates the supremacy of RF and ANNs in terms of classifying the phishing websites for dataset 2 while using the original attributes. Keeping in mind that dataset 2 has 30 different features and contains only 11000 distinct entries, still our employed ML models are achieving state of the art performance. In study [24], the authors achieved an accuracy of 95% using ANNs, however that study was focusing the classification of phishing emails and not websites. In the current work for classification of phishing websites, the ANN performed slightly higher by achieving accuracy of 97% for dataset 1 and 96% for dataset 2.

Table 1. Classifier performance over dataset 1 using all features

Classifier	Accuracy	Specificity	Precision	Recall	F1 Score
D- Tree	95.73%	95.48%	95.61%	95.98%	95.80%
SVM	94.37%	93.59%	93.83%	95.13%	94.48%
RF	97.87%	98.45%	98.47%	97.30%	97.88%
NB	82.17%	96.49%	95.22%	68.20%	79.48%
KNN	94.00%	95.68%	95.64%	92.36%	93.97%
ANN	97.83%	97.91%	97.96%	97.76%	97.86%

Table 2. Classifier performance on dataset 2 using all features

Classifier	Accuracy	Specificity	Precision	Recall	F1 Score
D- Tree	95.30%	94.65%	95.71%	95.82%	95.77%
SVM	92.58%	90.40%	92.22%	94.62%	93.40%
RF	95.96%	94.51%	95.67%	97.12%	96.39%
NB	60.48%	99.80%	99.44%	28.95%	44.85%
KNN	93.10%	92.21%	93.76%	93.81%	93.78%
ANN	95.90%	95.93%	96.71%	95.87%	96.29%

Table 3. Classifier performance on dataset 3 using all features

Classifier	Accuracy	Specificity	Precision	Recall	F1 Score
D-Tree	91.63%	93.72%	87.44%	87.44%	87.44%
SVM	89.66%	92.24%	84.48%	84.48%	84.48%
RF	92.94%	94.70%	89.41%	89.41%	89.41%
NB	89.33%	92.0%	83.99%	83.99%	83.99%
KNN	91.30%	93.47%	86.95%	86.95%	86.95%
ANN	90.48%	92.86%	85.71%	85.71%	85.71%

The results achieved by the employed classification algorithms on dataset 3 are shown in Table 3. It can be observed that the outcomes from all ML algorithms are slightly lower than the performances in case of dataset 1 and dataset 2. The highest accuracy obtained by any classifier on dataset 3 is 92.94% which is (4%) lower than what is observed in case of dataset 1 and dataset 2. However, there are some factors that should be noted in this case. Primarily, the dataset 3 is small (i.e. only 1353 distinct instances) with small number of distinguishing features (9 attributes) as compared to more than 10000 instances for datasets 1 and dataset 2. Likewise, feature set in later case are 48 (in dataset 1) and 30 (in dataset 2). Furthermore, dataset 3 is multi-class problem (3-classes) as compared to bi-class problem in case of dataset 1 and 2. This makes the classification task more challenging specifically when training data is limited as well. These factors influence the overall accuracy of our ML models and hence indicating relatively low performance in this case which was expected. More specifically, for a multiclass dataset, there needs to be sufficiently large number of distinct instances which the classifier can study and then try to make predictions. If a dataset is small as well as multiclass, the performances are expected to be mediocre.

From statistical results presented in Tables 1, 2 and 3, it can be seen that overall performance from all ML algorithms are quite satisfactory except NB which performed relatively low for the dataset 1 and 2, however indicated relatively better in case of dataset 3. One interesting aspect of the current study is the use of dataset 3 for phishing website classification. This dataset was not used by existing studies to the best of authors' knowledge. This may be due to the limited size of this dataset however, it might be helpful to investigate the classification performance using this dataset as it defines different attributes/features to other datasets. Furthermore, this dataset is multiclass as compared to dataset 1 and 2 which are bi-class dataset, hence, this study helps in getting better insights, as it is the only publicly available multiclass phishing websites dataset.

4.2 Classification Performance After Dimensionality Reduction Using PCA

Table 4, Table 5 and Table 6 present the statistical results achieved by different ML algorithms for the PCA-based dimension reduced datasets. It can be seen that ANN outperformed other classifiers when trained and tested over the PCA based dimension reduced Dataset 1. An accuracy of 97.13% is achieved while using first 30 components which contain the 95% variance of the overall PCs distributions. Dimensionality reduction is not previously employed in this specific domain of classification of phishing websites, this work is first of its kind and it is indeed getting good results while using reduced data and ANN in case of dataset 1.

Table 4. Classifier performance on dataset 1 after PCA

Classifier	Accuracy	Specificity	Precision	Recall	F1 Score
D-Tree	91.83%	91.29%	91.58%	92.00%	91.97%
SVM	93.97%	93.05%	93.33%	94.87%	94.09%
RF	94.90%	96.49%	96.46%	93.35%	94.88%
NB	78.37%	89.13%	86.49%	67.87%	76.06%
KNN	93.97%	95.61%	95.57%	93.36%	93.94%
ANN	97.13%	96.22%	96.48%	98.03%	97.19%

Similarly, first 18 PCs covers the 95% of overall components variance for the dataset 2 which originally consists of 30. Table 5 indicates that the classification performances are relatively lower than the Table 4 for the dataset 1 however, it is expected because the dataset 2 comprises of comparatively less features than dataset 1.

The overall performance is satisfactory though more specifically, we can see the balance between the sensitivity and specificity. This factor is very interesting because it validates the best compromise between true and false positives from a classifier.

Table 5. Classifier performance on dataset 2 after PCA

Classifier	Accuracy	Specificity	Precision	Recall	F1 Score
D- Tree	92.58%	91.12%	92.95%	93.75%	93.35%
SVM	92.43%	89.57%	91.89%	94.73%	93.29%
RF	93.79%	92.82%	94.26%	94.57%	94.41%
NB	90.50%	85.43%	89.01%	94.57%	91.70%
KNN	92.85%	91.80%	93.45%	93.70%	93.57%
ANN	94.33%	95.46%	96.25%	93.43%	94.82%

Table 6 shows the summary of statistical results performed by the aforementioned classifiers while trained and tested over the PCA based dimension reduced dataset 3. Similar to previous results, the outcomes indicated PCA to be handy on dataset 3 as well. By training on data produced by PCA, K-neighbors and ANN, performed even better than training on the whole datasets having all the features.

Table 6. Classifier performance on dataset 3 after PCA

Classifier	Accuracy	Specificity	Precision	Recall	F1 Score
D- Tree	90.31%	92.73%	85.47%	85.47%	85.47%
SVM	89.16%	91.87%	83.74%	83.74%	83.74%
RF	90.15%	92.61%	85.22%	85.22%	85.22%
NB	89.00%	91.75%	83.50%	83.50%	83.50%
KNN	92.12%	94.09%	88.18%	88.18%	88.18%
ANN	91.13%	93.35%	86.70%	86.70%	86.70%

Table 7 shows the top 10 ranked features within the three datasets identified by the PCA based on attribute loadings in components as described earlier (Sect. 3.3). It can be observed in Table 7 as well as Fig. 2 that the most important features in dataset 2 for instance, are the *'having-sub-dmian'* and *'age-of-domain'* while *'request-url'* and *'popUpWindow'* in dataset 3. The investigation of such distinguishing features would be helpful for domain experts and research community in this domain to further explore the varying combinations of only top-ranked features within the various datasets that might be helpful for further optimization of cyber security applications.

Table 7. Top-ranked features identified within three datasets using PCA

Feature rank	Features from dataset 1	Features from dataset 2	Features from dataset 3
1	RandomString	having_Sub_Domain	Request_URL
2	DomainInPaths	age_of_domain	popUpWidnow
3	NumUnderscore	Page_Rank	URL_of_Anchor
4	RightClickDisabled	Prefix_Suffix	SSLfinal_State
5	ExtFavicon	web_traffic	URL_Length
6	NumPercent	Statistical_report	having_IP_Address
7	NumSensitiveWords	having_At_Symbol	SFH
8	EmbeddedBrandName	SFH	web_traffic
9	TildeSymbol	Redirect	age_of_domain
10	SubmitInfoToEmail	Google_Index	

5 Conclusion and Future Work

This manuscript aims a comprehensive analysis of various ML algorithms to classify the fishing websites using multiple datasets. The study investigated the RF and ANN outperform other algorithms while tested over multiple datasets. We further conducted experiments on various datasets with and without dimension reductions using PCA and compared the performances of the state-of-the-art ML algorithms. The statistical results indicated the vital role of PCA specifically for eliminating the irrelevant features from the original datasets while not affecting the classification accuracy. The study further utilized the attribute loading-based ranking of various features within different datasets resulting some overlapping attributes within multiple datasets (e.g. web-traffic). The outcome might be useful for furthering the research within the domain of cyber security. For instance, it would be helpful to investigate the formation of a dataset consisting the composite of feature set identified as significant in this study and then use the ML techniques to classify the more complex problems (i.e. adversarial attacks) in this domain. Ensemble model can be utilized to enhance the classification accuracy specifically, in multi-class phishing website detection. Likewise, the existing datasets such as dataset 3 can further be extended that might be helpful to improve the classification performance.

References

1. What is phishing | Attack techniques & scam examples | Imperva, Imperva (2016). https://www.imperva.com/learn/application-security/phishing-attack-scam/. Accessed 12 June 2019
2. Sheng, S., Wardman, B., Warner, G., Cranor, L., Hong, J., Zhang, C.: An empirical analysis of phishing blacklists. In: Conference on Email and Anti-Spam (2009). https://doi.org/10.1184/R1/6469805.v1
3. Jain, A.K., Gupta, B.B.: Phishing detection: analysis of visual similarity based approaches. Secur. Commun. Netw. (2017). https://doi.org/10.1155/2017/5421046

4. Boser, B.E., Guyon, I.M., Vapnik, V.N.: A training algorithm for optimal margin classifiers. In: Proceedings of the Fifth Annual Workshop on Computational Learning Theory (1992). https://doi.org/10.1145/130385.130401
5. Quinlan, J.R.: "Induction of decision trees", readings in machine learning. Mach. Learn. **1**, 81–106 (1986). https://doi.org/10.1007/BF00116251
6. Breiman, L.: Random forests. Mach. Learn. **45**, 5–32 (2001). https://doi.org/10.1023/A:1010933404324
7. John, G.H., Langley, P.: Estimating continuous distributions in Bayesian classifiers. In: Proceedings of the Eleventh Conference on Uncertainty in Artificial Intelligence, pp. 338–345 (1995). https://arxiv.org/abs/1302.4964.
8. Altman, N.S.: An introduction to kernel and nearest-neighbor nonparametric regression. Am. Stat. **46**(3), 175–185 (1992). https://doi.org/10.1080/00031305.1992.10475879
9. Rosenblatt, F.F.: Princples of neurodynamics. Perceptions and the theory of brain mechanisms. Am. J. Psychol. (1963). https://doi.org/10.2307/1419730
10. Pearson, K.F.R.S.: On lines and planes of closest fit to systems of points in space. London Edinburgh Dublin Philos. Mag. J. Sci. **2**, 559–572 (1901). https://doi.org/10.1080/14786440109462720
11. Khan, W., Ansell, D., Kuru, K., Bilal, M.: Flight guardian: autonomous flight safety improvement by monitoring aircraft cockpit instruments. J. Aerospace Inf. Syst. AIAA **15**, 203–214 (2018)
12. Khan, W., Kuru, K.: An intelligent system for spoken term detection that uses belief combination. IEEE Intell. Syst. **32**, 70–79 (2017)
13. Khan, W., Badii, A.: Pathological gait abnormality detection and segmentation by processing the hip joints motion data to support mobile gait rehabilitation. J. Res. Med. Sci. **07**, 1–9 (2019)
14. Khan, W., Hussain, A., Khan, B., Shamsa, T.B., Nawaz, R.: Novel framework for outdoor mobility assistance and auditory display for visually impaired people. In: 12th International Conference on the Developments in eSystems Engineering (DeSE2019: Robotics, Sensors, Data Science and Industry 4.0.) (2019)
15. O'Shea, J., Crockett, K., Khan, W., Kindynis, P., Antoniades, A., Boultadakis, G.: Intelligent deception detection through machine based interviewing. In: International Joint Conference on Neural Networks (IJCNN) (2018)
16. Kuru, K., Khan, W.: Novel hybrid object-based non-parametric clustering approach for grouping similar objects in specific visual domains. Appl. Soft Comput. **62**, 667–701 (2018)
17. Dilek, S., Çakır, H., Aydın, M.: Applications of artificial intelligence techniques to combating cyber-crimes: a Review (2015). https://arxiv.org/abs/1502.03552
18. Qadir, H., Khalid, O., Khan, M.U., Khan, A.U., Nawaz, R.: An optimal ride sharing recommendation framework for carpooling services. IEEE Access **06**, 62296–62313 (2018). https://doi.org/10.1109/ACCESS.2018.2876595
19. Davis, J.: Phishing Attacks on the Rise, 25% Increase in Threats Evading Security, HealthITSecurity (2019). https://healthitsecurity.com/news/phishing-attacks-on-the-rise-25-increase-in-threats-evading-security
20. Ibrahim, D., Hadi, A.: Phishing websites prediction using classification techniques. In: International Conference on New Trends in Computing Sciences (ICTCS) (2017). https://doi.org/10.1109/ictcs.2017.38
21. Mohammad, R.M., McCluskey, T.L., Thabtah, F.: UCI Machine Learning Repository, Irvine, CA: University of California, School of Information and Computer Science (2012). https://archive.ics.uci.edu/ml/datasets/phishing+websites. Accessed 16 June 2019
22. Zhang, N., Yuan, Y.: Phishing detection using neural network (2012). https://cs229.stanford.edu/proj2012/ZhangYuan-PhishingDetectionUsingNeuralNetwork.pdf

23. Metrics and scoring: quantifying the quality of predictions — scikit-learn 0.22.1 documentation, Scikit-learn.org. https://scikit-learn.org/stable/modules/model_evaluation.html
24. Mohammad, R., McCluskey, L., Thabtah, F.: Intelligent rule-based phishing websites classification. IET Inf. Secur. **8**(3), 153–160 (2014). https://doi.org/10.1049/iet-ifs.2013.0202
25. Karnik, R., Bhandari, D.G.M.: Support vector machine based malware and phishing website detection (2016). https://pdfs.semanticscholar.org/ffea/603ec9f33931c9de630ba1a6ac71924f1539.pdf?_ga=2.226066713.262761491.1579621617-1102774226.1578838444
26. Babagoli, M., Aghababa, M.P., Solouk, V.: Heuristic nonlinear regression strategy for detecting phishing websites. Soft. Comput. **23**(12), 4315–4327 (2018). https://doi.org/10.1007/s00500-018-3084-2
27. Sahingoz, O.K., Buber, E., Demir, O., Diri, B.: Machine learning based phishing detection from urls (2019). https://doi.org/10.1016/j.eswa.2018.09.029
28. Tahir, M.A.U.H., Asghar, S., Zafar, A., Gillani, S.: A hybrid model to detect phishing sites using supervised learning algorithms (2016). https://doi.org/10.1109/CSCI.2016.0214
29. Chang, H.L., Dong, H.K., LEE, L.J.: Heuristic based approach for phishing site detection using URL features. In: Third International Conference on Advances in Computing, Electronics and Electrical Technology - CEET (2015). https://doi.org/10.15224/978-1-63248-056-9-84
30. Tan, C.L.: Phishing Dataset for Machine Learning: Feature Evaluation, Mendeley Data, v1 (2018). https://doi.org/10.17632/h3cgnj8hft.1. Accessed 16 June 2019
31. Abdelhamid, N.: UCI Machine Learning Repository, Irvine, CA: University of California, School of Information and Computer Science (2016). https://archive.ics.uci.edu/ml/datasets/Website+Phishing. Accessed 16 June 2019
32. Scikit-learn: machine learning in Python — scikit-learn 0.22.1 documentation, Scikit-learn.org. https://scikit-learn.org/stable/
33. Home - Keras Documentation, Keras.io. https://keras.io/
34. NumPy. https://numpy.org/
35. Python Data Analysis Library, Pandas.pydata.org. https://pandas.pydata.org/.

A Survey of Vision-Based Road Parameter Estimating Methods

Yan Wu[✉], Feilin Liu, Linting Guan, and Xinneng Yang

College of Electronics and Information Engineering, Tongji University, Shanghai 201804, China

{yanwu,1933048,glinting,1830836}@tongji.edu.cn

Abstract. Intelligent vehicles need to acquire real-time information on the road through sensors, calculate the limit of car speed and angular speed, so as to provide safety for the control decision. We argue that the road conditions such as snow, ice or humidity pose a major threat to driving safety. We divide the current methods of estimating the road parameters based on the visual sensor, as friction coefficient estimation method, road curvature estimation method and the road slope estimation method. The significance of various methods to intelligent driving, the current research status, as well as scientific difficulties are discussed in detail. Finally we discuss the possible research directions, including establish large-scale open data set, road status prediction methods under multi-task constraints and online learning mechanisms.

Keywords: Road parameter estimation · Visual sensor · Intelligent vehicle

1 Introduction

According to the U.S. Department of Transportation, about 22% of vehicle crashes occur each year, and about 16% of the casualties are weather-related. In addition, most accidents occur in wet road conditions, 73% of which occur on wet roads, and 17% on snow or sleet [1], which shows that road conditions do greatly affect driving safety, so real-time perception of road surface conditions is critical to the safe driving of cars. Human drivers perceive current road conditions through visual system, then adjusting the speed of the car according to the current road surface friction, road curvature and road slope. In order to ensure the driving safety of intelligent vehicle in various road surfaces and weather conditions, the intelligent driving system must obtain real-time information of the road through the sensor, e.g. predict the friction coefficient of the current road, road curvature and road slope and other basic parameters, so as to calculate the maximum driving speed and angular speed of the vehicle on the current road surface, and provide safety for the path planning and vehicle control system. This paper firstly introduces the method of road parameter estimation based on visual sensor, then summarizes the method of road surface perception and parameter estimation, finally discusses the current scientific progress of friction coefficient estimation, road curvature estimation and road slope estimation.

D.-S. Huang and P. Premaratne (Eds.): ICIC 2020, LNAI 12465, pp. 314–325, 2020.
https://doi.org/10.1007/978-3-030-60796-8_27

2 Road Friction Estimation

Road friction, as a traffic parameter which can directly influence vehicle braking distance, has been extensively researched in the field of transportation and autonomous driving. Autonomous driving and assisted driving systems can significantly benefit from real-time prediction of road friction, as the driving style of vehicles can be timely adjusted according to the road surface condition, thereby avoiding potential traffic accidents. Up to now, traditional method of estimating road friction is mainly based on vehicle response and tire dynamics [2]. Such methods can directly calculate the road adhesion coefficient by tire deformation, noise response and vehicle slip rate during braking, which demonstrate cost advantages as it can directly reuse the vehicle's inherent sensors. However, in practical applications, such methods have a disadvantage of lacking predictive ability, thus can only calculate the road friction of past driving area and cannot provide upcoming road surface information for the vehicle's decision-making system.

On the other hand, although road friction estimation method based on visual sensors requires additional hardware, it has a stronger predictive ability and thus can be a basic module of advanced autonomous driving applications. Such methods estimate road friction by establishing a model between road friction and related road parameters. Specifically, this process can be divided into two stages. Firstly, the type of road surface is predicted based on the image taken by a front camera on the vehicle. Secondly, a mapping function of road surface type to friction parameter values is established based on prior knowledge. In this way, we get a specific friction value.

This section mainly introduces the road friction estimation method based on visual sensors. According to the stages of constructing a model, existing works in this field are divided into road surface prediction and friction parameter prediction. The method overview is shown below in Fig. 1. Road surface prediction includes road material classification, road state classification, and mixed classification; friction parameter prediction includes adhesion coefficient estimation and other friction-related parameter estimation. The specific models used here include traditional machine learning models and deep learning models, usually combined with dynamics methods to achieve higher prediction accuracy.

2.1 Road Surface Prediction

Road Material Classification. Road material is closely related to road friction value. Common road material (including asphalt, soil, gravel, cobblestones, and grass) can be directly classified from an input image. Therefore, the classification of upcoming road material can be the first step of road friction estimation.

Kim et al. proposed a road material classification model [3] based on hand-crafted features and Bayesian classifiers in 2009, which divides the road surface into four different materials. The model first uses an over-segmentation algorithm to segment areas of the same material in an image, and then enhance contrast of the image to reduce the similarity of adjacent areas. Finally, the Bayesian classifier is used to classify road surface and achieve 95.38% classification accuracy. In contrast, preprocessing input images without contrast enhancement will result in a drop in accuracy to 90.55%.

Similarly, Rateke et al. proposed a CNN-based model [4] to classify road surface into asphalt, paved, and unpaved material in 2019. Due to the lack of appropriate public dataset, the authors construct a new dataset-the RTK dataset by collecting images from suburbs of Brazil using low-cost cameras, so it contains more unpaved road than previous ones. In terms of model design, the authors choose a simple 6-layer CNN for classification. And the lower part of the image which potentially contains more road information is used as RoI input. The model is trained on RTK, KITTI and CaRINA datasets, and the final accuracy on the three datasets reaches 95.7%, 93.1%, and 98.3%, respectively.

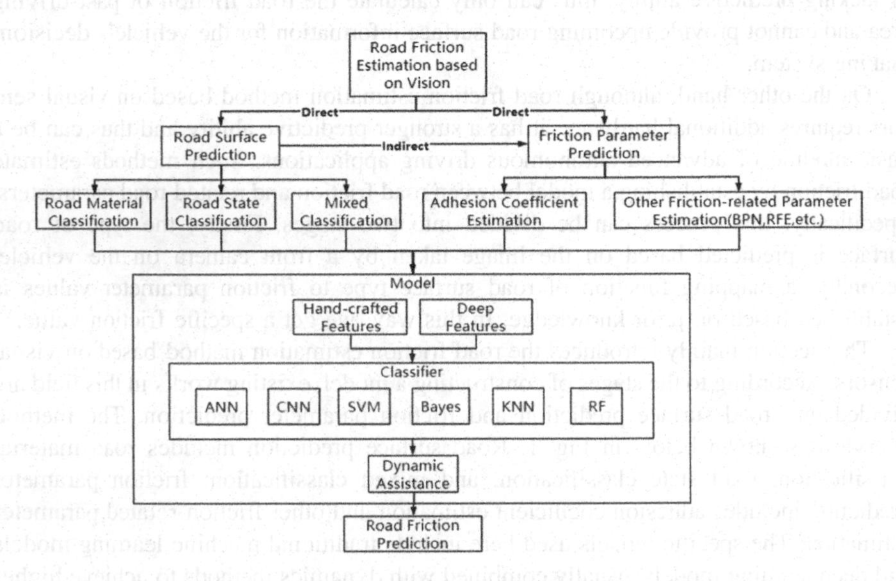

Fig. 1. Overview of vision-based road friction estimation methods.

Road State Classification. Road surface state is another important factor that affects road friction greatly. Unlike road material, road surface state tends to change more easily with weather conditions. Common road state such as dry, wet, water, ice, and snow can result in significantly different road friction and vehicle braking distance. Therefore, predict road surface state correctly could lead to a precise road friction estimation.

Qian et al. proposed a road state classification method [3] based on traditional machine learning algorithm in 2016. This method first learns a distribution of road area in the dataset, and then fix that area as the RoI input of the model. The model first uses MR8 filter and K-means cluster to extract luminance-invariant features, and then augment them with luminance-relative features based on pixel deviation histogram. A naive Bayes classifier is used to complete the classification. This method achieves

80%, 68% and 46% accuracy in two/three/five-classes classification respectively. The ablation experiment shows that manual selection of RoI can improve the accuracy by nearly 20%.

In order to get higher accuracy, Almazan et al. improved this work [6]. With other parts of the model basically remain unchanged, they mainly enhance the RoI prediction module. Additional geometric constraint is introduced by calculating vanishing point and horizontal line. Combined with the spatial priors learned from the dataset, a better RoI prediction result is obtained. With the same hand-crafted features being used, a simpler RBF SVM is chosen for classification, and a higher accuracy of 86%, 80% and 52% in two/three/five-classes classification respectively is achieved, which shows a great overall improvement.

Zhao et al. proposed another road state classification algorithm based on traditional machine learning models [7] in 2017. The feature extraction part of the model uses hand-crafted features as well, with SVM being used as a classifier. What makes it different is the parameter optimization algorithm and evaluation metric used here. Grid Search and PSO are used to optimize SVM parameter. And model accuracy is evaluated by calculating the ratio of correctly classified image grids to total 5x9 image grids. Finally, the single-state accuracy exceeds 90%, while the multi-state accuracy exceeds 85%. The SVM model based on PSO optimization algorithm is significantly better than grid search, which leads to an accuracy increase by more than 10%.

For winter road state with more snow, Pan et al. collected data on a highway in Canada and constructed a new dataset to estimate the amount of snow [8]. The dataset divides road state into 5 categories according to snow coverage condition. The model uses VGG16 pre-trained on ImageNet as a baseline classifier with a full image input, which is then compared to traditional machine learning models and a VGG16 model without pre-training. The result shows that the pre-trained VGG16 achieves the highest accuracy of 90.4%, 87.3% and 78.5% in two/three/five-classes classification respectively. Pre-training on ImageNet brings a 2% accuracy improvement.

Mixed Classification. Since the material and the state of the road surface both significantly affect road friction value, an intuitive idea is to construct a dataset containing both of them to train a mixed classification model.

Nolte et al. proposed a CNN-based mixed road surface classification model [9] in 2018. Considering that there exist lots of reusable data in public traffic datasets, they select and label images from multiple public datasets in a mixed way. After that, appropriate RoI is manually selected and resized to 224x224 as the input of the model. The paper compares performance of ResNet50 and InceptionV3 in road classification. Experiment shows that ResNet50 achieves 92% accuracy, 2% higher than InceptionV3. And selection of RoI is very important, which improves accuracy by 10% than simply using the whole image as input.

Similarly, Busch et al. select road images from existing traffic datasets to form a mixed dataset [10]. InceptionV3, GoogLeNet and SqueezeNet are used to compare the influence of different architectures on classification accuracy. In addition, considering the importance of RoI selection module in the previous work, this paper compares the effect of different RoI shapes on accuracy. Unexpectedly, the SqueezeNet architecture with the whole image input reaches the highest F1 accuracy of 95.36%, while different network architecture has little influence on accuracy.

2.2 Friction Parameter Prediction

Road Adhesion Coefficient Estimation. Road adhesion coefficient is defined as the ratio between ground adhesion force and tire normal force, which approximately equals to road friction coefficient when adhesion reaches the maximum. As it can directly change a car's braking distance, adhesion coefficient has been modelled in many different ways. The traditional dynamics-based method estimates adhesion coefficient with longitudinal response of the tire. Since no additional hardware is required and high accuracy can be ensured, such methods have always been the mainstream of adhesion coefficient estimation and have been widely studied. However, such methods also have inherent shortcomings like lacking predictive ability. On the other hand, camera-based adhesion coefficient estimation method is not as accurate as the former, but with more predictive ability and better real-time performance. Therefore, an intuitive idea is to combine both methods to solve the dilemma between speed and accuracy, improving the reliability of road adhesion coefficient estimation. In camera-based methods, dynamic models are always combined with vision predictions to achieve faster adhesion coefficient estimation with higher accuracy.

Xiong et al. proposed a method for calculating adhesion coefficient with dynamics and aided visual information [11] in 2019. This model uses color moment and GLCM to extract features and then use SVM to classify it into dry or wet asphalt, which achieves 92.47%/88.39% classification accuracy on dry/wet asphalt. Then, adhesion coefficients of dry/wet asphalt are specified as 0.85 and 0.6 based on statistical data, which are then used as approximate initial values of the dynamic model to calculate the final adhesion coefficient. As a result, the convergence speed of the hybrid estimator is obviously faster than pure dynamics model. And the prediction accuracy is also closer to ground truth, especially on road where dynamics pattern is not obvious.

Sabanovic et al. propose a method to estimate adhesion coefficient with a similar idea [12] in 2020. What makes the method different is that dynamic model is only needed during training. Road surface is first classified into six categories with AlexNet. After that, an adhesion coefficient-slip ratio curve is fitted for each road surface based on data collected by the vehicle in real time. In this way, an end to end adhesion coefficient estimation is achieved with only visual information. In addition, the paper combines the system with the ABS model to reduce the braking distance of the vehicle by predicting the adhesion coefficient in advance. Combined with the ABS model, the vehicle braking distance is reduced by up to 18%.

Other Friction-Related Parameter Estimation. In addition to road adhesion coefficient estimation, there are also some works manage to estimate other friction-related parameters to achieve a similar prediction effect. Road friction estimate RFE, friction level μ and anti-skid level BPN all belong to this category. With only visual information, prediction models can only roughly estimate their values in a coarse-grained manner. Therefore, this type of method has many similarities with previous road surface classification methods in data collection and implementation details.

Roychowdhury et al. proposed a multi-stage RFE estimation method [1] in 2018. First, the CNN-based method is used to classify roads into four categories: dry, wet, slush, and snow. Among them, dry roads can be directly considered to have high RFE. Then, manual segmentation is applied to other images to divide them into 15 trapezoid blocks, which are then stretched into rectangular bird's eye blocks according to perspective projection. Finally, the probability of each image block being dry/wet is predicted separately by the model, and the average value is calculated to comprehensively estimate the RFE. This method uses SqueezeNet to achieve the best accuracy of 97.36% in road classification, and 89.5% in RFE prediction.

The paper published by Jonnarth in 2018 explores the effect of network architecture, data distribution, and the use of simulated data on estimation of friction level μ [13]. VGG, ResNet, and DenseNet are used to classify the road surface into high friction ($0.2 <= \mu < 0.6$) or low friction ($\mu < 0.2$) level. The dataset used in this project consists of 37,000 real images and 54,029 simulated images, which is quite sufficient for a classification task. The trained model finally reached a prediction accuracy up to 90%. And the following conclusions are summarized: 1) the prediction accuracy between different network architectures has little difference; 2) the span of dataset is more important than its size, which significantly affects model performance; 3) there is a certain gap between simulated image and real image, and the use of simulated image does not lead to a noticeable performance improvement.

Du et al. published a paper on rapid estimation of road anti-skid level BPN from the perspective of anti-skid performance of autonomous vehicles [14] in 2019. The author combines CNN and hand-crafted features to propose a deep convolutional neural network-TLDKNet based on domain knowledge. The domain knowledge mentioned here refers to LBP (Local Binary Pattern), GMM (Gaussian Mixed Model) and GLCM (Gray-Level Co-occurrence Matrix), three texture features that have been proven to be strongly related to road anti-skid performance. The model combines the convolutional layer of VGG16 with three texture features to form a 4-branch feature extraction network. The features of different branches are merged together to classify the anti-skid level BPN. As a result, this model divides BPN into three levels: high (BPN > 57), medium (47 < BPN < 57), and low (BPN < 47), with a final accuracy of 90.67% and 80% in two/three-classes classification achieved respectively.

2.3 Existing Problems in Road Friction Estimation

Despite so many excellent works, there are still many problems in the current road friction estimation method based on vision.

1. There are no unified public benchmarks for road surface classification and road friction estimation. Many studies are based on the data collected by researchers themselves for model training and accuracy evaluation, which makes comparison between models very difficult. We need a unified and effective accuracy metric to evaluate the performance of different models.

2. An efficient image pre-processing method is needed to eliminate redundant background information and extract accurate road features. Although the context information can help to identify the state of road surface, it has undoubtedly a negative impact on the estimation of road material and friction parameters.
3. The road friction estimation based on vision is conduct in a coarse-grained manner, which can only be roughly classified into several value levels. And most accurate road friction estimation methods heavily rely on dynamic assistance, which greatly limits the use of the model.

3 Road Curvature Estimation

Intelligent vehicles need to perceive and predict the surrounding environment information in real time during autonomous driving, in which the curvature of the road can assist in predicting the direction of the lane, thus providing important help to the automatic control system of the vehicle. At present, the mainstream road curvature prediction algorithms are road curvature estimates based on lane detection [15–19], road curvature estimates based on GPS trajectories [20], and road curvature estimates based on vehicle dynamics [21]. Since the purpose of this article is to introduce the method of estimating road parameters based on visual sensors, in this section we mainly introduce the method of road curvature estimation based on lane detection.

3.1 Road Curvature Estimation Based on Lane Detection

Tsai et al. first proposed the use of road image information to calculate the curvature of the road [15], their proposed algorithm is divided into four steps: 1) use the vehicle camera to obtain the road image in front of the vehicle, and get the road curve edge through image processing, 2) convert the curve edge from the image coordinates to the world coordinate system through the inverse projection transformation (IPM) [22], 3) calibrate camera parameters, 4) calculate the radius and center point of the curve from the curve point of the world coordinate system. The road curvature estimate proposed by Seo et al. is also composed of four steps [17, 18], inspired by the prior of road is parallel in the world coordinate system, they adjusted the image processing steps: 1) convert the forward view of the road image from the image coordinate system to the world coordinate system, 2) detect the lane line from the world coordinate system bird's eye view, 3)fit lane using a random sampling consistent (RANSAC) algorithm [23], 4) calculate the lane curvature by sampling data points from the lane. Hu et al. used the continuous characteristics of images collected by the vehicle camera to estimate the error variance of the lane, and filter the data points with high variance, so as to improve the accuracy and robustness of the road curvature. The main technical details involved in road curvature estimation are described below [18].

Inverse Perspective Transformation. Because the camera's optical axis intersects with road surface, this causes distortion of the road surface information projected onto the image plane, such as the two parallel lanes on the road surface intersect in the image plane. The inverse perspective transformation projects pixels on the image coordinate system into the world coordinate system.

In fact, by collecting the coordinates of the four points on the road plane and the corresponding image plane under the world coordinate system, the corresponding single-entitlement matrix H can be calculated to transfer the image pixels from the image coordinate system to the world coordinate system.

$$[\mathrm{X\ Y\ 1}]^T = H^{-1}[uv\,1]^T \tag{1}$$

Lane Detection. Because the color of lane is usually white or yellow, it can be used as a priori to capture information about the location of the lane using a color threshold or gradient threshold. Because the lane are concentrated in the x-axis within a certain range, it is possible that the pixel distribution peak on the x-axis is likely to be the base point of the lane line, so the peak point can be used as the base point of the lane line, then the sliding windows are used to detect lane points, finally the straight lane line or arc lane line is fitted by the Hough Transform or RANSAC algorithm. The effect is shown in Fig. 2.

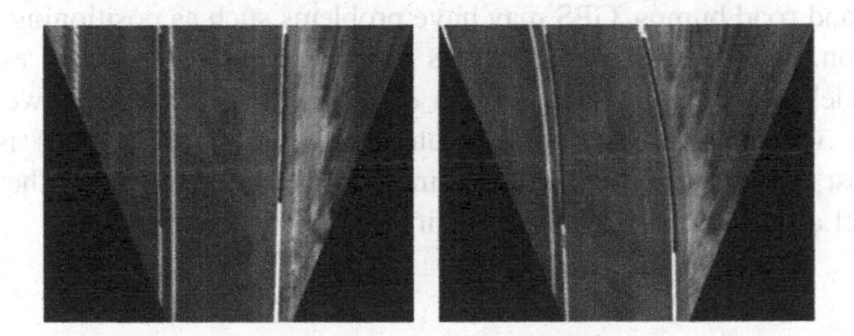

Fig. 2. Lane detection [18].

Road Curvature Calculation. The starting point, end point, and the center point are collected from the lane line on the picture to fit the radius of the arc's circle.

The curvature of the road κ is defined as the inverse of the arc radius, so the curvature of the road can be calculated directly.

$$\kappa = \frac{1}{R} \tag{2}$$

3.2 Problems of Current Visual-Based Road Curvature Method

The current vision-based road curvature study relies on robust lane detection, however, the results of lane detection will be subject to many conditions, such as poor visual images in bad weather conditions, road snow cover lane, lane wear or lane are severely blocked by cars, the current lane detection method will get poor performance in the above environment, so establish a large-scale road data set covering the above-mentioned situations will greatly promote the development of related methods.

4 Road Slope Estimation

As a key information to ensure driving safety and an important parameter of the electric control system of power transmission and chassis, the road slope can significantly improve the vehicle motion control performance if the road slope can be accurately estimated in real time. Road slope can be divided into lateral road slope and longitudinal road slope. Lateral road slope refers to the slope in the direction of the road crossing, while longitudinal road slope refers to the slope in the direction of the road moving forward. There are two types of longitudinal road slope: uphill and downhill. For the problem of road slope estimation, most studies focus on the estimation of longitudinal road slope [25, 26, 28, 29]. The methods of longitudinal road slope estimation are mainly divided into sensor-based [25] and model-based [26]. The sensor-based methods utilize additional sensors on the vehicle, such as inclination displacement sensors, accelerometers, GPS, etc. These methods are limited by the sensors used, which have good accuracy but have deficiencies. For example, the inclination displacement sensor is susceptible to the impact of body longitudinal acceleration and road bumps, GPS may have problems such as positioning error, signal loss and so on. The model-based methods estimate the road slope by establishing a dynamic model and obtaining the known data on the CAN bus. However, how to decouple the vehicle status parameters and road resistance changes is a difficult problem. Most of the lateral road slope estimation methods depend on the accuracy of the tire model and the road adhesion coefficient [27].

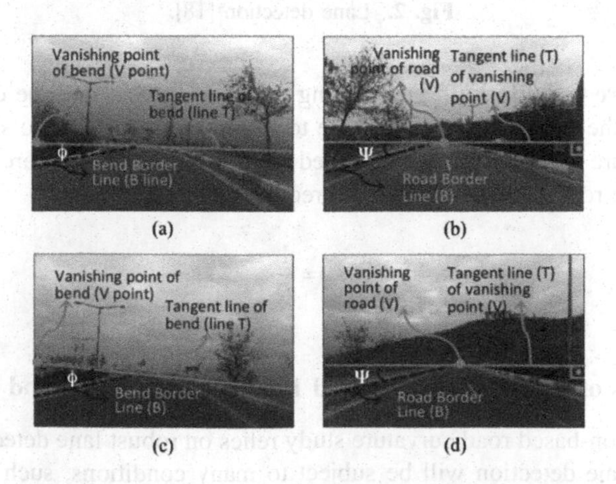

Fig. 3. Uphill and downhill. (a) and (b) refer to the images taken by the forward and backward cameras when going downhill. (c) and (d) refer to the images taken by the forward and backward cameras when going uphill [28].

4.1 Vision-Based Road Slope Estimation

Visual-based road slope estimation provides slope information before the vehicle enters a road with different slopes, which gives the driver or intelligent control system enough time to adopt the correct control strategy. However, the study of road slope estimation based on visual sensors is still in its preliminary stage. In [28], researchers combine forward and backward monocular cameras to classify longitudinal road slopes using geometric clues of the image, and divide road slopes into uphill and downhill. Figure 3 shows the images taken by the forward and backward cameras when going uphill and downhill. If $\phi > \psi$, the image represents an uphill. On the contrary, the image represents a downhill. If $\phi = \psi$, it means flat. In [29], researchers present three methods to estimate road slope from instant road images obtained from a front monocular camera. In the geometry-based method, they estimate the road slope using 2D road line derived from 3D road line and pin-hole camera model. In the local features-based method, they adopt the SIFT (scale-invariant feature transform) local features between two consecutive images. In the covariance-based method, they use 2D road line as feature to train multi-layer perceptron.

4.2 Limitation of Research on Vision-Based Road Slope Estimation

There are few studies on the estimation of road slope by using visual sensors. Relevant works are mainly conducted in the environment with good weather and high visibility, which cannot guarantee the robustness of the model in the snow and ice environment with bad weather or at night. In addition, these studies use only road lines or local features as inputs, and do not use whole images, resulting in that less information is available. Currently, there is no publicly available road slope image dataset, which hinders the development of road slope estimation using deep convolutional neural networks. In conclusion, the road slope information provided by visual sensors is insufficient to accurately estimate the road slope in real time, and its development is limited due to the lack of relevant dataset.

5 Conclusion

This paper mainly combs recent literature of road parameter estimation based on visual sensor, and classifies them into road friction estimation method, road curvature estimation method and road slope estimation method. The method of road friction estimation includes two part: road surface prediction and friction parameter estimation. In addition, the relevant progress of the road slope estimation method and road curvature estimation method are summarized in detail. This paper has reference value to the design of intelligent vehicle planning module in various road surfaces and weather conditions.

Human driver and intelligent vehicle are more prone to accident in snowy or icy environment, so road parameter estimation is particularly important in this situation. However, not much attention has been paid to this filed. Thus, future research can be expanded in the following directions:

1. Collect and label road parameter data sets for large-scale snow and ice roads. One reason for the lack of road parameters research under snow and ice surface is the lack of relevant open data sets, so collecting and labeling road parameter data sets in snow and ice environment can promote scientific progress in this area.
2. Road parameter prediction method under multi-task constraint. Intelligent vehicle usually requires simultaneous sensing of lane, coefficient of friction, travelable area and road curvature, these tasks have a mutually constrained relationship, so multitasking constraint road parameter prediction is worth studying.
3. Study how to estimate the vehicle's motion status through the redundant sensor data and the dynamic model, especially the current vehicle speed and corner speed, so as to assist the robustness of road parameter prediction and assist to the online update of prediction model is a very promising direction.

Acknowledgments. This work was supported by the National Natural Science Foundation of China (No. U19A2069).

References

1. Roychowdhury, S., Zhao, M., Wallin, A., Ohlsson, N., Jonasson, M.: Machine learning models for road surface and friction estimation using front-camera images. In: 2018 International Joint Conference on Neural Networks (IJCNN), pp. 1–8. IEEE, Rio, Brazil (2018)
2. Yuan, C.C., Zhang, L.F., Cheng, L.: Summary and prospect of development of road coefficient identification methods (in Chinese). Mach. Build. Autom. **47**(2), 1–4 (2018)
3. Kim, J., Kim, D., Lee, J., Lee, J., Joo, H., Kweon, I.S.: Non-contact terrain classification for autonomous mobile robot. In: 2009 IEEE International Conference on Robotics and Biomimetics (ROBIO), pp. 824–829. IEEE, Guilin (2009)
4. Rateke, T., Justen, K.A., Wangenheim, A.V.: Road surface classification with images captured from low-cost camera-road traversing knowledge (RTK) dataset. Revista de Informática Teórica e Aplicada **26**(3), 50–64 (2019)
5. Qian, Y., Almazan, E.J., Elder, J.H.: Evaluating features and classifiers for road weather condition analysis. In: 2016 IEEE International Conference on Image Processing (ICIP), pp. 4403–4407. IEEE, Phoenix (2016)
6. Almazan, E.J., Qian, Y., Elder, J.H.: Road segmentation for classification of road weather conditions. In: Hua, G., Jégou, H. (eds.) ECCV 2016. LNCS, vol. 9913, pp. 96–108. Springer, Cham (2016). https://doi.org/10.1007/978-3-319-46604-0_7
7. Zhao, J., Wu, H., Chen, L.: Road surface state recognition based on SVM optimization and image segmentation processing. J. Adv. Transp. **2017**(1), 1–21 (2017)
8. Pan, G., Fu, L., Yu, R., Muresan, M.I.: Winter road surface condition recognition using a pre-trained deep convolutional neural network. arXiv preprint arXiv:1812.06858 (2018)
9. Nolte, M., Kister, N., Maurer, M.: Assessment of deep convolutional neural networks for road surface classification. In: 2018 21st International Conference on Intelligent Transportation Systems (ITSC), pp. 381–386. IEEE, Orlando (2018)
10. Busch, A., Fink, D., Laves, M.-H., Ziaukas, Z., Wielitzka, M., Ortmaier, T.: Classification of road surface and weather-related condition using deep convolutional neural networks. In: Klomp, M., Bruzelius, F., Nielsen, J., Hillemyr, A. (eds.) IAVSD 2019. LNME, pp. 1042–1051. Springer, Cham (2020). https://doi.org/10.1007/978-3-030-38077-9_121

11. Xiong, L., Jin, D., Leng, B., Yang, X., Wu, L.H.: Road friction estimation method for distributed driving electric vehicle based on machine vision assistance (in Chinese). J. Tongji Univ. (Natl. Sci.) **47**(S1), 99–103 (2019)
12. Šabanovič, E., Žuraulis, V., Prentkovskis, O., Skrickij, V.: Identification of road-surface type using deep neural networks for friction coefficient estimation. Sensors **20**(3), 612 (2020)
13. Jonnarth, A.: Camera-based friction estimation with deep convolutional neural networks. Uppsala University, Uppsala, Sweden (2018)
14. Du, Y., Liu, C., Song, Y., Li, Y., Shen, Y.: Rapid estimation of road friction for anti-skid autonomous driving. IEEE Trans. Intell. Transp. Syst. **2019**(1), 1–10 (2019)
15. Tsai, Y., Wu, J., Wang, Z., Hu, Z.: Horizontal roadway curvature computation algorithm using vision technology. Comput. Aided Civil Infrastruct. Eng. **25**(2), 78–88 (2010)
16. Nelson, W.L.: Continuous-curvature paths for autonomous vehicles. In: IEEE International Conference on Robotics & Automation, pp. 1260–1264. IEEE, Scottsdale (1989)
17. Seo, D., Jo, K.H.: Inverse perspective mapping based road curvature estimation. In: IEEE/SICE International Symposium on System Integration, pp. 480–483. IEEE, Tokyo (2014)
18. Seo, D., Jo, K.H.: Road curvature estimation for autonomous vehicle. In: IEEE/SICE Proceedings of the Society of Instrument and Control Engineers Annual Conference, pp. 1745–1749. IEEE, Hangzhou (2015)
19. Hu, Z.Z., Zhang, L., Bai, D.F., Zhao, B.: Computation of road curvature from a sequence of consecutive in-vehicle images (in Chinese). J. Transp. Syst. Eng. Inf. Technol. **16**(1), 38–4563 (2016)
20. Ai, C.B., Tsai, Y.C.: Automatic horizontal curve identification and measurement method using GPS data. J. Transp. Eng. **141**(2), 04014078 (2015)
21. Dahmani, H., Chadli, M., Rabhi, A., Hajjaji, A.: Vehicle dynamics and road curvature estimation for lane departure warning system using robust fuzzy observers: experimental validation. Veh. Syst. Dyn. **53**(8), 1135–1149 (2015)
22. Bertozzi, M., Broggi, A.: GOLD: a parallel real-time stereo vision system for generic obstacle and lane detection. IEEE Trans. Image Process. **7**(1), 62–81 (1998)
23. Fischler, M.A.: Random sample consensus: a paradigm for model fitting with applications to image analysis and automated cartography. Commun. ACM **24**, 726–740 (1981)
24. Cáceres Hernández, D., Hoang, V.-D., Jo, K.-H.: Methods for vanishing point estimation by intersection of curves from omnidirectional image. In: Nguyen, N.T., Attachoo, B., Trawiński, B., Somboonviwat, K. (eds.) ACIIDS 2014. LNCS (LNAI), vol. 8397, pp. 543–552. Springer, Cham (2014). https://doi.org/10.1007/978-3-319-05476-6_55
25. Yong, W., Guan, H., Wang, B., Lu, P.: Identification algorithm of longitudinal road slope based on multi-sensor data fusion filtering (in Chinese). J. Mech. Eng. **54**(14), 116–124 (2018)
26. Jiang, S., Wang, C., Zhang, C., Bai, H., Xu, L.: Adaptive estimation of road slope and vehicle mass of fuel cell vehicle. eTransportation **2**, 100023 (2019)
27. Guan, X., Jin, H., Duan, C., Lu, P.: Estimation of lateral slope of vehicle driving road (in Chinese). J. Jilin Univ. **49**(6), 1802–1809 (2019)
28. Karaduman, O., Eren, H., Kurum, H., Celenk, M.: Road-geometry-based risk estimation model for horizontal curves. IEEE Trans. Intell. Transp. Syst. **17**(6), 1617–1627 (2016)
29. Ustunel, E., Masazade, E.: Vision-based road slope estimation methods using road lines or local features from instant images. IET Intell. Transp. Syst. **13**(10), 1590–1602 (2019)

Intelligent Fault Diagnosis

The TE Fault Monitoring Based on IPCR of Adjustable Threshold

Aihua Zhang(✉), Chengcong Lv, and Zhiqiang Zhang

College of Engineering, Bohai University, Jinzhou 121013, China
jsxinxi_zah@163.com

Abstract. The The algorithm of Improved Principal Component Regression (IPCR) judges whether there is a quality related fault in Tennessee Eastman (TE) process with T^2-statistics. Because the threshold value is never changed, there will be the problem of false alarm and missing alarm. To solve this problem, an adjustable threshold IPCR algorithm is proposed. Firstly, the IPCR model is built with normal data and the threshold of traditional T^2-statistics is obtained. In the online detection, the new threshold is calculated according to the fixed threshold and the exponentially weighted moving average of statistics, and the new threshold is used for fault detection. Finally, the simulation results in TE process show that this method can effectively enhance the detection results in TE process.

Keywords: Fault detection · IPCR · TE · Adjustable threshold

1 Introduction

Because the process monitoring based on multivariate statistics does not need complex mathematical model, it has been widely studied. In addition, the rapid development of sensors in recent years also promotes the development of process monitoring based on multivariate statistics. The basic theory of process monitoring method based on multivariate statistics involves principal component analysis (PCA), partial least squares (PLS) [1–7]. Most of the algorithms based on PCA can't determine whether the monitoring results are related to the quality. For example, if the monitored variables have been changed but the quality we care about doesn't be changed, we can ignore this kind of abnormality if it doesn't bring any other loss. The theory based on PLS usually links the monitoring results with quality, which can reduce some unnecessary alarms, improve industrial production and reduce production costs.

Algorithms similar to pls theory include multiple linear regression (MLR) [8–11], pls [12, 13], canonical variable analysis (CVA) [14], principal component regression (PCR) [15], etc.

These monitoring techniques usually use square prediction error (SPE) (also known as Q Statistics) and Hotelling's T^2 statistics draw control charts. The thresholds are fixed. If the statistics exceed the thresholds, an alarm will occur. The fixed threshold is defined based on a certain empirical distribution. It is also based on the balance of the relationship between the false alarm rate and the missed alarm rate. Therefore, it can not be better adjusted for the current situation, which will lead to the problem of false

D.-S. Huang and P. Premaratne (Eds.): ICIC 2020, LNAI 12465, pp. 329–338, 2020.
https://doi.org/10.1007/978-3-030-60796-8_28

alarm and missed alarm, so the monitoring results are not very ideal. In order to solve this problem, an IPCR algorithm with adjustable threshold is proposed, which can establish a relationship with quality and adjust the threshold at the same time. The adjustment of threshold value is based on the improved exponential weighted moving average (EWMA). This method can reduce the noise pollution, and the adjustable threshold can better measure the change of the system, which is more helpful to make a more reasonable decision on whether the system is healthy. Finally, it is verified in TE system that this method can obviously reduce the false alarm rate.

2 IPCR Algorithm

The IPCR algorithm is improved by PCA algorithm. As mentioned above, the traditional PCA algorithm can not establish a relationship between the monitoring results and the quality we care about, which will lead to the false alarm rate. The improved IPCR algorithm makes up for this disadvantage, and the results also show that IPCR can distinguish whether quality related faults occur or not. Where X $(n \times m)$ is the process variable and Y $(n \times l)$ is the quality variable. The specific IPCR algorithm is as follows.

First, decompose X according to PCA

$$X = \hat{X} + \tilde{X} = TP^T + \tilde{X} \tag{1}$$

where T is the score matrix, P is the load matrix, $\hat{X}$ is the main element part, and $\tilde{X}$ is the residual part. Then make T and Y do least square regression to get the load matrix W of Y.

$$W^T = \left(T^T T\right)^{-1} T^T Y \tag{2}$$

Get the coefficients M for Y and X

$$\hat{Y} = TW^T = XPQ^T = XM \tag{3}$$

In order to decompose X more thoroughly, the coefficient matrix MM^T is decomposed by SVD, and the following results are obtained.

$$MM^T = \left[P_M \tilde{P}_M\right] \begin{bmatrix} \Lambda_M & 0 \\ 0 & 0 \end{bmatrix} \begin{bmatrix} P_M^T \\ \tilde{P}_M \end{bmatrix} \tag{4}$$

$$\Pi_M = P_M P_M^T \tag{5}$$

$$T_{re} = XP_M \tag{6}$$

$$\hat{X} = X\Pi_M = T_{re} P_M^T \tag{7}$$

where $\hat{X}$ is extremely related to Y, T_{re} is the score matrix of $\hat{X}$.

On-line detection: t_{re} is obtained from Eq. (4).

$$t_{re} = P_M^T x \tag{8}$$

$$\hat{x}^T \hat{x} = x^T P_M P_M^T x = t_{re}^T t_{re} \tag{9}$$

Thus, we can determine the T^2 statistic of $\hat{X}$

$$T = t_{re}^T \left(\frac{T_{re}^T T_{re}}{N-1} \right)^{-1} t_{re} \tag{10}$$

If the confidence limit is set to α, the threshold of T_{re}^2 is as follows

$$T_\alpha^2 = \frac{A(N^2-1)}{N(N-m)} F_{A,N-A,\alpha} \tag{11}$$

where $F_{A,M-A,\alpha}$ is F-distribution with A and $M - A$ degrees of freedom, A is the number of latent variables

Finally, according to the threshold to determine whether the quality related fault occurs or not.

- $T \geq T_\alpha^2$
$\Rightarrow$ A quality related failure has occurred;
- $T < T_\alpha^2$,
$\Rightarrow$ No quality related failure occurred;

3 Improved EWMA Algorithm to Adjust Threshold

In order to overcome the shortcomings of using fixed threshold method in process monitoring, the paper [16] applies EWMA control scheme to process monitoring of PCA, and achieves good monitoring effect. However, although PCA can monitor the situation of the system, it is unable to distinguish whether quality related faults occur or not. This paper combines the idea of adjusting threshold with IPCR to make up for it. The formula of control line given in article [16] is as follows, T_α^2 is the traditional fixed threshold value, λ is the weight, and H is the window length. If there are alarm samples in the previous time, this adaptive threshold can get very small value or even show negative value due to the cumulative effect of fault samples in the previous time, which may lead to an increase in false positives. Therefore, a minimum threshold value of $\frac{T_\alpha^2}{2}$ is set.

$$t_i > \max\left\{ \frac{\left(T_\alpha^2 \sum_{j=1}^h \lambda^j - \sum_{j=1}^{h-1} \lambda^j t_{i-h+j} \right)}{\lambda^h}, \frac{T_\alpha^2}{2} \right\} \tag{12}$$

If λ is too small to be close to 1, although the false alarm can be reduced, the detection delay time will be increased. If λ is too large, it will increase the weight of the nearest sample. Although it can achieve the purpose of fast detection, it can not reduce the false alarm rate. The window length h can affect the calculation time. As time goes on, the later the data weight coefficient is smaller and even can be ignored. Therefore, h cannot and does not need to be too large. The specific values of h and λ still need to be determined in the specific environment.

4 IPCR with Adjustable Threshold

The EWMA method can be integrated into the IPCR algorithm, which can improve the accuracy of monitoring. By adjusting the parameters, it can reduce the false alarm rate and improve the detection rate. The specific algorithm steps are as follows.

The flow chart describes the detailed, shown in Fig. 1.

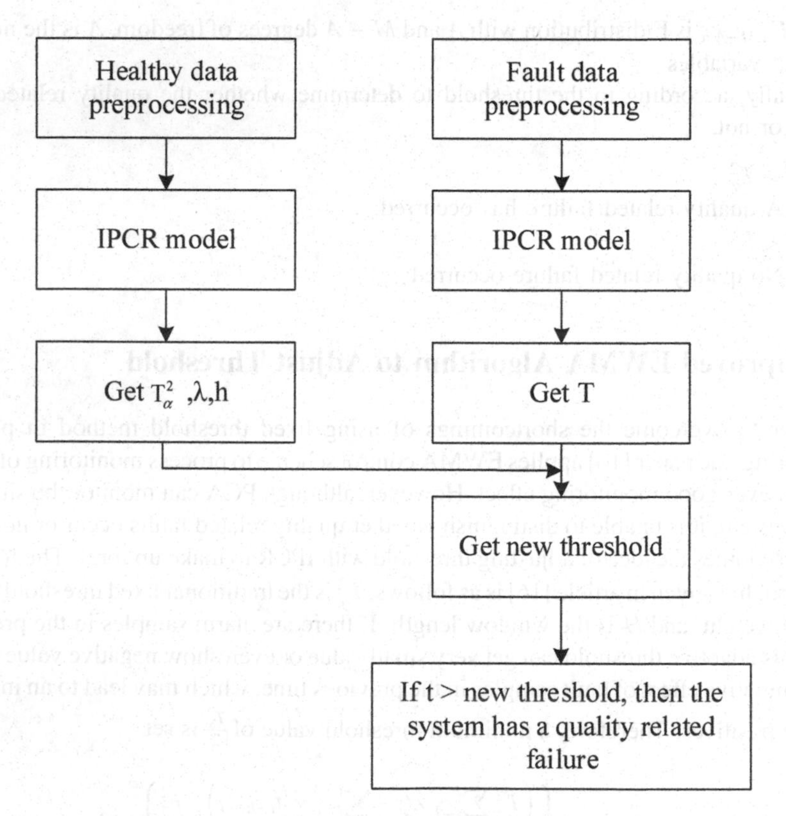

Fig. 1. Flow chart

Step 1 standardized processing of normal data
Step 2 processed data is brought into IPCR for modeling
Step 3 determines the confidence level α and get the threshold T_{α}^{2}
Step 4 determines the weight coefficient λ and window length h according to the quality related fault of normal data in the algorithm without alarm.
Step 5 In the online test, the test sample is brought into the IPCR model to obtain the statistic t
Step 6 The new threshold is determined according to T_{α}^{2} in Step 3 and T in step 5. Then the new threshold is used to judge whether the system has quality related faults. When the statistic t exceeds the threshold value, it means that there is a quality related fault; otherwise there is no quality related fault.

5 TE Process Monitoring Based on Adjustable Threshold IPCR

TE system comes from real chemical process. It is a chemical model widely used to develop research and evaluate process control technology and monitoring methods. The large number of literatures cited its data to study algorithm optimization, process monitoring and fault diagnosis. TE process includes 12 operation variables and 41 measurement variables. It is composed of 5 main operation units, including reactor, condenser, vapor-liquid separator, circulation compressor and product desorption tower. The TE chemical model is shown in Fig. 2. The reaction equation is as follows. The products g and H and the by-product F are liquid, and the rest of the reactants are gas

$$\begin{cases} A + C + D \rightarrow G \\ A + C + E \rightarrow H \\ A + E \rightarrow F \\ 2D \rightarrow 2F \end{cases} \tag{13}$$

There are 15 known faults in TE simulation, among which the quality related faults are (1 2 6 8 10 12 13) and the undisputed quality independent faults are (3 4 9 11 14 15). The fault types are given in Table 1, and the specific TE process can be referred to in reference [15], which will not be described here.

In this paper, 22 continuous variables and 11 manipulation variables are selected as input variables X, and the concentration of G in pipeline 11 is selected as quality variable Y. there are 500 samples in the training set and 960 samples in each test set. The fault is added after the 160th sample, and the IPCR model is established under the confidence level α = 99%. The values of h and λ are determined with reference to [16] and the current situation, h = 100, λ = 1.02.

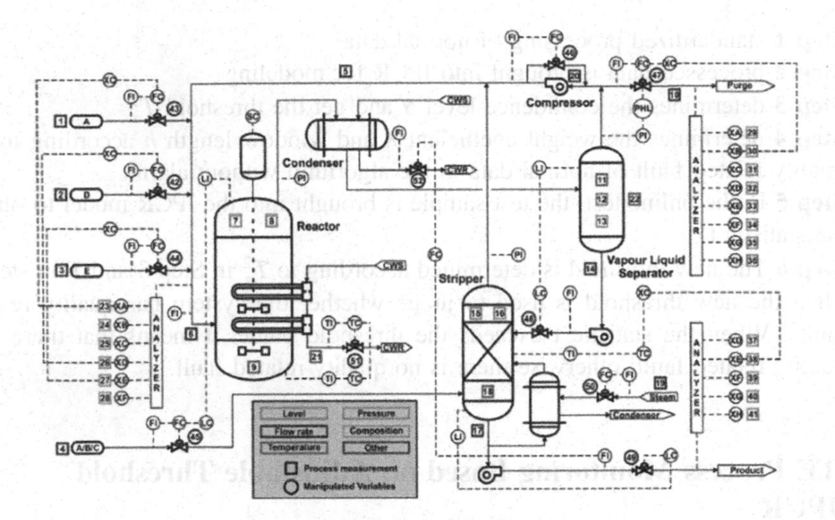

Fig. 2. TE mode

Table 1. The type of fault

Fault	Type	Fault	Type
1	step	8	Random
2	step	9	Random
3	step	10	Random
4	step	11	Random
5	step	12	Random
6	step	13	drift
7	step	14	sticking

Since the window length is set to 100, the threshold value of the first 100 t-statistics is still a fixed threshold T_{α}^{2}. FDR is the fault detection rate, FDR $= \frac{f}{F}$, where f is the detected fault sample and F is the total failure sample. In this simulation, PCA, IPCR and the methods proposed in this paper to compare with it. Through Table 2 and Table 3, it can get that the method proposed in this paper are better than IPCR in quality related faults and quality independent faults, especially in unrelated fault detection, the false alarm rate is very low. Almost no false alarm in faults 9, 11, 14. The PCA algorithm will alarm whether the quality related fault or the quality independent fault, which makes it impossible to distinguish whether the quality related fault occurs or not.

Table 2. Detection Rate of TE quality related Faults

Fault number	IPCR	PCA	Proposed method
3	13.63%	52.38%	1.50%
4	11.00%	70.13%	3.58%
9	7.50%	51.13%	0.00%
11	10.25%	74.38%	0.00%
14	10.00%	100.00%	0.00%
15	10.50%	44.00%	1.03%

In order to let readers see the monitoring result, the monitoring diagrams of different types of faults are listed in Figs. 3, 4, 5 and 6, in which faults 1 and 13 are quality related faults, while faults 9 and 14 are quality independent faults. During the monitoring, the quality related fault is alarmed, and the quality independent fault is not alarmed, which conforms to the fault diagnosis logic.

Table 3. Detection Rate of TE quality unrelated Faults

Fault number	IPCR	PCA	Proposed method
1	90.63%	99.25%	99.00%
2	88.38%	98.63%	85.55%
6	99.25%	99.88%	98.50%
8	68.88%	100.00%	73.25%
10	46.00%	73.75%	50.00%
12	84.13%	99.75%	87.63%
13	90.38%	97.88%	90.75%

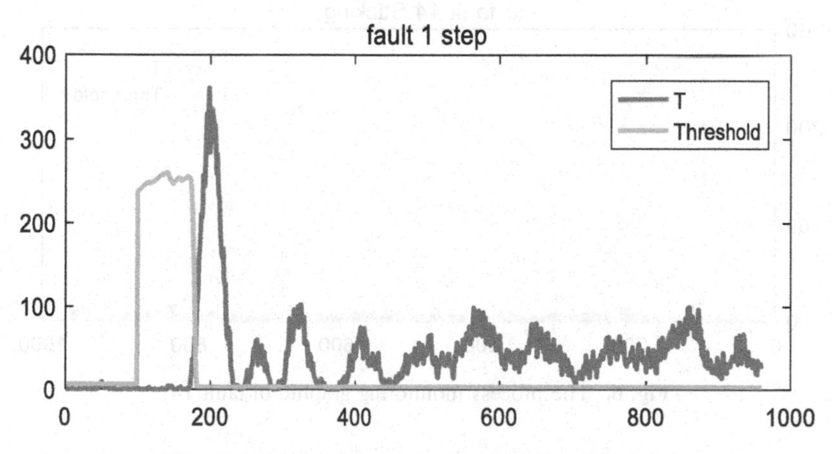

Fig. 3. The process monitoring graphic of fault 1

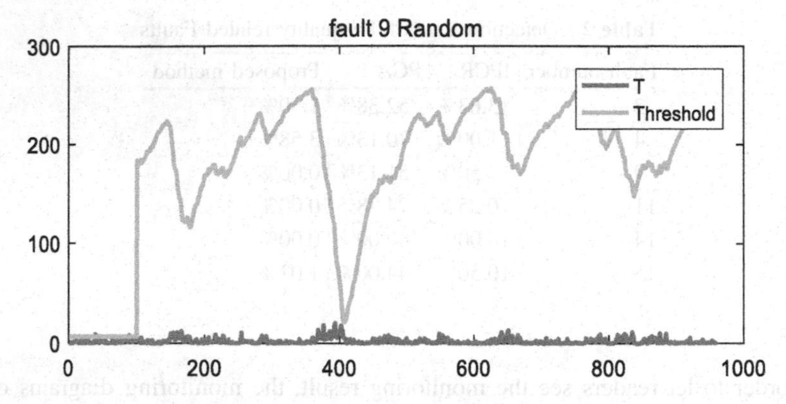

Fig. 4. The process monitoring graphic of fault 9

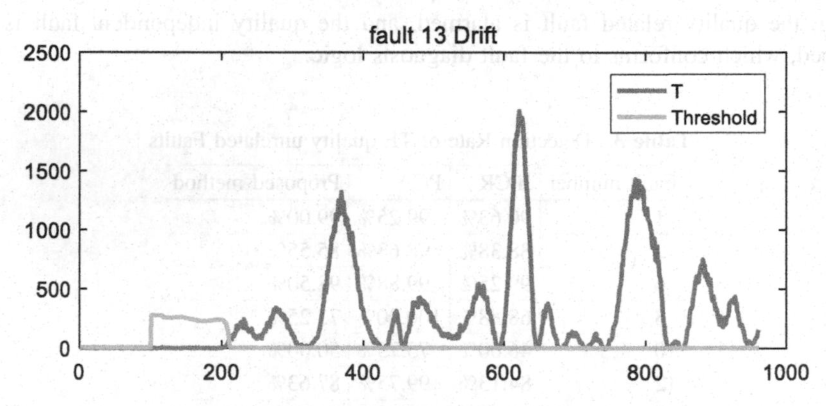

Fig. 5. The process monitoring graphic of fault 13

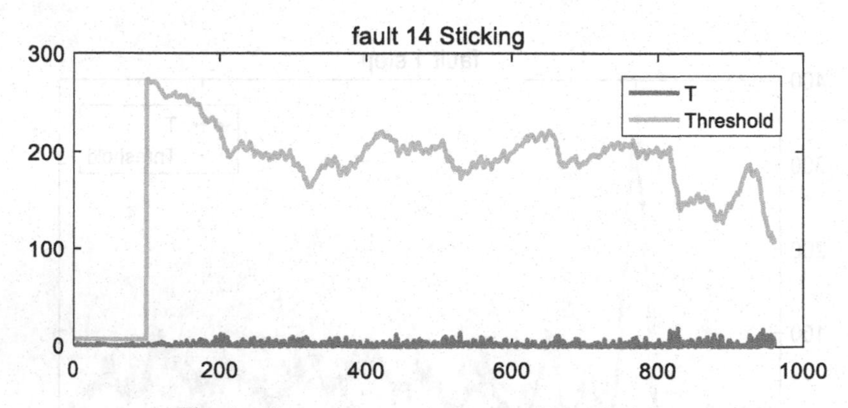

Fig. 6. The process monitoring graphic of fault 14

6 Conclusion

Due to the problem of some false alarms and missing alarms in the fault monitoring of TE chemical process by IPCR, an adjustable threshold IPCR algorithm is proposed, which greatly reduces the false alarm rate of quality independent fault monitoring, and ensuring the detection rate of quality related fault. Because this method considers the statistical results before detection in the current detection, the detection results are more accurate and stable. Besides, this method is simple, and does not need to change the IPCR algorithm, and too many mathematical formulas to be pushed to work. Whether the model is successful mainly depends on the adjustment parameter λ. Therefore, in future research, we can find a better λ through theory and practice.

Acknowledgements. This works is partly supported by the Natural Science Foundation of Liaoning, China under Grant 2019MS008, Education Committee Project of Liaoning, China under Grant LJ2019003.

References

1. Li, W., Yue, H.H., Valle-Cervantes, S., et al.: Recursive PCA for adaptive process monitoring. J. Process Control **10**(5), 471–486 (2000)
2. Chine, W., Mellit, A., Lughi, V., Malck, A., Sulligoi, G., Pavan, A.M.: A novel fault diagnosis technique for photovoltaic systems based on artificial neural networks. Renew. Energy **90**, 501–512 (2016)
3. Qin, S.J.: Statistical process monitoring: basics and beyond. J. Chem. **17**(8–9), 480–502 (2003)
4. Lv, C., Zhang, A., Zhang, Z.: A MIIPCR fault detection strategy for TEP. IEEE Access **7**, 18749–18754 (2019)
5. Qin, S.J.: Survey on data-driven industrial process monitoring and diagnosis. Ann. Rev. Control **36**(2), 220–234 (2012)
6. Wang, Y., Ma, X., Qian, P.: Wind turbine fault detection and identification through PCA-based optimal variable selection. IEEE Trans. Sustain. Energy **9**(4), 1627–1635 (2018)
7. Seghouane, A., Shokouhi, N., Koch, I.: Sparse principal component analysis with reserved sparsity pattern. IEEE Trans. Image Process. **28**(7), 3274–3285 (2019)
8. Li, B., Morris, A.J., Martin, E.B.: Generalized partial least squares regression based on the penalized minimum norm projection. Chem. Intell. Lab. Syst. **72**(1), 21–26 (2004)
9. Ergon, R.: Reduced PCR/PLSR models by subspace projections. Chem. Intell. Lab. Syst. **81** (1), 68–73 (2006)
10. Ding, S.X., Yin, S., Peng, K., et al.: A novel scheme for key performance indicator prediction and diagnosis with application to an industrial hot strip mill. IEEE Trans. Ind. Inform. **9**(4), 2239–2247 (2013)
11. Yin, S., Ding, S.X., Haghani, A., et al.: A comparison study of basic data-driven fault diagnosis and process monitoring methods on the benchmark Tennessee Eastman process. J. Process Control **22**(9), 1567–1581 (2012)
12. MacGregor, J.F., Jaeckle, C., Kiparissides, C., et al.: Process monitoring and diagnosis by multiblock PLS methods. AIChE J. **40**(5), 826–838 (1994)
13. Shen, Y.: Study on modifications of PLS approach for process monitoring. **44**(1), 12389–12394 (2011)

14. Russell, E.L., Chiang, L.H., Braatz, R.D.: Fault detection in industrial processes using canonical variate analysis and dynamic principal component analysis. Chem. Intell. Lab. Syst. **51**(1), 81–93 (2000)
15. Sun, C., Hou, J.: An improved principal component regression for quality-related process monitoring of industrial control systems. IEEE Access **5**, 21723–21730 (2017)
16. Bakdi, A., Kouadri, A.: A new adaptive PCA based thresholding scheme for fault detection in complex systems. Chem. Intell. Lab. Syst. **162**, 83–93 (2017)

Fuzzy Theory and Algorithms

Notes on Supervisory Control of Fuzzy Discrete Event Systems

Chongqing Lin[1] and Daowen Qiu[2](✉)

[1] Guangzhou Sport University, Guangzhou 510500, China
[2] Institute of Computer Science Theory, School of Data and Computer Science, Sun Yat-sen University, Guangzhou 510006, China
issqdw@mail.sysu.edu.cn

Abstract. Since the supervisory control of fuzzy discrete event systems (fuzzy DESs) was established in 2005, there have been meaningful development. The main contributions of the paper include two points: First we establish another supervisory control theorem of fuzzy DES. Here our purpose is to achieve the objective set (specifications) $\tilde{K}$ (instead of its prefix-closure language), and we use two supervisors to control the fuzzy DES, and $\tilde{K}$ can be between the controlled languages generated by the two supervisors. Also, a test algorithm is described to check whether or not the fuzzy controllability condition holds. Second, we further show two fundamental properties of the largest fuzzy sub-language and the s-mallest prefix-closed fuzzy controllable superlanguage of a given fuzzy language, and two equivalence characterizations of the largest fuzzy sublanguage and the smallest prefix-closed fuzzy controllable superlanguage are given.

Keywords: Fuzzy discrete event systems · Supervisory control · Fuzzy finite automata · Fuzzy controllability condition

1 Introduction

Discrete event systems (DESs) are formally dynamical systems whose states are discrete and the evolutions of its states are driven by the occurrence of events [1]. DESs have been applied to many real-world systems, such as traffic systems, manufacturing systems, smart grids systems, and logistic (service) systems, etc.

Supervisory Control Theory (SCT) is a basic and important subject in DESs [1]. Briefly, a DES is modeled as the generator (an automaton) of a formal language, and certain events (transitions) can be disabled by an external controller. The idea is to construct this controller so that the events it currently disables depend on the past behavior of the DES in a suitable way.

However, for some practical systems, the systems designers usually have not a precise picture of these systems in the stage of system modeling, and thus it is difficult to use the crisp DESs model to characterize the behaviors of such systems. For example, in a biomedical system, it is hard to present an exact definition for the "poor" state of one's health. In order to efficiently handle the vagueness, subjectivity and uncertainty in DESs, Lin and Ying [2] initialed the study of *fuzzy DESs* in 2002.

D.-S. Huang and P. Premaratne (Eds.): ICIC 2020, LNAI 12465, pp. 341–352, 2020.
https://doi.org/10.1007/978-3-030-60796-8_29

In 2004, Qiu [3] and Cao etc. [4] studied the supervisory control of fuzzy DESs. Then many scholars have developed fuzzy DESs concerning observability [5–11] diagnosis [12–14], and predictability [15]. In particular, these theories have been successfully applied to many practical applications, such as robot control [16–19], decision supporting [20, 21], uncertainty handling [22], etc.

In [3], the supervisory control of fuzzy DESs with fuzzy states and fuzzy events was established, and a test algorithm was designed for checking the fuzzy controllability condition that decides the existence of supervisor. This algorithm can also be used to check the controllability condition in crisp DESs. In addition, some fundamental properties related to the controllable languages were presented [3].

In supervisory control of (fuzzy) DESs [1–15], for the given set (specifications), say *K*, that belongs to the language generated (unnecessarily marked) by a finite automaton modeling the DES, a supervisor is required to achieve the prefix-closure of *K*. Here, in fuzzy DESs, our goal is to design two supervisors such that *K* is between the two controlled languages. Therefore, we establish another supervisory control theorem of fuzzy DESs, and also this result is new in crisp DESs. Furthermore, by using the method in [3] we present a test algorithm to check whether or not the fuzzy controllability condition in the theorem holds. In addition, there are still important properties regarding the largest fuzzy sublanguage and the smallest prefix-closed fuzzy controllable superlanguage to be studied. So, we give two equivalence characterization of the largest fuzzy sublanguage and the smallest prefix-closed fuzzy controllable superlanguage.

The remainder of the paper is organized as follows: In Sect. 2, we recall the supervisory control theory of fuzzy DESs and related notations and results that will be used in the paper. Then, in Sect. 3, we prove another supervisory control theorem of fuzzy DESs, and by virtue of the method of [3], a test algorithm is described for checking the fuzzy controllability condition. Sect. 4 is focused on the properties related to the fuzzy controllability languages, and we demonstrate two equivalence characterization of the largest fuzzy sublanguage and the smallest prefix-closed fuzzy controllable superlanguage for a given fuzzy subset. Finally, we conclude the paper with a short summary and mentioning some possible problems for further study.

2 Preliminaries

In this section, we recall the supervisory control of fuzzy DESs and relative properties, and the details are referred to [2, 3].

A fuzzy state is represented as a vector $[a_1, a_2, \cdots, a_n]$ that stands for the possibility distributions over crisp states, i.e., $a_i \in [0, 1]$ represents the possibility that the system is in the *i*th crisp state, (i = $1, 2, \cdots$, n). Similarly, a fuzzy event is denoted by a matrix $\tilde{\sigma} = [a_{ij}]_{n \times n}$, where $a_{ij} \in [0, 1]$ means the possibility of system transferring from the *i*th crisp state to the *j*th crisp state when event σ occurs, and *n* is the number of all possible crisp states.

Definition 1. *A fuzzy finite automaton is formally defined as a fuzzy system*

$$\tilde{G} = \left(\tilde{Q}, \tilde{\Sigma}, \tilde{\delta}, \tilde{q}_0, \tilde{Q}_m\right),$$

where $\tilde{Q}$ is the set of some state vectors (fuzzy states) over crisp state set; $\tilde{q}_0 \subseteq \tilde{Q}$ is the initial fuzzy state; $\tilde{Q}_m \subseteq \tilde{Q}$ is also a set of fuzzy states over Q, standing for the marking states; $\tilde{\Sigma}$ is the set of matrices (fuzzy events); $\tilde{\delta} : \tilde{Q} \times \tilde{\Sigma} \to \tilde{Q}$ is a transition function which is defined by $\tilde{\delta}(\tilde{q}, \tilde{\sigma}) = \tilde{q} \odot \tilde{\sigma}$ for $\tilde{q} \in \tilde{Q}$ and $\tilde{\sigma} \in \tilde{\Sigma}$, where $\odot$ denotes the max-product [3] or max-min [3] operation in fuzzy set theory: for $n \times m$ matrix $A = [a_{ij}]$ and $m \times k$ matrix $B = [b_{ij}]$, then $A \odot B = \left[c_{ij}\right]_{n \times n}$, where $c_{ij} = \max_{l=1}^{m} a_{il} \times b_{lj}$ after max-product operation, or $c_{ij} = \max_{l=1}^{m} \min\{a_{il}, b_{lj}\}$ after max-min operation.

Remark 1. The transition function $\tilde{\delta}$ can be naturally extended to $\tilde{Q} \times \tilde{\Sigma}^*$ in the usual manner:

$$\tilde{\delta}(\tilde{q}, \lambda) = \tilde{q}, \quad \tilde{\delta}(\tilde{q}, \tilde{s}\tilde{\sigma}) = \tilde{\delta}\left(\tilde{\delta}(\tilde{q}, \tilde{s}), \tilde{\sigma}\right),$$

where $\tilde{\Sigma}^*$ is the Kleene closure of $\tilde{\Sigma}$, ϵ denotes the empty string, $\tilde{q} \in \tilde{Q}$, $\tilde{\sigma} \in \tilde{\Sigma}$ and $\tilde{s} \in \tilde{\Sigma}^*$. Moreover, $\tilde{\delta}$ can be regarded as a partial transition function in practice.

$\tilde{\Sigma}^k$ is used to denote all string of fuzzy events with the length of k, i.e.,

$$\tilde{\Sigma}^k = \{\tilde{\sigma}_1\tilde{\sigma}_2 \cdots \tilde{\sigma}_k : \tilde{\sigma}_i \in \tilde{\Sigma},\ i = 1, 2, \cdots k\}. \tag{1}$$

Especially, $\tilde{\Sigma}^0 = \{\epsilon\}$.

The fuzzy languages generated and marked by $\tilde{G}$, denoted by $\mathcal{L}_{\tilde{G}}$ and $\mathcal{L}_{\tilde{G},m}$, respectively, are defined as a function from $\tilde{\Sigma}^*$ ($\tilde{\Sigma}^*$ represents the set of all strings of fuzzy events from $\tilde{\Sigma}$) to [0,1] as follows: For any $\tilde{\sigma}_1\tilde{\sigma}_2 \cdots \tilde{\sigma}_k \in \tilde{\Sigma}^*$ where $\tilde{\sigma}_i \in \tilde{\Sigma},\ i = 1, 2, \cdots, k$,

$$\mathcal{L}_{\tilde{G}}(\tilde{\sigma}_1\tilde{\sigma}_2 \ldots \tilde{\sigma}_k) = \max_{i=1}^{n} \tilde{q}_0 \circ \tilde{\sigma}_1 \circ \tilde{\sigma}_2 \circ \ldots \circ \tilde{\sigma}_k \circ \bar{s}_i^T, \tag{2}$$

$$\mathcal{L}_{\tilde{G},m}(\tilde{\sigma}_1\tilde{\sigma}_2 \ldots \tilde{\sigma}_k) = \sup_{\tilde{q} \in \tilde{Q}_m} \tilde{q}_0 \circ \tilde{\sigma}_1 \circ \tilde{\sigma}_2 \circ \ldots \circ \tilde{\sigma}_k \circ \tilde{q}^T, \tag{3}$$

where $\bar{s}_i^T$ is the transpose of $\bar{s}_i$, and $\bar{s}_i$ is as indicated above, i.e., $\bar{s}_i = [0 \cdots 1 \cdots 0]$ where 1 is in the ith place. From Eqs. (1) and (2) it follows that for any $\tilde{s} \in \tilde{\Sigma}^*$ and any $\tilde{\sigma} \in \tilde{\Sigma}$,

$$\mathcal{L}_{\tilde{G},m}(\tilde{s}\tilde{\sigma}) \leq \mathcal{L}_{\tilde{G}}(\tilde{s}\tilde{\sigma}) \leq \mathcal{L}_{\tilde{G}}(\tilde{s}). \tag{4}$$

Each event $\tilde{\sigma} \in \tilde{\Sigma}$ is associated with a degree of controllability, so, the un- controllable set $\tilde{\Sigma}_{uc}$ and controllable set $\tilde{\Sigma}_c$ are two fuzzy subsets of $\tilde{\Sigma}$, i.e., $\tilde{\Sigma}_{uc}$, $\tilde{\Sigma}_c \in$

$\mathcal{F}(\tilde{\Sigma})$ (in this paper, $\mathcal{F}(X)$ denotes the family of all fuzzy subsets of X), and satisfy: For any $\tilde{\sigma} \in \tilde{\Sigma}$,

$$\tilde{\Sigma}_{uc}(\tilde{\sigma}) + \tilde{\Sigma}_c(\tilde{\sigma}) = 1. \tag{5}$$

A sublanguage of $\mathcal{L}_{\tilde{G}}$ is represented as $\tilde{K} \in \mathcal{F}(\tilde{\Sigma}^*)$ satisfying $\tilde{K} \tilde{\subseteq} \mathcal{L}_{\tilde{G}}$. In this paper, $\tilde{A} \tilde{\subseteq} \tilde{B}$ stands for $\tilde{A}(\tilde{\sigma}) \leq \tilde{B}(\tilde{\sigma})$ for any element $\tilde{\sigma}$ of domain. A supervisor S of fuzzy DES $\tilde{G}$ is defined as a function:

$$\tilde{S} : \tilde{\Sigma}^* \rightarrow \mathcal{F}(\tilde{\Sigma}),$$

where for each $\tilde{s} \in \tilde{\Sigma}^*$ and each $\tilde{\sigma} \in \tilde{\Sigma}$, $\tilde{S}(\tilde{s})(\tilde{\sigma})$ represents the possibility of fuzzy event $\tilde{\sigma}$ being enabled after the occurrence of fuzzy event string $\tilde{s}$, and $\min\{\tilde{S}(\tilde{s})(\tilde{\sigma}), \mathcal{L}_{\tilde{G}}(\tilde{s}\tilde{\sigma})\}$ is interpreted to be the degree to which string $\tilde{s}\tilde{\sigma}$ is physically possible and fuzzy event $\tilde{\sigma}$ is enabled after the occurrence of fuzzy event string $\tilde{s}$.

$\tilde{S}$ is usually required to satisfy that for any $\tilde{s} \in \tilde{\Sigma}^*$ and $\tilde{\sigma} \in \tilde{\Sigma}$,

$$\min\{\tilde{\Sigma}_{uc}(\tilde{\sigma}), \mathcal{L}_{\tilde{G}}(\tilde{s}\tilde{\sigma})\} \leq \tilde{S}(\tilde{s})(\tilde{\sigma}). \tag{6}$$

This condition is called the *fuzzy admissibility condition* for supervisor $\tilde{S}$ of fuzzy DES $\tilde{G}$.

The fuzzy controlled system by $\tilde{S}$, denoted by $\tilde{S}/\tilde{G}$, is also a fuzzy DES and the languages $\mathcal{L}_{\tilde{S}/\tilde{G}}$ and $\mathcal{L}_{\tilde{S}/\tilde{G},m}$ generated and marked by $\tilde{S}/\tilde{G}$ respectively are defined as follows: For any $\tilde{s} \in \tilde{\Sigma}^*$ and each $\tilde{\sigma} \in \tilde{\Sigma}$,

$$\mathcal{L}_{\tilde{S}/\tilde{G}}(\epsilon) = 1, \quad \mathcal{L}_{\tilde{S}/\tilde{G}}(\tilde{s}\tilde{\sigma}) = \min\{\mathcal{L}_{\tilde{S}/\tilde{G}}(\tilde{s}), \mathcal{L}_{\tilde{G}}(\tilde{s}\tilde{\sigma}), \tilde{S}(\tilde{s})(\tilde{\sigma})\};$$
$$\mathcal{L}_{\tilde{S}/\tilde{G},m} = \mathcal{L}_{\tilde{S}/\tilde{G}} \tilde{\cap} \mathcal{L}_{\tilde{G},m},$$

where symbol $\tilde{\cap}$ is Zadeh fuzzy AND operator, i.e., $(\tilde{A} \tilde{\cap} \tilde{B})(x) = \min\{\tilde{A}(x), \tilde{B}(x)\}$.

We give a notation concerning prefix-closed property in the sense of fuzzy DESs. For any $\tilde{s} \in \tilde{\Sigma}^*$,

$$pr(\tilde{s}) = \{\tilde{t} \in \tilde{\Sigma}^* : \exists \tilde{r} \in \tilde{\Sigma}^*, \tilde{t}\tilde{r} = \tilde{s}\}. \tag{7}$$

For any fuzzy language $\mathcal{L}$ over $\tilde{\Sigma}^*$, its prefix-closure $pr(\mathcal{L}) : \tilde{\Sigma}^* \rightarrow [0, 1]$ is defined as:

$$pr(\mathcal{L})(\tilde{s}) = \sup_{\tilde{s} \in pr(\tilde{t})} \mathcal{L}(\tilde{t}). \tag{8}$$

So $pr(\mathcal{L})(\tilde{s})$ denotes the possibility of string $\tilde{s}$ belonging to the prefix-closure of $\mathcal{L}$. The two controllability theorems concerning fuzzy DESs is as follows.

Theorem 1. *Let a fuzzy DES be modeled by fuzzy finite automaton* $\tilde{G} = \left(\tilde{Q}, \tilde{\Sigma}, \tilde{\delta}, \tilde{q}_0\right)$. *Suppose fuzzy uncontrollable subset* $\tilde{\Sigma}_{uc} \in \mathcal{F}(\tilde{\Sigma})$, *and fuzzy legal subset* $\tilde{K} \in \mathcal{F}(\tilde{\Sigma}^*)$ *that satisfies:* $\tilde{K} \subseteq \mathcal{L}_{\tilde{G}}$, *and* $\tilde{K}(\epsilon) = 1$. *Then there exists supervisor* $\tilde{S} : \tilde{\Sigma}^* \to \mathcal{F}(\tilde{\Sigma})$, *such that* $\tilde{S}$ *satisfies the fuzzy admissibility condition Eq. (6) and* $\mathcal{L}_{\tilde{S}/\tilde{G}} = pr(\tilde{K})$ *if and only if for any* $\tilde{s} \in \tilde{\Sigma}^*$ *and any* $\tilde{\sigma} \in \tilde{\Sigma}$,

$$\min\{pr(\tilde{K})(\tilde{s}), \tilde{\Sigma}_{uc}(\tilde{\sigma}), \mathcal{L}_{\tilde{G}}(\tilde{s}\tilde{\sigma})\} \leq pr(\tilde{K})(\tilde{s}\tilde{\sigma}), \tag{9}$$

where Eq. (9) is called fuzzy controllability condition of $\tilde{K}$ *with respect to* $\tilde{G}$ *and* $\tilde{\Sigma}_{uc}$.

Theorem 2. *Let a fuzzy DES be modeled by fuzzy automaton* $\tilde{G} = \left(\tilde{Q}, \tilde{\Sigma}, \tilde{\delta}, \tilde{q}_0, \tilde{Q}_m\right)$, *and let* $\tilde{\Sigma}_{uc} \in \mathcal{F}(\tilde{\Sigma})$ *be the fuzzy uncontrollable subset of* $\tilde{\Sigma}$. *Suppose fuzzy language* $\tilde{K} \subseteq \mathcal{L}_{\tilde{G},m}$ *satisfying* $\tilde{K}(\epsilon) = 1$ *and* $pr(\tilde{K}) \subseteq \mathcal{L}_{\tilde{G},m}$. *Then there exists a nonblocking supervisor* $\tilde{S}$ *for* $\tilde{G}$ *such that* $\tilde{S}$ *satisfies the fuzzy admissibility condition Eq. (6), and*

$$\mathcal{L}_{\tilde{S}/\tilde{G},m} = \tilde{K} \text{ and } \mathcal{L}_{\tilde{S}/\tilde{G}} = pr(\tilde{K})$$

if and only if $\tilde{K} = pr(\tilde{K}) \tilde{\cap} \mathcal{L}_{\tilde{G},m}$ *and the fuzzy controllability condition of* $\tilde{K}$ *with respect to* $\tilde{G}$ *and* $\tilde{\Sigma}_{uc}$ *holds, i.e., Equation (9) holds.*

3 Supervisory Control of Fuzzy DESs with Two Supervisors

In this section, we first prove another supervisory control of fuzzy DESs, and then give a test algorithm for checking the fuzzy controllability condition.

3.1 Supervisory Control Theorem of Fuzzy DESs

In Theorem 1, if we require that the fuzzy controlled system $\tilde{S}/\tilde{G}$ approaches to the fuzzy legal subset $\tilde{K} \in \mathcal{F}(\tilde{\Sigma}^*)$, instead of $\mathcal{L}_{\tilde{S}/\tilde{G}} = pr(\tilde{K})$, then the fuzzy controllability condition will be changed. Also, we can pose the problem that if the fuzzy controllability condition is changed to an extent, then what the fuzzy controlled system will be. More exactly, we have the following result.

Theorem 3. *Let a fuzzy DES be modeled by fuzzy finite automaton* $\tilde{G} = \left(\tilde{Q}, \tilde{\Sigma}, \tilde{\delta}, \tilde{q}_0\right)$. *Suppose fuzzy uncontrollable subset* $\tilde{\Sigma}_{uc} \in \mathcal{F}(\tilde{\Sigma})$, *and fuzzy legal subset* $\tilde{K} \in \mathcal{F}(\tilde{\Sigma}^*)$ *that satisfies:* $\tilde{K} \subseteq \mathcal{L}_{\tilde{G}}$, *and* $\tilde{K}(\epsilon) = 1$. *If for any* $\tilde{s} \in \tilde{\Sigma}^*$ *and any* $\tilde{\sigma} \in \tilde{\Sigma}$,

$$\min\left\{\tilde{K}(\tilde{s}), \tilde{\Sigma}_{uc}(\tilde{\sigma}), \mathcal{L}_{\tilde{G}}(\tilde{s}\tilde{\sigma})\right\} \leq \tilde{K}(\tilde{s}\tilde{\sigma}), \tag{10}$$

then there exists two supervisors $\tilde{S}:\tilde{\Sigma}^*\rightarrow\mathcal{F}(\tilde{\Sigma})$ $(i=1,2)$, *such that*

$$\mathcal{L}_{\tilde{S}_1/\tilde{G}}\subseteq\tilde{K}\subseteq pr(\tilde{K})\subseteq\mathcal{L}_{\tilde{S}_2/\tilde{G}}\subseteq\mathcal{L}_{\tilde{G}}. \tag{11}$$

Proof. We define $\tilde{S}_1$: $\tilde{\Sigma}^*\rightarrow\mathcal{F}(\tilde{\Sigma})$ as: For any $\tilde{s}\in\tilde{\Sigma}^*$ and any $\tilde{\sigma}\in\tilde{\Sigma}$,

$$\tilde{S}(\tilde{s})(\tilde{\sigma})=\begin{cases}\min\{\tilde{\Sigma}_{uc}(\tilde{\sigma}),\mathcal{L}_{\tilde{G}}(\tilde{s}\tilde{\sigma})\}, & \text{if } \tilde{\Sigma}_{uc}(\tilde{\sigma})\geq\tilde{K}(\tilde{s}\tilde{\sigma}),\\ \tilde{K}(\tilde{s}\tilde{\sigma}), & \text{otherwise.}\end{cases} \tag{12}$$

We define $\tilde{S}_2$:$\tilde{\Sigma}^*\rightarrow\mathcal{F}(\tilde{\Sigma})$ as: For any $\tilde{s}\in\tilde{\Sigma}^*$ and any $\tilde{\sigma}\in\tilde{\Sigma}$,

$$\tilde{S}(\tilde{s})(\tilde{\sigma})=\begin{cases}\min\{\tilde{\Sigma}_{uc}(\tilde{\sigma}),\mathcal{L}_{\tilde{G}}(\tilde{s}\tilde{\sigma})\}, & \text{if } \tilde{\Sigma}_{uc}(\tilde{\sigma})\geq pr(\tilde{K})(\tilde{s}\tilde{\sigma}),\\ pr(\tilde{K})(\tilde{s}\tilde{\sigma}), & \text{otherwise.}\end{cases} \tag{13}$$

Clearly, $\tilde{S}_1$ and $\tilde{S}_2$ satisfy the fuzzy admissibility condition. Next our purpose is to show that for any $\tilde{s}\in\tilde{\Sigma}^*$,

$$\mathcal{L}_{\tilde{S}_1/\tilde{G}}(\tilde{s})\leq\tilde{K}(\tilde{s})\leq pr(\tilde{K})(\tilde{s})\leq\mathcal{L}_{\tilde{S}_2/\tilde{G}}(\tilde{s})\leq\mathcal{L}_{\tilde{G}}(\tilde{s}). \tag{14}$$

We only prove $\mathcal{L}_{\tilde{S}_1/\tilde{G}}(\tilde{s})\leq\tilde{K}(\tilde{s})$ and $pr(\tilde{K})(\tilde{s})\leq\mathcal{L}_{\tilde{S}_2/\tilde{G}}(\tilde{s})$ since the other inequalities are immediate. First we prove the first one. Proceed by induction for the length of $\tilde{s}$. If $|\tilde{s}|=0$, i.e., $\tilde{s}=\epsilon$, then $\mathcal{L}_{\tilde{S}/\tilde{G}}(\epsilon)=1=\tilde{K}(\epsilon)$. Suppose that $\mathcal{L}_{\tilde{S}_1/\tilde{G}(\tilde{s})}\leq\tilde{K}(\tilde{s})$ holds true for any $\tilde{s}\in\tilde{\Sigma}^*$ with $|\tilde{s}|\leq k-1$. Then our aim is to prove that it holds for any $\tilde{t}\in\tilde{\Sigma}^*$ with $|\tilde{t}|=k$. Let $\tilde{t}=\tilde{s}\tilde{\sigma}$ where $|\tilde{s}|=k-1$. Then with the assumption of induction, and the definition of $\mathcal{L}_{\tilde{S}_1/\tilde{G}}$, we have

$$\begin{aligned}\mathcal{L}_{\tilde{S}_1/\tilde{G}}(\tilde{s}\tilde{\sigma})&=\min\left\{\mathcal{L}_{\tilde{S}/\tilde{G}}(\tilde{s}),\mathcal{L}_{\tilde{G}}(\tilde{s}\tilde{\sigma}),\tilde{S}(\tilde{s})(\tilde{\sigma})\right\}\\&\leq\min\{\tilde{K}(\tilde{s}),\mathcal{L}_{\tilde{G}}(\tilde{s}\tilde{\sigma}),\tilde{S}(\tilde{s})(\tilde{\sigma})\}.\end{aligned}$$

By means of the definition $\tilde{S}(\tilde{s})(\tilde{\sigma})$, if $\tilde{\Sigma}_{uc}(\tilde{\sigma})\geq\tilde{K}(\tilde{s}\tilde{\sigma})$, then

$$\begin{aligned}\mathcal{L}_{\tilde{S}/\tilde{G}}(\tilde{s}\tilde{\sigma})&\leq\min\{\tilde{K}(\tilde{s}),\mathcal{L}_{\tilde{G}}(\tilde{s}\tilde{\sigma}),\tilde{\Sigma}_{uc}(\tilde{\sigma})\}\\&\leq\tilde{K}(\tilde{s}\tilde{\sigma});\end{aligned}$$

if $\tilde{\Sigma}_{uc}(\tilde{\sigma})<\tilde{K}(\tilde{s}\tilde{\sigma})$, then

$$\begin{aligned}\mathcal{L}_{\tilde{S}/\tilde{G}}(\tilde{s}\tilde{\sigma})&\leq\min\{\tilde{K}(\tilde{s}),\mathcal{L}_{\tilde{G}}(\tilde{s}\tilde{\sigma}),\tilde{K}(\tilde{s}\tilde{\sigma})\}\\&\leq\tilde{K}(\tilde{s}\tilde{\sigma});\end{aligned}$$

Therefore, we have shown $\mathcal{L}_{\tilde{S}/\tilde{G}}(\tilde{s}\tilde{\sigma}) \leq \tilde{K}(\tilde{s}\tilde{\sigma})$.

Now we prove the second inequality: $pr(\tilde{K})(\tilde{s}) \leq \mathcal{L}_{\tilde{S}_2/\tilde{G}}(\tilde{s})$. Similarly, proceed by induction for the length of $\tilde{s}$. If $|\tilde{s}| = 0$, i.e., $\tilde{s} = \epsilon$, then $pr(\tilde{K})(\epsilon) = \mathcal{L}_{\tilde{S}_2/\tilde{G}}(\epsilon) = 1$.

Suppose that $pr(\tilde{K})(\tilde{s}) \leq \mathcal{L}_{\tilde{S}_2/\tilde{G}}$ holds true for any $\tilde{s} \in \tilde{\Sigma}^*$ with $|\tilde{s}| \leq k-1$.

Then next we prove that it holds for any $\tilde{t} \in \tilde{\Sigma}^*$ with $|\tilde{t}| = k$.

Indeed, as [3] verifies that $\tilde{K} \subseteq \mathcal{L}_{\tilde{G}}$ implies $pr(\tilde{K}) \subseteq \mathcal{L}_{\tilde{G}}$, i.e., for any $\tilde{s} \in \tilde{\Sigma}^*$,

$$pr(\tilde{K})(\tilde{s}) = \sup_{\tilde{t} \in \tilde{\Sigma}^*} \tilde{K}(\tilde{s}\tilde{t}) \leq \sup_{\tilde{t} \in \tilde{\Sigma}^*} \mathcal{L}_{\tilde{G}}(\tilde{s}\tilde{t}) = \mathcal{L}_{\tilde{G}}(\tilde{s}). \tag{15}$$

Due to $pr(\tilde{K})(\tilde{s}\tilde{\sigma}) \leq pr(\tilde{K})(\tilde{s})$ and $pr(\tilde{K})(\tilde{s}\tilde{\sigma}) \leq \mathcal{L}_{\tilde{G}}(\tilde{s}\tilde{\sigma})$, we have

$$\begin{aligned} pr(\tilde{K})(\tilde{s}\tilde{\sigma}) &\leq \min\{pr(\tilde{K})(\tilde{s}), \mathcal{L}_{\tilde{G}}(\tilde{s}\tilde{\sigma})\} \\ &\leq \min\left\{\mathcal{L}_{\tilde{S}_2/\tilde{G}}(\tilde{s}), \mathcal{L}_{\tilde{G}}(\tilde{s}\tilde{\sigma})\right\}. \end{aligned} \tag{16}$$

Furthermore, if $\tilde{\Sigma}_{uc}(\tilde{\sigma}) \geq pr(\tilde{K})(\tilde{s}\tilde{\sigma})$, then by combining the definition of $\tilde{S}_2(\tilde{s})$, we have

$$\begin{aligned} pr(\tilde{K})(\tilde{s}\tilde{\sigma}) &\leq \min\left\{\mathcal{L}_{\tilde{S}_2/\tilde{G}}(\tilde{s}), \mathcal{L}_{\tilde{G}}(\tilde{s}\tilde{\sigma}), \tilde{\Sigma}_{uc}(\tilde{\sigma})\right\} \\ &= \min\left\{\mathcal{L}_{\tilde{S}_2/\tilde{G}}(\tilde{s}), \mathcal{L}_{\tilde{G}}(\tilde{s}\tilde{\sigma}), \tilde{S}_2(\tilde{s})(\tilde{\sigma})\right\}. \\ &= \mathcal{L}_{\tilde{S}_2/\tilde{G}}(\tilde{s}\tilde{\sigma}); \end{aligned}$$

if $\tilde{\Sigma}_{uc}(\tilde{\sigma}) < pr(\tilde{K})(\tilde{s}\tilde{\sigma})$, then with the definition of $\tilde{S}_2(\tilde{s})$ we have $\tilde{S}_2(\tilde{s})(\sigma) = pr(\tilde{K})(\tilde{s}\tilde{\sigma})$, and we can obtain that

$$\begin{aligned} pr(\tilde{K})(\tilde{s}\tilde{\sigma}) &\leq \min\left\{\mathcal{L}_{\tilde{S}_2/\tilde{G}}(\tilde{s}), \mathcal{L}_{\tilde{G}}(\tilde{s}\tilde{\sigma}), \tilde{S}_2(\tilde{s})(\tilde{\sigma})\right\} \\ &= \mathcal{L}_{\tilde{S}_2/\tilde{G}}(\tilde{s}\tilde{\sigma}). \end{aligned}$$

Therefore we have verified that $pr(\tilde{K})(\tilde{s}\tilde{\sigma}) = \mathcal{L}_{\tilde{S}_2/\tilde{G}}(\tilde{s}\tilde{\sigma})$ holds for any $\tilde{s} \in \tilde{\Sigma}^*$ and $\tilde{\sigma} \in \tilde{\Sigma}$ with $|\tilde{s}| = k-1$, and the proof is completed. □

Remark 2. As we are aware, the above result is also new in crisp DESs. In addition, an appropriate limit to derive the fuzzy controllability condition is worth considering.

3.2 The Decidability of the Fuzzy Controllability Condition Eq. (10)

In crisp DESs, checking the controllability condition can be finished clearly in polynomial time if the controlled specifications (usually represented by *pr*(*K*)) is generated by a finite automaton [1]. However, in fuzzy DESs, it is much com- plicated due to the infinity of the number of fuzzy states. However, if fuzzy DESs are modeled as max-min automata, then [3] presented a polynomial-time algorithm to test the fuzzy controllability condition. The method is new and different from the crisp case and also applies to crisp case.

By means of the ideas in [3], now we describe a test algorithm to check the controllability condition Eq. (10). We here use max-min automaton $\tilde{G} = \left(\tilde{Q}, \tilde{\Sigma}, \tilde{\delta}, \tilde{q}_0\right)$ to model fuzzy DES, and suppose there is a max-min fuzzy automaton $\tilde{H} = \left(\tilde{Q}_1, \tilde{\Sigma}, \tilde{\gamma}, \tilde{p}_0\right)$ to generate $\tilde{K}$. For any $\tilde{s} \in \tilde{\Sigma}^*$ and any $\tilde{\sigma} \in \tilde{\Sigma}$, denote

$$L(\tilde{G}, \tilde{G}, \tilde{s}, \tilde{\sigma}) = \min\{\tilde{K}(\tilde{s}), \tilde{\Sigma}_{uc}(\tilde{\sigma}), \mathcal{L}_{\tilde{G}}(\tilde{s}\tilde{\sigma})\}; \tag{17}$$

$$\tilde{K}(\tilde{s}) = \mathcal{L}_{\tilde{H}}(\tilde{s}) = [\tilde{p}_0 \odot \tilde{s}]; \tag{18}$$

$$\mathcal{L}_{\tilde{G}}(\tilde{s}\tilde{\sigma} = [\tilde{q}_0 \odot \tilde{s} \odot \tilde{\sigma}]) \tag{19}$$

By virtue of the method in [3], with polynomial time we can search for the set of all different fuzzy state pairs as:

$$\{(\tilde{q}_0 \odot \tilde{s}_i, \tilde{p}_0 \odot \tilde{s}_i) : i = 1, 2, \ldots, m, \tilde{s}_i \in \tilde{\Sigma}^*\}. \tag{20}$$

The finiteness of the above set of all different fuzzy state pairs was proved in [3]. So, for each $\tilde{s}_i \in \tilde{\Sigma}^*$ above (i = 1, 2, …, m), and for $\tilde{\sigma} \in \tilde{\Sigma}$, we check whether the following inequality holds:

$$L(\tilde{G}, \tilde{G}, \tilde{s}_i, \tilde{\sigma}) \leq \mathcal{L}_{\tilde{G}}(\tilde{s}_i\tilde{\sigma}). \tag{21}$$

The procedure is finite and we can finish with at most $O\left(m|\tilde{\Sigma}|\right)$ steps to decide whether or not the fuzzy controllability condition (10) holds.

4 Properties of Controllability of Fuzzy DESs

In this section, we deal with the largest fuzzy sublanguage and the smallest prefix-closed fuzzy controllable superlanguage of a given fuzzy subset in fuzzy DESs.

Definition 2. *Let $\tilde{K}$ and $\tilde{M}$ be fuzzy languages over set $\tilde{\Sigma}$ of fuzzy events, and $\tilde{K} \subseteq = \tilde{M} = pr(\tilde{M})$. Suppose that $\tilde{\Sigma}_{uc} \in \mathcal{F}(\tilde{\Sigma})$ denotes a fuzzy subset of uncontrollable events. Then $\tilde{K}$ is said to be controllable with respect to $\tilde{M}$ and $\tilde{\Sigma}_{uc}$ if for any $\tilde{s} \in \tilde{\Sigma}^*$ and any $\tilde{\sigma} \in \tilde{\Sigma}$,*

$$\min\{pr(\tilde{K})(\tilde{s}), \tilde{\Sigma}_{uc}(\tilde{\sigma}), \tilde{M}(\tilde{s}\tilde{\sigma})\} \leq pr(\tilde{K})(\tilde{s}\tilde{\sigma}). \tag{22}$$

Denote by $C\left(\tilde{M}, \tilde{\Sigma}_{uc}\right)$ the set of all those being controllable respect to $\tilde{M}$ and $\tilde{\Sigma}_{uc}$, that is,

$$C(\tilde{M}, \tilde{\Sigma}_{uc}) = \{\tilde{L} \in F(\tilde{\Sigma}^*) : \tilde{L} \text{ is controllable with respect to } \tilde{M} \text{ and } \tilde{\Sigma}_{uc}\}. \quad (23)$$

Denote

$$K(\tilde{K})^{<} = \{\tilde{L} \subseteq \tilde{K} : \tilde{L} \in C(\tilde{M}, \tilde{\Sigma}_{uc})\}, \quad (24)$$

$$k(\tilde{K})^{>} = \{\tilde{L} \in \mathcal{F}(\tilde{\Sigma}^*) : \tilde{K} \subseteq \tilde{L} \subseteq \tilde{M} \text{ and } pr(\tilde{L}) = \tilde{L} \text{ and } \tilde{L} \in C(\tilde{M}, \tilde{\Sigma}_{uc})\}, \quad (25)$$

$$\tilde{K}^{<} = \tilde{\bigcup}_{\tilde{L} \in \mathcal{K}(\tilde{K})^{<}} \tilde{L}, \text{ and } \tilde{K}^{>} = \bigcap_{\tilde{L} \in K(\tilde{K})^{>}} \tilde{L}. \quad (26)$$

First we give a characterization of the smallest prefix-closed fuzzy controllable superlanguage $\tilde{K}^{>}$.

Proposition 1. *Let $\tilde{K}$ and $\tilde{M}$ be fuzzy languages over set $\tilde{\Sigma}$ of fuzzy events, and $\tilde{K} \subseteq = \tilde{M} = pr(\tilde{M})$. Suppose that $\tilde{\Sigma}_{uc} \in \mathcal{F}(\tilde{\Sigma})$ denotes a fuzzy subset of uncontrollable events. Then for any $\tilde{s} \in \tilde{\Sigma}^*$, we have*

$$\tilde{K}^{>}(\tilde{s}) = \min\{(pr(\tilde{K})\tilde{\Sigma}^*_{uc})(\tilde{s}), \tilde{M}(\tilde{s})\}, \quad (27)$$

where

$$(pr(\tilde{K})\tilde{\Sigma}^*_{uc})(\tilde{s}) = \max_{\tilde{s}_1\tilde{s}_2=\tilde{s}} \min\{pr(\tilde{K})(\tilde{s}_1), \tilde{\Sigma}^*_{uc}(\tilde{s}_2)\}, \quad (28)$$

and $\tilde{\Sigma}^*_{uc}(\tilde{s}_2) = min_{1 \le i \le k} \tilde{\Sigma}_{uc}(\tilde{\sigma}_i)$ if $\tilde{s}_2 = \tilde{\sigma}_1\tilde{\sigma}_2 \cdots \tilde{\sigma}_k$, and it is 1 if $\tilde{s}_2 = \epsilon$.

Proof. Firstly we prove

$$\tilde{K}^{>}(\tilde{s}) \le \min\{(pr(\tilde{K})\tilde{\Sigma}^*_{uc})(\tilde{s}), \tilde{M}(\tilde{s})\}. \quad (29)$$

We define a fuzzy language $\tilde{K}' \subseteq \tilde{M}$ over $\tilde{\Sigma}$: For any $\tilde{t} \in \tilde{\Sigma}^*$, we set

$$\tilde{K}'(\tilde{t}) = \min\{(pr(\tilde{K})\tilde{\Sigma}^*_{uc})(\tilde{t}), \tilde{M}(\tilde{t})\}. \quad (30)$$

With the definition above, we have

$$\tilde{K}(\tilde{t}) \le \tilde{K}'(\tilde{t}) \le \tilde{M}(\tilde{t}), \quad (31)$$

$$\tilde{K}'(\tilde{t}) = pr(\tilde{K}')(\tilde{t}), \quad (32)$$

and for any $\tilde{\sigma} \in \tilde{\Sigma}$,

$$\min\{\tilde{K}'(\tilde{t}),\tilde{\Sigma}_{uc}(\tilde{\sigma}),\tilde{M}(\tilde{t}\tilde{\sigma})\}\leq\tilde{K}'(\tilde{t}\tilde{\sigma}). \quad (33)$$

Inequality (31) is immediate since $\tilde{K}\subseteq\tilde{M}$. For proving Inequality (32), we need a property: for any two fuzzy languages $\tilde{K}_1$, $\tilde{K}_2$, if $pr(\tilde{K}_1)=\tilde{K}_1$ and $pr(\tilde{K}_2)=\tilde{K}_2$, then $pr(\tilde{K}_1\cap\tilde{K}_2)=pr(\tilde{K}_1)\cap pr(\tilde{K}_2)$. Indeed, we note $\tilde{M}=pr(\tilde{M})$ and $pr(pr(\tilde{K})\tilde{\Sigma}^*_{uc})=pr(\tilde{K})\tilde{\Sigma}^*_{uc}$. Indeed, for any $\tilde{t}\in\tilde{\Sigma}^*$,

$$pr(pr(\tilde{K})\tilde{\Sigma}^*_{uc})(\tilde{t})=\sup_{\tilde{e}\in\tilde{\Sigma}}(pr(\tilde{K})\tilde{\Sigma}^*_{uc})(\tilde{t}\tilde{e}) \quad (34)$$

$$\leq(pr(\tilde{K})\tilde{\Sigma}^*_{uc})(\tilde{t}). \quad (35)$$

Next we prove Inequality (33).

$$\min\{\tilde{K}'(\tilde{t}),\tilde{\Sigma}_{uc}(\tilde{\sigma}),\tilde{M}(\tilde{t}\tilde{\sigma})\} \quad (36)$$

$$=\min\{\min\{(pr(\tilde{K})\tilde{\Sigma}^*_{uc})(\tilde{t}),\tilde{M}(\tilde{t})\},\tilde{\Sigma}_{uc}(\tilde{\sigma}),\tilde{M}(\tilde{t}\tilde{\sigma})\} \quad (37)$$

$$\leq\tilde{K}'(\tilde{t}\tilde{\sigma}). \quad (38)$$

So, $\tilde{K}'\in\mathcal{K}(\tilde{K}^>)$, and therefore Inequality (29) holds, due to the definition of $\tilde{K}^>$, i.e., $\tilde{K}^>=\bigcap_{\tilde{L}\in\mathcal{K}(\tilde{K})^>}\tilde{L}$.

Secondly, we prove

$$\tilde{K}^>(\tilde{s})\geq\min\{(pr(\tilde{K})\tilde{\Sigma}^*_{uc})(\tilde{s}),\tilde{M}(\tilde{s})\}. \quad (39)$$

For any $\tilde{L}\in\mathcal{K}(\tilde{K})^>$, then $pr(\tilde{K})\cap\tilde{M}\subseteq\tilde{M}$, and therefore we can further obtain

$$(pr(\tilde{K})\tilde{\Sigma}^*_{uc})\cap\tilde{M}\subseteq\tilde{M}. \quad (40)$$

So, for any $\tilde{s}\in\tilde{\Sigma}^*$, we have $\tilde{K}'(\tilde{s})\subseteq\tilde{L}(\tilde{s})$, and therefore, $\tilde{K}'(\tilde{s})\leq\tilde{K}^>(\tilde{s})$. The proof is completed. □

Now we give a characterization of the largest fuzzy sublanguage in fuzzy DESs.

Proposition 2. *Let a fuzzy DES be modeled by fuzzy automaton* $\tilde{G}=\left(\tilde{Q},\tilde{\Sigma},\tilde{\delta},\tilde{q}_0,\tilde{Q}_m\right)$, *and let* $\tilde{\Sigma}_{uc}\in\mathcal{F}(\tilde{\Sigma})$ *be the fuzzy uncontrollable subset of* $\tilde{\Sigma}$. *Suppose fuzzy language* $\tilde{K}\subseteq\mathcal{L}_{\tilde{G},m}$ *satisfying*

$$\tilde{K}=pr(\tilde{K})\cap L_{\tilde{G},m}, \quad (41)$$

then

$$\tilde{K}^<=pr\left(\tilde{K}^<\right)\cap L_{\tilde{G},m}. \quad (42)$$

Proof. We set $\tilde{K}' = pr(\tilde{K}^{<}) \cap \mathcal{L}_{\tilde{G},m}$. First, it is clear that $\tilde{K}^{<} \subseteq \tilde{K}'$ since $\tilde{K}^{<} \subseteq pr(\tilde{K}^{<})$ and $\tilde{K}^{<} \subseteq \tilde{K} \subseteq \mathcal{L}_{\tilde{G},m}$.

Next we prove $\tilde{K} \subseteq \tilde{K}^{<}$. From $\tilde{K}^{<} \subseteq \tilde{K}'$ it follows $pr(\tilde{K}^{<}) \subseteq pr(\tilde{K}')$.

Clearly, $\tilde{K} = pr(\tilde{K}^{<}) \cap \mathcal{L}_{\tilde{G},m} \subseteq pr(\tilde{K}^{<})$. Hence $pr(\tilde{K}') \subseteq pr(\tilde{K}^{<})$. So, we have

$$pr(\tilde{K}) = pr(\tilde{K}^{<}). \tag{43}$$

Since $pr(\tilde{K}^{<}) \in C(\tilde{G}, \Sigma_{uc})$, we have $pr(\tilde{K}') \in C(\tilde{G}, \Sigma_{uc})$, as well. Therefore, with $pr(\tilde{K}^{<}) \subseteq pr(\tilde{K})$, we conclude that

$$\tilde{K}' \subseteq pr(\tilde{K}) \cap \mathcal{L}_{\tilde{G},m} = \tilde{K}. \tag{44}$$

The proof is completed. □

5 Conclusion

In this paper, we have tried to give another supervisory control theorem of fuzzy DESs with two supervisors, and the given fuzzy specifications can be between two controlled languages achieved by the two supervisors respectively. As we are aware, this result is also new in crisp DESs. Also, we have presented a test algorithm to check the fuzzy controllability condition. The largest fuzzy sublanguage and the smallest prefix-closed fuzzy controllable superlanguage for a given fuzzy subset are two important controllable languages, so we have given two equivalence characterizations for the two controllable languages.

The fuzzy supervisory control and related issues of networked fuzzy discrete event systems (for the crisp case we can refer to, e.g., [23]) are worthy of further consideration, and we would like to study it in future.

Acknowledgement. The authors would like to thank the anonymous referees for important comments. This work is supported in part by the National Natural Science Foundation of China (Grant Nos. 61876195, 61572532), and the Natural Science Foundation of Guangdong Province of China (Grant No. 2017B030311011).

References

1. Cassandras, C.G., Lafortune, S.: Introduction to Discrete Event Systems. Springer, New York (2008). https://doi.org/10.1007/978-0-387-68612-7
2. Lin, F., Ying, H.: Modeling and control of fuzzy discrete event systems. IEEE Trans. Syst. Man and Cybern. Part B **32**, 408–415 (2002)
3. Qiu, D.: Supervisory control of fuzzy discrete event systems: a formal approach. IEEE Trans. Syst. Man and Cybern. Part B **35**, 72–88 (2005)
4. Cao, Y., Ying, M.: Supervisory control of fuzzy discrete event systems. IEEE Trans. Syst. Man and Cybern. Part B **35**, 366–371 (2005)

5. Liu, F., Qiu, D.: Decentralized supervisory control of fuzzy discrete event systems. Europe J. Control **14**, 234–243 (2008)
6. Qiu, D., Liu, F.: Fuzzy discrete-event systems under fuzzy observability and a test algorithm. IEEE Trans. Fuzzy Syst. **17**, 578–589 (2009)
7. Cao, Y., Ying, M., Chen, G.: State-based control of fuzzy discrete event systems. IEEE Trans. Syst. Man Cybern. Part B **37**, 410–424 (2007)
8. Lin, F., Ying, H.: State-feedback control of fuzzy discrete-event systems. IEEE Trans. Syst. Man Cybern. Part B **40**, 951–956 (2010)
9. Nie, M., Tan, W.: Theory of generalized fuzzy discrete-event systems. IEEE Trans. Fuzzy Syst. **23**, 98–110 (2015)
10. Deng, W., Qiu, D.: Supervisory control of fuzzy discrete event systems for simulation equivalence. IEEE Trans. Fuzzy Syst. **23**, 178–192 (2015)
11. Deng, W., Qiu, D.: Bi-fuzzy discrete event systems and their supervisory control theory. IEEE Trans. Fuzzy Syst. **23**, 2107–2121 (2015)
12. Liu, F., Qiu, D.: Diagnosability of fuzzy discrete-event systems: a fuzzy approach. IEEE Trans. Fuzzy Syst. **17**, 372–384 (2009)
13. Kilic, E.: Diagnosability of fuzzy discrete event systems. Inf. Sci. **178**, 858–870 (2008)
14. Deng, W., Qiu, D.: State-based decentralized diagnosis of bi-fuzzy discrete event systems. IEEE Trans. Fuzzy Syst. **25**, 854–867 (2017)
15. Benmessahel, Bilal, Touahria, Mohamed, Nouioua, Farid: Predictability of fuzzy discrete event systems. Discrete Event Dyn. Syst. **27**(4), 641–673 (2017). https://doi.org/10.1007/s10626-017-0256-7
16. Huq, R., Mann, G., Gosine, R.: Behavior-modulation technique in mobile robotics using fuzzy discrete event system. IEEE Trans. Robot. **22**, 903–916 (2006)
17. Schmidt, K., Boutalis, Y.: Fuzzy discrete event systems for multiobjective control: Framework and application to mobile robot navigation. IEEE Trans. Fuzzy Syst. **20**, 910–922 (2012)
18. Liu, R., Wang, Y., Zhang, L.: An FDES-based shared control method for asynchronous brain-actuated robot. IEEE Trans. Syst. Man Cybern. Part B **46**, 1452–1462 (2016)
19. Jayasiri, A., Mann, G., Gosine, R.: Generalizing the decentralized control of fuzzy discrete event systems. IEEE Trans. Fuzzy Syst. **20**, 699–714 (2012)
20. Lin, F., et al.: Decision making in fuzzy discrete event systems. Inf. Sci. **177**, 3749–3763 (2007)
21. Ying, H.: A self-learning fuzzy discrete event system for HIV/AIDS treatment regimen selection. IEEE Trans. Syst. Man Cybern. Part B **37**, 966–979 (2007)
22. Du, X., Ying, H., Lin, F.: Theory of extended fuzzy discrete-event systems for handling ranges of knowledge uncertainties and subjectivity. IEEE Trans. Fuzzy Syst. **17**, 316–328 (2009)
23. Lin, F.: Control of networked discrete event systems: Dealing with communication delays and losses. SIAM J. Control Optim. **52**, 1276–1298 (2014)

Kernel Methods and Supporting Vector Machines

A Multi-class Classification Algorithm Based on Geometric Support Vector Machine

Yuping Qin[1(✉)], Xueying Cheng[2], and Qiangkui Leng[3]

[1] College of Engineering, Bohai University, Jinzhou 121013, China
qlq888888@sina.com

[2] College of Mathematics and Physics, Bohai University, Jinzhou 121013, China

[3] College of Information Science and Technology, Bohai University, Jinzhou 121013, China

Abstract. A multi-class classification algorithm based on geometric support vector machine (SVM) is proposed. For each class of training samples, a convex hull is constructed in the sample space using the Schlesinger-Kozinec (SK) algorithm. For a sample to be classified, the class label is determined according to the convex hull in which it is located. If this sample is in more than one convex hull, or is not in any convex hull, the nearest neighbor rule is further employed. Subsequently, its class label is identified by the class of centroid closest to the sample. The experimental results show that compared with the existing multi-class SVM methods, the proposed algorithm can improve the classification accuracy.

Keywords: Multi-class classification · Support vector machine · Convex hull

1 Introduction

With the rapid development of computer technology, the sources of information to be managed, such as WebPages, news and database, are increasing rapidly, the need for multi-class classification is growing in real applications. Therefore, multi-class classification is one of the core issues in the field of machine learning [1–6]. Some research results has been applied to pattern recognition [7], image processing [8], text classification [9], and etc.

Support vector machine (SVM) [10] is a well-known arning method based on statistical learning theory. It exhibits many unique advantages when solving the problem of pattern recognition with small-scaled samples, nonlinear and high dimension. But SVM was originally designed for binary classification. How to effectively extend binary classification to multi-classification is an ongoing research issue.

The existing multi-class SVM mainly includes two kinds. One is to combine the parameter solutions of multiple hyperplanes into one optimization problem, which can be finally solved to achieve multi-class classification, such as qp-mc-sv and lp-mc-sv methods [11]. However, due to the high computational complexity and the difficult implementation, the training speed is very low, especially when the number of classes is large. In addition, the classification accuracy is not expected [12], so it is not

D.-S. Huang and P. Premaratne (Eds.): ICIC 2020, LNAI 12465, pp. 355–364, 2020.
https://doi.org/10.1007/978-3-030-60796-8_30

commonly used in practice. The second strategy is to decompose the multi-class problem into multiple two-class problems, and the resulting two-class classifiers are then combined to achieve multi-class classification, including one against rest(1-a-r) [13, 14], one againet one(1-a-1) [15, 16], and directed acyclic graphs support vector machine (DAGSVM) [17, 18] methods. All three methods have good performance and have been applied in practice, but the 1-a-r and 1-a-1 methods have inseparable regions. The 1-a-1 method requires a large number of sub-classifiers to be constructed and all sub-classifiers must be calculated when classifying. If a sub-classifier is not normalized, the entire classification system will tend to overlearning. The DAGSVM method solves the problem of inseparable regions, and it does not necessarily calculate all sub-classifiers. However, the position of each sub-classifier in the directed acyclic graph also has a greater impact on the classification effect.

A kernel nearest neighbor convex hull (KNNCH) classifier is proposed in [19], which first maps samples to high-dimensional feature space by kernel function. Then, it calculates the distances from the testing sample to convex hulls of all classes. The distances are used for classifying according to the nearest neighbor principle. Nevertheless, the computational complexity of this method is high, and it is not suitable for large data sets.

The intuitive geometric meaning of SVM is to achieve the separation of the convex hulls of the two-class training samples at the maximum margin. The optimal hyperplane is the hyperplane that maximizes the margin [20]. The literature [21] gives a soft Schlesinger-Kozinec (SK) algorithm for calculating the optimal hyperplane, which has the advantages of high calculation accuracy and easy application. Based on the geometric interpretation of SVM and SK algorithm, this paper proposes a multi-class SVM classification algorithm for large-scale data sets.

2 Geometric Interpretation of SVM and SK Algorithm

2.1 Geometric Interpretation of SVM

Given two finite sets $X, Y \subset R^n$, if they are linearly separable, a hyperplane can be computed by the nearest point pair between their convex hulls. Actually it is the vertical bisector connecting the two closest points. This hyperplane is called the hard-margin SVM, as shown in Fig. 1.

2.2 SK Algorithm

The SK algorithm is a typical method for solving the nearest points between convex hulls. The algorithm is described as follows:

Input: Two finite sets $X, Y \subset R^n$, precision parameter ε.

Output: A linear discriminant function $f(x) = w^* \cdot x + b$.

Step 1: Pick $x^* \in X, y^* \in Y$.

Step 2: Compute mx, my, and m according to Eq. (1), Eq. (2), and Eq. (3), respectively. If $\|x^* - y^*\| - m < \varepsilon$ is satisfied, goto Step 4. Otherwise, goto Step 3.

$$mx = \min\{\frac{(x_i - y^*) \cdot (x^* - y^*)}{\|x^* - y^*\|}|x_i \in X\} \quad (1)$$

$$my = \min\{\frac{(y_j - x^*) \cdot (y^* - x^*)}{\|x^* - y^*\|}|y_j \in Y\} \quad (2)$$

$$m = \min\{mx, my\} \quad (3)$$

Step 3: If $mx \le my$, calculate λ by Eq. (4) and then update x^* by Eq. (5). Otherwise, calculate μ by Eq. (6) and then update x^* by Eq. (7). Goto Step 2.

$$\lambda = \min\{1, \frac{(x^* - y^*) \cdot (x^* - x_t)}{\|x^* - x_t\|^2}\} \quad (4)$$

where $x_t \subseteq X$ is the sample with the smallest index satisfying the Eq. (1).

$$x^* = x^*(1 - \lambda) + \lambda x_t \quad (5)$$

$$\mu = \min\{1, \frac{(y^* - x^*) \cdot (y^* - y_t)}{\|y^* - y_t\|^2}\} \quad (6)$$

where $y_t \subseteq Y$ is the sample with the smallest index satisfying the Eq. (2).

$$y^* = y^*(1 - \mu) + \mu y_t \quad (7)$$

Step 4: Compute $f(x) = w^* \cdot x + b$, where $w^* = x^* - y^*$ and $b = (\|y^*\|^2 - \|x^*\|^2)/2$.

The SK algorithm first picks the samples x^* and y^* from X and Y respectively, and then it finds the nearest point x_t or y_t to the vector $x^* - y^*$. The nearest point is obtained by calculating mx and my. Taking mx as an example, we suppose that x_t is found to meet the condition, namely:

$$mx = \frac{(x_t - y^*) \cdot (x^* - y^*)}{\|x^* - y^*\|} \quad (8)$$

Let θ denote the angle of two vectors. According to the definition of inner product, we can simplify Eq. (8) to get Eq. (9):

$$mx = \frac{(x_t - y^*) \cdot (x^* - y^*)}{\|x^* - y^*\|} = \frac{\|x_t - y^*\| \cdot \|x^* - y^*\| \cdot \cos\theta}{\|x^* - y^*\|} = \|x_t - y^*\| \cdot \cos\theta \quad (9)$$

Furthermore, we can get the geometric meaning of mx by Eq. (9), as shown in Fig. 2.

From Fig. 2, we can see that mx is the projection length of the vector $x_t - y^*$ on the vector $x^* - y^*$, where $x_t \subseteq X$ is the point with the smallest projection length. θ is the angle between the vectors $x_t - y^*$ and $x^* - y^*$. In fact, $x_t \subseteq X$ is the sample that causes $x^* - x_i$ to be the largest project length on $x^* - y^*$ (see the indicator p_1 in Fig. 2). The purpose of Step 2 in the SK algorithm is to find a sample between x_t and y_t that makes the projection length larger to verify the stop condition. If the stop condition is satisfied, the algorithm stops and the classification hyperplane $f(x)$ is obtained; Otherwise, local adjustments and updates need to be performed. The updated rules are as follows: If $m = mx$, y^* is fixed and $x^*(x^* = (1 - \lambda)x^* + \lambda x_t)$ is updated. If $m = my$, x^* is fixed and $y^*(y^* = (1 - \mu)y^* + \mu\ y_t)$ is updated. The value of λ (or μ) is determined by Eq. (4) (or Eq. (6)), which is to ensure that the updated distance between x^* and y^* is the smallest.

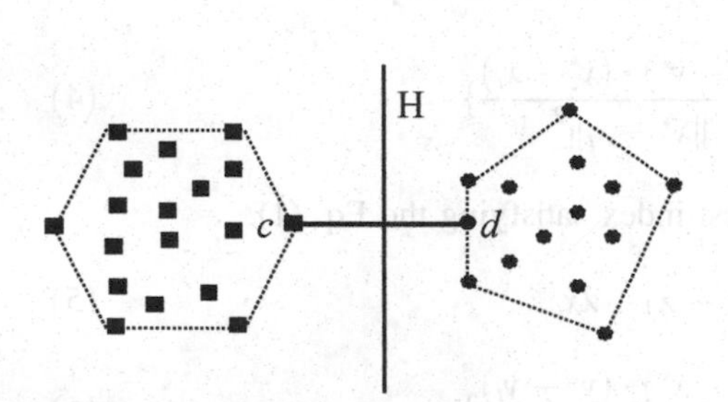

Fig. 1. Geometric interpretation of SVM

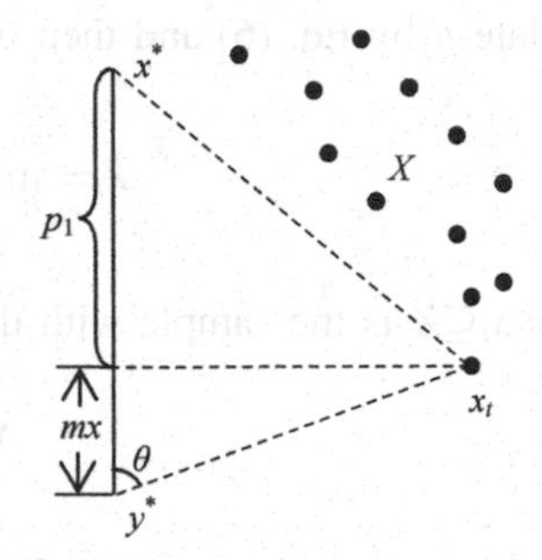

Fig. 2. The geometric meaning of mx

3 Multi-class Classification Algorithm

3.1 Training Algorithm for Generating Convex Hulls

Assume that a given set of training samples $X = \{x_i, y_i\}_{i=1}^{l}$, where $x_i \in R^n$, $y_i \in \{1, 2, \cdots, N\}$. l is the number of samples, and N is the number of classes. We use $X_m = \{x_i^m, y_i^m\}_{i=1}^{l_m}$ to represent the samples of the m-th classes with the number l_m.

The convex hull construction algorithm is as follows:

Step 1: Use the samples of the m-th classes $x_m (1 \leq m \leq N)$ as the positive class, and the rest samples $X - X_m$ as the negative class.
Step 2: Compute the convex hull of X_m (denote as $CH(X_m)$) by SK algorithm, and find the samples of the negative class that are not in $CH(X_m)$ to form a sample set C_m.
Step 3: Find the nearest point to $CH(X_m)$ from C_m, and construct a hyperplane, which will cut off some of the points. Then, repeat this process in the remaining samples of C_m until C_m is empty.
Step 4: Construct a convex classifier by using these resulting hyperplanes.
Step 5: Repeat Step 1 to Step 4 to obtain N convex classifiers.

3.2 Multi-class Classification Algorithm

For a sample x to be classified, its class label is determined according to Eq. (10):

$$f_{H_m^i}(x) = w_m^i \cdot x + b_m^i \; (i = 1, 2, \cdots, N_m) \tag{10}$$

where N_m is the number of hyperplanes in the m-th classifier, and $H_m^i (i \leq i \leq N_m)$ is the i-th hyperplane in the m-th classifier.

If $f_{H_m^i}(x) \geq 0$ for each hyperplane H_m^i, it is determined that x belongs to the m-th class. The classification process is described as follows:

Step 1: Determine whether x is within the m-th $(m = 1, 2, \cdots, N)$ convex classifier according to Eq. (12);
Step 2: If x belongs only to the m-th classifier, determine its class label as the m-th class, and go to step 5; Otherwise, go to step 3;
Step 3: If x is within more than one classifier, then determine its class label by Eq. (13), and goto Step 5. $d_m(x)$ represents the distance from x to all centroids, and its formula is Eq. (12). The centroid is calculated by Eq. (11). If x is not within any classifier, goto Step 4.

$$\alpha_m = \frac{1}{l} \sum_{i=1}^{l_m} \varphi(x_m^i) \tag{11}$$

$$[d_m(x)]^2 = \|\varphi(x) - \alpha_m\|^2 \tag{12}$$

$$class = \arg\min_m d_m(x) \tag{13}$$

Step 4: Calculate $\alpha_m (m = 1, 2, \cdots, N)$ and $d_m(x)$ by Eq. (11) and Eq. (12) respectively. Determine the class label of x by Eq. (13).
Step 5: Stop.

Table 1. Datasets used in the experiments

Name	Abbr.	#Class	#Training	#Testing	#Dim
Iris	Iri.	3	75	75	4
Wine	Win.	3	88	90	13
Vehicle	Veh.	4	422	424	18
Sub21578	Sub.	5	598	298	1000
Segment	Seg.	7	1155	1155	19
Letter	Let.	26	15000	5000	16
Sensorless	Sen.	11	29249	29260	48
Mnist	Mni.	10	60000	10000	780

4 Experimental Results and Analysis

The experiment uses 8 standard datasets to evaluate the performance of the proposed method. The datasets are shown in Table 1, where the four datasets including “iris”, “wine”, “Sensorless”, and “Mnist” are from UCI [22], the three datasets including “vehicle”, “segment”, and “letter” are from Statlog [23], and the dataset “sub21578” is from Reuters-21578 [24].

We use *accuracy* (*accu*) and macro-averaging *precision* (*P*), *recall* (*R*), F_1 value as evaluation indicators.

$$accu(accuracy) = \frac{TP_A + TN_A}{TP_A + TN_A + FP_A + FN_A} \tag{14}$$

$$P(precision) = \frac{TP_A}{Tp_A + FP_A} \tag{15}$$

$$R(recall) = \frac{TP_A}{TP_A + FN_A} \tag{16}$$

$$F_1 = \frac{2 \times P \times R}{P + R} \tag{17}$$

where TP_A, TN_A, FP_A, FN_A denote true positives, true negatives, false positives, false negatives for given class A, respectively.

We compare the proposed method with 1-a-1 SVM, 1-a-r SVM, and Kostin’s decition tree (KDT for short) [25]. 1-a-1 SVM is implemented using the method recommended in Ref. [12] with Radial Basis Function (RBF) $K(x,y) = e^{-\gamma\|x-y\|^2}$, where $\gamma = 1/dimensionality$. 1-a-r SVM is implemented by Liblinear [26] with linear kernel function. The penalty parameter *C* of SVM is set to 1. KDT is an approximately geometric method, which is based on the tree division of the subregion centroids. The precision parameter ε is set to 10^{-3}. The machine environment is I5-6500 CPU 3.20 GHz, 8 GB RAM, and windows 8.1 operating system.

Table 2 provides the accuracies, precision, recall and F_1 value. Table 3 shows the training time and testing time.

From Table 2, we can get that the proposed method has higher accuracy, precision, and F_1 value than 1-a-1 SVM on 5 datasets (i.e., “Iris”, “Vehicle”, “Sub21578”, “Segment”, “Letter”, and “Sensorless”), than 1-a-r SVM on 5 datasets (i.e., “Iris”, “Segment”, “Letter”, “Sensorless”, and “Mnist”), and than KDT on 6 datasets (i.e., “Iris”, “Wine”, “Segment”, “Letter”, “Sensorless”, and “Mnist”). In terms of recall rate, the proposed method performs better than 1-a-1 SVM, 1-a-r SVM, and KDT on 5 datasets (i.e., “Iris”, “Vehicle”, “Segment”, “Letter”, and “Sensorless”), 6 datasets (i.e., “Iris”, “Sub21578”, “Segment”, “Letter”, “Sensorless”, and “Mnist”), and 7 datasets (i.e., “Iris”, “Wine”, “Sub21578”, “Segment”, “Letter”, “Sensorless”, and “Mnist”), respectively. It can be seen that the larger the size of datasets and the more the number of classes, the more obvious advantage this proposed method has. The reason for obtaining higher accuracy is due to a tighter convex surrounding of the classification area.

Table 2. Comparison of the proposed method with 1-a-1 SVM, 1-a-r SVM and KDT on Accuracy (%), Precision (%), Recall Rate (%) and F_1 Value (%).

Name		Iri.	Win.	Veh.	Sub.	Seg.	Let.	Sen.	Mni.
Proposed method	*Accu.*	97.18	97.78	70.52	63.42	92.90	86.38	76.16	92.38
	P	97.53	97.59	69.41	68.38	92.94	86.87	75.82	92.32
	R	96.83	98.15	70.83	66.53	92.90	86.43	75.49	92.30
	F_1	97.05	97.82	69.77	67.21	92.79	86.43	75.65	92.31
1-a-1 SVM	*Accu.*	94.37	100.00	70.28	60.85	90.82	82.28	75.14	94.46
	P	94.54	100.00	68.52	62.91	90.80	83.28	74.70	94.39
	R	93.90	100.00	70.70	68.27	90.82	82.31	74.71	94.40
	F_1	94.10	100.00	68.10	64.71	90.72	82.50	74.70	94.39
1-a-r SVM	*Accu.*	85.92	98.89	74.76	64.77	91.26	66.44	68.52	91.64
	P	85.40	98.92	73.75	69.10	91.26	66.39	68.20	91.59
	R	85.40	99.07	75.03	66.38	91.26	66.73	67.65	91.60
	F_1	85.40	98.98	74.08	67.62	91.21	65.23	67.92	91.59
KDT	*Accu.*	90.67	91.11	70.99	64.22	86.58	72.98	71.73	87.47
	P	91.53	91.15	69.61	68.39	86.77	76.16	72.26	87.87
	R	90.67	91.76	71.49	48.96	86.58	72.98	71.73	87.48
	F_1	91.10	91.45	70.54	57.07	86.67	74.54	71.99	87.67

Table 3. Comparison of the proposed method with 1-a-1 SVM,1-a-r SVM and KDT on the training and testing time (seconds).

Name		Iri.	Win.	Veh.	Sub.	Seg.	Let.	Sen.	Mni.
Proposed method	*Training*	0.004	0.019	0.872	4.789	4.593	64.250	342.617	2807.564
	Testing	0.001	0.001	0.011	1.282	0.023	0.640	1.878	11.649
1-a-1 SVM	*Training*	0.001	0.001	0.037	0.260	0.069	14.083	69.050	171.705
	Testing	0.002	0.007	0.050	0.137	0.166	11.748	35.247	115.240
1-a-r SVM	*Training*	0.004	0.003	0.005	0.007	0.007	0.103	9.835	24.893
	Testing	0.001	0.004	0.004	0.005	0.004	0.007	0.027	0.025
KDT	*Training*	0.001	0.004	0.009	0.021	0.020	2.922	4.468	19.210
	Testing	0.001	0.002	0.013	0.008	0.007	0.010	0.127	1.854

However, since the proposed method contains multiple hyperplanes, it is slower than the other three methods. Involving the testing process, it also takes more time than 1-a-1 SVM, 1-a-r SVM, and KDT.

Next, we discuss the problem of determining the precision parameters ε. ε indicates the stop condition for the proposed method. Figure 3 provides the accuracies varying with the parameter ε. Its value is taken from {1, 0.5, 0.1, 0.05, 0.01, ..., 0.00005, 0.00001}. From Fig. 3, we can see a general trend in which the accuracies gradually

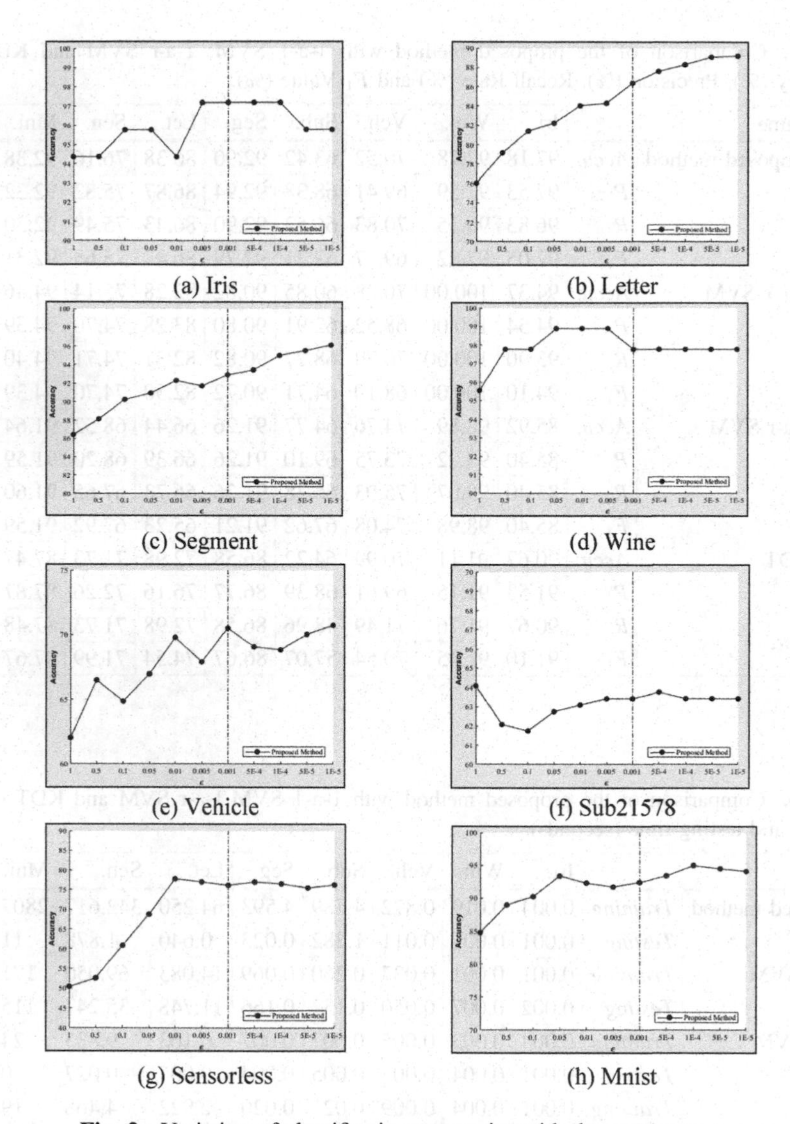

(a) Iris (b) Letter (c) Segment (d) Wine (e) Vehicle (f) Sub21578 (g) Sensorless (h) Mnist

Fig. 3. Variation of classification accuracies with the parameter ε

increase as ε becomes smaller. Meanwhile, the training time also should be taken into account. Therefore, the value of ε is set to 10^{-3}.

5 Conclusion

In this paper, we proposed a multi-class SVM classification algorithm. It is described in two stages, the training stage and the test stage. We conducted a series of experiments on 8 selected datasets. The experimental result shows that the proposed method generally has better accuracy, precision and recall rate than 1-a-1 SVM, 1-a-r SVM and KDT. And the effect is more obvious on theses datasets with larger size and more samples. To a certain extent, it solves the problem that the existing algorithm is not suitable to large-scale classification scenes. As further research work, we will study how to design a faster integration method of convex classifier and how to get fewer hyperplanes with better performance.

Acknowledgements. This work is supported by the National Natural Science Foundation of China under Grant 61602056, "Xingliao Yingcai Project" of Liaoning, China under Grant XLYC1906015, Natural Science Foundation of Liaoning, China under Grant 20180550525 and 201601348, Education Committee Project of Liaoning, China under Grant LZ2016005.

References

1. Mohammad, A.B., Gholam, A.M., Ehsanollah, K.: A subspace approach to error correcting output codes. Pattern Recogn. Lett. **34**(1), 176–184 (2013)
2. Forestier, G., Wemmert, C.: Semi-supervised learning using multiple clusterings with limited labeled data. Inf. Sci. **361**(C), 48–65 (2016)
3. Lee, Y., Lee, J.: Binary tree optimization using genetic algorithm for multiclass support vector machine. Expert Syst. Appl. **42**(8), 3843–3851 (2015)
4. Lai, S., Xu, L., Liu, K., Zhao, J.: Recurrent convolutional neural networks for text classification. In: Proceeding of 29th AAAI Conference on Artificial Intelligence, pp. 2267–2273. AAAI, Menlo Park (2015)
5. Liu, S.M., Chen, J.H.: A multi-label classification based approach for sentiment classification. Expert Syst. Appl. **42**(3), 1083–1093 (2015)
6. Xu, H., Yang, W., Wang, J.: Hierarchical emotion classification and emotion component analysis on Chinese micro-blog posts. Expert Syst. Appl. **42**(22), 8745–8752 (2015)
7. Omid, D., Bin, M.: Discriminative feature extraction for speech recognition using continuous output codes. Pattern Recogn. Lett. **33**(13), 1703–1709 (2012)
8. Gu, Y., Jin, Z., Chiu, S.C.: Active learning combining uncertainty and diversity for multi-class image classification. Iet Comput. Vis. **9**(3), 400–407 (2015)
9. Wu, Q., Tan, M., Song, H., Chen, J.: ML-FOREST: a multi-label tree ensemble method for multi-label classification. IEEE Trans. Knowl. Data Eng. **28**(10), 2665–2680 (2016)
10. Vapnik, V.: An overview of statistical learning theory. IEEE Trans. Neural Netw. **10**(5), 988–999 (1999)
11. Weston, J., Watkins, C.: Support vector machines for multi-class pattern recognition. In: Proceeding of 7th European Symposium on Artificial Neural Networks, Computational Intelligence and Machine Learning, pp. 219–224. IEEE, Piscataway (1999)
12. Hsu, C., Lin, C.J.: A comparison of methods for multi-class support vector machines. IEEE Trans. Neural Netw. **13**(2), 415–425 (2002)
13. Bennett, K.P.: Combining support vector and mathematical programming methods for classification. In: Advances in Kernel Methods: Support Vector Learning, pp. 307–326. MIT Press, Massachusetts (1999)
14. Xu, J.H.: An extended one-versus-rest support vector machine for multi-label classification. Neurocomputing **74**(17), 3114–3124 (2011)
15. Krebel, U.G.: Pairwise classification and support vector machines. In: Advances in Kernel Methods: Support Vector Learning, pp. 255–268. MIT Press, Massachusetts (1999)

16. Chen, Z.J., Jiang, G., Cai, Y.: Research of secondary subdivision method for one-versus-one multi-classification algorithm based on SVM. Transducer Microsystem Technol. **32**(4), 44–47 (2013)
17. Platt, J., Cristianini, N., Shawe-Taylor, J.: Large margin DAGs for multiclass classification. In: Advances in Neural Information Processing Systems, pp. 547–553. MIT Press, Massachusetts (2000)
18. Shen, J., Jiang, Y., Zou, L.: DAG-SVM multi-class classification based on nodes selection optimization. Comput. Eng. **41**(6), 143–146 (2015)
19. Miao, Z., Gandelin, M.H., Baozong, Y.: Fourier transform based image shape analysis and its Application to flower recognition. In: Proceeding of 6th International Conference on Signal Processing, pp. 1087–1090. IEEE, Piscataway (2002)
20. Lu, S.X., Wang, X.Z.: Margin and duality in support vector machines. J. Hebei Univ. (Nat. Sci. Edn.) **27**(5), 449–452 (2007)
21. Franc, V., Hlaváč, V.: An iterative algorithm learning the maximal margin classifier. Pattern Recogn. **36**(9), 1985–1996 (2003)
22. Frank, A., Asuncion, A.: UCI Machine Learning Repository (2010). http://archive.ics.uci.edu/ml
23. Brazdil, P., Gama, J.: Statlog Datasets (1999). http://www.liacc.up.pt/ml/old/statlog/datasets.html
24. Lewis, D.: Reuters-21578 text categorization test collection, Distribution 1.0, AT&T Labs-Research (1997)
25. Kostin, A.: A simple and fast multi-class piecewise linear pattern classifier. Pattern Recogn. **39**(11), 1949–1962 (2006)
26. Fan, R.E., Chang, K.W., Hsieh, C.J.: LIBLINEAR: a library for large linear classification. J. Mach. Learn. Res. **9**(9), 1871–1874 (2008)

Machine Learning

A Network Embedding-Based Method for Predicting miRNA-Disease Associations by Integrating Multiple Information

Hao-Yuan Li[1], Zhu-Hong You[2](✉), Zheng-Wei Li[1], Ji-Ren Zhou[2], and Peng-Wei Hu[2]

[1] School of Computer Science and Technology, China University of Mining and Technology, Xuzhou 221116, China
[2] Xinjiang Technical Institutes of Physics and Chemistry, Chinese Academy of Sciences, Urumqi 830011, China
zhuhongyou@ms.xjb.ac.cn, zhuhongyou@gmail.com

Abstract. MicroRNAs (miRNAs) play important roles in various human complex diseases. Therefore, identifying miRNA-disease associations is deeply significant for pathological progress, diagnosis, and treatment of complex diseases. However, considering the expensive and time-consuming of traditional biological experiments, more and more attentions have been paid on developing computational methods for predicting miRNA-disease associations (MDAs). In this paper, we propose a novel network embedding-based method for predicting miRNA-disease associations by integrating multiple information. Firstly, we constructed a multi-molecular associations network by integrating five known molecules and the associations among them. Then, the behavior features of miRNAs and diseases are extracted by the network embedding model Laplacian Eigenmaps. Finally, Random Forest classifier is trained to predict associations between miRNAs and diseases. As a result, the proposed method achieved outstanding performance on the HMDD V3.0 dataset by using five-fold cross validation, whose average AUC could be reached 0.9317. The promising results demonstrate that the proposed model is a reliable model for the prediction of potential miRNA-disease associations.

Keywords: miRNA-disease association · Network embedding · Random forest · Laplacian eigenmaps · Complex disease

1 Introduction

MicroRNAs (miRNAs) are a train of small (20–25 nucleotides) non-coding RNAs, which play a significant role in posttranscriptional negative regulation of target gene expression [1–3]. Nevertheless, it has also recently been indicated that miRNAs also could be positive regulators in many important biological progresses according to some cases [4–6]. In previous years, as lin-4 and let-7 were firstly discovered [7–9], many experimental methods and computational models started to investigate numerous of miRNAs [10–12]. Which influence many critical biological processes, including cell diffusion [13], growth [14], divergence [15], and death [16]. In addition, accumulating

D.-S. Huang and P. Premaratne (Eds.): ICIC 2020, LNAI 12465, pp. 367–377, 2020.
https://doi.org/10.1007/978-3-030-60796-8_31

reports have demonstrated that miRNAs play major roles in the biological disease research particularly in the pathology, diagnose and therapy [17]. Therefore, it is very important for biological and pathology field to predict miRNA-disease associations [18–20].

Nowadays, increasing computational methods have been put forward for identifying the relationship among miRNAs and disease [21–31]. For instance, Xuan *et al.* developed HDMP to predict miRNAs-diseases association by weighted *k* similar nodes [32]. Chen *et al.* developed RWRMDA by combining known miRNAs–diseases associations and feature similarity information of miRNAs–miRNAs to build miRNA functional similarly network [33]. Xu *et al.* demonstrated that target miRNA disordered with the change of special disease and based on this surmise presented the MTDN [34]. You *et al.* integrated biological network to propose PBMDA, which can fully utilize topological information of heterogeneous network by path [35]. Wang *et al.* presented a novel algorithm of LMTRDA combined with the NLP to obtain the feature information of the miRNA sequence [36]. In recent years, these proposed computational methods have largely compensated for the lack of time-consuming and costly in traditional biological experiment [37–39]. However, these methods may not be comprehensive enough to predict miRNA-disease associations [40–45]. Therefore, this paper proposed a novel network embedding-based method for predicting miRNA-disease associations by integrating multiple information.

2 Materials and Methods

2.1 Human miRNA-Disease Associations

HMDD v3.0 database (Human microRNA Disease Database) have collected 16427 verified miRNA-disease associations between 1023 miRNAs and 850 diseases [46]. The adjacency matrix $G(i,j)$ was described the miRNA-disease associations. If miRNA $m(i)$ have been verified to related with disease $d(j)$, the $G(i,j)$ would be equal to 1, otherwise 0.

2.2 Molecular Association Network

In this work, a heterogeneous network was constructed by integrating various known human molecules and relationship among them. The complex heterogeneous network combined five type nodes (miRNA, disease, lncRNA, protein, drug). Compared with the previous methods, the network contains three other molecules, which improves the efficiency and accuracy of predicting miRNA-disease associations. After preprocessing the data, the detail data source and the amount of the associations are shown in Table 1.

Table 1. The number of different types of associations in molecular associations network

Association	Database	Amount of relationships
miRNA-disease	HMDD [47]	16427
miRNA-protein	miRTarBase [48]	4944
drug-protein	DrugBank [49]	11107
lncRNA-disease	LncRNADisease [50] LncRNASNP2 [51]	1264
miRNA-lncRNA	lncRNASNP2 [51]	8374
lncRNA-protein	LncRNA2Target [52]	690
drug-disease	CTD [53]	18416
protein-protein	STRING [54]	19237
protein-disease	DisGeNET [55]	25087
Total	#N/A	105546

2.3 MiRNA Sequence Information

Attribute of the node is represented by the sequences of miRNA integrated from miRbase. MiRNA sequences are converted into numerical vectors by k-mers [56]. In order to convenient, miRNA sequences in this paper is converted to 64 ($4 \times 4 \times 4$) dimensional vector by using 3-mer to represent attribute of miRNA (e.g. AACUG to AAC, ACU, CUG).

2.4 Disease Semantic Similarity

National Library of Medicine (NLM) proposed a comprehensive system, Medical Subject Headings (Mesh) [57], to classify disease. In this system, disease could be converted into relevant Directed Acyclic Graph (DAG) [58] by MeSH of itself. The relationship between two different disease could be represented by a directed edge pointing from a parent node to a child node, such as DAG(D) = (D, N(D), E(D)), where N(D) is the ancestor node set of D including D, and E(D) indicates the edge set of all relationships. The contribution of disease term T to the semantic value of disease D is as the formula:

$$\begin{cases} D_D(T) = 1 & if\ T = D \\ D_D(T) = max\{\theta * D_D(T') | T' \in children\ of\ T\} & if\ T \neq D \end{cases} \tag{1}$$

Where θ is the semantic contribution factor, the contribution value of D to itself is set as 1. Therefore, we can obtain the sum DV(D) of D:

$$DV(D) = \sum\nolimits_{T \in N_D} D_D(T) \tag{2}$$

Based on the assumption that two diseases sharing more parts of their DAGs should hold higher similarity, we can obtain the semantic similarity among the diseases *a* and b by the following formula:

$$S(a,b) = \frac{\sum_{T \in N_a \cap N_b} (D_a(T) + D_b(T))}{DV(a) + DV(b)} \tag{3}$$

For convenience, we set the disease semantic similarity feature to 64 dimensions.

2.5 Laplacian Eigenmaps

Considering the entire network is large and complexity, some previous network methods need high time complexity or high space complexity, this work adopted Laplacian Eigenmaps (LE) [59–61] to globally represent the behavior information of nodes. LE could nonlinearly reduce dimensionality and has the characteristics of preserving locality and natural connecting with clustering, which makes this method has a good application in this paper.

A complex heterogeneous graph could be obtained from the constructed multi-molecular network. Then, we extract the behavior feature of miRNA and disease from the adjacency matrix of the graph through LE. In order to facilitate the calculation, we set the behavior feature dimension to 64, which is consistent with the attribute feature dimension.

LE regards dimensionality reduction as a high-dimensional to low-dimensional mapping, which should make connected points closer. Constructing a graph with adjacency matrix W to reconstruct the local structural features of the data manifold. If y_i is the point mapped from x_i, the objective function of LE is as the formula:

$$min \sum_{ij} (y_i - y_j)^2 W_{ij} \tag{4}$$

Based on the assumption of connected points closer, it is a matter to calculate the weights of relationship among each point. Here we used the thermonuclear function to evaluate the weights. When x_i is connected with x_j, we could define the weights of them as:

$$\omega_{ij} = e^{-\frac{\left\|x_i - x_j\right\|^2}{t}} \tag{5}$$

According above all, we could obtain a simpler objective function through calculate as following:

$$\begin{aligned} & \sum_{ij} (y_i - y_j)^2 W_{ij} \\ & = \sum_{ij} \left(y_i^2 + y_j^2 - 2y_i y_j\right) W_{ij} \\ & = \sum_{i} y_i^2 D_{ii} + \sum_{j} y_j^2 D_{jj} - 2\sum_{ij} y_i y_j W_{ij} \\ & = 2trace\left(Y^T DY\right) - 2trace\left(Y^T WY\right) \\ & = 2trace\left(Y^T LY\right) \end{aligned} \tag{6}$$

Here W is the adjacency matrix of graph, and $D\left(D_{ii} = \sum_j W_{ij}\right)$ is the measure matrix of graph. $L(L = D - W)$ is the Laplacian Matrix. Object function of Laplacian Eigenmaps could be expressed as following:

$$min\ trace\left(Y^T LY\right), \qquad s.t. Y^T DY = 1 \tag{7}$$

2.6 Random Forest

As a newly emerging and highly flexible machine learning algorithm, Random Forest (RF) has broad application prospects. Random forest is a tree-shaped classifier. It constructs a classification regression decision tree without pruning. The input is a two-dimensional matrix that determines the growth process of a single tree. The input of the forest uses the majority voting method. We constructed the training set based the random selected four fifths of HMDD database. And the test set was the another fifth. In this paper, we set the *number of trees = 99.*

2.7 Extract the Feature Descriptor

In this work, every node is defined contain its attributes and behavior (decided by its relationship with other nodes) information. According to the above, miRNA sequence has been converted into 3-mers as 64 ($4 \times 4 \times 4$) dimensional vector. A complete feature descriptor was formed by integrating the miRNA behavior information, disease behavior information, disease semantic similarity information and miRNA sequence information. Random Forest classifier is trained to classify association between miRNAs and disease. The algorithm is an effective classifier based on multiple CART (Classification and Regression Tree) [62–64]. Each sample in the training set is represented by the previously combined 256-dimensional vector. For each tree, the training set is sampled with replacement from the whole training set. In this work, we set the n_estimators = 99. The flow chart of the model has shown in Fig. 1.

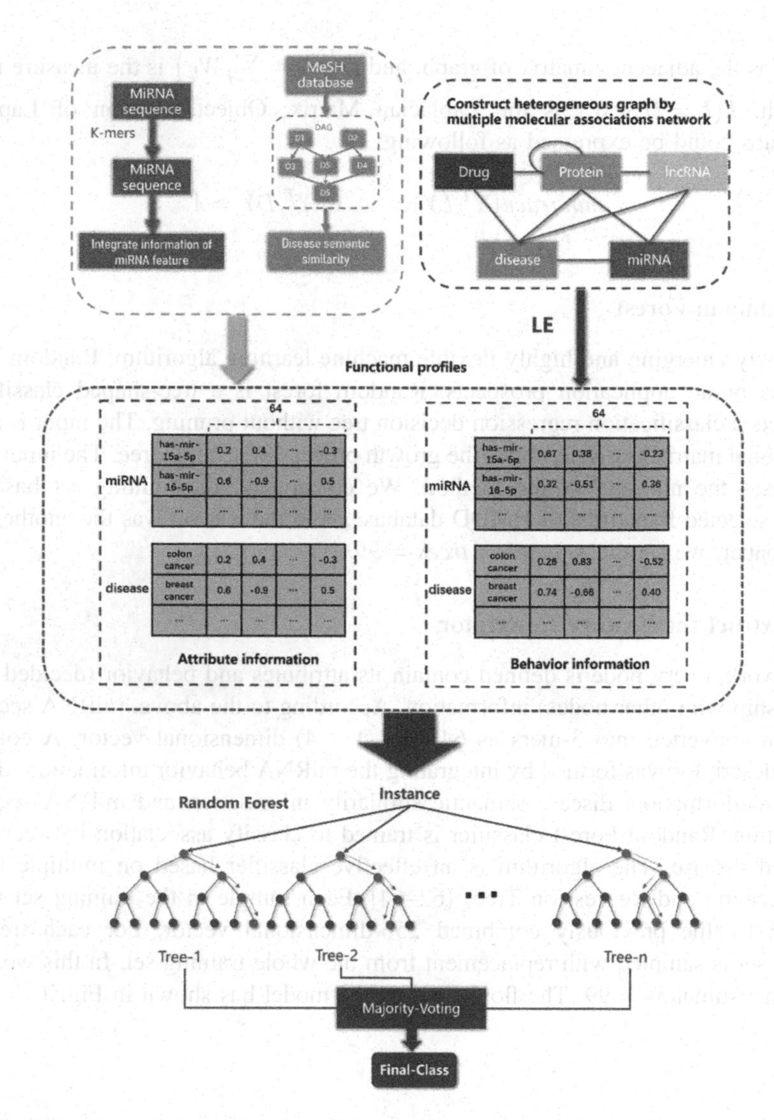

Fig. 1. Flowchart of the proposed model to predict potential miRNA-disease associations

3 Experimental Results

In this paper, we evaluate the predictive performance of the proposed model by implementing the five-fold cross validation based on known database HMDD v3.0. We divided all data into random five subsets of same size, and one of them is regarded as test set, others by integrating as training sets. By five times of the above operation, five sets of training and test data were generated and each pair of them has no intersection. In particular, for avoids the leakage of test information, only training data would construct the network and produce the behavior information at every validation.

For more comprehensive assessment the result of five-fold validation, we adopted a set of evaluation criteria including accuracy (Acc.), specificity (Spec.), precision (Prec.), Matthews Correlation Coefficient (MCC) and area under curve (AUC). The result of average Acc, Spec, Prec, MCC and AUC are respectively 0.8659, 0.8806, 0.8770, 0.7320 and 0.9317. Their standard deviations are 0.0022, 0.0051, 0.0043, 0.0043 and 0.0027 respectively. The detail performance of model in different evaluation criteria is shown in Table 2 and Fig. 2.

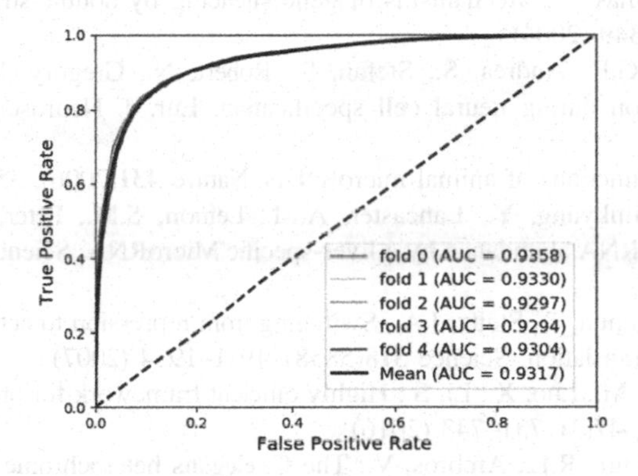

Fig. 2. The ROC curves performed by the proposed model on HMDD V3.0

Table 2. Five-fold cross validation results performed by the proposed model on HMDD V3.0

Datasets	Acc.	Spec.	Prec.	MCC	AUC
HMDD	0.8659	0.8806	0.8770	0.7320	0.9317

4 Conclusion

In this work, we presented a network embedding-based method for predicting miRNA-disease associations by integrating multiple information. This model combined miRNA, disease and other three related molecular (drug, protein, lncRNA) to construct a complex heterogeneous network. As a result, the model obtained average AUC of 0.9317 in the five-fold cross validation based on the HMDD v3.0 dataset. The prediction performance of our method is obviously better than that of existing methods. There are some reasons why the proposed method achieved high and reliable performance. Firstly, different from previous works, we constructed a more comprehensive heterogeneous network of molecular association. It is worth noting that we used both behavior and attribute features to form the feature descriptor of each node. In addition, an effective and concise model, Laplacian Eigenmaps (LE), was adopted to extract the behavior information from the complex molecular association network. The model is suitable for processing large amounts of nonlinear high-dimensional data. The prediction performance of MANMDA would improve with the increase of biological data and more effective feature extraction methods in the future work.

Funding. This work is supported by the Xinjiang Natural Science Foundation under Grant 2017D01A78.

Conflict of Interest. The authors declare that they have no conflict of interest.

References

1. Gunter, M., Thomas, T.: Mechanisms of gene silencing by double-stranded RNA. Nature **431**(7006), 343–349 (2004)
2. Lena, S., Anja, G.F., Andrea, S., Stefan, S., Robert, N., Gregory, W.F.: Regulation of miRNA expression during neural cell specification. Eur. J. Neurosci. **21**(6), 1469–1477 (2015)
3. Victor, A.: The functions of animal microRNAs. Nature **431**(7006), 350–355 (2004)
4. Jopling, C.L., Minkyung, Y., Lancaster, A.M., Lemon, S.M., Peter, S.: Modulation of hepatitis C virus RNA abundance by a liver-specific MicroRNA. Science **309**(5740), 1577–1581 (2005)
5. Shobha, V., Yingchun, T., Steitz, J.A.: Switching from repression to activation: microRNAs can up-regulate translation. Science **318**(5858), 1931–1934 (2007)
6. You, Z.H., Zhou, M., Luo, X., Li, S.: Highly efficient framework for predicting interactions between proteins, **47**(3), 731–743 (2016)
7. Lee, R.C., Feinbaum, R.L., Ambros, V.: The C. elegans heterochronic gene encodes small RNAs with antisense complementarity to lin-14
8. Reinhart, B.J., et al.: The 21-nucleotide let-7 RNA regulates developmental timing in Caenorhabditis elegans. Nature **403**(6772), 901–906 (2000)
9. Wightman, B., Ha, I., Ruvkun, G.: Posttranscriptional regulation of the heterochronic gene lin-14 by lin-4 mediates temporal pattern formation in C. elegans. Cell **75**(5), 855–862 (1993)
10. Ana, K., Sam, G.J.: miRBase: integrating microRNA annotation and deep-sequencing data. Nucleic Acids Res. **39**(Database issue), D152 (2011)
11. Yi, H.C., You, Z.H., Guo, Z.H., Huang, D.S., Chan, K.C.: Learning representation of molecules in association network for predicting intermolecular associations (2020)
12. Guo, Z.H., You, Z.H., Yanbin, W. and Yi, H.C.: Biomarker2vec: attribute-and behavior-driven representation for multi-type relationship prediction between various biomarkers, 849760 (2019)
13. Cheng, A.M., Byrom, M.W., Jeffrey, S., Ford, L.P.: Antisense inhibition of human miRNAs and indications for an involvement of miRNA in cell growth and apoptosis. Nucleic Acids Res. **33**(4), 1290–1297 (2005)
14. Xantha, K., Victor, A.: Developmental biology. Encountering microRNAs in cell fate signaling. Science **310**(5752), 1288–1289 (2005)
15. Miska, E.A.: How microRNAs control cell division, differentiation and death. Curr. Opin. Genet. Dev. **15**(5), 563–568 (2005)
16. Xu, P., Guo, M., Hay, B.A.: MicroRNAs and the regulation of cell death. Trends Genet. **20** (12), 617–624 (2004)
17. Esquela-Kerscher, A., Slack, F.J.: Oncomirs—microRNAs with a role in cancer. Nat. Rev. Cancer **6**(4), 259 (2006)
18. Chen, Z.H., Li, L.P., He, Z., Zhou, J.R., Li, Y., Wong, L.: An improved deep forest model for predicting self-interacting proteins from protein sequence using wavelet transformation, **10**, 90 (2019)

19. Chen, Z.H., You, Z.H., Li, L.P., Wang, Y.B., Wong, L., Yi, H.C.: Prediction of self-interacting proteins from protein sequence information based on random projection model and fast Fourier transform, **20**(4), 930 (2019)
20. Chen, Z.-H., et al.: Prediction of drug–target interactions from multi-molecular network based on deep walk embedding model, **8**, 338 (2020)
21. Zheng, K., You, Z.-H., Wang, L., Zhou, Y., Li, L.-P., Li, Z.-W.: MLMDA: a machine learning approach to predict and validate MicroRNA–disease associations by integrating of heterogenous information sources. J. Transl. Med. **17**(1), 1–14 (2019). https://doi.org/10.1186/s12967-019-2009-x
22. Chen, X., et al.: WBSMDA: within and between score for MiRNA-disease association prediction. Sci. Rep. **6**(1), 21106 (2016)
23. Zheng, K., You, Z.-H., Wang, L., Li, Y.-R., Wang, Y.-B., Jiang, H.-J.: MISSIM: improved miRNA-disease association prediction model based on chaos game representation and broad learning system. In: Huang, D.-S., Huang, Z.-K., Hussain, A. (eds.) ICIC 2019. LNCS (LNAI), vol. 11645, pp. 392–398. Springer, Cham (2019). https://doi.org/10.1007/978-3-030-26766-7_36
24. Jiang, Q., et al.: Prioritization of disease microRNAs through a human phenome-microRNAome network. BMC Syst. Biol. **4**(Suppl 1), S2 (2010)
25. Mørk, S., Pletscher-Frankild, S., Palleja Caro, A., Gorodkin, J., Jensen, L.J.: Protein-driven inference of miRNA-disease associations. Bioinformatics **30**(3), 392 (2014)
26. Chen, X., Yan, G.Y.: Semi-supervised learning for potential human microRNA-disease associations inference. Sci. Rep. **4**, 5501 (2014)
27. Chen, X., Yan, C.C., Zhang, X., You, Z.-H., Huang, Y.-A., Yan, G.-Y.: HGIMDA: heterogeneous graph inference for miRNA-disease association prediction. Oncotarget **7**(40), 65257 (2016)
28. Chen, X., Huang, L.: LRSSLMDA: Laplacian regularized sparse subspace learning for MiRNA-disease association prediction. PLoS Comput. Biol. **13**(12), e1005912 (2017)
29. Chen, X., Xie, D., Wang, L., Zhao, Q., You, Z.-H., Liu, H.: BNPMDA: bipartite network projection for MiRNA–disease association prediction. Bioinformatics **34**(18), 3178–3186 (2018)
30. Li, J.-Q., Rong, Z.-H., Chen, X., Yan, G.-Y., You, Z.-H.: MCMDA: matrix completion for MiRNA-disease association prediction. Oncotarget **8**(13), 21187 (2017)
31. Wang, M.N., You, Z.H., Li, L.P., Wong, L., Chen, Z.H., Gan, C.Z.: GNMFLMI: graph regularized nonnegative matrix factorization for predicting LncRNA-MiRNA interactions, **8**, 37578–37588 (2020)
32. Xuan, P., et al.: Correction: prediction of microRNAs associated with human diseases based on weighted k most similar neighbors. PLoS ONE **8**(9), e70204 (2013)
33. Chen, X., Liu, M.X., Yan, G.Y.: RWRMDA: predicting novel human microRNA-disease associations. Mol. BioSyst. **8**(10), 2792–2798 (2012)
34. Xu, C., et al.: Prioritizing candidate disease miRNAs by integrating phenotype associations of multiple diseases with matched miRNA and mRNA expression profiles. Mol. BioSyst. **10** (11), 2800–2809 (2014)
35. You, Z.H., et al.: PBMDA: a novel and effective path-based computational model for miRNA-disease association prediction. PLoS Comput. Biol. **13**(3), e1005455 (2017)
36. Wang, L., et al.: LMTRDA: using logistic model tree to predict MiRNA-disease associations by fusing multi-source information of sequences and similarities. PLoS Comput. Biol. **15**(3), e1006865 (2019)
37. Chen, X., Yin, J., Qu, J., Huang, L.: MDHGI: matrix decomposition and heterogeneous graph inference for miRNA-disease association prediction. PLoS Comput. Biol. **14**(8), e1006418 (2018)

38. Chen, X., Wu, Q.-F., Yan, G.-Y.: RKNNMDA: ranking-based KNN for MiRNA-disease association prediction. RNA Biol. **14**(7), 952–962 (2017)
39. Wang, M.-N., You, Z.-H., Wang, L., Li, L.-P., Zheng, K.J.N.: LDGRNMF: LncRNA-disease associations prediction based on graph regularized non-negative matrix factorization (2020)
40. Guo, Z.-H., Yi, H.-C., You, Z.-H.: Construction and comprehensive analysis of a molecular associations network via lncRNA-miRNA-disease-drug-protein graph (2019)
41. Wang, L., You, Z.H., Li, Y.M., Zheng, K., Huang, Y.A.: GCNCDA: a new method for predicting circrna-disease associations based on graph convolutional network algorithm, **16** (5), e1007568 (2020)
42. Guo, Z.H., You, Z.H., Yi, H.C.: Integrative construction and analysis of molecular association network in human cells by fusing node attribute and behavior information, **19**, 498–506 (2020)
43. Wang, L., You, Z.H., Huang, Y.A., Huang, D.S., Chan, K.C.: An efficient approach based on multi-sources information to predict circRNA–disease associations using deep convolutional neural network, **36**(13), 4038–4046 (2020)
44. Chen, Z.H., You, Z.H., Zhang, W.B., Wang, Y.B., Cheng, L., Alghazzawi, D.: Global vectors representation of protein sequences and its application for predicting self-interacting proteins with multi-grained cascade forest model, **10**(11), 924 (2019)
45. Guo, Z.H., You, Z.H., Huang, D.S., Yi, H.C., Chen, Z.H., Wang, Y.B.: A learning based framework for diverse biomolecule relationship prediction in molecular association network, **3**(1), 1–9 (2020)
46. Huang, Z., et al.: HMDD v3. 0: a database for experimentally supported human microRNA–disease associations. Nucleic Acids Res. **47**(D1), D1013–D1017 (2018)
47. Huang, Z., et al.: HMDD v3.0: a database for experimentally supported human microRNA-disease associations. Nucleic Acids Res. **41**, D1013–D1017 (2018)
48. Chou, C.-H., et al.: miRTarBase update 2018: a resource for experimentally validated microRNA-target interactions. Nucleic Acids Res. **46**, D296–D302 (2017)
49. Wishart, D.S., et al.: DrugBank 5.0: a major update to the DrugBank database for 2018. Nucleic Acids Res. **46**(D1), D1074 (2018)
50. Chen, G., et al.: LncRNADisease: a database for long-non-coding RNA-associated diseases. Nucleic Acids Res. **41**(D1), D983–D986 (2013)
51. Miao, Y.R., Liu, W., Zhang, Q., Guo, A.Y.: lncRNASNP2: an updated database of functional SNPs and mutations in human and mouse lncRNAs. Nucleic Acids Res. **46**, D276–D280 (2018)
52. Cheng, L., et al.: LncRNA2Target v2.0: a comprehensive database for target genes of lncRNAs in human and mouse. Nucleic Acids Res. **47**, D140–D144 (2019)
53. Davis, A.P., et al.: The comparative toxicogenomics database: update 2019. Nucleic Acids Res. **47**, D948–D954 (2019)
54. Szklarczyk, D., et al.: The STRING database in 2017: quality-controlled protein–protein association networks, made broadly accessible. Nucleic Acids Res. **45**, gkw937 (2017)
55. Janet, P., et al.: DisGeNET: a comprehensive platform integrating information on human disease-associated genes and variants. Nucleic Acids Res. D833–D839 (2017)
56. Pan, X., Shen, H.-B.: Learning distributed representations of RNA sequences and its application for predicting RNA-protein binding sites with a convolutional neural network. Neurocomputing **305**, 51–58 (2018)
57. Lipscomb, C.E.: Medical subject headings (MeSH). Bull. Med. Libr. Assoc. **88**(3), 265 (2000)
58. Kalisch, M., Buehlmann, P.: Estimating high-dimensional directed acyclic graphs with the PC-algorithm. J. Mach. Learn. Res. **8**(2), 613–636 (2012)

59. Belkin, M., Niyogi, P.: Laplacian eigenmaps for dimensionality reduction and data representation (2003)
60. Belkin, M., Niyogi, P.: Laplacian eigenmaps and spectral techniques for embedding and clustering. In: Advances in Neural Information Processing Systems, pp. 585–591 (2002)
61. Belkin, M., Niyogi, P.: Convergence of Laplacian eigenmaps. In: Advances in Neural Information Processing Systems, pp. 129–136 (2007)
62. Chen, X., Wang, C.-C., Yin, J., You, Z.-H.: Novel human miRNA-disease association inference based on random forest. Mol. Ther.-Nucleic Acids **13**, 568–579 (2018)
63. Qi, Y.: Random forest for bioinformatics. In: Zhang, C., Ma, Y. (eds.) Ensemble Machine Learning, pp. 307–323. Springer, Boston (2012). https://doi.org/10.1007/978-1-4419-9326-7_11
64. He, Y., et al.: A support vector machine and a random forest classifier indicates a 15-miRNA set related to osteosarcoma recurrence. OncoTargets Ther. **11**, 253 (2018)

BP Neural Network-Based Deep Non-negative Matrix Factorization for Image Clustering

Qianwen Zeng[1], Wen-Sheng Chen[1,2,3](✉), and Binbin Pan[1,2]

[1] College of Mathematics and Statistics, Shenzhen University, Shenzhen, People's Republic of China
{chenws,pbb}@szu.edu.cn

[2] Guangdong Key Laboratory of Media Security, Shenzhen University, Shenzhen, People's Republic of China

[3] Shenzhen Key Laboratory of Advanced Machine Learning and Applications, Shenzhen 518060, People's Republic of China

Abstract. Deep non-negative matrix factorization (DNMF) is a promising method for non-negativity multi-layer feature extraction. Most of DNMF algorithms are repeatedly to run single-layer NMF to build the hierarchical structure. They have to eliminate the accumulated error via fine-tuning strategy, which is, however, too time-consuming. To deal with the drawbacks of existing DNMF algorithms, this paper proposes a novel deep auto-encoder using back-propagation neural network (BPNN). It can automatically yield a deep non-negative matrix factorization, called BPNN based DNMF (BP-DNMF). The proposed BP-DNMF algorithm is empirically shown to be convergent. Compared with some state of the art DNMF algorithms, experimental results demonstrate that our approach achieves superior clustering performance and has high computing efficiency as well.

Keywords: Deep Non-negative Matrix Factorization (DNMF) · Back-Propagation Neural Network (BPNN) · Image clustering

1 Introduction

Non-negative matrix factorization (NMF) [1] aims to find two factors W and H such that $X \approx WH$, where X is an image data matrix, W and H are non-negative, called basis-image matrix and feature matrix, respectively. NMF can learn parts-based image data representation and has exhibited its ability to handle classification and clustering tasks [2–5]. Nevertheless, NMF and its variants are merely single-layer decomposition methods and thus cannot uncover the underlying hierarchical-feature structure of the data. Empirical results in deep learning indicate that the multi-layer feature-based approaches outperform the shallow-layer based learning methods. Therefore, some researchers presented deep NMF models based on single-layer NMF algorithms. Cichocki et al. [6] proposed a multi-layer NMF algorithm for blind source separation. They adopted a single-layer sparse NMF to generate a deep NMF structure under the iterative rules $H_{i-1} \approx W_i H_i, i = 1, \cdots, L$, where $H_0 = X$. The final decomposition is obtained as $X \approx W_1 W_2 \cdots W_L H_L$. But this DNMF algorithm has a large error of

D.-S. Huang and P. Premaratne (Eds.): ICIC 2020, LNAI 12465, pp. 378–387, 2020.
https://doi.org/10.1007/978-3-030-60796-8_32

reconstruction and its performance is negatively affected. Recently, there are some improved DNMF schemes have been proposed by using the fine-tuning technique. For instance, Lyu et al. [7] extended a single-level orthogonal NMF to a deep architecture. Their update rules are $W_{i-1} \approx W_i H_i, i = 1, \cdots, L$, where $W_0 = X$. The final deep decomposition is $X \approx W_1 H_1 H_2 \cdots H_L$. This orthogonal DNMF uses a fine-tuning step to reduce the total error of factorization and shows its effectiveness in facial image clustering. Trigeorgis et al. [8] suggested a semi-DNMF model with deep decomposition form $X^{\pm} \approx W_1^{\pm} W_2^{\pm} \cdots W_L^{\pm} H_L^{\pm}$, where '$\pm$' means the matrix has no limitation on the sign of entry, '$+$' denotes the matrix is non-negative. Semi-DNMF model is also solved via two stages, namely pre-training and fine-tuning, and can learn hidden representation for clustering and classification on facial images. Similar deep NMF models, such as [9, 10], have been proposed for hyperspectral unmixing, clustering of human facial images. It can be seen that most DNMF approaches need to decrease the entire reconstruction error of the models using the fine-tuning tactic. However, that leads to high computational complexity. Moreover, none of the existing DNMF algorithms acquire hierarchical-feature structure using deep neural networks (DNN) and are unable to make use of the advantage of DNN for clustering.

To address the problems of singer-layer NMF based DNMF methods, this paper proposes a novel BPNN based DNMF (BP-DNMF) approach. We exploit RBF on labeled original image data to obtain a block diagonal similarity matrix which is used as the input of BPNN. Meanwhile, the original data are set as the ground-truth target of the network. Our model can be viewed as a deep auto-encoder. Especially, our auto-encoder automatically yields a DNMF with a deep hierarchical structure for image-data representation. The proposed BP-DNMF approach has high computing efficiency because it directly avoids the fine-tuning step. The experiments on facial images reveal that our BP-DNMF algorithm has fast convergence speed. Finally, evaluated results on facial-image clustering indicate that our method achieves competitive performance.

The rest of this paper is organized as follows. In Sect. 2, we briefly introduce the idea of the DNMF algorithm. The proposed BP-DNMF approach is given in Sect. 3. Experimental results, involving convergence and clustering, are reported in Sect. 4. The final section draws the conclusions.

2 The Framework of Deep NMF

This section will briefly introduce the framework of deep NMF model. Most of DNMF algorithms generate hierarchical-feature structure by recursively utilizing certain single-layer NMF and obtain the following deep factorization:

$$X \approx W_1 W_2 \cdots W_L H_L. \tag{1}$$

The initial decomposition (1) is called the pretraining stage, which however has a large reconstruction error caused by the accumulated error. Hence, it is necessary to reduce the entire error of the DNMF model using a fine-tuning strategy. In detail, all matrices acquired at the previous stage are slightly adjusted to minimize the following objective function:

$$C_{deep} = \frac{1}{2} \|X - W_1 W_2 \cdots W_L H_L\|. \tag{2}$$

The update rules of the fine-tuning stage are derived using gradient descent method and shown as below:

$$W_i \leftarrow W_i \otimes \frac{(\Psi_{i-1}^T X H_{t-1}^T)}{(\Psi_{i-1}^T \Psi_{i-1} W_i H_i H_i^T)}, H_i \leftarrow H_i \otimes \frac{(\Psi_i^T X)}{(\Psi_i^T \Psi_i H_i)}, \tag{3}$$

where $\Psi_{i-1} = W_1 W_2 \cdots W_i, i = 1, 2, \cdots, L$, and Ψ_0 denotes the identity matrix.

3 Proposed BP-DNMF Approach

This section will present an auto-encoder based on BP neural network. The proposed auto-encoder is capable of automatically creating a deep non-negative matrix factorization on image data and thus avoids the high computational complexity of the fine-tuning stage. The proposed BP-DNMF is finally applied to hierarchical feature extraction and image clustering.

3.1 Auto-encoder

Let $X = [X_1, X_2, \cdots, X_c]$ be a training data matrix, where $X_i = \left[x_1^i, x_2^i, \cdots, x_{n_i}^i\right]$ is the *ith* class data matrix, c is the number of class, and the total number of the data is $n = \sum_{i=1}^{c} n_i$. The proposed auto-encoder is composed of two parts: data to similarity matrix and similarity matrix to data.

Data X to Similarity Matrix H: we exploit radial basis function (RBF) on training data to generate a block diagonal similarity matrix H according to the criterion that if two data belong to the same class, they have high similarity, otherwise their similarity is low. In detail, the similarity matrix $H = diag(H_1, H_2, \cdots, H_c) \in R^{m \times n}$ and $H_i = \left(H_{sl}^i\right) \in R^{n_i \times n_i}, i = 1, \cdots, c$, where $H_{sl}^i = k\left(x_s^i, x_s^l\right)$ and $k(x, y) = \exp\left(\frac{\|x-y\|^2}{t}\right)$ with t > 0. It can be seen that matrix H possesses good clustering feature of data X.

Similarity Matrix H to data X: we establish and optimize the structure of a multi-layer BP neural network using input H and target X. The structure of BPNN is firstly determined by setting the number of layer L and the number of neurons in each layer etc. The weight matrices $W_i (i = 1, \cdots, L)$ are initialized such that their entries obey normal distribution $N(0, 1)$. The activation function and bias are respectively set to $f(x) = p^{\frac{1}{L}} \cdot x (p > 0)$ and zero. Consequently, the loss function of the network can be expressed as $E = \frac{1}{2} \|X - p \cdot W_L \cdot W_{L-1} \cdots W_1 H\|_F^2$. The weight matrices are updated using gradient descent method. The optimal procedure of our BPNN is as follows.

1. *Forward pass*

- Build the structure of a deep neural network, including the total number of layers L and the number of neurons in each layer;
- Set the input and the output target of network $a_0 = H_j$ and $X_j (j = 1, 2, \cdots, n)$ respectively, where H_j and X_j are the jth column of H and X;
- Initialize the weight matrices $W_i (i = 1, 2, \cdots, L)$. such that their entries obey normal distribution $N(0, 1)$ and set bias $b_i = 0$;
- For the ith layer, calculate its input $z_i = W_i \cdot a_{i-1}$ and output $a_i = f(z_i) = p^{\frac{1}{L}} \cdot z_i, i = 1, \cdots, L$.

2. *Back pass*

- For input H_j and target $X_j (j = 1, 2, \cdots, n)$, the loss energy $E_j = \frac{1}{2} \|X_j - a_L\|_F^2$;
- $\delta_L = \frac{\partial E_j}{\partial z_L} = p^{\frac{1}{L}} \cdot (a_L - X_j)$;
- $\delta_i = \frac{\partial E_j}{\partial z_i} = p^{\frac{1}{L}} \cdot (W_{i+1}^T \delta_{i+1}), i = L - 1, \cdots, 2, 1$;
- $\frac{\partial E}{\partial W_i} = \delta_i a_{i-1}^T, i = 1, \cdots, L$;
- $W_i \leftarrow P\left[W_i - r \cdot \frac{\partial E}{\partial W_i}\right], i = 1, \cdots, L$, where p[·] is a gradient projection operator defined by $P[W] = \max\{W, 0\}$;
- Let $\frac{dE}{dp} = 0$ and get $p = \frac{tr(V^T X)}{tr(V^T V)}$ to update parameter p, where $V = W_L \cdots W_1 H$.

After training the network, we have that $X_j \approx a_L$, where

$$a_L = f(W_L a_{L-1}) = p^{\frac{1}{L}} W_L a_{L-1} = p \cdot W_L W_{L-1} \cdots W_1 H_j.$$

Hence, our BPNN based deep NMF (BP-DNMF) is acquired as follows:

$$X \approx a_L = p \cdot W_L W_{L-1} \cdots W_1 H. \tag{4}$$

3.2 Hierarchical Feature Extraction

Assume y is a query sample and h_i is its hidden feature on the ith layer, where $i = 1, \cdots, L$. Then we can calculate the feature h_i via the following formula:

$$h_{L-i+1} = (W_L \cdots W_i)^+ y, i = 1, \cdots, L, \tag{5}$$

where A^+ denotes the pseudo-inverse of matrix A.

3.3 Application to Image Clustering

The hierarchical feature extracted by our BP-DNMF algorithm will be applied to image clustering. The algorithm is shown below:

– Training Step

Step 1. Compute the similarity matrix H on training image-data matrix X using radial basis function. Set up the structure of BPNN. Give the error bound ε and the number of maximum iteration I_{max}. Initialize parameter P and the weight matrices $W_i(i = 1, \cdots, L)$.
Step 2. Update the matrices W_i and parameter p according to the rules mentioned in Back Pass stage.
Step 3. If the total loss function $E \leq \varepsilon$ or the iterative number attains I_{max}, then stop the iteration and output the weight matrices $W_i, i = 1, \cdots, L$. Otherwise, go to *Step 2*.

– Clustering Step

Step 4. For the pending clustering data set $Y = \{y_1, y_2, \cdots, y_m\}$, calculate the *ith* layer feature vectors h_i^k of sample y_k using formula (5), $k = 1, \cdots, m$.
Step 5. For fixed i, cluster the *ith* layer feature vector set $\{h_i^1, \cdots, h_i^m\}(i = 1, \cdots, L)$ using K-means algorithm.
Step 6. Compare clustering results with the actual label of data Y and evaluate the experiment effects.

4 Experimental Results

In this section, we will evaluate the clustering performance of our BP-DNMF on two facial image databases, namely Yale database and FERET database. The single-layer and multi-layer NMF algorithms, such as NMF [1], Multi-NMF [6, 7], and Semi-DNMF [8], are for comparisons. For the proposed BP-DNMF model, we set the number of network layer $L = 3$, the maximum number of iteration $I_{max} = 500$, and the learning rate $r = 1e{-}4$. For all deep NMF approaches, we use K-means algorithm and choose the highest layer feature H3 for clustering. The experiments on each database are run 10 times. The average clustering accuracy (AC) and normalized mutual information (NMI) are recorded. The higher values of AC and NMI mean better clustering performance. Finally, empirical convergence and computational efficiency are discussed.

4.1 Facial Image Databases

Yale database contains 15 people, each individual possesses 11 images which are taken in different situation such as the lighting condition, with/without glasses and facial expression. Figure 1 shows 11 images of one person from Yale face database.

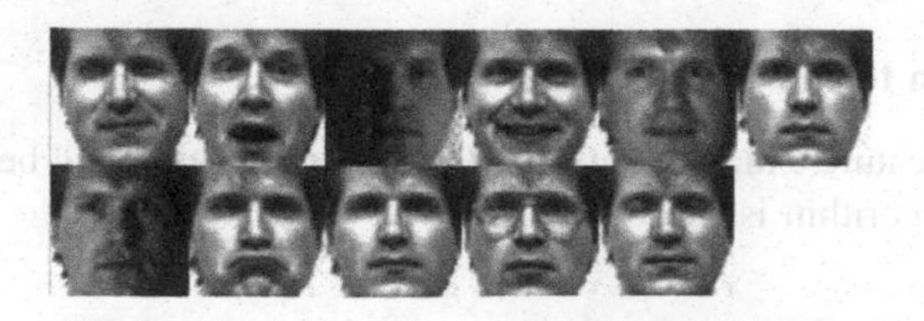

Fig. 1. Images of one person from Yale database

While for FERET database, it involves 720 face images from 120 people. This database consists of four different sets, namely, Fa, Fb, Fc, and duplicate. The size of each image is 112 × 92 and six images of one person from FERET database are shown in Fig. 2.

Fig. 2. Images of one person from FERET database

4.2 Results on Yale Database

We randomly select 6 images from each individual for training while the rest images are for testing. Let k (ranges from 3 to 15) be the number of clusters. All of the compared algorithms are conducted in the same experimental conditions. Two indices, namely AC and NMI, are adopted for clustering evaluation. Their average results are recorded and tabulated in Table 1 and Table 2, respectively. Figure 3 shows the line chart of the results. It can be seen that our BP-DNMF achieves the best clustering performance.

Table 1. Mean accuracy (%) versus Clustering Numbers (CN) on Yale database

CN	3	6	9	12	15
BP-DNMF	**64.00**	**63.00**	**59.56**	**57.00**	**58.93**
Multi-NMF [6]	50.67	44.33	36.22	33.83	32.00
ODNMF [7]	59.33	51.00	42.00	37.67	36.00
Semi-DNMF [8]	61.33	54.33	48.89	47.83	45.33
NMF [1]	51.33	41.33	34.44	33.17	33.20

Table 2. Mean NMI (%) versus Clustering Numbers (CN) on Yale database

CN	3	6	9	12	15
BP-DNMF	**40.79**	**65.38**	**67.18**	**69.19**	**72.34**
Multi-NMF [6]	19.66	38.14	39.87	43.29	46.28
ODNMF [7]	36.36	48.96	50.08	50.76	51.95
Semi-DNMF [8]	34.67	50.77	53.83	57.54	58.65
NMF [1]	20.82	33.15	38.18	42.74	47.10

4.3 Results on FERET Database

The experimental settings on FERET database are similar to those of Yale data set. Three images from each person are randomly selected for training and the remainder images of each individual are used for testing. The number of clusters k increases from 20 to 120 with gap 20. The clustering results are respectively recorded in Table 3 (AC) and Table 4 (NMI), which are plotted in Fig. 4. We see that the proposed BP-DNMF approach greatly surpasses all of the compared approaches.

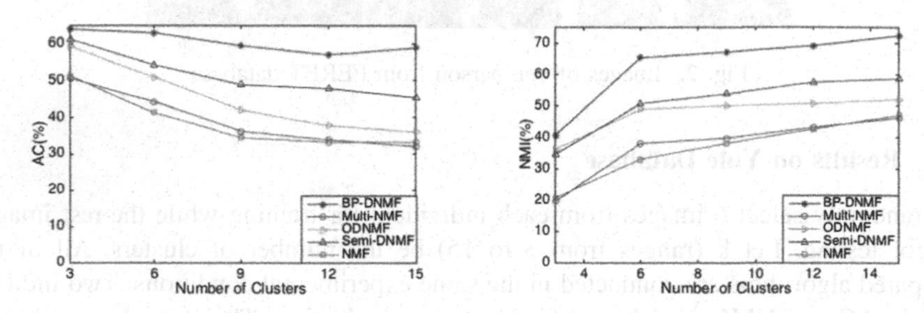

Fig. 3. The clustering performance of feature H_3 on Yale data set

Table 3. Mean accuracy (%) versus Clustering Numbers (CN) on FERET database

CN	20	40	60	80	100	120
BP-DNMF	**56.67**	**54.75**	**54.33**	**53.12**	**52.47**	**51.32**
Multi-NMF [6]	38.33	36.33	34.22	33.80	33.40	33.12
ODNMF [7]	40.17	37.25	34.67	34.32	34.00	33.56
Semi-DNMF [8]	51.67	49.67	45.94	43.32	42.00	41.14
NMF [1]	52.67	51.25	49.17	48.73	48.20	47.75

Table 4. Mean NMI (%) versus Clustering Numbers (CN) on FERET database

CN	20	40	60	80	100	120
BP-DNMF	**77.20**	**80.47**	**81.94**	**82.36**	**83.72**	**84.59**
Multi-NMF [6]	62.39	67.35	69.15	71.01	72.04	73.51
ODNMF [7]	64.78	69.39	71.13	73.23	74.41	75.21
Semi-DNMF [8]	71.67	74.93	74.81	74.51	74.72	75.21
NMF [1]	69.48	77.47	77.29	78.00	78.62	79.36

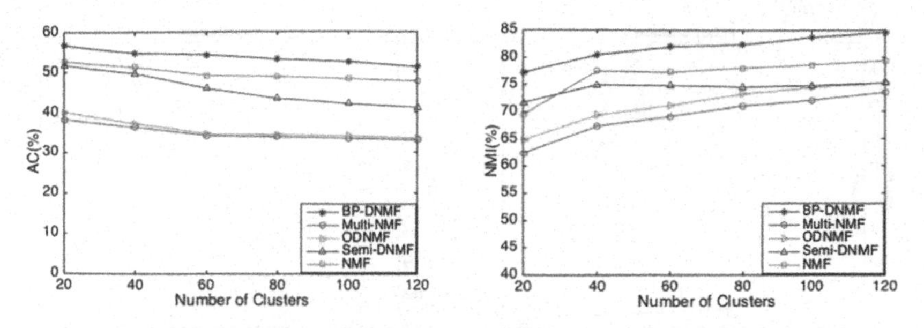

Fig. 4. The clustering performance of feature H_3 on FERET data set

4.4 Computational Efficiency

This subsection will compare the running time of each deep model, including Multi-NMF [6], ODNMF [7], Semi-DNMF [8], and our BP-DNMF algorithms. For all compared algorithms, the experimental settings are the same on the deep decomposition. We choose 6 images of each people from Yale data set and 3 images of each individual from FERET data set to form the data matrix X respectively. The decomposition times are tabulated in Table 5. It can be seen from Table 5 that the running times of our BP-DNMF are 209.57 s and 289.92 s on Yale and FERET databases respectively, while Multi-NMF, Semi- DNMF, ODNMF run for 291.94 s, 669.38 s and 763.40 s on Yale database, 298.04 s, 1056.94 s and 833.99 s on FERET database, respectively. Due to lack of fine-tuning stage, Multi-NMF is faster than Semi-DNMF and ODNMF. This implies that the fine-tuning stage is very time-consuming. We also see that our method achieves the best computational efficiency among the compared methods.

4.5 Convergence Analysis

We will give an empirical convergence of our BP-DNMF algorithm on Yale and FERET databases. The total error against the iteration number is plotted in Fig. 5, from which we can observe that the total loss monotonously decreases as the number of iteration increases. It empirically verifies the convergence of the proposed BP-DNMF algorithm.

Table 5. Running Time (seconds) on Yale and FERET Database

Method	Yale	FERET
BP-DNMF	209.57	289.92
Multi-NMF	291.94	298.04
Semi-DNMF	669.38	1056.94
ODNMF	763.40	833.99

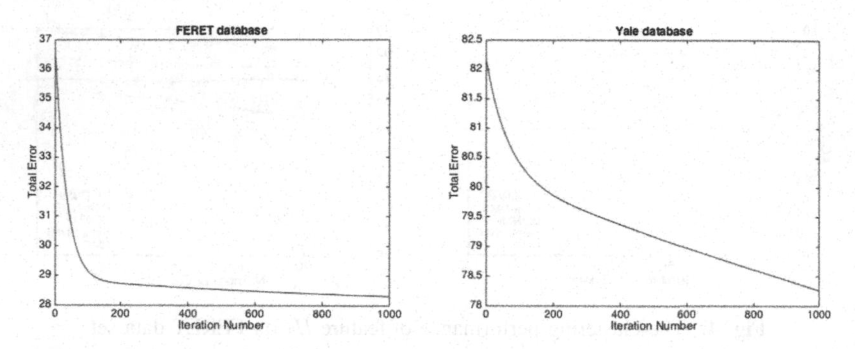

Fig. 5. Convergence curve of BP-DNMF on FERET and Yale Database

5 Conclusions

The existing deep non-negative matrix factorization methods generate their hierarchical features by repeatedly implementing single-layer NMF. These DNMF methods have to reduce the total error of reconstruction via the fine-tuning stage, which is, however, very time-consuming. To solve the problem of existing DNMF algorithms, this paper comes up with a novel deep non-negative matrix factorization approach based on an auto-encoder, which is constructed using back-propagation neural network. Our algorithm is evaluated on both image clustering and computational efficiency. Experimental results have shown that our BP-DNMF approach surpasses the compared state of the art DNMF methods.

Acknowledgements. This work was supported by the National Natural Science Foundation of China under Grant 61272252 and the Interdisciplinary Innovation Team of Shenzhen University. We would like to thank Yale University and the US Army Research Laboratory for the contributions of Yale database and FERET database, respectively.

References

1. Lee, D.D., Seung, H.S.: Learning the parts of the objects by non-negative matrix factorization. Nature **401**, 788–791 (1999)
2. Xu, W., Liu, X., Gong, Y.: Document clustering based on non-negative matrix factorization. In: Proceedings of 26th Annual International on ACM SIGIR Conference, pp. 267–273 (2003)
3. Kim, H., Park, H.: Sparse non-negative matrix factorizations via alternating non-negativity-constrained least squares for microarray data analysis. Bioinformatics **23**(12), 1495–1502 (2007)
4. Ding, C.H., Li, T., Jordan, M.T.: Convex and semi-non-negative matrix factorization. IEEE Trans. Pattern Anal. Mach. Intell. **32**(1), 45–55 (2020)
5. Cai, D., He, X., Han, J., Huang, T.S.: Graph regularized non-negative matrix factorization for data representation. IEEE Trans. Pattern Anal. Mach. Intell. **33**(8), 1548–1560 (2011)
6. Cichcki, A., Zdunek, R.: Multilayer non-negative matrix factorization. Electron. Lett. **42** (16), 947–948 (2006)

7. Lyu, B.S., Xie, K., Sun, W.J.: A deep orthogonal non-negative matrix factorization method for learning attribute representations. In: Liu, D., Xie, S., Li, Y., Zhao, D., El-Alfy, E.S. (eds.) Neural Information Processing. Lecture Notes in Computer Science, vol. 10639, pp. 443–452. Springer, Cham (2017). https://doi.org/10.1007/978-3-319-70136-3_47
8. Trigeorgis, G., Bousmalis, K., Zafeiriou, S., Schuller, B.W.: A deep matrix factorization method for learning attribute representations. IEEE Trans. Pattern Anal. Mach. Intell. **39**(3), 417–429 (2017)
9. Fang, H., Li, A., Xu, H., Wang, T.: Sparsity constrained deep non-negative matrix factorization for hyperspectral unmixing. IEEE Geosci. Remote Sens. Lett. **15**, 1–5 (2018)
10. Yu, J., Zhou, G., Cichocki, A., Xie, S.: Learning the hierarchical parts of objects by deep non-smooth non-negative matrix factorization. IEEE Access **6**, 58096–58105 (2018)

Parameters Selection of Twin Support Vector Regression Based on Cloud Particle Swarm Optimization

Xiuxi Wei, Huajuan Huang(✉), and Weidong Tang

College of Artificial Intelligence, Guangxi University for Nationalities, Nanning 530006, China
hhj-025@163.com

Abstract. Twin Support Vector Regression (TSVR), a novel regressor, obtaining faster learning speed than classical support vector regression (SVR), has attracted the attention of many scholars. Similar to SVR, TSVR is also sensitive to its parameters. Therefore, how to select the suitable parameters has become an urgent problem for TSVR. In this paper, a parameters selection version for TSVR, termed parameters selections of twin support vector regression based on cloud particle swarm optimization (TSVR-CPSO), is proposed. Using the characteristics of randomness and stable tendency of normal cloud model, the inertia weight of PSO can be generated by the basic cloud generator of cloud model. To do so, we can improve the diversity of population for PSO, thus greatly improve the ability of diagnosis to avoid falling into local optimal. Based on the above idea, the cloud particle swarm optimization (CPSO) model is constructed. At last, CPSO is used to search the optimal combination of TSVR parameters. Simulations show that the proposed algorithm is an effective way to search the TSVR parameters and has good performance in nonlinear function estimation.

Keywords: Twin Support Vector Regression · Cloud model · Particle swarm optimization · Parameters selection

1 Introduction

Support Vector Machines (SVM) is known as a new generation learning system based on statistical learning theory [1–3]. Because of its profound mathematical theory, SVM has played excellent performance on many real-world predictive data mining applications such as text categorization, medical and biological information analysis [4–10].

As for support vector regression (SVR), similar to SVM, its training cost also is expensive. In the spirit of TSVMs, in 2010, Peng [11] introduced a new nonparallel plane regression, termed as the twin support vector regression (TSVR). TSVR also aims at generating two nonparallel functions such that each function determines the ε-insensitive down- or up- bounds of the unknown regressor. Similar to TSVMs, TSVR only need to solve a pair of smaller QPPs, instead of solving the large one in SVR. Furthermore, the number of constraints of each QPP in STVR is only half of the classical SVR, which makes TSVR work faster than SVR. In order to further improve

D.-S. Huang and P. Premaratne (Eds.): ICIC 2020, LNAI 12465, pp. 388–399, 2020.
https://doi.org/10.1007/978-3-030-60796-8_33

the performance of TSVR, some improved algorithms have been proposed. For example, in 2010, Peng [12] proposed a primal version for TSVR, termed primal TSVR (PTSVR). PTSVR directly optimized the QPPs in the primal space based on a series of sets of linear equations. Experimental results showed that PTSVR is effective. In 2012, in order to address the shortcoming of TSVR, Chen et al. [13] developed a novel SVR algorithm termed as smooth TSVR (STSVR) by introducing a smooth technique. The effectiveness of STSVR had been demonstrated via experiments on synthetic and real-word benchmark datasets. In the same year, Shao et al. [14] proposed a new regressor called ε-twin support vector regression (ε-TSVR) based on TSVR. ε-TSVR determined a pair of ε-insensitive proximal functions by solving two related SVM-type problems. Experimental results showed that the proposed method had remarkable improvement of generalization performance with short training time, etc.

As a new machine learning methods, there are still many places needing to be perfect for TSVR Specially, the learning performance and generalization ability of TSVR is very dependent on its parameters selection. If the choice is unreasonable, it will be very difficult to approach superiorly. However, the current research on this aspect is very little. At present, for the parameters selection, the grid search method is commonly used. However, the search time of this method is too long, especially dealing with the large dataset. In order to solve this problem, one solving algorithm called twin support vector regression based on cloud particle swarm optimization (TSVR-CPSO) is proposed in this paper. Firstly, in order to improve the performance of Particle Swarm Optimization (PSO), we use clod model to generate the inertia weight of PSO and then present the cloud PSO (CPSO) model. Because cloud model has the characteristics of randomness and stable tendency, it can improve the diversity of population for PSO, thus greatly improve the ability of diagnosis to avoid falling into local optimal. Finally, we use CPSO model to select the TSVR parameters. The experimental results show the effectiveness and stability of the proposed method.

The paper is organized as follows: In Sect. 2, we briefly introduce the basic theory of TSVR and the analysis of its parameters. In Sect. 3, TSVR-CPSO algorithm is introduced and analyzed. Computational comparisons are done in Sect. 4 and Sect. 5 gives concluding remarks.

2 Twin Support Vector Regression and Its Parameters

2.1 Twin Support Vector Regression

Let $A \in R^{l \times n}$ denote the input sample matrix, whose row vectors $A_i = (A_{i1}, A_{i2}, \cdots, A_{in})$, $i = 1, 2, \cdots, l$ are the training samples. Also let $Y = (y_1, y_2, \cdots, y_l)^T$ denote the output vector, in which y_i, $i = 1, 2, \cdots, l$ are the corresponding response values. We will discuss the problem formulations and their dual problems of SVR and TSVR respectively as follows.

Similar to TSVMs, TSVR would generate two nonparallel functions around the data points.

For the linear case, TSVR aims at finding a pair of nonparallel functions

$$f_1(x) = w_1^T x + b_1 \tag{1}$$

$$f_2(x) = w_2^T x + b_2, \tag{2}$$

such that each function determines the ε-insensitive down- or up- bounds regressor. The two functions are obtained by solving the following QPPs:

$$\begin{aligned} \min \ & \frac{1}{2}\|Y - e\varepsilon_1 - (Aw_1 + eb_1)\|^2 + C_1 e^T \xi \\ s.t. \ & Y - (Aw_1 + eb_1) \geq e\varepsilon_1 - \xi, \ \xi \geq 0 \end{aligned} \tag{3}$$

$$\begin{aligned} \min \ & \frac{1}{2}\|Y + e\varepsilon_2 - (Aw_2 + eb_2)\|^2 + C_2 e^T \eta \\ s.t. \ & (Aw_2 + eb_2) - Y \geq e\varepsilon_2 - \eta, \ \eta \geq 0 \end{aligned} \tag{4}$$

where, $C_1, C_2 > 0$; $\varepsilon_1, \varepsilon_2 > 0$ are the parameters, ξ, η are the slack vectors and e is the vector of ones of appropriate dimensions.

Introducing the lagrangian multiplier vectors α and γ considering the KKT conditions, the dual QPPs of (9) and (10) can be obtained as follows:

$$\begin{aligned} \max \ & -\frac{1}{2}\alpha^T G(G^T G)^{-1} G^T \alpha + f^T G(G^T G)^{-1} G^T \alpha - f^T \alpha \\ s.t. \ & 0 \leq \alpha \leq C_1 e \end{aligned} \tag{5}$$

$$\begin{aligned} \max \ & -\frac{1}{2}\gamma^T G(G^T G)^{-1} G^T \gamma - h^T G(G^T G)^{-1} G^T \gamma + h^T \gamma \\ s.t. \ & 0 \leq \gamma \leq C_2 e \end{aligned} \tag{6}$$

where, $G = [A \quad e]$, $f = Y - \varepsilon_1$ and $h = Y + \varepsilon_2 e$.

After optimizing (11) and (12), we can obtain the regression function of TSVR as follows:

$$f(x) = \frac{1}{2}(f_1(x) + f_2(x)) = \frac{1}{2}(w_1 + w_2)^T x + \frac{1}{2}(b_1 + b_2) \tag{7}$$

where, $[w_1 \ b_1]^T = (G^T G)^{-1} G^T (f - \alpha)$, $[w_2 \ b_2]^T = (G^T G)^{-1} G^T (h + \gamma)$.

For the nonlinear case, TSVR considers the following kernel-generated functions:

$$f_1(x) = K(x^T, A^T)w_1 + b_1, \quad f_2(x) = K(x^T, A^T)w_2 + b_2 \tag{8}$$

Similarly, solving (14) can be obtained by dealing with the following QPPs:

$$\begin{aligned} \min \quad & \frac{1}{2}\left\|Y - e\varepsilon_1 - (K(A,A^T)w_1 + eb_1)\right\|^2 + C_1 e^T \xi \\ s.t. \quad & Y - (K(A,A^T)w_1 + eb_1) \geq e\varepsilon_1 - \xi, \ \ \xi \geq 0 \end{aligned} \tag{9}$$

$$\begin{aligned} \min \quad & \frac{1}{2}\left\|Y + e\varepsilon_2 - (K(A,A^T)w_2 + eb_2)\right\|^2 + C_2 e^T \eta \\ s.t. \quad & (K(A,A^T)w_2 + eb_2) - Y \geq e\varepsilon_2 - \eta, \ \ \eta \geq 0 \end{aligned} \tag{10}$$

According to the KKT conditions, the dual problems of (15) and (16) are as follows:

$$\begin{aligned} \max \quad & -\frac{1}{2}\alpha^T H(H^T H)^{-1} H^T \alpha + f^T H(H^T H)^{-1} H^T \alpha - f^T \alpha \\ s.t. \quad & 0 \leq \alpha \leq C_1 e \end{aligned} \tag{11}$$

$$\begin{aligned} \max \quad & -\frac{1}{2}\gamma^T H(H^T H)^{-1} H^T \gamma - h^T H(H^T H)^{-1} H^T \gamma + h^T \gamma \\ s.t. \quad & 0 \leq \gamma \leq C_2 e \end{aligned} \tag{12}$$

where, $H = [K(A,A^T) \quad e]$. After optimizing (17) and (18), we can obtain the augmented vectors for $f_1(x)$ and $f_2(x)$, which are

$$[w_1 \ b_1]^T = (H^T H)^{-1} H^T (f - \alpha), \quad [w_2 \quad b_2]^T = (H^T H)^{-1} H^T (h + \gamma) \tag{13}$$

Then the regression function of nonlinear TSVR is constructed as follows:

$$f(x) = \frac{1}{2}(f_1(x) + f_2(x)) = \frac{1}{2}K(x^T, A)(w_1 + w_2) + \frac{1}{2}(b_1 + b_2) \tag{14}$$

2.2 Analysis the Penalty Parameters of TSVR

The role of penalty parameters c_1 and c_2 is to adjust the ratio between the confidence range with the experience risk in the defining feature, so that the generalization ability of TSVR can achieve the best state. The values of c_1 and c_2 smaller expresses the punishment on empirical error smaller. Do it this way, the complexity of TSVR is smaller, but its fault tolerant ability is worse. The values of c_1 and c_2 are greater, the data fitting degree is higher, but its generalization capacity will be reduced. From the above analysis, we can know that the parameters selection is very important for TSVR.

3 Cloud PSO Model

3.1 Cloud Theory

Li Deyi [15] proposed the cloud model which can be used to transform the qualitative linguistic values to quantitative numerical value. It is based on the traditional fuzzy mathematics and probability-statistics. Cloud model theory has been applied in the field of data mining [16], intrusion detection [17], intelligent control [18], and reliability evaluation [19] successfully. The basic theory of cloud model is introduced as follows.

Let U be a quantitative domain, C be a qualitative concept in U. If quantitative value x $(x \in U)$ is a stochastic implementation of concept C, the certainty degree $u(x) \in [0, 1]$ of x to qualitative concept C is a random number with stable tendency:

$$u : U \to [0, 1] \qquad \forall x \in U, x \to u(x)$$

The distribution of x in domain U is called cloud. Each x is called a cloud droplet [20].

In the above definition, the mapping from U to the interval [0, 1] is a one-point to multi-point transition, which shows the uncertainty by integrating fuzziness and randomness of an element belonging to a term in U. So the degree of membership of u to [0, 1] is a probability distribution rather than a fixed value, which is different from the fuzzy logic.

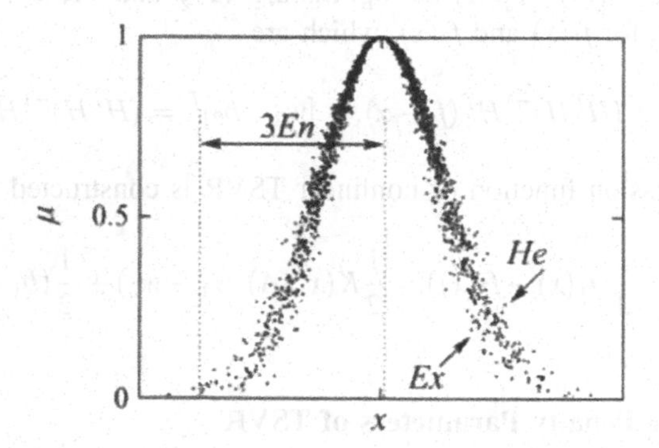

Fig. 1. Illustration of the three digital characteristics of a normal cloud

The numerical characters of cloud are described by expected value, entropy and hyper entropy, as $C = (Ex, En, He)$. The expected value denotes the expectation of cloud droplet in domain distribution, which determines the center of the cloud. The entropy reflects the uncertainty measure of qualitative concept, which determines the range of the cloud. So a normal cloud is described as Fig. 1.

3.2 Algorithm of the Basic Cloud Generator

Algorithm 1 Basic normal cloud generator algorithm
Input: $\{Ex, En, He\}$, n
Output: n cloud drops
For $i = 1 : n$
$Enn = rand(En, He)$
$x_i = rand(Ex, Enn)$
$u_i = e^{\frac{-(x_i - Ex)^2}{2(Enn)^2}}$
Drop (x_i, u_i)

If x_0 is given, we would use the following algorithm, which is called X-conditional cloud generator algorithm.

Algorithm 2 X-conditional cloud generator algorithm
Input: $\{Ex, En, He\}$, n, x_0
Output $\{(x_0, \mu_1), \cdots, (x_0, \mu_n)\}$
For $xi = 1 : n$
$En' = rand(En, He)$
$\mu_i = e^{\frac{-(x_0 - Ex)^2}{2(En')^2}}$
Drop (x_0, μ)

3.3 Cloud PSO Algorithm

3.3.1 The Basic Algorithm of Cloud PSO

In order to improve the diversity of population for PSO, thus greatly improve the ability of diagnosis to avoid falling into local optimal, in this paper, we use cloud model to generate the inertia weight of PSO. Based on this idea, we propose cloud PSO algorithm (CPSO). The basic principle of CPSO is described as follows.

For PSO, set the size of the particles is N, the fitness value of the i particle denoted X_i is f_i, the average fitness value is $f_{avg} = \frac{1}{N}\sum_{i=1}^{N} f_i$, the average of the fitness values whose value is better than f_{avg} is denoted by f'_{avg}, the average of the fitness values whose value is worse than f_{avg} is denoted by f''_{avg} and the fitness value of the best particle is denoted as $f_{\min}$.

In our algorithm, the whole of particles is divided into three populations. Each population uses different strategies to generate the inertia weight of PSO. Three different strategies for generating the inertia weight of PSO are shown as follows.

(1) When f_i is better than f'_{avg}, we use this strategies:

The particles whose fitness values are worse than f'_{avg}, is considered the best particles. Because these particles have been closer to the global optimum position, so we should use smaller value of inertia weight of PSO, which can speed up the convergence rate. In this case, we set the value of inertia weight of PSO is 0.2.

(2) When f_i is better than f''_{avg}, but is worse than f'_{avg}, we use this strategies:

These particles are the general group in the whole. The inertia weight values of these particles are generated by X-conditional cloud generator as follows.

Algorithm 3 The inertia weight values of PSO generator algorithm

$Ex = f_{avg}'$

$En = (f_{avg}' - f_{min}) / c_1$, where c_1 is the controlling parameter.

$He = En / c_2$, where c_2 is the controlling parameter.

$En' = normrnd(En, He)$

$w = 0.9 - 0.5 \times e^{\frac{-(f_i - Ex)^2}{2(En')^2}}$, where w is the inertia weight values of PSO.

With the fitness of particle reduced, by mathematical limit theorem we can know that

$0 < e^{\frac{-(f_i - Ex)^2}{2(En')^2}} < 1$, which can ensure the results $w \in [0.4, 0.9]$.

(3) When f_i is worse than f''_{avg}, we use this strategies:

These particles are the worst group in the whole. So we set $w = 0.9$.

3.3.2 The Parameters Selection of CPSO

As we know, the value of En affects the steep degree of the normal cloud model. According to the principle of "3 En" [15], for the language value of the domain U, the quantitative contribution to the linguistic value of 99.74% drops in c_1. A larger En, the horizontal width of cloud cover is more. Combination of speed and precision of the algorithm, we set $c_1 = 2.9$.

He determines the discrete degree of the cloud droplet. *He* is too small, which will lose the "random" to a certain extent. *He* is too large, which will lose "stable tendency". So in this paper, we set $c_2 = 10$.

3.3.3 The Algorithm Process of CPSO

Algorithm 4 The algorithm process of CPSO

Step1: Initialize PSO, including the parameters, the speed, the position, the best value of each particle p_i and the best global value p_g.

Step2: For each particle X_i, do the following operations.

(1) According to the above strategy, the inertia weight value of each particle is generated. Then according to (15) and (16), update the speed and position of each particle X_i.

(2) Calculate the fitness value f_i of X_i.

(3) If f_i is better than the fitness value of p_i, update the current position of X_i as p_i.

(4) If f_i is better than the fitness value of p_g, update the current position of X_i as p_g.

Step 3: Determine whether the algorithm meets the termination condition, if it has met the condition, please implement Step 4, or implement Step 2.

Step 4: Output p_g.

4 Experiment Results and Analysis

In order to verify the efficiency of TSVR-CPSO, compared with three algorithms, that is, SVR, TSVR and TSVR-PSO, we conduct two experiments on one nonlinear function and five benchmark datasets from the UCI machine learning repository, where TSVR-PSO means that parameters of TSVR based on PSO. In all algorithms in this paper, we only consider their nonlinear case and we use the gauss kernel function $K(x, x_i) = \exp(-\frac{\|x-x_i\|^2}{2\sigma^2})$ as their kernel function. For these datasets, the regression results are obtained by using 10-fold cross-validation. For SVR and TSVR, their parameters are selected over the range $\{2^i | i = -7, \cdots, 7\}$ using cross validation method. The environments of all experiments are in Intel (R) Core (TM) 2Duo CUP E4500, 2G memory and MATLAB 7.11.0. The parameter values of PSO are as follows: The number of particles $N = 50$, are the positive constants $c_1 = c_2 = 1.5$.

4.1 The First Experiment

The nonlinear function $z = \frac{(\sin x-1)^2}{8} + \frac{(\sin y-2)^2}{9}$, where $x \in [0, 3]$, $y \in [0, 3]$,is usually used to test the performance of the regression method. In this section, we use this function to test the fitness ability of TSVR-CPSO. In the interval, we randomly select 40 (x_i, y_i) as the training samples and 200 (x_i, y_i) as the testing samples. We can obtain the parameters of TSVR using TSVR-CPSO as follows. We get the penalty parameters of TSVR $c_1 = c_2 = 270.56$ and the kernel parameter $\sigma = 2.7845$. The comparison results are shown as Table 1. Figure 2 is the practical model of this nonlinear function and Fig. 3 is the approximation model of TSVR-CPSO.

Table 1. The comparison results of SVR, TSVR, TSVR-PSO and TSVR-CPSO

Algorithm	SSE	SSE/SST	SSR/SST
SVR	0.0524 ± 0.00226	0.0061 ± 0.00024	0.9526 ± 0.0327
TSVR	0.0489 ± 0.00124	0.0049 ± 0.00021	0.9528 ± 0.0315
TSVR-PSO	0.0481 ± 0.00156	0.0041 ± 0.00038	0.9528 ± 0.0321
TSVR-CPSO	0.0325 ± 0.00135	0.0021 ± 0.00029	0.9657 ± 0.0346

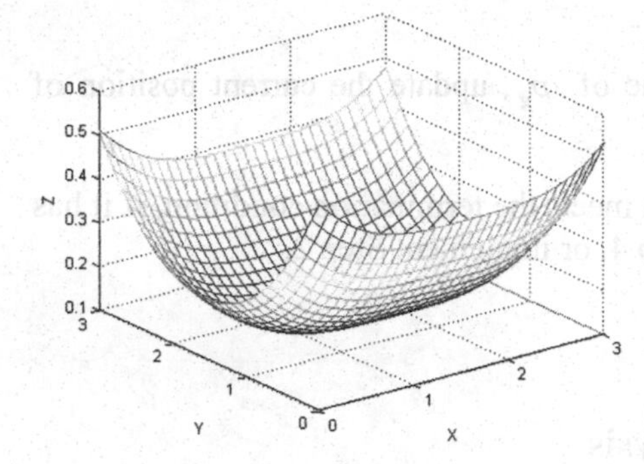

Fig. 2. The actual function model

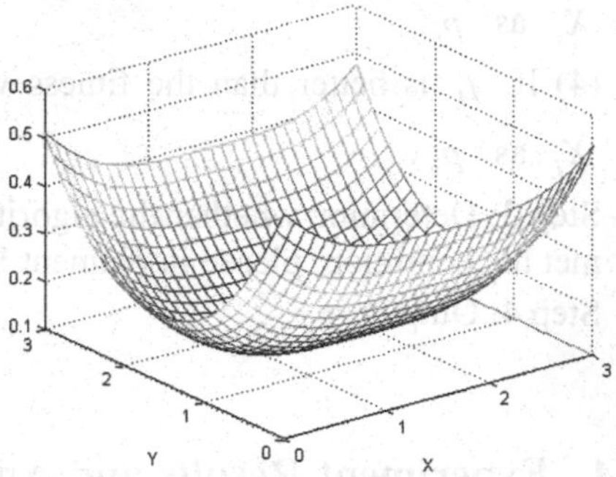

Fig. 3. The fitting function model of TSVR-CPSO

From Table 1, we can see that TSVR-CPSO can obtain the smallest SSE, SSE/SST and the largest SSR/SST than other algorithms. This results show that TSVR-CPSO owns better regression performance than other three algorithms. Figure 3 indicates that the fitting capacity of TSVR-CPSO is perfect.

4.2 The Second Experiment

For further evaluation, five benchmark datasets are tested in this section, which include Diabetes, Boston Housing, Auto-Mpg, Machine CPU, Servo. These datasets are usually used to validate the performances of regression methods. Table 2 shows the average results of SVR, TSVR, TSVR-PSO and TSVR-CPSO with 15 independent runs on five benchmark datasets. Figure 4 and Fig. 5 are the fitness curves of PSO and CPSO searching the optimal parameters for dealing with the Diabetes dataset respectively. Figure 6 and Fig. 7 are the fitness curves of PSO and CPSO searching the optimal parameters for dealing with the Boston Housing dataset respectively. In our experiments, the fitness function is expressed as follows.

Fitness function = $\sum_{i=1}^{n} (y(i) - y'(i))^2$, where n is the number of samples, is the real value and is predictive value.

Table 2. The comparison results of SVR, TSVR, TSVR-PSO and TSVR-CPSO on UCI dataset

Dataset	Algorithm	SSE	SSE/SST	SSR/SST
Diabetes (433)	SVR	0.45090.0526	0.51920.5673	0.60140.0174
	TSVR	0.40060.1573	0.47780.5954	0.64960.1147
	TSVR-PSO	0.32840.1984	0.42620.5806	0.71040.0415
	TSVR-CPSO	0.30690.0027	0.39340.2016	0.78240.0246
Boston Housing (50614)	SVR	0.40560.0174	0.12870.0352	0.90560.1547
	TSVR	0.40520.1523	0.12780.0348	0.97890.1276
	TSVR-PSO	0.39780.2544	0.12360.0364	1.00570.1026
	TSVR-CPSO	0.36630.4523	0.12320.0357	1.00680.0359
Auto-Mpg (3928)	SVR	0.12470.1578	0.11410.0314	0.98750.0145
	TSVR	0.12470.1413	0.10640.0528	0.98740.0074
	TSVR-PSO	0.09650.1238	0.10250.0424	0.98980.0012
	TSVR-CPSO	0.08260.0173	0.10230.0407	0.99890.0019
Machine CPU (2099)	SVR	0.10240.1742	0.10490.0741	0.96780.0052
	TSVR	0.10850.1643	0.10190.0121	0.96980.0048
	TSVR-PSO	0.08280.1215	0.11120.0741	0.97090.0044
	TSVR-CPSO	0.07840.1357	0.10270.0754	0.97650.0048
Servo (1674)	SVR	0.25420.0547	0.14150.0089	0.95420.0085
	TSVR	0.24840.1209	0.14120.0076	0.95750.0049
	TSVR-PSO	0.22460.0145	0.13640.0068	0.97490.0012
	TSVR-CPSO	0.20540.0143	0.11550.0028	0.98450.0019

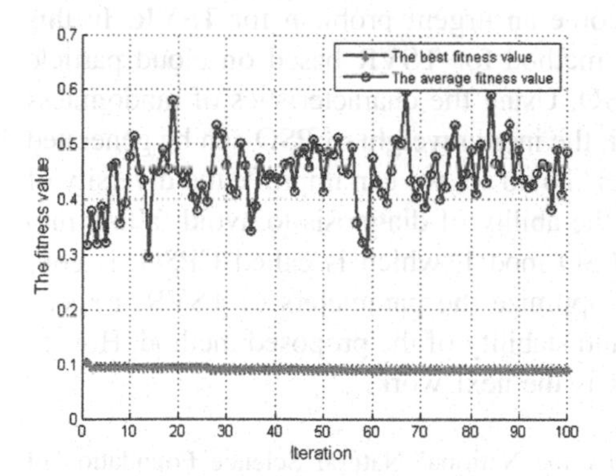

Fig. 4. The fitness curves of PSO searching the optimal parameters for dealing with diabetes

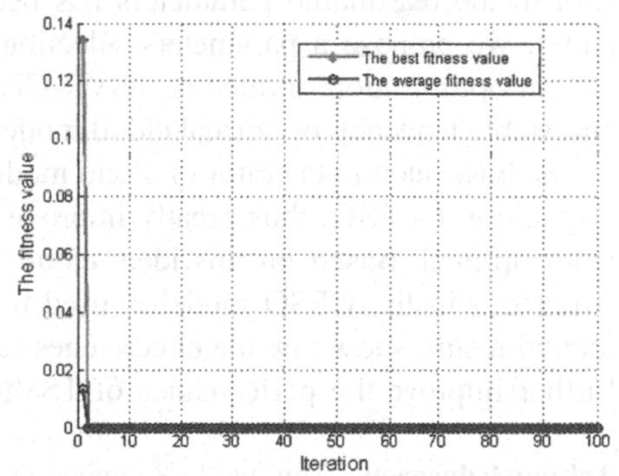

Fig. 5. The fitness curves of CPSO searching the optimal parameters for dealing with diabetes

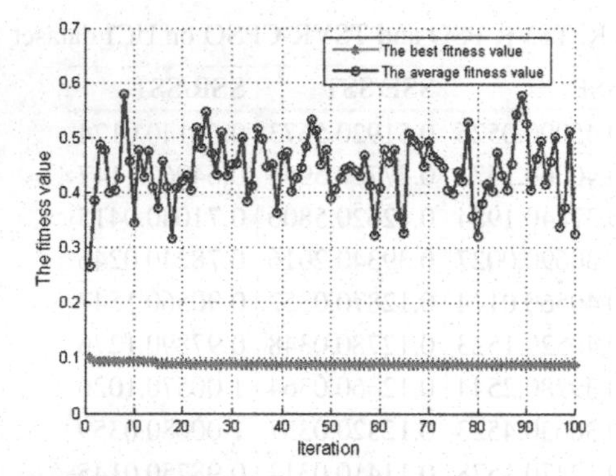

Fig. 6. The fitness curves of PSO searching the optimal parameters for dealing with Boston Housing

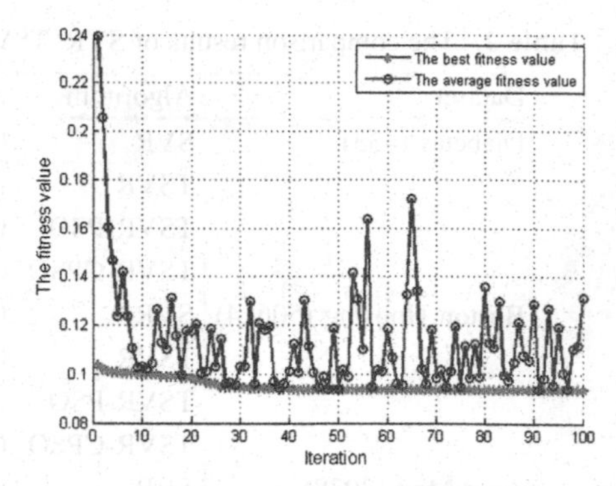

Fig. 7. The fitness curves of CPSO searching the optimal parameters for dealing with Boston Housing

From Table 2, we can see that the regression performance of TSVR-CPSO is relatively better than other algorithms. It indicates the optimization ability of CPSO is better than PSO. From Fig. 5 and Fig. 7, we can visually see that the optimization ability of CPSO is very strong.

5 Conclusion

Similar to SVR, the performance of TSVR is very dependent on its parameters. How to choose the reasonable parameters has become an urgent problem for TSVR. In this paper, we propose a parameters selection method for TSVR based on cloud particle swarm optimization, termed as TSVR-CPSO. Using the characteristics of randomness and stable tendency of normal cloud model, the inertia weight of PSO can be generated by the basic cloud generator of cloud model. To do so, we can improve the diversity of population for PSO, thus greatly improve the ability of diagnosis to avoid falling into local optimal. Based on this idea, cloud PSO model, which is called CPSO, is constructed. Finally, CPSO model is used to optimize the parameters of TSVR. Experimental results show that the effectiveness and stability of the proposed method. How to further improve the performance of TSVR is the next work.

Acknowledgement. This work is supported by the National Natural Science Foundation of China (61662005). Guangxi Natural Science Foundation (2018GXNSFAA294068, 2017GXNSFAA198008); Basic Ability Improvement Project for Young and Middle-aged Teachers in Colleges and Universities in Guangxi (2019KY0195); Research Project of Guangxi University for Nationalities (2019KJYB006).

References

1. Vapnik, V.N.: The Nature of Statistical Learning Theory. Springer, New York (1995). https://doi.org/10.1007/978-1-4757-2440-0
2. Liu, X., Jin, J., Weining, W., Herz, F.: A novel support vector machine ensemble model for estimation of free lime content in cement clinkers. ISA Trans. **99**, 479–487 (2020)
3. Borrero, L.A., Guette, L.S., Lopez, E., Pineda, O.B., Castro, E.B.: Predicting toxicity properties through machine learning. Procedia Comput. Sci. **170**, 1011–1016 (2020)
4. Liu, G., Chen, L., Zhao, W.: Internal model control of permanent magnet synchronous motor using support vector machine generalized inverse. IEEE Trans. Ind. Inf. **9**(2), 890–898 (2013)
5. Tang, X., Ma, Z., Hu, Q., Tang, W.: A real-time arrhythmia heartbeats classification algorithm using parallel delta modulations and rotated linear-kernel support vector machines. IEEE Trans. Bio-Med. Eng. **67**(4), 978–986 (2020)
6. Jayadeva, Reshma, K., Chandra, S.: Twin support vector machines for pattern classification. IEEE Trans. Pattern Anal. Mach. Intell. **29**(5), 905–910 (2007)
7. Fung, G., Mangasarian, O.L.: Proximal support vector machine classifiers. In: Proceedings of 7th ACM SIFKDD International Conference on Knowledge Discovery and Data Mining, pp. 77–86 (2001)
8. Mangasarian, O.L., Wild, E.W.: Multisurface proximal support vector machine classification via generalized eigenvalues. IEEE Trans. Pattern Anal. Mach. Intell. **28**(1), 69–74 (2006)
9. Mello, A.R., Stemmer, M.R., Koerich, A.L.: Incremental and decremental fuzzy bounded twin support vector machine. Inf. Sci. **526**, 20–38 (2020)
10. Zhang, X.S., Gao, X.B., Wang, Y.: Twin support vector machine for MCs detection. J. Electron. (China) **26**(3), 318–325 (2009)
11. Peng, X.: TSVR: an efficient twin support vector machine for regression. Neural Netw. **23**, 365–372 (2010)
12. Peng, X.: Primal twin support vector regression and its sparse approximation. Neurocomputing **73**, 2846–2858 (2010)
13. Chen, X., Yang, J., Liang, J.: Smooth twin support vector regression. Neural Comput. Appl. **21**, 505–513 (2012). https://doi.org/10.1007/s00521-010-0454-9
14. Sheykh Mohammadi, F., Amiri, A.: TS-WRSVM: twin structural weighted relaxed support vector machine. Connect. Sci. **31**(3), 215–243 (2019)
15. Liu, Y., Li, D., Zhang, G., et al.: Atomized feature in cloud based evolutionary algorithm. J. Electron. **37**(8), 1651–1658 (2009)
16. Li, D., Di, K., et al.: Mining association rules with linguistic cloud models. J. Softw. **11**, 143–158 (2000)
17. Rastogi, R., Saigal, P., Chandra, S.: Angle-based twin parametric-margin support vector machine for pattern classification. Knowl.-Based Syst. **139**, 64–77 (2018)
18. Dai, C., Zhu, Y., et al.: Cloud method based genetic algorithm and its applications, algorithm. J. Electron. (China) **35**(7), 1419–1424 (2007)
19. Dai, C., Zhu, Y., et al.: Adaptive genetic algorithm based on cloud theory. Control Theory Appl. **24**(4), 646–650 (2007)
20. Fu, Q., Cai, Z., Wu, Y.: A novel hybrid method: genetic algorithm based on asymmetrical cloud model. In: 2010 International Conference on Artificial Intelligence and Computational Intelligence (2010)

A MapReduce-Based Parallel Random Forest Approach for Predicting Large-Scale Protein-Protein Interactions

Bo-Ya Ji[1,2,3], Zhu-Hong You[1(✉)], Long Yang[1,2,3], Ji-Ren Zhou[1], and Peng-Wei Hu[1]

[1] Xinjiang Technical Institutes of Physics and Chemistry, Chinese Academy of Sciences, Urumqi 830011, China
zhuhongyou@ms.xjb.ac.cn
[2] University of Chinese Academy of Sciences, Beijing 100049, China
[3] Xinjiang Laboratory of Minority Speech and Language Information Processing, Urumqi, China

Abstract. The protein-protein interactions (PPIs) play an important part in understanding cellular mechanisms. Recently, a number of computational approaches for predicting PPIs have been proposed. However, most of the existing methods are only suitable for relatively small-scale PPIs prediction. In this study, we propose a MapReduce-based parallel Random Forest model for predicting large-scale PPIs using only proteins sequence information. More specifically, the Moran autocorrelation descriptor is firstly used to extract the local features from protein sequence. Then, the MapReduce-based parallel Random Forest model is utilized to perform PPIs prediction. In the experiment, the proposed method greatly reduces the required time to train the model, while maintaining the high accuracy in the prediction of potential PPIs. The promising results demonstrate that our method can be used as an efficient tool in the field of large-scale PPIs prediction, which greatly reduces the required training time and has high prediction accuracy.

Keywords: Protein-protein interactions · MapReduce · Random forest · Protein sequence

1 Introduction

The protein-protein interactions (PPIs) are critical for the growth, development, differentiation, and apoptosis of biological cells. In recent years, the high-throughput proteomics technology opens up new prospects for PPIs system identification [1–3]. However, these methods always face problems such as long training time, high cost, and low accuracy. In addition, most of the current calculation methods also required prior knowledge about proteins for PPI prediction [4, 5]. In recent years, there have been many researches to solve these limitations [6–21]. Today's high-throughput sequencing technology can sequence millions of protein molecules at a time, so sequence data is now growing very fast, which motivated researchers to develop new efficient parallel training methods for large-scale protein sequence data [18, 22–35]. For

D.-S. Huang and P. Premaratne (Eds.): ICIC 2020, LNAI 12465, pp. 400–407, 2020.
https://doi.org/10.1007/978-3-030-60796-8_34

example, Collobert *et al.* [36] proposed a novel parallelization method. They firstly used different subsets of the training set to train different SVM classifiers and then integrated the different classifiers into the final classifier. Zanghirati *et al.* [37] advanced a parallel decomposition technique, which uses Message Passing Interface (MPI) decomposition technology to improve training efficiency by decomposing a large problem into multiple small quadratic programming problems. Besides, the MapReduce is a programming framework for distributed computing programs. It integrates user-written business logic code and native default components into a complete distributed computing program. Mahout is an open-source project of the Apache Software Foundation (ASF), which provides a parallel implementation of classic machine learning algorithms. Thence, in this work, we put forward a MapReduce-based parallel Random Forest model for predicting large-scale protein-protein interactions only utilizing the proteins sequence information.

2 Methods and Materials

2.1 Dataset

For the positive training dataset, we downloaded the human PPI datasets in the article of Pan *et al.* [38]. After removing the duplicate interactions and self-interactions, the final positive datasets were obtained with a total of 36,630 pairs of PPI data from 9630 different human proteins. For the negative training dataset, we combined the negative data from the 57.3 version of the Swiss-Prot database [39] and the article of Smallowski *et al.* [40]. Moreover, we removed some sequences according to the following four principles when selecting the negative dataset in the Swiss-Prot database: (1) the protein sequence annotated with uncertain sub-cell location terms. (2) the protein sequence annotated with multi-locations. (3) the protein sequence annotated with "fragment". (4) sequences with fewer than 50 amino acid residues. Finally, the final negative dataset contains 36,480 pairs of non-interacting protein pairs. The final training data set is obtained by integrating the positive and negative datasets, which contains a total of 73110 pairs of protein-protein interactions. Then we randomly divide it into a training data set and a test data set, in which the training data set occupies 4/5, and the other 1/5 is the test data set.

2.2 Extraction of Physicochemical Properties of Proteins

In this work, we utilize six physicochemical properties of amino acids to encode protein sequences, which are respectively polarity (P1), solvent-accessible surface area (SASA), hydrophobicity (H), volumes of side chains of amino acids (VSC), net charge index of side chains (NCISC) and polarizability (P2). In this way, each amino acid residue can be converted into the numerical value based on its physicochemical properties. After that, we normalized them to unit standard deviation (SD) and zero mean according to the Equations:

$$P'_{ij} = \frac{P_{ij} - \bar{P}_J}{S_j}, (i = 1, 2, \ldots, 6; j = 1, 2, \ldots 20) \quad (1)$$

$$\bar{P}_J = \frac{\sum_{i=1}^{20} P_{ij}}{20} \quad (2)$$

$$S_j = \sqrt{\frac{\sum_{i=1}^{20} \left(P_{ij} - \bar{P}_J\right)^2}{20}} \quad (3)$$

where $\bar{P}_J$ is the mean of j_{th} descriptor over the 20 amino acids, P_{ij} is the j_{th} descriptor value for i_{th} amino acid and S_j is the corresponding SD.

2.3 Convert Protein Feature Vectors into Uniform Matrices

In this article, the Moran autocorrelation (MA) descriptor tool was used to convert the protein feature vectors into uniform matrices. It converts the protein feature vectors while taking into account the distribution of amino acid properties and the effect of residue proximity in the amino acid sequence, and describes the level of correlation between two protein sequences based on the specific physicochemical properties of amino acids. In detail, the Moran autocorrelation descriptor (MA) can be defined as follows:

$$MA(d) = \frac{1}{N-d} \sum_{j=1}^{N-d} \left(P_j - \bar{P}\right)\left(P_{j+d} - \bar{P}\right) \Big/ \frac{1}{N} \sum_{j=1}^{N} \left(P_j - \bar{P}\right)^2 \quad (4)$$

where P_j is the property value of the j_{th} amino acid, N is the length of protein sequence, P_{j+d} is the property value of the $(j+d)_{th}$ amino acid, d = 1, 2,…, 30 is the distance between the residual and its neighbor, $\bar{P}$ is the average of the considered property P along the protein sequence as follows:

$$\bar{P} = \sum_{j=1}^{N} \frac{P_j}{N} \quad (5)$$

2.4 The Random Forest Algorithm in Mahout

The Random Forest [41] is a supervised learning algorithm, which is an integrated learning algorithm based on Decision Trees. This algorithm is easy to implement and has a low computational cost while showing amazing performance in classification and regression. Mahout [42] is an open-source project of the Apache Software Foundation (ASF) that provides a parallel implementation of classic machine learning algorithms to help developers create smart applications faster and easier. The Random Forest algorithm in Mahout has been parallelized based on MapReduce.

2.5 MapReduce Model

MapReduce is a software architecture for parallel processing of large data sets proposed by Google to solve the problem of massive data calculation. It is inspired by functional languages and targets various practical tasks accompanying large data sets. It is written and maintained by experts in parallel programming to ensure the robustness and optimization of the system. The user only needs to pay attention to the data processing function without paying attention to the details of parallelism. Figure 1 below shows the overview of the MapReduce model:

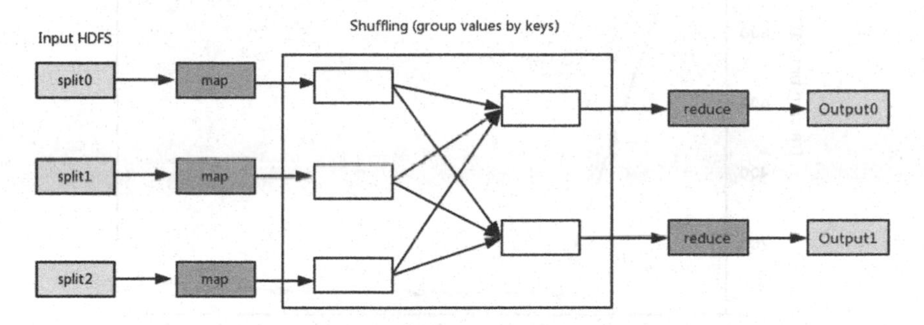

Fig. 1. Overview of the MapReduce framework

2.6 Hadoop Overview

Hadoop [43] is a distributed system infrastructure developed by the Apache Foundation. It enables users to develop distributed programs without knowing the details of the distributed foundation, to use the cluster function for high-speed calculation and storage. This distributed cluster platform includes the Hadoop Distributed File System (HDFS) and MapReduce. HDFS is designed to reliably store large data sets and transfer them to applications with high bandwidth. It uses a master-slave structure model composed of one NameNode (master node) and several DataNodes (slave nodes). NameNode is mainly used to manage the file system and client access to files. DataNode is mainly used to manage stored data. MapReduce is designed to distribute storage and compute tasks among different servers, allowing resources to be expanded as needed.

3 Experimental Results

We prepared 20000 PPI pairs as training data for comparative experiments. More specifically, we firstly utilized the Random Forest algorithm to train and predict training samples in the single-machine state (without using MapReduce) and recorded the training time and accuracy parameters. Secondly, we uploaded the large-scale PPI training set to HDFS on the Hadoop platform and then used the Random Forest algorithm in Mahout for training and prediction. Thirdly, we divided the training samples into 80, 160, 320, 640 parts by changing the largest block of data that can be processed in each map node of MapReduce. Figure 2 and 3 respectively shows the training time and accuracy of 20000 training pairs in two different modes. As can be seen from Fig. 2 and 3, our method can greatly reduce the training time required to process large-scale PPI datasets and has no significant loss in training accuracy.

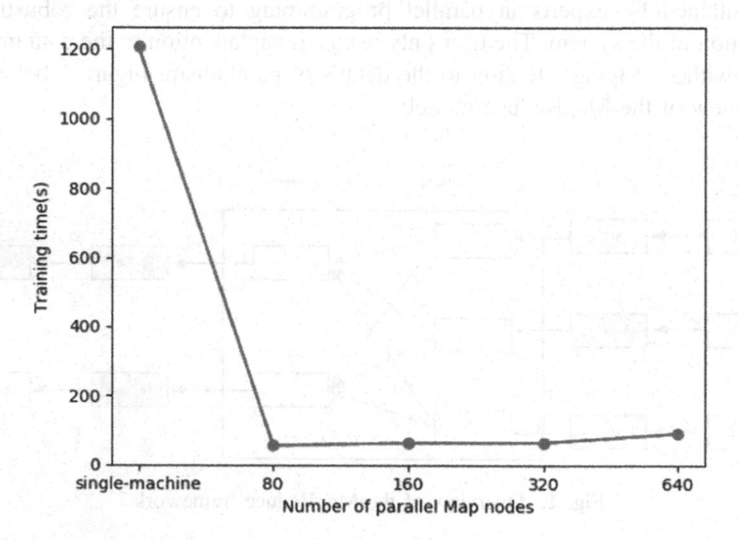

Fig. 2. Comparison of training times with single-machine Random Forest and MapReduce-based distributed Random Forest for PPIs prediction

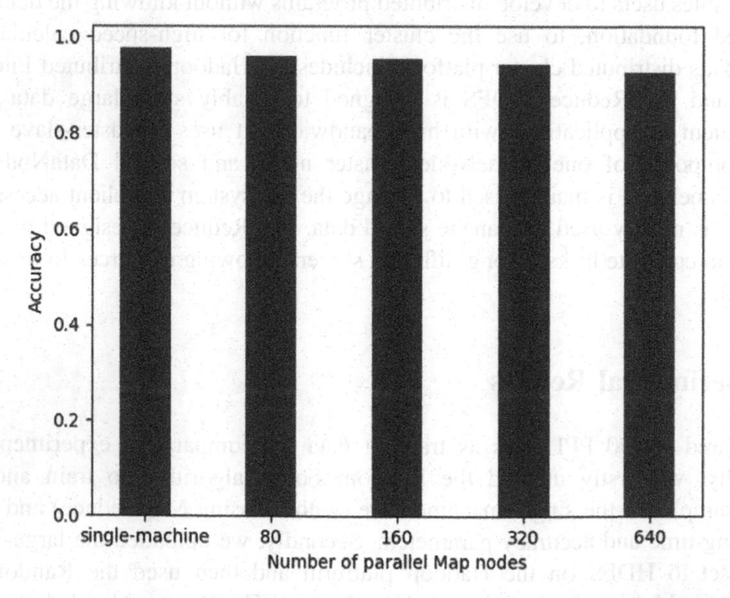

Fig. 3. Comparison of accuracy with single-machine Random Forest and MapReduce-based distributed Random Forest for PPIs prediction

4 Conclusions

Since the development of genome sequencing projects in recent years has provided large-scale protein sequence information, there is an increasing need to develop advanced methods that utilize the large-scale sequence information of proteins to predict potential PPIs. In this paper, we developed a MapReduce-based parallel Random Forest model to predict potential PPIs only using proteins sequence information. The experimental results prove that the proposed model can greatly accelerate the training speed of large-scale PPIs prediction and has high prediction accuracy.

Funding. This work is supported by the Xinjiang Natural Science Foundation under Grant 2017D01A78.

Conflict of Interest. The authors declare that they have no conflict of interest.

References

1. Krogan, N.J., et al.: Global landscape of protein complexes in the yeast Saccharomyces cerevisiae. Nature **440**, 637–643 (2006)
2. Ito, T., Chiba, T., Ozawa, R., Yoshida, M., Hattori, M., Sakaki, Y.: A comprehensive two-hybrid analysis to explore the yeast protein interactome. Proc. Natl. Acad. Sci. **98**, 4569–4574 (2001)
3. Gavin, A.-C., et al.: Functional organization of the yeast proteome by systematic analysis of protein complexes. Nature **415**, 141–147 (2002)
4. Wang, L., You, Z.-H., Li, L.-P., Yan, X., Zhang, W.: Incorporating chemical sub-structures and protein evolutionary information for inferring drug-target interactions. Sci. Rep. **10**, 1–11 (2020)
5. Wang, Y., You, Z., Li, L., Chen, Z.: A survey of current trends in computational predictions of protein-protein interactions. Front. Comput. Sci. **14**, 144901 (2020). https://doi.org/10.1007/s11704-019-8232-z
6. Zhu, H.-J., You, Z.-H., Shi, W.-L., Xu, S.-K., Jiang, T.-H., Zhuang, L.-H.: Improved prediction of protein-protein interactions using descriptors derived from PSSM via gray level co-occurrence matrix. IEEE Access **7**, 49456–49465 (2019)
7. Chen, Z.-H., You, Z.-H., Li, L.-P., Wang, Y.-B., Wong, L., Yi, H.-C.: Prediction of self-interacting proteins from protein sequence information based on random projection model and fast Fourier transform. Int. J. Mol. Sci. **20**, 930 (2019)
8. You, Z.-H., Lei, Y.-K., Gui, J., Huang, D.-S., Zhou, X.: Using manifold embedding for assessing and predicting protein interactions from high-throughput experimental data. Bioinformatics **26**, 2744–2751 (2010)
9. You, Z.-H., Huang, W.-Z., Zhang, S., Huang, Y.-A., Yu, C.-Q., Li, L.-P.: An efficient ensemble learning approach for predicting protein-protein interactions by integrating protein primary sequence and evolutionary information. IEEE/ACM Trans. Comput. Biol. Bioinf. **16**, 809–817 (2018)
10. Wang, L., You, Z.-H., Huang, D.-S., Zhou, F.: Combining high speed ELM learning with a deep convolutional neural network feature encoding for predicting protein-RNA interactions. IEEE/ACM Trans. Comput. Biol Bioinform. (2018)

11. Zhu, L., Deng, S.-P., You, Z.-H., Huang, D.-S.: Identifying spurious interactions in the protein-protein interaction networks using local similarity preserving embedding. IEEE/ACM Trans. Comput. Biol. Bioinf. **14**, 345–352 (2015)
12. Wang, Y., et al.: Predicting protein interactions using a deep learning method-stacked sparse autoencoder combined with a probabilistic classification vector machine. Complexity **2018** (2018)
13. Huang, Y.-A., You, Z.-H., Gao, X., Wong, L., Wang, L.: Using weighted sparse representation model combined with discrete cosine transformation to predict protein-protein interactions from protein sequence. BioMed Res. Int. **2015** (2015)
14. Wang, L., et al.: Using two-dimensional principal component analysis and rotation forest for prediction of protein-protein interactions. Sci. Rep. **8**, 1–10 (2018)
15. Li, L.-P., Wang, Y.-B., You, Z.-H., Li, Y., An, J.-Y.: PCLPred: a bioinformatics method for predicting protein–protein interactions by combining relevance vector machine model with low-rank matrix approximation. Int. J. Mol. Sci. **19**, 1029 (2018)
16. Guo, Z.-H., Yi, H.-C., You, Z.-H.: Construction and comprehensive analysis of a molecular association network via lncRNA–miRNA–disease–drug–protein graph. Cells **8**, 866 (2019)
17. Guo, Z.-H., You, Z.-H., Wang, Y.-B., Yi, H.-C., Chen, Z.-H.: A learning-based method for lncRNA-disease association identification combing similarity information and rotation forest. iScience **19**, 786–795 (2019)
18. Guo, Z.-H., You, Z.-H., Huang, D.-S., Yi, H.-C., Chen, Z.-H., Wang, Y.-B.: A learning based framework for diverse biomolecule relationship prediction in molecular association network. Commun. Biol. **3**, 1–9 (2020)
19. Guo, Z.-H., You, Z.-H., Yi, H.-C.: Integrative construction and analysis of molecular association network in human cells by fusing node attribute and behavior information. Mol. Ther.-Nucleic Acids **19**, 498–506 (2020)
20. Chen, Z.-H., Li, L.-P., He, Z., Zhou, J.-R., Li, Y., Wong, L.: An improved deep forest model for predicting self-interacting proteins from protein sequence using wavelet transformation. Front. Genet. **10**, 90 (2019)
21. Chen, Z.-H., You, Z.-H., Li, L.-P., Wang, Y.-B., Qiu, Y., Hu, P.-W.: Identification of self-interacting proteins by integrating random projection classifier and finite impulse response filter. BMC Genom. **20**, 1–10 (2019)
22. Wang, Y.-B., You, Z.-H., Li, X., Jiang, T.-H., Cheng, L., Chen, Z.-H.: Prediction of protein self-interactions using stacked long short-term memory from protein sequences information. BMC Syst. Biol. **12**, 129 (2018). https://doi.org/10.1186/s12918-018-0647-x
23. You, Z.-H., Li, X., Chan, K.C.: An improved sequence-based prediction protocol for protein-protein interactions using amino acids substitution matrix and rotation forest ensemble classifiers. Neurocomputing **228**, 277–282 (2017)
24. Wang, L., et al.: An ensemble approach for large-scale identification of protein-protein interactions using the alignments of multiple sequences. Oncotarget **8**, 5149 (2017)
25. Wang, Y.-B., et al.: Predicting protein–protein interactions from protein sequences by a stacked sparse autoencoder deep neural network. Mol. BioSyst. **13**, 1336–1344 (2017)
26. Huang, Y.-A., You, Z.-H., Chen, X., Yan, G.-Y.: Improved protein-protein interactions prediction via weighted sparse representation model combining continuous wavelet descriptor and PseAA composition. BMC Syst. Biol. **10**, 485–494 (2016)
27. Huang, Y.-A., You, Z.-H., Li, X., Chen, X., Hu, P., Li, S., Luo, X.: Construction of reliable protein–protein interaction networks using weighted sparse representation based classifier with pseudo substitution matrix representation features. Neurocomputing **218**, 131–138 (2016)

28. An, J.Y., Meng, F.R., You, Z.H., Chen, X., Yan, G.Y., Hu, J.P.: Improving protein–protein interactions prediction accuracy using protein evolutionary information and relevance vector machine model. Protein Sci. **25**, 1825–1833 (2016)
29. You, Z.-H., Chan, K.C., Hu, P.: Predicting protein-protein interactions from primary protein sequences using a novel multi-scale local feature representation scheme and the random forest. PLoS ONE **10**, e0125811 (2015)
30. You, Z.-H., et al.: Detecting protein-protein interactions with a novel matrix-based protein sequence representation and support vector machines. BioMed Res. Int. **2015** (2015)
31. Zheng, K., You, Z.-H., Li, J.-Q., Wang, L., Guo, Z.-H., Huang, Y.-A.: iCDA-CGR: identification of circRNA-disease associations based on Chaos Game Representation. PLoS Comput. Biol. **16**, e1007872 (2020)
32. Zheng, K., You, Z.-H., Wang, L., Zhou, Y., Li, L.-P., Li, Z.-W.: MLMDA: a machine learning approach to predict and validate MicroRNA–disease associations by integrating of heterogenous information sources. J. Transl. Med. **17**, 260 (2019). https://doi.org/10.1186/s12967-019-2009-x
33. Zheng, K., You, Z.-H., Wang, L., Zhou, Y., Li, L.-P., Li, Z.-W.: Dbmda: A unified embedding for sequence-based mirna similarity measure with applications to predict and validate mirna-disease associations. Mol. Ther.-Nucleic Acids **19**, 602–611 (2020)
34. Wang, M.-N., You, Z.-H., Wang, L., Li, L.-P., Zheng, K.: LDGRNMF: LncRNA-Disease associations prediction based on graph regularized non-negative matrix factorization. Neurocomputing (2020)
35. Wang, M.-N., You, Z.-H., Li, L.-P., Wong, L., Chen, Z.-H., Gan, C.-Z.: GNMFLMI: Graph regularized nonnegative matrix factorization for predicting LncRNA-MiRNA interactions. IEEE Access **8**, 37578–37588 (2020)
36. Collobert, R., Bengio, S., Bengio, Y.: A parallel mixture of SVMs for very large scale problems. In: Advances in Neural Information Processing Systems, pp. 633–640 (Year)
37. Zanghirati, G., Zanni, L.: A parallel solver for large quadratic programs in training support vector machines. Parallel Comput. **29**, 535–551 (2003)
38. Pan, X.-Y., Zhang, Y.-N., Shen, H.-B.: Large-Scale prediction of human protein − protein interactions from amino acid sequence based on latent topic features. J. Proteome Res. **9**, 4992–5001 (2010)
39. Bairoch, A., Apweiler, R.: The SWISS-PROT protein sequence database and its supplement TrEMBL in 2000. Nucleic Acids Res. **28**, 45–48 (2000)
40. Smialowski, P., et al.: The Negatome database: a reference set of non-interacting protein pairs. Nucleic Acids Res. **38**, D540–D544 (2010)
41. Liaw, A., Wiener, M.: Classification and regression by randomForest. R News **2**, 18–22 (2002)
42. Solanki, R., Ravilla, S.H., Bein, D.: Study of distributed framework hadoop and overview of machine learning using apache mahout. In: 2019 IEEE 9th Annual Computing and Communication Workshop and Conference (CCWC), pp. 0252–0257. IEEE (2019)
43. Shvachko, K., Kuang, H., Radia, S., Chansler, R.: The hadoop distributed file system. In: 2010 IEEE 26th Symposium on Mass Storage Systems and Technologies (MSST), pp. 1–10. IEEE (2010)

Feature Extraction and Random Forest to Identify Sheep Behavior from Accelerometer Data

Natasa Kleanthous[1(✉)], Abir Hussain[1(✉)], Wasiq Khan[1], Jenny Sneddon[2], and Alex Mason[3]

[1] Department of Computer Science, Liverpool John Moores University, Liverpool, UK
N.K.Orphanidou@2015.ljmu.ac.uk, {A.Hussain,W.Khan}@ljmu.ac.uk
[2] Natural Sciences and Psychology, Liverpool John Moores University, Liverpool, UK
J.C.Sneddon@ljmu.ac.uk
[3] Faculty of Science and Technology, Norwegian University of Life Sciences, As, Norway
alex.mason@nmbu.no

Abstract. Sensor technologies play an essential part in the agricultural community and many other scientific and commercial communities. Accelerometer signals and Machine Learning techniques can be used to identify and observe behaviours of animals without the need for an exhaustive human observation which is labour intensive and time consuming. This study employed random forest algorithm to identify grazing, walking, scratching, and inactivity (standing, resting) of 8 Hebridean ewes located in Cheshire, Shotwick in the UK. We gathered accelerometer data from a sensor device which was fitted on the collar of the animals. The selection of the algorithm was based on previous research by which random forest achieved the best results among other benchmark techniques. Therefore, in this study, more focus was given to feature engineering to improve prediction performance. Seventeen features from time and frequency domain were calculated from the accelerometer measurements and the magnitude of the acceleration. Feature elimination was utilised in which highly correlated ones were removed, and only nine out of seventeen features were selected. The algorithm achieved an overall accuracy of 99.43% and a kappa value of 98.66%. The accuracy for grazing, walking, scratching, and inactive was 99.08%, 99.13%, 99.90%, and 99.85%, respectively. The overall results showed that there is a significant improvement over previous methods and studies for all mutually exclusive behaviours. Those results are promising, and the technique could be further tested for future real-time activity recognition.

Keywords: Accelerometer data · Animal activity recognition · Feature extraction · Machine learning · Random forest · Sheep behaviour · Signal processing

D.-S. Huang and P. Premaratne (Eds.): ICIC 2020, LNAI 12465, pp. 408–419, 2020.
https://doi.org/10.1007/978-3-030-60796-8_35

1 Introduction

Sheep play an essential role in our society as they are kept for meat, wool, as well as pasture management. According to research conducted, sheep are shown to be as effective as herbicides in controlling winter weed, as well as insecticides [1, 2]. In order to manage the land they graze, human observation is the traditional mean of monitoring the distribution of the animals, which is a time consuming and labour intensive process. Thus, development of smart devices is essential for efficient monitoring and controlling of the animals' distribution on the pasture.

Automated monitoring of animals also allows early detection of illness, particularly lameness; present in an estimated 80% of UK flocks [3–6]. Furthermore, evidence showed that the reduced activity or decreased food intake of the animal might be an indicator of disease. Therefore, computerized monitoring of animals in real-time has become a pressing requirement in sheep production systems. Using insight from automatic monitoring capability can offer sufficient knowledge of the animal's welfare and food intake, and the decision making of the land and animal managers can be made more efficiently.

Accelerometers are widely used with machine learning techniques to identify animal behaviour such as cattle [7–14], horses [15], sharks [16], goats [17, 18] and other domesticated or wild animals. However, in this study, we focused only on previous research that involves sheep behaviour in order to identify challenges concerned with this type of animal and be able to compare between previous studies. Additionally, we aim to improve prediction performance of the activities of the animals.

The reminder of this paper is organized as follows. Section 2 consists of background information. Section 3 provides information about the materials and methodology, while Sect. 4 demonstrates results and discussions. Section 5 includes the conclusion and the future work.

2 Background

Marais et al. [19] developed a device capable of collecting accelerometer signals at 100 Hz from a collar. The authors extracted features using 5.12 s windows and applied linear discriminant analysis (LDA) and quadratic discriminant analysis (QDA) using 10 features to classify five common behaviours of the animals (lying, standing, walking, running and grazing). LDA and QDA achieved an overall accuracy of 87.1% and 89.7%, respectively. Discriminant analysis was tested by Giovanetti et al. to classify grazing, ruminating, and resting of sheep using a 60 s window and accelerometer data. The algorithm yielded an overall accuracy of 93.0% and k coefficient of 89.0% [20].

Nadimi et al. [21] classified five mutually exclusive behaviours (grazing, lying, walking, standing, and others) with 76.2% success rate. Additionally, they classified two behaviours (grazing and lying) with a success rate of 83.5% using the Nguyen–Widrow method and the Levenberg– Marquardt back-propagation algorithm. Compared to similar studies, the authors showed significant improvement of the designed system.

Kamminga et al. compared several machine learning algorithms to detect five mutually exclusive behaviours using data gathered from goats and sheep using accelerometer, gyroscope, and magnetometer signals. The best results were obtained using a 1 s window and Deep neural networks with a 94% accuracy [22]. The same dataset were used by Kleanthous et al. that tested multilayer perceptron, random forests, extreme gradient boosting, and k-Nearest neighbors to classify sheep and goat behaviour [23]. The best results achieved using random forest algorithm and classified grazing, lying, scratching or biting, standing, and walking with an overall accuracy of 96.47% and kappa value of 95.41%. The authors conducted another experiment and they gathered accelerometer and gyroscope data from more sheep using smartphones to test the performance of random forest and their previous method using a smaller sample rate; 10 Hz [24]. The technique proved successful and they achieved accuracy and kappa value of 96.43%, and 95.02%, respectively by using only accelerometer features.

Mansbridge et al. collected accelerometer and gyroscope signals from sensors attached to the ear and collar of sheep at 16 Hz [25]. Various machine learning algorithms were tested using multiple features from the signals. Random forest yielded the highest results using 39 feature characteristics and a 7 s window, achieving accuracy of 92% and 91% for collar and ear data, respectively.

Barwick et al. were also interested in applying machine learning to describe sheep behaviour using accelerometers to evaluate the effectiveness by placing accelerometers on different parts of the body; ear, collar, leg [26]. The authors applied QDA and best results obtained from the ear acceleration data, at 94%, 96% and 99% for grazing, standing, and walking, respectively.

Walton et al. evaluated sampling frequency (8, 16, and 32 Hz), window size (3, 5, and 7 s) and sensor position (ear and collar) to classify sheep behaviour using random forests [27]. Their results suggested that the 16 Hz sampling frequency and a 7 s window offer benefits concerning battery energy and it has the potential to be used for real-time monitoring system. The authors achieved results of 91%–93% accuracy and F-score of 88%–95%.

Alvarenga et al. [28] evaluated the performance of decision trees for accelerometer data obtained from sheep. The algorithm was validated for 3, 5 and 10 s epochs. The best results in terms of accuracy were achieved for the 5 s epoch having accuracy of 85.5%.

The sheep activity was also evaluated from Le Roux et al. [29] The authors developed an energy-aware feature and classifier selection technique for low-power sensor applications to minimize the energy consumed and also minimizing the accuracy loss of the classifier. The sheep data that they used included accelerometer signals and the authors were able to achieve a reduction in energy consumption while achieving an accuracy of 88.4% for classification of five behaviours. The authors, in a previous study, also evaluated sheep behaviour based on accelerometer data [30]. The classification algorithm achieved an accuracy of 82.4% for standing, walking, grazing, running and lying behavioural classes.

Guo et al. gathered signals from an IMU (Inertial Measurement Unit) sensors at a 20 Hz sampling rate and compared the grazing behaviour results according to different sward surface heights [31]. The authors applied Linear discriminant analysis on several datasets which they consisted of three sward surface heights. Overall, they achieved accuracy over 95% with the best results achieved using a 10 s window having accuracy of 98.2%. The authors showed that the IMU sensors are capable of providing robust information on the grazing behaviour of the animal despite the sward surface heights.

Decandia et al. [32] evaluated the performance of canonical discriminant analysis (CDA), and discriminant analysis (DA) to distinguish between three behaviours of sheep; grazing, ruminating, and others. The authors aimed to identify the window which provides the best algorithm performance and they evaluated windows of 5, 10, 30, 60, 120, 180 and 300 s from accelerometer signals sampled at 62.5 Hz. Best results were achieved with the 30 s epoch having accuracy and kappa value of 89.7% and 80%, respectively. Vazquez et al. [33] aimed to develop a combined online (k-means) and offline (k nearest neighbors) algorithm, which deals with concept drift to deal with three behaviours of sheep. The combined algorithm produced results with average accuracies of 85.18%, average specificities of 82.84%, and an average recall of 57.82%.

All of the abovementioned studies involve the use of Machine Learning techniques to identify sheep behaviour at pasture, however there is still a need for improvement of the prediction accuracy. The aim of our study is to significantly improve our previously tested method [24] by expanding the feature set and decreasing the sliding window to 5 s. In our experiment, we focused on four behaviours; grazing, walking, scratching, and inactive. For the experiment, we used only accelerometer data sampled at 12.5 Hz, which was previously demonstrated adequate and did not compromise the battery life of the device [27].

3 Materials and Methods

This section describes the materials and methods used to examine the performance of Random Forest algorithm regarding the classification of four mutually exclusive behaviours of sheep; grazing, walking, scratching, and inactive. Figure 1, shows the process followed to conduct the study.

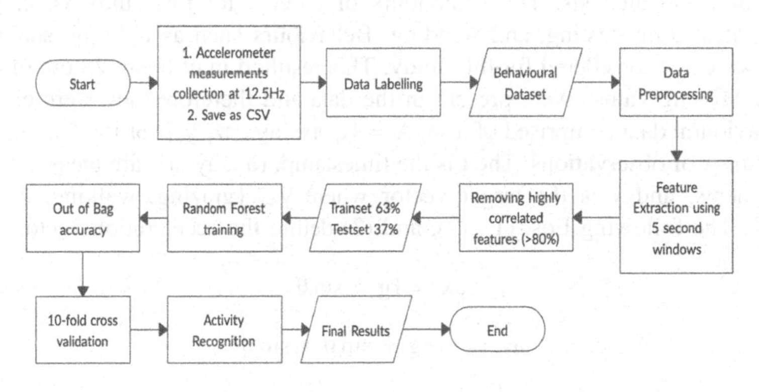

Fig. 1. Methodology

3.1 Animals, Location, and Sensor Device

This study was conducted in July-August 2019 in Cheshire Shotwick (OS location 333781,371970), UK. Eight Hebridean ewes between the ages of 5–12 years were fitted with a sensor device collar. The animals were free to use a paddock of 1500 m^2 area size and had access to grass and water all the time. The Senior Research Officer and LSSU Manager of Liverpool John Moores University approved the protocol of the experiment (approval AH_NKO/2018-13).

The MetamorionR® [34] wearable device was used for the current experiment. The sensor device collects motion and environmental data, however for this experiment we only used accelerometer measurements. The device weights 0.3 oz and its dimensions are of 36 mm × 27 mm × 10 mm with the case. Additionally, a 60 mAH MicroUSB rechargeable li-po battery powers it. For this study, we used only accelerometer measurements at a sample rate of 12.5 Hz. The device logged and saved the data on its offboard memory as a CSV file.

3.2 Data Collection and Annotation

The animals were fitted with collars, which had the device attached in a nonfixed position to have a more generalised algorithm performance independent of the sensor orientation and position. The animals were video recorded during the morning, afternoon or night, and one observer was present each time. At the end of each day, the CSV file was saved for later use. Once all the recordings were completed with a total of 40 h of recorded behaviours, the accelerometer readings were time synchronised with the video recordings for behavioural annotation. For animal behaviour annotation, we used ELAN_5.7_AVFX Freeware tool [35] and manually labeled the behaviours as grazing, walking, scratching, and inactive.

3.3 Data Preprocessing

After the data annotation, all the CSV files were merged and imported in Rstudio® for visualization and analysis. The behaviours of interest for this study were: grazing, walking, scratching, resting, and standing. Behaviours such as fighting, shaking, and rubbing were not considered for this study. This resulted in utilising 28 out of 40 h for analysis. Missing values were present in the data and therefore they were eliminated. The behavioural data comprised of a set $A = \{t_i, ax_i, ay_i, az_i, y_i\}$ for $i = 1, .., n$, where n is the number of observations. The t is the timestamp, (ax, ay, az) are the accelerometer measurements, and y is the target vector where $y \in$ {grazing, walking, scratching, inactive}. The following Eqs. (1), (2), and (3) define the acceleration vector:

$$ax = 1g * \sin\theta \tag{1}$$

$$ay = -1g * \sin\theta * \sin\phi \tag{2}$$

$$az = 1g * \cos\theta \qquad (3)$$

Where θ is the angle between az relative to gravity, ϕ is the angle of ax relative to ground, and g is the gravitational constant where 1 g = 9.81 m/s^2.In this step we extracted the magnitude of the acceleration (4):

$$\text{Magnitude} = \sqrt{ax^2 + ay^2 + az^2} \qquad (4)$$

3.4 Feature Extraction, Feature Importance, and Dimensionality Reduction

A total of 17 features were calculated from the x, y, z, and magnitude of the acceleration signals for each activity resulting in a total of 68 newly created features (i.e. 17 features $\times$ 4 activities). Those features include the mean, standard deviation, root mean square, root mean square velocity, energy, sum of changes, mean of changes, absolute and squared integrals, madogram [36], peak frequency, peak to peak value, kurtosis and skewness, zero crossing, crest factor, and signal entropy. The features were extracted using a 5 s sliding window. Having a greater window in a real-time classification could provoke mislabeling because the animal might exhibit more than one behaviour in a short time interval; therefore, a 5 s window is considered sufficient.

The distributions for first four principal components (PCs) with respect to target class, original attributes and corresponding impacts of the target classes within the dataset are represented in Fig. 2(a) and (b). These figures also indicate the non-linearity of the problem specifically in terms of first four PCs covering the highest variances ($\sim$65%) within the overall principal components. Though, there is a small degree of overlap between all activities. However, this was expected since the head movements of the animal might exhibit similar patterns in some instances. Furthermore, the plots help to understand the corresponding influence of the features within the datasets on the classification of animal behaviours (i.e. 4 target classes). For instance, in Fig. 2(a) the madogram of the magnitude & the madogram of ay measurement have a clear impact on class 'inactive' as compared to root mean square velocity which influence the 'scratching and grazing' classes.

We used the most commonly used dimensionality reduction technique PCA [37] to identify the most significant attributes/features within the dataset set and eliminating the unnecessary features. In other words, PCA can be used to transform a large dataset containing large number of features/variables to a lower dimension which still holds most of the information contained in the original high dimensional dataset. One of the important properties of PCA is the attribute loadings on the principal components that can also be used for the identification of attribute importance within the original dataset.

The correlation coefficient between the dataset attributes is represented by the principal components' loadings (i.e. obtained through PCA). The component rotations provide the maximized sum of variances of the squared loadings. The absolute sum of component rotations gives the degree of importance for the corresponding attributes in dataset. Figure 3 shows the feature significance score within the original dataset which

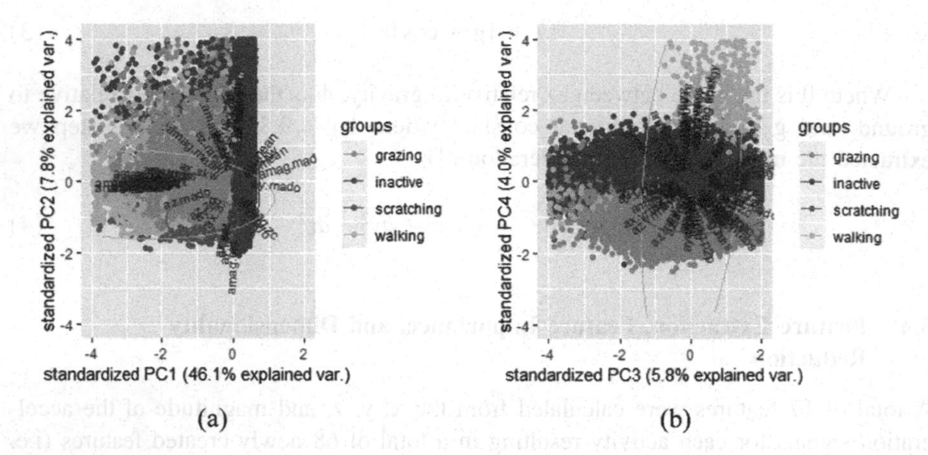

Fig. 2. (a) First two PCA components' distributions; (b) 3rd and 4th components' distributions within the PCA components

is calculated through the PCs loadings. There are variations in the importance measure of features which can be used to identify and hence remove the unnecessary features from the dataset. For instance, 'madogram' of z, and x axis are indicated the top-ranked

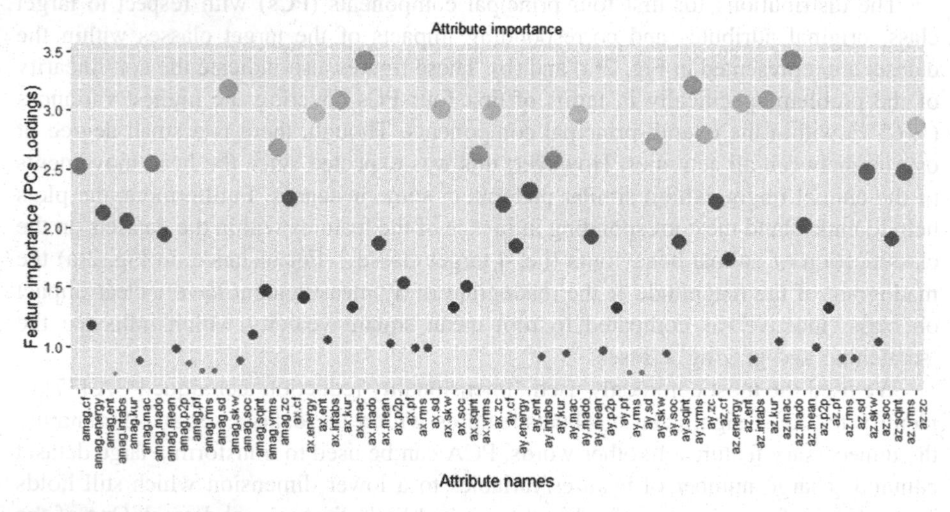

Fig. 3. Measure of feature importance within the dataset using principal components loading

variables compared to magnitude 'integrals' and 'rms' of the ay axis which are indicated the least important variables within the original dataset.

To further investigate the features/attributes within the dataset, we used the correlation coefficients. The correlated features with correlation above 80% were removed and the remaining features are in agreement with the feature importance ranking indicated by the PCA. Therefore, we eliminated our features from 17 to 9. The remaining features are the mean, crest factor, root mean square velocity, skewness, kurtosis, madogram, zero crossing rate, squared integrals, and signal entropy.

3.5 Classification

The classification algorithm selected to evaluate our dataset and test the activity prediction performance of the animals was the Random Forest as it was proved successful in our previous studies, as well as other studies concerned with animal behaviour [9–11, 14, 24]. Random forest [38] is an ensemble method which consists of a combination of decision trees which are dependent on random values. All trees are sampled independently with the same distribution. The classification decision is then made based on the majority of votes from each tree.

To estimate the performance of the algorithm, we evaluated the model with the Out of bag (OOB) accuracy. The idea behind OOB was since the decision trees are learning from a subset of the dataset (63%), then we have unseen data (37%) to be used for evaluation. This method is a good estimate of the ability of the model to generalize on unseen data [39]. We then recursively evaluated the performance using sensitivity, specificity, accuracy, and kappa value quality measures by means of 10-fold cross validation. The results are presented in the next section.

4 Results and Discussion

The performance of random forest is presented in Table 1. The four behaviours are classified correctly at a high rate. The overall accuracy of the algorithm is 99.43% with kappa value of 98.66%. Additionally, the f1-score is between 91.53%-99.90%. The lowest F1-score is resulted from scratching, and the highest from inactive behaviour.

Table 1. Random forest performance on unseen data

	Activities			
	Grazing	Walking	Scratching	Inactive
Sensitivity	98.26%	98.66%	99.87%	99.86%
Specificity	99.91%	99.60%	99.92%	99.84%
F1-score	98.97%	94.64%	91.53%	99.90%
Balanced accuracy	99.08%%	99.13%	99.90%	99.85%
Overall accuracy: 99.43%, Kohen's Kappa value: 98.66%				

The sensitivities of all behaviours are between 98.26% to 99.87%. Also, the specificities are between 99.60% to 99.92%. Scratching was misclassified only once with grazing, while grazing was misclassified with scratching and walking in some cases. The same is valid with walking as it was misclassified with grazing and scratching.

Only limited cases misclassified inactive behaviour with the other behaviours However, the misclassification is limited, and consequently, the results showed high accuracy, sensitivity, and specificity in all 4 cases.

In this study, we noted that the movements the animals are conducting while they graze can sometimes have similarities with the walking and scratching behaviour. Additionally, resting and standing provide a similar pattern of the acceleration signals, because of the animals' inactive state and this was also noted by Barwick et al. [26]. On the other hand, while the animals scratch or bite, the activity is detected easily as the magnitude changes markedly. While the animals are ruminating, the head movements are relatively small, and stationary compared with grazing and it does not interfere with the correct classification of the activity they perform. From the results, we noted that 5 s windows can provide a very good activity pattern representation and therefore could be suggested that this size is adequate. However, Decandia et al. [32], conducted experiments with various window sizes, such as 5, 10, 30, 60, 120, 180 and 300 s, and they identified that the best performance was obtained from a 30 s window having sensitivity 94.8% for grazing, 80.4% for ruminating, and 92.3% for other behaviours. Though, the two studies cannot be compared because the ML model applied, the selection of features, and the position of the sensor is different. On the other hand, a 5 s window achieved best performance in a study of Alvarenga et al. [28] when they compared 3, 5, and 10 s windows. The authors achieved an overall accuracy of 85.50% with Decision Trees and 5 s windows which exhibited higher accuracy in comparison with 3 and 10 s windows. However, the variety of feature combinations, ML techniques, sample rate, and window size used in previous and the current study show that there is still need for further investigation and there is no clear indication yet on the technique that is more suitable to be used for sheep activity recognition.

5 Conclusion and Future Work

This study was focused on detecting four mutually exclusive behaviours of interest to the animal health and production industry. Data was collected from eight Hebridean ewes located in Cheshire Shotwick, UK. Accelerometer signals were collected from a sensor which was attached on the collar of each animal. A total of 28 h was used to test the performance of random forest to detect each behaviour. The behaviours of interest were the grazing, walking, scratching, and inactive. To test the algorithm, 17 features were extracted from the x, y, z, and magnitude of the acceleration signal resulting in 68 newly created variables. We then removed features with higher than 80% correlation and eliminated the features to 9. The evaluation of the random forest algorithm was then assessed using out-of-bag (OBB) estimate which is empirically proven that is as accurate as using a test set of the same size as the training set [39].

The results were very high for all the activities having accuracies of 99.08% for grazing, 99.13% for walking, 99.90% for scratching, and 99.85% for inactive. The overall accuracy and kappa value were 99.43% and 98.66%. The results showed that there is an important improvement over previous methods. The technique can be further tested and used for online activity recognition system and be part of a multi-functional smart device for monitoring and controlling animal behaviour and position.

In future work, we will use GPS coordinates to track the position of the animals and monitor the land they mostly graze. The implementation of such a device can be used as an intelligent assistant to provide valuable information regarding the food intake of the animals and their activities during the day, which can improve the decision making of the land managers. Such information can contribute to the animal's welfare, pasture utilisation and overall farm and animal decision management approach.

Acknowledgement. We would like to thank the Douglas Bomford Trust [40] for the funding support of this study.

References

1. Umberger, S.H.: Sheep grazing management
2. Doran, M.P., Hazeltine, L., Long, R.F., Putnam, D.H.: Strategic grazing of alfalfa by sheep in California's Central Valley (2010)
3. Winter, A.C.: Lameness in sheep. Small Ruminant Res. **76**, 149–153 (2008)
4. Barwick, J., Lamb, D., Dobos, R., Schneider, D., Welch, M., Trotter, M.: Predicting lameness in sheep activity using tri-axial acceleration signals. Animals **8**, 1–16 (2018)
5. Al-Rubaye, Z., Al-Sherbaz, A., McCormick, W.D., Turner, S.J.: The use of multivariable wireless sensor data to early detect lameness in sheep (2016)
6. Gougoulis, D.A., Kyriazakis, I., Fthenakis, G.C.: Diagnostic significance of behaviour changes of sheep: a selected review. Small Ruminant Res. **92**, 52–56 (2010)
7. González, L.A.A., Bishop-Hurley, G.J.J., Handcock, R.N.N., Crossman, C.: Behavioral classification of data from collars containing motion sensors in grazing cattle. Comput. Electron. Agric. **110**, 91–102 (2015)
8. Robert, B., White, B.J., Renter, D.G., Larson, R.L.: Evaluation of three-dimensional accelerometers to monitor and classify behavior patterns in cattle. Comput. Electron. Agric. **67**, 80–84 (2009)
9. Rahman, A., Smith, D.V., Little, B., Ingham, A.B., Greenwood, P.L., Bishop-Hurley, G.J.: Cattle behaviour classification from collar, halter, and ear tag sensors. Inf. Process. Agric. **5**, 124–133 (2018)
10. Smith, D., et al.: Behavior classification of cows fitted with motion collars: Decomposing multi-class classification into a set of binary problems. Comput. Electron. Agric. **131**, 40–50 (2016)
11. Dutta, R., et al.: Dynamic cattle behavioural classification using supervised ensemble classifiers. Comput. Electron. Agric. **111**, 18–28 (2015)
12. Andriamandroso, A.L.H., et al.: Development of an open-source algorithm based on inertial measurement units (IMU) of a smartphone to detect cattle grass intake and ruminating behaviors. Comput. Electron. Agric. **139**, 126–137 (2017)
13. Riaboff, L., et al.: Evaluation of pre-processing methods for the prediction of cattle behaviour from accelerometer data. Comput. Electron. Agric. **165**, 104961 (2019)
14. Vázquez Diosdado, J.A., et al.: Classification of behaviour in housed dairy cows using an accelerometer-based activity monitoring system. Anim. Biotelemetry **3**, 15 (2015). https://doi.org/10.1186/s40317-015-0045-8
15. Gutierrez-Galan, D., et al.: Embedded neural network for real-time animal behavior classification. Neurocomputing **272**, 17–26 (2018)
16. Hounslow, J.L.L., et al.: Assessing the effects of sampling frequency on behavioural classification of accelerometer data. J. Exp. Mar. Biol. Ecol. **512**, 22–30 (2019)

17. Navon, S., Mizrach, A., Hetzroni, A., Ungar, E.D.: Automatic recognition of jaw movements in free-ranging cattle, goats and sheep, using acoustic monitoring. Biosys. Eng. **114**, 474–483 (2013)
18. Kamminga, J.W., Le, D.V., Meijers, J.P., Bisby, H., Meratnia, N., Havinga, P.J.M.: Robust sensor-orientation-independent feature selection for animal activity recognition on collar tags. Proc. ACM Interact. Mobile Wearab. Ubiquit. Technol. **2**, 1–27 (2018)
19. Marais, J., et al.: Automatic classification of sheep behaviour using 3-axis accelerometer data (2014)
20. Giovanetti, V., et al.: Automatic classification system for grazing, ruminating and resting behaviour of dairy sheep using a tri-axial accelerometer. Livest. Sci. **196**, 42–48 (2017)
21. Nadimi, E.S., Jørgensen, R.N., Blanes-Vidal, V., Christensen, S.: Monitoring and classifying animal behavior using ZigBee-based mobile ad hoc wireless sensor networks and artificial neural networks. Comput. Electron. Agric. **82**, 44–54 (2012)
22. Kamminga, J.W., Bisby, H.C., Le, D.V., Meratnia, N., Havinga, P.J.M.: Generic online animal activity recognition on collar tags. In: Proceedings of the 2017 ACM International Joint Conference on Pervasive and Ubiquitous Computing and Proceedings of the 2017 ACM International Symposium on Wearable Computers on - UbiComp 2017, pp. 597–606. ACM, New York (2017)
23. Kleanthous, N., et al.: Machine learning techniques for classification of livestock behavior. In: Cheng, L., Leung, A.C.S., Ozawa, S. (eds.) ICONIP 2018. LNCS, vol. 11304, pp. 304–315. Springer, Cham (2018). https://doi.org/10.1007/978-3-030-04212-7_26
24. Kleanthous, N., Hussain, A., Mason, A., Sneddon, J.: Data science approaches for the analysis of animal behaviours. In: Huang, D.-S., Huang, Z.-K., Hussain, A. (eds.) ICIC 2019. LNCS (LNAI), vol. 11645, pp. 411–422. Springer, Cham (2019). https://doi.org/10.1007/978-3-030-26766-7_38
25. Mansbridge, N., et al.: Feature selection and comparison of machine learning algorithms in classification of grazing and rumination behaviour in sheep. Sensors (Switzerland) **18**, 1–16 (2018)
26. Barwick, J., Lamb, D.W., Dobos, R., Welch, M., Trotter, M.: Categorising sheep activity using a tri-axial accelerometer. Comput. Electron. Agric. **145**, 289–297 (2018)
27. Walton, E., et al.: Evaluation of sampling frequency, window size and sensor position for classification of sheep behaviour. R. Soc. Open Sci. **5**, 171442 (2018)
28. Alvarenga, F.A.P., Borges, I., Palkovič, L., Rodina, J., Oddy, V.H., Dobos, R.C.: Using a three-axis accelerometer to identify and classify sheep behaviour at pasture. Appl. Anim. Behav. Sci. **181**, 91–99 (2016)
29. le Roux, S.P., Wolhuter, R., Niesler, T.: Energy-aware feature and model selection for onboard behavior classification in low-power animal borne sensor applications. IEEE Sens. J. **19**, 2722–2734 (2019)
30. Le Roux, S., Wolhuter, R., Niesler, T.: An overview of automatic behaviour classification for animal-borne sensor applications in South Africa (2017)
31. Guo, L., Welch, M., Dobos, R., Kwan, P., Wang, W.: Comparison of grazing behaviour of sheep on pasture with different sward surface heights using an inertial measurement unit sensor. Comput. Electron. Agric. **150**, 394–401 (2018)
32. Decandia, M., et al.: The effect of different time epoch settings on the classification of sheep behaviour using tri-axial accelerometry. Comput. Electron. Agric. **154**, 112–119 (2018)
33. Vázquez-Diosdado, J.A., Paul, V., Ellis, K.A., Coates, D., Loomba, R., Kaler, J.: A combined offline and online algorithm for real-time and long-term classification of sheep behaviour: Novel approach for precision livestock farming. Sensors (Switzerland) **19**, 3201 (2019)
34. Mbientlab Inc.: MetaMotionR – MbientLab. https://mbientlab.com/metamotionr/

35. ELAN - The Language Archive. https://tla.mpi.nl/tools/tla-tools/elan/
36. Gneiting, T., Ševčíková, H., Percival, D.B.: Estimators of fractal dimension: Assessing the roughness of time series and spatial data. Stat. Sci. **27**, 247–277 (2012)
37. Wold, S., Esbensen, K., Geladi, P.: Principal component analysis. Chemometr. Intell. Lab. Syst. **2**, 37–52 (1987)
38. Breiman, L.: Random forests. Mach. Learn. **45**, 5–32 (2001)
39. Breiman, L.: Out-of-bag estimation, Technical report, pp. 1–13 (1996)
40. The Douglas Bomford Trust. https://www.dbt.org.uk/

Multi-core Twin Support Vector Machines Based on Binary PSO Optimization

Huajuan Huang and Xiuxi Wei(✉)

College of Artificial Intelligence, Guangxi University for Nationalities, Nanning 530006, China
weixiuxi@163.com

Abstract. How to select the suitable parameters and kernel model is a very important problem for Twin Support vector Machines (TWSVM). In order to solve this problem, one solving algorithm called binary PSO for optimizing the parameters of multi-core Twin Support Vector Machines (BPSO-MTWSVM) is proposed in this paper. Firstly, introducing multiple kernel functions, the twin support vector machines based on multi-core is constructed. This strategy is a good way to solve the kernel model selection. However, it has added three adjustable parameters. In order to solve the parameters selection problem which contain TWSVM parameters and multi-core model parameters, binary PSO (BPSO) is introduced. BPSO is an optimization algorithm who has strong robustness and good global searching ability. Finally, compared with the classical TWSVM the experimental results show that BPSO-MTWSVM has higher classification accuracy.

Keywords: Multi-core · Binary PSO · Twin support vector machines · Parameter optimization

1 Introduction

Support Vector Machines (SVM) is known as a new generation learning system based on statistical learning theory [1]. Because of its profound mathematical theory, SVM has played excellent performance on many real-world predictive data mining applications such as text categorization, medical and biological information analysis [2–4], etc.

One of the main challenges for the traditional SVM is the high computational complexity. The training cost of $O(n^3)$, where n is the total size of the training data, is too expensive. In order to improve the computational speed, Jayadeva et al. [5] proposed a new machine learning method called Twin Support Vector Machines (TWSVM) for the binary classification in the spirit of proximal SVM [6, 7] in 2007. TWSVM would generate two non-parallel planes, such that each plane is closer to one of the two classes and is as far as possible from the other. In TWSVM, a pair of smaller sized quadratic programming problems (QPP) are solved, whereas SVM solves a single QPP problem. Furthermore, in SVM, the QPP problem has all data points in the constraints, but in TWSVM they are distributed in the sense that patterns of class −1 give the constraints of the QP used to determine the hyperplane for class 1, and vice-

D.-S. Huang and P. Premaratne (Eds.): ICIC 2020, LNAI 12465, pp. 420–431, 2020.
https://doi.org/10.1007/978-3-030-60796-8_36

versa. This strategy of solving two smaller sized QP problems, rather than one larger QP problem, makes the computational speed of TWSVM approximately 4 times faster than the traditional SVM. Because of its excellent performance, TWSVM has been applied to many areas such as speaker recognition [8], medical detection [9–11], etc. At present, many improved TWSVM algorithms have been proposed. For example, in 2010, Kumar et al. [12] brought the prior knowledge into TWSVM and least square TWSVM and then got two improved algorithms. Experimental results showed the proposed algorithms were effective. In 2011, Yu et al. [13] adding the regularization method into the TWSVM model, proposed the TWSVM model based on regularization method. This method ensured that the proposed model was the strongly convex programming problem. In 2012, Xu et al. [14] proposed a twin multi-class classification support vector machine. Experimental results demonstrated the proposed algorithm was stable and effective.

As a new machine learning method, there are still many places needing to be perfect for TWSVM. Specially, the learning performance and generalization ability of TWSVM is very dependent on its parameters and kernel model selection. If the choice is reasonable, it will be very difficult to approach superiorly. However, the current research on this aspect is very little. At present, the kernel model selection adopts the random or experimental method. These methods are blindness and time consuming. For the parameters selection, the grid search method is commonly used. However, the search time of this method is too long, especially in dealing with the large dataset. In order to solve this problem, one solving algorithm called binary PSO for optimizing the parameters of multi-core Twin Support Vector Machines (BPSO-MTWSVM) is proposed in this paper. Firstly, in view of the blindness of the kernel model selection for TWSVM, one kernel function with good generalization ability and the other kernel function with good learning ability is combined, formed a mixed kernel model with the more excellent performance. Secondly, because of the limitation of the traditional selection method for TWSVM, we use binary PSO algorithm which has fast global searching ability to select the TWSVM parameters and the mixed kernel parameters, so that we would obtain the optimal parameters combination. Finally, the experimental results show the effectiveness and stability of the proposed method.

The paper is organized as follows: In Sect. 2, we briefly introduce the basic theory of TWSVM and the analysis of its parameters. In Sect. 3, BPSO-MTWSVM algorithm is introduced and analyzed in detail. Computational comparisons on UCI datasets are done in Sect. 4 and Sect. 5 gives concluding remarks.

2 Background

2.1 Twin Support Vector Machines

Consider a binary classification problem of classifying m_1 data points belonging to class +1 and m_2 data points belonging to class −1. Then let matrix A in $R^{m_1 \times n}$ represent the data points of class +1 while matrix B in $R^{m_2 \times n}$ represent the data points of class −1. Two nonparallel hyper-planes of the linear TWSVM can be expressed as follows.

$$x^T w_1 + b_1 = 0 \text{ and } x^T w_2 + b_2 = 0 \tag{1}$$

The target of TWSVM is to generate the above two nonparallel hyper-planes in the n-dimensional real space R^n, such that each plane is closer to one of the two classes and is as far as possible from the other. A new sample point is assigned to class +1 or −1 depending upon its proximity to the two nonparallel hyper-planes. The linear classifiers are obtained by solving the following optimization problems.

$$\begin{aligned} \min_{w^{(1)},b^{(1)},\xi^{(2)}} \ & \tfrac{1}{2}\left\|Aw^{(1)} + e_1 b^{(1)}\right\|^2 + c_1 e_2^T \xi^{(2)} \\ s.t. \ & -(Bw^{(1)} + e_2 b^{(1)}) \geq e_2 - \xi^{(2)}, \\ & \xi^{(2)} \geq 0. \end{aligned} \tag{2}$$

$$\begin{aligned} \min_{w^{(2)},b^{(2)},\xi^{(1)}} \ & \tfrac{1}{2}\left\|Bw^{(2)} + e_2 b^{(2)}\right\|^2 + c_2 e_1^T \xi^{(1)} \\ s.t. \ & (Aw^{(2)} + e_1 b^{(2)}) \geq e_1 - \xi^{(1)}, \\ & \xi^{(1)} \geq 0. \end{aligned} \tag{3}$$

where c_1 and c_2 are penalty parameters, $\xi^{(1)}$ and $\xi^{(2)}$ are slack vectors, $A = [x_1^{(1)}, x_2^{(1)}, \ldots, x_{m1}^{(1)}]^T$, $B = [x_1^{(1)}, x_2^{(1)}, \ldots, x_{m1}^{(1)}]^T$, e_1 and e_2 are the vectors of ones of appropriate dimensions. $x_j^{(i)}$ represents the jth sample of the ith class. Introducing the Lagrange variables α and β, the dual problems of (2) and (3) can be expressed as follows:

$$\begin{aligned} \max_{\alpha} \ & e_2^T \alpha - \tfrac{1}{2}\alpha^T G(H^T H)^{-1} G^T \alpha \\ s.t. \ & 0 \leq \alpha \leq c_1 e_2 \end{aligned} \tag{4}$$

$$\begin{aligned} \max_{\beta} \ & e_1^T \beta - \tfrac{1}{2}\beta^T H(G^T G)^{-1} H^T \beta \\ s.t. \ & 0 \leq \beta \leq c_2 e_1 \end{aligned} \tag{5}$$

where, $H = [A \quad e_1]$, $G = [B \quad e_2]$. Defining $u_i = [(w^{(i)})^T \quad b^{(i)}]^T$, $i = 1, 2$, the solution becomes:

$$u_1 = -(H^T H)^{-1} G^T \alpha, \ u_2 = (G^T G)^{-1} H^T \beta \tag{6}$$

To judge a new sample belonging to which class, we should find this sample is closer to which class. We can calculate the distance of a sample from a class by (7).

$$f(x) = \arg \min_i (d_i(x)) \tag{7}$$

where,

$$d_i(x) = \frac{\left|x^T w^{(i)} + b^{(i)}\right|}{\|w^{(i)}\|_2}, \quad i = 1,2. \tag{8}$$

For the nonlinear case, the two nonparallel hyper-planes of TWSVM based on kernel can be expressed as follows:

$$K(x^T, C^T)w^{(1)} + b^{(1)} = 0, \; K(x^T, C^T)w^{(2)} + b^{(2)} = 0 \tag{9}$$

where, $C = [A^T, B^T]^T$. So the optimization problem of nonlinear TWSVM can be expressed as follows.

$$\begin{aligned} &\min_{w^{(1)}, b^{(1)}, \xi^{(2)}} \tfrac{1}{2}\left\|K(A, C^T)w^{(1)} + e_1 b^{(1)}\right\|^2 + c_1 e_2^T \xi^{(2)} \\ &s.t. \quad -(K(B, C^T)w^{(1)} + e_2 b^{(1)}) \geq e_2 - \xi^{(2)}, \\ &\xi^{(2)} \geq 0. \end{aligned} \tag{10}$$

$$\begin{aligned} &\min_{w^{(1)}, b^{(1)}, \xi^{(2)}} \tfrac{1}{2}\left\|K(B, C^T)w^{(2)} + e_2 b^{(1)}\right\|^2 + c_2 e_1^T \xi^{(1)} \\ &s.t. \quad (K(A, C^T)w^{(2)} + e_1 b^{(2)}) \geq e_1 - \xi^{(1)}, \\ &\xi^{(1)} \geq 0. \end{aligned} \tag{11}$$

According to the Lagrange theorem, the dual problems of (10) and (11) can be expressed by (12) and (13).

$$\begin{aligned} &\max_{\alpha} \; e^T \alpha - \tfrac{1}{2}\alpha^T R(S^T S)^{-1} R^T \alpha \\ &s.t. \quad 0 \leq \alpha \leq c_1 e_2 \end{aligned} \tag{12}$$

$$\begin{aligned} &\max_{\alpha} \; e^T \beta - \tfrac{1}{2}\beta^T S(R^T R)^{-1} S^T \beta \\ &s.t. \quad 0 \leq \beta \leq c_2 e_1 \end{aligned} \tag{13}$$

where, $S = [K(A, C) \quad e_1]$, $R = [K(B, C) \quad e_2]$. Defining $v_i = [(w^{(i)})^T \quad b^{(i)}]^T$, $i = 1,2$, the solution becomes:

$$v_1 = -(S^T S)^{-1} R^T \alpha, \; v_2 = (R^T R)^{-1} S^T \beta \tag{14}$$

2.2 Analysis the Penalty Parameters of TWSVM

The role of penalty parameters c_1 and c_2 is to adjust the ratio between the confidence range with the experience risk in the defining feature, so that the generalization ability of TWSVM can achieve the best state. The values of c_1 and c_2 smaller express the punishment on empirical error smaller. Do it this way, the complexity of TWSVM is smaller, but its fault tolerant ability is worse. The values of c_1 and c_2 are greater, the data fitting degree is higher, but its generalization capacity will be reduced. From the above analysis, we can know that the parameters selection are very important for TWSVM.

2.3 Construction Multiple Kernel Functions

The mercer theorem is the theory of the traditional kernel function construction, which is showed as follows:

Theorem 1: When $g(x) \in L_2(R^N)$ and $k(x,x') \in L_2(R^N \times R^N)$, if

$$\iint k(x,x')g(x)g(x')dxdy \geq 0 \tag{15}$$

There is $k(x,x') = (\Phi(x) \cdot \Phi(x'))$.

According to the Theorem 1, the properties of the kernel function can be easily proved as follows:

Theorem 2: Let k_1 and k_2 is the kernel function defined in the $X \times X$. At the same time, $a \in R^+, f$ is a real function on X, $\phi : X \to R^N$, k_3 is a kernel function defined in the $R^N \times R^N$, B is a semi positive definite symmetric matrix of $n \times n$. The following functions are kernel functions.

(1) $k(x,z) = k_1(x,z) + k_2(x,z)$
(2) $k(x,z) = ak_1(x,z)$

After selecting a kernel function, the learning model also has been identified. As we know, the performance of a learning model is decided by the learning ability and the generalization ability. Generally, the kernel functions can be classified into two types: global kernel function and local kernel function. However, for the global kernel function, its generalization ability is strong when the learning ability is weak. On the contrary, for the local kernel function, it has strong learning ability but its generalization ability is weak. In view of the respective characteristic of the global and local kernel function, if the two type of kernel functions are mixed into a hybrid kernel function, which will be able to achieve the good classification performance. Based on the above ideas, we will construct a multiple kernel functions as follows.

As we know, the Sigmoid function is the one of the commonly used global kernel function in TWSVM.

$$k(x,x_i) = \tanh(p_1(x \bullet x_i) + p_2) \tag{16}$$

For the Sigmoid function, the sketch map of the testing point 0.1 when $p_1 = 100$, p_2 taking different values is shown as Fig. 1. From Fig. 1 we can see that the Sigmoid function has good generalization ability in the appropriate parameters because of having a role for the near test point and far across the data points. But the learning ability in the test point is not obvious, which means it's learning ability is not only strong. After several experiments, we find that it is appropriate when $p_1 = 100, p_2 \geq 3$.

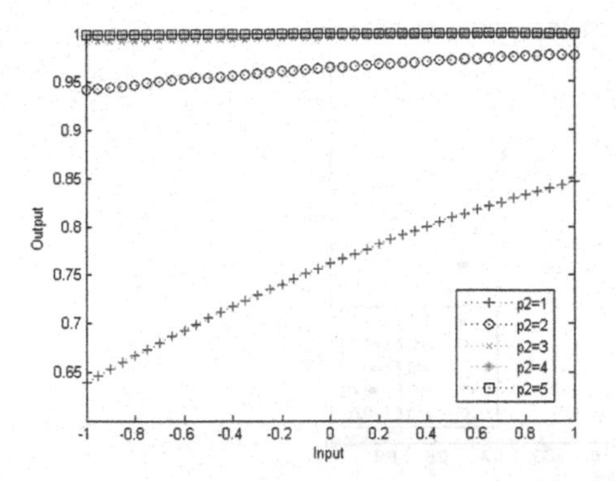

Fig. 1. The curve of sigmoid kernel function in function in test point 0.1

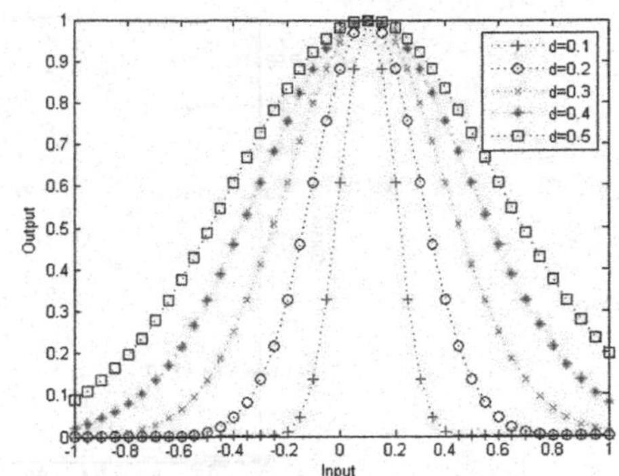

Fig. 2. The curve of Gaussian kernel test point 0.1

Gauss Radial Basis Function is the one of the commonly used local kernel function in TWSVM.

$$K(x, x_i) = \exp(-\frac{\|x - x_i\|^2}{2\sigma^2}) \tag{17}$$

For Gauss Radial Basis Function, the sketch map of the testing point 0.1 when σ taking different values is shown as Fig. 2 (d replaced by σ in Fig. 2). From Fig. 2 we can see that the Gauss Radial Basis Function has good learning ability because of only having a role for the near test point, but its generalization ability is weak. Therefore, if the Sigmoid function and the Gauss Radial Basis function is mixed to generate a new multiple kernel functions, which can have better learning ability and better generalization ability. From Fig. 2 we can also see that the value of σ smaller, the performance of Gauss Radial Basis Function better. Generally, the range of the σ value is 0.1 ~ 1.

Based on the above idea, a new mixed function is proposed:

$$k(x, x_i) = a\tanh(p_1(xgx_i) + p_2) + b\exp(-\frac{\|x - x_i\|^2}{2\sigma^2}) \tag{18}$$

Theorem 3: The Eq. (13) is a kernel function.

Therefore, $k(x, x_i) = k_5(x, x_i) = a\tanh(p_1(x, x_i) + p_2) + b\exp(-\frac{\|x - x_i\|^2}{2\sigma^2})$ is a kernel function.

Where, a and b is the proportion coefficient of the two kernel function in multiple kernel functions. Generally, let $0 \leq a, b \leq 1$, $a + b = 1$.

For the multiple kernel functions, the sketch map of the testing point 0.1 when a, b taking different values is shown as Fig. 3.

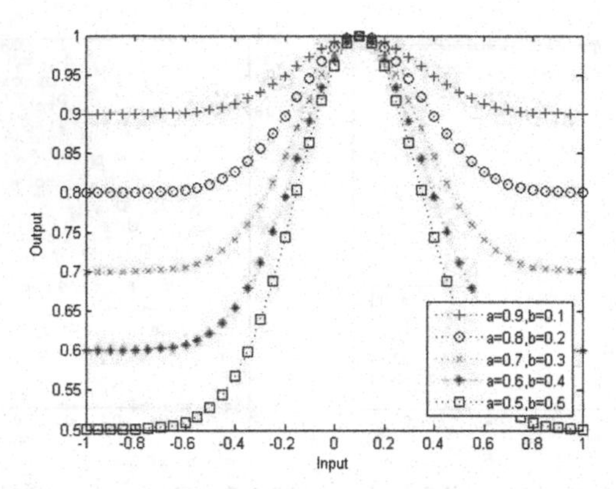

Fig. 3. The curve of multiple kernel functions in test point 0.1

2.4 Analysis the Parameters of Multiple Kernel Functions

After introducing the mixed kernel, TWSVM have added three adjustable parameters which contain the weight of multiple kernel functions a, the Gauss kernel function parameter σ and the polynomial kernel parameter q. According to the properties of kernel function, the value of a is between $0 \sim 1$. The multiple kernel functions is closer to the polynomial kernel function when $a \to 0$. On the contrary, The multiple kernel functions is closer to the Gauss kernel function when $a \to 1$. Therefore, it is very important to select a. If the choice is not appropriate, it may make the performance of multiple kernel functions below the single one, thus losing the advantage of multiple kernel functions. σ and q is the kernel parameters, which also take important role in the performance of multiple kernel functions. At present, there are two selection methods of kernel parameters. One is the random method and the other is cross validation method. The random method is that the kernel parameters are randomly given and then the value of kernel parameters are constantly adjusted until getting a satisfactory precision. In view of lack of adequate theoretical basis, the random method has certain blindness. The cross validation method tests a range of kernel parameters individually to find the optimal value using traversal approach. Generally, this method can find the best values, but its time complexity is relatively high.

After the above analysis, in this paper, binary PSO algorithm which has fast global searching ability is used to select the TWSVM parameters and the mixed kernel parameters.

2.5 Binary PSO Algorithm

Particle swarm optimization (PSO) [15] is in principle a much simpler algorithm. It operates on the principle that each solution can be represented as a particle denoted by x_i in a swarm. A population of particles is randomly generated initially. Then a swarm of particles moves through the problem space, with the moving velocity of each particle represented by a velocity vector v_i. Each particle keeps track of its own best position, which is associated with the best fitness it has achieved so far in a vector p_i.

Furthermore, the best position among all the particles obtained so far in the population is kept track of as p_g. In addition to this global version, another local version of PSO keeps track of the best position among all the topological neighbors of a particle.

At each time step t, by using the individual best position, $p_i(t)$, and global best position, $p_g(t)$, a new velocity for particle i is updated by

$$v_i(t+1) = v_i(t) + c_1\phi_1(p_i(t) - x_i(t)) + c_2\phi_2(p_g(t) - x_i(t)), \tag{19}$$

where c_1 and c_2 are positive constants, ϕ_1 and ϕ_2 are uniformly distributed random numbers in [0, 1]. Based on the updated velocities, each particle changes its position according to the following:

$$x_i(t+1) = x_i(t) + v_i(t+1) \tag{20}$$

Based on (19) and (20), the population of particles tends to cluster together with each particle moving in a random direction. PSO is considered as the simplest swarm intelligence optimization algorithm [17]. In order to adapt to the application of PSO in discrete problems, Kennedy [18] proposed a binary PSO (BPSO) algorithm in 1997. In BPSO, each particle is encoded as a binary vector.

According to the above procedures, we conduct our BPSO-based model selection for multi-core TWSVM.

2.6 The Algorithm Steps of BPSO-MTWSVM

The algorithm steps of BPSO-MTWSVM is as follows:

Step1: Select the training dataset and the testing dataset.
Step2: Preprocessing the dataset.
Step3: Constructe the mixed kernel function.
Step4: Select the optimal parameters using BPSO algorithm.
Step5: Train the multi-core TWSVM using the optimal parameters.
Step6: Predict the testing dataset.
Step7: Output the classification accuracy.

3 Experimental Results and Analysis

In order to verify the efficiency of BPSO-MTWSVM, meanwhile, in order to compare the performance of three algorithms, that is, SVM, TWSVM and BPSO-MTWSVM, we conduct experiments on seven benchmark datasets from the UCI machine learning repository. The environments of all experiments are in Intel (R) Core (TM) 2Duo CUP E4500, 2G memory and MATLAB 7.11.0. The parameter values of BPSO are as follows: $D = 5$, $iter_{\max} = 30$, $N_0 = 10$, $s_{\max} = 5$, $s_{\min} = 1$, $n = 3$, $\sigma_{init} = [1, 0.1, 1, 1, 0.5]$, $\sigma_{final} = [0.1, 0.1, 0.1, 0.1, 0.1]$. In BPSO algorithm, the accuracy in the sense of CV is used for the fitness of BPSO. Therefore, the fitness value closer to 100, the obtained parameters closer to the optimal value. The experiment results of BPSO-MTWSVM are shown as Table 1. Furthermore, the comparisons of BPSO-MTWSVM and other algorithms are shown as Table 2. In order to more objectively test the performance of each algorithm, we test each dataset 20 times independently. And the values of Table 1 and Table 2 are the average values. Figure 4 and Fig. 5 are the fitness curves of BPSO searching the optimal parameters for dealing with the Australian dataset and Breast-cancer dataset respectively. Figure 6 represents the classification results on seven UCI dataset by three algorithms.

Table 1. The classification results of BPSO-MTWSVM

Dataset	The optimal parameter values					Training accuracy (%)	Testing accuracy (%)
	c_1	c_2	a				
Australian	1.4	5.2	0.8	12.0	36.4	98.71.2	87.74.1
Breast-cancer	2.9	15.7	0.5	85.8	12.0	83.25.3	69.14.4
Heart	45.9	54.3	0.6	92.8	52.0	91.56.5	84.67.3
Pima	54.7	1.5	0.8	4.2	12.8	94.74.2	82.02.4
Votes	83.1	2.5	0.4	17.1	3.7	99.30.2	96.62.6
Sonar	6.2	8.7	0.8	2.2	60.1	94.61.3	90.14.9
CMC	62.5	69.2	0.9	36.7	96.0	88.84.5	77.27.0

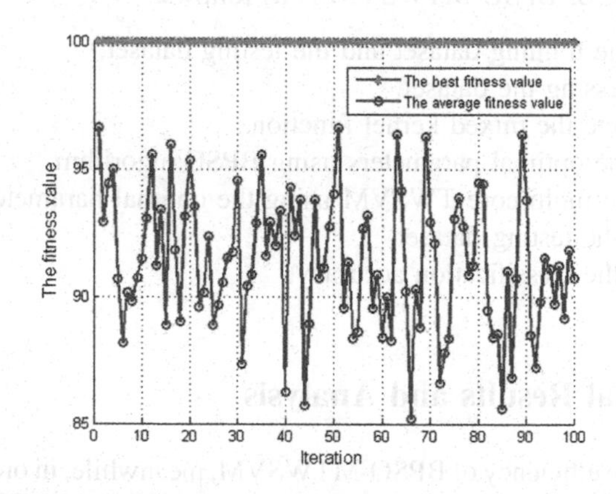

Fig. 4. The fitness curves of BPSO searching the optimal parameters for dealing with the Australian dataset

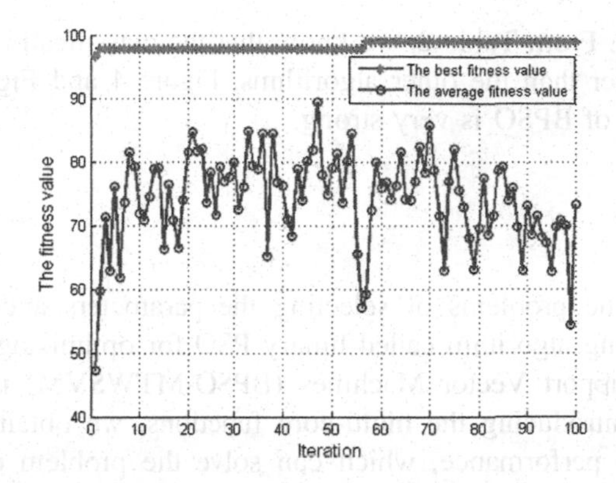

Fig. 5. The fitness curves of BPSO searching the optimal parameters for dealing with the Breast-cancer dataset

Table 2. Testing accuracy comparisons of BPSO-MTWSVM and other algorithms

Dataset	BPSO-MTWSVM	TWSVM	SVM
Australian	**87.7 ± 4.1**	84.8 ± 2.1	85.9 ± 2.2
Breast-cancer	**69.1 ± 4.4**	64.4 ± 3.9	65.4 ± 4.5
Heart	**84.6 ± 7.3**	81.9 ± 4.3	82.2 ± 6.7
Pima	**82.0 ± 2.4**	73.8 ± 6.0	76.6 ± 2.4
Votes	**96.6 ± 2.6**	95.0 ± 4.2	95.9 ± 2.2
Sonar	**90.1 ± 4.9**	89.5 ± 3.4	88.1 ± 9.7
CMC	**77.2 ± 7.0**	73.6 ± 9.9	68.0 ± 2.2

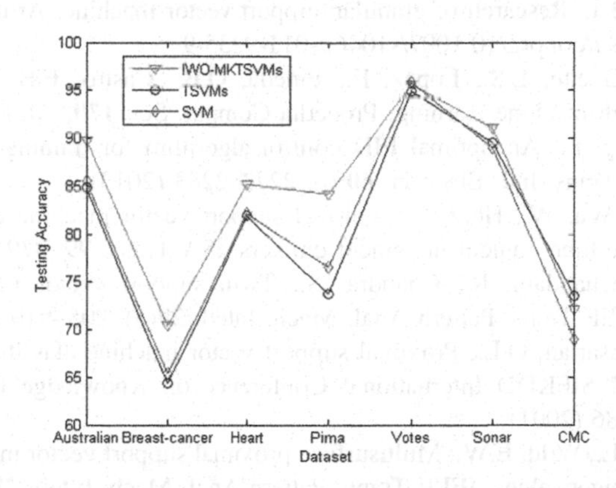

Fig. 6. The classification results on seven UCI dataset by three algorithms

From Table 1, we can see that the training accuracy and testing accuracy of BPSO-MTWSVM is relatively high. Meanwhile, Table 1 lists the optimal parameters using BPSO for searching. Table 2 is the testing accuracy comparisons of BPSO-MTWSVM,

TWSVM and SVM. From Table 2, we know that the classification results of BPSO-MTWSVM are better than the other algorithms. Figure 4 and Fig. 5 shows that the optimization ability of BPSO is very strong.

4 Conclusions

In order to solve the problems of selecting the parameters and kernel model for TWSVM, one solving algorithm called binary PSO for optimizing the parameters of multi-core Twin Support Vector Machines (BPSO-MTWSVM) is proposed in this paper. Firstly, by introducing the multi-core functions, we obtain a kind of kernel function with good performance, which can solve the problem of selecting kernel function in TWSVM. Secondly, in view of the good optimization ability of binary PSO (BPSO) algorithm, it is used to optimize the parameters containing the TWSVM parameters and the multi-core parameters. Finally, the experimental results show the effectiveness and stability of the proposed method. How to further improve the performance of BPSO is the next work.

Acknowledgement. This work is supported by the National Natural Science Foundation of China (61662005). Guangxi Natural Science Foundation (2018GXNSFAA294068); Basic Ability Improvement Project for Young and Middle-aged Teachers in Colleges and Universities in Guangxi (2019KY0195); Research Project of Guangxi University for Nationalities (2019KJYB006).

References

1. Ding, S.F., Qi, B.J.: Research of granular support vector machine. Artif. Intell. Rev. **38**(1), 1–7 (2012). https://doi.org/10.1007/s10462-011-9235-9
2. Borrero, L.A., Guette, L.S., Lopez, E., Pineda, O.B., Castro, E.B.: Predicting toxicity properties through machine learning. Procedia Comput. Sci. **170**, 1011–1016 (2020)
3. Jing, X.J., Cheng, L.: An optimal PID control algorithm for training feedforward neural networks. IEEE Trans. Ind. Electron. **60**(6), 2273–2283 (2013)
4. Liu, X., Jin, J., Wu, W., Herz, F.: A novel support vector machine ensemble model for estimation of free lime content in cement clinkers. ISA Trans. **99**, 479–487 (2020)
5. Jayadeva, Khemchandani, R., Chandra, S.: Twin support vector machines for pattern classification. IEEE Trans. Pattern Anal. Mach. Intell. **29**(5), 905–910 (2007)
6. Fung, G., Mangasarian, O.L.: Proximal support vector machine classifiers. In: Proceedings of the 7th ACM SIFKDD International Conference on Knowledge Discovery and Data Mining, pp. 77–86 (2001)
7. Mangasarian, O.L., Wild, E.W.: Multisurface proximal support vector machine classification via generalized eigenvalues. IEEE Trans. Pattern Anal. Mach. Intell. **28**(1), 69–74 (2006)
8. Tang, X., Ma, Z., Hu, Q., Tang, W.: A real-time arrhythmia heartbeats classification algorithm using parallel delta modulations and rotated linear-kernel support vector machines. IEEE Trans. Bio-Med. Eng. **67**(4), 978–986 (2020)
9. Sheykh Mohammadi, F., Amiri, A.: TS-WRSVM: twin structural weighted relaxed support vector machine. Connection Sci. **31**(3), 215–243 (2019)

10. Huang, H.J., Ding, S.F., Shi, Z.Z.: Primal least squares twin support vector regression. J. Zhejiang Univ.-Sci. C-Comput. Electron. **14**(9), 722–732 (2013). https://doi.org/10.1631/jzus.CIIP1301
11. Xu, Y.T., Wang, L.S.: A weighted twin support vector regression. Knowl.-Based Syst. **33**, 92–101 (2012)
12. Kumar, M.A., Khemchandani, R., Gopal, M., Chandra, S.: Knowledge based least squares twin support vector machines. Inf. Sci. **180**(23), 4606–4618 (2010)
13. Ye, Q.L., Zhao, C.X., Chen, X.B.: A feature selection method for TWSVM via a regularization technique. J. Comput. Res. Dev. **48**(6), 1029–1037 (2011)
14. Xu, Y.T., Guo, R., Wang, L.S.: A twin multi-class classification support vector machine. Cogn. Comput. **5**(4), 580–588 (2012). https://doi.org/10.1007/s12559-012-9179-7
15. Ghosh, S.K., Biswas, B., Ghosh, A.: A novel approach of retinal image enhancement using PSO system and measure of fuzziness. Procedia Comput. Sci. **167**, 1300–1311 (2020)
16. Zheng, X., Gao, Y., Jing, W., Wang, Y.: Multidisciplinary integrated design of long-range ballistic missile using PSO algorithm. J. Syst. Eng. Electron. **31**(02), 335–349 (2020)
17. Wan, P., Zou, H., Wang, K., Zhao, Z.: Research on hot deformation behavior of Zr-4 alloy based on PSO-BP artificial neural network. J. Alloy. Compd. **826** (2020)
18. Kennedy, J., Eberhart, R.C.: A discrete binary version of the particle swarm algorithm. In: 1997 Proceedings of the World Multiconference on Systemics, Cybernetics and Informatics, Piscataway, NJ, pp. 4104–4109 (1997)

Multi-stage Hierarchical Clustering Method Based on Hypergraph

Yue Xi and Yonggang Lu(✉)

School of Information Science and Engineering, Lanzhou University, Lanzhou 730000, Gansu, China

ylu@lzu.edu.cn

Abstract. Clustering analysis is a data analysis technique, it groups a set of data points into multiple clusters with similar data points. However, clustering of high dimensional data is still a difficult task. In order to facilitate this task, people usually use hypergraphs to represent the complex relationships between high dimensional data. In this paper, the hypergraph is used to improve the representation of the complex high dimensional data, and a multi-stage hierarchical clustering method based on hypergraph partition and Chameleon algorithm is proposed. The proposed method constructs a hypergraph in the shared-nearest-neighbor (SNN) graph from the dataset and then employs a hypergraph partitioning method hMETIS to obtain a series of subgraphs, finally those subgraphs are merged to get the final clusters. Experiments show that the proposed method is better than Chameleon algorithm and the other four clustering methods when applied on four UCI datasets.

Keywords: Clustering · Hypergraph · Chameleon algorithm · UCI datasets

1 Introduction

Clustering analysis is one of the most frequently used data analysis methods in data mining. According to the attribute information of data or the relationships between data points, it groups data into multiple clusters, so that the data points are similar in the same cluster and are different between different clusters [1]. There are two kinds of traditional clustering methods: partitioning based clustering and hierarchical clustering [2]. For a given *k*, a partitioning based clustering algorithm first gives the initial *k* partitions and then changes the partitioning of data points through repeated iterations, so that the new partitioning scheme is better than the previous partitioning scheme, moves data points between different partitions until meeting certain criteria. A hierarchical clustering method constructs a hierarchical tree of clusters according to the similarity between data points. According to the way of a hierarchical tree formation, the hierarchical clustering method can be classified as being either Agglomerative methods or Divisive methods. The agglomerative hierarchical clustering methods treat each point as a single cluster and iteratively merge small clusters to larger clusters, until all data points are in one cluster or some conditions are satisfied. On the contrary, the divisive hierarchical clustering methods set all data points in one cluster. Then they divide the initial cluster into several smaller sub-clusters, and recursively partition those

D.-S. Huang and P. Premaratne (Eds.): ICIC 2020, LNAI 12465, pp. 432–443, 2020.
https://doi.org/10.1007/978-3-030-60796-8_37

sub-clusters into smaller ones until each cluster contains only one point or data points within a cluster are similar enough. The commonly used partitioning methods are *k*-means [3] and *k*-medoids [2], and the commonly used hierarchical clustering methods are DIANA algorithm and AGNES algorithm [2]. In addition, Chameleon [4], a dynamic multi-stage hierarchical clustering method, is also extensively used in the clustering.

In order to facilitate clustering, people usually use graphs to represent the relationships between data points. Spectral clustering is a kind of clustering method defined on the graph. It constructs the eigenvector space by calculating the first *k* eigenvalues and eigenvectors of the similarity matrix or Laplace matrix of the data, and clusters the eigenvectors in the eigenvector space by using *k*-means to get the clustering results [5]. Chameleon [4] is a multi-stage hierarchical clustering algorithm. In the first stage, a *k*-Nearest-Neighbor (*k*NN) graph constructed from data points is divided into a series of subgraphs by the graph partition algorithm. Each subgraph represents an initial sub-cluster. In the second stage, Chameleon method employs an agglomerative hierarchical clustering algorithm to merge sub-clusters again and again until the real cluster is found. And a dynamic modeling framework is used to determine the similarity between sub-clusters. Zhao and Karypis employed in document clustering (high dimensional data), and the method is similar to Chameleon algorithm [6]. Cao proposed an optimized Chameleon algorithm based on local features and grid structure [7]. Dong described a new improved method by introducing the recursive dichotomy, flood fill, the quotient of cluster density γ, and the first jump cutoff [8]. Barton and Bruna improved the internal cluster quality measure and put forward an improved graph-based method (Chameleon 2) [9].

In addition to graphs, hypergraphs are also used for clustering. Because the graphs can only represent the data points with pairwise relationships. In real-world problems, data are usually in the high dimensional space, and the relationships between data points are quite complex, using graphs to squeeze the complex relationships to pairwise ones will lead to the loss of information. However, in a hypergraph, a hyperedge can connect more than two vertices, which can be used to represent high order relationships between data. Zhou generalized spectral clustering, which transfers the original operations based on graphs to hypergraphs [10]. HMETIS [11] is a hypergraph partitioning algorithm that can be used to partition large-scale hypergraphs. Its advantages are high quality of hypergraph partition results and high speed. Wang proposed a dense subgraph merge method based on hypergraph partition and verified the effectiveness of the method on the handwritten digital datasets [12]. Kumar introduced a hypergraph null model that used to define a modularity function, proposed a refinement over clustering by iteratively reweighting cut hyperedges [13]. Veldt proposed a framework for local clustering in hypergraphs based on minimum cuts and maximum flows [14].

Most of the existing Chameleon method and its improved methods are based on the graph, which have a good performance on low dimensional data but cannot handle high dimensional data well. This may be because the normal graphs can only represent the pairwise relationships in the data and inevitably lead to the loss of information. However, hypergraphs can be used to represent high order relationships between data. Thus, in this paper, we improve the traditional hierarchical clustering method Chameleon by introducing the hypergraph and propose a multi-stage hierarchical clustering method based on hypergraph (MHCH), which discovers the clusters in the dataset through three stages: constructing the hypergraph, partitioning the hypergraph and merging subgraphs. Experiments on four UCI datasets show the effectiveness of the proposed method.

The rest of this paper is organized as follows. In Sect. 2, the multilevel graph partitioning methods and Chameleon algorithm are introduced. In Sect. 3, our method is proposed. Section 4 provides experimental results and Sect. 5 gives the conclusion.

2 Related Work

2.1 Graph Partition Method: METIS and HMETIS

METIS [15] is a software package for partitioning graphs developed in Karypis Lab. The algorithms in METIS are based on the multilevel graph partitioning. It consists of three phases: graph coarsening phase, initial partition phase, and the multilevel refinement (or uncoarsening) phase. In the coarsening phase, a series of small consecutive graphs are obtained by collapsing vertices and edges. In the initial partitioning phase, a bisection (*k*-way partitioning) scheme of the smallest graph is derived. In the uncoarsening phase, the partitioning scheme of the smallest graph is projected to the larger graphs successively, until the original graph is projected. The multilevel graph partitioning algorithm is illustrated in Fig. 1.

Then, Karypis Lab proposed a multilevel hypergraph partitioning algorithm hMETIS [11] based on METIS. And hMETIS can be directly applied to hypergraphs. In hMETIS, a lot of consecutive hypergraphs are constructed. A partitioning scheme of

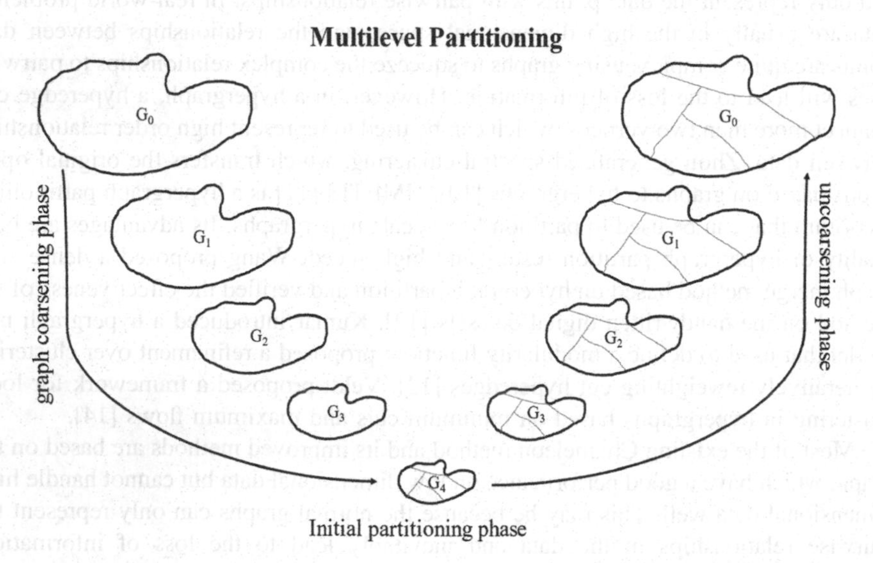

Fig. 1. The three phases of multilevel *k*-way graph partitioning.

the smallest hypergraph is derived and it is used to obtain the partitioning scheme of the original hypergraph by projecting and refining the partitioning scheme to the larger hypergraph. The algorithm can provide high-quality partition and fast operation in hypergraph partition. In view of the information characteristics of the massive data era and the characteristics of the data studied in this paper, we use this algorithm to study hypergraphs.

2.2 A Hierarchical Clustering: Chameleon

Chameleon [4] is a multi-stage hierarchical clustering algorithm. Chameleon first constructs a *k*NN graph from the dataset and then uses graph partitioning method to partition the *k*NN graph into a series of subgraphs, each subgraph can be regarded as a sub-cluster, and finally merges the subgraph. When merging sub-clusters, the similarity between a pair of sub-clusters is determined by observing the *Relative Interconnectivity* (*RI*) and *Relative Closeness* (*RC*) between a pair of sub-clusters. It selects cluster pairs with high *RC* and *RI* values to merge.

The *RI* (C_i, C_j) of two clusters C_i and C_j is the normalized absolute interconnectivity between C_i and C_j of its internal interconnectivity:

$$\mathrm{RI}(C_i, C_j) = \frac{|EC(C_i, C_j)|}{\frac{1}{2}\left(|EC(C_i)| + |EC(C_j)|\right)} \tag{1}$$

where $|EC\ (C_i, C_j)|$ is the sum of the weights of the edges that are cut off when cluster C is divided into C_i and C_j, it is used to evaluate the absolute interconnectivity between C_i and C_j Similarly, $|EC\ (C_i)|$ (or $|EC\ (C_j)|$) is the minimum weight sum of the cut edges that divide C_i (or C_j) into two roughly equal parts.

The *RC* (C_i, C_j) of two clusters C_i and C_j is the normalized absolute closeness between C_i and C_j of its internal closeness:

$$\mathrm{RC}(C_i, C_j) = \frac{|\bar{S}EC(C_i, C_j)|}{\frac{|C_i|}{|C_i| + |C_j|}|\bar{S}EC(C_i)| + \frac{|C_j|}{|C_i| + |C_j|}|\bar{S}EC(C_j)|} \tag{2}$$

where $|\bar{S}EC(C_i, C_j)|$ is the average weight of the edges that connect the C_i and C_j, Similarly, $|\bar{S}EC(C_i)|$ (or $|\bar{S}EC(C_j)|$) is the average weight of the cut edge that divides C_i (or C_j) into two roughly equal parts.

The formula of similarity function is:

$$\mathrm{S}(C_i, C_j) = \mathrm{RI}(C_i, C_j) \times \mathrm{RC}(C_i, C_j)^{\alpha} \tag{3}$$

where α is an user-specified parameter, if $\alpha > 1$, *RC* is more important, and when $\alpha < 1$,*RI* is more important.

3 Method

In this section, a multi-stage hierarchical clustering method based on hypergraph (MHCH) is proposed, which is an improved one based on Chameleon algorithm. The main idea of the proposed MHCH method is to construct a hypergraph from the shared-nearest-neighbor (SNN) graph of the dataset firstly, and then use a hypergraph partitioning method hMETIS [11] to partition the hypergraph into a series of relatively small subgraphs, finally the method of similarity in Chameleon algorithm is used to merge the subgraphs to get the final clusters. Figure 2 shows the flowchart of our algorithm. Compared with the traditional Chameleon algorithm, the hypergraph model of the dataset is employed in this paper, and hMETIS is used to partition hypergraph to get subgraphs. In Chameleon algorithm, a *k*NN graph is constructed from data, but it only uses local information between data points and inevitably leads to the loss of information. However, in the proposed method, substituting the hypergraphs for the graphs will avoid the disadvantages caused by using graphs in Chameleon algorithm.

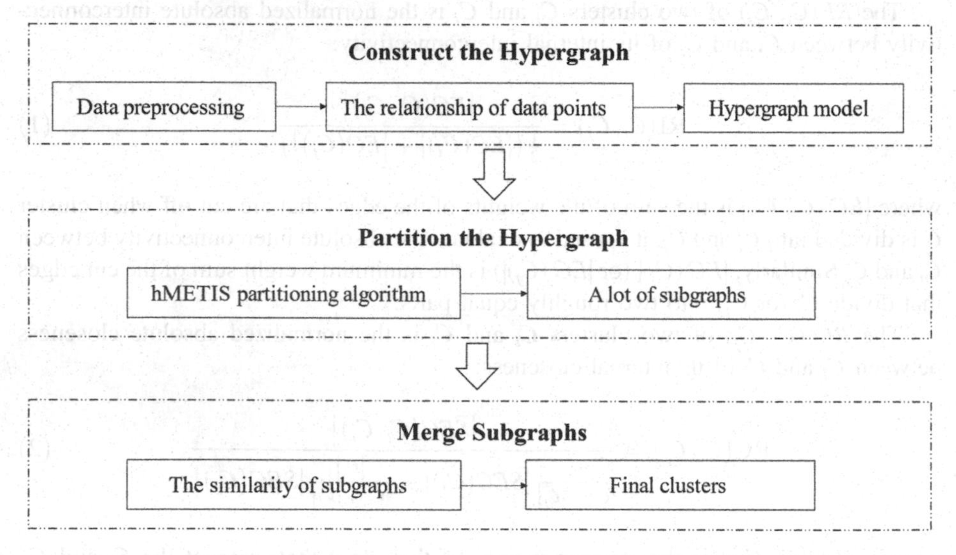

Fig. 2. Flowchart of the proposed MHCH algorithm.

3.1 Construct the Hypergraph

Data preprocessing is an indispensable part of the algorithm. The main purpose of this process is to better represent the data and improve the accuracy of the algorithm clustering. T-Distributed Stochastic Neighbor Embedding (t-SNE) [16] is used to pre-process the data, which is a method to reduce the high dimensional space to low-dimensional space. T-SNE models the distribution of the nearest neighbors of each data point in the original high dimensional space and transforms the distance between the data points in the high dimensional space into a conditional probability to express the similarity. It can catch the local structural features of a lot of high dimensional data, and also can reveal the global structure of clusters of different sizes.

A hypergraph G can be composed of a triple, $G = (V, E, w)$, V is a set of vertices, E is a set of hyperedges, and each hyperedge can be understood as an extension of an ordinary graph, which can connect more than two vertices, and w is the weight of the hyperedge. In this paper, a method of finding the maximum clique [17] is used to construct a hypergraph. This method first constructs a shared-nearest-neighbor (SNN) graph from the dataset and then finds the maximum clique in the SNN graph as a hyperedge [18].

First, finding the kNN of the data. And then the similarity between two data points is redefined according to how many kNNs are shared between the two data points. For dataset D, the number of SNN between any two data points, d_1 and d_2, is defined as:

$$\mathrm{SNN}(d_1, d_2) = |kNN(d_1) \cap kNN(d_2)| \tag{4}$$

When the number of SNN of d_1 and d_2 is greater than the threshold Sc, in the SNN graph, the vertex d_1 and the vertex d_2 are connected by an edge.

Then the hypergraph is constructed by searching for the maximum clique from the SNN graph. The maximum clique is the maximal connected subgraph of a graph. In [18], the method found maximum cliques by association mining. If the dataset is larger, the running time of the method will be very slow. Therefore, in our method, an improved Bron-Kerbosch [17] method is used to search for the maximum cliques, which runs $O(d(n-d)3^{d/3})$ times to find the maximum cliques, where n is the number of data points and d is the maximal degree of the data points in the graph.

Then, the all maximum cliques are regarded as the hyperedges of the hypergraph. The weight of a hyperedge E is defined as:

$$\mathrm{W(E)} = \frac{|E|(|E|-1)}{2} \tag{5}$$

Where $|E|$ is the number of data points in a hyperedge E.

3.2 Partition the Hypergraph

In the clustering algorithms based on the hypergraph model, the relationship between data points are mapped to a hypergraph. And in the hypergraphs, the clustering problem is considered as a hypergraph partition problem.

In the proposed method, hMETIS [11] is used to partition hypergraph into a series of subgraphs, each subgraph can be regarded as a sub-cluster. It consists of three phases: coarsening phase, initial partitioning phase, and uncoarsening and refinement phase.

Coarsening Phase. During the coarsening phase, a series of smaller successive hypergraphs is generated from the original hypergraph. Each hypergraph is constructed from the previous hypergraph by merging the vertices in the hyperedges.

Initial Partitioning Phase. In the initial partitioning phase, a bisection (or a k-way partitioning) scheme of the smallest hypergraph is obtained, so that it gets a little cut, and satisfies a balance constraint.

Uncoarsening and Refinement Phase. In the uncoarsening and refinement phase, the partition scheme of the smallest hypergraph is projected to the larger hypergraphs successively. A partitioning refinement algorithm [19] is utilized to reduce the cut set, and the partitioning quality is improved.

3.3 Merge Subgraphs

After obtaining subgraphs, the method of similarity in Chameleon algorithm is used to merge the subgraphs. The pseudocode for this phase is shown in Algorithm 1.

Algorithm 1. Merge Subgraphs

input C - the set of subgraphs
α- the parameter in similarity calculation formula
c - the number of clusters

output C - c clusters

1. **Repeat**
2. Computing the similarity matrix S for each pair of clusters in C using (1), (2) and (3)
3. $(C_i, C_j) = \underset{C_m, C_n \in C}{\operatorname{argmax}}[S(C_m, C_n)]$
4. $C_{new} \leftarrow C_i \cup C_j$
5. $C \leftarrow (C - \{C_i\} - \{C_j\}) \cup \{C_{new}\}$
6. **Until** $|C| = c$

4 Experimental Results

4.1 Datasets

Four UCI datasets [20] are employed in the experiment. The MNIST dataset contains 60000 grayscale images of handwritten digits, each of which has $28 \times 28 = 784$ pixels. The Semeion dataset contains 1593 handwritten digits from around 80 persons, which stretches in a rectangular box 16×16 in a grayscale of 256 values. The USPS dataset contains 11000 grayscale images, each image contains $16 \times 16 = 256$ pixels. The Statlog (Landsat Satellite) dataset includes 6435 multispectral instances of satellite images, each with 36 attributes. All these data are high dimensional datasets. 1000 images are randomly selected from MNIST, USPS, and Statlog datasets as the samples in this experiment.

4.2 Parameters Selection

The Selection of the Number of Subgraphs N_S. The number of subgraphs obtained by the graph partition method, N_S, has a direct impact on the clustering results. Therefore, to evaluate the effect of different Ns on clustering results, some N_S are used in the experiments, such as $N_S = 2c, 3c, 4c, \ldots$, where c is the number of real clusters. Then the proposed MHCH method is compared with the original Chameleon method on the four datasets. The results of this experiment are given in Fig. 3.

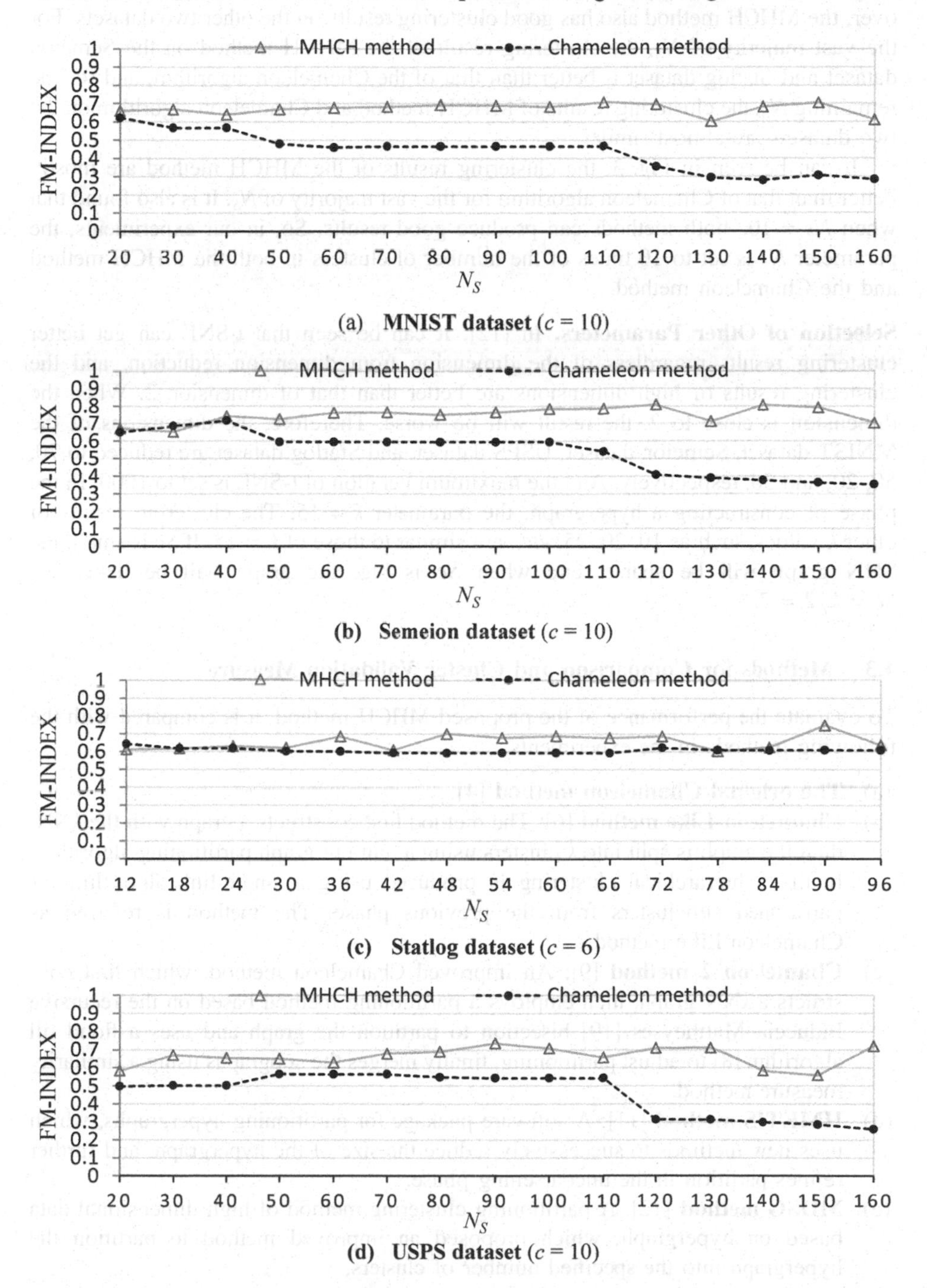

(a) **MNIST dataset** ($c = 10$)

(b) **Semeion dataset** ($c = 10$)

(c) **Statlog dataset** ($c = 6$)

(d) **USPS dataset** ($c = 10$)

Fig. 3. Clustering results of the MHCH method and Chameleon method on the four datasets.

As shown in Fig. 3, the MHCH method performs well on both MNIST dataset and USPS dataset. In the experiments on these two datasets, for each N_S, the clustering result of the MHCH method is preferable to that of the Chameleon algorithm. Moreover, the MHCH method also has good clustering results on the other two datasets. For the vast majority of N_S, the clustering result of the MHCH method on the Semeion dataset and Statlog dataset is better than that of the Chameleon algorithm, and for the remaining N_S, the clustering results of MHCH method and Chameleon algorithm on the two datasets are almost similar.

It can be seen in Fig. 3, the clustering results of the MHCH method are mostly better than that of Chameleon algorithm for the vast majority of N_S. It is also found that when $Ns = 10c$ both methods can produce good results. So, in our experiments, the parameter Ns is set to 10 times of the number of clusters in both the MHCH method and the Chameleon method.

Selection of Other Parameters. In [12], it can be seen that t-SNE can get better clustering results regardless of the dimension from dimension reduction, and the clustering results of high dimensions are better than that of dimension 2. When the dimension is close to 2, the result will be worse. Therefore, the dimensions of the MNIST dataset, Semeion dataset, USPS dataset, and Statlog dataset are reduced to 70, 50, 20, and 50, respectively. And the maximum iteration of t-SNE is set to 1000. In the phase of constructing a hypergraph, the parameter $k = 15$. The clustering results of other k values, such as 10, 20, 25, *etc.*, are similar to those of $k = 15$. If Sc is small, the SNN graph will be sparse, and when Sc is big, the graph will be dense, so $Sc = k/2 = 7$.

4.3 Methods for Comparison and Cluster Validation Measure

To evaluate the performance of the proposed MHCH method, it is compared with the following methods in the experiments:

(a) **The original Chameleon method** [4].
(b) **Chameleon-Like method** [6]: The method first constructs a graph with the kNN, then the graph is split into k clusters using a min-cut graph partitioning algorithm, finally a hierarchical clustering is produced using a single-link algorithm for partitioned subclusters from the previous phase. The method is referred as Chameleon-Like method.
(c) **Chameleon 2 method** [9]: An improved Chameleon method, which first constructs a kNN graph, then employs a partitioning method based on the recursive Fiduccia–Mattheyses [19] bisection to partition the graph and uses a flood fill algorithm [8] to adjust partitioning, finally merges the subgraphs using a similarity measure method.
(d) **HMETIS method** [11]: A software package for partitioning hypergraphs, which uses new methods to successively reduce the size of the hypergraph, and further refines partition in the uncoarsening phase.
(e) **MDSG method** [12]: A partitioning clustering method of high dimensional data based on hypergraph, which proposed an improved method to partition the hypergraph into the specified number of clusters.

The clustering results are evaluated using the Fowlkes–Mallows index [21] (FM-index), which ranges from 0 to 1. The larger the FM index, the better the clustering result.

4.4 Comparing the Proposed MHCH Method with Other Methods

Five clustering methods are used to compare with the proposed MHCH method on the four UCI datasets, which are the original Chameleon, Chameleon-Like, Chameleon 2, hMETIS, and MDSG. The results are given in Table 1.

Table 1. The FM-index of different clustering algorithms on four UCI datasets.

	MNIST	Semeion	Statlog	USPS
Chameleon	0.4662	0.5953	0.5933	0.5435
Chameleon-Like	0.4573	0.5248	0.6312	0.4256
Chameleon 2	0.6267	0.6591	0.6309	0.5672
HMETIS	0.5282	0.4228	0.4531	0.4423
MDSG	0.6703	0.6884	**0.7260**	0.5300
MHCH	**0.6829**	**0.7835**	0.6875	**0.7085**

It can be seen from Table 1 that the FM-indices of the MHCH method is greater than that of other methods on all datasets except Statlog dataset. For Statlog dataset, the FM-index of the MHCH method is the second best one.

From the FM-indices of Chameleon and its improved methods, it can be seen that the MHCH method is superior to other Chameleon methods, it may because the MHCH method is based on hypergraph, which can be used to represent high order relationships between data. The result of Chameleon 2 is the second-best, maybe that is due to this method uses a flood fill algorithm [8] to adjust partitions after partitioning graphs. And the result of Chameleon-Like method is the worst, this may be because the method uses a single-link algorithm to replace the complex merging phase in Chameleon algorithm.

From the FM-indices of hypergraph methods, it can be seen that the clustering result of the MHCH method is almost the best. For MNIST datasets, the FM-index of the MHCH and MDSG method is approximately similar. For Statlog dataset, the FM-index of MDSG method is preferable to that of the MHCH method and hMETIS. For all datasets, hMETIS method performs poorly compared to the other two clustering methods.

5 Conclusion

In this paper, a MHCH method is proposed by introducing hypergraphs in Chameleon algorithm, and it discovers the clusters in the dataset through three steps: constructing hypergraph, partitioning hypergraph, and merging subgraph. Experimental results show that the proposed MHCH method is superior to Chameleon algorithm and other clustering methods. From our experiments, it can be seen that the introduction of hypergraphs is helpful for the hierarchical clustering on the high dimensional datasets. In future work, we will study the improvement of the hypergraph partition algorithm and apply it to the proposed method.

Acknowledgments. This work is supported by the National Key R&D Program of China (Grants No. 2017YFE0111900, 2018YFB1003205).

References

1. Han, J., Kamber, M.: Data Mining: Concept and Technology. Machine Industry Press (2001)
2. Kaufman, L., Rousseeuw, P.J.: Finding Groups in Data: An Introduction to Cluster Analysis. Wiley, New York (2009)
3. MacQueen, J.: Some methods for classification and analysis of multivariate observations. In: Proceedings of the Fifth Berkeley Symposium on Mathematical Statistics and Probability, pp. 281–297 (1967)
4. Karypis, G., Han, E.H., Kumar, V.: Chameleon: hierarchical clustering using dynamic modeling. Computer **32**(8), 68–75 (1999)
5. Shi, J., Malik, J.: Normalized cuts and image segmentation. IEEE Trans. Pattern Anal. Mach. Intell. **22**(8), 888–905 (2000)
6. Zhao, Y., Karypis, G.: Hierarchical clustering algorithms for document datasets. Data Min. Knowl. Discov. **10**(2), 141–168 (2005). https://doi.org/10.1007/s10618-005-0361-3
7. Cao, X., Su, T., Wang, P., et al.: An optimized chameleon algorithm based on local features. In: Proceedings of the 2018 10th International Conference on Machine Learning and Computing, pp. 184–192 (2018)
8. Dong, Y., Wang, Y., Jiang, K.: Improvement of partitioning and merging phase in chameleon clustering algorithm. In: 2018 3rd International Conference on Computer and Communication Systems, pp. 29–32 (2018)
9. Barton, T., Bruna, T., Kordík, P.: Chameleon 2: an improved graph-based clustering algorithm. ACM Trans. Knowl. Discov. Data **13**(1), 10.1–10.27 (2019)
10. Zhou, D., Huang, J., Schölkopf, B.: Learning with hypergraphs: clustering, classification, and embedding. In: Proceedings of the Twentieth Annual Conference on Neural Information Processing Systems, pp. 1601–1608. MIT Press (2010)
11. Karypis, G., Aggarwal, R., Kumar, V., et al.: Multilevel hypergraph partitioning: applications in VLSI domain. IEEE Trans. Very Large Scale Integr. (VLSI) Syst. **7**(1), 69–79 (1999)
12. Wang, T., Lu, Y., Han, Y.: Clustering of high dimensional handwritten data by an improved hypergraph partition method. In: Huang, D.-S., Hussain, A., Han, K., Gromiha, M.M. (eds.) ICIC 2017. LNCS (LNAI), vol. 10363, pp. 323–334. Springer, Cham (2017). https://doi.org/10.1007/978-3-319-63315-2_28
13. Kumar, T., Vaidyanathan, S., Ananthapadmanabhan, H., et al.: Hypergraph clustering: a modularity maximization approach. arXiv preprint arXiv:1812.10869 (2018)
14. Veldt, N., Benson, A.R., Kleinberg, J.: Localized flow-based clustering in hypergraphs. arXiv preprint arXiv:2002.09441 (2020)
15. Karypis, G., Kumar, V.: METIS–unstructured graph partitioning and sparse matrix ordering system, version 2.0 (1995)
16. van der Maaten, L., Hinton, G.: Visualizing data using t-SNE. J. Mach. Learn. Res. **9**, 2579–2605 (2008)

17. Eppstein, D., Löffler, M., Strash, D.: Listing all maximal cliques in sparse graphs in near-optimal time. In: Cheong, O., Chwa, K.-Y., Park, K. (eds.) ISAAC 2010. LNCS, vol. 6506, pp. 403–414. Springer, Heidelberg (2010). https://doi.org/10.1007/978-3-642-17517-6_36
18. Hu, T., Liu, C., Tang, Y., et al.: High dimensional clustering: a clique-based hypergraph partitioning framework. Knowl. Inf. Syst. **39**(1), 61–88 (2014). https://doi.org/10.1007/s10115-012-0609-3
19. Fiduccia, C.M., Mattheyses, R.M.: A linear-time heuristic for improving network partitions. In: Papers on Twenty-Five Years of Electronic Design Automation, pp. 241–247. ACM (1988)
20. UCI Machine Learning Repository. http://archive.ics.uci.edu/ml/index.php. Accessed 11 Apr 2020
21. Fowlkes, E.B., Mallows, C.L.: A method for comparing two hierarchical clusterings. J. Am. Stat. Assoc. **78**(383), 553–569 (1983)

Knowledge Discovery and Data Mining

Discovery of Cancer Subtypes Based on Stacked Autoencoder

Bo Zhang[1], Rui-Fen Cao[1], Jing Wang[1,2], and Chun-Hou Zheng[1(✉)]

[1] School of Computer Science and Technology, Anhui University, Hefei, China
zhengch99@126.com

[2] School of Computer and Information Engineering, Fuyang Normal University, Fuyang, China

Abstract. The discovery of cancer subtypes has become one of the research hotspots in bioinformatics. Clustering can be used to divide the same cancer into different subtypes, which can provide a basis and guidance for precision medicine and personalized medicine, so as to improve the treatment effect. It was found that multi-omics clustering had better effect than single cluster of omics data. However, omics data is usually of high dimensionality and noisy, and there are some challenges in multi-omics clustering. In this paper, we first use a stacked autoencoder neural network to reduce the dimensionality of multi-omics data and obtain the feature representation of low dimension. Then the similarity matrix is constructed by scaled exponential similarity kernel. Finally, we use spectral clustering method to calculate the clustering results. The experimental results on three datasets show that our method is more effective than the traditional dimensionality reduction method.

Keywords: Cancer subtypes · Clustering · Stacked autoencoder · High dimensionality

1 Introduction

The development of deep sequencing and high-throughput technology enables people to get a lot of omics data, for instance, DNA expression, RNA expression, DNA methylation and so on [1, 2]. With lower costs and advances in science and technology, larger and more diverse sets of genomic data are available. The use of these omics data to identify cancer subtypes has become one of the hotspots in bioinformatics.

Human cancer is a heterogeneous disease driven by random somatic mutations and multiple mutations [2]. The discovery of cancer subtypes is an important direction in oncology. The same cancer is divided into different molecular subtypes, and patients with the same molecular subtypes have the same therapeutic response to a large extent. The discovery of cancer subtypes can provide a basis and guidance for accurate medicine and personalized treatment, thus improving the curative effect of cancer.

Clustering algorithms are often used for cancer clustering in single or multiple omics data. However, there may be a lot of noise in a single omics data that is not suitable for directly discovering new subtypes. For multi-omics data (miRNA expression, mRNA expression, DNA methylation), these data are both correlated and

D.-S. Huang and P. Premaratne (Eds.): ICIC 2020, LNAI 12465, pp. 447–454, 2020.
https://doi.org/10.1007/978-3-030-60796-8_38

mutually focused [3]. The similarity and complementarity of omics data can be used to cluster cancer patients.

In recent years, many integrated frameworks for cancer clustering using multi-omics data have been published [4]. These methods can be divided into several categories. The first categorie is early integration: in the early integration method, the main omics feature matrices are simply connected in series to form a matrix with multiple omics features, and then the resulting single matrix is clustered. LRACluster used four different omics data to analyze 11 types of cancer and further identify subtypes of those types [5]. The second categorie is intermediate integration, which mainly includes the integration of sample similarity, statistical modeling, and joint dimensionality reduction. iCluster is a multi-omics data integration probabilistic model derived from the gaussian latent variable model, which mainly applies integration and dimensionality reduction [6]. By applying k-means algorithm to joint latent variables, disease subtypes can be calculated [10, 11]. In iCluster, PCA is often used to decrease the dimensionality of high-dimensional statistical data. However, PCA method also has some shortcomings, for example, PCA is a linear subspace model, but in practical application, many data do not conform to the linear subspace model. Similarity network fusion (SNF) [7] is a method based on sample similarity network which is the fusion of multiple networks in an iterative method. The main innovation of stratification via subspace merging (VSM) [8] is the application of Grassmann manifold to merge the low-dimensional subspace of each omics data. The third categorie is late integration. This method mainly clusters each group of data. And then the unique clustering results are obtained by integrating the obtained clustering results.

In this paper, we propose a new method for cancer clustering using multi-omics data. This method is mainly composed of three steps. Firstly, we input the raw data into the stacked autoencoder to obtain the low dimensional features representation of the raw data [9]. Then, the low dimensional features of different omics were splicing together to form the patient-patient similarity network using scaled exponential similarity kernel function [4]. Finally, spectral clustering method is used to cluster the obtained similar networks. Compared with the previous methods of dimensionality reduction, our method uses stacked autoencoder to obtain a more meaningful potential representation of lower dimensionality. Compared with combining multi-omics of genome directly, our method is easier to mitigate the bias of measurement differences from multi-omics of genome during integration.

2 Methods

2.1 Methods Overview

The proposed model uses three groups of different omics data, including DNA methylation, mRNA expression and miRNA expression. We reduced the dimension of the data and then integrated the low-dimensional subspace data to verify the cancer subtypes. As shown in Fig. 1, firstly, the raw data(miRNA expression, mRNA expression, DNA methylation) were separately inputted into the stacked autoencoder to obtain their lower dimensional subspace representation matrixs (*H1, H2, H3*). Then, the low-dimensional features are integrated into a joint latent matrix H,and we can get the network similarity matrix U using scaled exponential similarity kernel. Finally, we used spectral clustering to identify candidate cancer subtypes.

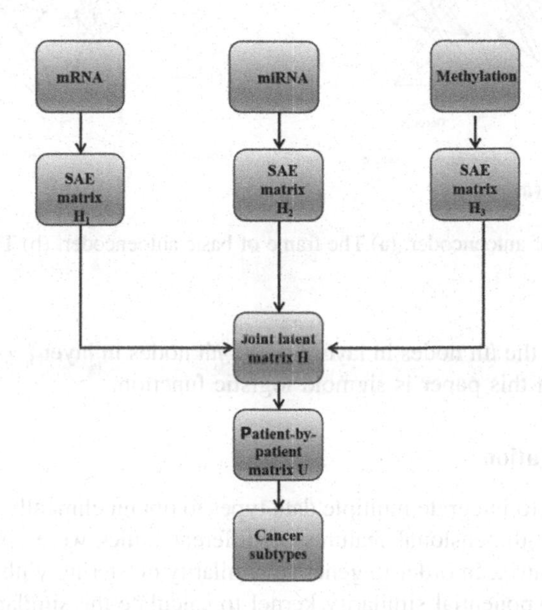

Fig. 1. Workflow for integrating multi-omics data based stacked autuencoder.

2.2 Stacked Autoencoder

Stacked autoencoder is composed of several sparse autoencoder. As shown in Fig. 2(a), autoencoder is mainly separated into three layer which consists of input layer, hidden layer and output layer. For stacked autoencoder, the output of each hidden layer of the sparse autoencoder is the input of the next autoencoder [15]. Figure 2(b) shows stacked autoencoder with three hidden layers. The loss function for sparse autoencoder can be defined as:

$$J_{SAE}(W,b) = \frac{1}{N}\sum_{r=1}^{N}\frac{1}{2}||x_r' - x_r||^2 + \beta\sum_{j=1}^{s_2} KL(\theta||\hat{\theta}_j) + \lambda\sum_{l=1}^{m_l}\sum_{i=1}^{s_l}\sum_{j=1}^{s_{l-1}}(W_{ij}^{(l)})^2 \quad (1)$$

The first term is the average sum-of-squares error of the discrepancy in dataset. The second term is the sparse constraint. The third term is $L2$ regularization term to prevent our network from overfitting. Here N means the amount of samples, x' means the output sample features after reconstruction, x represents the original sample features, W, b represents the parameters to be learned, which mean the weight matrix and bias parameter. β controls the weight of the sparsity penalty factor. s_2 is the number of hidden layer nodes. KL is Kullback-Leibler Divergence. θ is Sparsity parameter. m_l is the number of sparse autoencoder layers, s_l is the number of nodes in lth layer, and $w_{ij}^{(l)}$

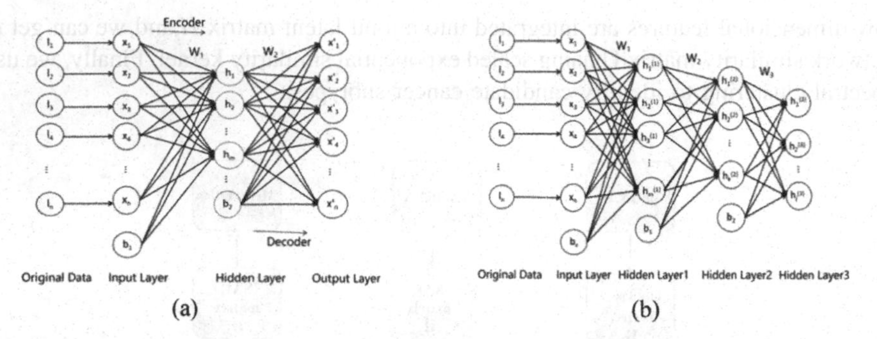

Fig. 2. The frame of autoencoder. (a) The frame of basic autoencoder. (b) The frame of stacked autoencoder.

is the weight from the ith nodes in layer l to the jth nodes in layer $l-1$. The activation function we use in this paper is sigmoid logistic function.

2.3 Data Integration

In order to be able to integrate multiple data types to obtain clinically significant cancer subtypes. The low-dimensional features of different omics were splicing together to obtain a feature matrix. In order to generate similarity clustering with strong similarity, we use a scaled exponential similarity kernel to calculate the similarity network:

$$U(i,j) = \exp(-\frac{\delta^2(x_i, x_j)}{\mu \upsilon_{ij}}) \tag{2}$$

$U(i,j)$ represents the similarity in patients. Where $\delta(x_i, x_j)$ means the Euclidean distance between x_i and x_j. μ is a superparameter between [0.3, 0.8]. It is generally believed that our local similarity is more representative than the remote similarity. υ_{ij} is used to eliminate scaling problems that cause local affinity. υ_{ij} is defined as:

$$\upsilon_{i,j} = \frac{avg(\delta(x_i, N_i)) + avg(\delta(x_j, N_j)) + \delta(i,j)}{3} \tag{3}$$

Where $avg(\delta(x_i, N_i))$ is the average from x_i to its neighbor N_i.

Finally, we do spectral clustering of the obtained similarity matrix $U(i,j)$ to obtain the clustering results.

3 Results

3.1 Comparison with Results of VSM and SNF

To verify the validity of this method, we applied our method on three cancer data sets from wang et al. and compared our method with SNF and VSM [7, 8]. We used three omics datasets (mRNA expression, DNA methylation and miRNA expression) for each cancer type. Types of cancer included glioblastoma multiforme (GBM) with 215 patients, lung squamous cell carcinoma (LSCC) with 106 patients and breast invasive carcinoma (BIC) with 105 patients. The details of data are showed in the Table 1. The Components represent the dimensions of each data type. In our method, the omics data for each cancer were fed separately into our stacked autoencoder. It is significant to pay attention to that since the dimension of each data type is different, we need to adjust the number of nodes in the hidden layer. For example, for the 12,042 mRNAs in GBM, we set the number of nodes as 1,000, 100, 50 in the hidden layer.

Table 1. The number of nodes in the hidden layer in different omic data.

Dataset	Data type	#Components	#Layers(3)
GBM	mRNA	12,042	1,000-100-50
	Methylation	1,491	500-200-50
	miRNA	534	256-128-50
LSCC	mRNA	12,042	1,000-100-50
	Methylation	23,074	2,000-200-50
	miRNA	354	256-128-50
BIC	mRNA	16,818	1,000-100-50
	Methylation	23,094	2,000-200-50
	miRNA	354	256-128-50

Cox log-rank p value of survival analysis is used as the evaluation criterion [12, 13]. When $p < 0.05$ is showed statistically significant difference between different groups of survival. The smaller the p value, the more significant the difference between the different groups. And the clustering results are better.

In our method, the loss function parameter is set to $\beta = 0.05$ and $\lambda = 0.1$. First, we input the original data into the stacked autoencoder to obtain the low dimensional features subspace of the raw data. Then, the low-dimensional features of different omics were splicing together to solve the patient-patient similarity network using scaled exponential similarity kernel. Finally, spectral clustering was used to cluster cancer patients. The experimental results are listed in Table 2. To ensure the comparability of results, we select the same number of clusters for each cancer type as SNF and VSM. As shown in the table, although our results performed slightly worse in BIC than VSM, they were superior in GBM and LSCC. Overall, our method provides a better survival significance than the other two methods.

3.2 A Case Study: Subtype Analysis in Breast Cancer (BRCA)

We further analyzed BRCA, and we obtained DNA methylation (22,533), mRNA expression (gene 20,100) and miRNA expression (gene 718) of 172 patients with primary breast Cancer from TCGA (Cancer Genome Atlas Research Network et al. 2012). For DNA methylation and mRNA expression we set number of hidden nodes to

Table 2. Comparison of Cox survival P-values with those from SNF and VSM.

Dataset	SNF	VSM	Our method
GBM (k = 3)	2.0×10^{-4}	4.3×10^{-3}	5.6×10^{-5}
LSCC (k = 4)	2.0×10^{-2}	1.6×10^{-2}	1.6×10^{-3}
BIC (k = 5)	1.1×10^{-3}	2.0×10^{-4}	1.4×10^{-3}

be 2,000, 200, and 50 from the bottom to the top in stacked autoencoder. For miRNA expression, we set 256, 128 and 50.

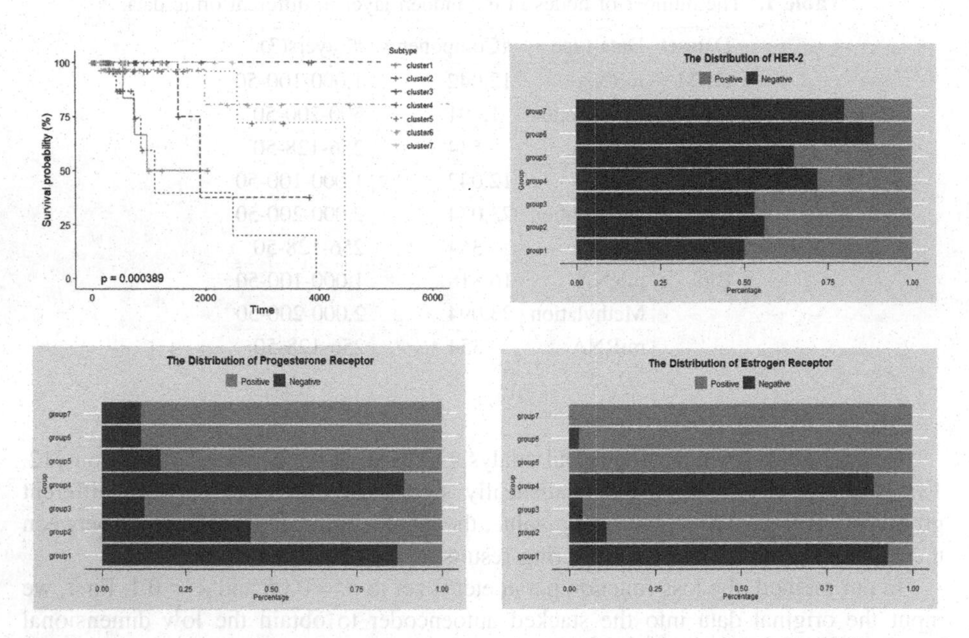

Fig. 3. Survival analysis of breast tumors and clinical features (PR, ER, HER-2) distribution of TCGA subtype of breast cancer.

We applied the proposed method to this dataset and divided the patient population into seven comprehensive subtypes. The choice of quantity depends on the survival p value. Figure 3 shows the overall survival curve obtained by integration clustering. The estimated p value < 0.001 (log-rank test). At the same time, we performed cluster analysis on the single omics data. The results of DNA methylation ($p = 0.2279$), mRNA expression ($p = 0.8434$), and miRNA expression ($p = 0.0911$) show that the clustering results had no significant prognosis. Which shows that the clusters can be more clearly separated by synthetic clustering. As shown in Fig. 3, subtype 4 and subtype 1 were negative in estrogen receptors. However, there were significant differences in survival. It is consistent with the clinical diagnosis that estrogen receptor negative includes at least two biologically distinct tumor subtypes. The two subtypes may need to be treated as separate diseases. The ER, PR and HER-2 states of subtype 6 and subtype 7 are similar, but the prognosis is different [14]. These differences may represent useful outcomes and, more importantly, therapeutic opportunities.

4 Discussion

With the development of deep sequencing and high-throughput technology, a huge range of omics data can be measured to cluster cancer subtypes. The effect of multi-omics clustering is better than that of a single-omics data to find the subtypes in the cluster. We propose a stacked autoencoder for dimensionality reduction of omics data in this paper. The resulting low-dimensional subspace is integrated into a patient-to-patient similarity matrix. Finally, we use spectral clustering to cluster the similarity matrix. Compared with the other two methods, the experimental results show that our method is better than the other two methods. Our approach can separate the clusters more clearly. The results also confirmed that estrogen receptor negative includes at least two biologically distinct tumor subtypes. In addition, although our method is able to identify biologically significant cancer subtypes, the analysis of the biological significance needs further study.

Acknowledgments. This work was supported by grants from the National Natural Science Foundation of China (Nos. U19A2064, 61873001), the Key Project of Anhui Provincial Education Department (No. KJ2017ZD01), and the Natural Science Foundation of Anhui Province (No. 1808085QF209).

References

1. Pollack, J.R., et al.: Microarray analysis reveals a major direct role of DNA copy number alteration in the transcriptional program of human breast tumors. Proc. Natl. Acad. Sci. U.S. A. **99**, 12963–12968 (2002)
2. Stratton, M.R., Campbell, P.J., Futreal, P.A.: The cancer genome. Nature **458**, 719–724 (2009)
3. Yang, Y., Wang, H.: Multi-view clustering: a survey. Big Data Min. Anal. **1**, 3–27 (2018)
4. Rappoport, N., Shamir, R.: Multi-omic and multi-view clustering algorithms: review and cancer benchmark. Nucleic Acids Res. **46**, 10546–10562 (2018)
5. Wu, D., Wang, D., Zhang, M.Q., Gu, J.: Fast dimension reduction and integrative clustering of multi-omics data using low-rank approximation: application to cancer molecular classification. BMC Genomics 16 (2015). Article number: 1022 https://doi.org/10.1186/s12864-015-2223-8
6. Shen, R., et al.: Integrative subtype discovery in glioblastoma using iCluster. PLoS ONE **7**, e35236 (2012)
7. Wang, B., Mezlini, A.M., Demir, F., Fiume, M.: Similarity network fusion for aggregating data types on a genomic scale. Nat. Methods **11**, 333–337 (2014)
8. Ding, H., Sharpnack, M., Wang, C., Huang, K., Machiraju, R.: Integrative cancer patient stratification via subspace merging. Bioinformatics **35**(10), 1653–1659 (2019)

9. Zabalza, J., Ren, J., Zheng, J., Zhao, H., Marshall, S.: Novel segmented stacked autoencoder for effective dimensionality reduction and feature extraction in hyperspectral imaging. Neurocomputing **185**, 1–10 (2015)
10. Zha, H., He, X., Ding, C., Ming, G., Simon, H.D.: Spectral relaxation for K-means clustering. In: Advances in Neural Information Processing Systems 14 (2001)
11. Ding, C., He, X.: Cluster structure of *K*-means clustering via principal component analysis. In: Dai, H., Srikant, R., Zhang, C. (eds.) PAKDD 2004. LNCS (LNAI), vol. 3056, pp. 414–418. Springer, Heidelberg (2004). https://doi.org/10.1007/978-3-540-24775-3_50
12. Mo, Q., et al.: Pattern discovery and cancer gene identification in integrated cancer genomic data. Proc. Natl. Acad. Sci. U.S.A. **110**, 4245–4250 (2013)
13. Hosmer, D.W., Lemeshow, S., May, S.: Applied survival analysis: regression modeling of time to event data. J. Stat. Plann. Infer. **95**, 173–175 (2000)
14. Subik, K., et al.: The expression patterns of ER, PR, HER2, CK5/6, EGFR, Ki-67 and AR by immunohistochemical analysis in breast cancer cell lines. Breast Cancer Basic Clin. Res. (2010)
15. Yang, G., Zheng, J., Shang, X., Li, Z.: A similarity regression fusion model for integrating multi-omics data to identify cancer subtypes. Genes **9**, 314 (2018)

A Meta Graph-Based Top-k Similarity Measure for Heterogeneous Information Networks

Xiangtao Chen(✉), Yonghong Jiang, Yubo Wu, Xiaohui Wei, and Xinguo Lu

Hunan University, Changsha 410082, China
xtchen2009@sina.cn, {yhj,Xh_wei}@hnu.edu.cn, 844529478@qq.com, hnluxinguo@126.com

Abstract. Studies have demonstrated that real-world data can be modeled as a heterogeneous information network (HIN) composed of multiple types of entities and relationships. Similarity search is a basic operation requiring many problems in HINs. Similarity measures can be used in various applications, including friend recommendation, link prediction, and online advertising. However, most existing similarity measures only consider meta path. Complex semantic meaning cannot be expressed through meta path. In this paper, we study the similarity search problem of complex semantics meaning between two HIN objects. In order to solve the problem, we use meta graphs to express the semantic meaning between objects. The advantage of meta graphs is that it can describe the complex semantic meaning between two HIN objects. And we first define a new meta graph-based relation similarity measure, GraphSim, which is to measure the similarity between objects in HINs, then we propose a similarity search framework based on GraphSim. The experiments with real-world datasets from DBLP demonstrated the effectiveness of our approach.

Keywords: Heterogeneous information network · Similarity search · Meta graph

1 Introduction

A heterogeneous information network is a type of logical network that typically consists of a large number of multiple types and interconnected entities. Interconnections in heterogeneous networks often represent different types of relationships, such as bibliographic networks [7], disease information network, and social media network [2]. In recent years, data mining communities [10, 11, 15, 16] have become increasingly interested in the study of heterogeneous information networks. Many interesting and practically important research questions can be performed on heterogeneous information networks, where the similarity measure is a basic work. It is very important to provide effective search functions in heterogeneous information networks. Links play a vital role, and it is difficult to fully express the latent semantic relationships of links in the network. In particular, we are interested in providing similarity search functions for objects that are from the same type and multiple semantics. For example, in a

D.-S. Huang and P. Premaratne (Eds.): ICIC 2020, LNAI 12465, pp. 455–466, 2020.
https://doi.org/10.1007/978-3-030-60796-8_39

bibliographic network, in a given multiple semantic meaning of relation (e.g., two authors have published papers in the same venue and have also mentioned the same topic), a user may be interested in the (top-k) most similar authors for a given author, or the most similar venues for a given venue, and so on.

In order to calculate the similarity between two objects in a heterogeneous information network, neighborhood-based measures such as Jaccards coefficient and common neighborhoods are proposed [8]. Other theoretical measures based on random walks between objects include Personalized PageRank [1], SimRank [5], and SCAN [18]. These measures do not take into account object and edge type information in the HIN. In order to process this information, the concept of the meta-path [12] was proposed. A meta-path is a series of object types that define a composite relationship between a start type and an end type. Considering that the bibliographic network extracted from DBLP which includes author (A), paper (P), term (T) and conference (C) and other types of objects and the different semantics behind the meta-path, we can list two different the meta-path: (a) CPAPC, because the different papers published by the same author at two different conferences indicated two conferences are related, and (b) CPTPC, indicating that the papers published at the two conferences contain the same subject (T). Based on the meta-path, several similarity measures have been proposed, such as PathSim, RelSim, and JoinSim [12, 14, 17]. These similarity measures have proven to be better than similarity measures that do not consider object and edge type information. However, a meta path fails to capture complex semantic relation, We know that connecting two objects has different meta paths, which contain different semantics, which can lead to different similarities. Meta-paths can not express more than two kinds of complex semantics. For the above two meta-paths, we can use meta-graphs to express two complex semantics. Meta-graph CP {A, T} PC, describes that if two meetings are similar, the paper at the conference is written by the same author and the paper has the same terminology, in order to capture complex semantic relationships, we introduce a based meta graph similarity framework for objects of the same type in a heterogeneous network. Our experiments also show that meta graph are more effective than meta paths.

Under the proposed meta graph based similarity framework, based on the specific instances in the given meta graph, a new method is proposed to define the similarity measure between two objects. We propose a new similarity measure, GraphSim, which captures the more subtle semantics of similarity between peer entities in heterogeneous information networks. Given a meta graph, GraphSim computes the similarity between objects using the adjacency matrix of a given meta graph.

Compared to PathSim [12] and JoinSim [17], the similarity calculated by GraphSim is more accurate because it captures complex semantic relationships. However, it still involves complex matrix multiplication of Top-k search function and more complex matrix multiplication for the reason that the meta graph in GraphSim contains a directed acyclic graph. Besides, in order to support processing the fast online query of large-scale networks, we propose a method that initially implements the directed acyclic graph matrix multiplication in the meta graph, and next implements the adjacency matrix multiplication according to the given meta graph, and then connects them online to obtain longer meta graph similarity. We proposed a baseline method GraphSim which calculates the similarity between the query object x and all candidate objects y of the same type.

The contributions of this paper are summarized below.

1. It studies the similarity search in heterogeneous information networks, which is an increasingly important issue because of the proliferation of linked data and its broad application.
2. It proposes a new framework based on the similarity of meta graphs and a new definition of similarity measures, GraphSim, which captures more potential similarity semantics between peers in the network.
3. Our experiments demonstrate GraphSim achived higher validity in terms of similarity measurements than PathSim and JoinSim.

This paper mainly focuses on the similarity search based on meta-graph in heterogeneous information networks. There are a total of five chapters in the paper, which are arranged as follows:

The first chapter introduces the basic theoretical knowledge and related concepts of heterogeneous information network, and then expounds the similarity algorithm and its literature review, which is analyzed by the methods of feature-based and link-based measurement. Finally, the similarity algorithms based on meta-path are analyzed, and the characteristics of these methods are briefly analyzed.

In the second chapter, the definitions of heterogeneous information network, network pattern, meta-path, meta-graph and relational matrix are given, and a similarity search algorithm GraphSim based on meta-graph is proposed. The basic concepts related to GraphSim and the relation matrix of meta-graph are given. This algorithm mainly aims at the problem that the existing similarity search algorithms do not consider complex semantics.

The third chapter describes the first k similarity search algorithm GraphSim for online query objects.

The fourth chapter describes the experimental scheme, the experimental results and analysis, and the analysis of the time complexity of the algorithm.

The fifth chapter is the summary and prospect. The main work of this paper is summarized, and the future research and future plans are prospected.

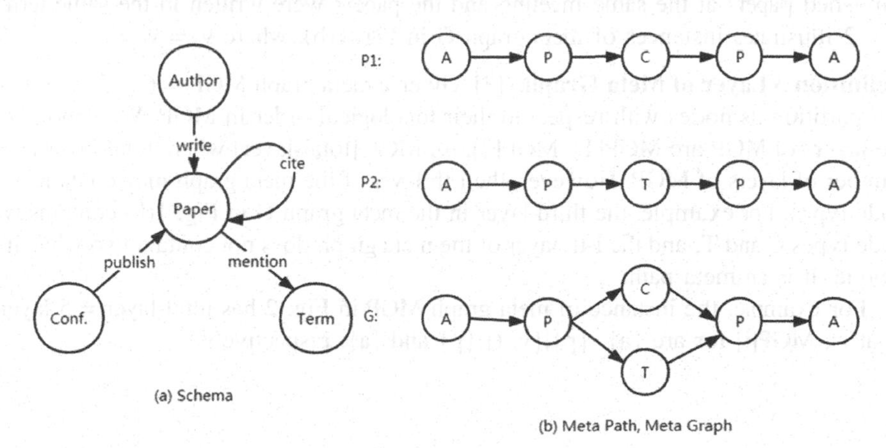

Fig. 1. (a) is a document network schema. (b) p1: the relevant meaning is the relationship between the two authors who published the paper at the same meeting. p2: The relevant meaning is that two authors use the same term to write a paper. G: The relevant meaning is that the two authors published papers at the same conference, and the papers were written in the same terminology.

2 Problem Definition

2.1 Heterogeneous Information Network

The heterogeneous information network can be abstracted into a directed graph G = (V, E), which contains the entity type mapping function $\varphi : V \rightarrow A$ and the link type mapping function $\phi : E \rightarrow L$, where each entity $v \in V$ belongs to a specific entity type (v) A, each link e E belongs to a specific relationship type ϕ (e) $\in L$, and both entities and links have the characteristics of attributes. If the number of entity types | A| > 1 or the number of relationship types |L| > 1, the information network is a heterogeneous information network [12].

Definition 1 HIN Schema [12]. The network mode is a template of the heterogeneous information network G = (V, E), denoted as T_G = (A, L), where the directed graph vertex represents the entity A, and the edge type represents the relationship L. Figure 1 (a) shows an example of a bibliography network schema. It expresses all possible link types between objects.

Meta path [12] is defined on the mode T_G = (A, L). It can be represented by the object type as P = (A_1, A_2, ···, A_n) and the path instance of P is p = (a_1, a_2, ···, a_n), where we use lower-case letters to represent the object type. For example, Fig. 1(b) p1: the physical meaning of the meta-path APCPA is the relationship between two authors who wrote papers published at the same conference. Figure 1(b) p2: the relevance meaning of APTPA is that two authors use the same term to write papers.

The meta graph designed to capture complex relationship between two HIN objects, is defined as follows.

Definition 2 Meta Graph [3]. A meta graph MGP is a directed acyclic graph, with a single source object v_s and a single target object v_t, defined on a schema T_G = (A, L). It can be denoted as MGP = (V, E, v_s, v_t), where V is a set of objects and E is a set of edges. For example, the physical meaning of Fig. 1(b) meta graph G is that two authors published papers at the same meeting and the papers were written in the same terms. Fig. 2 illustrates instances of meta graph G in Fig. 1(b), where $v_s = v_t = a$.

Definition 3 Layer of Meta Graph. [3]. Given a meta graph MGP = (V, E ,v_s, v_t), we can partition its nodes with respect to their topological order in MGP. We denote that the layers of MGP are MGP[1], MGP[2], ···, MGP [total-layer] where total-layer is the number of layers of MGP. However, the i-th layer of the meta graph may contains two node types. For example, the third layer in the meta graph G in Fig. 1(b) contains two node types C and T, and the i-th layer of the meta graph does not contain two when it is a node, it is an meta path.

For example, the instance of meta graph MGP in Fig. 2 has total-layer = 5 layers. That is, MGP[i] for are {a}, {p},{v, t},{p} and {a}, respectively.

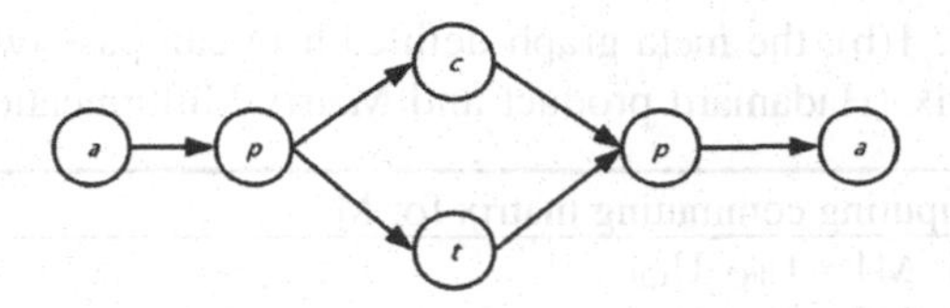

Fig. 2. Intance of meta graph for Fig. 1(b)

We use the meta graph MGP = MGP[1] MGP[2] ··· {MGP[i], MGP[i′]} ··· MGP[l] to find similar pairs of objects in MGP[i] based on the similarity of their meta graph in MGP[l].

2.2 GraphSim: Graph-Based Similarity Measure

Similarity measures PathSim and JoinSim have been proposed in [12, 17] to capture semantics, but they are unable to capture complex semantics in the network. For example, PathSim can only compute the similarity of a given path, indicating that it can only capture one semantic. JoinSim can integrate various semantics behind the path into the similarity measure, but it can only capture one semantic. Therefore, we need to define a new measure of similarity. To integrate the complex semantics of the meta graph into our similarity measure, we further use the matrix form to describe the composite relationship between the start type and the end type of the meta graph, as shown below.

Definition 4 Relation Matrix. Given an information network G = (V, E), and its network schema T_G, a relation matrix M for a meta graph MGP is defined as $M = U_{MGP1MGP2}U_{MGP2MGP3}\cdots(U_{MGPi-1MGPi}U_{MGPiMGPi+1}) \odot (U_{MGPi'-1MGPi'}U_{MGPi'MGPi'+1})\cdots U_{MGPl-1MGPl}$, where $U_{MGPiMGPj}$ is the adjacency matrix between type MGPi and type MGPj. M (i, j) represents the number of graph instances between node xi ∈ MGP1 and node yj ∈ MGPl under meta graph MGP, where M (i, j) ∈ N.

For a meta graph, it can be a bit complicated because it can contain multiple nodes in the meta layer. For example, for G in Fig. 1(b), there are two ways to implement a meta graph, namely (A, P, C, P, A) and (A, P, T, P, A), the third layer contains two nodes in the meta graph. Note that P represents the entity type paper in the HIN. The path here (A, P, C, P, A), (P, C, P) means that if both papers are published in the same C (Conference), then they have some similarities. Similarly, in (A, P, T, P, A), (P, T, P) means that if two papers contain the same T (term), they also have some similarities. When there are multiple ways to flow from the source node to the target node, we should define the logic. When there are two paths, we can allow the process to pass either path, or we can constrain the process to satisfy both. By analyzing the former strategy, we found that it is similar to simply splitting the meta graph into multiple meta-paths. Therefore, we choose the latter, which requires more matrix operations than simple matrix multiplication, i.e. Hadamard product or element product. Algorithm 1 describes an algorithm for calculating the similarity based on the meta graph G in Fig. 1(b), where is the Hadamard product. After obtaining Csr, the entire relation matrix M is more easily obtained by multiplication of matrix sequences. In fact, not limited to M in Fig. 1(b), the meta graph defined here can pass two operations on the corresponding matrix (Hadamard product and Matrix Multiplication) to calculate.

Algorithm 1 Computing commuting matrix for M

1: Compute M1: $M1 = U_{PC} \cdot U_{CP}$.
2: Compute M2: $M2 = U_{PT} \cdot U_{TP}$.
3: Compute C_{sr}: $C_{sr} = M1 \odot M2$.
4: Compute M: $M = U_{AP} \cdot C_{sr} \cdot U_{PA}$.

Definition 5 GraphSim: A Meta Graph-Based Similarity Measure. Given a meta graph MGP, the definition of GraphSim between objects x and y is as follows:

$$s(x, y) = \frac{|g_{x \to y} : g_{x \to y} \in MGP|}{\sqrt{g_{x \to x}} \times \sqrt{g_{y \to y}} : g_{x \to x}, g_{y \to y} \in MGP} \quad (1)$$

where $g_{x \to y}$ represents the instance of meta graph between x and y. Besides, $g_{x \to x}$ and $g_{y \to y}$ represent similar meanings to $g_{x \to y}$.

To understand how this new measure works, Fig. 3 is a toy HIN, We use this toy HIN for GraphSim calculations, and use the meta graph G in Fig. 1(b) to calculate the correlation between a1 and a2. GraphSim generates similarity scores: $s(a1, a2) = 1/(\sqrt{2} \times \sqrt{2}) = 0.5$

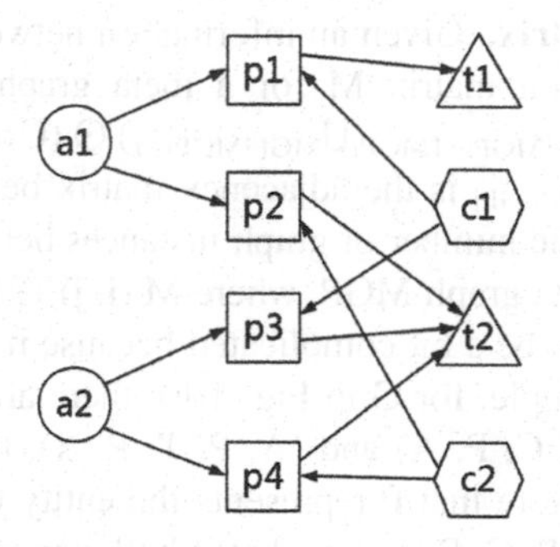

Fig. 3. A toy HIN to understand how GraphSim calculations works

Calculating the GraphSim value between two objects in a meta graph involves matrix multiplication, and we have defined the relationship matrix in Definition 4. Given the meta graph MGP = MGP[1] MGP[2] ··· {MGP[i], MGP[i′]} ··· MGP[l], object $x_i \in$ MGP[1] and $y_j \in$ MGP[l] under meta graph MGP. GraphSim between computing object x_i and object y_j is available $s(x, y) = M(i, j)/(\sqrt{M(i, i)} \times \sqrt{M(j, j)})$ to calculate, where M is the relation matrix of the meta graph MGP.

Definition 6 Top-k Similarity Search Under GraphSim. Given an heterogeneous information network G and the network schema T_G, given the meta graph MGP = MGP[1] MGP[2] ⋯{MGP[i], MGP[i′]}⋯MGP[l], the top-k similarity search of the object $x_i \in$ MGP[1] is found to be sorted the k objects of type MGP[l] such that $s(x_i, x_j) \geq s(x_i, x_j')$, for any x_j' not in the returned list and x_j in the returned list, where $s(x_i, x_j)$ is defined for Definition 5.

3 Online Single Meta Graph Calculation

This section describes an efficient top-k GraphSim similarity search for online queries. Which returns the exact top-k result for a given query.

3.1 GraphSim Algorithm Framework

Given the meta graph MGP = MGP[1] MGP[2] ⋯ {MGP[i], MGP[i′]} ⋯ MGP[2] MGP[1], divide it into two meta graphs respectively MGP1 = MGP[1] MGP[2] ⋯ {MGP[i], MGP[i′]} MGP[i + 1] and MGP2 = MGP[i + 1] ⋯ MGP[l], the relationship matrix of the meta graph MGP1 is $M1 = U_{MGP1MGP2}U_{MGP2MGP3} \cdots (U_{MGPi-1MGPi}U_{MGPiMGPi+1}) \odot (U_{MGPi'-1MGPi'}U_{MGPi'MGPi'+1})$ and meta graph MGP2 is $M2 = U_{MGPlMGPl-1}U_{MGPl-2MGPl-3} \cdots U_{MGPi+2MGPi+1}$, the relationship matrix of the meta graph MGP is M = M1M2. For example, given the meta graph G in Fig. 1, it is divided into two parts: MGP1 = (AP{C, T}P) and MGP2 = (PA), and the relationship matrix corresponding to the subgraph MGP1 is $M1 = U_{AP}((U_{PT}U_{TP}) \odot (U_{PC}U_{CP}))$, the relation matrix corresponding to the subgraph MGP2 is $M2 = U_{PA}$, the commuting matrix of the meta graph MGP is M = M1M2.

Let n be the number of objects in MGP[1], and the object types of MGP[1] and MGP [l] are the same. However, materializing the relationship matrix of all meta graphs is not realistic because its spatial complexity ($o(n^2)$) makes it impossible to store the similarity matrix M of each meta graph. In order to avoid the above extreme cases, we only materialize the relationship matrix M1 of meta graph MGP1 and the relationship matrix M2 of meta graph MGP2, and calculate the top-k result by connecting MGP1 and MGP2 to MGP online. Therefore $s(x_i, x_j) = \frac{\sum_{k=1}^{|MGP1[n]|} M1(i,k)M2(j,k)}{\sqrt{\sum_{k=1}^{|MGP1[n]|} M1(i,k)M2(i,k)}\sqrt{\sum_{k=1}^{|MGP1[n]|} M1(j,k)M2(j,k)}}$, node $x_i, x_j \in$ MGP[1].

Algorithm 2 describes the basic framework of the GraphSim algorithm. Given the meta graph MGP = MGP[1] MGP[2] ⋯ {MGP[i], MGP[i′]} ⋯ MGP[1], and steps 2 count the number of path instances of x_iMGP[2] ⋯ {MGP[i], MGP[i′]} ⋯ x_i, steps 5 count the number of path instances of x_jMGP[2] ⋯ {MGP[i], MGP[i′]} ⋯ x_j, steps 6 count the number of path instances of x_iMGP[2] ⋯ {MGP[i], MGP[i′]} ⋯ x_j, x_i, $x_j \in$ MGP[1], and MGP[1] and MGP[l] are of the same objects type. Note: For query node x_i, it must be similar to its own node, so it can be added directly to the search results (see Algorithm 2, Step 1).

3.2 Time Complexity Analysis of the Method

Algorithm 1, calculate the relation matrix M of meta graph MGP, the number of execution is a constant, so the time complexity of this part is O (1). Algorithm 2, GraphSim algorithm, the time complexity of this method is O (n), Where n is the number of all nodes in MGP[1]. Generally speaking, the time complexity of GraphSim method is O (n).

Algorithm 2 GraphSim algorithm

Input: HIN G ,meta graph MGP, commuting matrix M1,M2, Parameter k, query node x_i

1: R ← $\{x_i\}$;

2: $\sum_{k=1}^{|MGP1[n+1]|} M1(i,k)M2(i,k)$

3: N ← all type in MGP[1];

4: **for** each $x_j \in$ MGP[1] \ x_i **do**

5: $\sum_{k=1}^{|MGP1[n+1]|} M1(j,k)M2(j,k)$

6: $\sum_{k=1}^{|MGP1[n+1]|} M1(i,k)M2(j,k)$

7: **if** | R |< k **then**

8: add x_j to R

9: **else if** $s(x_i, x_j) \geq s(xi, x'_j)$ **then**

10: update R with x_j;

11: **end if**

12: **end for**

13: return R;

Output: top-k node R

---

4 Experiments

This chapter will conduct related experiments in the Digital Bibliography and Library Project (DBLP) [6] data set. Through experiment, the similarity search algorithm GraphSim proposed in this paper is compared with the traditional meta path-based similarity search method. Verify the effectiveness of GraphSim.

4.1 Experimental Setting

Experimental environment: The processor is Intel Core i5-3230 M CPU @ 2.60 GHz, RAM 4 GB, operating system is Windows 10.

In the experiment, the dataset used was a subset of the selection in the DBLP network, i.e., DBLP-4_Area [9], including major conferences in four research areas: databases, data mining, artificial intelligence, and information retrieval. In this data set, there are 4 research areas, 2000 authors, 4,366 papers, 20 conferences and 5,081 topics.

We compare our relevance metrics with other representative meta-path metrics (i.e., JoinSim [17], and PathSim [12]). These measures use the meta paths and meta graph shown in Fig. 4.

4.2 Effectiveness

Given the meta graph G and the meta-paths P_1 and P_2 in Fig. 4, GraphSim is used in the meta graph in Fig. 4, and PathSim and JoinSim are applied to the meta-paths P_1 and P_2 Fig. 4. Then analyze the three algorithms separately.

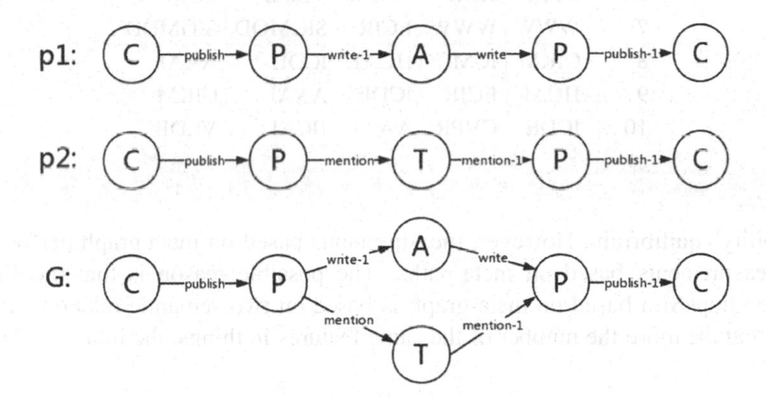

Fig. 4. Meta paths and meta graph used in experiments

First, Table 1 gives an example of a laboratory data set that analyzes a given meta graph and meta-path. It can be seen from the query PKDD that the results of the three algorithms are conferences with similar scale and reputation to PKDD, but the rankings are different. PathSim and JoinSim tend to have meetings with the same author set, such as IJCAI, in meta-path P_1. PathSim and JoinSim, on the other hand, tend to have meetings that contain the same terms, such as CIKM, in meta-path P_2. The result of GraphSim is basically a meeting with two kinds of semantic relations, which can be seen from the table, it is like a combination of two kinds of semantic correlation under two algorithms.

Next, in our effectiveness experiment, we performed the relevance ranking task as follows. We first use four levels to mark the relevance of each pair of meetings in the DBLP: 0 for "non-correlated", 1 for "some related", 2 for "comparatively related", and 3 for "very relevant". We consider the level and scope of the meeting when tagging. For example, SIGMOD and VLDB have a correlation score of 3 because they are highly correlated. We use the meta graph G and the two meta-paths P_1, P_2 shown in Fig. 4. We then evaluate the quality of the returned ranking list w.r.t. Using the different measures of the standardized discount cumulative gain (nDCG) [4], which is a common measure of ranking quality, the bigger the better.

The nDCGs results are shown in Table 2. We can observe that on the meta-path based measurement, the first meta path P_1 = CPAPC of the PathSim metric produces better results than the second meta path P_2 = CPTPC, but JoinSim is just the opposite. In general, the nDCG values of JoinSim are higher, which may be better for the quality

Table 1. An example of a similar measure in the experimental data set for querying "PKDD"

Rank	PathSim		JoinSim		GraphSim
	P1	P2	P1	P2	G
1	PKDD	PKDD	PKDD	PKDD	PKDD
2	SDM	ICDM	KDD	KDD	ICDM
3	KDD	SDM	SDM	ICDM	KDD
4	ICDM	CIKM	ICML	SDM	SDM
5	ICML	PODS	ICDM	CIKM	ICML
6	ECIR	KDD	WWW	VLDB	ECIR
7	WWW	WWW	ECIR	SIGMOD	SIGMOD
8	CIKM	ICML	IJCAI	ICDE	AAAI
9	IJCAI	ECIR	ICDE	AAAI	CIKM
10	ICDE	CVPR	AAAI	IJCAI	VLDB

of visibility equilibrium. However, measurements based on meta graph perform better than measurements based on meta-paths. The possible reason is that the similarity measure GraphSim based on meta-graph is based on two semantic relations. It is well known that the more the number of the same features in things, the more similar things will be.

Table 2. Ranking quality nDCG and clustering accuracy NMI

Metric	PathSim		JoinSim		GraphSim
	P1	P2	P1	P2	G
nDCG	0.9621	0.9273	0.9638	0.9645	0.9684
NMI	0.6272	0.6560	0.6701	0.6453	0.6771

Finally, to further evaluate the quality of similarity search, we use Normalized Mutual Information (NMI) [13] to calculate clustering accuracy. Applying three algorithms to the cluster, PathSim and JoinSim cluster the conference in the meta-paths P_1 and P_2 of Fig. 4, and GraphSim clusters the conference in the meta graph G of Fig. 4. Different similarity matrices are obtained by using PathSim, JoinSim and GraphSim, and k-means is used for clustering. The NMI is then used to evaluate the clustering results. Table 2 shows the clustering accuracy obtained by each algorithm. It can be seen that for the clustering result of the meta path, the PathSim algorithm has an advantage in the meta path P_2, and the clustering result of JoinSim in the meta path P_1 is better. Then we can see that GraphSim has the best performance over all measures. Because GraphSim metrics can capture complex semantics, GraphSim can be applied to clustering.

5 Conclusion

In this paper, we propose a novel similarity search method based on meta graph. Based on meta graph, we introduce a relevance framework on heterogeneous information networks, which can express complex relevance of two objects. Moreover, we propose a new similarity measure, GraphSim, under this framework, which produces overall better similarity qualities than the existing measures based on meta path. Experiments on real datasets demonstrate the effectiveness of our methods.

In the future, we will study the method of automatically learning the meta graph from HIN. We will also examine the use of meta graphs in different applications, such as citation recommendations and bioinformatics.

Acknowledgements. This work was supported in part by the National Natural Science Foundation of China under Grant 61873089, Grant 61572180, and in part by the China National Key R&D Program during the 13th Five-year Plan Period under Grant 2018YFC0910405.

References

1. Chakrabarti, S.: Dynamic personalized pagerank in entity-relation graphs. In: The Web Conference, pp. 571–580 (2007)
2. Han, J.: Mining heterogeneous information networks by exploring the power of links. In: Gama, J., Costa, V.S., Jorge, A.M., Brazdil, P.B. (eds.) DS 2009. LNCS (LNAI), vol. 5808, pp. 13–30. Springer, Heidelberg (2009). https://doi.org/10.1007/978-3-642-04747-3_2
3. Huang, Z., Zheng, Y., Cheng, R., Sun, Y., Mamoulis, N., Li, X.: Meta structure: computing relevance in large heterogeneous information networks, pp. 1595–1604 (2016)
4. Jarvelin, K., Kekalainen, J.: Cumulated gain-based evaluation of IR techniques. ACM Trans. Inf. Syst. **20**(4), 422–446 (2002)
5. Jeh, G., Widom, J.: SimRank: a measure of structural-context similarity, pp. 538–543 (2002)
6. Ji, M., Sun, Y., Danilevsky, M., Han, J., Gao, J.: Graph regularized transductive classification on heterogeneous information networks. In: Balcázar, J.L., Bonchi, F., Gionis, A., Sebag, M. (eds.) ECML PKDD 2010. LNCS (LNAI), vol. 6321, pp. 570–586. Springer, Heidelberg (2010). https://doi.org/10.1007/978-3-642-15880-3_42
7. Ley, M.: The DBLP computer science bibliography: evolution, research issues, perspectives. In: Laender, A.H.F., Oliveira, A.L. (eds.) SPIRE 2002. LNCS, vol. 2476, pp. 1–10. Springer, Heidelberg (2002). https://doi.org/10.1007/3-540-45735-6_1
8. Libennowell, D., Kleinberg, J.M.: The link-prediction problem for social networks. J. Assoc. Inf. Sci. Technol. **58**(7), 1019–1031 (2007)
9. Meng, C., Cheng, R., Maniu, S., Senellart, P., Zhang, W.: Discovering meta-paths in large heterogeneous information networks. In: The Web Conference, pp. 754–764 (2015)
10. Shi, C., Li, Y., Zhang, J., Sun, Y., Yu, P.S.: A survey of heterogeneous information network analysis. IEEE Trans. Knowl. Data Eng. **29**(1), 17–37 (2015)
11. Sun, Y., Han, J.: Mining heterogeneous information networks: a structural analysis approach. Sigkdd Explor. **14**(2), 20–28 (2013)
12. Sun, Y., Han, J., Yan, X., Yu, P.S., Wu, T.: PathSim: meta path-based top-k similarity search in heterogeneous information networks. Very Large Data Bases **4**(11), 992–1003 (2011)
13. Sun, Y., Han, J., Zhao, P., Yin, Z., Cheng, H., Wu, T.: RankClus: integrating clustering with ranking for heterogeneous information network analysis, pp. 565–576 (2009)

14. Wang, C., et al.: RelSim: relation similarity search in schema-rich heterogeneous information networks, pp. 621–629 (2016)
15. Wang, S., Xie, S., Zhang, X., Li, Z., He, Y.: Coranking the future influence of multiobjects in bibliographic network through mutual reinforcement. ACM Trans. Intell. Syst. Technol. **7** (4), 1–28 (2016)
16. Xiang, L., Zhao, G., Li, Q., Hao, W., Li, F.: TUMK-ELM: a fast unsupervised heterogeneous data learning approach. IEEE Access **6**, 35305–35315 (2018)
17. Xiong, Y., Zhu, Y., Yu, P.S.: Top-k similarity join in heterogeneous information networks. IEEE Trans. Knowl. Data Eng. **27**(6), 1710–1723 (2015)
18. Xu, X., Yuruk, N., Feng, Z., Schweiger, T.A.J.: SCAN: a structural clustering algorithm for networks, pp. 824–833 (2007)

Joint Deep Recurrent Network Embedding and Edge Flow Estimation

Gaoyuan Liang[1], Haoran Mo[2(✉)], Zhibo Wang[3,6], Chao-Qun Dong[4], and Jing-Yan Wang[5]

[1] Heriot Watt University, Dubai Campus, Dubai International Academic City, 294345 Dubai, United Arab Emirates
[2] Innopolis University, Innopolis, Russia
m.haoran.cq@gmail.com
[3] Merck & Co., Inc., Kenilworth, NJ, USA
[4] School of Economics and Management, Harbin University of Science and Technology, Harbin, China
[5] New York University Abu Dhabi, Abu Dhabi, United Arab Emirates
[6] College of Engineering and Computer Science, University of Central Florida, Orlando, USA

Abstract. The two most important tasks of network analysis are network embedding and edge flow estimation. The network embedding task seeks to represent each node as a continuous vector, and the edge flow estimation seeks to predict the flow direction and amount along each edge given some known flows of edges. In the past works, they are always studied separately, while their inner connection are completely ignored. In this paper, we fill this gap by building a joint learning framework for both node embedding and flow amount learning. We firstly use a long short-term memory network (LSTM) model to estimate the embedding of a node from its neighboring nodes' embeddings, meanwhile we use the same LSTM model with a multi-layer perceptron (MLP) to estimate a value of the node which presents its importance over the network. The node value is further used to regularize the edge flow learning, so that for each node the balance of flowing-in and flowing-out reach the node value. We simultaneously minimize the reconstruction error of neighborhood LSTM for each node, the approximation error of node value, and the consistency loss between node value and its conjunctive edge flow values. Experiments show the advantage of the proposed algorithm over benchmark datasets.

Keywords: Deep learning · Long short-term memory network · Network embedding · Edge flow learning

1 Introduction

In network analysis, edge flow estimation is a problem of learning the directed flow through each edge connecting two nodes on a graph [4, 10]. Given a graph with a set of nodes and edges, for some edges, we already know the flowing amounts along them, but

H. Mo—Co first author.

D.-S. Huang and P. Premaratne (Eds.): ICIC 2020, LNAI 12465, pp. 467–475, 2020.
https://doi.org/10.1007/978-3-030-60796-8_40

for the remaining edges, the flows are still unknown. It is important to estimate these missing flows from the known flows of edges, and the graph structure. The edge flow estimation has a various range of applications in the real-world, including the areas of transportation control, innovation management, cyber-security, etc. In this paper, we study the problem of edge flow learning by utilizing the network structure and the existing edge flows. This is a semi-supervised learning problem, however, different from the traditional graph-based semi-supervised problem, the object of each data point is an edge, instead of a node. We propose a novel semi-supervised edge flow learning algorithm by exploring the graph structure and the unknown edge flows. To present the graph structure, we use the network embedding technology which represents each node as a low-dimensional continues vector [9], and further use the embedding vectors to approximate a node value to regularize the flows of the node's conjunctive edges.

Despite the wide applications of edge flow learning over networks, there are a limited number of works in this field. Jia et al. [10] proposed a novel edge flow learning method by regularizing the flows of the edges according to the balance of each node's incoming and outgoing flows. Meanwhile, the flows of labeled edges are constraint to the known flows. The assumption of this method is that for each node, the incoming flow should be equal to the outgoing flow for its conjunctive edges. Meanwhile, network embedding aims to represent the nodes of a graph to embedding vectors for the purpose of node classification, link prediction, etc. There are lots of works done in the field of network embedding. For example, Grover and Leskovec [9] proposed a method to learn continuous feature representations for nodes in graphs, named node2vec. It learns the embedding vectors of nodes so that the likelihood of preserving network neighborhoods of nodes can be maximized. To learn from the neighborhood of each node, a biased random walk procedure is developed to efficiently explores diverse neighborhoods. Recently, deep learning has been applied to the graph embedding problems [5, 6, 15, 16]. Zhu et al. [18] proposed a Deep Variational Network Embedding in Wasserstein Space (DVNE) by learning a Gaussian distribution in the Wasserstein space to represent the nodes of graphs. This representation can both reserve the network structure and measure the uncertainty of nodes. The 2-Wasserstein distance is used to measure the similarity between the distributions, so that the transitivity in the network can be preserved. Tu et al. [14] proposed to represent the nodes in a graph by using a popular recursive network model, namely the layer normalized Long Short-Term Memory (ln-LSTM) [12]. The nodes are firstly presented as embedding vectors, and then the neighboring nodes of each node is used as input of a LSTM model, and the output of the model is the embedding vector of the node itself. Moreover, the authors proposed to use a multi-layer perceptron (MLP) model [2] to project the ln-LSTM outputs to the degree of the node. The embedding vectors are jointly learned by minimizing the reconstruction errors the nodes, and the reconstruction errors of the degrees of the nodes. Zhang et al. [17] developed the and arbitrary-order proximity preserved network embedding method, which is based on the singular value decomposition (SVD) framework [8]. The theorem of eigen-decomposition reweighting has been proven to reveal the intrinsic relationship between proximities of different orders. The scalable eigen-decomposition solution is proposed to derive the embedding vectors.

Although there are many existing works of network embedding and also some work for flow estimation over network, however, the current solutions are of the following shortages,

- The existing flow estimation method imposes a strong assumption to balance the flows into/out of a node [10]. It assumes that the amount flowing into a node is equal to the amount flowing out from the same node. However, this assumption does not always hold in all cases.
- The node natural is critical for the flowing in and out amount. This fact has been ignored by all the existing network imbedding and flow estimation methods.
- There is a gap between the network embedding and the edge flow estimation. Accord to our last knowledge, up to now, there is not work done to learn network node embedding vectors for the purpose of edge flow estimation, vice versa. However, according to our observation, the node embedding is a good reflection of the nature of the node, which is critical to estimate the flows of the edge connected to itself. Thus, we believe the edge flow estimation can be improved by the network embedding technology. Furthermore, the edge flow information is also a good guide for the node embedding learning.

To solve the above issues, we propose a novel joint learning framework for both network embedding and edge flow estimation. In this framework, we propose the concept of node value to present the nature of a node. This node value will be the measure of the neighboring edge flowing over the node, and also reflects the value mount attached to this node. The function of the node value is that it cannot only guide the learning of node embedding vector, so that the values can be estimated from the embedding vectors, but also guide the learning of the edge flows. We firstly use a LSTM model to approximate the embedding vector of a node from its neighborhood, then use the embedding vector of each node to estimate the node value, and finally estimate the edge flows according to the nodes connected to the edge. In learning process, we learn the embedding vectors and value of the nodes, and the edge flows together by modeling them into one single unified objective and solving them in an iterative algorithm. In this way, we learn meaningful embeddings, values, and edge flows simultaneously.

2 Proposed Method

Assume we have a graph composed a set of nodes, and a set of edges among the nodes, we denote it as $G = \{\mathcal{V}, \mathcal{E}\}$. $\mathcal{V} = \{1, \cdots, n\}$ is the set of n nodes, and $\mathcal{E} = \left\{\varepsilon_k = (i,j)|_{i,j=1}^{n}\right\} \in \mathcal{V} \times \mathcal{V}$ is the set of edges, where ε_k is the k-th edge between the i-th and j-th nodes. Since all the edges are undirected, without loss of generalization, we impose $i<j$ for all edges, $\varepsilon_k = (i,j) \in \mathcal{E}$. For a portion of the edges, $\varepsilon_k \in \mathcal{E}_L$, we have a flow value for each edge ε_k defined as $\bar{f}_k \in \mathbb{R}$, where the amplitude measures the amount of the flow, and the sign indicators the direction of the flow. $\bar{f}_k > 0$ if the flow is from i-th node to the j-th node, and $\bar{f}_k < 0$. For the other edges, the flow values are unknown. The learning problem of semi-supervised learning problem is to learn the flow values for the edges which are not included in $\mathcal{E}_L$, and we define a set of unlabeled edges as $\mathcal{E}_U$ for these edges. To this end, we define a vector of the flow values of the edges as $\boldsymbol{f} = [f_1, \cdots, f_{|\mathcal{E}|}]^\top \in \mathbb{R}^{|\mathcal{E}|}$, where each element is the flow value of one edge in

$\mathcal{E}$, and the edge flow learning problem is converted to the problem of learning of f from the graph structure of G, subject to $f_k = \bar{f}_k, \forall e_k \in \mathcal{E}_L$.

To learn the flow values of edges, we propose to embed each node $i \in \mathcal{V}$ to a embedding vector of d dimensions, and estimate the overall out flow amount δ_i of this node from its embedding vector. Moreover, we use this flow amount to regularize the flow values learning for its conjunctive edges, so that the sum of the flow amount of the node is close to the summation of the flow values of its conjunctive edges. Moreover, we regularize the embedding function parameters and flow values so that the solution is unique and simple.

- **Recursive node embedding.** The embedding vectors of the nodes are denoted as $\boldsymbol{x}_1, \cdots, \boldsymbol{x}_n$, where $\boldsymbol{x}_i \in \mathbb{R}^d$ is the embedding vector of the i-th node. To learn the embedding vectors, we propose to reconstruct a nodes' embedding vector from a sequence of its neighbors' vectors. The set of neighboring nodes of node i is denoted as

$$\mathcal{N}_i = \{j|(i,j) \in \mathcal{E} \ or \ (j,i) \in \mathcal{E}\}. \tag{1}$$

To create such a neighboring node sequence, we sort the neighboring nodes of the i-th node according to their flow contributions to the i-th node, i.e., the mount of value flowing to the i-th node, denoted as c_i^j, which will be defined in the following section. The sorting is conducted to nodes $j \in \mathcal{N}_i$ according to their contribution values c_i^j descending, so that the most contributing neighboring nodes are ranked at the top, and the least one at the bottom. The sorted sequence of embedding vectors of nodes in $\mathcal{N}_i$ is denoted as

$$\mathcal{S}_i = \{\boldsymbol{x}_{i1}, \cdots, \boldsymbol{x}_{i|\mathcal{N}_i|}\}. \tag{2}$$

To represent the neighborhood structure, we learn a deep recursive neural network to calculate the representation vector for $\mathcal{S}_i$. To this end, we apply the ln-LSTM model as the recursive model. In this model, it slides a LSTM Cell function g over each timestamp $\boldsymbol{x}_t \in \mathcal{S}_i$ sequentially to calculate the output vector $\boldsymbol{h}_t$ of the node from its embedding vector and the previous output vector, $\boldsymbol{h}_{t-1}$,

$$\boldsymbol{h}_t = g(\boldsymbol{x}_t, \boldsymbol{h}_{t-1}; \theta), t = 1, \cdots, |\mathcal{N}_i|, \tag{3}$$

where θ is the parameter set of the LSTM cell function. The output vector of the last timestamp is the output of the recursive model,

$$LSTM(\boldsymbol{x}_{i1}, \cdots, \boldsymbol{x}_{i|\mathcal{N}_i|}; \theta) = \boldsymbol{h}_{|\mathcal{N}_i|} \tag{4}$$

- **Node value estimation from recursive embedding.** To describe such a node-wise value, we define a slack variable $\phi_i \in \mathbb{R}$ for each node. This value is an estimation of the impact and importance of the node, essentially determined by the nature of the node, and it also impacts the flowing of the edges. Since we use the LSTM output $LSTM(\boldsymbol{x}_{i1}, \cdots, \boldsymbol{x}_{i|\mathcal{N}_i|}; \theta)$ to approximate the value amount of the i-th node, we apply a single layer neural network to it to estimate ϕ_i,

$$\phi_i \leftarrow \varphi(LSTM(\boldsymbol{x}_{i1}, \cdots, \boldsymbol{x}_{i|\mathcal{N}_i|}; \theta)) \tag{5}$$

where $\varphi(\boldsymbol{x}) = \sigma(\boldsymbol{w}^\top \boldsymbol{h})$ is the single layer neural network with a rectified linear unit (ReLU) activation function σ.

- **Node value regularized edge flow learning.** To learn the flow value of each edge $\varepsilon_k \in \mathcal{E}, f_k$ we use the node value amount to regularize the edge flow. For each node at any moment of the flowing process, the amount of value ϕ_i hold by this node is equal to the balance of incoming and outgoing values from the conjunctive edges. Given a node $i \in \mathcal{V}$, we define its conjunctive edges as $\mathcal{R}_i = \{\varepsilon_k | \varepsilon_k \in \mathcal{E}, \varepsilon_k = (u, v), u = i \ or \ v = j\}$. We further divide $\mathcal{R}_i$ to a subset of edges with node i as the larger index of the edge, $\mathcal{R}_i^+ = \{\varepsilon_k | \varepsilon_k \in \mathcal{E}, \varepsilon_k = (u, i)\}$, and another subset of edges with node i as the smaller index of the edge, $\mathcal{R}_i^- = \{\varepsilon_k | \varepsilon_k \in \mathcal{E}, \varepsilon_k = (i, v)\}$, so that $\mathcal{R}_i = \mathcal{R}_i^+ \cup \mathcal{R}_i^-$. Since the positive flowing of edge is directed from the smaller index node to the larger index node, while the negative flowing of edges is directed from the larger index node to the smaller index node, the overall flow to a node can be calculated as

$$\phi_i = \sum_{k \in \mathcal{R}_i^+} f_k - \sum_{k \in \mathcal{R}_i^-} f_k = \sum_{k=1}^{|\mathcal{E}|} \tau_{ik} f_k, \tag{6}$$

where

$$\tau_{ik} = \begin{cases} +1, \ if \ k \in \mathcal{R}_i^+ \\ -1, \ if \ k \in \mathcal{R}_i^- \\ 0, \ otherwise. \end{cases} \tag{7}$$

We define a vector of node values as $\boldsymbol{\phi} = [\phi_1, \cdots, \phi_n]^\top \in \mathbb{R}^n$, and a matrix of flow-node contributions mapping, $\Phi = [\tau_{ik}] \in \{+1, -1, 0\}^{n \times |\mathcal{E}|}$. We rewrite (6) as a matrix form as

$$\boldsymbol{\phi} = \Phi \boldsymbol{f}. \tag{8}$$

This constrain imposes that for each node, the flowing-in value amount is consistent to its own node value.

To build the overall learning problem, we consider the three problems of network embedding, node value estimation from embeddings, and edge flow approximation regularized by the node values together, and formulate the following minimization problem,

$$
\begin{aligned}
\min_{f,\phi,x_i|_{i=1}^n,\theta,w} \quad & \{\sum_{i=1}^{n} \|x_i - LSTM(x_{i1}, \cdots, x_{i|\mathcal{N}_i|}; \theta)\|_F^2 \\
& + \lambda_1 \sum_{i=1}^{n} \|\phi_i - \varphi(LSTM(x_{i1}, \cdots, x_{i|\mathcal{N}_i|}; \theta))\|_F^2 \\
& + \lambda_2 \|\phi - \Phi f\|_F^2 \\
& + \lambda_3 (\|f\|_F^2 + \|\phi\|_F^2 + \sum_{i=1}^{n} \|x_i\|_F^2 + \|\theta\|_F^2 + \|w\|_F^2)\}, \\
s.t. \quad & f_k = \bar{f}_k, \forall k : \varepsilon_k \in \mathcal{E}_L.
\end{aligned} \tag{9}
$$

In the objective, the first term is the reconstruction error of the embedding vectors over the recursive model, the second term is the approximation error of the node values from the neighborhood embedding vectors, the third term is the consistency loss of the node value and the flows of the conjunctive edges, and the last term is the squared ℓ_2 normalization term to obtain the unique simplest solution. To solve the problem of (9), we adopt the alternate optimization strategy. In an iterative algorithm, we solve one parameter while fixing the other parameters by using the alternating direction method of multipliers (ADMM) algorithm [3].

3 Experiments

In this section, we experimentally study the performance and properties of the proposed algorithm, namely Embedding to Flow (E2F).

3.1 Benchmark Data Set

In the experiments, we used three benchmark network data sets, which are listed as follows

- the **Minnesota road network** data set, which contains 2642 nodes and 3303 edges, and each node is an intersection, while each edge is a road [7],
- the **US power grid network** of KONECT, which has 4941 nodes and 6593 nodes, where each node is an individual consumer, and each edge is a transmission line [11],
- the **water irrigation network** of Balerma, Spain, which has 447 nodes and 454 edges, while in this case, each node is a water supplies or hydrants, and each edge is a water pipe [13].

3.2 Experimental Setting

To conduct the experiments, given a network, we firstly split the set of edges to a labeled set and an unlabeled set. For the labeled edges, the amount of flow is given as known input of the model, and for the unlabeled edges, the model is supposed to predict the amount of flows from the network structure and the known flows. We use the 10-fold cross validation protocol for the labeled/unlabeled set splitting. We firstly split the entire set of edges to 10 folds of the same size, and then use each fold as a labeled set, while the remaining 9 folds as unlabeled. For each fold, we train the model and predict the unlabeled edges' flows. The overall performance is measured by the Pearson correlation coefficient between the predicted edge flows and the ground truth edge flows of the unlabeled edges [1].

3.3 Experimental Results

Comparison to State-of-the-Arts

We firstly compare the proposed edge flow learning algorithm against the other methods of the same function, including,

- the FlowSSL algorithm, which is a graph-based semi-supervised & active learning for edge flows proposed by Jia et al. [10],
- the LineGraph algorithm, which performs a line-graph transformation of the network, and then uses a standard vertex-based semi-supervised learning method to learn the edge flows. It is a baseline of [10].

The comparison results are shown in Fig. 1. In all four networks experiments, the E2F algorithm stably and significantly outperforms the other two methods. This is a strong evidence of necessary to use network embedding technology to enhance the performance of edge flow estimation.

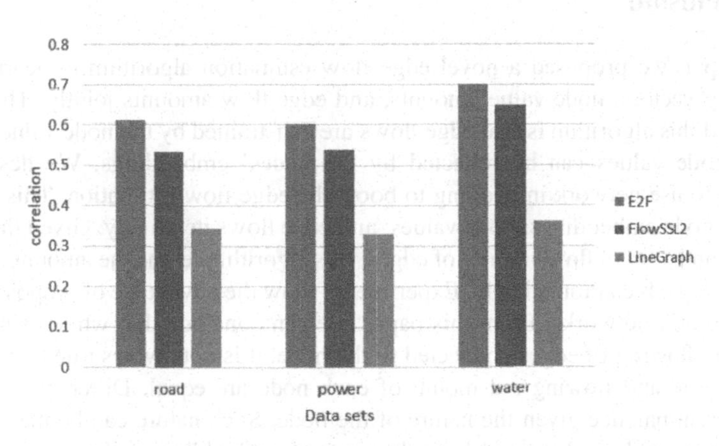

Fig. 1. Comparison results over state-of-the-arts.

Sensitivity to Parameter Changes

In our algorithm, there are three trade-off parameters, λ_1, λ_2 and λ_3. We study how the performance changes with different values of λ_1, λ_2 and λ_3, and plot the curves of correlation measures in Fig. 2. From this figure, we have the following observations.

- For all the four benchmark data sets, when λ_1 is increased from 0.1 to 1, the correlation is also increasing. But when it is larger than 1, the performance's change is not significant. This indicates that the node reconstruction error plays an important role in the edge flow estimation process, which is the reason at a larger value of λ_1 in a certain range can make the performance better.

- When the value of λ_2 is increasing from 0.1 to 100, in most cases, the correlation is increasing, except the case of water supply network with λ_2 increase from 10 to 100. Since λ_2 is the weight of loss term of edge flow learning from the node value, this observation means that the node value is critical for the edge flow estimation.
- The proposed algorithm is very stable to the change of λ_3, the weight of ℓ_2 norm of the parameters. When the value of λ_3 changes, the correlation does not change too much.

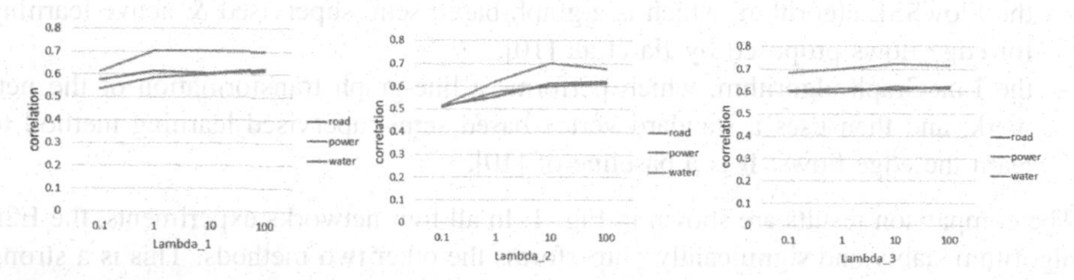

Fig. 2. Curves of performance with different values of trade-off parameters.

4 Conclusion

In this paper, we proposed a novel edge flow estimation algorithm by learning node embedding vectors, node value amounts, and edge flow amounts jointly. The philosophy behind this algorithm is that edge flows are constrained by the node value balances, and the node values can be reflected by the nodes' embeddings. We designed the algorithm to use network imbedding to boost the edge flow estimation. This algorithm learns the node embeddings, node values, and edge flows iteratively. Given the network structure, and a set of flow amount of edges, this algorithm learns the amount of flows of the remaining edge automatically. Experiments show the advantage of proposed method over real-world networks. From this paper, we can conclude that while using node to balance the flowing of edges connected to the node, it is not always true to assume that the flowing-in and flowing-out mount of each node are equal. Different node should have different balance given the nature of the node. Such nature can be measured by a node value variable and learned from the network embedding methods.

References

1. Benesty, J., Chen, J., Huang, Y., Cohen, I.: Pearson correlation coefficient. In: Cohen, I., Huang, Y., Chen, J., Benesty, J. (eds.) Noise Reduction in Speech Processing. STSP, vol. 2, pp. 1–4. Springer, Heidelberg (2009). https://doi.org/10.1007/978-3-642-00296-0_5
2. Bourlard, H., Kamp, Y.: Auto-association by multilayer perceptrons and singular value decomposition. Biol. Cybern. **59**(4-5), 291–294 (1988). https://doi.org/10.1007/BF00332918

3. Boyd, S., Parikh, N., Chu, E., Peleato, B., Eckstein, J.: Distributed optimization and statistical learning via the alternating direction method of multipliers. Found. Trends® Mach. Learn. **3**(1), 1–122 (2011)
4. Even, S., Tarjan, R.E.: Network flow and testing graph connectivity. SIAM J. Comput. **4**(4), 507–518 (1975)
5. Geng, Y., et al.: Learning convolutional neural network to maximize pos@ top performance measure. In: ESANN 2017 - Proceedings, pp. 589–594 (2016)
6. Geng, Y., et al.: A novel image tag completion method based on convolutional neural transformation. In: Lintas, A., Rovetta, S., Verschure, P.F.M.J., Villa, A.E.P. (eds.) ICANN 2017. LNCS, vol. 10614, pp. 539–546. Springer, Cham (2017). https://doi.org/10.1007/978-3-319-68612-7_61
7. Gleich, D.F., Saunders, M.: Models and algorithms for pagerank sensitivity. Stanford University, Stanford (2009)
8. Golub, G.H., Reinsch, C.: Singular value decomposition and least squares solutions. In: Bauer, F.L. (ed.) Linear Algebra. HDBKAUCO, vol. 2, pp. 134–151. Springer, Heidelberg (1971). https://doi.org/10.1007/978-3-662-39778-7_10
9. Grover, A., Leskovec, J.: node2vec: scalable feature learning for networks. In: Proceedings of the 22nd ACM SIGKDD International Conference on Knowledge Discovery and Data Mining, pp. 855–864. ACM (2016)
10. Jia, J., Schaub, M.T., Segarra, S., Benson, A.R.: Graph-based semi-supervised & active learning for edge flows. arXiv preprint arXiv:1905.07451 (2019)
11. Kunegis, J.: KONECT: the Koblenz network collection. In: Proceedings of the 22nd International Conference on World Wide Web, pp. 1343–1350. ACM (2013)
12. Liwicki, M., Graves, A., Fernàndez, S., Bunke, H., Schmidhuber, J.: A novel approach to on-line handwriting recognition based on bidirectional long short-term memory networks. In: Proceedings of the 9th International Conference on Document Analysis and Recognition, ICDAR 2007 (2007)
13. Reca, J., Martínez, J.: Genetic algorithms for the design of looped irrigation water distribution networks. Water Resour. Res. **42**(5) (2006)
14. Tu, K., Cui, P., Wang, X., Yu, P.S., Zhu, W.: Deep recursive network embedding with regular equivalence. In: Proceedings of the 24th ACM SIGKDD International Conference on Knowledge Discovery & Data Mining, pp. 2357–2366. ACM (2018)
15. Zhang, G., et al.: Learning convolutional ranking-score function by query preference regularization. In: Yin, H., et al. (eds.) IDEAL 2017. LNCS, vol. 10585, pp. 1–8. Springer, Cham (2017). https://doi.org/10.1007/978-3-319-68935-7_1
16. Zhang, G., Liang, G., Su, F., Qu, F., Wang, J.-Y.: Cross-domain attribute representation based on convolutional neural network. In: Huang, D.-S., Gromiha, M.Michael, Han, K., Hussain, A. (eds.) ICIC 2018. LNCS (LNAI), vol. 10956, pp. 134–142. Springer, Cham (2018). https://doi.org/10.1007/978-3-319-95957-3_15
17. Zhang, Z., Cui, P., Wang, X., Pei, J., Yao, X., Zhu, W.: Arbitrary-order proximity preserved network embedding. In: Proceedings of the 24th ACM SIGKDD International Conference on Knowledge Discovery & Data Mining, pp. 2778–2786. ACM (2018)
18. Zhu, D., Cui, P., Wang, D., Zhu, W.: Deep variational network embedding in wasserstein space. In: Proceedings of the 24th ACM SIGKDD International Conference on Knowledge Discovery & Data Mining, pp. 2827–2836. ACM (2018)

An Effective Multi-label Classification Algorithm Based on Hypercube

Yuping Qin[1(✉)], Xueying Cheng[2], Xiangna Li[3], and Qiangkui Leng[4]

[1] College of Engineering, Bohai University, Jinzhou 121013, China
qlq888888@sina.com
[2] College of Mathematics and Physics, Bohai University, Jinzhou 121013, China
[3] Beijing Guo Dian Tong Network Technology Co., Ltd, Beijing 100761, China
[4] College of Information Science and Technology, Bohai University, Jinzhou 121013, China

Abstract. To solve the problem of multi-label classification, a classification algorithm based on hypercube is proposed. For each label in the training sample set, a minimum hypercube containing all its samples is constructed in the sample space. In classification, the labels of the sample to be classified are determined according to the hypercubes to which the sample belongs. If the sample to be classified is not in any hypercube, the magnification factor of the hypercube containing the sample to be classified is calculated respectively, and the label is determined according to the magnification factor. The algorithm avoids the influence of unbalanced data and has good scalability. The experimental results show that the algorithm has faster training speed and classification efficiency, and has higher classification accuracy. The larger the training set size and the more labels, the more obvious the effect.

Keywords: Classification · Multi-label · Hypercube · Amplification factor

1 Introduction

Multi-label refers to a data instance associated with multiple class labels. For example, a document can cover multiple topics, an image can be annotated with multiple tags, and a single gene can be associated with several functional categories, and etc. Multi-label classification is to classify an instance into a group of labels. Multi-label classification has attracted people's attention due to the rapid growth of the application field, and has become a research hotspot in the field of machine learning [1–5].

At present, the main research results of multi-label classification include problem transformation strategy, support vector machine, neural network, decision tree and K-nearest neighbor [6–9]. These methods have been successfully applied to text classification [10, 11], image recognition [12, 13], genomics [14, 15], and emotional classification [16, 17].

Most of the existing multi-label classification methods transform the multi-label classification problem into multiple binary classification problems. If the size of the data set is large, the training speed is slow, and the training time required brings great difficulties to practical application. If the number of labels in the dataset is large, the

D.-S. Huang and P. Premaratne (Eds.): ICIC 2020, LNAI 12465, pp. 476–483, 2020.
https://doi.org/10.1007/978-3-030-60796-8_41

classification efficiency is relatively low, and its complex calculation can not meet the actual needs of real-time classification. If the data of different labels in the data set is unbalanced, the classification accuracy will be affected. The unbalance problem is a difficult problem to solve in multi-label classification. In addition, the scalability and inheritance of these methods are poor. If new label samples are added, the classifier needs to be retrained. Therefore, the methods based binary classifications are not very applicable in many cases.

On the basis of SVM, a single-valued classification method named one-class SVM, also known as support vector domain description, is proposed in literature [18]. The idea is to describe the set of data by calculating the minimum hypersphere containing a set of data, and the minimum hypersphere is used as the classifier of One-class problem. Based on one-class SVM, a hypersphere support vector machine multi-label classification algorithm is proposed in literature [19]. This algorithm trains a hypersphere for each label sample. When classifying, the labels of the sample to be classified are determined by the hyperspheres to which it belongs. Because the distribution of samples is mostly convex and hyperellipsoid with different directions, a multi-label classification algorithm based on hyperellipsoid support vector machine is proposed in literature [20]. This algorithm describes the sample set of each label with a hyperellipsoid. When classifying, the labels of the sample to be classified are determined according to the hyperellipsoids to which it belongs.

The method based on one-class SVM effectively solves the problem of multi-label classification, and is not affected by unbalanced samples, because the domain range of each SVM is determined by only one class of samples. At the same time, the method has inheritance and extensibility. When a new label sample is added, the classifier independent of the new label need not be retrained. However, for large-scale data sets, the training time of this method is too long and the number of support vectors is too large. For practical applications which require high real-time performance, the classifier trained on large-scale data sets is usually not available.

In order to solve the problem of training speed and classification efficiency of large-scale data sets, based on the advantages of data convex hull description, this paper proposes a multi-label classification algorithm based on hypercube. For a subset of training samples with the same label, a minimum hypercube is constructed in the sample space to enclose all the samples, so that the samples with the same label are bounded by a hypercube. For the sample to be classified, the labels are determined according to the hypercubes to which it belongs.

The rest of this paper is organized as follows. The construction method of hypercube is given in Sect. 2. The multi-label classification algorithm based on hypercube is elaborated in detail in Sect. 3. The experimental results and analysis on the standard dataset are given in Sect. 4. Finally, the conclusion is drawn.

2 Construction of Hypercube

Given a sample set with the same label $X = \{x_i\}_{i=1}^{l}$, where $x_i \in R^n$, l is the number of the samples. Constructing a minimal hypercube enclosing all samples in the feature space, which is denoted as $HC = (\alpha, \beta, o)$, where α is the maximum vertex, β is the minimum vertex, and o is the center. The solid line in Fig. 1 is the smallest hypercube that surrounds the same class of samples. The maximum vertex α is calculated according to formula (1). The minimum vertex β is calculated according to formula (2). The center o is calculated according to formula (3).

$$\alpha_i = \max\{x_{ij}|j = 1, 2, \cdots, l\} \tag{1}$$

$$\beta_i = \min\{x_{ij}|j = 1, 2, \cdots, l\} \tag{2}$$

$$o_i = (\alpha_i + \beta_i)/2 \tag{3}$$

At this time, any sample $\boldsymbol{x_i}$ in the data set $\boldsymbol{X}$ satisfies formula (4).

$$\beta_i \leq x_{ij} \leq \alpha_i, j = 1, 2, \cdots, n \tag{4}$$

If a new sample x is added to X and the sample point is outside the hypercube, the minimum hypercube containing x can be obtained by amplification factor. Firstly, the shortest distance $d_i(x)(i = 1, 2, \cdots n)$ from $\boldsymbol{x_i}$ to α_i and β_i is calculated according to formula (5), and then the amplification factor $F(x)$ of the hypercube is calculated according to formula (6). The smallest hypercube containing x obtained by magnification factor is shown as dotted line in Fig. 1.

$$d_i(x) = \begin{cases} \min(|x_i - \alpha_i|, |x_i - \beta_i|) & x_i < \beta_i \ or\ x_i > \alpha_i \\ 0 & \beta_i \leq x_i \leq \alpha_i \end{cases} \tag{5}$$

$$F(x) = \max\left\{1 + \frac{d_i(x)}{\beta_i - o_i}|i = 1, 2, \cdots, n\right\} \tag{6}$$

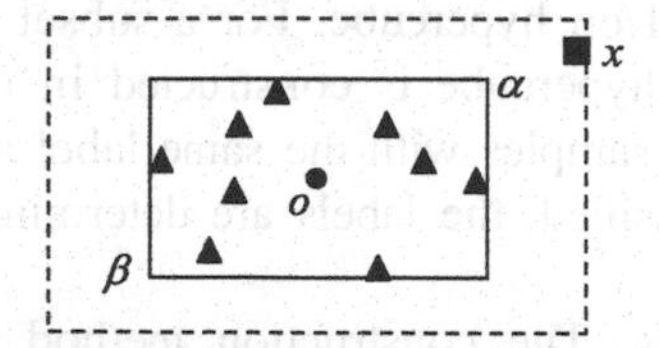

Fig. 1. Minimum hypercube

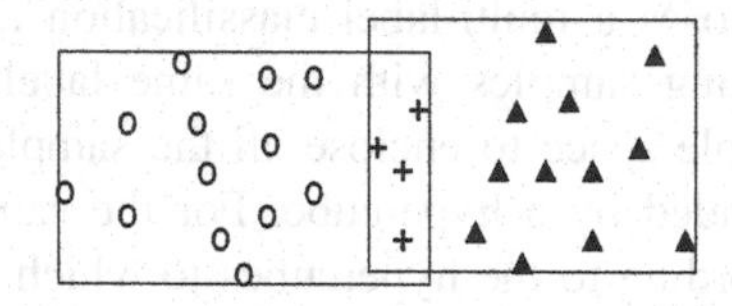

Fig. 2. Minimum hypercube domain division

3 Multi-label Classification Algorithm

The multi-label classification algorithm based on hypercube consists of training and classification. The training process is to construct the smallest hypercubes enclosing the same label samples in feature space, and delimit the different label samples by the hypercubes. Since a sample may correspond to multiple different labels, i.e. it has a label set, the constructed hypercubes are not independent of each other. Multiple hypercubes will cross each other. The samples in the intersection area have the labels represented by the intersecting hypercubes. The division of hypercube domain in feature space is shown in Fig. 2. The classification process is to calculate the region of the sample to be classified in the feature space, and its labels are determined by the hypercubes in which it belongs. The flow chart of multi-label classification algorithm based on hypercube is shown in Fig. 3.

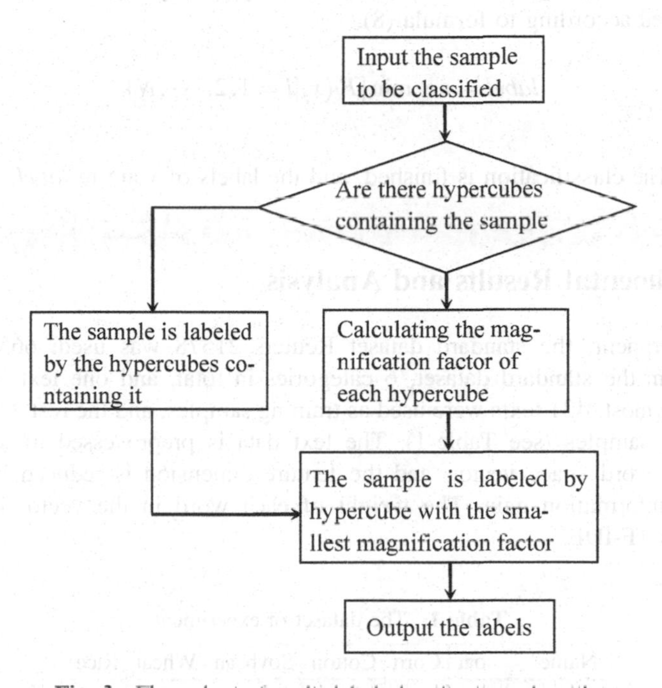

Fig. 3. Flow chart of multi-label classification algorithm

Given the multi-label sample set $A\{x_i, E_i\}_{i=1}^{l}$, where $x_i \in R^n$, $E_i = \{y_{ij}\}_{j=1}^{p}, y_{ij} \in \{1, 2, 3, \cdots, N\}$, l is the number of samples in the sample set A, N is the number of labels in sample set A, $p(1 \leq p \leq N)$ is the number of labels for sample x_i. $A^m = \{x_i^m\}_{i=1}^{l_m}$ is a subset of samples with label $m(1 \leq m \leq N)$, where l_m is the number of samples in the sample set A^m.

For each sample subset $A^m(1 \leq m \leq N)$, Constructing hypercube $HC_m = (\alpha_m, \beta_m, o_m)$ in feature space according to formula (1), formula (2) and formula (3). The multi-label classifier formula (7) is obtained.

$$C = (HC_1, HC_2, \cdots, HC_N) \tag{7}$$

For the sample x to be classified, the classification process is specifically described as follows:

Step 1: $label = \phi$, $m = 1$.
Step 2: Determine whether x is in the hypercube HC_m according to formula (4), if $x \in HC_m$, go to step 3, otherwise go to step 4.
Step 3: Add m to *label* and go to step 4.
Step 4: $m = m+1$, if $m \leq N$, go to step 2, otherwise go to step 5.
Step 5: If $label \neq \phi$, go to step 7, otherwise, go to step 6.
Step 6: The amplification factor $F_i(x)$ of each hypercube $HC_i(i = 1, 2, \cdots N)$ is calculated according to formula (5) and (6) respectively, and then the label of $\boldsymbol{x}$ is determined according to formula (8).

$$label = \arg\min_{l}\{F_l(x)|i = 1, 2, \cdots, N\} \tag{8}$$

Step 7: The classification is finished, and the labels of x are in *label*.

4 Experimental Results and Analysis

In the experiment, the standard dataset Reuters 21578 was used. 665 texts were selected from the standard dataset, 6 categories in total, and one text belongs to 3 categories at most. 431 texts were used as training samples, and the rest 234 texts were used as test samples (see Table 1). The text data is preprocessed to form a high-dimensional word space vector, and the feature dimension is reduced by using the method of information gain. The weight of each word in the vector is calculated according to TF-IDF.

Table 1. The dataset of experiment

Name	oat	Corn	Cotton	Soybean	Wheat	Rice
#Class	1	2	3	4	5	6
#Training	9	168	44	79	204	44
#Testing	5	84	22	40	101	23

In order to compare the performance of the algorithm, the hypercube method and the hypersphere method are used to carry out classification experiments on the same dataset. The kernel function of hypersphere method is radial basis function $\kappa(x, y) = e^{-\gamma\|x-y\|2}$, where $\gamma = 0.001$. System parameter $v = 0.6$.

In the experiment, the average precision, the average recall and the average F_1 value are used as the evaluation indexes, which are defined as formula (9), formula (10) and formula (11), respectively.

$$\text{Average Precision}(AP) = \frac{1}{n}\sum\frac{N_c}{N_a} \tag{9}$$

$$\text{Average Recall}(AR) = \frac{1}{n}\sum \frac{N_c}{N_r} \tag{10}$$

$$\text{Average F}_1\ (AF) = \frac{1}{n}\sum \frac{2 \times AP \times AR}{AP + AR} \tag{11}$$

Where, N_r is the actual number of labels of a test sample, N_a is the number of labels of a sample test result, and N_c is the correct number of labels of a sample test result. If n is the number of sample with the same number of labels, formula (9), formula (10) and formula (11) are called micro average precision (MIAP), micro average recall (MIAR) and micro average F_1 value (MIAF) respectively. If n is the total number of test sample, formula (9), formula (10) and formula (11) are called macro average precision (MAAP), macro average recall (MAAR) and macro average F_1 value (MAAF) respectively.

The experimental environment is i5-6500 CPU 3.20ghz, 8 GB memory, windows 8.1 operating system.

Table 2 shows the micro average precision, micro average recall and micro average F_1 values of the two algorithms. Table 3 shows the macro average precision, macro average recall and macro average F_1 values of the two algorithms. Table 4 shows the training time and classification time of the two algorithms.

Table 2. Comparison of MIAP, MIAR and MIAF

Algorithm	#Class	MIAP(%)	MIAR(%)	MIAF(%)
Hypersphere	1	71.34	73.91	72.14
	2	83.33	55.32	63.93
	3	100.00	50.00	65.00
Hypercube	1	70.85	74.57	71.73
	2	78.52	70.37	72.26
	3	66.67	68.33	67.67

It can be seen from the experimental results that the precision of the hypercube algorithm is slightly lower than that of the hypersphere algorithm. This is because the region of the hypercube of the same kind of sample is larger than that of the hypersphere, if there are noise points, the area of hypercube is larger, which makes the cross area of different types of hypercube increase, and leads to the decrease of precision. The training speed of the hypercube algorithm has been significantly improved compared to the hypersphere algorithm. This is because the hypersphere training is a convex quadratic programming (QP) problem, and the time complexity of the standard QP optimization algorithm is $O(n^3)$ (n is the number of samples), with the increase of the number of training samples, the memory space required for QP solution increases rapidly, and the solution time also increases significantly. While the hypercube construction is only to find the maximum and minimum value of each dimension, and the time complexity is $O(nm)$ (n is the number of samples, m is the dimension). The

Table 3. Comparison of MAAP, MAAR and MAAF

Algorithm	MAAP(%)	MAAR(%)	MAAF(%)
Hypersphere	78.38	77.92	77.52
Hypercube	76.22	78.41	77.24

Table 4. Comparison of training time and testing time

Algorithm	Training time(ms)	Testing time(ms)
Hypersphere	159	108
Hypercube	78	51

classification speed of the hypercube algorithm has also been significantly improved compared to the hypersphere algorithm. This is because the classification calculation of hypersphere algorithm is to solve the distance between the sample to be classified and the center of hypersphere, and all the support vectors participate in the calculation, the more support vectors are, the more complex the calculation is, the longer the classification time is. Therefore, the classification calculation of the hypercube algorithm is to determine whether each dimension of the sample to be classified is in the interval of the corresponding dimension of the hypercube, which is a simple comparison calculation.

5 Conclusion

For the problem of multi-label classification, a hypercube classifier is designed. In the process of training hypercube, only the maximum and minimum values of each dimension need to be calculated in the feature space. The calculation of training process is simple, the imbalance influence of training samples is avoided, and the classifier has expansibility. In the process of classification, it only needs to calculate the interval of each dimension of the sample to be classified, and the complexity of the classification process is low. The algorithm not only has higher precision and recall, but also has faster classification speed and training speed. The larger the sample set and the more labels, the more obvious the advantage of the algorithm, which effectively solves the problem of multi-label classification of large-scale dataset. However, this method is greatly influenced by outliers and sample distribution. In the further work, one is to design algorithm to remove the abnormal data in the multi-label dataset, and the other is to use kernel function to increase the density of samples in the feature space.

Acknowledgement. This work is supported by the National Natural Science Foundation of China under Grant 61602056, "Xingliao Yingcai Project" of Liaoning, China under Grant XLYC1906015, Natural Science Foundation of Liaoning, China under Grant 20180550525 and 201601348, Education Committee Project of Liaoning, China under Grant LZ2016005.

References

1. Madjarov, G., Kocev, D., Gjorgjevikj, D.: An extensive experimental comparison of methods for multi-label learning. Pattern Recogn. **45**(9), 3084–3104 (2012)
2. Zhang, M.L., Zhou, Z.H.: A review on mutil-label learning algorithms. IEEE Trans. Knowl. Data Eng. **26**(8), 1819–1837 (2014)
3. Gibaja, E., Ventura, S.: A tutorial on multi-label learning. ACM Comput. Surv. **47**(3), 1–38 (2015)
4. Yeh, C.K., Wu, W.C., KO, W.J.: Learning deep latent space for multi-label classification. In Proceedings of the Thirty-First AAAI Conference on Artificial Intelligence, AAAI, Menlo Park, pp. 2838–2844 (2017)
5. Luo, F.F., Guo, W.Z., Yu, Y.L.: A multi-label classification algorithm based on kernel extreme learning machine. Neurocomput. **260**, 313–320 (2017)
6. Feng, P., Qin, D., Ji, P.: Multi-label learning algorithm with SVM based association. High Technol. Lett. **25**(1), 97–104 (2019)
7. Zhuang, N., Yan, Y., Chen, S.: Multi-label learning based deep transfer neural network for facial attribute classification. Pattern Recogn. **80**, 225–240 (2018)
8. Cai, Z., Zhu, W.: Feature selection for multi-label classification using neighborhood preservation. IEEE/CAA J. Automatica Sinica **5**(1), 320–330 (2018)
9. Prati, R.C., Charte, F., Herrera, F.: A first approach towards a fuzzy decision tree for multilabel classification. In: Proceeding of IEEE Conference on Fuzzy Systems, Piscataway, pp. 1–6. IEEE (2017)
10. Brucker, F., Benites, F., Sapozhnikova, E.: Multi-label classification and extracting predicted class hierarchies. Pattern Recogn. **44**(3), 724–738 (2011)
11. Agrawal, S., AgrawaL, J., Kaur, S.: A comparative study of fuzzy PSO and fuzzy SVD-based RBF neural network for multi-label classification. Neural Comput. Appl. **29**(1), 245–256 (2018)
12. Zhang, M.L., Zhou, Z.H.: ML-KNN: a lazy learning approach to multi-label learning. Pattern Recogn. **40**(7), 2038–2048 (2007)
13. Wang, P., Zhang, A.F., Wang, L.Q.: Image automatic annotation based on transfer learning and multi-label smoothing strategy. J. Comput. Appl. **38**(11), 3199–3203 (2018)
14. Chou, K.C., Shen, H.B.: Cell-PLoc 2.0: an improved package of web-servers for predicting subcellular localization of proteins in various organisms. Nat. Sci. **2**(10), 1090–1103 (2010)
15. Guan, R.C., Wang, X., Yang, M.Q.: Multi-label deep learning for gene function annotation in cancer pathways. Sci. Rep. **8**(1), 267 (2018)
16. Tax, D., Duin, R.: Outliers and data descriptions. In: Proceeding of the 7th Annual Conference of the Advanced School for Computing and Imaging, Betascript Publishing Beau Bassin, pp. 234–241 (2001)
17. Trohidis, K., Tsoumakas, G., Kalliris, G.: Multi-label classification of music into emotions. In: Proceeding of the Ninth International Conference on Music Information Retrieval, Drexel University, Philadelphia, pp. 325–330 (2008)
18. Tomar, D., Agarwal, S.: Multi-label classifier for emotion recognition from music. In: Nagar, A., Mohapatra, D.P., Chaki, N. (eds.) Proceedings of 3rd International Conference on Advanced Computing, Networking and Informatics. SIST, vol. 43, pp. 111–123. Springer, New Delhi (2016). https://doi.org/10.1007/978-81-322-2538-6_12
19. Qin, Y.P., Wang, X.K., Wang, C.L.: An incremental learning algorithm for multi-class sample. Control Decis. **24**(1), 137–140 (2009)
20. Qin, Y.P., Chen, Y.D., Wang, C.L.: A new multi-label text classification algorithm. Comput. Sci. **38**(11), 204–205 (2011)

Using Self Organizing Maps and K Means Clustering Based on RFM Model for Customer Segmentation in the Online Retail Business

Rajan Vohra[1(✉)], Jankisharan Pahareeya[2], Abir Hussain[1], Fawaz Ghali[1], and Alison Lui[3]

[1] Department of Computer Science, Liverpool John Moores University, Liverpool, UK
{r.vohra,a.hussain,f.ghali}@ljmu.ac.uk
[2] Rustamji Institute of Technology, BSF Academy, Tekanpur, Gwalior, India
talkto.pahariya@gmail.com
[3] School of Law, Liverpool John Moores University, Liverpool, UK
a.lui@ljmu.ac.uk

Abstract. This work based on the research of Chen et al. who compiled sales data for a UK based online retailer for the years 2009 to 2011. While the work presented by Chen et al. used k means clustering algorithm to generate meaningful customer segments for the year 2011, this research utilised 2010 retail data to generate meaningful business intelligence based on the computed RFM values for the retail data set. We benchmarked the performance of k means and self organizing maps (SOM) clustering algorithms for the filtered target data set. Self organizing maps are utilized to provide a framework for a neural networks computation, which can be benchmarked to the simple k means algorithm used by Chen et al.

Keywords: Online retail data · RFM model · K means clustering · Self-organizing maps · Business intelligence

1 Introduction

According to the retail and e-commerce sales figures put out by e-marketer, the total online retail sales for 2019 was 106.46 billion pounds representing 22.3% of total retail sales. This is expected to grow to 139.24 billion pounds in 2023 representing 27.3% of total retail sales in the UK. The percentage of mobile commerce using smart phones expected to rise from 58.9% in 2019 to 71.2% in 2023. According to the office for National Statistics, UK, clothes or sports goods account for 60% of the goods purchased online in Great Britain in 2019. The other key goods or services are: House hold goods representing 49%, holiday accommodation with 44%, travel arrangements with 43% and tickets for events 43%. Retailers are interested in gaining business intelligence about their customers. This can represent the buying patterns, expenditure, repeat purchases, longevity of association and high profit customer segments. In addition sales pattern by region, season and time are key components of such knowledge. This enables design of suitable marketing campaigns and the discovery of new patterns in

D.-S. Huang and P. Premaratne (Eds.): ICIC 2020, LNAI 12465, pp. 484–497, 2020.
https://doi.org/10.1007/978-3-030-60796-8_42

the sales data which were previously unknown to the retailer. Chen et al. [1] have analysed the data set for an online retailer for the year 2011. Using k means clustering they have derived meaningful customer segments, and then used decision tree based rule induction to build decision rules that represent gained business intelligence. This work based on the work of Chen et al. by using simple k means clustering [12] and self-organizing maps [15] to perform clustering for the 2010 retail data set.

RFM (Recency, Frequency, and Monetary) is a model to analyse the shopping behavior of a customer. Recency represents the duration in time since the last purchase while frequency represents the number of purchases made in a given time period and monetary denotes the amount spent by a customer in the given time period which in the analysis performed is the calendar year 2010. Dogan et al. [2] used RFM computations and k means clustering to segment customers of a sports retail company based in Turkey to design a customer loyalty card system based on this analysis. The k means clustering analysis performed by Dogan et al. has used the retail data set designed by Hu & Yeh [6]. Sarvari et al. [7] used RFM analysis on a global food chain data set. They used k means clustering and association rule mining for segmenting customers and buying patterns. They highlighted the importance of assigning weights to RFM values. Yeh et al. [8] added time since first purchase to the basic RFM model which improved the predictive accuracy of the RFM model. Our analysis uses time in months to indicate the first purchase made by a customer in the time period under consideration – for year 2010 in our analysis. Wei et al. [9] have discussed comprehensively a review of the RFM model including its scoring scheme, applications especially in customer segmentation, merits and demerits, along with how RFM model can be extended to perform a more comprehensive analysis by adding other variables like churn and also incorporating call centre data. Customer segmentation using Neural networks is demonstrated for the global tourist business by Bloom [10]. Holmbom et al. [11] used self-organizing maps to cluster customers for portfolio analysis in order to determine profitability for target marketing purposes – in this study they used both demographic data as well as product profiles. Vellido et al. used self-organizing maps for segmenting online customer data [13]. Self-organizing maps can be visualized using the U matrix (unified distance matrix) which displays the Euclidean distance between neurons, according to ultsch [14]. While Kiang et al. [17] used self-organizing maps to discover interesting segments of customers in telecommunication service providers data sets. Although Chen et al. published their retail analytics paper in 2012 [1], the associated data set was uploaded on the UCI machine learning repository only in September 2019 [5]. Chen et al. have analysed the data set for an online retailer for the year 2011. We want to analysed for the online retailer for the year 2010 data which Chen et al. did not do. We also want to explore the potential of neural network that's why have chosen SOM on account of its Neural networks framework because of its robust architecture and lesser sensitivity to Noise in the input data set.

We did our study in two phases:

Phase 1: Cluster Profiling.

Phase 2: Performance benchmarking.

these two phases discus in details in research methodology Sect. 2.

The data set pertains to the operations of a small online retailer based in the UK. Chen et al. used RFM model and k means clustering to derive a new segmentation of customers of this online retailer. After profiling the clusters generated, decision tree based rule induction was used to derive decision rules representing business intelligence gained. We use k means and self organizing maps as well as RFM values computed to perform clustering to obtain new customer segments. While Chen et al. used 2011 data, we use the data set for 2010, from the same data source. We have profiled the clusters generated and compare the performance of k means and self organizing maps based on certain parameter values generated during execution. The remainder of this paper is organized as follows. Section 2 discusses the proposed research methodology while Sect. 3 shows the utilized data set, Sect. 4 describes the pre processing steps performed on the data set while Sect. 5 describes Simulation results and discussion. The final section concludes the paper.

2 Research Methodology

The original retail data set consists of 5,25,461 records for 2009–10 and 5,41,910 records for 2010–11. We determined 3940 distinct customer ids in the year 2010. There were 151 outliers which yielded a total of 3879 records in the filtered data set which was used for clustering computations. The first step is Data preparation and pre processing. The second step is to generate the target data set which has the distinct customer ids and the computed RFM values for each customer id. Once the target data set is generated we perform the cluster analysis using both the k means and self organizing maps. This analysis is performed using WEKA version 3.8.3. In order to get the target data set from the raw data set we have used the Excel data set in conjunction with MS Access. After the generation of the target data set, removal of outliers and normalization was done to obtain the final filtered data set which was the input for the two clustering algorithms- the K means and Self organizing maps.

The Computations in this study occur in two phases:

Phase 1: Cluster Profiling.
Phase 2: Performance benchmarking.

In Cluster Profiling we start by setting K = 4 for generating the cluster profiles, for both k means and self organizing maps. The Objective is to demonstrate the generation of cluster profiles for each of these techniques. The four clusters generated for each technique represent knowledge gained from the analysis- Distribution of data instances across clusters, Total monetary value represented by each cluster, Mean values of R, F, M, and FP for each cluster and mean spending per customer for the cluster. This gives a detailed profile of the generated clusters.

In Performance benchmarking, we compare the performance of K means and SOM by the following parameters: Execution Time, Number of iterations, Space complexity and Time complexity. The value of K is now varied for K = 2, 4 and 6. The corresponding values of these parameters are computed, tabulated and bar charts are drawn for Execution time and Number of iterations. These computed values benchmark the performance of these two clustering techniques. The Time and Space Complexity are computed for each.

The methodology adopted is this study different from Chen et al. due to the factors of introducing a neural networks computational framework in the form of SOM in addition to using K means clustering, Analysis of the Data set not studied by Chen et al. (2010), and implementing performance benchmarking of the two clustering techniques for the data set. The detailed steps contained in Data preparation and target data set generation are described in the next two sections.

3 The Data Set

The online retail data set was uploaded on 21st September 2019 on the UCI machine learning, repository. The original data set is processed suitably to create a Target Data set which is then analysed to generate the clusters. While the original data set contained 11 attributes we selected six attributes for starting the data preparation as shown in Table 1.

Table 1. Data attributes for our ML algorithms

Name	No. of Digits	Description
Invoice Number	6	Identifies each transaction uniquely
Item code	5	Identifies each product uniquely
Quantity	Numeric	The quantity per item purchased by a customer
Unit price	Numeric	Price per unit of an item
Invoice Date	Date	Date and Time of each transaction
Customer id	5 digit	Uniquely identifies each distinct customer

The Customer id is used instead of the Post code as Post code is subject to the data protection laws of the UK.

4 Data Pre Processing

The next step is to create a number of variable for our machine learning (ML) algorithms including the Amount which is calculated as the Quantity in to Unit Price. The Amount is computed for each distinct customer id for the country = UK. We then computed the number of distinct customer ids in our 2010 data set. Next segregate Date & Time components of the Invoice Date data so that distinct date & time values can be obtained for the transactions in the data set. Considering only the UK transactions we delete records with no customer id and also any missing records. Three aggregate variables have been created including recency (r), frequency (f) and monetary (m). These have the following interpretation:

Recency (r): Measures the recency of the transactions made by any customer value is in months.
Frequency (f): Measures the frequency of the purchases made by a customer over a time period in our case the year 2010.
Monetary (m): Measures the total amount spent by a customer across transactions over the year 2010.

First Purchase: Time in months since the beginning of 2010 when the first purchase is made by a customer. Accordingly the Target Data set consists of the following five attributes: Customer id, Recency, Frequency, Monetary and First Purchase. The work flow of our approach is shown in Fig. 1, while Algorithm 1 illustrated the proposed methodology.

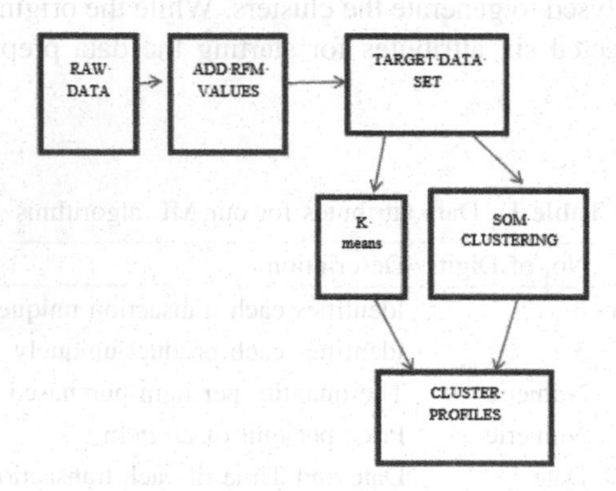

Fig. 1. The work flow

There are two computational tasks in this study – Cluster profiling and Performance benchmarking. The Cluster Profiling is done using simple K means and SOM clustering with K set at 4, and uses the target data set which consists of five attributes – namely: Cust_id, R,F,M and FP.

Performance bench marking for both the techniques is done using 4 key parameters of Execution time, Number of iterations, Space complexity and Time complexity with K varying from 2 through 4 to 6. The computed values of these computations are tabulated in Table 10.

Algorithm 1: Our proposed methodology for the analysis of online retail data.
Let X represents a set of retail data for online shopping customers where
X = {Invoice, Item code, Description, Quantity, Price, Invoice Data, Customer ID}
Let $C \subset X$, a set of customers with a number of f transactions.
$C = \{c \mid c \text{ has } f > 0\}$
$\forall c \in C, \exists$ r and m $\Rightarrow$ m is the monetary and r is the recency.
now add r, f and m to the X, X become X1 for online shopping customers where
X1 = {Customer ID, Monetary, Frequency, Recency, First Purchase Month}
$\forall c \in C, \exists$ outlier removal of X1.
$\forall c \in C, \exists$ normalization of X1.

Let ML to be our machine learning set
ML = {K means, SOM}
∀ml ∈ ML, find c cluster using, r, f, m and fp where fp is the first purchase

Algorithm 2: Our proposed methodology for Performance benchmarking for the K means and Self organizing maps.
Let X1 represents a set of retail data for online shopping customers where
X1 = {Customer ID, Monetary, Frequency, Recency, First Purchase Month}
For cluster Ck (k = 2, 4, 6)
calculate Ck from algorithm 1
∀ml∈ ML, find E, N, S and T

where E is Execution time
where N is No of iterations
where S is Space complexity
where T Time complexity.

Many machine learning algorithms are sensitive to the range and distribution of attribute values in the input data. Outliers in input data can skew and mislead the training process of machine learning algorithms resulting in longer training times, less accurate models and ultimately poorer results. so we did outliers removal based on interquartile ranges. We determined 3940 distinct customer ids in the year 2010. There were 151 outliers which yielded a total of 3879 records in the filtered data set which was used for clustering computations. we also did normalization, The goal of normalization is to change the values of numeric columns in the dataset to use a common scale, without distorting differences in the ranges of values or losing information. Normalization is also required for some algorithms to model the data correctly. We normalized data followed by outlier removal using weka tool. Normalization and outliers removal are provided in Fig. 2. Red data point is showing outliers in Fig. 2 (Figs. 3, 4, and 5).

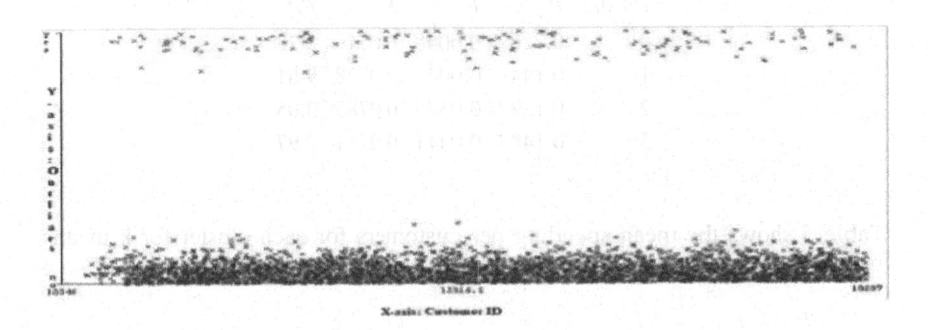

Fig. 2. Determining outliers

5 Simulation Results and Discussion

In this section, the simulation results for utilising K- mean and Self - organising map are presented. We utilised our filtered target data set for the analysis of the data. Table 2 shows the no of instances in each clusters for k means.

Table 2. K means (k = 4)

Cluster	Number of instances	%
0	748	20
1	1274	34
2	885	23
3	882	23

Table 3 shows the distribution of monetary value across the clusters for k means. Total Monetary for all clusters:285.024757.

Table 3. Total monetary value

Cluster	*Total Monetary by cluster*	*%*
0	53.807791	18.88
1	92.789915	32.56
2	69.51373	24.38
3	68.913321	24.17

Table 4 shows the RFM value computed during K means clustering.

Table 4. RFM values for K-Means clustering

Cluster	*R*	*F*	*M*	*FP*
0	0.728	0.0048	0.0719	3.28
1	0.1449	0.007	0.0728	9.61
2	0.1292	0.0373	0.0785	3.05
3	0.1483	0.0411	0.0781	2.97

Table 5 shows the mean spending per customers for each cluster for k means.

Table 5. Mean spending per customer for k-means

Cluster	*Mean spending per customer*
0	0.0719
1	0.0728
2	0.0785
3	0.0781

As seen above we get the following result:

According to the total monetary value the highest monetary value is in Cluster 1, the second highest in cluster 2 and the lowest in cluster 0. The highest mean spending per customer is in cluster 2, the second highest in cluster 3 and the lowest in cluster 0. In this case

Cluster 0: Lowest by monetary value and mean spending per customer. Low recency and low frequency.
Cluster 1: Highest group by monetary value but the second lowest by mean spending per customer. High recency and higher frequency.
Cluster 2: The second highest group by monetary value and the highest group by mean spending per customer. high recency and higher frequency.
Cluster 3: Similar to cluster 2 in reference to monetary value and second highest group by mean spending per customer. High recency and medium frequency.

For the Self Organizing Maps we have utilised similar to K-means 4 clusters. Table 6 shows the no of instances in each clusters for SOM.

Table 6. SOM (k = 4)

Cluster	*Instances*	*%*
0	695	18
1	966	25
2	808	21
3	1320	35

Table 7 shows the distribution of monetary value across the clusters for SOM.

Table 7. Total monetary value

Cluster	Monetary	%
0	50.648962	17.78
1	70.868476	24.86
2	105.24944	20.44
3	105.24944	36.92

Total monetary value of clusters: 285.024757.

Table 8 Shows the mean spending per customers for each cluster for SOM.

Table 8. Mean spending per customer for SOM

Cluster	*Mean spending per customer*
0	0.0729
1	0.0734
2	0.0721
3	0.0797

Table 9 shows the RFM value computed during SOM clustering.

Table 9. RFM values for SOM clustering

Cluster	*R*	*F*	*M*	*FP*
0	0.0807	0.0059	0.0729	10.61
1	0.2065	0.0124	0.0734	7.32
2	0.7098	0.0058	0.0721	3.24
3	0.11	0.0462	0.0797	2.23

According to the total monetary value per cluster the highest monetary value is in cluster 3 which also has the highest mean spending per customer. The second highest monetary value is in cluster 1 which also has the second highest mean spending per customer. The lowest monetary value is in cluster 0 which has the lowest mean spending per customer described as follows

Cluster 0: lowest group in monetary value and lowest by mean spending per customer. High recency and low frequency.
Cluster 1: The second highest group in terms of monetary value and also by mean spending per customer. High recency and higher frequency than cluster 0.
Cluster 2: The second lowest group by monetary value and the lowest by mean spending per customer. Low recency and low frequency.
Cluster 3: The highest group by monetary value and also by the mean spending per customer. High recency and high frequency.

This completes the cluster profiles for the Self organizing maps with 4 clusters.

The following cluster plots show the visual assignment of RFM for k means and som.

Frequency Plot for k means and som. X- axis represent Frequency and Y-axis represent different cluster assignment.

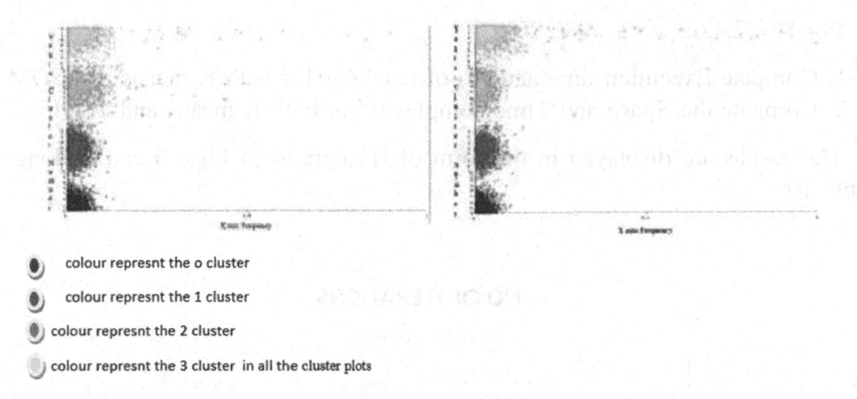

Fig. 3. Frequency plot

Monetary plot for k means and som. X- axis represent Monetary and Y- axis represent different cluster assignment.

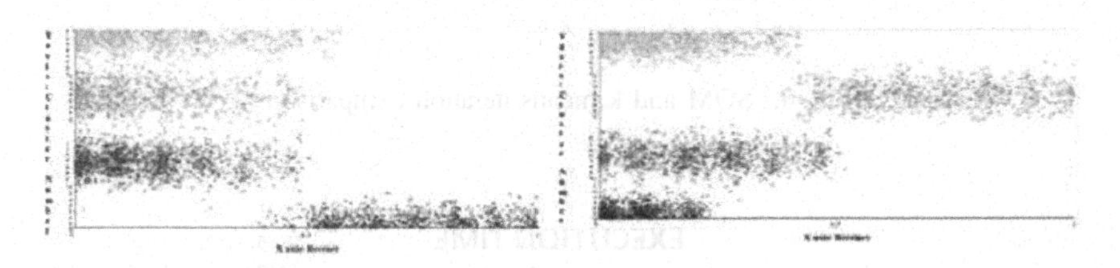

Fig. 4. Monetary plot

Recency plot for k means and SOM. X- axis represent Recency and Y- axis represent different cluster assignment.

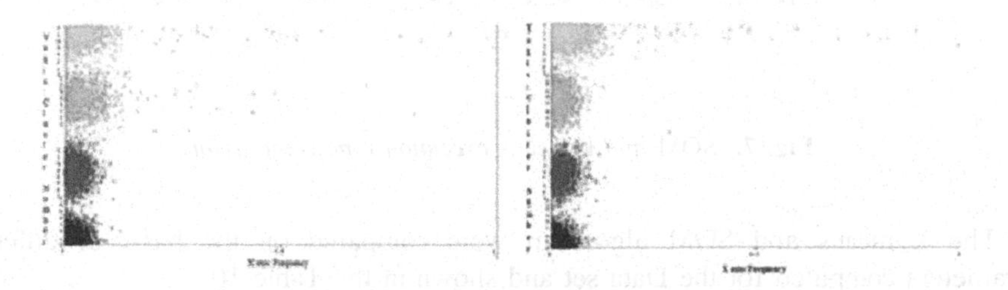

Fig. 5. Recency plot

Comparison between K means and self organizing maps (SOM).

While we have demonstrated Cluster profiling and the related computations for K = 4, in the case of both K means and self organizing maps, we now proceed to bench mark the performance of these two clustering algorithms.

For K = 2,4,6:

1. Compute Execution time and No of iterations for both K means and SOM.
2. Compute the Space and Time complexity for both K means and SOM.

The results are displayed in the form of Histograms in Figs. 6 and 7 along with Table 10.

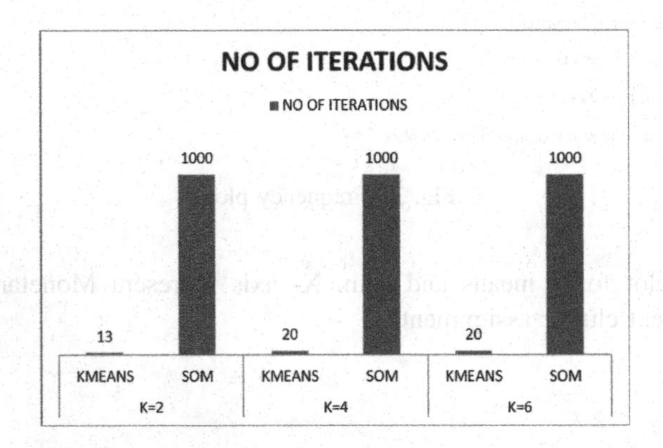

Fig. 6. SOM and k means iteration comparison

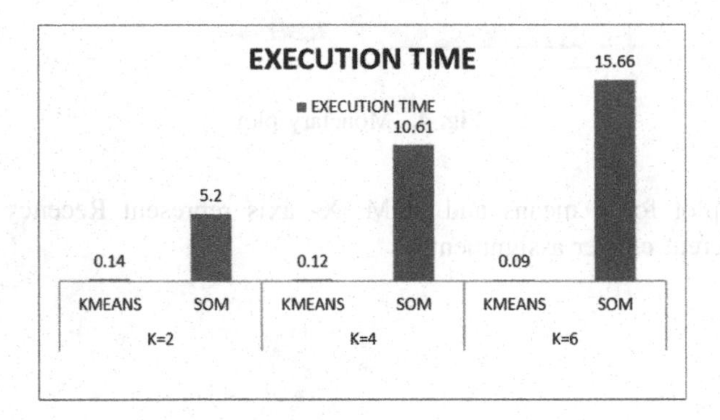

Fig. 7. SOM *and* k-means *execution time comparison*

The k means and SOM algorithm were compared on the basis of different parameters computed for the Data set and shown in the Table 10.

Linkage to Chen et al. and Differences

Chen et al. [1] analysed the same retail data set for 2011 and used k means clustering to segment the customers of the online retail store. While they tried with k = 3, 4 and 5, it concludes that the results obtained for k = 5, have a clearer understanding of the target data set than the results for k = 3 and k = 4 (Table 11).

Table 10. Comparison between K means and Self Organizing Maps

	K = 2		K = 4		K = 6	
	KMEANS	SOM	KMEANS	SOM	KMEANS	SOM
Execution Time	0.14 s	5.2 s	0.12 s	10.61 s	0.09 s	15.66 s
No of Iterations	13	1000	20	1000	20	1000
Space Complexity	O(18955)	O(7582)	O(18965)	O(7586)	O(18975)	O(7590)
Time Complexity	O(492570)	O(15156000)	O(985140)	O(30312000)	O(1477010)	O(45468000)

Table 11. K means Clustering results of *Chen et al.*

Cluster	*Instances*	%
1	527	14
2	636	17
3	1748	47
4	627	17
5	188	5

The relative contributions of these clusters to Monetary have also been described in the paper and there after cluster profiling in terms of r, f and m values has been done. The approach taken by this paper is to demonstrate the clustering of the customers by both K means and self organizing maps and thereafter to profile the clusters obtained in terms of r, f and m values. However it has been done for 2010. Also the performance of K means and Self Organizing Maps has been benchmarked and compared for k = 2, 4 and 6 as seen in Figs. 9 and 10 and Table 10, showing the computed values of time and space complexity for both these clustering algorithms.

Also in the case of the Self organizing maps, N = H * W, where N is the number of clusters, H is the height of the lattice and W the width of the lattice. Thus we have ensured that the comparison is done for 2, 4 and 6 clusters for both K means and Self Organizing maps. In the case of the SOM, this results in the creation of a 2*1,2*2 and 2*3 lattice, facilitating further computations. This facilitates the comparison of the two algorithms for the same parameters. Accordingly we chose K means algorithm as it is simple and popular amongst practitioners and was used by Chen et al. alto do the analysis. It allows us to compare results obtained with those of Chen et al. for the k means algorithm. In addition self organizing maps were chosen to give a Neural Networks perspective and frame work due to its robust architecture and lesser sensitivity to noise in the input data. It allows us to compare the working of the clustering algorithm using self organizing maps for the same data set. We can also gain insights on the computations done using a Neural Networks frame work and then compare the results obtained from these two key techniques.

6 Conclusions

This research paper based on the work done by Chen et al. [1] who used K means clustering to obtain a segmentation of customers for an online retailer. While the concerned data set was up loaded on the UCI Machine learning repository on the 21st of September 2019, the analysis by Chen et al. covered the retail data set for 2011. It uses the RFM model to construct a Target data set containing distinct post codes. There after decision tree based rule induction was used to obtain decision rules representing customer specific business intelligence. Clear and stable results representing the under lying data set were obtained for k = 5. In this paper we selected the data set for 2010 and then obtained the number of distinct customers who did transactions with the online retailer over the year 2010. We constructed the target data set and performed Normalization and removal of outliers. We performed K means clustering for k = 4 and then used Self organizing maps with number of clusters = 4. The clusters obtained were profiled in terms of their RFM values and the mean spending per customer. The highest and lowest monetary value clusters were identified. The K means and Self organizing maps clustering algorithms were compared for their performance for 2, 4 and 6 clusters and the results were tabulated in the histograms of Fig. 9 and 10, along with Table 10, depicting the time and space complexity for the two clustering algorithms. In reference to this research paper, further work can be done in identifying buying patterns of customers in terms of items purchased (association rules). Also a buyer loyalty program can be designed based on the buying choices made by customers and there after high value customers can be identified. The design of such a loyalty based card membership can increase the popularity and visibility of the retailer in terms of their business operations. Finally, advanced techniques of machine learning such as Deep learning can be used to design new computational architectures and obtain new results. Fuzzy learning techniques can also be used to determine which paradigm to select to obtain better and more accurate results with greater efficiency and speed.

References

1. Chen, D., Sain, S., Guo, K.: Data mining for the online retail industry: a case study of RFM model-based customer segmentation using data mining. J. Database Mark. Customer Strategy Manage. **19**(3), 197–208 (2012) https://doi.org/10.1057/dbm.2012.17
2. Dogan, O., Ayçin, E., Bulut, Z.: Customer segmentation by using rfm model and clustering methods: a case study in retail industry. Int. J. Contemp. Econ. Adm. Sci. **8**(1), 1–19 (2018)
3. https://Emarketer.com/content/uk-ecommerce-2019, read on 14 November 2019
4. https://Statista.com/statistics/275973/types-of-goods-purchased-online-in-great-britain/, read on 14 November 2019
5. https://archive.ics.uci.edu/ml/index.php, online_retail II, This is the data set used by Chen et al, which has also been used in this paper. The Data set was uploaded on 21 September 2019
6. Hu, Y.-H., Yeh, T.-W.: Discovering valuable frequent patterns based on RFM analysis without customer identification information. J. Knowl. Based Syst. **61**, 76–88 (2014). https://doi.org/10.1016/j.knosys.2014.02.009

7. Sarvari, P.A., Ustundag, A., Takci, H.: Performance evaluation of different customer segmentation approaches based on RFM and demographics analysis. Kybernetes. **45**(7), 1129–1157 (2016)
8. Yeh, I.C., Yang, K.J., Ting, T.M.: Knowledge discovery on RFM model using Bernoulli sequence. Expert Syst. Appl. **36**, 5866–5871 (2008)
9. Wei, J.-T., Lin, S.-Y., Hsin-Hung, W.: A review of the application of RFM Model. Afr. J. Bus. Manage. **4**(19), 4199–4206 (2010)
10. Bloom, J.Z.: Market segmentation – a neural network application. Ann. Tourism Res. **32**(1), 93–111 (2005)
11. Holmbom, A.H., Eklund, T., Back, B.: Customer portfolio analysis using the som. Int. J. Bus. Inf. Syst. **8**(4), 396–412 (2011)
12. Kohonen, T.: Self Organizing Maps. Springer Verlag, Berlin (2001)
13. Vellido, A., Lisboa, P.J.G., Meehan, K.: Segmentation of the online shopping market using Neural networks. Expert Syst. Appl. **17**(4), 303–314 (1999)
14. Ultsch, A.: Self organized feature maps for monitoring and knowledge aquisition of a chemical process. In: Gielen, S., Kappen, B. (eds.) ICANN 1993, pp. 864–867. Springer, London (1993). https://doi.org/10.1007/978-1-4471-2063-6_250
15. Miljkovic.: Brief overview of Self organizing maps. In: Proceedings of 40th International conference on information and communication technology, electronics and micro electronics (MIPRO), IEEE (2017)
16. https://www.cs.waikato.ac.nz/ml/weka/
17. Kiang, M.Y., Hu, M.Y., Fisher, D.M.: An extended self-organizing map network for market segmentation—a telecommunication example. Decis. Support Syst. **42**, 36–47 (2006)

An Adaptive Seed Node Mining Algorithm Based on Graph Clustering to Maximize the Influence of Social Networks

Tie Hua Zhou, Bo Jiang, Yu Lu, and Ling Wang(✉)

Department of Computer Science and Technology, School of Computer Science, Northeast Electric Power University, Jilin, China
smile2867ling@neepu.edu.cn

Abstract. Recently, the issue of maximizing the influence of social networks is a hot topic. In large-scale social networks, the mining algorithm for maximizing influence seed nodes has made great progress, but only using influence as the evaluation criterion of seed nodes is not enough to reflect the quality of seed nodes. This paper proposes an Out-degree Graph Clustering algorithm (OGC algorithm) to dynamically select the out-degree boundary to optimize the range of clustering. On this basis, we propose an Adaptive Seed node Mining algorithm based on Out-degree (ASMO algorithm). Experiments show that our algorithm keeps the balance between the cost and benefit of seed node mining, and greatly shortens the running time of seed node mining.

Keywords: Social network · Influence maximization · Seed node mining · Adaptive algorithm

1 Introduction

Social networks contain a lot of valuable data, and popular topics in politics, economics, culture and other fields are generated and spread every day. For example, Rafael Prieto Curiel et al. (2020) collected millions of tweets from the 18 largest Spanish-speaking countries/regions in Latin America in 70 days and analyzed them to find out people's fear of crime [1]. Therefore, social network research is a hot topic.

In recent years, social network influence maximization algorithms have made great progress in greedy algorithms and heuristic algorithms. In the study of influence maximization based on greedy thought, Kempe et al. proved for the first time in 2003 that the optimal solution can be effectively approximated to 63% by using sub-modules of aggregate functions [2]. Rezvan Mohamadi et al. (2017) proposed a new information diffusion model CLIM, which considers the continuous state of each node instead of the discrete state [3]. Bhawna Saxena et al. (2019) proposed a UACRank algorithm to identify initial adopters, and fully considered user behavior when calculating user influence potential [4]. The CELF algorithm (2007) proposed by Jure Leskovec et al. uses the sub-module of the aggregate function to greatly reduce the evaluation range of node influence diffusion, making the selection time of nodes 700 times faster than that of the greedy algorithm [5]. The advantage of the greedy algorithm is that the seed

D.-S. Huang and P. Premaratne (Eds.): ICIC 2020, LNAI 12465, pp. 498–509, 2020.
https://doi.org/10.1007/978-3-030-60796-8_43

nodes obtained by searching have a strong influence, but the greedy algorithm needs to visit all nodes when looking for each seed node. So the running time of the greedy algorithm is not fast enough.

In the research of heuristic algorithm, Chen Wei et al. (2009) proposed a degree discount algorithm, but the degree discount algorithm is only suitable for independent cascade models [6]. Zhang Dayong et al. (2019) proposed the CumulativeRank method and proved the accuracy and stability of the algorithm on the SIR model [7]. Meng Han et al. (2016) proposed a framework to explore a part of the community, and then explore the true changes of the network by considering the divide and conquer technology of the community [8]. Tang Youze et al. (2014) proposed a two-stage influence maximization algorithm TIM +, and proved through experiments that it is better than the greedy algorithm of Kempe et al. [9]. Tang Youze et al. (2015) further proposed an IMM algorithm based on TIM+, which can be extended to a wider range of diffusion models [10]. Liu Dong et al. (2017) proposed the LIR algorithm and proved that the running time of the algorithm is hundreds of times faster than that of the greedy algorithm [11]. László Hajdu et al. (2018) proposed a method of maximizing community infection based on a greedy algorithm, and increased the solvable network scale [12]. In terms of the efficiency of finding seed nodes, heuristic algorithms are much higher than greedy algorithms. But the influence of the seed nodes searched by the heuristic algorithm is not high.

In this paper. We propose an adaptive seed node mining algorithm based on graph clustering. In this algorithm, we combine the high efficiency of the heuristic algorithm and the high accuracy of the greedy algorithm. In addition, we also propose a new seed node evaluation method that takes into account the influence of seed nodes and the time spent in finding seed nodes.

2 Out-Degree Graph Clustering

2.1 Motivation

Research on maximizing influence in social networks has practical value in election canvassing, public opinion control, brand marketing, etc. Therefore, maximizing the influence of social networks has always been a hot research topic. But at present, the algorithm for maximizing the influence is to specify the Top-k size to mine seed nodes, regardless of the benefits of seed nodes and the cost of mining seed nodes. Therefore, we propose an Out-degree Graph Clustering algorithm (OGC algorithm) to dynamically select the out-degree boundary to optimize the range of clustering. Based on the OGC algorithm, we proposed an Adaptive Seed node Mining algorithm based on Out-degree (ASMO algorithm). Experiments show that the ASMO algorithm maintains the balance between the cost and benefit of seed node mining, and greatly shortens the running time of seed node mining.

2.2 OGC Algorithm

In this section, we will use an example to describe the detailed process of OGC algorithm. The symbols used in this paper and their meanings are given in Table 1 below.

Table 1. Symbols used in the paper.

Notation	Description
G	Social network graph
V	Represent node set in social network graph
E	Represent edge set in social network graph
D	Out-degree set of nodes
dv_i	The out-degree of node v_i
S	Represents a seed node set
C	Set representing cluster center
Cost(S)	Cumulative cost of mining seed nodes
Gain(S)	Cumulative benefits from mining seed nodes
B(S)	Influence gain per unit cost
e	Preselected seed set size. In this paper, *e* = 3 is used by default
r	*r* is the clustering granularity. It controls the size of clusters

In order to improve the efficiency of mining seed nodes, we first proposed the graph clustering algorithm OGC based on out-degree, and then performed seed node mining on the results of the OGC algorithm. *r* is the clustering parameter. The following example shows the influence on clustering results when *V* = {*v1*, *v2*,..., *v15*}, clustering parameter *r* = {2, 3, 4} for graph G (*V*, *E*).

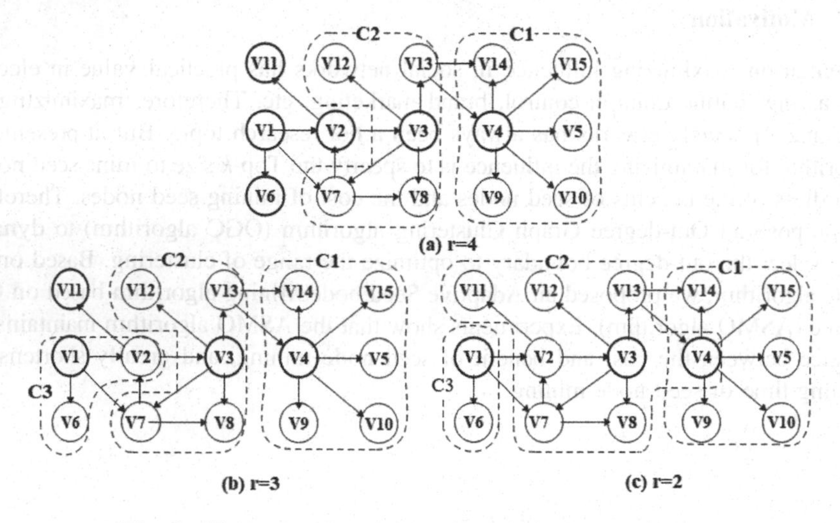

Fig. 1. Shows the clustering results under different *r* values.

For Fig. 1 (a) (b) (c), it can be seen from $r = 4$ and $r = 3$ that the larger r is, the more isolated cluster centers are. In $r = 3$, since C_2 completes before C_3, the neighbor node V_3 of V_2 satisfies $dv3 \geq 3$, and V_3 is not in the cluster of V_2. In the case of $r = 2$, because the clusters in the clustering results are not independent of each other, when r is small, the clusters generated after may contain the cluster centers generated before.

The input of the OGC algorithm is the social network graph G, the node set D sorted in descending order according to the degree of node out, and the parameter r that controls the size of the cluster. The OGC algorithm traverses an unvisited node v from D each time as a cluster center and marks the node v as visited. Then traverse all the nodes pointed to by v and add the nodes whose out-degree is greater than r to the set where v is located. Then iterate through the nodes pointed to by the visited nodes until their out-degrees are less than r. At this time, a cluster with v as the cluster center is generated. Then traverse the next unvisited node from D and repeat the above operation. Until all nodes in D have been visited, the OGC algorithm ends.

The detailed steps of OGC algorithm are as follows:

Algorithm 1: OGC(*G, D, r*)

Input Graph G = (N, E), Descending Order Out-degree D, Clustering parameter r
Output Cluster Result C

1. C, Visited $\leftarrow \emptyset$
2. **for** $D \neq \emptyset$ **do**
3. Queue, Result $\leftarrow \emptyset$
4. center $\leftarrow$ D.pop(0)
5. **If** center in visited **do**
6. Continue
7. Queue, Result $\leftarrow$ center
8. **for** Queue $\neq \emptyset$ **do**
9. $v_i \leftarrow$ Queue.pop(0)
10. **If** $dv_i \geq r$ and v_i not in visited **do**
11. Visited, Result $\leftarrow v_i$
12. Queue $\leftarrow$ unvisited neighbor of v_i
13. **else if** v_i is center **do**
14. Visited $\leftarrow v_i$
15. C.append(***result***)
16. Return Cluster Result C

3 Seed Node Mining Algorithm for Maximizing Influence

3.1 Iterative Process of ASMO Algorithm

ASMO algorithm does not need to specify the size of the seed set. The algorithm automatically selects the seed set that meets the requirements according to the termination conditions B(S). S is the set of seed nodes, which is initially empty. With the iteration of ASMO algorithm, new nodes will be added to *S*. B(*S*) is the decreasing function of seed set *S*. The following will prove this conclusion:

Formula (1) is to calculate the influence gain of each node v:

$$f(S+v) - f(S) = \Delta Gain(v) \tag{1}$$

Let the seed set $S = \{v_1, v_2, \ldots, v_n\}$, where *1, 2,…, n* represents the joining order of nodes. For example, for every two adjacent nodes *vi, vj* and $1 \leq i < j \leq n$, according to formula (1), there are:

$$f(S+v_i) - f(S) = \Delta Gain(v_i) \tag{2}$$

$$f(S+v_j) - f(S) = \Delta Gain(v_j) \tag{3}$$

Note: in formula (3), $S = S + v_i$

According to the submodules of set function, we can get:

$$f(S+v_i) - f(S) \geq f(S+v_i+v_j) - f(S+v_i) \tag{4}$$

Then, we can get the gain relation between *vi* and *vj*. As follows:

$$\Delta Gain(vi) \geq \Delta Gain(vj) \tag{5}$$

Therefore, we define the gain function Gain(*S*) of seed node as follows:

$$Gain(S) = \Delta Gain(v_1) + \Delta Gain(v_2) + \ldots + \Delta Gain(v_i) + \Delta Gain(v_j) + \ldots + \Delta Gain(v_n) \tag{6}$$

$$S = \{v_1, v_2, \ldots, v_i, v_j, \ldots, v_n\}$$

$$\Delta Gain(v_1) > \Delta Gain(v_2) > \ldots > \Delta Gain(v_i) > \Delta Gain(v_j) > \ldots > \Delta Gain(v_n) \tag{7}$$

$\Delta Cost(v)$ is the average time, The function Cost(*S*) of cost is defined as follows:

$$Cost(S) = \Delta Cost(v_1) + \Delta Cost(v_2) + \ldots + \Delta Cost(v_i) + \Delta Cost(v_j) + \ldots + \Delta Cost(v_n) \tag{8}$$

$$S = \{v_1, v_2, \ldots, v_i, v_j, \ldots, v_n\}$$

$$\Delta Cost(v_1) \approx \Delta Cost(v_2) \approx \ldots \approx \Delta Cost(v_i) \approx \Delta Cost(v_j) \approx \ldots \approx \Delta Cost(v_n) \tag{9}$$

To sum up, the cumulative gain Gain(*S*) and the cumulative cost Cost(*S*) are two increasing functions of the seed node set S. We define the gain rate as:

$$B(S) = \frac{Gain(S)}{Cost(S)} \tag{10}$$

3.2 Principle of ASMO Algorithm

According to the above description, Algorithm 2 gives the detailed process of adaptive algorithm ASMO. The input of the ASMO algorithm is the social network graph G and the result C of the OGC algorithm. ASMO algorithm first establishes the preselected seed set and initializes the seed set *S* (3-4). Calculate cumulative gain Gain(*S*), cumulative cost Cost(*S*) and gain rate B(*S*) (5-9). Finally, the algorithm judges whether to stop seed node mining by iterative termination conditions and returns the seed set *S*. (10-15).

Algorithm 2: ASMO(*G*, *C*)

Input: Graph $G = (N, E, W)$, Clustering Result *C*
Output: Seeds Set *S*

1. $S \leftarrow \emptyset$
2. $Pre \leftarrow \emptyset$
3. **For** i $\leftarrow$**0 to e do**
4. Pre.append(*C*.pop(0))
5. **For** pre $\neq \emptyset$ **do**
6. **If** Pre.size() $< e$ **and** $C \neq \emptyset$ **do**
7. Pre.append(C.pop(0))
8. Gain($S + v$), Cost($S + v$) $\rightarrow$ According to Formula(6), Formula (8)
9. B($S + v$) $\leftarrow$ Gain($S + v$) / Cost($S + v$)
10. **If** B($S + v$) > 1 do
11. $S \leftarrow S + v$
12. Continue
13. **Else**
14. break
15. Return Seed Node Set *S*

4 Evaluation

4.1 Data Set

The experimental environment is windows 10, CPU: Inter (R) Core (TM) i5 CPU 2.20 GHz, memory 8 GB, programming language: Python. All the data sets we use come from the Stanford Social Network [16]. Details of these data are shown in Table 2. We use CELF [5] and IMRank [13] algorithms to compare with ASMO algorithm. In the aspect of seed nodes influence: we compare the influence of three algorithms in independent cascade model and linear threshold model. In the aspect of the time spent in finding seed nodes: we only collect the running time of CELF algorithm and ASMO algorithm in two information diffusion models. Because IMRank algorithm is based on the adjacency matrix rather than the information diffusion models to find seed nodes, we use time complexity to compare the running time of IMRank and ASMO. We use *dict* data structure in Python to build social network model, such as $G = \{A: \{B: 0.27, C: 0.58\} \ldots \}$. Indicates that node *A* points to *B* and *C*, where the weight of *AB* edge is 0.27, and that of *AC* edge is 0.58.

Table 2. Data set.

Name	Nodes	Edges	Average clustering coefficient
Email-Eu-core	1005	25571	0.3994
Wiki-Vote	7115	103689	0.1409

4.2 Experimental Results and Discussion

4.2.1 The Influence of *R* on OGC Algorithm

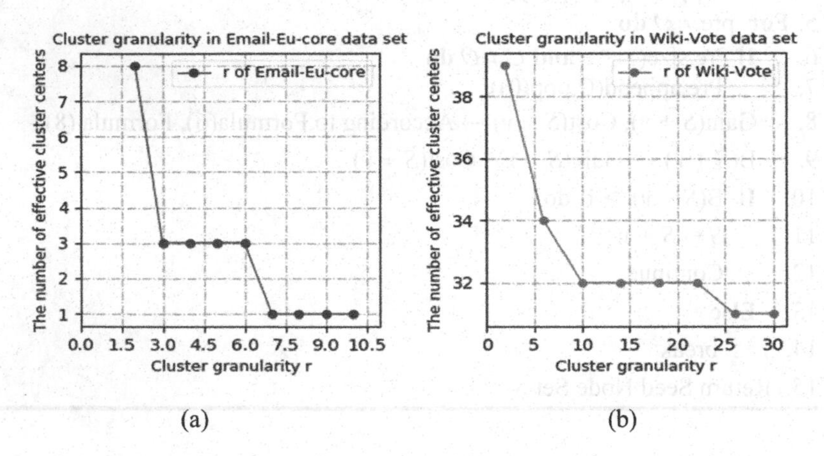

Fig. 2. (a) (b) Shows the relationship between *r* and the effective cluster centers on Email-Eu-core and Wiki-Vote.

Table 3. Limit value *r* for clustering.

Properties	Email-Eu-core		Wiki-Vote	
	$r = 1$	$r \geq 7$	$r = 1$	$r \geq 774$
Cluster center	41	42	4739	5490
Effective clustering center	27	1	64	1
Total nodes	1005	1005	7115	7115

Figure 2 (a) and (b) show the effect of r value on the size of cluster center in two different data sets. r is the out-degree of the node. The larger r, the smaller the cluster center. We call the cluster center with the ability of Influence diffusion as the effective cluster center. Obviously, when r is greater than or equal to 1, the cluster center is the effective cluster center. Table 3 shows the minimum and maximum values of r on the two datasets. When the effective cluster center is 1, only one cluster has data. After many experiments, we get that when $r \geq 7$ and $r \geq 774$, we reach the upper limit of clustering center of Email-Eu-core dataset and Wiki-Vote dataset.

4.2.2 The Effect of B (*S*) on the Number of Seeds

In this section, we use experiments to analyze the iterative stopping conditions of the ASMO algorithm, its influence in social networks, and the running time. In the experiment, we use the IC (independent cascade) model [14] and the LT (linear threshold) model [15]. Among them, the LT model is a value accumulation model, and the IC model is a probability model.

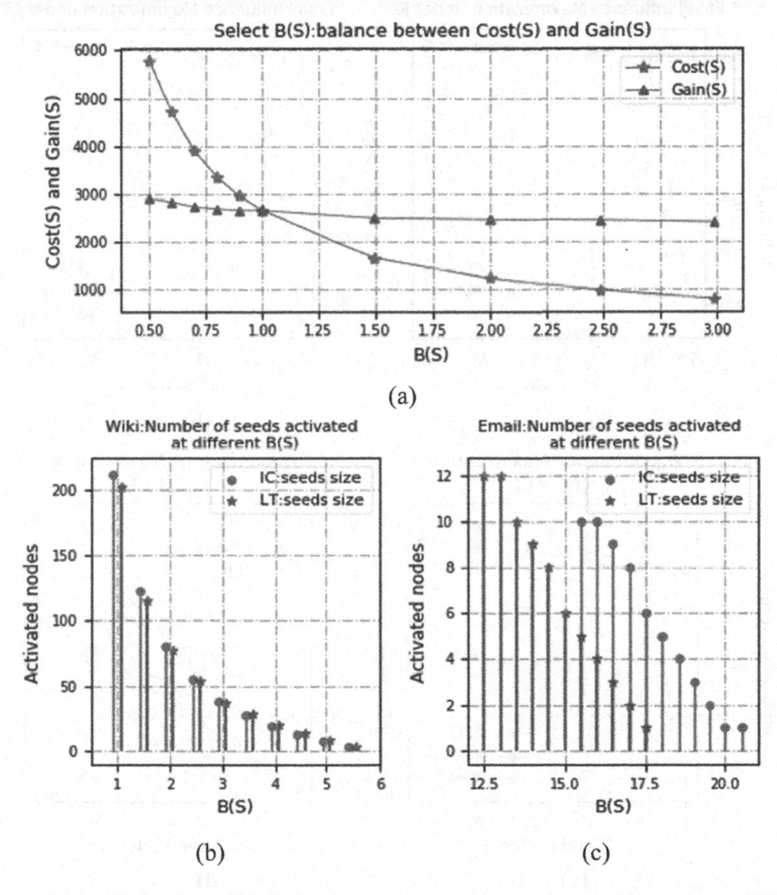

Fig. 3. (a) (b) (c) The relationship between cumulative cost and cumulative gian on Wiki-Vote data set and the influence of B (*S*) on the number of seeds on Email-Eu-core and Wiki-Vote data sets.

Figure 3 (a), (b) and (c) are experiments on the iterative termination condition B (S). Figure 3(a) shows that when the Wiki-Vote data set B (S) is one, we get the largest total revenue. Figure 3 (b) and (c) shows the number of nodes activated by different values of B (S). In Fig. 3(c), when B (S) < 16 in IC model and B (S) < 13 in LT model, the number of seed nodes is constant. This is because the average clustering coefficient of Email-Eu-core network is high and the network scale is small, which leads to better performance of B (S). However, the social networks we encounter are large and complex, so in most cases, they will show the performance on Fig. 3 (b).

4.2.3 Comparison of ASMO Algorithm with CELF and IMRank Algorithm Under the Same Number of Seed Nodes

In Fig. 4 (a) and (b), we compare the influence of the three algorithms on the Email-Eu-core dataset. In the two models, when there are more than 10 seed nodes, the influence

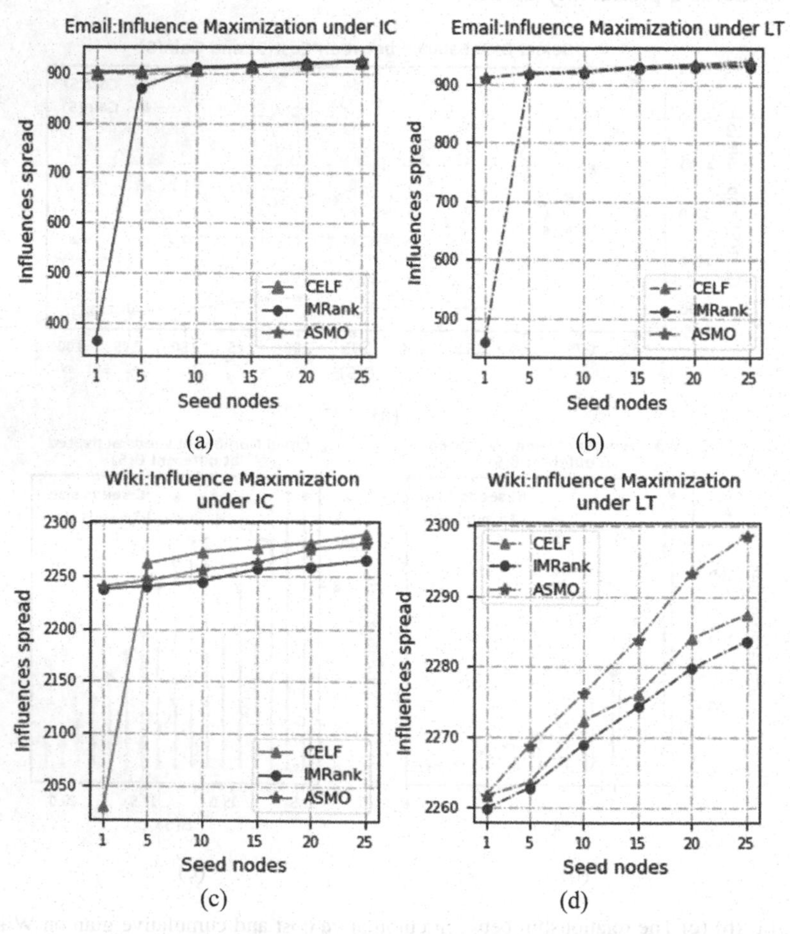

Fig. 4. (a) (b) (c) (d) In the Email-Eu-core and Wiki-Vote datasets, the performance of the three algorithms is compared when searching for the same number of nodes in IC and LT models.

of the three algorithms is roughly same. However, in IC model and LT model, when the number of seed nodes is less than 10 and 5, IMRank is not as good as ASMO algorithm. Figure 4 (c) and (d) show the influence comparison of the three algorithms on the Wiki-Vote dataset. In IC model, when the number of seed nodes is more than 5, ASMO algorithm is not as good as CELF algorithm, but better than IMRank algorithm. In LT model, the performance of ASMO algorithm is much better than CELF and IMRank algorithm under the same number of nodes. To sum up, compared with CELF and IMRank algorithms, the influence of seed nodes mined by ASMO algorithm is similar to CELF and IMRank algorithm in small data sets, but ASMO algorithm performs better in LT model for larger datasets.

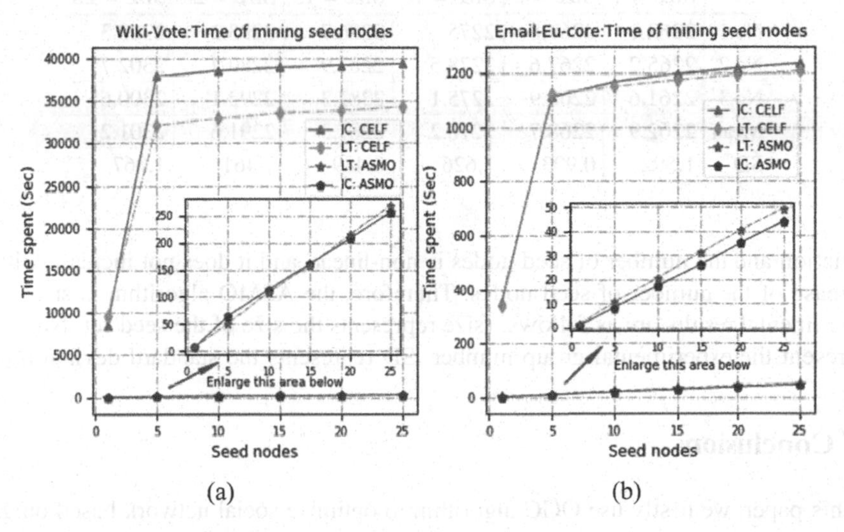

Fig. 5. (a) (b). Running time of CELF and ASMO mining different seed sets on IC and IT models on Email-Eu-core and Wiki-Vote datasets.

The time complexity of the ASMO algorithm is *O(nlogn)*, and the running time is shown in Fig. 5. In Fig. 5 (a), ASMO algorithm of both IC model and LT model on Email-Eu-core dataset is about 24 times faster than CELF algorithm. In Fig. 5 (b), the running time of ASMO algorithm on Wiki-Vote dataset is 140 and 160 times faster than that of CELF algorithm in LT model and IC model.

4.3 Stability Evaluation of ASMO Algorithm

We use standard deviation to evaluate the stability of the ASMO algorithm. We compare the influence of different numbers of seed sets on the Wiki-Vote social network. We conducted three independent experiments and calculated the mean and standard deviation of different seed sets under the three experimental conditions. Table 4 and Table 5 correspond to the experimental results under the IC model and the LT model, respectively. Under these two models, the relationship between the standard

Table 4. The stability analysis of ASMO algorithm under IC model

	Size = 1	Size = 5	Size = 10	Size = 15	Size = 20	Size = 25
No.1	2240.1	2242.4	2253.1	2263	2271	2280.6
No.2	2239.2	2245.8	2255.1	2266.6	2269.8	2283.9
No.3	2238.6	2248.7	2255.8	2265	2268.9	2277.6
Mean	2239.3	2245.6	2254.6	2264.8	2269.9	2280.7
SD	0.616	2.574	1.144	1.472	0.860	2.572

Table 5. The stability analysis of ASMO algorithm under LT model.

	Size = 1	Size = 5	Size = 10	Size = 15	Size = 20	Size = 25
No.1	2262.1	2269.8	2275	2286.7	2293.2	2300.3
No.2	2265.2	2267.6	2278.5	2287.9	2289.7	2302.7
No.3	2261.6	2268.9	2275.1	2287.3	2292.1	2300.6
Mean	2262.9	2268.7	2276.2	2287.3	2291.6	2301.2
SD	1.592	0.903	1.626	0.489	1.461	1.067

deviation and the number of seed nodes is non-linear, and it does not increase with the increase of the number of seed nodes. Therefore, the ASMO algorithm is stable. The experimental results are as follows: (size represents the size of the seed set. No. 1, 2, 3 represent the experimental group number. SD represents the standard deviation.)

5 Conclusions

In this paper, we firstly use OGC algorithm to optimize social network based on graph clustering before seed mining, then use ASMO algorithm to mine seed nodes for each update clusters, which can guarantee the time performance. In terms of seed quality, ASMO algorithm uses greedy strategy and set function submodular feature to ensure that the seed node selected from each iteration process is the most influential node at present. So, ASMO algorithm can keep good quality of the seed from low cost and high influence. Finally, appropriate value of parameter *e* can reduce the running time of the algorithm exponentially, especially for large-scale graph environment. However, if the value of e is too small, the influence of the seed node cannot be expanded.

Acknowledgement. This work was supported by the National Natural Science Foundation of China (No. 61701104), and by the Science and Technology Development Plan of Jilin Province, China (No.20190201194JC).

References

1. Curiel, R.P., Cresci, S., Muntean, C.I., Bishop, S.R.: Crime and its fear in social media. Palgrave Commun. **6**(57), 5–9 (2020)

2. Kempe, D., Kleinberg, J., Tardos, É.: Maximizing the spread of influence through a social network. In: 9th ACM SIGKDD international conference on knowledge discovery and data mining, KDD'03, pp. 137–146 (2003)
3. MohamadiBaghmolaei, R., Mozafari, N., Hamzeh, A.: Continuous states latency aware influence maximization in social networks. AI Commun. **30**(2), 99–116 (2017)
4. Saxena, B., Kumar, P.: A node activity and connectivity-based model for influence maximization in social networks. Soc. Netw. Anal. Min. **9**(1), 1–16 (2019). https://doi.org/10.1007/s13278-019-0586-6
5. Leskovec, J., et al.: Cost-effective outbreak detection in networks. In: Proceedings of the 13th ACM SIGKDD international conference on knowledge discovery and data mining, KDD'07, pp. 420–429 (2007)
6. Chen, W., Wang, Y., Yang, S.: Efficient influence maximization in social networks. In: Proceedings of the 15th ACM SIGKDD Conference on Knowledge Discovery and Data Mining, KDD'09, pp. 199–208 (2009)
7. Zhang, D., Wang, Y., Zhang, Z.: Identifying and quantifying potential super-spreaders in social networks. Sci. Report. **9**(14811), 1–10 (2019)
8. Han, M., et al.: Influence maximization by probing partial communities in dynamic online social networks. Trans. Emerg. Telecommun. Technol. **28**(4), 5–13 (2016)
9. Youze, T., Xiaokui, X., Yanchen, S.: Influence maximization: near-optimal time complexity meets practical efficiency. In: ACM SIGMOD International Conference on Management of Data, pp. 75–86. ACM (2014)
10. Tang, Y., Shi, Y., Xiao, X.: Influence maximization in near-linear time: a martingale approach. In: ACM SIGMOD International Conference on Management of Data, pp. 1539–1554. ACM (2015)
11. Liu, D., et al.: A fast and efficient algorithm for mining top-k nodes in complex networks. Sci. Report **7**(43330), 1–7 (2017)
12. Hajdu, L., et al.: Community based influence maximization in the independent cascade model. Federated Conference on Computer Science and Information Systems, pp. 237–243 (2018)
13. Cheng, S., et al.: IMRank: influence maximization via finding self-consistent ranking. In: SIGIR, pp. 475–484 (2014)
14. Goldenberg, J., Libai, B., Muller, E.: Talk of the network: a complex systems look at the underlying process of word-of-mouth. Mark. Lett. **12**(3), 211–223 (2001). https://doi.org/10.1023/A:1011122126881
15. Rogers, E.M.: Diffusion of Innovations, 5th ed Paperback – August 16 (2003) http://snap.stanford.edu/data/

Wavelet-Based Emotion Recognition Using Single Channel EEG Device

Tie Hua Zhou, Wen Long Liang, Hang Yu Liu, Wei Jian Pu, and Ling Wang(✉)

Department of Computer Science and Technology, School of Computer Science, Northeast Electric Power University, Jilin, China
smile2867ling@neepu.edu.cn

Abstract. Using computer technology to recognize emotion is the key to realize high-level human-computer interaction. Compared with facial and behavioral, physiological data such as EEG can detect real emotions more efficiently to improving the level of human-computer interaction. Because of the traditional EEG equipment is complex and not portable enough, the single channel EEG device is cheap and easy to use that has attracted our attention. In this paper, the main goal of this study is to use a single channel EEG device to acquire the EEG signal, which has been decomposed to corresponding frequency bands and features have been extracted by the Discrete Wavelet Transforms (DWT). Then, classify three different emotional states data so as on to achieve the purpose of emotion recognition. Our experimental results show that three different emotional states include positive, negative and neutral can be classified with best classification rate of 92%. Moreover, using the high-frequency bands, specifically gamma band, has higher accuracy compared to using low-frequency bands of EEG signal.

Keywords: EEG · DWT · Emotion recognition · Single channel

1 Introduction

Electroencephalogram (EEG) is one of the most effective tools to measure brain activity. In recent years, recognition of emotions by physiological signals mainly with EEG signals has attracted researchers' attention [1]. There are many neurons in the brain and when the neurons are active, they will generate current flows and wave patterns locally, that is known as the wave of the brain (brain waves) [2]. Different brain active state have different brain waves, these brain waves are classified into 5 classification indicating different conditions [3]. The classification is shown in Table 1. Using physiological data can really reflect brain activity, so this paper will use EEG data for this research.

Emotion is a psychological and physiological state related to various feelings, thoughts and behaviors. Researchers have put forward different views and theories in the field of emotional research. At present, there are two main theories of emotion: discrete emotion model and the bi-dimensional emotion model [4]. The bi-dimensional model is frequently used in the literature. For example, DEAP data set is used for

D.-S. Huang and P. Premaratne (Eds.): ICIC 2020, LNAI 12465, pp. 510–519, 2020.
https://doi.org/10.1007/978-3-030-60796-8_44

Table 1. Brain waves classification

Types of wave	Frequency range	Short description
Delta (δ)	0.5 – 4 Hz	Adult slow-wave sleep
Theta (θ)	4 – 7 Hz	Deep relaxation, drowsiness
Alpha (α)	8 – 15 Hz	Relaxation, closing the eyes
Beta (β)	16 – 31 Hz	Active thinking, focus
Gamma (γ)	>32 Hz	Regional learning

emotion recognition in [5], and study the influence of music on emotion in [6] under bi-dimensional emotion model. The experimental study was evaluated from two dimensions of valence and arousal. But there are still many literatures which use the discrete emotion model. Such as [7], they studied EEG Correlates of Ten Positive Emotions under discrete emotion model. Ten positive emotions were evaluated respectively. This paper also uses the discrete emotion model. In addition, out of various mother wavelets are analyzed and compared in [6], Daubechies (db4) was selected as a mother wavelet since it had high SNR value (51.37 dB), db4 is also used as the mother wavelet in this paper.

Traditionally, EEG devices are multi-channel, and some devices even reach 256 channels. DEAP dataset is a most commonly used dataset for emotion analysis, which still uses 32 EEG channels [8]. Although the accuracy of multichannel EEG devices is higher, the preparation time is too long and the portability is not enough. The latest development of single channel, dry electrode EEG sensor technology has aroused the interest of researchers, because it has higher availability. It can provide the possibility for use in informal environment such as home or outdoor environment, such as NeuroSky's Single-Channel EEG Sensor for Drowsiness Detection in [9]. This study analyzed the difference of positive and negative emotions based on 32 channels EEG signals, two kinds of signals are successfully classified in [10]. Emotion recognition based on 8 channels EEG signals, these studies utilizing wavelet coefficients in [11]. This paper focused on emotion recognition using single-channel EEG signals.

The rest of this paper is organized as follows: Sect. 2 describes the experimental process and motivation. Section 3 describes the materials and methods (including algorithm, acquisition equipment, data acquisition, noise reduction, feature extraction and classification) used in this paper. In Sect. 4, present the experimental results and evaluation. It is followed by the conclusion in Sect. 5.

2 Overall Design

2.1 Motivation

Different from the traditional medical wet sensor which needs conductive adhesive, the device which Mindlink device of MindAsset as a single channel device can be directly connected to the dry contact when it is used in this paper. The device is so convenient to use that transmits the data wirelessly via a Bluetooth connection, people can use it in any place and environment, so more and more people begin to pay attention to and study it, which is also trend of development in the future.

At present, this equipment has been widely used in health, education, research, entertainment and other industries, many applications and games have been developed about the device, such as Mental Fruit Bomb which is the real time first duel players brainwave game on a smart phone and tablet platform. For the study and comprehension of the EEG signal, the analysis of the frequency bands is widely used. In addition, it is worth noting that emotions are becoming the focus of people's attention. Emotion was synonymous to all mankind, studying emotions can help some patients with mental illness such as depression, and also can help people better manage and understand their mental health. So, this paper will focus on the above two issues, focusing on the feasibility of using single channel EEG device for emotion recognition.

2.2 Framework

In this study, we use a single channel EEG device to collect EEG data. General process as follows: First, we stimulate emotions by watching video, and then we preprocess, extract feature and classify the collected data segments, finally judge the emotional state according to the classification results (see Fig. 1). The details of the experiment are introduced in the next section.

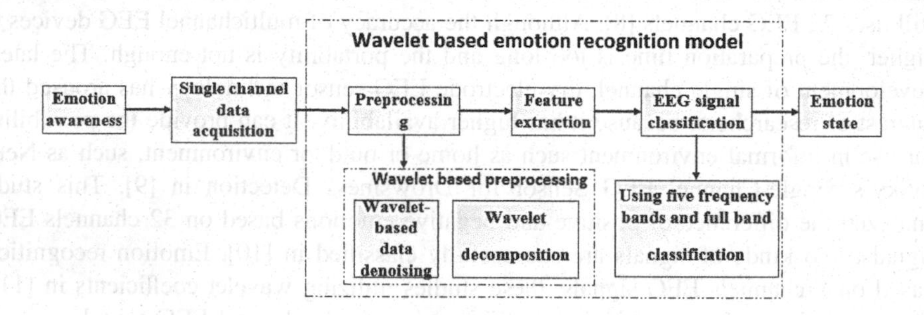

Fig. 1. Overall flowchart

3 Wavelet-Based Emotion Classification Model

3.1 Wavelet-Based Emotion Recognition Algorithm

Participants were asked to sit quietly before and during the experiment. Participants were also asked to avoid extra actions such as head movement and blinking, in order to obtain the minimum interference EEG signal during the experiment. Participants watched Internet videos that stimulated neutral, positive and negative emotions and recorded the stimulation position. Then segmented EEG signals into *60 s* emotional labeled data include emotion stimulation point. The EEG data is read, saved and processed using Matlab R2018b.

To acquire the EEG signal we used MindAsset's MindLink. The sensor samples neuronal activities with a frequency up to 512 Hz and outputs EEG data at 1 Hz frequency, the samples frequency range is 3 Hz to 100 Hz. The position of single channel point is at the *Fp1* of 10–20 system (see Fig. 2) [12]. Under better detection environment and less interference, the output value of the raw EEG signal will be between − 300 μV and + 300 μV. However, the value of muscle electrical interference signal caused by blinking, eyebrow lifting and head swinging will be less than − 1000 μV or more than + 1000 μV. So in order to get better raw EEG signal, if the value is less than − 300 μV or more than + 300 μV, they will be discarded.

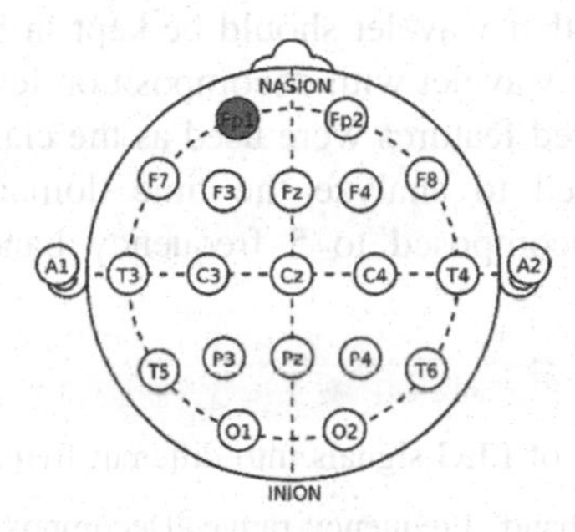

Fig. 2. The position of *Fp1*.

Although we try to choose better data in the process of data acquisition, there were still be noise in the data, so it is necessary to de-noising for the next work. In this study, using wavelet denoising techniques based on thresholding to removal noise. Daubechies (db4) was selected as a mother wavelet. The db4 mother wavelet with decomposition level of 6 was used to remove the noise, and extremum threshold estimation rule was selected in the threshold settings (see Fig. 3).

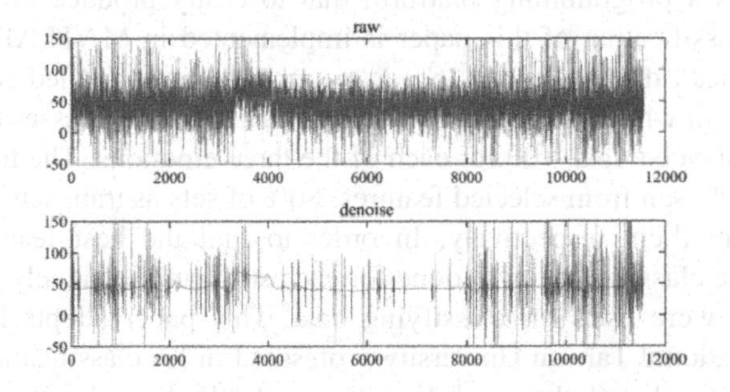

Fig. 3. Using wavelet to reduce noise of EEG signal.

Wavelet transform has lower time resolution and higher frequency resolution at low frequency, while it has higher time resolution and lower frequency resolution at high frequency, which is in line with the characteristics of slow change of low frequency signal and fast change of high frequency signal. Therefore, wavelet analysis is more suitable for analyzing non-stationary EEG signals than Fourier transform and short-time

Fourier transform. According to the principle of wavelet transform, the samples frequency range is 3 Hz to 100 Hz, so the frequency range of the subband is as follows, where fs is the sampling frequency:

$$\left[3, \frac{fs}{2^5}\right], \left[\frac{fs}{2^5}, \frac{fs}{2^4}\right], \left[\frac{fs}{2^4}, \frac{fs}{2^3}\right], \left[\frac{fs}{2^3}, \frac{fs}{2^2}\right], \left[\frac{fs}{2^2}, \frac{fs}{2}\right], \left[\frac{fs}{2}, 100\right] \quad (1)$$

In this study, the DWT was used to extract features from EEG signal. In the current research, selected same mother wavelet should be kept in both denoising and decomposition. So the db4 mother wavelet with decomposition level of 6 was used to feature extraction. Then, the acquired features were used as the classifier inputs. In this paper, wavelet transform was used to analyze the time domain feature of EEG signal. The EEG signals were decomposed to 5 frequency bands by DWT as shown in Table 2.

Table 2. Decomposition of EEG signals into different frequency bands using DWT.

Frequency band	Frequency range	Decomposition level
Delta (δ)	3 – 4 Hz	D6
Theta (θ)	4 – 7 Hz	D5
Alpha (α)	7 – 13 Hz	D4
Beta (β)	13 – 25 Hz	D3
Gamma (γ)	25 – 50 Hz	D2
Noises	50 – 100 Hz	D1

MATLAB is a programming platform that to easily produce time series visualizations. The classification of this paper is implemented in MATLAB(R 2018b). We have an imbalanced data collection. So as to improve it we applied some methods to balance the data. In which we randomly copy the data of minor classes to balance them, there are 300 balanced data, 100 for each of the three emotions. The train and test sets were randomly chosen from selected features, 80% of sets as train sets and 20% of sets as test sets from them respectively. In order to find the best feature for emotion classification, the classification was done for each sub-band separately and also the full frequency band were used for classifying data. This paper adopts LIBSVM which developed by National Taiwan University professor Lin for classification [13, 14]. The data was normalized and the used Kernel was RBF. In order to choose the best parameters, we used the grid method on the training data. A 10-fold cross validation was used to enhance the authenticity of the classifier outputs. In addition, we uses KNN and K-Means algorithm as comparison. In order to get best accuracy value, we repeated the classification for data.

3.2 WLER Algorithm Description

In this paper, we purposed wavelet based emotion recognition algorithm (WLER) to achieve the purpose of emotion recognition by a single channel EEG device. Firstly, the collected EEG data is input, and then the input of the classifier is obtained by wavelet denoising and wavelet decomposition. Finally, the best accuracy and average accuracy of the classifier is obtained and output by cross validation method. The time complexity of the algorithm is O (n^2). The algorithm of WLER is as follows:

Algorithm: WLER algorithm

Input: raw EEG data
Output: maximum classification accuracy and average classification accuracy
1: import raw EEG data to Matlab
2: *X ← raw EEG data; TPTR ← minimaxi; SORH ← h;*
SCAL ← sln; N ← 6; wname ← db4
3: **for** each of raw data ∈ dataset **do**
4: denoised data = wavelet denoising (*X, 'TPTR', 'SORH', 'SCAL', N, 'wname'*)
5: **end for**
6: *X ← denoised data; N ← 6; wname ← db4*
7: **for** each of denoised data ∈ dataset **do**
8: frequency bands = discrete wavelet transforms (*X, N, 'wname'*)
9: **end for**
10: **for** each of frequency bands ∈ 5 frequency bands and all frequency band **do**
11: *-t* = *2* % The used Kernel was RBF
12: **for** 10 fold cross validation **do** %meshgrid method
13: get best parameter *-c* and parameter *-g*
14: **end for**
15: cmd = [*' -t',' -c ', ' -g*]
16: model = libsvmtrain (*train_label, train_matrix, cmd*)
17: classification accuracy = libsvmpredict (*test_label, test_matrix, model*)
18: **end for**
19: get maximum accuracy and calculate average accuracy
20: **return** maximum classification accuracy and average classification accuracy

4 Evaluation Analysis

In this paper, using a single channel EEG device to collect our own data, and then use this data set to carry out our experiment. As a comparison, we use K-means method in [15] and KNN method in [16]. Import the raw EEG data to IBM SPSS software for some simple analysis. After observed the bar chart of EEG data, we found that the raw EEG data accorded with the normal distribution. Then use one-way ANOVA for analysis, amplitude as de-pendent list, emotion state as factor, and the significance level $P < 0.05$ between groups (see Table 3). Further analysis found the significance level $P > 0.05$ of negative emotion and positive emotion (see Table 4). The results show that only use ANOVA cannot distinguish these three emotions. So we need to use the classifier to classification the EEG data of three emotional states in next work.

Table 3. ANOVA of the raw EEG data of three emotional states.

	Sum of squares	df	Mean square	F	Sig
Between Groups	5074.976	2	2537.488	3.113	.045
Within Groups	9437932.273	11577	815.231		
Total	9443007.249	11579			

Table 4. Multiple comparisons of the raw EEG data of three emotional states.

(I)state	(J)state	Mean difference(I-J)	Std. error	Sig	Lower bound	Upper bound
Negative	Neutral	−1.505*	.740	.042	−2.96	−.05
	Positive	−.303	.799	.704	−1.87	1.26
Neutral	Negative	1.505*	.740	.042	.05	2.96
	Positive	1.202*	.602	.046	.02	2.38
Positive	Negative	.303	.799	.704	−1.26	1.87
	Neutral	−1.202*	.602	.046	−2.38	−.02

The mean difference is significant at the 0.05 level.

We spliced the EEG signals of negative, neutral and positive emotions one by one, *180* s each, *540* s in total. After wavelet de-noising and wavelet decomposition, waveforms of 5 frequency bands are obtained. In some samples, we have found that activate more for positive emotions than negative emotions in the beta and gamma bands, neutral emotions have lower alpha responses, and the negative emotion have significant higher delta responses and higher gamma responses (see Fig. 4). A sample is shown in the figure below, this rule can be observed.

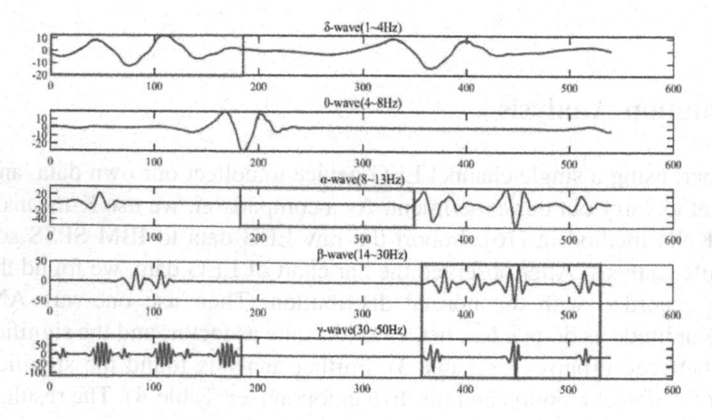

Fig. 4. Wavelet decomposition.

Our method relies on just EEG signals and is not dependent on the other physiological signals, but three kinds of EEG signals are successfully classified. The raw EEG signal is denoised by wavelet threshold method using mother wavelet db4, then the EEG signal was decomposed to 5 frequency bands by DWT using mother wavelet db4. Next, the 5 frequency bands as features were input to the classifier. We compared the accuracy of the raw signal and the denoised signal; the accuracy after denoising is significantly higher than that without denoising. So we use the denoised signal for classification and the classification results are shown in Fig. 5.

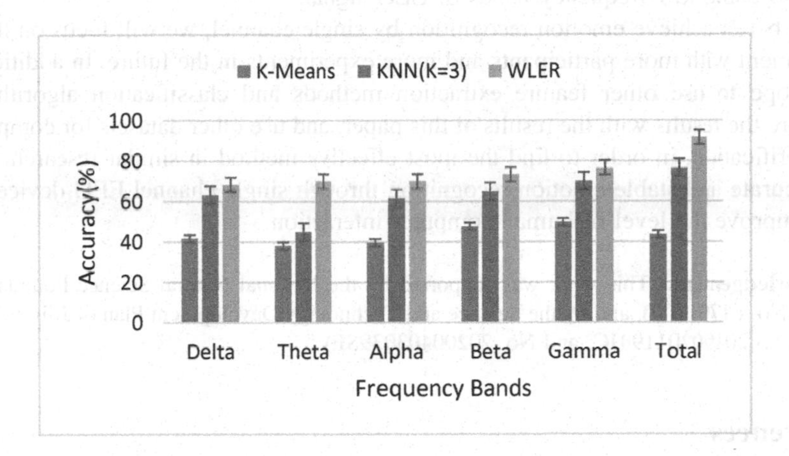

Fig. 5. The accuracy rate comparison using five frequency bands and full band.

Figure 5 shows that the best classification accuracy of WLER, K-NN and K-means is 91.68%, 76.67% and 49.67%, respectively. The maximum classification accuracy achieved was 91.68% and average classification accuracy achieved was 84.67% using the WLER on the full frequency band. This value exceeds the results of most studies, whether for those who use a considerable number of channels for measurement or those who use a reduced number of channels. In addition, we also compare the classification results of different band features in emotional state classification. The results show that high band has better classification results than low band; gamma band and full frequency band have the best classification results. We also used SJTU Emotion EEG Dataset (SEED) to test [17]. The accuracy rates of five bands (delta, theta, alpha, beta, gamma) and total is 54.5%, 58.8%, 60.6%, 54.3%, 54.5%, 54.5%, respectively. Our data set corresponds to an emotional label every 60 s, and the SEED dataset corresponds to an emotional label every 240 s. However, different data sets tend to conform to their own emotional labels by our proposed method, which are closely related to the corresponding data sets themselves and different acquisition equipment. In fact, we only select the EEG data with obvious emotional performance with a label in 60 s over our own dataset, so as to get more accurate wavelet features and achieve better results.

5 Conclusion

Multi-channel is most used for feature extraction and classifier training, while our experiment only uses single channel. The research shows that the effectiveness of using a commercial device such as MindAsset's MindLink for emotion recognition is checked. The three kinds of emotions have been successfully classified and obtained better results using our WLER algorithm by single channel EEG signals. Moreover, using the high-frequency bands, specifically gamma band, has higher accuracy compared to using low-frequency bands of EEG signal.

To better achieve emotion recognition by single channel, we will focus on this the experiment with more participants and more experiments in the future. In addition, we also hope to use other feature extraction methods and classification algorithms to compare the results with the results of this paper, and use other datasets for comparison and verification, in order to find the most effective method in similar research. Strive for accurate and stable emotion recognition through single channel EEG devices, and then improve the level of human-computer interaction.

Acknowledgement. This work was supported by the National Natural Science Foundation of China (No. 61701104), and by the Science and Technology Development Plan of Jilin Province, China (No.20190201194JC, and No. 20200403039SF).

References

1. Soleymani, M., Asghari-Esfeden, S., Fu, Y., Pantic, M.: Analysis of eeg signals and facial expressions for continuous emotion detection. IEEE Trans. Affect. Comput. **7**(1), 17–28 (2016)
2. Wan Ismail, W.O.A.S., Hanif, M., Mohamed, S.B., Hamzah, N., Rizman, Z.I.: Human emotion detection via brain waves study by using electroencephalogram (EEG). Int. J. Adv. Sci. Eng. Inf. Technol. **6**(6), 1005–1011 (2016)
3. Wikipedia. https://en.wikipedia.org/wiki/Electroencephalography. Accessed 15 Apr 2020
4. Mohammadi, Z., Frounchi, J., Amiri, M.: Wavelet-based emotion recognition system using EEG signal. Neural Comput. Appl. **28**(8), 1985–1990 (2016)
5. Zhuang, Ning., Zeng, Y., Tong, L., Zhang, C., Zhang, H., Yan, B.: Emotion recognition from EEG signals using multidimensional information in EMD domain. BioMed Res. Int. **2017** (2017)
6. Balasubramanian, G., Kanagasabai, A., Mohan, J., Seshadri, N.P.G.: Music induced emotion using wavelet packet decomposition—An EEG study. Biomed. Signal Process. Control **42**, 115–128 (2018)
7. Hu, X., et al.: EEG correlates of ten positive emotions. Front. Hum. Neurosci. **11**, 26 (2017)
8. Koelstra, S., et al.: DEAP: a database for emotion analysis; using physiological signals. IEEE Trans. Affect. Comput. **3**(1), 18–31 (2012)
9. Patel, K., Shah, H., Dcosta, M., Shastri, D.: Evaluating neurosky's single-channel EEG Sensor for drowsiness detection. In: Stephanidis, C. (ed.) HCI 2017. CCIS, vol. 713, pp. 243–250. Springer, Cham (2017). https://doi.org/10.1007/978-3-319-58750-9_35
10. Li, J.: Analysis of positive and negative emotions based on EEG signal. In: Proceedings of Joint 2016 International Conference on Artificial Intelligence and Engineering Applications (AIEA 2016), Wuhan Zhicheng Times Cultural Development Co, pp. 170–174 (2016)

11. Momennezhad, A.: EEG-based emotion recognition utilizing wavelet coefficients. Multimedia Tools Appl. **77**, 27089–27106 (2018)
12. Wikipedia. https://en.wikipedia.org/wiki/10–20_system_(EEG). Accessed 15 Apr 2020
13. Chang, C.C., Lin, C.-J.: LIBSVM: a library for support vector machines. ACM Trans. Intell. Syst. Technol. **2**(3), 1–27 (2011)
14. Lu, H.B., Wang, J.L.: Optimization of carbon content in fly ash of utility boiler based on LIBSVM and intelligent algorithm. J. Northeast Dianli Univ. **34**(1), 16–20 (2014)
15. Gurudath, N., BryanRiley, H.: Drowsy driving detection by EEG analysis using wavelet transform and k-means clustering. Procedia Comput. Sci. **34**, 400–409 (2014)
16. Li, M., Xu, H., Liu, X., Lu, S.: Emotion recognition from multichannel EEG signals using K-nearest neighbor classification. Technol. Health Care: Official J. Eur. Soc. Eng. Med. **26** (S1), 509–519 (2018)
17. Zheng, W.-L., Lu, B.-L.: Investigating critical frequency bands and channels for EEG-based emotion recognition with deep neural networks. IEEE Trans. Auton. Mental Dev. (IEEE TAMD) **7**(3), 162–175 (2015)

Dense Subgraphs Summarization: An Efficient Way to Summarize Large Scale Graphs by Super Nodes

Ling Wang, Yu Lu, Bo Jiang, Kai Tai Gao, and Tie Hua Zhou(✉)

Department of Computer Science and Technology, School of Computer Science, Northeast Electric Power University, Jilin, China
thzhou@neepu.edu.cn

Abstract. For large scale graphs, the graph summarization technique is essential, which can reduce the complexity for large-scale graphs analysis. The traditional graph summarization methods focus on reducing the complexity of original graph, and ignore the graph restoration after summarization. So, in this paper, we proposed a graph Summarization method based on Dense Subgraphs (DSS) and attribute graphs (dense subgraph contains cliques and quasi cliques), which recognizes the dense components in the complex large-scale graph and converts the dense components into super nodes after deep sub-graph mining process. Due to the nodes in the dense component are closely connected, our method can easily achieve the lossless reduction of the summarized graph. Experimental results show that our method performs well in execution time and information retention, and with the increase of data, DSS algorithm shows good scalability.

Keywords: Super nodes · Quasi-cliques · Graph summarization · Dense subgraph mining

1 Introduction

Most information can be represented and analyzed by graphs, such as social networks, protein connection networks, knowledge maps (including but not limited to these) and these graphs will contain billions of the nodes and edges. Mining, management, maintenance, and other operations on these large-scale graphs will greatly increase the execution time of these algorithms, and it will be difficult to maintain these large-scale graphs. So, we need to summarize some of the structures in these large-scale graphs into super nodes and compress a certain amount of edges to achieve the simplification of the original graph.

For the graph summarization problem, there are many existing methods. Graph summarization methods based on grouping technology are: Grass [1], Coarse Net [2], UDS [3], etc. A more novel approach is UDS [3] method, which uses a method called zero loss encoding, summarizes its Graph according to this encoding method. Artificially set a loss threshold to control the loss of the summarized graph relative to the original graph. However, this method is difficult for lossless restoration. This method

D.-S. Huang and P. Premaratne (Eds.): ICIC 2020, LNAI 12465, pp. 520–530, 2020.
https://doi.org/10.1007/978-3-030-60796-8_45

focuses on the summarization of the graph, and does not consider the relative retention of most of the original information on the original graph.

There are also some summary methods based on MDL (The Minimum Description Length Principle) principle, such as the method proposed in [4], to compress bipartite graphs. The bipartite graph cannot represent the dense relationship between the same part of the node set. In the process of summarizing the graph, most of the important dense components in the original graph will be greatly broken. There will be many self-loops in the summarized graph, and the execution time is too long. There are some other summarization methods such as [5], whose goal is to influence; [6] whose goal of summarization is visualization; [7] whose summary goal is that entity resolution and so on. They are different from our goal.

The performance of the storage, execution algorithms [4], analysis [8–10], and processing of the summarized post-graph is better than that of original graph. However, the summarization of graph will cause several problems relative to the original, including: 1. The summarized graph may introduce false edges; 2. The finished post-graph may not be restored to the original without loss; 3. The summarized graph may damage the structure and information of the original graph.

In this paper, we propose a Graph Summarization method based on the Dense Subgraph (Cliques and Quasi-cliques) (DSS). DSS is to summarize the dense components in graph by considering the main ideas of attribute graphs and overlapping graphs [11], to summarize the dense components into super nodes and connect them with super edges. In this way, Firstly, we can not only retain most of the structure in the original graph, but also summary the original graph; Secondly, our method also records the missing information in the summarized graph, so the summarized graph can be restored without loss.

As for clique, there are many methods for complete clique mining, such as enumeration methods, quickly heuristic enumeration methods, boundary approximation method [12, 13] and so on. For the problem of maximal quasi-clique mining, the most classic method is Quick [14] which is an enumeration method. The principle of this method is to use the DFS solution space tree and many new pruning strategies to find possible quasi-cliques; However, this algorithm still takes too long to execute and it is inefficient. Enumerate top-k algorithm [15] is based on the idea of Quick algorithm, and its main principle is based on the concept of kernel. However, in this method, it must first find all possible candidate kernels, this process is very time-consuming. In our method, our main purpose is to summarize the given graph, so that we can reduce the requirements of the quasi cliques; we set some conditions to limit the number of kernel search to a certain extent, thereby speeding up the overall quasi-cliques mining time. The modified version we call L-Enum.

2 Motivation

The graph summarization technique is to summarize a given graph into a new graph with relatively few nodes and edges. With the improvement of the summary effect, the structure of the original graph will be greatly destroyed in the summarized graph. This makes the results of analyzing the graph data on the summarized graph produce a large error relative to the original graph.

Figure 1 is the process of the summarization algorithm which is based on 2-hop. It can be seen from the summary result (b) that most of the important information in the original graph is destroyed. Such as the set $\{a, b, c\}$, $\{a, e, g\}$, $\{a, d, g\}$ with strong connection has been destroyed in the summarized graph. This not only results in large errors in the analysis of data and the execution of the algorithm on the summarized graph, but also makes it impractical to use the summarized graph to represent the original data.

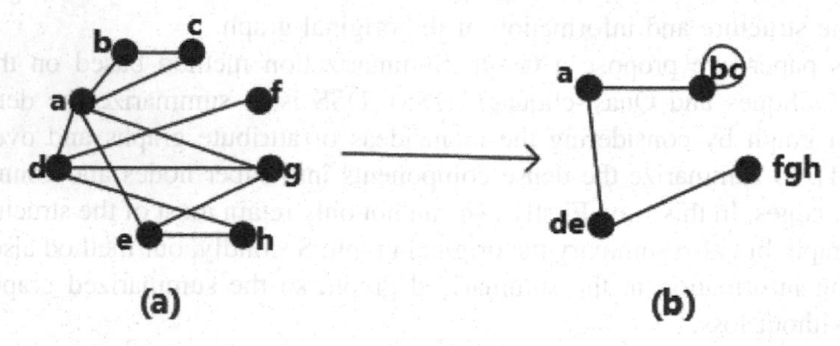

Fig. 1. Graph summarization algorithm based on 2-hop: (b) is the summarized graph for (a)

Therefore, we propose a graph summarization method (DSS) based on dense subgraphs. This method summarizes dense components into super nodes and connects these super nodes with super edges. It can not only realize the summarization of the graph, but also retain the important structure of the original graph.

3 Dense Subgraph Summarization

Given a graph, the graph Summarization method based on Dense Subgraphs (DSS) summarizes the dense subgraph of the graph: cliques and quasi cliques into super nodes and connects them by super edges. Meanwhile, the nodes contained in each super node are recorded by using the attribute graph. In this way, not only the important structure of the original diagram is retained in the summarized graph, but also the purpose of summarization is achieved.

3.1 Definition

Given an original Graph $G = (N, E)$, we use the property Graph SG to represent the summarized Graph of G. $SG = (SN, SE)$, where the set SN represents the super nodes set in the summarized graph. $SN = \{s_1, s_2, \ldots, s_n\}$, each super node s_i has two attribute fields: *node_set* and *loss_edges*. *Node_set* according to the set of super node contains all the nodes (the collection may come from k-clique or γ- quasi-clique of vertices), and $SN \subseteq N$. The set SE represents the super edges between the super nodes in the connection summarized graph (SE may contain the old edges in the original graph or the newly extended edges), $SE = \{se_1, se_2, \ldots, se_n\}$, $se_i = (s_j, s_k)$.

Definition1. Super Node Set $SN = \{s_1, s_2, \ldots, s_n\}$ which contains all the super nodes. Super Node $S_i = (idx, node_set, loss_edges)$ in SN, where idx is the index of the super node.
Definition2. Super Edge Set $SE = \{se_1, se_2, \ldots, se_n\}$ which contains all the super edges. Super Edge $Se_i = ((Sn_j, Sn_k), connected_set)$ in SE, where *connected_set* is the overlap of each two super nodes.

3.2 L-Enum Top-K Quasi Clique Algorithm

Currently, there are many methods for complete clique mining, such as enumeration methods, quickly heuristic enumeration methods, boundary approximation method and so on. Any complete cliques mining technique can be used here to get the compete cliques set we want.

As for maximal quasi clique mining, that is a classic NP-Hard problem for it. Since we only pay attention to the summarization of the original graph, and the requirement of quasi clique can be slightly reduced. So, we modified the kernel structure search process based on the Enumerate top-k quasi clique algorithm, so that the whole mining process can be mined more quickly in the basis of the original algorithm (The optimized version called L-Enum).

Enumerate top-k quasi clique algorithm's main idea is that a $\gamma - quasiclique$ usually contains a smaller but denser subgraph (there is a $\gamma' - quasiclique$ where $\gamma' > \gamma$). The kernel is the smaller and denser subgraph: $\gamma' - quasiclique$. The process for L-Enum which is based on Enumerate top-k quasi clique algorithm is as follows:

Step1. L-Enum sets the parameters γ, γ' and k, where $\gamma' > \gamma$ and k which must be big enough. After repeated experiments, we concluded that when $\gamma = 0.8$, $\gamma' = 0.9$, our algorithm performed well in execution time and summary rate.
Step2. L-Enum searches the Kernels that is $\gamma' - quasiclique$ by Quick algorithm. In the process of Enumerate top-k quasi clique algorithm, it will search all the Kernels in the given Graph that is a time-consuming process. L-Enum limits the number of Kernels in this process by any monotonically increasing function. In this paper, the function is as follow:

$$f = k * \log(k + kp)/2 + k + kp \quad (1)$$

Where k_p can be set manually. L-Enum is insensitive to setting $k\,and\,kp$, When $k\,and\,kp$ have different values, the effect of DSS is stable.

Step3. L-Enum expands the Kernels found into $\gamma - quasiclique$ which finds the maximal quasi cliques that contain each Kernel.

Step4. L-Enum can quickly get the set of quasi cliques that is needed in the graph summarization process of DSS.

4 Structural Optimization of Large-Scale Graphs

4.1 Super Nodes Transformation

The super node transformation is based on the overlap graph method of hypergraph. Such as (a) (b) (c) in Fig. 2. Among them (a) includes the nodes $\{a, b, c, d, e\}$, (b) includes the nodes $\{d, e, f, g\}$, then we can express it in the form of (c) in the overlapping graph (The $\{d, e\}$ is the overlapping part.). Based on this overlapping idea, we replace the complete cliques and quasi cliques we found above with corresponding super nodes. As shown in Fig. 3, suppose we find two dense components: complete clique $\{a, b, c, d, e\}$ and quasi clique $\{d, e, f, g, h\}$.

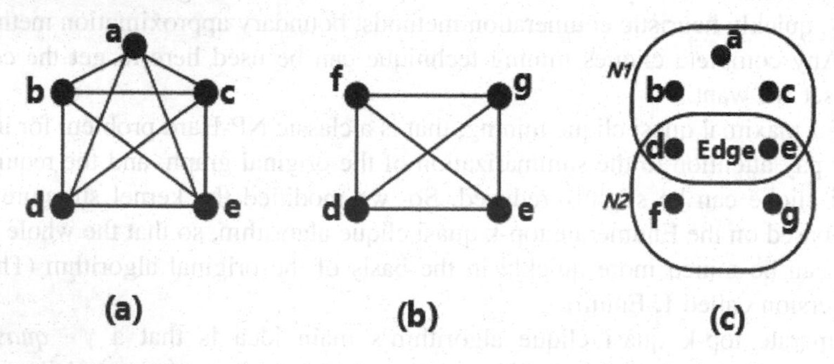

Fig. 2. Subgraph (a): $N1 = \{a, b, e, d, e\}$; Subgraph (b): $N2 = \{d, e, f, g\}$; Then merge (a) and (b) into an overlap graph (c)

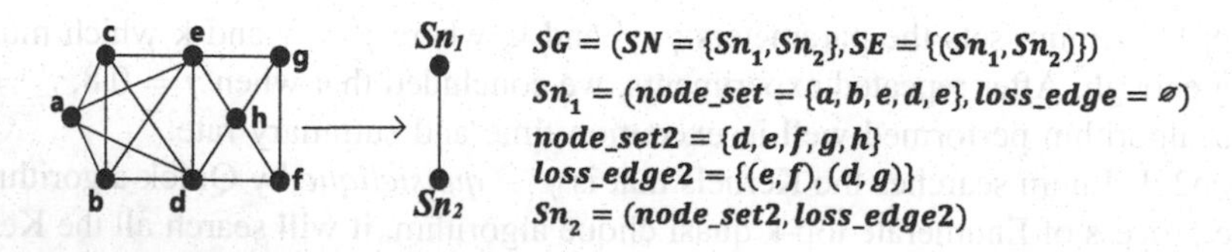

Fig. 3. DSS method: there are two dense components: complete clique $\{a, b, c, d, e\}$ and quasi clique $\{d, e, f, g, h\}$. These two components are summarized into two super nodes and a super edge.

(a) DSS summarize the complete clique into super node, and each super node records all the nodes in the original graph it contains; (b) For the quasi clique situation, DSS also use the above method to summarize the quasi cliques as the complete cliques. However, it needs to add an additional attribute records the quasi clique Loss-Edges in each super node. This attribute records the false edges introduced by the quasi clique when it acts as a super node. Through this method, it can make DSS restore on the node without loss.

4.2 Super Edges Transformation

DSS initializes each edge in the original graph to be replaced with a super edge. When super nodes are replaced, it deletes the nodes in the original graph contained in each super node and delete all connected edges between these nodes. At the same time, the edges that are connected to the deleted nodes (that is, neighbors of the deleted nodes except the edges between the deleted nodes) will also be deleted accordingly. These neighbor nodes need to be connected to this super node. For all the newly added super nodes in the summarized graph, DSS uses the overlap between these super nodes as the basis for connecting them. Each super edge will define an attribute domain: connect-set, whose set records the overlap between two super nodes.

However, this will cause redundant edges between the super nodes (if redundant edges are not processed, the number of edges in the summarized graph will exceed the number of edges in the original graph in some case. This violates our original intention). Such as, suppose the three super nodes: *{1, 2, 3}*, *{1, 2, 3}*, *{1, 2, 5}*. The same overlapping part *{1, 2}* exists between these three super nodes. Theoretically, it needs to connect these super nodes with 3 super edges which only need two super edges to represent the connection relationship between these three super nodes in fact. For the processing if redundant edges, we set up a tracker T when connecting the super nodes. For each super node, first making a connection of the overlapping part. At the same time, the tracker T will record the information of all connected edges of this node (i.e. the overlapping part). When the following super edge connection is performed, if the super edge already exists in the tracker T, the connection of this super edge is skipped. In this way, DSS realizes the processing of redundant edges.

4.3 DSS Algorithm

Graph Summarization method based on Dense Subgraphs (DSS) summarizes the dense subgraphs in the graph into super nodes, which can retain important structures in the summarized graph. DSS can quickly discover dense subgraphs in large-scale graphs through L-Enum that can efficiently summarize the large-scale graphs.

The DSS algorithm details are shown below:

```
Algorithm: DSS(G)
     Input: A given Graph G = (N, E)
     Output: A Summarized Graph SG = (SN, SE)
 1   for any n ∈ N, e ∈ E do
 2       SN = SN ∪ n, SE ∪ e
 3   SG = (SN, SE)
 4   cliques ← FindMaximalClique(G)
 5   quasi_cliques ← L-Enum (γ=0.8, G) /* 0.8 quasi cliques in graph G*/
 6   Dense Component: DC = cliques + quasi_cliques
 7   for each of the dense components dc ∈ DC do
 8       node_set = dc /* All nodes contained in the dense component*/
 9       loss_edge = edges in clique that made up by nodes in dc –edges in dc
10       sn = (node_idx, node_set, loss_edge)
11       SN = SN ∪ sn; SN = SN - node_set /* Delete se at the same time */
12   Define the tracker: T
13   sorts SN by node_idx
14   for first_idx in range (0, node_idx − 1) do
15       for second_idx in range (first_idx + 1, node_idx) do
16           if isconnected (T, first_idx, second_idx) do
17              Connect these two super nodes
18              updates T with intersection of two super nodes'node_set
19           else continue
20   Return Summarized graph SG
```

5 Experiment

The experiments were done on a 2.2 GHz Intel(R) Core (TM) laptop with 16 GB main memory, that were implemented with Python in Windows 10.

Dataset. The data sets of our experiment come from Stanford Large Network Dataset Collection [16]. The network graphs used in our experiments are all undirected and unweighted graphs. The real data sets are **Ca–GrQc**, **Ca– HepPh** and **Ca-AstroPh**. The details of them are shown in Table 1. We compared the experiment with DSS using the Greedy algorithm [4] and VoG algorithm [17] in terms of original graph retention and execution time.

Table 1. Datasets details

Data sets	Nodes	Edges	Average clustering coefficient	Number of triangles
Ca–GrQc	5242	14496	0.5296	48260
Ca– HepPh	12008	118521	0.6115	3358499
Ca-AstroPh	18772	198110	0.6306	1351441

Preprocessing of Raw Data. For these three data sets, there are two nodes in each which represent two nodes on one edge. We first load each edge into graphs and then remove the duplicate data. Finally, delete the self-cycles in the graphs.

We use the following formula to measure the retention rate and compressibility of the original graph:

$$LD = [\sqrt{}(Forward(SN)/N) + Gama * \sqrt{}(Forward(SE)/E)]/2 \quad (2)$$

Among them: $Forward(SN)$ represents the reserved nodes in the summarized graph relative to the original graph, and $Forward(SN) = SN - SN \cap N$; $Forward(SE)$ represents the reserved edges in the summarized graph, and $Forward(SE) = SE \cap E$. Gama is the ratio of edges effects, we set $Gama = 1$. LD includes the node retention degree and edge retention degree in the summarized graph, and we take the average to synthesize the two parts. Besides, $Forward(SN)$ and $Forward(SE)$ represent the compressibility of the original graph, LD measures both the retention of the original information and the compression of the summarization graph.

On the one hand, for the experimental results of Fig. 4 and Fig. 5, it can be seen that our method is not as good as the Greedy and VoG method within small dataset. However, after our summary method, the compression rate of the nodes and edges of the original graph is above 50%; As the number of nodes and edges in the graph increases, and the dense components in the graph increase, Dss compresses large-scale graph effect is similar to Vog. Our algorithm shows good scalability.

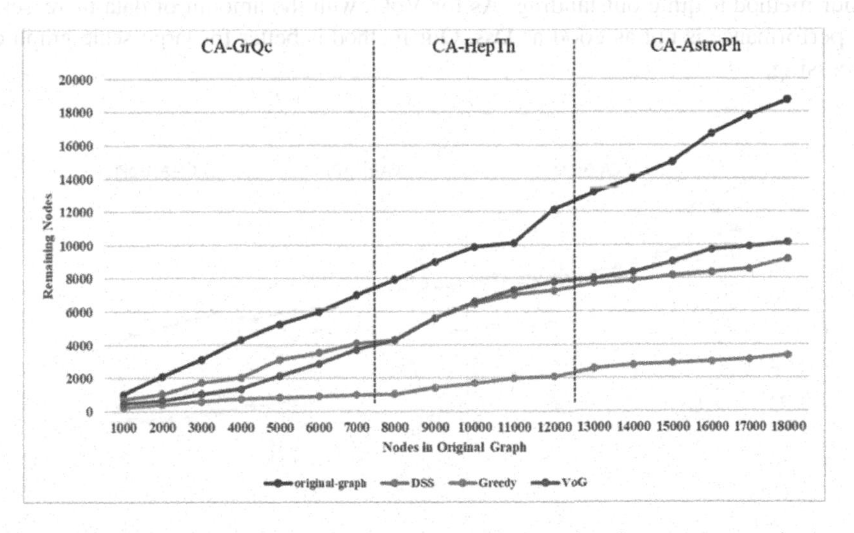

Fig. 4. The number of remaining nodes in the graph after the summary method

Here we explain: our summary method sacrifices a certain degree of summary effect to retain most of the important information in the original picture. (Collect the dense components into super nodes and save them).

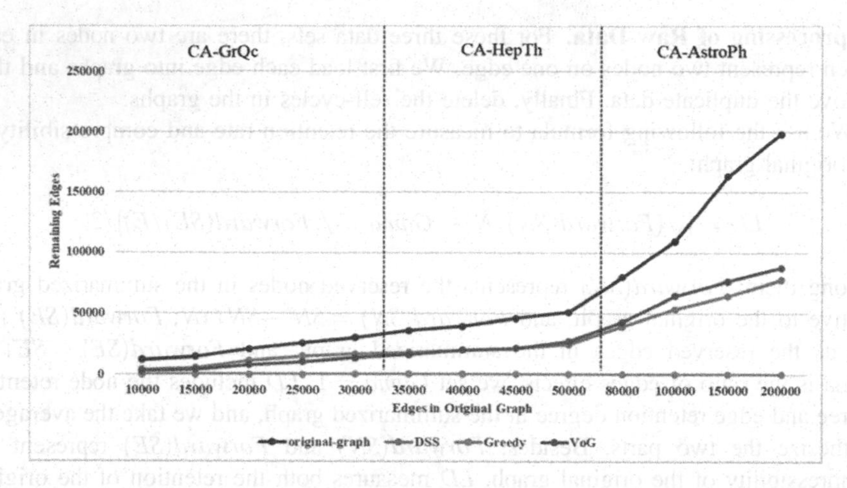

Fig. 5. The number of remaining edges in the graph after the summary method

On the other hand, as shown in Fig. 6, the experimental results show that although the Greed algorithm's summary effect is very good. However, after Greed summarization, most of the structure in the original graph has been destroyed, making it impossible to analyze information, execute algorithms and etc. in the already summarized graph. On the contrary, our summary method improves the retention of the information in the original graph on the basis of a certain summary effect. And the LD of our method is quite outstanding. As for VoG, with the amount of data increases, its LD performance is not as good as Dss. Our method is better for large-scale graph data processing.

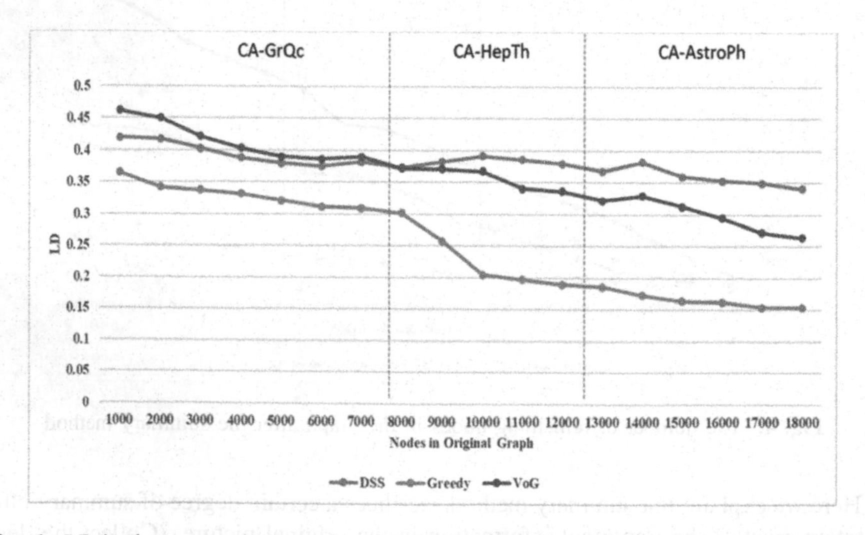

Fig. 6. LD in the summarized graph relative tohe original graph after the summary method.

Because each of the greedy and VoG methods summarizes the data of the original graph, the execution time of these two methods is shorter when processing small sets of data. However, Dss is to extract the dense components in the graph which are easy to be mined, and then realize the summary of the original graph, thus reducing the processing time of the original graph. In Fig. 7, as the data set size increases, the execution time of our method is significantly lower than that of the Greedy and VoG algorithm. For large data sets, the scalability of our algorithm is good.

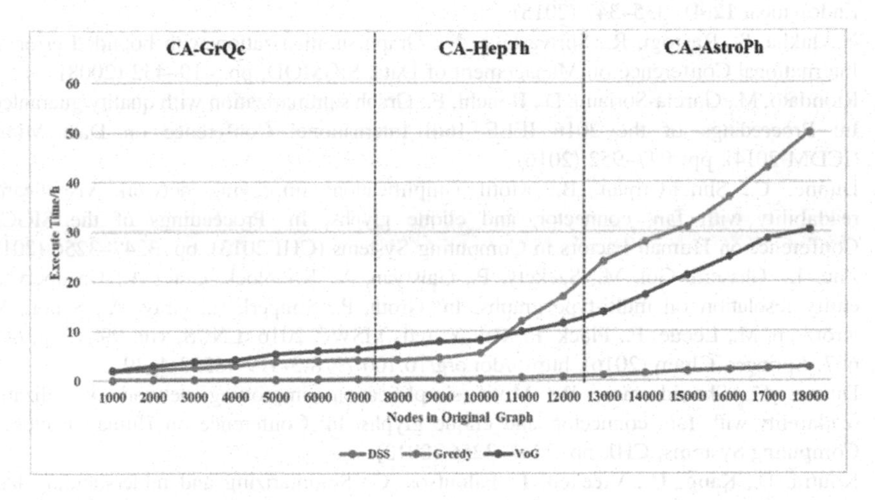

Fig. 7. The comparison between the execution time of these two algorithms

6 Conclusion

In this paper, we propose a summary method based on dense subgraphs and attribute graphs. This method sacrifices a certain degree of summary effect to improve the retention of the original graph information in the summarized graph. The experimental part shows that although DSS is not as good as the Greed algorithm in terms of aggregation effect, its aggregation effect has reached more than half of the original graph. The biggest advantage of DSS is that its execution time is much faster than Greed and VoG algorithm. However, on the one hand, the dense components found by DSS in order to speed up the execution time are low-level, and it does not exhaust all the quasi-cliques in the graph; on the other hand, DSS only considers the two denseness of cliques and quasi-cliques Ingredients, without considering other dense ingredients. Therefore, in future work, we will focus on the application of other dense components in summarization technology; at the same time, improve the algorithm to find as many quasi-cliques as possible to improve the summarization effect.

Acknowledgement. This work was supported by the National Natural Science Foundation of China (No. 61701104), and by the Science and Technology Development Plan of Jilin Province, China (No.20190201194JC, and No.20200403039SF).

References

1. LeFevre, K., Terzi, E.: GraSS: graph structure summarization. In: Proceedings of the SIAM International Conference on Data Mining, pp. 454–465 (2010)
2. Purohit, M., Prakash, B.A., Kang, C., Zhang, Y., Subrahmanian, V.S.: Fast influence-based coarsening for large networks. In: Proceedings of the ACM Conference on Knowledge Discovery and Data Mining (KDD'2014), pp. 1296–1305 (2014)
3. Kumar, K.A., Efstathopoulos, P.: Utility-driven graph summarization. Proc. VLDB Endowment **12**(4), 335–347 (2018)
4. Navlakha, S., Rastogi, R., Shrivastava, N.: Graph summarization with bounded error. In: International Conference on Management of Data, SIGMOD, pp. 419–432 (2008)
5. Riondato, M., García-Soriano, D., Bonchi, F.: Graph summarization with quality guarantees. In: Proceedings of the 2016 IEEE 16th International Conference on Data Mining (ICDM'2014), pp. 947–952 (2016)
6. Dunne, C., Shneiderman, B.: Motif simplification: improving network visualization readability with fan, connector, and clique glyphs. In: Proceedings of the SIGCHI Conference on Human Factors in Computing Systems (CHI'2013), pp. 3247–3256 (2013)
7. Zhu, L., Ghasemi-Gol, M., Szekely, P., Galstyan, A., Knoblock, Craig A.: Unsupervised entity resolution on multi-type graphs. In: Groth, P., Simperl, E., Gray, A., Sabou, M., Krötzsch, M., Lecue, F., Flöck, F., Gil, Y. (eds.) ISWC 2016. LNCS, vol. 9981, pp. 649–667. Springer, Cham (2016). https://doi.org/10.1007/978-3-319-46523-4_39
8. Dunne, C., Shneiderman, B.: Motif simplification: improving network visualization readability with fan, connector, and clique glyphs. In: Conference on Human Factors in Computing Systems, CHI, pp. 3247–3256 (2013)
9. Koutra, D., Kang, U., Vreeken, J., Faloutsos, C.: Summarizing and understanding large graphs. Stat. Anal. Data Min. **8**(3), 183–202 (2015)
10. Li, C., Baciu, G., Wang, Y.: Modulgraph: modularity-based visualization of massive graphs. SIGGRAPH Asia 2015 Visualization in High Performance Computing, pp. 1–4. (2015)
11. Meng, J., Tu, Y.-C.: Flexible and feasible support measures for mining frequent patterns in large labeled graphs. In: SIGMOD'17: Proceedings of the 2017 ACM International Conference on Management of Data, pp. 391–402 (2017)
12. Verma, A., Butenko, S.: Network clustering via clique relaxations: a community based. Graph Partitioning Graph Clustering **588**, 129 (2013)
13. Bron, C., Kerbosch, J.: Algorithm 457: finding all cliques of an undirected graph. Commun. ACM **16**(9), 575–577 (1973)
14. Liu, G., Wong, L.: Effective pruning techniques for mining quasi-cliques. In: Daelemans, W., Goethals, B., Morik, K. (eds.) ECML PKDD 2008. LNCS (LNAI), vol. 5212, pp. 33–49. Springer, Heidelberg (2008). https://doi.org/10.1007/978-3-540-87481-2_3
15. Sanei-Mehri, S.-V., Das, A., Tirthapura, S.: Enumerating top-k quasi-cliques. In: IEEE International Conference on Big Data (Big Data) (2018)
16. http://snap.stanford.edu/data/
17. Koutra, D., Kang, U., Vreeken, J., Faloutsos, C.: VOG: summarizing and understanding large graphs. Stat. Anal. Data Min. **8**(3) (2014)

Uncertainty of Multi-granulation Hesitant Fuzzy Rough Sets Based on Three-Way Decisions

Hong Wang(✉) and Huanhuan Cheng

College of Mathematics and Computer Science, Shan'xi Normal University, Linfen, Shanxi 041004, People's Republic of China
whdw218@163.com

Abstract. Three-way decisions, whose extensive application finds relevance in risk decision making, have become an indispensable tool for handling uncertainty information. This paper investigates the decision-theoretic rough sets approach in the framework of multi-granulation hesitant fuzzy approximation space. Primarily, a basic theoretical framework has been developed by combining decision-theoretic with multi-granulation rough sets using three-way decisions. Thereafter, two types of double parameter rough membership degree of a hesitant fuzzy set have been constructed based on the multi-granulation decision-theoretic hesitant fuzzy rough sets, and their basic properties and relationship are discussed. Further, a modified entropy has been constructed.

Keywords: Multi-granulation hesitant fuzzy rough sets · Rough membership degree · Uncertainty measure

1 Introduction

The rough set has become an important tool for dealing with uncertain, imprecise and incomplete information [1]. This has been rectified by generalizing the method of Pawlak to decision-theoretic rough sets (DTRS) by utilizing Bayesian decision procedure [2]. The salient features of the DTRS model are conditional probability and loss function, which play a vital role in determining the thresholds from the given cost function. DTRSs can be used to compute the required thresholds from the given cost functions on the basis of minimum Bayesian decision cost procedure. Various generalizations of DTRSs models have been proposed, such as variable precision rough sets [3], 0.5-probabilistic rough sets [4], among others. As the generalization of the two-way decisions, the theory of three-way decisions has been proposed in a later work [5].

In the three-way decisions theory, a domain is divided into three disjointed parts that correspond to the three regions of the Pawlak's rough set [6]. Considering the minimum risk, we can generate three corresponding decision rules. According to the three-way decisions, we can construct rules for the acceptance from the positive region, rejection from the negative region, and non-commitment from the boundary region. Thus, the elements, whose decision cannot be made immediately need to be discussed further. However, the information in real-life applications are often fuzzy, which

D.-S. Huang and P. Premaratne (Eds.): ICIC 2020, LNAI 12465, pp. 531–541, 2020.
https://doi.org/10.1007/978-3-030-60796-8_46

necessitates an intensive study on fuzziness in fuzzy approximation. A novel method to attribute reduction for fuzzy rough sets has been proposed [7]. The triangular fuzzy decision-theoretic rough sets have been investigated in a later work [8]. Meanwhile, different extensions of the fuzzy set have been developed, such as intuitionistic [9], interval-valued intuitionistic [10], and hesitant fuzzy sets [11]. The hesitant fuzzy sets (HFSs) as a generalized set, have been intensely discussed with respect to the domain of decision-making. We take into account the combination of entropy and conditional entropy and propose a double parameter uncertainty measure in multi-granulation hesitant fuzzy approximation. To keep the decision of elements in the positive or negative regions unchanged, a novel reduction has been developed by employing the proposed uncertainty in consistent decision system. The rest of this paper has been organized as follows. Certain related notions and results are reviewed in Sect. 2. In Sect. 3, we define rough membership degrees, and certain ensuing results of considerable interest are presented here. Section 4 concludes the paper.

2 Preliminaries

2.1 Fuzzy Logic Operators and Typical Hesitant Fuzzy Sets

Definition 2.1.1 [14]. An implicator is a function φ: $[0, 1]^2 \longrightarrow [0, 1]$ satisfying $\varphi(1, 0) = 0$ and $\varphi(1,1) = \varphi(0, 1) = \varphi(0, 0) = 1$. An implicatory φ is called left monotonic (resp. right monotonic) if for every $\alpha \in [0, 1]$, $\varphi(.,\alpha)$ is decreasing (resp. $\varphi(.,\alpha)$ is increasing). If φ is both left monotonic and right monotonic, that is called hybrid monotonic. For all $x,y \in [0, 1]^2$, φ satisfies $x \leqq y \Leftrightarrow \varphi(x, y) = 1$; then it follows the confinement principle (CP principle).

Definition 2.1.2 [11]. Let $X = \{x_1, x_2,\ldots, x_n\}$be a finite and nonempty universe of discourse and H be the set of all finite nonempty subsets of unite [0, 1]; a hesitant fuzzy set A on X is a function: h_A:X- > H which can be expressed as follows:

$$A = \{\langle x, h_A(x)\rangle | x \in X\} \tag{1}$$

where $h_A(x)$ denotes the fuzzy membership degrees of the element $x \in X$ to the set A. For convenience, $h = h_A(x)$ is called a hesitant fuzzy element (HFE) and $l(h_A(x))$ is the number of values in a HFE $h_A(x)$. The set of all the hesitant fuzzy sets on X is denoted as HFS(X).

In the following parts, for simplicity, all the values in each $h_A(x)$ are arranged in increasing order.

Given a HFE h and a HFS A, some operators are defined as follows [25]:

(1) Lower bounder: $h^-(x) = \min h(x)$,

(2) Upper bounder: $h^+(x) = \max h(x)$,

(3) $|h| = \sum_{i=1}^{n} h^i(x)$.

Definition 2.1.3 [11]. let U be a universe of discourse, an HF relation $\Re$ on U is an HFS in $U \times U$, $\Re$ is expressed by:

$$\Re = \{\langle (x,y), h(x,y) \rangle | x, y \in U\} \tag{2}$$

where $h_{\Re}(x,y) : U \times U \rightarrow [0, 1]$ is a set of some values in [0, 1]. The family of all HF relations on $U \times U$ is denoted by HFR (U).

Definition 2.1.4 [15]. (HM inclusion measure for HFE) Let h_1, $h_2 \in (H, \leqq_{vH})$. A real number $Inc \leqq_{vH} (h_1, h_2) \in [0, 1]$ is called the inclusion measure between h_1 and h_2, if $Inc \leqq_{vH} (h_1, h_2)$ satisfies the following properties:

(TI1) $0 \leq Inc_{\leq \mathrm{vH}}(h_1, h_2) \leq 1$

(TI2) if $h_1 \leq_{\mathrm{vH}} h_2$, then $Inc_{\leq \mathrm{vH}}(h_1, h_2) = 1$

(TI3) if $h = 1$, then $Inc_{\leq \mathrm{vH}}(h, h^c) = 0$

(TI4) if $h_1 \leq_{\mathrm{vH}} h_2$, for any THFE h_3 $Inc_{\leq \mathrm{vH}}(h_3, h_1) \leq Inc_{\leq \mathrm{vH}}(h_3, h_2)$ and $Inc_{\leq \mathrm{vH}}(h_2, h_3) \leq Inc_{\leq \mathrm{vH}}(h_1, h_3)$

When the partially ordered set $(H, \leqq_{vH})$ is replaced by $(THF(X), \subseteq_{vHF})$, then the inclusion between any two hesitant fuzzy sets can be defined in the same way.

Definition 2.1.5 [15]. Let $h_1 = \{h_1^1, h_1^2, \ldots, h_1^{l_1}\}$, $h_2 = \{h_2^1, h_2^2, \ldots, h_2^{l_2}\}$ be two HFEs, the order $\leqq_{vH}$ between h_1 and h_2 is defined as follows:

$$h_1 \leq_{\mathrm{vH}} h_2 \text{ iff } \begin{cases} h_1^i \leq h_2^i, & i = 1, 2, \ldots, l_1, \quad l_1 \leq l_2 \\ h_1^{l_1 - l_2 + i} \leq h_2^i, & i = 1, 2, \ldots, l_2, \quad else \end{cases} \tag{3}$$

For any two HFSs A and B, $A \subseteq_{vHF} B$ iff $h_A(x) \leqq_{vH} h_B(x)$, $\forall x \in X$. The ordered set is denoted as $(HF(X), \subseteq_{vHF})$.

Definition 2.1.6 [15]. Let h_1 and h_2 be two HFEs, φ be an implicator which satisfies the hybrid monotonicity and CP principle, then $Inc\leqq_{vH}$ (h_1, h_2) is an HM inclusion measure for HFEs under the partial order $\leqq_{vH}$.

$$Inc_{\leq_{\mathrm{vH}}}(h_1, h_2) = \begin{cases} \bigwedge_i \varphi(h_1^i, h_2^i), & i = 1, 2, \ldots, l_1, \quad if\ l_1 \leq l_2 \\ \bigwedge_i \varphi(h_1^{l_1 - l_2 + i}, h_2^i), & i = 1, 2, \ldots, l_2, \quad else \end{cases} \tag{4}$$

Definition 2.1.7 [15]. Let $A \in (HFS(X), \subseteq)$ be HFSs, $\Re$ is a hesitant fuzzy relations on U and φ be an implicator which satisfies the hybrid monotonicity and CP principle, then the inclusion measure $Inc([x]_{\Re}, A)$ can be defined as follows:

$$Inc([x]_{\Re}, A) = \frac{\sum_{i=1}^{n} [x]_{\Re}^{+}(x_i) Inc_{\leq \mathrm{vH}}([x]_{\Re}(x_i), h_A(x_i))}{\sum_{i=1}^{n} [x]_{\Re}^{+}(x_i)} \tag{5}$$

For convenience, in the following, we let $\varphi = \varphi\Delta$, where $\varphi\Delta(x, y) = 1$ for $x \leqq y$ and $\varphi\Delta(x, y) = y/x$ otherwise, based on T_P.

2.2 Multi-granulation Hesitant Fuzzy Decision Rough Set

Definition 2.2.1. Suppose $(U, \Re_i)$ is a multigranulation hesitant fuzzy approximation space, $\Re = \{R_1, R_2, \ldots, R_m\}$ is a multigranulation structure on U, $\forall X \in HF(U)$, $x \in U$, the novel membership is defined as:

$$Inc([x]_{\Re}, X) = \frac{\sum_{i=1}^{n} [x]_{\Re}^{+}(x_i) Inc \leq_{VH} ([x]_{\Re}(x_i), h_X(x_i))}{\sum_{i=1}^{n} [x]_{\Re}^{+}(x_i)} \tag{6}$$

Proposition 2.2.1. Suppose $(U, \Re_i)$ is a multigranulation hesitant fuzzy approximation space, $\Re = \{R_1, R_2, \ldots, R_m\}$ is a multigranulation structure on U, $\forall T \subseteq U$, $x \in U$, then $Inc([x]_{\Re}, T) + Inc([x]_{\Re}, \sim T) = 1$.

Proof According to T is crisp, $h_T(x_i) = 1$ or $h_T(x_i) = 0$. If $h_T(x_i) = 1$, we have $[x]_{\Re}(x_i) \leq_{VH} h_T(x_i)$, thus $Inc \leq_{VH}([x]_{\Re}(x_i), h_T(x_i)) = 1$. If $h_T(x_i) = 0$, according to the Gaines implicator: $\varphi\Delta(x, y) = 1$ for $x \leq y$ and $\varphi\Delta(x, y) = y/x$ otherwise, based on T_P, thus $Inc \leq_{VH}([x]_{\Re}(x_i), h_T(x_i)) = 0$, we can get the same results. So,

$$\begin{aligned} & Inc([x]_{\Re}, T) + Inc([x]_{\Re}, \sim T) \\ & = \frac{\sum_{i=1}^{n} [x]_{\Re}^{+}(x_i) Inc \leq_{VH}([x]_{\Re}(x_i), h_T(x_i))}{\sum_{i=1}^{n} [x]_{\Re}^{+}(x_i)} + \frac{\sum_{i=1}^{n} [x]_{\Re}^{+}(x_i) Inc \leq_{VH}([x]_{\Re}(x_i), h_{\sim T}(x_i))}{\sum_{i=1}^{n} [x]_{\Re}^{+}(x_i)} \\ & = \frac{\sum_{i=1}^{n} [x]_{\Re}^{+}(x_i) \left(Inc \leq_{VH}([x]_{\Re}(x_i), h_T(x_i)) + Inc \leq_{VH}([x]_{\Re}(x_i), h_{\sim T}(x_i)) \right)}{\sum_{i=1}^{n} [x]_{\Re}^{+}(x_i)} = 1 \end{aligned} \tag{7}$$

Definition 2.2.2. Suppose $(U, \Re_i)$ is a multigranulation hesitant fuzzy approximation space, $\Re = \{R_1, R_2, \ldots, R_m\}$ is a multigranulation structure on U, the novel degree are defined as: $\forall X \in HF(U)$, $x \in U$,

$$\omega_X^{\Re}(x) = \max_{i=1}^{m} Inc([x]_{\Re}, X) \tag{8}$$

Remark 2.2.1. Because of $\omega_X^{\Re}(x) + \omega_{\sim X}^{\Re}(x) \neq 1$, so we let the rough membership degree of an element $x \in U$ in X is defined as:

$$\frac{\omega_X^{\Re}(x)}{\omega_X^{\Re}(x)+\omega_{\sim X}^{\Re}(x)} \tag{9}$$

2.3 Bayesian Decision Procedure Based on Type-1 Multi-granulation Hesitant Fuzzy Decision Rough Set

In the Bayesian decision procedure, a finite set of states can be written as $\Omega = \{\omega_1, \omega_2, \ldots, \omega_s\}$, and a finite set of r possible actions can be denoted by $A = \{a_1, a_2, \cdots, a_r\}$. Let $P(\omega_i|x)$ be the conditional probability of an object x being in state ω_j given that the object is described by x. Let $\lambda(a_i|\omega_j)$ denote the loss, or cost, for taking action a_i when the state is ω_j, the expected loss associated with taking action a_i is given by $R(a_i|x) = \sum_{j=1}^{s} \lambda(a_i|\omega_j)P(\omega_i|x)$. In classical rough set theory, the approximation operators partition the universe into three disjoint classes *POS(A)*, *NEG(A)*, *BND(A)*. Through using the conditional Probability $P(X|[x])$, the Bayesian decision procedure can decide how to assign x into these three disjoint regions. The expected losses of each action for object $x \in U$ are defined by formulas:

$$\Re(P|X) = \lambda_{PP}\omega_T^{\Re}(x) + \lambda_{PN}\omega_{\sim T}^{\Re}(x) \tag{10}$$

$$\Re(B|X) = \lambda_{BP}\omega_T^{\Re}(x) + \lambda_{BN}\omega_{\sim T}^{\Re}(x) \tag{11}$$

$$\Re(N|X) = \lambda_{NP}\omega_T^{\Re}(x) + \lambda_{NN}\omega_{\sim T}^{\Re}(x) \tag{12}$$

The Bayesian decision procedure suggests the following minimum-risk decision rules:

(1) If $\Re(P|X) \le \Re(B|X)$ and $\Re(P|X) \le \Re(N|X)$, then $x \in POS^{\omega}(T)$;
(2) If $\Re(B|X) \le \Re(P|X)$ and $\Re(B|X) \le \Re(N|X)$, then $x \in BND^{\omega}(T)$;
(3) If $\Re(N|X) \le \Re(B|X)$ and $\Re(N|X) \le \Re(P|X)$, then $x \in NEG^{\omega}(T)$.

Now let us consider a special kind of loss function (C_0):

$$\lambda_{PP} \le \lambda_{BP} < \lambda_{NP},\ \lambda_{NN} \le \lambda_{BN} < \lambda_{PN} \tag{13}$$

Under condition (C_0), if $\omega_T^{\Re}(x) = 0$, then $x \in NEG^{\omega}(T)$; If $\omega_{\sim T}^{\Re}(x) = 0$, then $x \in POS^{w}(T)$;

Further according to Bayesian decision procedure, we have

$$\begin{aligned} &\Re(P|X) \le \Re(N|X) \\ &\Leftrightarrow \lambda_{PP}\omega_T^{\Re}(x) + \lambda_{PN}\omega_{\sim T}^{\Re}(x) \le \lambda_{NP}\omega_T^{\Re}(x) + \lambda_{NN}\omega_{\sim T}^{\Re}(x) \\ &\Leftrightarrow \frac{\omega_T^{\Re}(x)}{\omega_{\sim T}^{\Re}(x)} \ge \frac{\lambda_{PN}-\lambda_{NN}}{\lambda_{NP}-\lambda_{PP}} \end{aligned} \tag{14}$$

$$\begin{aligned} &\Re(P|X) \leq \Re(B|X) \\ &\Leftrightarrow \lambda_{PP}\omega_T^{\Re}(x) + \lambda_{PN}\omega_{\sim T}^{\Re}(x) \leq \lambda_{BP}\omega_T^{\Re}(x) + \lambda_{BN}\omega_{\sim T}^{\Re}(x) \\ &\Leftrightarrow \frac{\omega_{\sim T}^{\Re}(x)}{\omega_T^{\Re}(x)} \geq \frac{\lambda_{BP} - \lambda_{PP}}{\lambda_{PN} - \lambda_{BN}} \end{aligned} \tag{15}$$

Let $f(x,T) = \frac{\omega_T^{\Re}(x)}{\omega_{\sim T}^{\Re}(x)}$, then the decision rules can be equivalently rewritten as follows:

(P1) If $f(x,T) \geq \frac{\lambda_{PN}-\lambda_{NN}}{\lambda_{NP}-\lambda_{PP}}$ and $\frac{1}{f(x,T)} \leq \frac{\lambda_{BP}-\lambda_{PP}}{\lambda_{PN}-\lambda_{BN}}$, decide $x \in POS^{\omega}(T)$;
(B1) If $\frac{1}{f(x,T)} > \frac{\lambda_{BP}-\lambda_{PP}}{\lambda_{PN}-\lambda_{BN}}$ and $f(x,T) > \frac{\lambda_{BN}-\lambda_{NN}}{\lambda_{NP}-\lambda_{BP}}$, decide $x \in BND^{\omega}(T)$;
(N1) If $f(x,T) < \frac{\lambda_{PN}-\lambda_{NN}}{\lambda_{NP}-\lambda_{PP}}$ and $f(x,T) \leq \frac{\lambda_{BN}-\lambda_{NN}}{\lambda_{NP}-\lambda_{BP}}$, decide $x \in NEG^{\omega}(T)$;

If the loss function still satisfies the following conditions: $\lambda_{PP} < \lambda_{BP} < \lambda_{NP}$, $\lambda_{NN} < \lambda_{BN} < \lambda_{PN}$, then the decision rules an simplified as follows:

(P2) If $f(x,T) \geq \frac{\lambda_{PN}-\lambda_{NN}}{\lambda_{NP}-\lambda_{PP}}$ and $f(x,T) \geq \frac{\lambda_{PN}-\lambda_{BN}}{\lambda_{BP}-\lambda_{PP}}$, decide $x \in POS^{\omega}(T)$;
(B2) If $f(x,T) < \frac{\lambda_{PN}-\lambda_{BN}}{\lambda_{BP}-\lambda_{PP}}$ and $f(x,T) > \frac{\lambda_{BN}-\lambda_{NN}}{\lambda_{NP}-\lambda_{BP}}$, decide $x \in BND^{\omega}(T)$;
(N2) If $f(x,T) < \frac{\lambda_{NN}-\lambda_{PN}}{\lambda_{PP}-\lambda_{NP}}$ and $f(x,T) \leq \frac{\lambda_{BN}-\lambda_{NN}}{\lambda_{NP}-\lambda_{BP}}$, decide $x \in NEG^{\omega}(T)$.

If $\frac{\lambda_{PN}-\lambda_{BN}}{\lambda_{BP}-\lambda_{PP}} \geq \frac{\lambda_{BN}-\lambda_{NN}}{\lambda_{NP}-\lambda_{BP}}$, thus we can obtain the decision rules:

(P3) If $f(x,T) \geq \frac{\lambda_{PN}-\lambda_{BN}}{\lambda_{BP}-\lambda_{PP}}$, decide $x \in POS^{\omega}(T)$
(B3) If $\frac{\lambda_{BN}-\lambda_{NN}}{\lambda_{NP}-\lambda_{BP}} < f(x,T) < \frac{\lambda_{PN}-\lambda_{BN}}{\lambda_{BP}-\lambda_{PP}}$ and decide $x \in BND^{\omega}(T)$;
(N3) If $f(x,T) \leq \frac{\lambda_{BN}-\lambda_{NN}}{\lambda_{NP}-\lambda_{PP}}$, decide $x \in NEG^{\omega}(T)$.

Let us set $\alpha = \frac{\lambda_{PN}-\lambda_{BN}}{\lambda_{BP}-\lambda_{PP}}$, $\beta = \frac{\lambda_{BN}-\lambda_{NN}}{\lambda_{NP}-\lambda_{BP}}$, thus decision rules (P3)–(N3) can be equivalently rewritten as follows:

(P4) If $f(x,T) \geq \alpha$, decide $x \in POS^{\omega}(T)$;
(B4) If $\beta < f(x,T) < \alpha$, decide $x \in BND^{\omega}(T)$;
(N4) If $f(x,T) \leq \beta$, decide $x \in NEG^{\omega}(T)$.

Then the decision-theoretic rough set in multi-granulation fuzzy decision system can be listed as follows:

$$\begin{aligned} \underline{\omega}(T) &= \{x : x \in POS^{\omega}(T)\} = \{x : f(x,T) \geq \alpha \vee \omega_{\neg T}^{\Re}(x) = 0\} \\ \overline{\omega}(T) &= \{x : x \in U - NEG^{\omega}(T)\} = \{x : f(x,T) > \beta\} \end{aligned} \tag{16}$$

3 Uncertainty of Multi-granulation Hesitant Fuzzy Rough Sets Based on Three-Way Decisions

From the three-way decision viewpoint, if the element in the positive region and negative region, we can immediately make certain decision. Conversely, if the element in the boundary region, we cannot immediately make decision or need to defer decision -making. On the basis of this, we can define an uncertainty measure based on the decision rules listed in Sect. 2.

3.1 Type-*I* α, β-Rough Membership Degree

We first define a novel type-$I\alpha, \beta$ uncertainty based on three-way decision model in the multi-granulation hesitant fuzzy sets.

The model of rough membership an element $x \in U$ in X should satisfy:

(1) When $x \in POS^{\omega}(X)$, then $\Re_{\omega}^{\alpha,\beta}(X)(x) = 1$,

(2) When $x \in NEG^{\omega}(X)$, then $\Re_{\omega}^{\alpha,\beta}(X)(x) = 0$.

So, we can obtain that the uncertainty is caused by the boundary.

Definition 3.1.1. Suppose $(U, \Re_i)$ is a multigranulation hesitant fuzzy approximation space, $\Re = \{\Re_1, \Re_2, \cdots, \Re_m\}$ is a multigranulation structure on membership degree of an element $x \in U$ in X is defined as:

$$\Re_{\omega}^{\alpha,\beta}(X)(x) = \begin{cases} 1, & \frac{1}{f(x,X)} \leq \frac{1}{\alpha}, \\ \frac{\omega_X^{\Re}(x)}{\omega_X^{\Re}(x) + \omega_{\sim X}^{\Re}(x)}, & \beta < f(x,X) < \alpha, \\ 0, & f(x,X) \leq \beta. \end{cases} \tag{17}$$

Proposition 3.1.1. Suppose $(U, \Re_i)$ is a multi-granulation hesitant fuzzy approximation space, $\Re = \{\Re_1, \Re_2, \cdots, \Re_m\}$ is a set of hesitant fuzzy relations, then

$$\forall X, Y \in HF(U), X \subseteq_{VHF} Y \Rightarrow \Re_{\omega}^{\alpha,\beta}(X) \subseteq \Re_{\omega}^{\alpha,\beta}(Y)$$

Proof. Suppose $X \subseteq_{VHF} Y$ then $\forall x \in U, h_X(x_i) \leq_{VH} h_Y(x_i)$, we have

$$\begin{aligned} \max_{i=1}^{m} Inc([x]_{\Re}, X) &= \max_{i=1}^{m} \frac{\sum_{i=1}^{n} [x]_{\Re}^{+}(x_i) Inc \leq_{VH}([x]_{\Re}(x_i), h_X(x_i))}{\sum_{i=1}^{n} [x]_{\Re}^{+}(x_i)} \\ &\leq \max_{i=1}^{m} \frac{\sum_{i=1}^{n} [x]_{\Re}^{+}(x_i) Inc \leq_{VH}([x]_{\Re}(x_i), h_Y(x_i))}{\sum_{i=1}^{n} [x]_{\Re}^{+}(x_i)} = \max_{i=1}^{m} Inc([x]_{\Re}, Y) \end{aligned} \tag{18}$$

Similarly, $\max_{i=1}^{m} Inc([x]_{\Re}, \sim X) \geq \max_{i=1}^{m} Inc([x]_{\Re}, \sim Y)$, Then, we have

$$\begin{aligned}\max_{i=1}^{m} Inc([x]_{\Re}, \sim X) = 0 \Rightarrow \max_{i=1}^{m} Inc([x]_{\Re}, \sim Y) \\ = 0, \max_{i=1}^{m} Inc([x]_{\Re}, X) \\ = 1 \Rightarrow \max_{i=1}^{m} Inc([x]_{\Re}, Y) = 1\end{aligned} \tag{19}$$

If $f(x, X) \geq \alpha \Rightarrow f(x, Y) \geq \alpha$, thus $\Re_{\omega}^{\alpha,\beta}(X)(x) = 1 \Rightarrow \Re_{\omega}^{\alpha,\beta}(Y)(x) = 1$.
If $\beta < f(x, X) < \alpha$, then $\beta < f(x, X)$.

$$\Re_{\omega}^{\alpha,\beta}(X)(x) = \frac{\omega_X^{\Re}(x)}{\omega_X^{\Re}(x) + \omega_{\sim X}^{\Re}(x)} \Rightarrow \begin{cases} 1, & f(x, Y) > \alpha, \\ \frac{\omega_Y^{\Re}(x)}{\omega_Y^{\Re}(x) + \omega_{\sim Y}^{\Re}(x)}, & \beta < f(x, Y) < \alpha. \end{cases} \tag{20}$$

And $\frac{\omega_X^{\Re}(x)}{\omega_X^{\Re}(x) + \omega_{\sim X}^{\Re}(x)} \leq \frac{\omega_Y^{\Re}(x)}{\omega_Y^{\Re}(x) + \omega_{\sim Y}^{\Re}(x)}$.
If $f(x, Y) < \beta$, then $\Re_{\omega}^{\alpha,\beta}(X)(x) = 0 \Rightarrow \Re_{\omega}^{\alpha,\beta}(Y)(x) \geq 0$.
Thus, $\Re_{\omega}^{\alpha,\beta}(X) \subseteq \Re_{\omega}^{\alpha,\beta}(Y)$.

Proposition 3.1.2. Suppose $S = (U, \Re_i)$ is a multi-granulation hesitant fuzzy approximation space, $\Re = \{\Re_1, \Re_2, \cdots, \Re_m\}$ is a set of hesitant fuzzy relations, in which $\Re_i$ is generated by $A_i \subseteq AT$.

(1) If $\alpha_2 > \alpha_1$, then $\Re_{\omega}^{\alpha_2,\beta}(X) \subseteq \Re_{\omega}^{\alpha_1,\beta}(X)$.
(2) If $\beta_2 > \beta_1$, then $\Re_{\omega}^{\alpha,\beta_2}(X) \subseteq \Re_{\omega}^{\alpha,\beta_1}(X)$.

Proof. It is easy to proof by the Definition 3.1.1.

3.2 Type-*I* η, ξ-Rough Membership Degree

In order to compare it with the three decision models that are common to us, we can do a simple transformation. $\eta = \frac{\lambda_{PN} - \lambda_{BN}}{\lambda_{PN} - \lambda_{BN} + \lambda_{BP} - \lambda_{PP}}$, $\xi = \frac{\lambda_{BN} - \lambda_{BN}}{\lambda_{BN} - \lambda_{BN} + \lambda_{NP} - \lambda_{BP}}$.

If $\eta > \xi$, we have

$$f(x, X) \geq \frac{\lambda_{PN} - \lambda_{BN}}{\lambda_{BP} - \lambda_{PP}} \Leftrightarrow \frac{1}{f(x, X)} \leq \frac{\lambda_{PP} - \lambda_{BP}}{\lambda_{BN} - \lambda_{PN}} = \frac{1}{\eta} - 1 \Leftrightarrow f(x, X) \geq \frac{\eta}{1 - \eta} \tag{21}$$

$$f(x, X) \leq \frac{\lambda_{BN} - \lambda_{NN}}{\lambda_{NP} - \lambda_{BP}} \Leftrightarrow \frac{1}{f(x, X)} \leq \frac{\lambda_{BP} - \lambda_{NP}}{\lambda_{NN} - \lambda_{BN}} = \frac{1}{\xi} - 1 \Leftrightarrow f(x, X) \leq \frac{\xi}{1 - \xi} \tag{22}$$

Definition 3.2.1. Suppose $(U, \Re_i)$ is a multi-granulation hesitant fuzzy approximation space, $\Re = \{\Re_1, \Re_2, \cdots, \Re_m\}$ is a multi-granulation structure on U, the Type-$I\alpha, \beta$ rough set membership degree defined as follows:

$$\Re_{\omega}^{\eta,\xi}(X)(x)\begin{cases}1, & \frac{1}{f(x,X)}\leq\frac{\eta}{1-\eta},\\ \frac{\omega_X^{\Re}(x)}{\omega_X^{\Re}(x)+\omega_{\sim X}^{\Re}(x)}, \frac{1}{f(x,X)}>\frac{\eta}{1-\eta}, & f(x,X)>\frac{\xi}{1-\xi},\\ 0, & f(x,X)\leq\frac{\xi}{1-\xi}.\end{cases} \tag{23}$$

Proposition 3.2.1. Suppose $(U,\Re_i)$ is a multi-granulation hesitant fuzzy approximation space, $\Re=\{\Re_1,\Re_2,\cdots,\Re_m\}$ is a set of hesitant fuzzy relations, then $\forall X\in HF(U), \Re_{\omega}^{\eta,1-\eta}(`\sim X)=\Re_{\omega}^{\eta,1-\eta}(X)$.

Proof. Since $\forall x\in U$.

$$\Re_{\omega}^{\eta,1-\eta}(X^c)(x)=\begin{cases}1, & \frac{1}{f(x,X)}\leq\frac{1-\eta}{\eta},\\ \frac{\omega_X^{\Re}(x)}{\omega_X^{\Re}(x)+\omega_{X^C}^{\Re}(x)}, \frac{1-\eta}{\eta}<f(x,X)<\frac{\eta}{1-\eta}, \\ 0, & f(x,X)\leq\frac{1-\eta}{\eta}.\end{cases} \tag{24}$$

and

$$\Re_{\omega}^{\eta,1-\eta}(X^c)(x)=\begin{cases}1, & f(x,X)\leq\frac{1-\eta}{\eta},\\ \frac{\omega_{X^C}^{\Re}(x)}{\omega_X^{\Re}(x)+\omega_{X^C}^{\Re}(x)}, \frac{1-\eta}{\eta}<f(x,X)<\frac{\eta}{1-\eta}, \\ 0, & \frac{1}{f(X,x)}\leq\frac{1-\eta}{\eta}.\end{cases} \tag{25}$$

thus,

$$\Re_{\omega}^{\eta,1-\eta}(X^c)(x)+\Re_{\omega}^{\eta,1-\eta}(X)(x)=\begin{cases}1, & f(x,X)\leq\frac{\eta}{1-\eta},\\ 1, \frac{1-\eta}{\eta}<f(x,X)<\frac{\eta}{1-\eta}, \\ 1, & \frac{1}{f(X,x)}\leq\frac{1-\eta}{\eta}.\end{cases} \tag{26}$$

Definition 3.2.2. Suppose $(U,\Re_i)$ be a multi-granulation hesitant fuzzy approximation system, Type-I η,ξ-uncertainty measure of a hesitant fuzzy rough set in $(U,\Re_i)$, denoted by $F\Re_{\omega}^{\eta,\xi}(X)$ is defined as follows:

$$\begin{aligned}F\Re_{\omega}^{\eta,\xi}(X)=\frac{k}{-n}\sum_{i=1}^{n}(&\Re_{\omega}^{\eta,\xi}(X)(x_i)\ln\Re_{\omega}^{\eta,\xi}(X)(x_i)\\ &+(1-\Re_{\omega}^{\eta,\xi}(X)(x_i))\ln(1-\Re_{\omega}^{\eta,\xi}(X)(x_i))\end{aligned} \tag{27}$$

Proposition 3.2.2. Let $(U,\Re_i)$ be a multi-granulation hesitant fuzzy approximation space, where $\Re=\{\Re_1,\Re_2,\cdots,\Re_m\}$ is a set of hesitant fuzzy relations. If $\xi=1-\eta$, X $F\Re_{\omega}^{\eta,\xi}(X)=0$ and $\sim X$ are two crisp, then $F\Re_{\omega}^{\eta,1-\eta}(X)=0$.

Proof. Since X is a crisp and definable set, $\forall x \in X$, we have

$$\underline{\Re}^{\omega,\eta}(X) = X, \overline{\Re}^{\omega,1-\eta}(\sim X) = (\sim X) \cdot x \in X \Leftrightarrow x \in \underline{\Re}^{\omega,\eta}(X) \Leftrightarrow x \in \omega_X^{\Re} \geq \eta \quad (28)$$

$$x \cap \sim X = \emptyset \Leftrightarrow x \cap \bar{\Re}^{\omega,\eta}(\sim X) = \emptyset \Leftrightarrow \omega_{\sim X}^{\Re} \leq 1 - \eta \quad (29)$$

thus, we conclude that $\frac{1}{f(x,X)} = \frac{\omega_{\sim X}^{\Re}(x)}{\omega_X^{\Re}(x)} \leq \frac{1-\eta}{\eta}$, according to the Definition 3.2.1, we obtain that $\Re_{\omega}^{\eta,1-\eta}(X)(x) = 1$

Similarity,$\forall x \cap X = \emptyset$, from $\underline{\Re}^{\omega,\eta}(\sim X) = \sim X, \bar{\Re}^{\omega,1-\eta}(X) = (X)$. We have $f(x,X) = \frac{\omega_X^{\Re}(x)}{\omega_{\sim X}^{\Re}(x)} \leq \frac{1-\eta}{\eta}$, according to the Definition 3.2.1, we can obtain that $\Re_{\omega}^{\eta,1-\eta}(X)(x) = 0$, So $\forall x \in U, (\Re_{\omega}^{\eta,\xi}(X)(x_i) \ln \Re_{\omega}^{\eta,\xi}(X)(x_i) + (1 - \Re_{\omega}^{\eta,\xi}(X)(x_i)) \ln(1 - \Re_{\omega}^{\eta,\xi}(X)(x_i)) = 0$ thus, $F\Re_{\omega}^{\eta,1-\eta}(X) = 0$

Proposition 3.2.3. Let $S = (U, \Re)$ be a multi-granulation hesitant fuzzy approximation space, where $\Re = \{\Re_1, \Re_2, \cdots, \Re_m\}$ is a set of hesitant fuzzy relations. If $\xi = 1 - \eta, \forall X \subseteq U$, then $\mathrm{F}\Re_{\omega}^{\eta,1-\eta}(\sim X) = \mathrm{F}\Re_{\omega}^{\eta,1-\eta}(X)$.

Proof $\forall x \in HF(U)$, according to the Proposition 3.2.1, we have $\Re_{\omega}^{\eta,1-\eta}(\sim X)(x) = 1 - \Re_{\omega}^{\eta,1-\eta}(X)(x)$. Then

$$\begin{aligned} &\mathrm{F}\Re_{\omega}^{\eta,1-\eta}(\sim X) \\ &= \frac{k}{-n}\sum_{i=1}^{n}(\Re_{\omega}^{\eta,1-\eta}(\sim X)(x_i) \ln \Re_{\omega}^{\eta,1-\eta}(\sim X)(x_i) + (1 - \Re_{\omega}^{\eta,1-\eta}(\sim X)(x_i)) \ln(1 - \Re_{\omega}^{\eta,1-\eta}(\sim X)(x_i)) \\ &= \frac{k}{-n}\sum_{i=1}^{n}(1 - \Re_{\omega}^{\eta,1-\eta}(\sim X)(x_i)) \ln(1 - R_{\omega}^{\eta,1-\eta}(X)(x_i)) + \Re_{\omega}^{\eta,1-\eta}(X)(x_i)) \ln \Re_{\omega}^{\eta,1-\eta}(X)(x_i) \\ &= F\Re_{\omega}^{\eta,1-\eta}(X) \end{aligned} \quad (30)$$

4 Conclusions

In this paper, we propose the membership degree of an object with respect to a hesitant fuzzy set in a single granulation fuzzy rough set model. By using the maximal and minimal membership degrees of an object with respect to a hesitant fuzzy set, we have given two types of Multi-granulation hesitant fuzzy decision-theoretic rough sets. Thereafter, we have discussed the decision-theory of Type-1 Multi-granulation hesitant fuzzy decision-theoretic rough set, using the method of three-way decisions. The Type-2 Multi-granulation hesitant fuzzy decision-theoretic rough set is similar to the Type-1 rough set model. Finally, we study the reduction and uncertainty measure of Multi-granulation hesitant fuzzy decision-theoretic rough sets.

References

1. Pawlak, Z.: Rough Sets: Theoretical Aspects of Reasoning About Data. Kluwer Academic Publishers, Dordrecht (1991)
2. Wong, S.K.M., Ziarko, W.: Comparison of the probabilistic approximate classification and the fuzzy set model. Fuzzy Sets Syst. **21**, 357–362 (1987)
3. Ziarko, W.: Variable precision rough sets model. Int. J. Comput. Inf. Sci. **46**(1), 39–59 (1993)
4. Skowron, A., Stepaniuk, J.: Tolerance approximation spaces. Fundamental Informaticae **27**, 245–253 (1996)
5. Yao, Y.Y., Wong, S.K.M.: A decision theoretic framework for approximating concepts. Int. J. Man-Machine Stud. **37**, 793–809 (1992)
6. Hu, B.Q.: Three-ways decisions space and three-way decisions. Inf. Sci. **281**, 21–52 (2014)
7. Liang, J.Y., Chin, K.S., Dang, C., Yam, R.C.M.: A new method for measuring uncertainty and fuzziness in rough set theory. Int. J. Gen. Syst. **31**, 331–342 (2002)
8. Liang, D.C., Liu, D., Pedrycz, W., Hu, P.: Triangular fuzzy decision-theoretic rough sets. Int. J. of Approximata Reasoning **54**, 1087–1106 (2013)
9. Atanassov, K.: Intuitionistic fuzzy sets. Fuzzy Sets Syst. **20**, 87–96 (1986)
10. Atanassov, K., Gargov, G.: Interval-valued intuitionistic fuzzy sets. Fuzzy Sets Syst. **31**, 343–349 (1989)
11. Torra, V.: Hesitant fuzzy sets. Int. J. Intell. Syst. **25**, 529–539 (2010)
12. Lin, G.P., Liang, J.Y., Qian, Y.H.: Multigranulation rough sets: From partition to covering. Inf. Sci. **241**, 101–118 (2013)
13. Liu, C., Pedrycz, W.: Decision-theoretic rough set approaches to multi-covering approximation spaces based on fuzzy probability measure. J. Intell. Fuzzy Syst. **34**, 1917–1931 (2018)
14. Mas, M., Monserrat, M., Torrens, J., Trillas, E.: A survey on fuzzy implicator functions. IEEE Trans. Fuzzy Syst. **15**(6), 1107–1121 (2007)
15. Zhang, H.Y., Yang, S.H.: Inclusion measures for typical hesitant fuzzy sets, the relative similarity measure and fuzzy entropy. Soft. Comput. **4**, 1–11 (2016)

WGMFDDA: A Novel Weighted-Based Graph Regularized Matrix Factorization for Predicting Drug-Disease Associations

Mei-Neng Wang[1], Zhu-Hong You[2,3](✉), Li-Ping Li[2], Zhan-Heng Chen[2,3], and Xue-Jun Xie[1]

[1] School of Mathematics and Computer Science, Yichun University, Yichun Jiangxi 336000, China

[2] Xinjiang Technical Institutes of Physics and Chemistry, Chinese Academy of Sciences, Urumqi 830011, China

zhuhongyou@ms.xjb.ac.cn

[3] University of Chinese Academy of Sciences, Beijing 100049, China

Abstract. Identification of drug-disease associations play an important role for expediting drug development. In comparison with biological experiments for drug repositioning, computational methods may reduce costs and shorten the development cycle. Thus, a number of computational approaches have been proposed for drug repositioning recently. In this study, we develop a novel computational model WGMFDDA to infer potential drug-disease association using weighted graph regularized matrix factorization (WGMF). Firstly, the disease similarity and drug similarity are calculated on the basis of the medical description information of diseases and chemical structures of drugs, respectively. Then, weighted *K*-nearest neighbor is implemented to reformulate the drug-disease association adjacency matrix. Finally, the framework of graph regularized matrix factorization is utilized to reveal unknown associations of drug with disease. To evaluate prediction performance of the proposed WGMFDDA method, ten-fold cross-validation is performed on Fdataset. WGMFDDA achieves a high AUC value of 0.939. Experiment results show that the proposed method can be used as an efficient tool in the field of drug-disease association prediction, and can provide valuable information for relevant biomedical research.

Keywords: Drug-disease association · Graph regularization · Matrix factorization · *K*-nearest neighbor

1 Introduction

New drug research and development is still a time-consuming, high-risky and tremendously costly process [1–4]. Although the investment in new drug research and development has been increasing, the number of new drugs approved by the US Food and Drug Administration (FDA) has remained limited in the past few decades [5–7]. Therefore, more and more biomedical researchers and pharmaceutical companies are paying attention to the repositioning for existing drugs, which aims to infer the new

D.-S. Huang and P. Premaratne (Eds.): ICIC 2020, LNAI 12465, pp. 542–551, 2020.
https://doi.org/10.1007/978-3-030-60796-8_47

therapeutic uses for these drugs [8–11]. For example, Thalidomide, and Minoxidil, were repositioned as a treatment to insomnia and the androgenic alopecia, respectively [12–15]. In other words, drug repositioning is actually to infer and discover potential drug-disease associations [16].

Recently, some computational methods have been presented to identify associations of drugs with diseases, such as deep walk embedding [17, 18], rotation forest [19–22], network analysis [23–25], text mining [26, 27] and machine learning [28–31], etc. Martínez *et al.* proposed a new approach named DrugNet, which performs disease-drug and drug-disease prioritization by constructing a heterogeneous network of interconnected proteins, drugs and diseases [32]. Wang *et al.* developed a triple-layer heterogeneous network model called TL-HGBI to infer drug-disease potential associations [33]. The network integrates association data and similarity about targets, drugs and diseases. Luo *et al.* utilized Bi-Random walk algorithm and comprehensive similarity measures (MBiRW) to infer new indications for existing drugs [34]. In fact, predicting associations of drug with disease can be transformed into a recommendation system problem [35–38]. Luo *et al.* developed a drug repositioning recommendation system (DRRS) to identify new indications for a given drug [39]. In this work, we develop a novel computational model WGMFDDA, which utilizes graph regularized matrix factorization to infer the potential associations between drugs and diseases. The experiment results indicate that the performance of WGMFDDA is better than other compared methods.

2 Methods and Materials

2.1 Method Overview

To predict potential associations of drugs with diseases, the model of WGMFDDA consists of three steps (See Fig. 1): (1) we measure the similarity for drugs and diseases based on the collected dataset; (2) According to the weighted K-nearest neighbor profiles of drugs and diseases, the drug-disease association adjacency matrix is re-established; (3) the graph Laplacian regularization and Tikhonov (L_2) terms are incorporated into the standard Non-negative matrix factorization (NMF) framework to calculate the drug-disease association scores.

2.2 Dataset

In this study, we obtain the dataset (Fdataset) from Gottlieb *et al.* [40]. This dataset is used as the gold standard datasets for identifying drug-disease associations, which includes 1933 known associations between 313 diseases and 593 drugs [41, 42]. In order to more conveniently describe the drug-disease associations information, the drug-disease association adjacency matrix $Y^{n \times m}$ is constructed, where n and m are the number of drugs and diseases, respectively. The element $Y(i,j) = 1$ if drug r_i associated with disease d_j, otherwise $Y(i,j) = 0$. The similarities for drugs and diseases are obtained from the Chemical Development Kit (CDK) [43] based on SMILES [44] and MimMiner [45] based on the OMIM [41] database, respectively. In ten-fold cross-

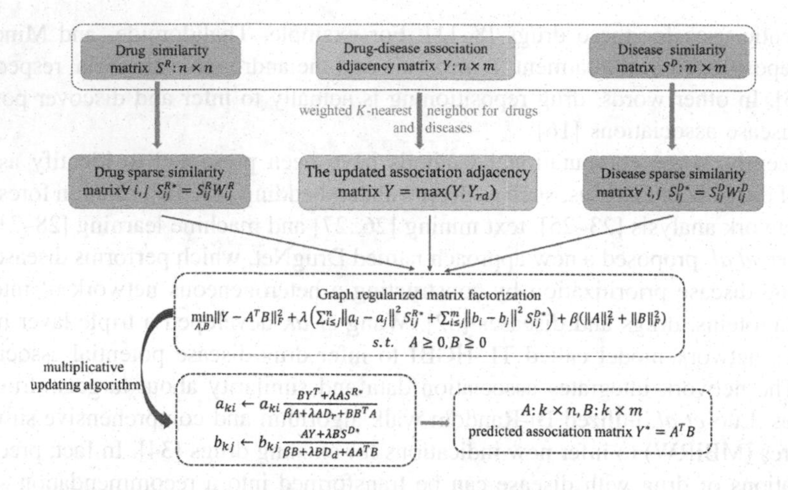

Fig. 1. Overview of the WGMFDDA framework.

validation experiments, all known associations are random divided into ten equal sized subsets, in which the training data set occupies 9/10, and the remaining partition is utilized as the test set.

2.3 Reformulate the Drug-Disease Association Adjacency Matrix

Let $R = \{r_1, r_2, \cdots, r_n\}$ and $D = \{d_1, d_2, \cdots, d_m\}$ are the set of n drugs and m diseases. $Y(r_i) = (Y_{i1}, Y_{i2}, \cdots, Y_{im})$ and $Y(d_j) = (Y_{1j}, Y_{2j}, \cdots, Y_{nj})$ are the ith row vector and jth column vector of matrix Y, respectively. $Y(r_i)$ and $Y(d_j)$ denote the interaction profiles of drugs and diseases, respectively. Since many drug-disease pairs with unknown associations (i.e. the value of these elements in Y is zero) may be potential true associations, this will affect prediction performance. In order to assign associated likelihood scores to drug-disease pairs with unknown associations, weighted K-nearest neighbor (WKNN) is implemented to calculate new interaction profiles of drugs and diseases [38, 46].

For each drug r_p (or disease d_q), the novel interaction profile can be calculated as follows:

$$Y_r(r_p) = \frac{1}{\sum_{1 \leq i \leq K} S^R(r_i, r_p)} \sum\nolimits_{i=1}^{k} a^{i-1} * S^R(r_i, r_p) Y(r_i) \tag{1}$$

or

$$Y_d(d_q) = \frac{1}{\sum_{1 \leq j \leq K} S^D(d_j, d_q)} \sum\nolimits_{j=1}^{k} a^{j-1} * S^D(d_j, d_q) Y(d_j) \tag{2}$$

$a \in [0, 1]$ denotes a decay term. S^R and S^D are the similarity matrices for drugs and diseases, respectively.

Subsequently, we define the updated association adjacency matrix Y as follows:

$$Y = \max(Y, Y_{rd}) \tag{3}$$

where

$$Y_{rd} = (Y_r + Y_d)/2 \tag{4}$$

2.4 WGMFDDA

The standard Nonnegative matrix factorization (NMF) aims to find two low-rank Nonnegative matrices whose product as more as possible to approximation to the original matrix [36, 47–49]. $Y \cong A^T B (k \leq \min(n, m))$, $A \in R^{k \times n}$ and $B \in R^{k \times m}$. To avoid overfitting, the graph Laplacian regularization and Tikhonov (L_2) terms are introduced into the standard NMF model. The objective function of WGMFDDA can be constructed as follows:

$$\begin{aligned} \min_{A,B} \left\| Y - A^T B \right\|_F^2 + \lambda \left(\sum\nolimits_{i \leq j}^{n} \left\| a_i - a_j \right\|^2 S_{ij}^{R*} + \sum\nolimits_{i \leq j}^{m} \left\| b_i - b_j \right\|^2 S_{ij}^{D*} \right) \\ + \beta \left(\|A\|_F^2 + \|B\|_F^2 \right) s.t. A \geq 0, B \geq 0 \end{aligned} \tag{5}$$

where $\|\cdot\|_F$ denotes the Frobenius norm. λ and β are the regularization parameters. a_j and b_j are jth column of matrices A and B, respectively. S^{R*} and S^{D*} denote the sparse similarity matrices for drugs and diseases, respectively.

According to the spectral graph theory, the p-nearest neighbor graph can preserve the intrinsic geometrical structure of the original data [46]. Therefore, p-nearest neighbors is utilized to construct the graphs S^{R*} and S^{D*}. The details are as follows:

$$W_{ij}^R = \begin{cases} 1 & i \in N_p(r_j) \& j \in N_p(r_i) \\ 0 & i \notin N_p(r_j) \& j \notin N_p(r_i) \\ & 0.5 \quad otherwise \end{cases} \tag{6}$$

where $N_p(r_i)$ and $N_p(r_j)$ denote the sets of p-nearest neighbors of r_i and r_j respectively. Then, we define the sparse matrix S^{R*} of drug as follows:

$$\forall i,j \quad S_{ij}^{R*} = S_{ij}^R W_{ij}^R \tag{7}$$

Similarly, the sparse matrix S^{D*} of disease can be expressed as follows:

$$\forall i,j \quad S_{ij}^{D*} = S_{ij}^D W_{ij}^D \tag{8}$$

The Eq. (5) can be written as:

$$\min_{A,B} \|Y - A^T B\|_F^2 + \lambda Tr\left(AL_r A^T\right) + \lambda Tr\left(BL_d B^T\right) + \beta\left(\|A\|_F^2 + \|B\|_F^2\right) \quad s.t. A \geq 0, B \geq 0 \tag{9}$$

Here, $L_r = D_r - S^{R*}$ and $L_d = D_d - S^{D*}$ are the graph Laplacian matrices for S^{R*} and S^{D*}, respectively. $D_r(i,i) = \sum_p S^{R*}_{ip}$ and $D_d(j,j) = \sum_q S^{D*}_{jq}$ are diagonal matrices, $Tr(\cdot)$ denotes the trace of matrix.

In order to optimize the objective function in Eq. (9), the corresponding Lagrange function $\mathcal{H}_f$ is defined as:

$$\mathcal{H}_f = Tr\left(YY^T\right) - 2Tr\left(YB^T A\right) + Tr\left(A^T BB^T A\right) + \lambda Tr\left(AL_r A^T\right) + \lambda Tr\left(BL_d B^T\right) + \beta Tr\left(AA^T\right) + \beta Tr\left(BB^T\right) + Tr\left(\Phi A^T\right) + Tr\left(\Psi B^T\right) \tag{10}$$

In which, $\Phi = \{\phi_{ki}\}$ and $\Psi = \{\psi_{kj}\}$ are Lagrange multipliers that constrain $a_{ki} \geq 0$ and $b_{kj} \geq 0$, respectively. We calculate $\frac{\partial \mathcal{H}_f}{\partial A}$ and $\frac{\partial \mathcal{H}_f}{\partial B}$ as follows:

$$\frac{\partial \mathcal{H}_f}{\partial A} = -2BY^T + 2BB^T A + 2\lambda AL_r + 2\beta A + \Phi \tag{11}$$

$$\frac{\partial H_f}{\partial B} = -2AY + 2AA^T B + 2\lambda BL_d + 2\beta B + \Psi \tag{12}$$

After using Karush–Kuhn–Tucker (KKT) conditions $\phi_{ki} a_{ki} = 0$ and $\psi_{kj} b_{kj} = 0$, the updating rules can be obtained as follows:

$$a_{ki} \leftarrow a_{ki} \frac{BY^T + \lambda AS^{R*}}{\beta A + \lambda AD_r + BB^T A} \tag{13}$$

$$b_{kj} \leftarrow b_{kj} \frac{AY + \lambda BS^{D*}}{\beta B + \lambda BD_d + AA^T B} \tag{14}$$

The predicted drug-disease association matrix is obtained by $Y^* = A^T B$. Generally, the larger the element value in predicted matrix Y^*, the more likely the drug is related to the corresponding disease.

3 Experimental Results

In this study, the model of WGMFDDA has six parameters that determine by grid search. The ROC curve and AUC value are widely used to evaluate the predictor [50–54]. WGMFDDA produces best AUC values when $P = 5$, $K = 5$, $a = 0.5$, $k = 160$, $\lambda = 1$ and $\beta = 0.02$. We implement ten-fold cross-validation (CV) experiments on the Fdataset and compare it with the previous methods: DrugNet [32], HGBI [33], MBiRW

[34] and DDRS [39]. To implement 10-CV experiment, all known drug-disease associations in Fdataset are random divided into ten equal sized subsets. the training data set occupies 9/10, while the remaining partition is utilized as the test set. As shown in Fig. 2 and Table 1, WGMFDDA achieves the AUC value of 0.939, while DrugNet, HGBI, MBiRW and DDRS are 0.778, 0.829, 0.917and 0.930, respectively. This result shows that compared with DDRS, MBiRW, HGBI and DrugNet, WGMFDDA obtains the best performance.

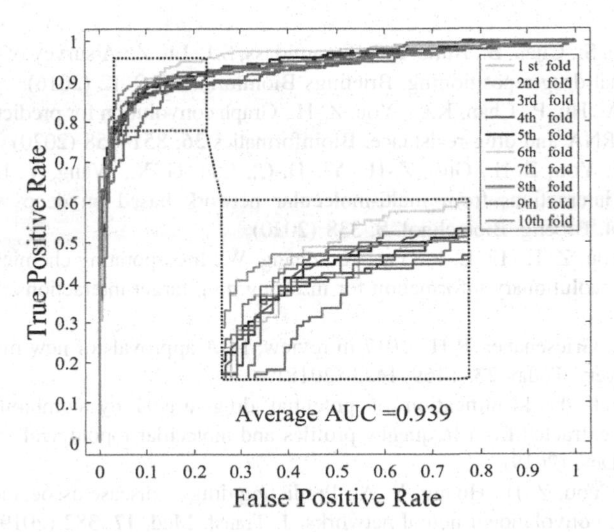

Fig. 2. The ROC curves of WGMFDDA on Fdataset under ten-fold cross-validation.

Table 1. The average AUC values of WGMFDDA and other compared methods on Fdataset.

Methods	DrugNet	HGBI	MBiRW	DDRS	WGMFDDA
AUC	0.778	0.829	0.917	0.930	0.939

4 Conclusions

The purpose of drug repositioning is to discover new indications for existing drugs. Compared to traditional drug development, drug repositioning can reduce risk, save time and costs. In this work, we present a new prediction approach, WGMFDDA, based on weighted graph regularized matrix factorization. The proposed method casts the problem of inferring the associations between drugs and diseases into a matrix factorization problem in recommendation system. The main contribution of our method is that a preprocessing step is performed before matrix factorization to reformulate the drug-disease association adjacency matrix. In ten-fold cross-validation, experiment results indicate that our proposed model outperforms other compared methods.

Acknowledgement. This work was supported in part by the NSFC Excellent Young Scholars Program, under Grants 61722212, in part by the Science and Technology Project of Jiangxi Provincial Department of Education, under Grants GJJ180830, GJJ190834.

Competing Interests
The authors declare that they have no competing interests.

References

1. Li, J., Zheng, S., Chen, B., Butte, A.J., Swamidass, S.J., Lu, Z.: A survey of current trends in computational drug repositioning. Briefings Bioinform. **17**, 2–12 (2016)
2. Huang, Y.-A., Hu, P., Chan, K.C., You, Z.-H.: Graph convolution for predicting associations between miRNA and drug resistance. Bioinformatics **36**, 851–858 (2020)
3. Chen, Z.-H., You, Z.-H., Guo, Z.-H., Yi, H.-C., Luo, G.-X., Wang, Y.-B.: Prediction of drug-target interactions from multi-molecular network based on deep walk embedding model. Front. Bioeng. Biotechnol. **8**, 338 (2020)
4. Wang, L., You, Z.-H., Li, L.-P., Yan, X., Zhang, W.: Incorporating chemical sub-structures and protein evolutionary information for inferring drug-target interactions. Sci. Rep. **10**, 1–11 (2020)
5. Kinch, M.S., Griesenauer, R.H.: 2017 in review: FDA approvals of new molecular entities. Drug Discovery Today **23**, 1469–1473 (2018)
6. Wang, L., et al.: Identification of potential drug–targets by combining evolutionary information extracted from frequency profiles and molecular topological structures. Chem. Biol. Drug Des. (2019)
7. Jiang, H.-J., You, Z.-H., Huang, Y.-A.: Predicting drug − disease associations via sigmoid kernel-based convolutional neural networks. J. Transl. Med. **17**, 382 (2019)
8. Hurle, M., Yang, L., Xie, Q., Rajpal, D., Sanseau, P., Agarwal, P.: Computational drug repositioning: from data to therapeutics. Clin. Pharmacol. Ther. **93**, 335–341 (2013)
9. Huang, Y.-A., You, Z.-H., Chen, X.: A systematic prediction of drug-target interactions using molecular fingerprints and protein sequences. Curr. Protein Pept. Sci. **19**, 468–478 (2018)
10. Wang, L., You, Z.-H., Chen, X., Yan, X., Liu, G., Zhang, W.: Rfdt: a rotation forest-based predictor for predicting drug-target interactions using drug structure and protein sequence information. Curr. Protein Pept. Sci. **19**, 445–454 (2018)
11. Li, Y., Huang, Y.-A., You, Z.-H., Li, L.-P., Wang, Z.: Drug-target interaction prediction based on drug fingerprint information and protein sequence. Molecules **24**, 2999 (2019)
12. Graul, A.I., et al.: The year's new drugs & biologics-2009. Drug News Perspect **23**, 7–36 (2010)
13. Sardana, D., Zhu, C., Zhang, M., Gudivada, R.C., Yang, L., Jegga, A.G.: Drug repositioning for orphan diseases. Briefings Bioinform. **12**, 346–356 (2011)
14. Zhang, S., Zhu, Y., You, Z., Wu, X.: Fusion of superpixel, expectation maximization and PHOG for recognizing cucumber diseases. Comput. Electron. Agric. **140**, 338–347 (2017)
15. Li, Z., et al.: In silico prediction of drug-target interaction networks based on drug chemical structure and protein sequences. Sci. Rep. **7**, 1–13 (2017)
16. Zheng, K., You, Z.-H., Wang, L., Zhou, Y., Li, L.-P., Li, Z.-W.: Dbmda: a unified embedding for sequence-based mirna similarity measure with applications to predict and validate mirna-disease associations. Mol. Therapy-Nucleic Acids **19**, 602–611 (2020)

17. Guo, Z., Yi, H., You, Z.: Construction and comprehensive analysis of a molecular association network via lncRNA–miRNA–disease–drug–protein graph. Cells **8**, 866 (2019)
18. Chen, Z., You, Z., Zhang, W., Wang, Y., Cheng, L., Alghazzawi, D.: Global vectors representation of protein sequences and its application for predicting self-interacting proteins with multi-grained cascade forest model. Genes **10**, 924 (2019)
19. You, Z.-H., Chan, K.C., Hu, P.: Predicting protein-protein interactions from primary protein sequences using a novel multi-scale local feature representation scheme and the random forest. PLoS ONE **10**, e0125811 (2015)
20. Guo, Z., You, Z., Wang, Y., Yi, H., Chen, Z.: A learning-based method for LncRNA-disease association identification combing similarity information and rotation forest. iScience **19**, 786–795 (2019)
21. Wang, L., et al.: Using two-dimensional principal component analysis and rotation forest for prediction of protein-protein interactions. Sci. Rep. **8**, 1–10 (2018)
22. You, Z., Li, X., Chan, K.C.C.: An improved sequence-based prediction protocol for protein-protein interactions using amino acids substitution matrix and rotation forest ensemble classifiers. Neurocomputing **228**, 277–282 (2017)
23. Oh, M., Ahn, J., Yoon, Y.: A network-based classification model for deriving novel drug-disease associations and assessing their molecular actions. PLoS ONE **9**, e111668 (2014)
24. Zheng, K., You, Z.-H., Li, J.-Q., Wang, L., Guo, Z.-H., Huang, Y.-A.: iCDA-CGR: identification of circRNA-disease associations based on Chaos Game Representation. PLoS Comput. Biol. **16**, e1007872 (2020)
25. Yi, H.-C., You, Z.-H., Guo, Z.-H.: Construction and analysis of molecular association network by combining behavior representation and node attributes. Front. Genet. **10**, 1106 (2019)
26. Yang, H., Spasic, I., Keane, J.A., Nenadic, G.: A text mining approach to the prediction of disease status from clinical discharge summaries. J. Am. Med. Inform. Assoc. **16**, 596–600 (2009)
27. Chen, Z.-H., Li, L.-P., He, Z., Zhou, J.-R., Li, Y., Wong, L.: An improved deep forest model for predicting self-interacting proteins from protein sequence using wavelet transformation. Front. Genet. **10**, 90 (2019)
28. Li, L., Wang, Y., You, Z., Li, Y., An, J.: PCLPred: a bioinformatics method for predicting protein-protein interactions by combining relevance vector machine model with low-rank matrix approximation. Int. J. Mol. Sci. **19**, 1029 (2018)
29. Li, S., You, Z.-H., Guo, H., Luo, X., Zhao, Z.-Q.: Inverse-free extreme learning machine with optimal information updating. IEEE Trans. Cybern. **46**, 1229–1241 (2015)
30. Zheng, K., You, Z.-H., Wang, L., Zhou, Y., Li, L.-P., Li, Z.-W.: MLMDA: a machine learning approach to predict and validate MicroRNA–disease associations by integrating of heterogenous information sources. J. Transl. Med. **17**, 260 (2019)
31. Yi, H.-C., You, Z.-H., Wang, M.-N., Guo, Z.-H., Wang, Y.-B., Zhou, J.-R.: RPI-SE: a stacking ensemble learning framework for ncRNA-protein interactions prediction using sequence information. BMC Bioinform. **21**, 60 (2020)
32. Martinez, V., Navarro, C., Cano, C., Fajardo, W., Blanco, A.: DrugNet: network-based drug–disease prioritization by integrating heterogeneous data. Artif. Intell. Med. **63**, 41–49 (2015)
33. Wang, W., Yang, S., Zhang, X., Li, J.: Drug repositioning by integrating target information through a heterogeneous network model. Bioinformatics **30**, 2923–2930 (2014)
34. Luo, H., Wang, J., Li, M., Luo, J., Peng, X., Wu, F.-X., Pan, Y.: Drug repositioning based on comprehensive similarity measures and bi-random walk algorithm. Bioinformatics **32**, 2664–2671 (2016)

35. You, Z., Wang, L., Chen, X., Zhang, S., Li, X., Yan, G., Li, Z.: PRMDA: personalized recommendation-based MiRNA-disease association prediction. Oncotarget **8**, 85568–85583 (2017)
36. Wang, M.-N., You, Z.-H., Li, L.-P., Wong, L., Chen, Z.-H., Gan, C.-Z.: GNMFLMI: graph regularized nonnegative matrix factorization for predicting LncRNA-MiRNA interactions. IEEE Access **8**, 37578–37588 (2020)
37. Huang, Y., You, Z., Chen, X., Huang, Z., Zhang, S., Yan, G.: Prediction of microbe-disease association from the integration of neighbor and graph with collaborative recommendation model. J. Transl. Med. **15**, 1–11 (2017)
38. Wang, M.-N., You, Z.-H., Wang, L., Li, L.-P., Zheng, K.: LDGRNMF: LncRNA-disease associations prediction based on graph regularized non-negative matrix factorization. Neurocomputing (2020)
39. Luo, H., Li, M., Wang, S., Liu, Q., Li, Y., Wang, J.: Computational drug repositioning using low-rank matrix approximation and randomized algorithms. Bioinformatics **34**, 1904–1912 (2018)
40. Gottlieb, A., Stein, G.Y., Ruppin, E., Sharan, R.: PREDICT: a method for inferring novel drug indications with application to personalized medicine. Mol. Syst. Biol. **7**, 496 (2011)
41. Hamosh, A., Scott, A.F., Amberger, J.S., Bocchini, C.A., McKusick, V.A.: Online Mendelian Inheritance in Man (OMIM), a knowledgebase of human genes and genetic disorders. Nucleic Acids Res. **33**, D514–D517 (2005)
42. Wishart, D.S., et al.: DrugBank: a comprehensive resource for in silico drug discovery and exploration. Nucleic Acids Res. **34**, D668–D672 (2006)
43. Steinbeck, C., Han, Y., Kuhn, S., Horlacher, O., Luttmann, E., Willighagen, E.: The Chemistry Development Kit (CDK): an open-source Java library for chemo-and bioinformatics. J. Chem. Inf. Comput. Sci. **43**, 493–500 (2003)
44. Weininge, D.: SMILES, a chemical language and information system. 1. Introduction to methodology and encoding rules. J. Chem. Inf. Comput. Sci. **28**, 31–36 (1988)
45. Van Driel, M.A., Bruggeman, J., Vriend, G., Brunner, H.G., Leunissen, J.A.: A text-mining analysis of the human phenome. Eur. J. Hum. Genet. **14**, 535–542 (2006)
46. Xiao, Q., Luo, J., Liang, C., Cai, J., Ding, P.: A graph regularized non-negative matrix factorization method for identifying microRNA-disease associations. Bioinformatics **34**, 239–248 (2018)
47. Huang, Y.-A., You, Z.-H., Li, X., Chen, X., Hu, P., Li, S., Luo, X.: Construction of reliable protein–protein interaction networks using weighted sparse representation based classifier with pseudo substitution matrix representation features. Neurocomputing **218**, 131–138 (2016)
48. Chen, Z.-H., You, Z.-H., Li, L.-P., Wang, Y.-B., Wong, L., Yi, H.-C.: Prediction of self-interacting proteins from protein sequence information based on random projection model and fast fourier transform. Int. J. Mol. Sci. **20**, 930 (2019)
49. Ji, B.-Y., You, Z.-H., Cheng, L., Zhou, J.-R., Alghazzawi, D., Li, L.-P.: Predicting miRNA-disease association from heterogeneous information network with GraRep embedding model. Sci. Rep. **10**, 1–12 (2020)
50. Chen, Z.-H., You, Z.-H., Li, L.-P., Wang, Y.-B., Qiu, Y., Hu, P.-W.: Identification of self-interacting proteins by integrating random projection classifier and finite impulse response filter. BMC Genom. **20**, 1–10 (2019)
51. Chen, X., Yan, C.C., Zhang, X., You, Z.-H.: Long non-coding RNAs and complex diseases: from experimental results to computational models. Briefings Bioinform. **18**, 558–576 (2017)

52. Jiao, Y., Du, P.: Performance measures in evaluating machine learning based bioinformatics predictors for classifications. Quant. Biol. **4**(4), 320–330 (2016). https://doi.org/10.1007/s40484-016-0081-2
53. Chen, X., Huang, Y.-A., You, Z.-H., Yan, G.-Y., Wang, X.-S.: A novel approach based on KATZ measure to predict associations of human microbiota with non-infectious diseases. Bioinformatics **33**, 733–739 (2017)
54. You, Z.-H., Huang, Z.-A., Zhu, Z., Yan, G.-Y., Li, Z.-W., Wen, Z., Chen, X.: PBMDA: a novel and effective path-based computational model for miRNA-disease association prediction. PLoS Comput. Biol. **13**, e1005455 (2017)

Natural Language Processing and Computational Linguistics

Word Embedding by Unlinking Head and Tail Entities in Crime Classification Model

Qinhua Huang(✉) and Weimin Ouyang

School of AI and Law, Shanghai University of Political Science and Law,
Shanghai 201701, China
{hqh, oywm}@shupl.edu.cn

Abstract. Word embedding is one of the natural language processing. It is designed to represent the entities and relations with vectors or matrix to make knowledge graph model. Recently many related models and methods were proposed, such as translational methods, deep learning based methods, multiplicative approaches. We proposed an embedding method by unlink the relation of head and tail entity representation when these two are the same entity. By doing so, it can free the relation space thus can have more representations. By comparing some typical word embedding algorithms and methods, we found there are tradeoff problem to deal with between algorithm's simplicity and expressiveness. After optimizing the parameter of our proposed embedding method, we test this embedding on the HMN model, a model used to generate auto-judge system in the law area. We carefully replaced the encoder part of the model using our embedding strategy, and tested the modified HMN model on real legal data set. The result showed our embedding method has some privileges on performance.

Keywords: Word embeddings · Machine learning · Crime classification

1 Introduction

The idea of using AI as the judge of law cases has a long history. In 1958, Lucien Mehi has purposed the automatic problem of AI world [16]. Afterwards, many researches were being developed. And in 1980s, the joint disciplines researches of AI and Law had come to its prosperity. In the early stages, people focused on developing AI with similar logic used in law, which emphasize the mechanism and tool of reasoning, representing, known a rule-based reasoning (RBR). This has the same idea with the NLP field. In 1980s researchers realized the complex attributes of natural language, and the law information retrieval method was put out. Experienced the basic stage of using keyword searching in AI, Carole Hafner proposed her novel method of semantic net representations. And many law expert system were build, but there were one flaw that matters, which mainly due to the openness nature of law suit predicate. While these reasoning systems based on rules continued to be developed, the other way using case to do reasoning appeared, which is called cased-based reasoning (CBR). Rissland developed the HYPO system, which was taken to be the first real CBR system in AI and Law. This research stream attracted many interests. To combine the benefits of

D.-S. Huang and P. Premaratne (Eds.): ICIC 2020, LNAI 12465, pp. 555–564, 2020.
https://doi.org/10.1007/978-3-030-60796-8_48

RBR and CBR, Rissland proposed the CABARET system, a reasoned using hybrid of CBR-RBR. CABARET can dynamically apply RBR and CBR, rather than serially call it in a fixed order. With all these efforts, it is broad consensus that automatically representation of law entity and concept searching are the core target of AI in law. But due to the limitation of data sparsity, most AI law models were far from success.

With the deep learning method developed in recent years, many researches are working on building legal AI models using NLP pre-trained models based on deep learning technique, trained and generated from big dataset of real legal documents. With the great progress made in the field of natural language processing, especially in the language model, such as ELMo, BERT, GPT-2, etc., the application of knowledge graph has been greatly developed. The legal artificial intelligence based on knowledge map can be used as a trial assistant in the ODR scene, providing accurate reference. Multitasking learning is a hot topic in the field of deep learning in the last two years, refers to training data from a number of related tasks, through sharing the representation of related tasks, can make the model can have a better generalization, Xin Zhou and so on, the study of the network transaction dispute cases of legal artificial intelligence application, combined with deep learning end-to-end and representation learning characteristics, through the legal judgment model and dispute resolution model joint training, to obtain end-to-end multi-task learning model. In addition, the study is important through joint learning, ODR can provide important legal data, is seen as an effective solution to the sparseness of legal data. The technology has been piloted in the Hangzhou Internet Court. Liu Zonglin and others noticed that the crime prediction and legal recommendation are important sub-tasks of legal judgment prediction, and proposed to use multi-task learning model to model the two tasks jointly, and to integrate the crime keyword information into the model, which can improve the accuracy of the model for these two tasks.

In the field of legal intelligence question-and-answer, especially legal advice, there is sometimes a need to answer questions that do not depend on a certain fact, and gayle McElvain has studied and established the West Plus system to provide question-and-answer capabilities by using IR and NLP technologies rather than relying on a structured knowledge base. This is a continuation of the idea of legal concept search. Due to the large degree of data openness, Chinese's study of automatic trial prediction is relatively active, limited by policy and other reasons, other languages are relatively few, Ilias Chalkidis and other 40, combined with the data of the European Court of Human Rights, set up a deep learning model based on the BERT language model, in the binary classification, multi-label classification, case importance prediction and other aspects of the traditional feature-based model, relatively speaking, the traditional feature model is more comprehensible, in addition, The study further discusses the bias of the model. Haoxi Zhong and others discuss the application of topological learning in trial prediction, in the actual legal trial, the trial is generally composed of several sub-tasks, such as application of laws, charges, fines, sentences, etc., based on the dependence between sub-tasks, the establishment of topological mapping, and thus the construction of trial models. In recent years, natural language processing has made great progress in the field of general reading comprehension, such as BERT model application, attention mechanism, Shangbang Long and so on, studied the problem of legal reading comprehension, according to the description of facts, complaints and laws, according to the judge's working mechanism, to give predictions, and thus to achieve automatic prediction. Wenmin Yang and other research based on the topological structure of sub-tasks, combined with the attention mechanism of word matching, proposed a deep learning framework, the model performance obtained to obtain a greater improvement.

The application of artificial intelligence in law is promising, and its important foundation is machine learning and natural language processing. At present, natural language processing technology is in the process of continuous development, has been or is being applied in the field of machine translation, reading comprehension, automated question-and-answer and auxiliary creation. Based on the development of NLP, Google first introduced a knowledge map in May 2012, and users will see smarter answers, i.e. more structured physical knowledge, in addition to getting links to search pages. Then the major search engine companies have followed up, such as China's Baidu "heart", so on the search dog "search cube" and so on. A typical large-scale knowledge map, such as Wikipedia, has more than 22 million terms, making it the sixth most visited site in the world. Freebase contains 39 million entities and 1.8 billion entity relationships. There are many similar knowledge maps. In addition to the large-scale general knowledge base, there is also a specialized domain knowledge base, the so-called vertical domain. Knowledge in vertical areas complements and strengthens large-scale knowledge bases, which in turn can be established. Unlike intelligent image processing technology in the field of artificial intelligence, natural language processing is closely related to scenes. The current large-scale mature commercial knowledge map focuses on the broad sense of web data knowledge, for a specific vertical areas can not be all-encompassing, which is the core interests of professional search engine companies and technical difficulty decision. High-quality training and testdata collection is difficult in specific vertical areas, data is not freely available in some areas, the number of users is relatively small, and user experience is difficult to collect. It is not difficult to find that in the legal field, the current advantage is the long-term policy of artificial intelligence at the national level, but the application of the scene driving force is still insufficient, the lack of high-quality training data and user stability of the general needs are important reasons.

At present, knowledge mapping technology is an important method to realize artificial intelligence. Knowledge map itself has the form of structural map, the current research of knowledge map mainly focuses on knowledge representation learning and knowledge map embedding, knowledge representation is moving towards a more up-level culture, intelligent and semantic direction. Some progress has been made in recent years in the knowledge map method, which is related to the complex reasoning of knowledge map event timing, reasoning and causality. The Markov logical network is combined with kgE to take advantage of logical rules, taking into account their uncertainties. The interpretability of knowledge representation, injection and reasoning is the key issue in the application of real legal artificial intelligence. In the field of knowledge mapping, knowledge migration is carried out using sparse vectors and explained by attention visualization. By using embedded path search to generate explanations of link predictions, the interpretation scheme of knowledge maps is explored. Further research is yet to be developed.

2 Related Works

In this section, we briefly listed the related models in knowledge graph embeddings. Due to the space limitation we mainly discuss two kinds of method. One the translational method. The other is tensor factorization method.

TransE firstly purposed projecting the entities into the same space, where the relation can be taken as a vector from head entity to tail entities. Formally, we have a triple (h, r, t), where $h, r, t \in \mathbb{R}^k$, h is the head entity vector, r is the relation vector and t is the tail entity vector. The TransE model represents the a relationship by a translation from head entity to tail entity, thus it holds $h + r \approx t$. By minimizing the score function $f(h, r, t) = \| h + r - t \|_2^2$, which means h + r is the closest to t in distance. This representation has very clear geometric meaning as it showed in Fig. 1.

TransH was proposed to address the issue of N-to-1, 1-to-N and N-to-N relations. It projected (h, r, t) onto a hyperplane of w_r, where w_r is the hyperplane normal vector of r. TransR noticed both TransE and TransH took the assumption that embeddings of entities and relations are represented in the same space $\mathbb{R}^k$. And relations and entities might have different semantic meaning. So TransE suggest project entities and relations onto different spaces in representation, respectively. The score function will be minimized by translating entity space into relation space.

There are some other models like Unstructured Model, which is a simplified TransE. It suppose that all r = 0; Structured Embedding, it adopted L_1 as its distance

Table 1. Entity and relation embedding models: embeddings and score functions

Model name	Embeddings	Score function s(h, r, t)
Neural Tensor Network (NTN)	$M_{r,1}, M_{r,2} \in \mathbb{R}^{k \times d}$, $b_r \in \mathbb{R}^k$	$u_r^\top g(h^\top M_r t + M_{r,1} h + M_{r,2} t + b_r)$
Latent Factor Model (LFM) [10]		$h^\top M_r t$
Semantic Matching Energy (SME)	M_1, M_2, M_3, M_4 are weight matrices, $\otimes$ is the Hadamard product, b_1, b_2 are bias vectors	$((M_1 h) \otimes (M_2 r) + b_1)\top((M_3 t) \otimes (M_4 r) + b_2)$
TranE [1]	$h, r, t \in \mathbb{R}^k$	$\| h + r - t \|$
TransH [5]	$h, t \in \mathbb{R}^k, w_r, d_r \in \mathbb{R}^k$	$\| (h - w_r^\top h w_r) + d_r - (t - w_r^\top t w_r) \|$
TransD	$\{h, h_p \in \mathbb{R}^k\}$ for entity h, $\{t, t_p \in \mathbb{R}^k\}$, for entity t, $\{r, r_p \in \mathbb{R}^d\}$ for relation r	$\| \left(h + h_p^\top h r_p\right) + r - \left(t + t_p^\top t r_p\right)) \|$
TransR [2]	$h, t \in \mathbb{R}^k, r \in \mathbb{R}^d, M_r \in \mathbb{R}^{k \times d}$, M_r is a projection matrix	$\| M_r h + r - M_r t \|$

measure since it has two relation-specific matrices for head and tail entities; Neural Tensor Network (NTN), which has some complexity that makes it only suit for small knowledge graphs. For the convenience of comparison, we listed the embeddings and score functions of some models in Table 1.

In Table 2, the constraints of each models are presented. As we should point out that with the models developed, the embeddings and constraints actually become more complicated. One thing is sure that if the model is more complicated, the computation cost goes higher. This problem should be carefully considered in related algorithm design.

Table 2. Entity and relation embedding models: constraints

Model name	Constraints
TranE	$h, r, t \in \mathbb{R}^k$
TransH	$h, t \in \mathbb{R}^k, w_r, d_r \in \mathbb{R}^k$
TransD	$\{h, h_p \in \mathbb{R}^k\}$ for entity h, $\{t, t_p \in \mathbb{R}^k\}$, for entity t, $\{r, r_p \in \mathbb{R}^d\}$ for relation r
TransR	$h, t \in \mathbb{R}^k, r \in \mathbb{R}^d, M_r \in \mathbb{R}^{k \times d}$, M_r is a projection matrix

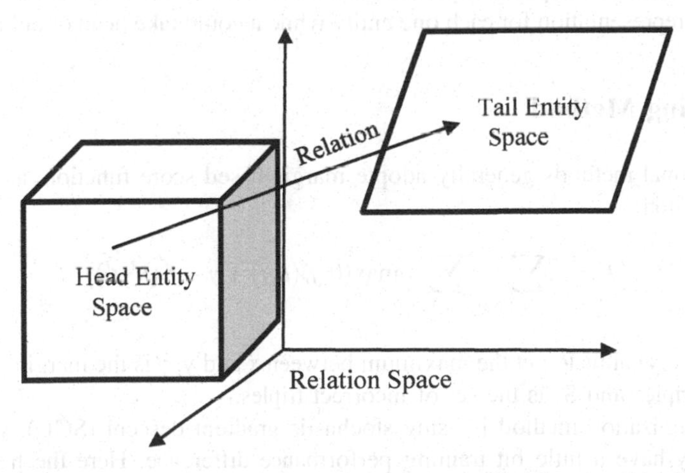

Fig. 1. A simple illustration of entity and relation spaces in embeddings model, where the space dimensions k, l and d might not be the same between any two spaces.

Also there are ways of freeing limitation on the entity embeddings. The main idea is to let head and tail embedding representation independent on each other. We give a possible implementation method here, as showed in Fig. 2.

For each triple (h, r, t), $h \in \mathbb{R}^k, t \in \mathbb{R}^l, whiler \in \mathbb{R}^d$. Here k, l and d can be different. For the sake of calculation, like in TransR and TransE, we project head entities and tail entities into relation space. The projected vectors of head entities and tail entities are defined as

$$h_r = hM_{hr}, t_r = tM_{tr} \tag{1}$$

Where $M_{hr} \in \mathbb{R}^{k\times d}, M_{tr} \in \mathbb{R}^{l\times d}$ are transition matrix.

Routinely the score function is thus defined as

$$f_r(h,t) = \| h_r + r - t_r \|_2^2 \tag{2}$$

And there are also constraints on the norms of embeddings h, r, t and the transition matrix. As it showed below

$$\forall h, r, t, \quad \| h \|_2 \leq 1, \| r \|_2 \leq 1, \| t \|_2 \leq 1, \| hM_{hr} \|_2 \leq 1, \| tM_{hr} \|_2 \leq 1 \tag{3}$$

The Canonical Polyadic method in link prediction take the head and tail entities by learning independently. After nalyzing the negative side of independency, the Simple model simplified the freedom limitation of Canonical Polyadic decomposition. It let the two kind entities learn dependently, while both of them have the same idea that using two embedding representation for each one entity while it could take head or tail position.

3 Training Method

The transitional methods generally adopte margin-based score function, as showed in follow equation.

$$L = \sum_{(h,r,t)\in S} \sum_{(h',r,t')\in S'} \max(0, f_r(h,t) + \gamma - f_r(h',t')) \tag{4}$$

where max(x, y) aims to get the maximum between x and y, γ is the margin, S is the set of correct triples and S' is the set of incorrect triples.

The optimization method is using stochastic gradient descent (SGD), while other method may have a little bit training performance difference. Here the h and t may come from different weight list for one entity.

The SimplE defined a function $f(h, r, t) = \langle h, r\ t\rangle$. The $\langle h, rt\rangle$ is defined to be $\langle h, rt\rangle\ \underline{\underline{def}}\ \sum_{j=1}^{d} h[j] * r[j] * t[j]$, for the brevity, $\langle h, r, t\rangle\ \underline{\underline{def}}\ (v \odot w) \cdot x$, where the $\odot$ is Hadamrd multiplication (element-wise), the $\cdot$ stands for dot product. The optimization object is to minimize each batch with $\sum$ softplus$(-l \cdot \langle h, rt\rangle + \lambda\theta)$, here λ is the model hyper-parameter and θ is the regularization factor, softplus $= \log(1 + e^x)$ is a softened relu function. In addition, the l is in $\{-1, 1\}$, which is the labeled result.

For the translational method, take TransE for example, to make the predication model work, the parameters mainly are $\{h^{k\times d}, r^{n\times d}\}$. If we free the ties between head entity and tail entity for an entity, a $t^{k \times d}$ also is needed to represent the tail entity, for simplicity, assuming the 3 types has same embedding dimension. Let us consider the size of weight parameters in SimplE. We can rewrite parameter spaces here, parameter $= \{h^{k \times d}, r^{n \times d}, r^{n \times d}, t^{k \times d}\}$. As a comparison, the TransE method has a smaller weight spaces in size. Roughly, the SimplE's weight size is about twice as the TransE's if we specify the two embedding size for the same dimension. It could be an important factor that can weaken TransE's expressiveness. If we simply try to increase TransE's embedding size to improve the expressiveness, it might cause problem of overfitting, which is not wanted. Generally there are no perfect theoretical result in deciding how the dimension d should be in this sort of problem. The parameter n, k are characters of the specified training data. Consider if the Knowledge Graph can grow, n, l, k can be very different. Thus it will generate performance impact on a trained model.

4 Combining Word Embedding in Multi-label Classification in Crime Classification

Consider a scenario in law using AI to judge. A law may have many articles. To decide a judgement of a fact violation of law, we can do the supervised learning from labeled data. In literature [13] a hierarchical matching network is proposed to build a training their model. Formally, let $F = \{f_1, f_2, \ldots f_{|F|}\}$ denote all the facts, $W_{f1} = \{w_1, w_2, \ldots, w_k\}$ denote all words in a fact f_1, $L = \{l_1, l_2, \ldots, l_{|L|}\}$ denote all the law, and $A = \{a_1, a_2, \ldots, a_{|A|}\}$ denote all the articles []. In HMN model, the mission is to evaluate a relevance score of a fact related to all laws and their articles.

In encoder layer, unlike original HMN model, we represent articles and facts by averaging word embedding representation instead of bag-of-words.Thus h(f) = AVG ($W_{f1}(1), W_{f2}(2), \ldots$,), and $V_f = [h_f(1), h_f(2), \ldots, h_f(n)]$. $H_f(t)$ is the representation at time t, which are obtained by GRU.

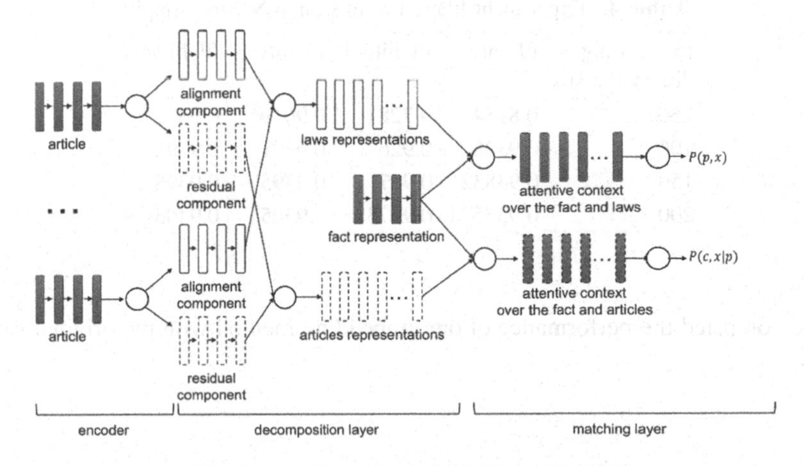

Fig. 2. An illustration of HMN model [13]

$$h_f(t) = GRU(v_t : t \in x, v_t \in V, h_f(t-1)). \quad (5)$$

We adopt the rest of HMN model. In decomposition layer, the similarity between law and article is measured by cosine metric. For the optimization method, we tried both Adam and random gratitude ascend.

5 Experiment and Result

Firstly, we evaluate our embedding ethods with two typical knowledge graphs, built with Freebase [8] and WordNet [9]. These two datasets are chosen in the same way with SimplE [13] and other embedding models. The statics of the datasets is showed in Table 3. This step is to fix the optimization of our embedding settings. To get the optimized dimension size of the embedding, we tested size from 50 to 200, and set 150 as the embedding parameter in HMN model test.

Table 3. Datasets statics

Dataset	#Rel	#Ent	#Train	#Valid	#Test
FB15k	1,345	14,951	483,142	50,000	59,071
WN18	18	40,943	141,442	5,000	5,000

Secondly, we conduct empirical experiments in HMN model on crime classification task. Here is the experimental settings in Table 4. We conduct our empirical experiments on the real-world legal datasets, i.e., Fraud and Civil Action dataset, which has 17160 samples. The Statistics of the dataset is described in Table 5. The learning rate, as suggested in HMN, is set to be 0.0005.

Table 4. Experiment filtered results on WN18 of SimplE

Embedding dimension size	fil_mrr	fil_hit@1	fil_hit@3	fil_hit@10
50	0.8154	0.7281	0.9019	0.9372
100	0.9326	0.926	0.9391	0.9392
150	0.93832	0.937	0.9395	0.9398
200	0.938574	0.9373	0.9395	0.9405

We compared the performance of our embedding method with the original BoW in HMN.

Table 5. Statistics of experiment data set [14]

Dataset	#Fact	#Laws	#Articles	AVG fact description size	AVG article definition size	AVG law set size perfact	AVG article set size perfact
Fraudand Civil Action	17,160	8	70	1,455	136	2.6	4.3

From Table 6. it can be found that our embedding strategy can have some positive effects on the performance. On performance of macro-P and Jaccard, our method can achieve higher value while using the original simple BoW strategy can have better performance in terms of macro-R and macro-F.

Table 6. Results and performance comparison

Dataset	Macro-P	Macro-R	Macro-F	Jaccard	Model
Fraud and Civil Action	65.2	30.6	43.3	67.5	HMN using BoW
	67.4	28.1	40.2	69.3	HMN using our embedding

6 Conclusion

In this paper, we researched the word embedding algorithms. After we carefully investigated the typical algorithms, TransE and SimplE, we optimized our embedding strategy by unlink the head and tail entity. By do so, the expression in entities has been richen and get more freedom. To illustrate the result of our embedding method, we adopt our embedding method to provide the input of encoder in HMN model, which was developed for the purpose of auto judge in AI and law area. By testing the performance on real data set, we can find our embedding method has some privileges in result. Our work is very preliminary and there still some problem remained to be settled. In the future, we will test on more dataset in terms of scale and variety, and try to deep optimize some strategy adopted by HMN to get better performance. Also more models are considered to be taken using our embedding strategy.

References

1. Bordes, A., Usunier, N., Garcia-Duran, A., Weston, J., Yakhnenko, O.: Translating embeddings for modeling multi-relational data. In: NIPS, pp. 2787–2795 (2013)
2. Lin, Y., Liu, Z., Sun, M., Liu, Y., Zhu, X.: Learning entity and relation embeddings for knowledge graph completion. In: AAAI, pp. 2181–2187 (2015)

3. Niu, X.-F., Li, W.-J.: ParaGraphE: a library for parallel knowledge graph embedding. arXiv: 1703.05614v3 (2017)
4. Recht, B., Re, C., Wright, S., Niu, F.: Hogwild: a lock-free approach to parallelizing stochastic gradient descent. In: NIPS, pp. 693–701 (2011)
5. Wang, Z., Zhang, J., Feng, J., Chen, Z.: Knowledge graph embedding by translating on hyperplanes. In: AAAI, pp. 1112–1119 (2014)
6. Xiao, H., Huang, M., Yu, H., Zhu, X.: From one point to a manifold: knowledge graph embedding for precise link prediction. In: IJCAI, pp. 1315– 1321 (2016)
7. Zhao, S.-Y., Zhang, G.-D., Li, W.-J.: Lock-free optimization for nonconvex problems. In: AAAI, pp. 2935–2941 (2017)
8. Miller, G.A.: Wordnet: a lexical database for english. Commun. ACM **38**(11), 39–41 (1995)
9. Bollacker, K., Evans, C., Paritosh, P., Sturge, T., Taylor, J.: Freebase: a collaboratively created graph database for structuring human knowledge. In: Proceedings of KDD, pp. 1247–1250 (2008)
10. Jenatton, R., Roux, N.L., Bordes, A., Obozinski, G.R.: A latent factor model for highly multi-relational data. In: Proceedings of NIPS, pp. 3167–3175 (2012)
11. Ji, G., He, S., Xu, L., Liu, K., Zhao, J.: Knowledge graph embedding via dynamic mapping matrix. In: ACL, pp. 687–696 (2015)
12. Singhal, A.: Introducing the knowledge graph: things, not strings. Google Official Blog. 16 May 2012. Accessed 6 Sep 2014
13. Kazemi, S.M., Poole, D.: Simple embedding for link prediction in knowledge graphs. In: Advances in Neural Information Processing Systems (2018)
14. Wang, P., Fan, Y., Niu, S., Yang, Z., Zhang, Y., Guo, J.: Hierarchical matching network for crime classification. In: Proceedings of the 42nd International ACM SIGIR Conference on Research and Development in Information Retrieval, SIGIR'2019, Paris, France (2019)
15. Wang, P., Yang, Z., Niu, S., Zhang, Y., Zhang, L., Niu, S.: Modeling dynamic pairwise attention for crime classification over legal articles. In: The 41st International ACM SIGIR Conference on Research and Development in Information Retrieval, SIGIR 2018, Ann Arbor, MI, USA, pp. 485–494 (2018)
16. Mehl, L.: Automation in the Legal World, Conference on the Mechanisation of Thought Processes held at Teddington. England (1958)

Recent Advances in Swarm Intelligence: Computing and Applications

A Novel Hybrid Bacterial Foraging Optimization Algorithm Based on Reinforcement Learning

Ben Niu[1,2], Churong Zhang[1], Kaishan Huang[2](✉), and Baoyu Xiao[1]

[1] College of Management, Shenzhen University, Shenzhen 518060, China
drniuben@gmail.com, churong97@163.com,
winsonxiao2019@163.com

[2] Great Bay Area International Institute for Innovation, Shenzhen University,
Shenzhen 518060, China
467778995@qq.com

Abstract. This paper proposes a novel hybrid BFO algorithm based on reinforcement learning (QLBFO), which combines Q-learning with the improved BFO operators. In the QLBFO algorithm, under the guidance of Q-learning mechanism, each bacterium has the chance to adaptively choose appropriate one from three chemotaxis mechanisms to adjust step size. In addition, to maintain the diversity of the whole bacterial population and promote the convergence speed of the algorithm, we also improved two operators. On the one hand, we add the learning communication mechanism in the chemotaxis operator, which can make the bacterium learn from the current best one during the searching process. On the other hand, to alleviate the premature problem, a novel mechanism is adopted into the process of elimination and dispersal for each bacterium. Finally, experimental results show that the proposed QLBFO performs better than four compared algorithms.

Keywords: Bacterial foraging optimization · Q-learning · Reinforcement learning · Chemotaxis

1 Introduction

The Optimization problem is one of the most common problems in academic research and industrial engineering practice [1]. With the progress of science and technology, the real problems have become more and more complex. At the same time, how to solve complex practical problems has become a hot research topic. To solve these problems, scholars have proposed a series of heuristic optimization algorithms, such as evolutionary algorithms, swarm intelligence optimization algorithms, and so on. Compared with the traditional optimization methods, the heuristic optimization algorithms can obtain the approximate optimal solution with better efficiency and effectiveness.

Swarm intelligence optimization algorithms are a kind of random search algorithms, which are inspired by the social behavior of biology as well as the natural foraging phenomenon. So far, scholars have proposed diverse swarm intelligence

D.-S. Huang and P. Premaratne (Eds.): ICIC 2020, LNAI 12465, pp. 567–578, 2020.
https://doi.org/10.1007/978-3-030-60796-8_49

algorithms including particle swarm optimization (PSO) [2], ant colony optimization algorithm (ACO) [3], and so on. Moreover, owing to the good robustness and the outstanding global search capability, in recent years, this kind of methods have been obtained widespread attention in the public and have been broadly applied into various practical fields, e.g. feature selection [4] and job-shop scheduling [5].

As a member of swarm intelligence optimization algorithm, Bacterial foraging optimization algorithm (BFO) [6] is inspired by the foraging activities of E. coli. It mainly consists of three operators: chemotaxis, reproduction, and elimination & dispersal. Although it has strong robustness, BFO uses three nested loops to perform these three operators, which requires a lot of time and cost. Consequently, some scholars proposed many variants of BFO by improving its parameters [9], restructuring its complex structure [10] and hybridizing it with other algorithms [5] to improve its performance. Because of the good optimization performance of BFO variants, they have been widely and successfully applied in various fields, such as image segmentation [7], job-shop scheduling [5], path planning [8], etc.

Reinforcement learning is a remarkable learning mechanism in which an agent interacts with trial and error in a dynamic environment, enabling the agent to have the ability of self-learning [11]. In recent years, scholars have combined swarm intelligence optimization algorithm with reinforcement learning and have verified the feasibility and effectiveness of the corresponding hybrid algorithms. For example, Samma & Lim et al. [12] proposed a memetic particle swarm optimization algorithm based on reinforcement learning which enabled each particle to select and perform five operations (exploration, convergence, high jump, low jump and fine tuning) driven by the reinforcement learning (RL) algorithm, and then verified that its performance was better than other variants of PSO algorithm. Besides that, Alipour & Razavi et al. [13] combined the genetic algorithm with the reinforcement learning algorithm to propose a hybrid heuristic algorithm and applied the proposed algorithm to solve the TSP problem.

In this paper, we propose a novel hybrid algorithm (QLBFO), which combined a typical reinforcement learning algorithm named Q-learning with the improved BFO operators. More detailly, QLBFO uses the Q-learning to select and execute three kinds of chemotaxis operators respectively for fixed chemotaxis step size (FC), linear-decreasing chemotaxis step size (LDC) and nonlinear-decreasing chemotaxis step size (NDC). Moreover, Q-learning plays an important role in selecting appropriate chemotaxis operators for each bacterium at different evaluation times. In addition, we improve the learning mechanism in the chemotaxis operator, and we generate new bacteria in the restricted environment space for elimination & dispersal operator, changing the previous way of generating new individuals in the whole search space, to improve the convergence speed of the algorithm. The contribution of this paper is not only to simplify the structure of BFO, but also to break through the standard BFO in which each bacterium adopts the same chemotaxis strategy. Moreover, few scholars combined Q-learning with BFO to propose a hybrid heuristic algorithm, so this paper is relatively novel.

The paper is organized as follows. The Sect. 2. introduces the standard BFO. In Sect. 3, we describe the Q-learning algorithm, and propose a novel hybrid algorithm (QLBFO) based on Q-learning and BFO. The experimental results and analyses are presented in Sect. 4. Conclusions and the future work are shown in Sect. 5.

2 Bacterial Foraging Optimization Algorithm

Bacterial foraging optimization algorithm (BFO) is a heuristic optimization algorithm designed by Passino in 2002, which is inspired by the foraging process of E. coli. Figure 1 shows the BFO's brief pseudo code that mainly consists of three operations: chemotaxis, reproduction, and elimination and dispersal [6].

2.1 Chemotaxis

Chemotaxis is the core operator of BFO. The operator is to simulate the behavior of E. coli to change its original position through flagella activity, mainly including two steps: tumbling and swimming. Moreover, the bacteria randomly choose one direction to tumble and then update its positions along the direction with a fixed step size. When the bacterium locates in a better position, it can have the chance to move forward along the same direction. Specifically, if the bacterium obtains the worse position or reaches the limitation of swimming number, it may trigger the signal to stop swimming. In general, the chemotaxis is the way of local search, which is convenient for bacteria to exploit the optimal solution of the search space.

$$\theta^i(j+1,k,l) = \theta^i(j,k,l) + C(i) * \frac{\Delta(i)}{\sqrt{\Delta^T(i)\Delta(i)}} \tag{1}$$

where $\theta^i(j+1,k,l)$ is the position of the i bacteria on the $j+1th$ chemotaxis, the kth replication, the lth elimination and dispersal step. $C(i)$ is the step length of the bacterium $i\Delta(i)$ represents the tumbling vector for the bacterium i in which all elements range from −1 to 1. $\frac{\Delta(i)}{\sqrt{\Delta^T(i)\Delta(i)}}$ describes a random direction for the bacterium i.

2.2 Reproduction

Reproduction greatly mirrors the main accept of survival of the fittest followed by Darwin' s theory of evolution Based on the measurement of the healthy degree, this operation is to calculate the cumulative fitness value of each bacterium in their lifecycle, and then all bacteria are sorted in ascending order according to cumulative value. Furtherly, the half of better bacteria with greater healthy degree would replace the worse one.

2.3 Elimination and Dispersal

The elimination and dispersal operator simulates that the bacterium immigrates to a new position because of environmental changes or the bacterium will die when confronting with harmful substances in the searching process. In the standard BFO, if the randomly generated value is less than or equal to the probability of P_{ed}, the bacterium will die and generate a new bacterium in the current search environment. This operator is helpful for the BFO to jump out of the local optimum and find a better solution.

```
Initialize population and parameters: S is the population size; P is the dimension of the
optimization problem. N_c is the maximum number of chemotaxis step; N_re is the maximum
number of reproduction step; N_ed is the maximum number of elimination & dispersal; N_s is
the swimming and tumbling maximum times .J^i is the fitness value of bacterium i after
tumbling or swimming.J^i_last is the fitness value of bacterium i before tumbling or swimming.
for l = 1:N_ed: elimination & dispersal loop
  for k = 1:N_rd: reproduction loop
    for j = 1:N_c: chemotaxis step
      for i= 1:S:
        Do chemotaxis operator:
        Tumble and move using equation (1): update θ the position of the i bacteria.
        Swimming: m = 0
          While m < N_s
            m = m + 1
            if J^i_last > J^i
                update θ using equation (1) and J^i(J^i_last = J^i)
            else
                m = N_s
            end
      end
    end
    Do reproduction operator
  end
  Do elimination & dispersal operator
end
```

Fig. 1. The BFO's brief pseudo code.

3 The Proposed Algorithm

In this paper, we proposed a new hybrid algorithm called QLBFO, which combined Q-learning with the improved BFO operators. The QLBFO is described as follows.

3.1 Q-Learning

Q-learning is a classical model-independent algorithm in reinforcement learning [14], consisting of five elements: agent, environment, reward, action, and state. The main idea of Q-learning reflects interactions between the process of exploratory and evaluation for each individual. After the agent perceives the state of the environment, it executes the action and applies it to the environment. When the environment receives the action, it gives rewards to the agent and updates the current state. The cycle repeats until the optimal result is found, as shown in Fig. 2.

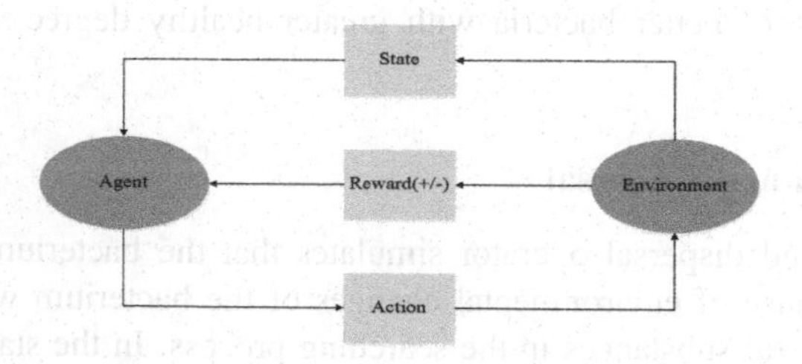

Fig. 2. The framework of Q-learning.

The specific steps of Q-learning are as follows:

- Step1: Initialize state denoted as $s_i \in [s_1, s_2, \ldots, s_n]$, action denoted as $a_i \in [a_1, a_2, \ldots, a_n]$, and generate a $n \times n$ matrix, named Q-table. Then set Q-table to zeros.
- Step 2: According to Q-table and state s_i, the agent selects the optimal action a_i to execute it.
- Step 3: The environment feeds the reward back to the agent. Whether the environment reward r_{i+1} is positive or not will depend on the action a_i performed by the agent.
- Step 4: Update cumulative rewards for executing action a_i in state s_i in the Q-table, using formula (2).

$$Q_{i+1}(s_i, a_i) = Q_i(s_i, a_i) + \alpha\left[r_{i+1} + \gamma\left(\max_a Q_i(s_{i+1}, a) - Q_i(s_i, a_i)\right)\right] \tag{2}$$

where $Q_i(s_i, a_i)$ is the cumulative reward after the execution of action a_i in the state s_i at time i. The sate $s_i \in [s_1, s_2, \ldots, s_n]$, and the action $a_i \in [a_1, a_2, \ldots, a_n]$. r_{i+1} is the reward for the environmental feedback after the execution of action a_i. Besides that, α is a learning factor and $\alpha \in [0, 1]$,while γ is the discount factor and belongs to $[0, 1]$.

- Step 5: Update the next state s_{i+1}, then judge whether the evaluation times are satisfied. If not, return to Step 2.

In addition, when α is closed to 1, the agent is more inclined to explore the unknown space, otherwise, while α is closed to 0, it will largely learn with the previous experience. When γ approaches 1, the agent heavily focus on the long-term reward, while γ is approached to 0, it means the agent is more concerned about the current short-term reward.

3.2 The Q-Learning Bacterial Foraging Algorithm Model

In the standard BFO, the structure involved in three fixed nesting order makes the convergence slow down, and the BFO is hard to find the optimal solution. More specifically, the bacterium updates its position only just with a fixed step size, which makes the significantly negative impact on the convergence speed and convergence accuracy.

To improve the constant step size, lots of scholars have incorporated the self-adaptation mechanisms into the standard BFO. For example, Niu et al. [9] proposed BFO-LDC and BFO-NDC to improve the standard BFO. However, to the best of our knowledge, in prior studies, the bacterium adjusts its positions just based on a self-adjustment chemotaxis mechanism. Moreover, our paper has the main purpose to discuss the effect of alternative selection from different strategies on the performance of the standard BFO. When the bacteria have the ability to choose proper strategies in terms of current situations, the algorithm might have the better performance in optimization.

As a result, in this paper, we propose a novel hybrid algorithm named QLBFO, which incorporates Q-learning into the BFO. The Q-table is a 3×3 matrix, where has three states and three action. What's more, each bacterium has its own Q-Table to update their position. According to the bacteria's environment, each bacterium can choose the appropriate actions from FC, LDC and NDC, as shown in Fig. 3. In addition, in order to simplify the algorithm's structure and make the algorithm more efficient to find the optimal solution, this paper only selected and improved two operators: chemotaxis, elimination and dispersal. The flowchart of QLBFO is shown in Fig. 4.

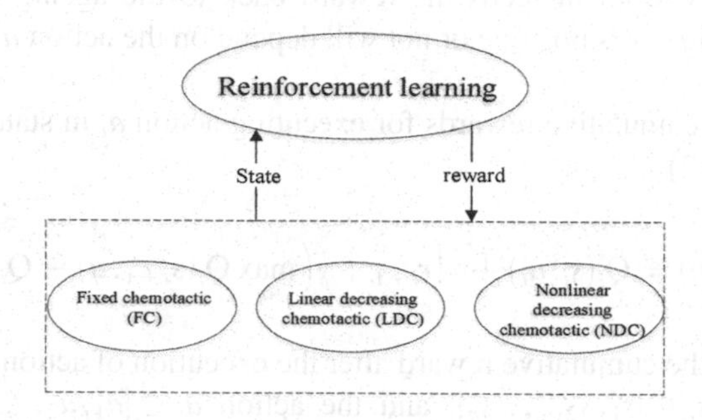

Fig. 3. The proposed QLBFO structure.

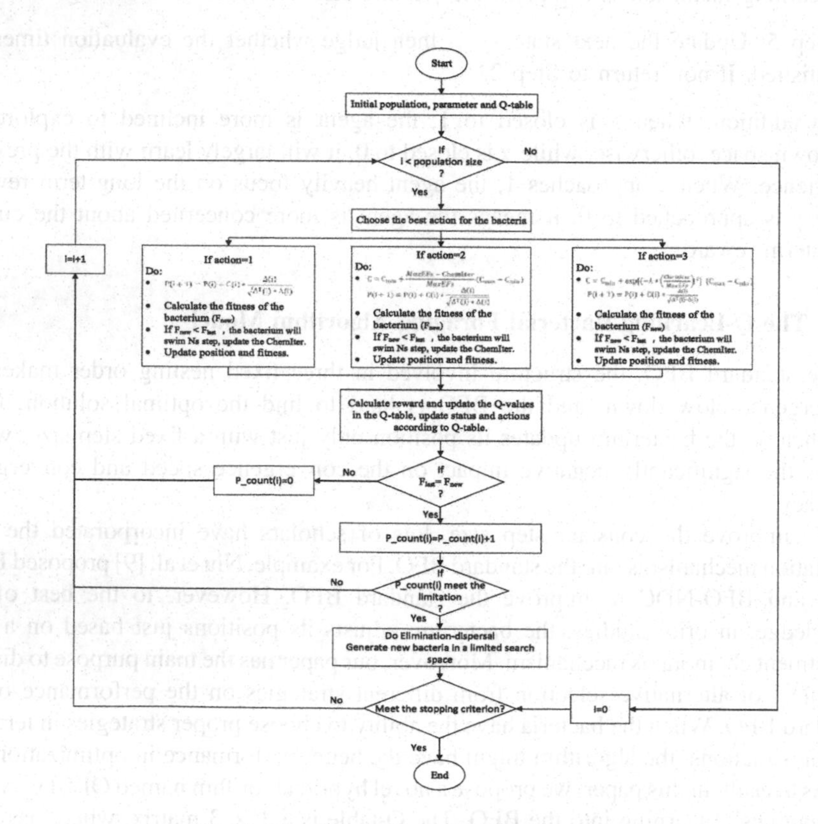

Fig. 4. The flowchart of QLBFO algorithm.

Interactive Mechanism Between Q-Learning and BFO Operator. Inspired by the integrated method of PSO and Q-learning [12], we take account of the bacteria in the BFO as the agents in the Q-learning. The environment is treated as a search space for bacteria. Moreover, the three chemotaxis operators (FC, LDC and NDC) are regarded as three actions and the state can be regarded as the current action. The reward is defined as follows:

$$reward = \left\{ \begin{array}{l} 1\ if\ F_{new} < F_{last} \\ -1\ if\ F_{new} \geq F_{last} \end{array} \right\} \tag{3}$$

where *reward* means that the agent gets the reward after taking an action. F_{new} is the updated fitness value. F_{last} is the original fitness value.

The interactive mechanisms between Q-learning and BFO operators are described as follows:

- Step1: Initialize the parameters and choose the optimal strategy as the first action among the three chemotaxis operators.
- Step 2: The bacteria obtain the optimal action to execute in the current state.
- Step 3: Calculate rewards using formula (3) and update Q-table using formula (2). It is noted that the learning factor adopts linear-decreasing strategy.

$$\alpha = \alpha_{min} + \frac{MaxFEs - CurrentFEs}{MaxFEs} \times (\alpha_{max} - \alpha_{min}) \tag{4}$$

where α is learning factor, α_{min} is the minimum value of the learning factor, α_{max} is the maximum value of the learning factor.*MaxFEs* is the maximum number of evaluations. *CurrentFEs* is the current evaluation.

- Step 4: The current state changes and the agent selects the next optimal action based on the Q-table.
- Step 5: If the maximum number of evaluations has been met, the algorithm will stop. Otherwise, it will return to Step 2.

Improved Chemotaxis Operator. In the standard BFO, the chemotaxis operator is to randomly select a direction for a bacterium, and there is no learning mechanism in it. In QLBFO, we add the learning mechanism [10, 15] into chemotaxis where the bacteria can learn from the best individual. To be specific, for the moving direction of bacteria, we not only consider its own randomly moving direction, but also consider the direction of the best bacterium. This method overcomes the shortcoming that the standard BFO is hard to converge quickly because of its randomly direction in tumbling and swimming. Thus, we introduce the learning mechanism to the chemotaxis, which may be helpful to improve the search efficiency of the QLBFO.

$$D(i) = \mathrm{w}* \frac{\Delta(i)}{\sqrt{\Delta^T(i)\Delta(i)}} \tag{5}$$

$$w = w_{min} + \frac{MaxFEs - CurrentFEs}{MaxFEs} \times (w_{max} - w_{min}) \quad (6)$$

$$\theta(i) = \theta(i) + C(i) * D(i) + 1.5 * rand * (Gbest - \theta(i)) \quad (7)$$

where $\theta(i)$ is the position of the bacterium i; $C(i)$ represents the chemotaxis step size of the bacterium i; $D(i)$ is the bacterium i's moving direction; *Gbest* is the position of the best bacterium, who has the minimum fitness in all bacteria;*rand* is a random value.

Improved Elimination and Dispersal Operator. As the environment change, the bacteria will die or immigrate, resulting in creating a new bacterium or an original one moves to a new position in the search environment. In QLBFO, we create new bacteria in a limited environment, that is, the upper and lower bounds of the bacterial generation space are defined by the minimum and maximum values of all bacterial position. Besides, the elimination and dispersal operator is activated by the counting number *P_count* when the bacterium's fitness has not change in *P_count* times. In a word, compared with the standard BFO's elimination and dispersal operator of moving or generating bacteria in the whole search area, the new strategy can help the algorithm achieve rapid convergence while maintaining diversity of the bacteria and avoiding the algorithm falling into the local search.

4 Experiments and Results

4.1 Experimental Parameter Settings

To verify the effectiveness of the proposed algorithm QLBFO, we choose two classical heuristic algorithms to compare, such as PSO, BFO. Moreover, we also compare with two variants of BFO, including BFO-LDC and BFO-NDC. To evaluate the proposed algorithm with four optimization algorithms, we use three unimodal functions, (Sphere, Rosenbrock and Sum of different powers) and three multimodal functions (Rastrigin, Griewank and Weierstrass).

The specific parameters of the experiment are set as follows: the number of runs is set to 30 and the maximum number of fitness evaluations is 300,000. The population size is set to 30 in PSO-w, BFO, BFO-LDC, BFO-NDC, and the dimension is set to 30.

In PSO with modified inertia weight (PSO-w) [15], w_{min} is 0.9, w_{max} is 1.2. Social and cognition learning factor $c_1$and c_2 are 2.

In BFO [6], the chemotaxis step C is 0.1, the number of chemotaxis N_c is 100, the maximum number of swimming N_s is 4, the number of reproduction N_{re} is 4 and the number of elimination and dispersal N_{ed} is 1. The probability of elimination and dispersal P_{ed} is 0.25.

In BFO-LDC and BFO-NDC, the minimum value of chemotaxis step size C_{min} is 0.01 and the maximum value of chemotaxis step size C_{max} is 0.02. $N_c, N_s, N_{re}, N_{ed}, P_{ed}$ are the same as BFO.

In QLBFO, the population size is 4, the discount factor γ is set to 0.8, the minimum value of learning factor α_{min} is set to 0.1, the maximum value of learning factor α_{max} is set to 1, $w_{min} = 0.4$, $w_{max} = 0.9$. *P_count* represents the signal of doing elimination and dispersal operator, $P_{count} = 200$. C_{min} C_{max} and P_{ed} are the same as BFO-LDC and BFO-NDC.

To ensure the fairness of comparison, it's noted that we use the number of fitness evaluations instead of generations as the termination condition.

4.2 Result and Analysis

Table 1 represents the experimental results of four comparison algorithms and QLBFO on six benchmark functions. The Fig. 5 shows the convergence graphs on six functions.

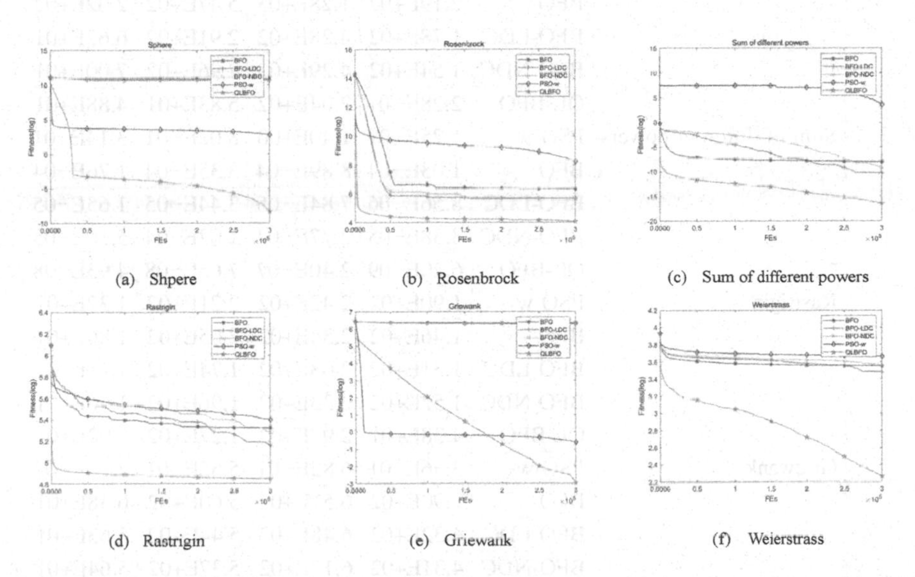

(a) Shpere (b) Rosenbrock (c) Sum of different powers

(d) Rastrigin (e) Griewank (f) Weierstrass

Fig. 5. The convergence graphs on six functions (a-c: unimodal functions; d-f: multimodal functions).

As shown in Table 1, we can observe that QLBFO is superior to other optimization algorithms on three unimodal functions. However, in the multimodal functions, comparing with other optimization algorithms, QLBFO's average results of 30 runs are better than other four algorithm. From these experimental results, the performance superiority of QLBFO can be mainly attributed to Q-learning and the improved operators that can well balance the exploration and exploitation during the process of finding the optimal results.

We can observe that the convergence graphs about these five algorithms on 30-D functions from Fig. 5, which shows that the QLBFO have the fastest convergence speed on four functions (Rosenbrock Sum of different powers Rastrigin Weierstrass). Although QLBFO on Sphere and Griewank have not obtained satisfactory convergence, it may sacrifice fast convergence to obtain the diversity of the population, indicating that the addition of Q-learning and improved operators play an important role in making the algorithm have more potential capability to explore better results.

Table 1. The comparison results of five algorithms on six functions.

No.	Function	Algorithm	Best	Worse	Mean	Std.
1	Sphere	PSO-w	1.05E+01	2.21E+01	1.62E+01	2.67E+00
		BFO	3.85E+04	6.25E+04	5.17E+04	6.81E+03
		BFO-LDC	3.27E+04	5.34E+04	4.43E+04	5.45E+03
		BFO-NDC	2.92E+04	5.23E+04	4.22E+04	5.61E+03
		QL-BFO	1.57E−04	4.37E−04	2.48E−04	6.71E−05
2	Rosenbrock	PSO-w	2.38E+03	6.15E+03	3.74E+03	9.36E+02
		BFO	2.19E+02	1.28E+03	5.47E+02	2.02E+02
		BFO-LDC	1.78E+02	4.28E+02	2.91E+02	6.62E+01
		BFO-NDC	1.54E+02	4.29E+02	2.96E+02	7.00E+01
		QL-BFO	2.28E−01	2.04E+02	5.83E+01	4.88E+01
3	Sum of different powers	PSO-w	1.25E−01	1.40E+00	8.02E−01	3.14E−01
		BFO	1.03E−04	8.89E−04	3.35E−04	1.76E−04
		BFO-LDC	8.56E−06	7.84E−05	3.44E−05	1.65E−05
		BFO-NDC	3.38E−05	2.77E−04	1.02E−04	5.10E−05
		QL-BFO	6.20E−09	2.40E−07	7.65E−08	4.93E−08
4	Rastrigin	PSO-w	1.90E+02	2.42E+02	2.21E+02	1.32E+01
		BFO	1.46E+02	2.34E+02	2.05E+02	1.86E+01
		BFO-LDC	1.31E+02	2.03E+02	1.74E+02	1.95E+01
		BFO-NDC	1.52E+02	2.33E+02	1.96E+02	1.96E+01
		QL-BFO	4.38E+01	2.97E+02	1.27E+02	5.92E+01
5	Griewank	PSO-w	3.36E−01	6.82E−01	5.52E−01	7.74E−02
		BFO	3.90E+02	6.57E+02	5.60E+02	6.38E+01
		BFO-LDC	4.32E+02	6.48E+02	5.44E+02	5.63E+01
		BFO-NDC	4.31E+02	6.17E+02	5.37E+02	5.64E+01
		QL-BFO	1.81E−05	1.17E+00	5.13E−02	2.12E−01
6	Weierstrass	PSO-w	3.71E+01	4.13E+01	3.89E+01	1.13E+00
		BFO	3.17E+01	3.72E+01	3.49E+01	1.20E+00
		BFO-LDC	2.98E+01	3.55E+01	3.25E+01	1.47E+00
		BFO-NDC	3.02E+01	3.76E+01	3.50E+01	1.49E+00
		QL-BFO	5.76E+00	1.75E+01	9.65E+00	2.45E+00

Note: "Best" in the fourth column means the minimum value of 30 runs for each algorithm, while "Worse" in the fifth column means the maximum value of 30 runs for each algorithm. "Mean" in the sixth column represents the average of experimental results of 30 runs for each algorithm. "Std." in the seventh column means the standard deviation of experimental results of 30 runs for each algorithm.

Furthermore, we employ the Friedman test on the termination condition to check significant differences among these algorithms. It can be observed from the Table 2 that our proposed algorithm (QLBFO) can improve the BFO performance significantly. QLBFO's performance differences are statistically significant on unimodal functions, while the average result of QLBFO is still superior to other algorithms on the multimodal functions although there is no statistically significant on Rastrigin and Griewank functions. We can confirm that RL strategy we adopted can select the appropriate BFO chemotaxis operator for each bacterium during the searching process, and the improved two BFO operators can further accelerate the convergence and enhance the diversity of the bacteria to obtain optimal results.

Table 2. The statistical test results of the Friedman test for average fitness of 30 runs among five algorithms.

Function	The statistical test results
Sphere	QLBFO ≈ PSO-w ≈ BFO-NDC ≈ BFO-LDC ≈ BFO
Rosenbrock	QLBFO ≫ BFO-LDC ≈ BFO-NDC ≫ PSO-w ≈ BFO
Sum of different powers	QLBFO ≫ BFO-LDC ≈ BFO-NDC > BFO > PSO-w
Rastrigin	QLBFO ≈ BFO-LDC ≈ BFO-NDC ≈ BFO ≈ PSO-w
Griewank	QLBFO ≈ PSO-w ≫ BFO-NDC ≈ BFO-LDC ≈ BFO
Weierstrass	QLBFO > BFO-LDC ≈ BFO ≈ BFO-NDC > PSO-w

Note: A >> B, A > B represent that A is significantly better than B with significant levels of 1%, 5% respectively.
A ≈ B means that although A is better than B, there is no significant difference between them (two algorithms adjacent to each other.).

5 Conclusion and Future Work

This paper proposes a novel hybrid BFO algorithm (QLBFO) based on the reinforcement learning mechanism. In QLBFO, using the Q-learning, the bacteria can choose the proper strategy to update their position among three different chemotaxis operators (FC, LDC, NDC). Moreover, we improved the chemotaxis and elimination & dispersal operators, this is, we introduce the information learning mechanism to the chemotaxis operator and change the region in which new bacteria are generated in the elimination & dispersal operator. Finally, the comparison experiments on six functions show that QLBFO outperforms its four compared algorithms. Moreover, the experiments of statistical test confirm the better performance of the QLBFO.

In the future, we will propose other variants of QLBFO to find better results and make algorithms become more robust. Moreover, we also can apply it or its variants to solve industrial engineering problems to further verify its effectiveness.

Acknowledgement. This study is supported by The National Natural Science Foundation of China (Grants Nos. 71971143), Guangdong Province Soft Science Project (2019A101002075), Guangdong Province Educational Science Plan 2019 (2019JKCY010) and Guangdong Province Bachelor and Postgraduate Education Innovation Research Project (2019SFKC46).

References

1. Lynn, N., Suganthan, P.N.: Ensemble particle swarm optimizer. Appl. Soft Comput. **55**, 533–548 (2017)
2. Eberhart, R., Kennedy, J.: Particle swarm optimization. In: the Proceedings of the IEEE international conference on neural networks, Citeseer (1995)
3. Dorigo, M., Di Caro, G.: Ant colony optimization: a new meta-heuristic. In: the Proceedings of the 1999 congress on evolutionary computation-CEC, pp. 1470–1477. IEEE (1999)
4. Gu, S., Cheng, R., Jin, Y.: Feature selection for high-dimensional classification using a competitive swarm optimizer. Soft. Comput. **22**(3), 811–822 (2018)
5. Vital-Soto, A., Azab, A., Baki, M.F.: Mathematical modeling and a hybridized bacterial foraging optimization algorithm for the flexible job-shop scheduling problem with sequencing flexibility. J. Manuf. Syst. **54**, 74–93 (2020)
6. Passino, K.M.: Biomimicry of bacterial foraging for distributed optimization and control. IEEE Control Syst. Mag. **22**(3), 52–67 (2002)
7. Pan, Y., Xia, Y., Zhou, T., Fulham, M.: Cell image segmentation using bacterial foraging optimization. Appl. Soft Comput. **58**, 770–782 (2017)
8. Hossain, M.A., Ferdous, I.: Autonomous robot path planning in dynamic environment using a new optimization technique inspired by bacterial foraging technique. Robot. Auton. Syst. **64**, 137–141 (2015)
9. Niu, B., Fan, Y., Wang, H., Li, L., Wang, X.: Novel bacterial foraging optimization with time-varying chemotaxis step. Int. J. Artif. Intell. **7**(A11), 257–273 (2011)
10. Niu, B., et al.: Coevolutionary structure-redesigned-based bacterial foraging optimization. IEEE/ACM Trans. Computational Biol. Bioinform. **15**(6), 1865–1876 (2017)
11. Kaelbling, L.P., Littman, M.L., Moore, A.W.: Reinforcement learning: a survey. J. Artif. Intell. Res. **4**, 237–285 (1996)
12. Samma, H., Lim, C.P., Saleh, J.M.: A new reinforcement learning-based memetic particle swarm optimizer. Appl. Soft Comput. **43**, 276–297 (2016)
13. Alipour, M.M., Razavi, S.N., Derakhshi, M.R.F., Balafar, M.A.: A hybrid algorithm using a genetic algorithm and multiagent reinforcement learning heuristic to solve the traveling salesman problem. Neural Comput. Appl. **30**(9), 2935–2951 (2018)
14. Watkins, C.J., Dayan, P.: Q-learning. Mach. Learn. **8**(3–4), 279–292 (1992)
15. Shi, Y., Eberhart, R.: A modified particle swarm optimizer. In: the 1998 IEEE international conference on evolutionary computation proceedings, pp. 69–73. IEEE (1998)

Improved Water Cycle Algorithm and K-Means Based Method for Data Clustering

Huan Liu[1], Lijing Tan[2(✉)], Luoxin Jin[2], and Ben Niu[1]

[1] College of Managment, Shenzhen University, Shenzhen 518060, China
[2] School of Management, Shenzhen Institute of Information Technology, Shenzhen 518172, China
mstlj@163.com

Abstract. K-means is a classical clustering method, but it is easy to fall into local optimums because of poor centers. Inspired by the good global search performance of Inter-Peer Communication Mechanism Based Water Cycle Algorithm (IPCWCA), three hybrid methods based on IPCWCA and K-means are presented in this paper, which are used to address the shortcoming of K-means and explore better clustering approaches. The hybrid methods consist of two modules successively: IPCWCA module and K-means module, which means that K-means module will inherit the best individual from IPCWCA module to start its clustering process. Compared with original K-means and WCA + K-means methods on eight datasets (including two customer segmentation datasets) based on SSE, accuracy and Friedman test, proposed methods show greater potential to solve clustering problems both in simple and customer segmentation datasets.

Keywords: Water Cycle Algorithm · IPCWCA · K-means · Clustering · Customer segmentation dataset · Friedman test

1 Introduction

In data mining, clustering is one of the most commonly used methods to divide a set of unlabeled data into related clusters. Clustering has no prior knowledge about data, which leads to it could acquire some hidden information in data. Among many clustering algorithms, K-means is the one of most popular methods for its high efficiency and simplicity, but K-means is prone to getting trapped into local optimums when having poor initial centroids [1].

Nature-inspired heuristic algorithms, such as Genetic Algorithm (GA) [2–4], Particle Swarm Optimization (PSO) [5–7], Ant Colony Optimization (ACO) [8, 9], which attract scholars to apply them in clustering problems, have good performance in data clustering. In this context, Water Cycle Algorithm (WCA) was proposed by Eskandar et al. [10], focusing on the processes of water cycle and how streams and rivers flow to sea.

In WCA, besides the main step of flow, evaporation and raining are also important portions, which help WCA escape from local optimization. To enhance the performance of WCA, many improvements of WCA are proposed. Chen et al. [11] presented

D.-S. Huang and P. Premaratne (Eds.): ICIC 2020, LNAI 12465, pp. 579–589, 2020.
https://doi.org/10.1007/978-3-030-60796-8_50

Hierarchical Learning WCA (HLWCA) to divide the solutions into collections with hierarchy differences to improve WCA's global searching ability. Al-Rawashdeh et al. [12] applied hybrid Water Cycle and Simulated Annealing to improve the accuracy of feature selection and to evaluate proposed Spam Detection. Bahreininejad [13] studied the impact of the Augmented Lagrange Method (ALM) on WCA and presented WCA-ALM algorithm to enhance convergence and solution quality. In 2019, an Inter-Peer Communication Mechanism Based Water Cycle Algorithm (IPCWCA) was presented by Niu et al. [14], aiming to utilize the information communication of inter-peer individuals to enhance the performance of whole WCA. According to IPCWCA, each stream and river need to learn and get information from one of their peers on some dimensions before flow step, which is also beneficial to improve population diversity.

In this paper, we try to combine IPCWCA with K-means and apply it to clustering analysis, including data clustering and customer segmentation. This kind of method can be divided into IPCWCA module and K-means module: IPCWCA module is executed at first to get a global best individual and then K-means module inherits this individual to continue its clustering process. SSE (sum of squared error) is adopted as fitness function to judge the performance of clustering. The smaller SSE is, the better; otherwise, the reverse. In addition, to compare the performances of the above algorithms from a statistical viewpoint, Friedman test is used in this paper.

The rest of the paper is organized as follows: Sect. 2, 3 and 4 introduce Water Cycle Algorithm, Inter-Peer Communication Mechanism Based Water Cycle Algorithm (IPCWCA), and K-means Algorithm respectively. Section 5 presents the series of WCA + K-means based methods in details. In Sect. 6, the experiment and results are discussed. In the final Sect. 7, conclusions of the work are presented.

2 Water Cycle Algorithm

Water Cycle Algorithm (WCA), simulating natural phenomenon of water cycle, is originally presented to address engineering optimization problems. WCA mainly consists of three steps: flow, evaporation and raining.

Specifically, WCA pays attention to the flow among streams, rivers and sea. It is noted that sea is the best individual in the whole population while rivers are some good individuals which are inferior to sea. Finally, the remaining individuals are considered as streams.

After flow, a stream's position will be updated, using

$$X_{Stream}(t+1) = X_{Stream}(t) + rand \times C \times (X_{Sea}(t) - X_{Stream}(t)) \tag{1}$$

$$X_{Stream}(t+1) = X_{Stream}(t) + rand \times C \times (X_{River}(t) - X_{Stream}(t)) \tag{2}$$

Then, if the fitness value of a stream is better than specific river's or sea's, exchange their roles.

A river's position is updated after flowing to the sea, using

$$X_{River}(t+1) = X_{River}(t) + rand \times C \times (X_{Sea}(t) - X_{River}(t)) \tag{3}$$

Similarly, if the river has better fitness value than sea, exchange their roles.

3 Inter-Peer Communication Mechanism Based Water Cycle Algorithm

In order to decrease information loss and enhance communication efficiency among individuals, an Inter-Peer Communication Mechanism Based Water Cycle Algorithm (IPCWCA) is presented.

Unlike original WCA, IPCWCA considers the relationship between inter-peer individuals, i.e. streams to streams, rivers to rivers. Besides learning from a higher level individual, a stream/river can acquire information from another stream/river before flow step in IPCWCA.

Peer of a stream or river is determined randomly, which helps to improve population diversity, using Eq. (4)–(5)

$$I_{Stream} = fix(rand * (S - Nsr)) + 1 \qquad I_{Stream} \neq X_i \tag{4}$$

$$I_{river} = fix(rand * (Nsr - S)) + 1 \qquad I_{river} \neq X_j \tag{5}$$

Where S is the number of individuals, Nsr is the total number of rivers and sea.

$$\begin{aligned} Position_{Stream}(1,d) &= Position_{Stream}(1,d) * gauss \\ gauss &= N(0, |Position_{I_{Stream}}(1,d)|) \end{aligned} \tag{6}$$

$$\begin{aligned} Position_{river}(1,d) &= Position_{river}(1,d) * gauss \\ gauss &= N(0, |Position_{I_{river}}(1,d)|) \end{aligned} \tag{7}$$

where “gauss” is a normal distribution with a mean of 0 and a variance of the I_{stream}'s or I_{river}'s dth dimension's absolute value. It is noted that the dimensions of learning between inter-peer is selected randomly instead of studying from all dimensions.

4 K-Means Algorithm

K-means is a well-known clustering method, which divides data vectors into K groups, usually adopting Euclidean metric to calculate the distance between data vectors and cluster centers.

First of all, K-means needs to select initial K centroids ($M = (M_1, M_2, \ldots, M_j, \ldots, M_K)$) and distribute each data vector to the cluster C_j ($j = 1, \ldots, K$) by Euclidean metric:

$$d(X_p, M_j) = \sqrt{\sum_{n=1}^{N_d} (X_{pn} - M_{jn})^2} \tag{8}$$

where X_p is the *p-th* data vector; M_j represents the *j-th* centroid; N_d is the dimension of data vector.

In K-means, it is important to recalculate cluster centroids, using:

$$M_j = \frac{1}{n_j} \sum_{X_p \in C_j} X_p \tag{9}$$

where n_j is the number of data vectors in cluster C_j.

5 WCA + K-Means Based Methods

5.1 The IPCWCA/WCA Module

In the module of WCA or IPCWCA, each individual is encoded as follows:

$$P_i = (M_{i1}, M_{i2}, \ldots M_{ij}, \ldots, M_{iK})$$

where K represents the number of clusters; M_{ij} is the *j-th* cluster centroid vector of the *i-th* individual in cluster C_{ij}. The fitness function is used to calculate the fitness value of each individual to data vectors, which can be described as:

$$SSE = \sum\nolimits_{J=1}^{K} \sum_{\forall X_p \in C_{ij}} d(X_p, M_{ij})^2 \tag{10}$$

where d is defined in Eq. (8); n_{ij} is the number of data vectors in cluster C_{ij}.

For clustering, inter-peer communication process is different from flow step in learning dimensions, which can be concluded to three versions: IPCWCA-1, IPCWCA-A, IPCWCA-R. IPCWCA-1 only gets information from the first category of a peer, IPCWCA-A studies from all of the peer's categories and IPCWCA-R learns randomly. Additionally, the dimension of learning in each category is random.

Take an example, there are thousands of four-dimensional data vectors that need to be divided into three categories. Therefore, each individual in population is a 3×4 matrix. Three potential learning methods from a peer are illustrated in Fig. 1.

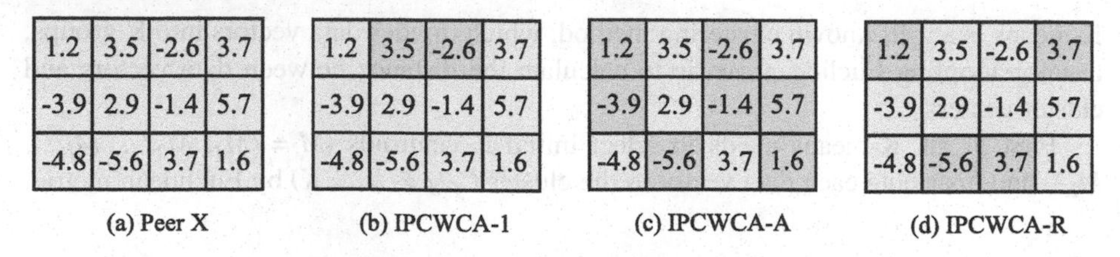

Fig. 1. Three potential learning methods from peer X

5.2 The K-Means Module

K-means module runs after WCA or IPCWCA module, acquiring initial cluster centroids from the best individual of last module and then searching for the final solution. Figure 2 shows the flowchart of IPCWCA + K-means.

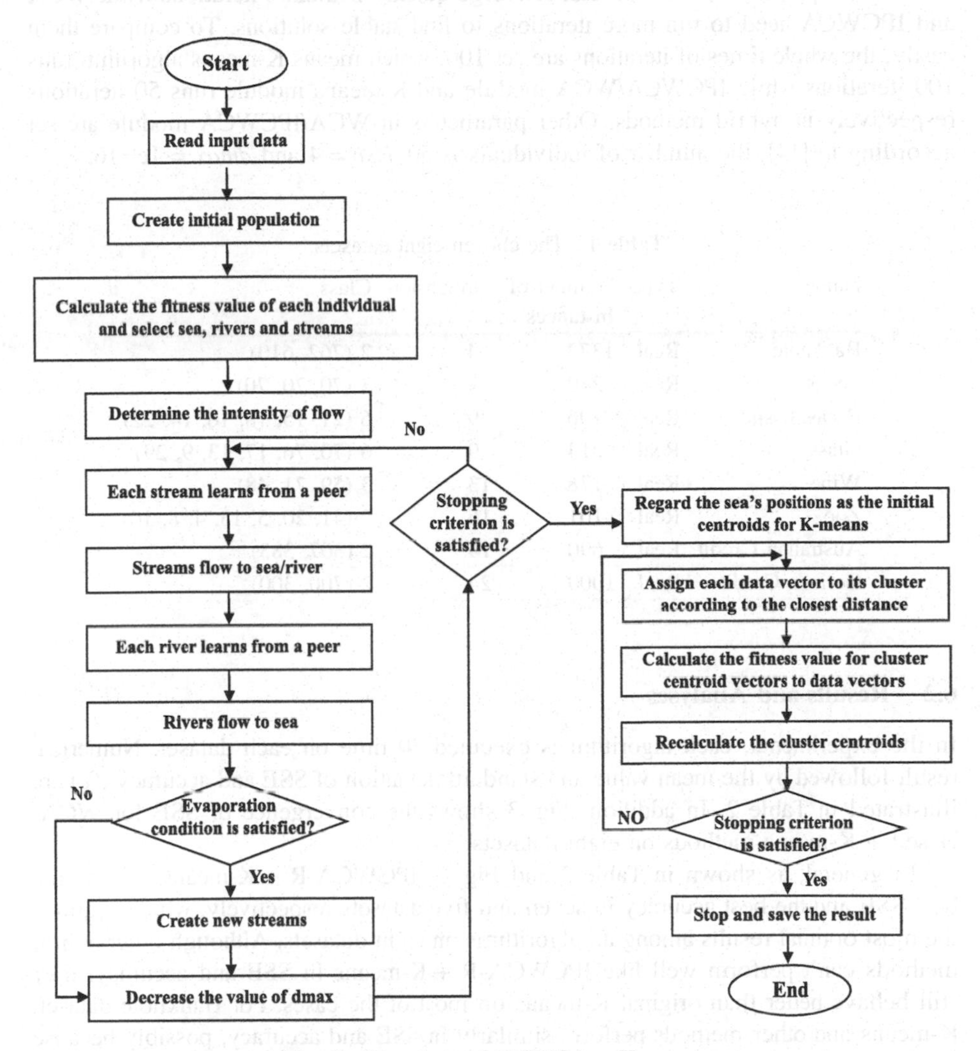

Fig. 2. Flowchart of IPCWCA + K-means

6 Experiments and Results

6.1 Datasets and Experiment Settings

In this section, eight datasets from UCI are selected to test the performance of the proposed algorithm, including six simple datasets for data clustering and two business datasets (Australian Credit and German Credit) for customer segmentation. The information of these datasets is described in Table 1. For the purpose of decreasing the negative effects of abnormal data points, all datasets are preprocessed by minimum and maximum normalization. Besides SSE, accuracy is also selected to test the performance of clustering.

In the experiments, K-means can converge quickly within 50 iterations while WCA and IPCWCA need to run more iterations to find stable solutions. To compare them easily, the whole times of iterations are set 100, which means K-means algorithm runs 100 iterations while IPCWCA/WCA module and K-means module runs 50 iterations respectively in hybrid methods. Other parameters in WCA/IPCWCA module are set according to [14], the number of individuals is 50, $Nsr = 4$ and $dmax = 1e{-}16$.

Table 1. The chosen eight datasets

Name	Type	Number of Instances	Dimension	Class
Banknote	Real	1372	4	2 (762, 610)
Seeds	Real	210	7	3 (70, 70, 70)
Breast tissue	Real	106	9	6 (21, 15, 18, 16, 14, 22)
Glass	Real	214	9	6 (70, 76, 17, 13, 9, 29)
Wine	Real	178	13	3 (59, 71, 48)
Zoo	Real	101	16	7 (41, 20, 5, 13, 4, 8, 10)
Australian Credit	Real	690	14	2 (307, 383)
German Credit	Real	1000	24	2 (700, 300)

6.2 Results and Analyses

In the experiments, each algorithm is executed 30 time on each dataset. Numerical result followed by the mean value and standard deviation of SSE and accuracy (%) are illustrated in Table 2. In addition, Fig. 3 shows the convergence of SSE for WCA-Based + K-means methods on eight datasets.

In general, as shown in Table 2 and Fig. 3, IPCWCA-R + K-means obtains the best SSE and the best accuracy in seven and five datasets respectively, which acquires the most optimal results among all algorithms on eight datasets. Although other hybrid methods can't perform well like IPCWCA-R + K-means in SSE and accuracy, they still behave better than original K-means on most of the cases. For Banknote dataset, K-means and other methods perform similarly in SSE and accuracy, possibly because Banknote dataset is simple with low dimensions, which makes K-means capable of solving this clustering problem well.

As for customer segmentation datasets, they have more instances and higher dimension. For Australian Credit dataset, three proposed methods acquire better results than K-means and WCA + K-means in SSE and accuracy, which indicates that the three hybrid methods are applicable to solve this clustering problem. In German Credit dataset, three proposed methods still get better SSE, but fail to acquire the best accuracy. Interestingly, on the customer segmentation Australian Credit dataset,

IPCWCA-A + K-means gets the optimal result, but on German Credit dataset, IPCWCA-R + K-means gets the optimal SSE value, which indicates that different scenarios may require different approaches and one algorithm may not find the best solution for all problems.

In order to compare the performances of the above algorithms from a statistical viewpoint, Friedman test is adopted in this paper. The Friedman test is a nonparametric statistical test of multiple group measures, which can be used to determine whether a set of algorithms have differences in performance. Null hypothesis H0 is proposed: There is not difference in the performance among these algorithms. The significance level in this testing hypothesis is $\alpha = 0.05$. We reject H0 when $T_F > F_\alpha$, where, T_F-value is given by

$$T_F = \frac{T_{\chi^2}(N-1)}{N(k-1) - T_{\chi^2}} \tag{11}$$

$$T_{\chi^2} = \frac{12N}{k(k+1)}\left(\sum\nolimits_{i=1}^{k} R_i^2 - \frac{k(k+1)^2}{4}\right) \tag{12}$$

T_F follows the F distribution with $k-1$ and $(k-1)$ $(N-1)$ degree of freedom. Where k and N are the number of algorithms and datasets respectively, i.e. $k = 5, N = 8$. T_{χ^2} is defined in Eq. (12), R_i is the ith algorithm's average rank value. As an unsupervised method without label information guidance, clustering performance is evaluated by SSE in this paper, i.e. the smaller the SSE, the better the clustering effect. Therefore, in this Friedman test, mean value of SSE acquired by compared algorithms on each dataset is used as evaluation indicator. Table 3 shows the aligned ranks of algorithms. By Eq. (11)-(12), the result of T_F-value is 4.852. Because $T_F > F_{0.05}$ (4,28) = 2.714, H0 is rejected, i.e. these algorithms have difference in performance.

To further explore how these algorithms are different, Nemenyi subsequent validation is used as follow. CD is the critical range of the algorithm's average rank- value difference, which is defined as

$$CD = q_\alpha \sqrt{\frac{k(k+1)}{6N}} \tag{13}$$

$q_\alpha = 2.728$ $(\alpha = 0.05)$ in this paper. Calculated by Eq. (13), the value of CD is equal to 2.157. Therefore, Friedman test pattern based on CD-value is shown in Fig. 4.

From the results of Friedman test, firstly, the null hypothesis H0 is rejected, which means these compared algorithms have difference in performance. Secondly, according to the rank-values of algorithms in Table 3, IPCWCA-R + K-means have the best average rank-value followed by IPCWCA-A +K-means, IPCWCA-1 +K-means, WCA + K-means and K-means. Compared with original K-means, three proposed methods acquire better average rank-values, which means the three algorithms have better performances.

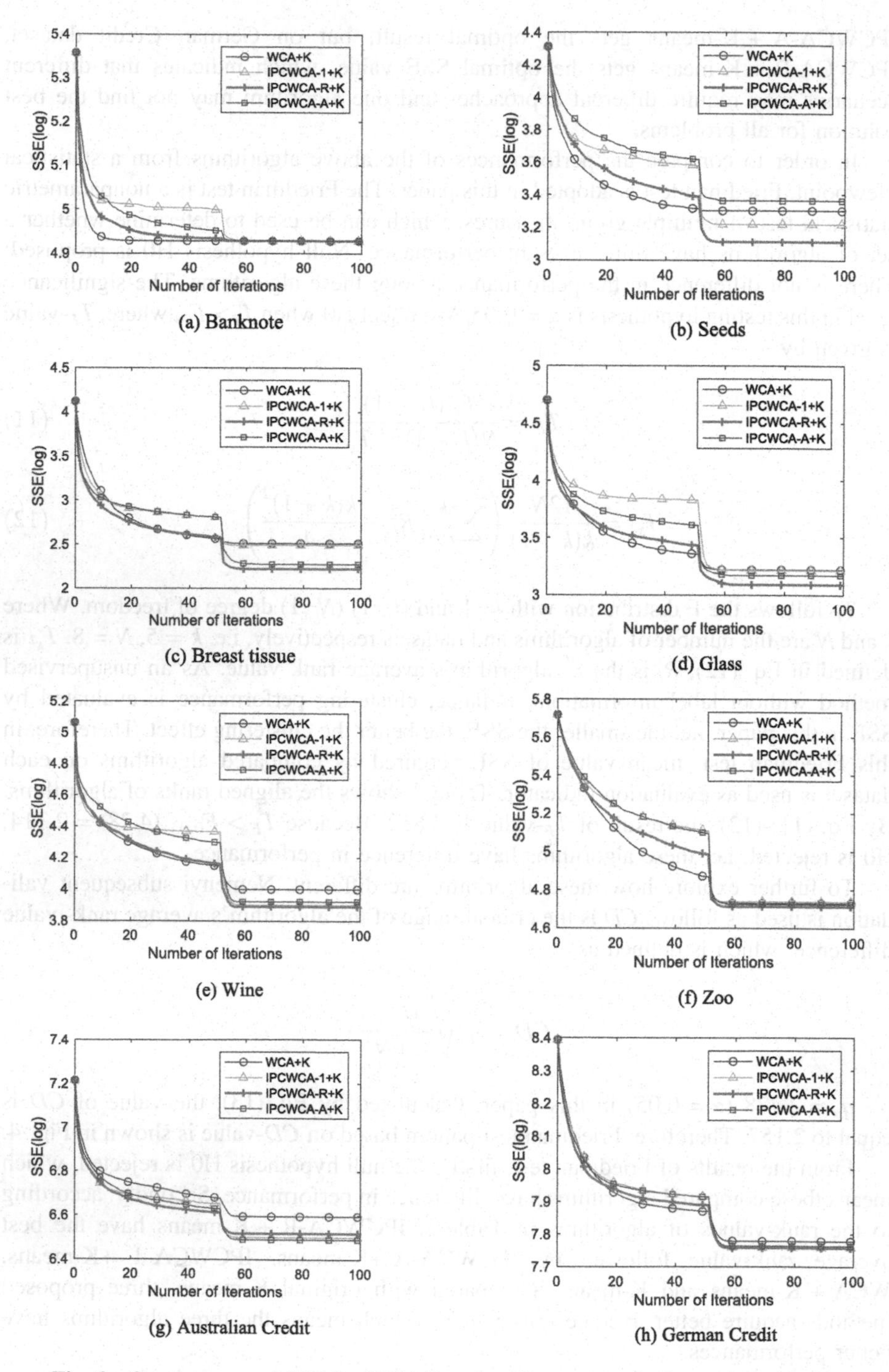

Fig. 3. Convergence of SSE for WCA-Based + K-means methods on eight datasets

Table 2. Numerical Results on Eight Datasets

Dataset			K-means	WCA+ K-means	IPCWCA-1 +K-means	IPCWCA-A + K-means	IPCWCA-R + K-means
Banknote	SSE	Mean	138.1455	138.1455	138.1455	138.1455	138.1455
	Accuracy	Std.	1.0489E−05	1.0119E−05	1.0650E−05	1.0500E−05	1.0119E−05
		Mean	57.4611	57.432	57.4636	57.4611	57.432
		Std.	0.1489	0.163	0.1453	0.1489	0.163
Seeds	SSE	Mean	27.2413	27.1411	25.0094	28.8453	22.4509
		Std.	12.882	6.372	5.501	6.4892	2.3349
	Accuracy	Mean	81.619	79.5079	83.381	76.5714	88.2222
		Std.	13.4696	11.4562	9.9151	11.8058	4.2535
Breast tissue	SSE	Mean	11.1032	12.3061	12.0672	9.7089	9.2067
		Std.	2.29	1.643	1.6703	2.0097	1.418
	Accuracy	Mean	43.8365	41.1635	40.5975	47.1698	48.9308
		Std.	5.505	5.1463	5.3164	4.9301	4.5999
Glass	SSE	Mean	31.1259	25.0278	24.9301	23.8359	21.9728
		Std.	2.5294E−14	1.7595	2.394	2.5055	2.1933
	Accuracy	Mean	45.7944	49.486	49.2835	49.8131	49.8287
		Std.	7.2269E−15	1.2536	1.5166	1.506	1.0303
Wine	SSE	Mean	49.0154	54.6125	49.4884	50.5332	48.9835
		Std.	2.8908E−14	7.5367	2.8424	4.748	0.0199
	Accuracy	Mean	93.2584	81.7978	93.5019	91.3109	94.7191
		Std.	4.3361E−14	16.6366	6.3194	10.5894	0.9401
Zoo	SSE	Mean	138.5768	113.3467	114.2121	112.6272	111.3844
		Std.	5.7815E−14	12.6859	10.7377	9.4395	6.6043
	Accuracy	Mean	65.3465	81.6172	82.8383	83.4323	86.1716
		Std.	0	6.1458	6.1549	3.8385	2.7206
Australian Credit	SSE	Mean	755.4022	732.5268	665.5452	658.8763	678.1709
		Std.	0	94.3183	22.2473	14.2846	31.0021
	Accuracy	Mean	62.6087	68.5217	79.1691	82.7826	73.3188
		Std.	2.8908E−14	12.0841	11.3949	8.5878	16.0789
German Credit	SSE	Mean	2526.4	2387.7	2342.5	2352.5	2341.2
		Std.	9.2504E−13	111.8499	39.3201	61.3984	38.945
	Accuracy	Mean	66.3	63.7633	60.9533	62.5033	61.7333
		Std.	2.8908E−14	3.3302	3.3991	3.1862	3.2184

Finally, as illustrated in Fig. 4, the performance of IPCWCA-R + K-means is significantly different from K-means' and WCA + K-means' performances, which proves that the results obtained on all the datasets with IPCWCA-R + K-means differ from the original K-means' and WCA + K-means' final results. In addition, it is noted that the three proposed methods have overlaps in Fig. 4, therefore, there is no significant difference in performance between them.

Table 3. Aligned ranks of algorithms

Dataset	K-means	WCA+ K-means	IPCWCA-1 + K-means	IPCWCA-A + K-means	IPCWCA-R + K-means
Banknote	3	3	3	3	3
Seeds	4	3	2	5	1
Breast tissue	3	5	4	2	1
Glass	5	4	3	2	1
Wine	2	5	3	4	1
Zoo	5	3	4	2	1
Australian Credit	5	4	2	1	3
German Credit	5	4	2	3	1
Average Rank-Value	4	3.875	2.875	2.75	1.5

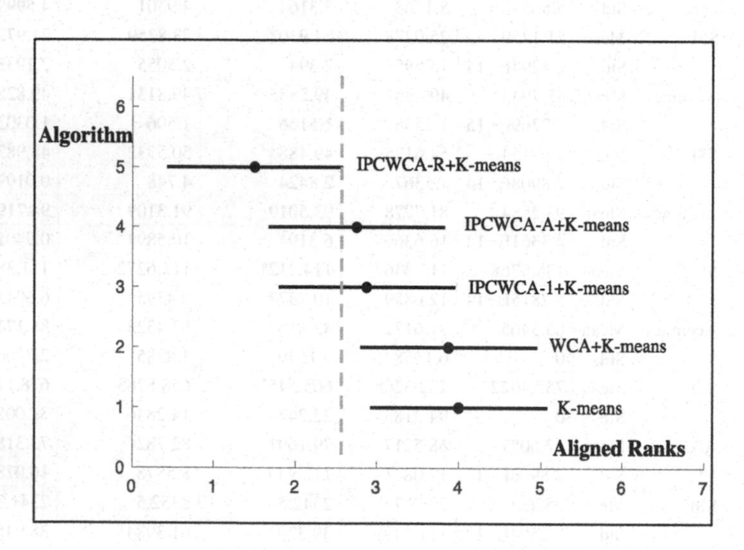

Fig. 4. Friedman Test Pattern

7 Conclusions and Further Work

Inspired by the good global search ability of IPCWCA, three hybrid clustering methods based on IPCWCA and K-means are presented and compared in this paper. Among these three proposed methods, according to SSE, accuracy and statistical analyses, IPCWCA-R + K-means behaved best in most of the datasets. However, in the dataset of customer segmentation Australian Credit dataset, IPCWCA-R + K-means cannot perform well, which indicates that different datasets may require different approaches. Fortunately, compared with original K-means and WCA + K-means, IPCWCA + K-means-based methods behave better in SSE and accuracy in most occasions and perform better in Friedman test.

In future research, we will continue to improve the proposed methods to solve different kinds of clustering problems, especially on high-dimensional data. In addition, customer segmentation problems will be studied more comprehensively, such as customer segmentation models, evaluation criterions and so on.

Acknowledgement. The work described in this paper was supported by Innovating and Upgrading Institute Project from Department of Education of Guangdong Province (2017GWTS CX038), Innovative Talent Projects in Guangdong Universities (2018GWQNCX143), Guangdong Province Soft Science Project (2019A101002075), Guangdong Province Educational Science Plan 2019 (2019JKCY010) and Guangdong Province Postgraduate Education Innovation Research Project (2019SFKC46).

References

1. Pollard, D.: A central limit theorem for K-means clustering. Ann. Probab. **10**(4), 919–926 (1982)
2. Dutta, D., Sil, J., Dutta, P.: Automatic clustering by multi-objective genetic algorithm with numeric and categorical features. Expert Syst. with Appl. **137**, 357–379 (2019)
3. Mustafi, D., Sahoo, G.: A hybrid approach using genetic algorithm and the differential evolution heuristic for enhanced initialization of the K-means algorithm with applications in text clustering. Soft. Comput. **23**(15), 6361–6378 (2019)
4. Gribel, D., Vidal, T.: HG-means: a scalable hybrid genetic algorithm for minimum sum-of-squares clustering. Pattern Recogn. **88**, 569–583 (2019)
5. Lai, D.T.C., Miyakawa, M., Sato, Y.: Semi-supervised data clustering using particle swarm optimisation. Soft. Comput. **24**(5), 3499–3510 (2020)
6. Janani, R., Vijayarani, S.: Text document clustering using spectral clustering algorithm with particle swarm optimization. Expert Syst. Appl. **134**, 192–200 (2019)
7. Liu, W.B., Wang, Z.D., Liu, X.H., Zeng, N.Y., Bell, D.: A novel particle swarm optimization approach for patient clustering from emergency departments. IEEE Trans. Evol. Comput. **23**(4), 632–644 (2019)
8. Menendez, H.D., Otero, F.E.B., Camacho, D.: Medoid-based clustering using ant colony optimization. Swarm Intell. **10**(2), 123–145 (2016)
9. Inkaya, T., Kayaligil, S., Ozdemirel, N.E.: Ant colony optimization based clustering methodology. Appl. Soft Comput. **28**, 301–311 (2015)
10. Eskandar, H., Sadollah, A., Bahreininejad, A., Hamdi, M.: Water cycle algorithm–A novel metaheuristic optimization method for solving constrained engineering optimization problems. Comput. Struct. **110–111**(10), 151–166 (2012)
11. Chen, C.H., Wang, P., Dong, H.C., Wang, X.J.: Hierarchical learning water cycle algorithm. Appl. Soft Comput. **86**, p. 105935 (2020) https://doi.org/10.1016/j.asoc.2019
12. Al-Rawashdeh, G., Mamat, R., Abd Rahim, N.H.B.: Hybrid water cycle optimization algorithm with simulated annealing for spam E-mail detection. IEEE Access. **7**, 143721–143734 (2019)
13. Bahreininejad, A.: Improving the performance of water cycle algorithm using augmented lagrangian method. Adv. Eng. Softw. **132**, 55–64 (2019)
14. Niu, B., Liu, H., Song, X.: An inter-peer communication mechanism based water cycle algorithm. In: Tan, Y., Shi, Y.H., Niu, B. (eds.) Advances in Swarm Intelligence. LNCS, vol. 11655, pp. 50–59. Springer, Chiang Mai (2019). https://doi.org/10.1007/978-3-030-26369-0_5

Information Security

The Research of Music AI in the Context of Information Security

Hui Sun(✉)

School of Music and Dance, Zhengzhou Normal University, Zhengzhou, China
253370721@qq.com

Abstract. In this article through the analysis and summary of the specialty characteristics of undergraduate and postgraduate students, we explore in detail the advantages and features of AI in music composition, performance and education and conclude that AI cannot replace humans, but it performs much better than humans in many aspects, and that in the future, the new music ecology of music AI + database + music education + social interaction will become an inevitable trend with the development of AI. The article also studies the "information security" factors in the application of music AI, and emphasizes that the "information security" includes not only the awareness and understanding of information system risks, but also the technical and operational specifications for preventing such risks. We put forward the idea that in order to survive and thrive in the digital age, students must consciously abide by the ethnical codes and laws and regulations in the information society and be responsible in publication, use and dissemination of information.

Keywords: Information security · Artificial intelligence · Music education · Music interaction

1 Introduction

Artificial intelligence, abbreviated as AI, was proposed and named at the 1956 Dartmouth conference. It is an interdisciplinary boundary discipline of natural science, social science, and technological science, and a new science and technology that aim to study human intelligence, and simulate, extend and expand it for applications in various fields [1].

Music AI is based on the artificial intelligence technology, analyzes the human music intelligence through big data, simulates the information process of human's sight, hearing, touch, feeling, thinking and reasoning, and constructs its own neural network and algorithm generation [2]. Finally, it can be applied to the music perception, cognition, research and creation by humans, and innovate the new music teaching model of "human-computer interaction".

With cloud space, big data, "Internet of music" and AI, it is an era evolving from information explosion towards "intelligence explosion". It changes the life of every one and exerts influence on the global economy, culture, education, etc. As far as the current situation is concerned, in the broad context of AI development, the research, promotion and application of music AI in the field of music education will be perfected

D.-S. Huang and P. Premaratne (Eds.): ICIC 2020, LNAI 12465, pp. 593–600, 2020.
https://doi.org/10.1007/978-3-030-60796-8_51

progressively. Specifically, both of AI technology and music education are benefited from each other. Traditional music education can be promoted and enlightened with the development of AI technology, where the format of education process, feedback of education quality and management of education schedule in music are all improved by introducing AI device and products. Moreover, the adoption of AI technology in music education can also expand the application fields and brings a potential advancement for AI technology.

In July 2015, the "artificial intelligence" was written into the Guiding Opinions of the State Council on Vigorously Advancing the "Internet Plus" Action. In May 2016, the National Development and Reform Commission and other four departments jointly promulgated the Three-Year Action Plan for "Internet +" Artificial Intelligence [3]. On March 5, 2017, Li Keqiang issued the Report on the Work of the Government for 2017 and proposed to accelerate the development of emerging industries, the most important part of which was the AI technology. This was the first time that "artificial intelligence" was written into the government report and fully reflected the importance attached by the government to the research and development of AI technology.

The author will focus on artificial intelligence, information security and music education, and apply the approaches of literature research, interdisciplinary research and comparative research to study the characteristics of AI and students at different school ages. This research gives the detailed and scientific reference materials with the hope to provide some new ideas for the development of AI and information security in the field of music education, so as to make AI better serve the music education of humans.

2 Traditional Music Perception for Infants and Young Children and Music Interaction in the Context of AI

The traditional music perception education for infants and young children is divided into two types: one is based on the daily life of infants and young children, and played as the background music surrounding listeners frequently for unconscious music edification. When they need the emotion regulation and spatial ambience before sleep, or pacification during sleep, the music should be quiet, soft and peaceful; when they play and become lively and active, the music should be upbeat with a certain rhythm. The synchronization of emotions and music is beneficial for their healthy growth physically and mentally. This type is called the functional music. Another type refers to the conscious and purposeful permeation of the basic music theory education in the daily music perception of infants and young children, such as training of pitch and rhythm, and beat imitation. Scientific practice has proven that the ages from 3 to 6 years old are very crucial to the memory ability of absolute pitch in music. The musical rhythm training is extremely helpful for the intellectual development and body coordination of infants and young children. Nowadays, multifarious music early education institutions for infants and young children mostly offer the unidirectional pitch and rhythm training, which is limited only to classroom interactions. The development of music technology can allow infants and young children to have more scientific, reasonable and enjoyable music perception. This type is called the skill-based music.

In recent years, Google, Sony, Baidu, Tencent, Alibaba and other large international and domestic companies are continuously enhancing their investment and R&D in the field of artificial intelligence. In 2011, a pet game app "My Talking TOM" was developed by the world-renowned mobile game developer Outfit7. TOM is a cute cat that you can interact by touching him, caressing his tail or tapping him slightly. What's important, he is a talking pet that can completely repeat what you have said in a humorous and funny voice. The machine has learned to listen, and consciously imitate and actively change the human voice. The design principle of "My Talking TOM" inspires and unlocks the new model of infant education.

Since 2016, the artificial intelligence was elevated to be a national strategy in China. Large companies designed and produced a series of humanoid intelligent robots, which could gradually become a member of the family. They had the knowledge of astronomy, geography, Chinese, mathematics, English, science, music, fine arts, etc. and could speak in various life scenarios. The machine had the ability of language recognition, and began to talk with humans through its own neural network combined with big data for analysis.

From the perspective of the music education for infants and young children, the traditional music perception training for infants and young children is gradually evolving towards a new AI music education model with active teaching and interactive communication by the robots. Music information retrieval (MIR) is an important music technology based on the musical acoustics. It extracts musical features by audio signal processing, and the various machine learning techniques in artificial intelligence are utilized in the back station [4]. MIR technology can extract massive audio information of digital music or automated technical analysis, and classify according to the unique features of every piece of music. Music AI can carry out the big data screening of early education resources, embed the abundant knowledge library with materials suitable for the music education of infants and young children in the back end, and form its own regulated and accurate early education system.

The robot can perceive the real life of infants and young children and their vocal emotions through the automatic language and speech recognition processing technology, and automatically recognize and play the functional music. The robot is just like a resident music teacher, who can accompany the daily life of infants and young children with specific pitch and rhythm according to their life habits, and progressively infuse the basic music theory education. The robot is a giant music library if there is music AI + Internet. Traditionally, keywords are used to search music on the Internet, but currently, AI can read the speech intentions of young children and parents, through its own neural network, interact with humans by voice and provide various music services as needed.

3 The Convenience of AI in Music Education

Today's music education in primary and secondary schools in China mainly comprises the following areas: music appreciation, theoretical study (basic music theory and music history), and play and performance skills (earning to play musical instruments, singing and chorus, band training). The current situation is that students love music, but they do not like the music courses. Consequently, music teachers continuously innovate the teaching models, such as increasing the multimedia teaching and the use of modern network information technology; and mobilizing students' teaching participation to the maximum extent by the research and application of the music education systems of Carl Orff, Kodaly Zoltan and Emile Jaques-Dalcroze [5]. Regarding the rapid development of AI in the new era, the author proposes to build and configure a "3D AI music classroom – musical scenario space for primary and secondary schools", which can greatly arouse the interest and enthusiasm of students to learn music, thus providing some exploration, research and thinking on how to realize the teaching in the new era with new concepts and new technology.

(i) Perceiving Music Stories in The "Musical Scenario Space"
The "musical scenario space for primary and secondary schools" applies the Dolby Atmosphere technology. The hardware configuration is Dolby Atmos speaker setup, which can be 5.1 or 7.1 for parallel processing. If the classroom has a large enough space, it can also add a pair of front widening speakers in 7.1 setup. The voice will come from all directions, including overhead, thus creating a clear, full and layered sound space. During the class, different characters, music, sound effects and events will surround the students in a detailed and deep three-dimensional space, so that students seem to instantly become part of the immersive music story scenario.

(ii) Performing, Adapting and Composing Music Stories in The "Musical Scenario Space"
In the "3D AI music classroom", the combination of voice assistant and intelligent 3D audio enables students to change from the typing on the screen to the convenient and fast mode of language command, listening, appreciation and creation. With the music information retrieval (MIR) technology, students can immediately call the music, sound effects and other materials they need from the massive music library. The music stories can be imitated or adapted or used to inspire new creative ideas, improve their ability to think proactively, and interact with music teaching in an instant and efficient manner.

(iii) The Rapid Development of New Things Naturally Needs to Improve Many Related Elements Synchronously
The biggest problem faced by the "3D AI music classroom – musical scenario space for primary and secondary schools" is the very few sources of music education videos with full audio atmosphere. There are laws for all things and everything has two sides. The occurrence of a new problem implies new ideas and solutions, which will give a new direction of learning and practice for the undergraduate music education.

4 Application of AI to Music Undergraduate Students

The traditional music education system of undergraduate students is important, but the incorporation of AI technology into the music education in the new era is also imperative. Music AI will provide a new multi-dimensional teaching practice platform for the music education of undergraduate students. We should also pay special attention to information security risks in music AI applications and enhance the awareness of prevention. Any security risks caused by human factors must be prevented in terms of "humans", and the commonly used prevention techniques include identity verification and authority management. For risks caused by software or hardware factors, the host system security technology and the network and system security emergency response technology should be applied for prevention. The risks caused by the data factors may be prevented by data encryption technology and data backup [6], etc.

(i) Application of AI in Composition

The traditional techniques and theory of composition have a modular pedagogical system, comprising harmony, texture and musical passage structure. Music AI can not only simulate the pedagogical system, but also have the enormous computing power of composition [7]. Orb Composer, the first intelligent musical composition software with AI, is of great significance. Now it has six basic music templates, namely Orchestral, Strings, Piano, Electro, Pop-Rock and Ambient, to help you pre-select a musical environment. Orb Composer can instantly create the specified style of musical composition by the following easy steps: A. to set the tempo, rhythm and tonality; B. to select the complete structure of composition determined by the block scheme, select chords and instruments (the preset basic music template may be used); and C. to select the automatic generation. Music AI enables the dream of "composition" of people who know a little about music to instantly come true. For those who have the ability of composition to some extent, the software can give them inspiration and make the personalized, professional and detailed revision of the works automatically generated.

(ii) Application of AI in Music Performance

The traditional music performance achieves the image building and formal expression by music. With the development of AI, the intervention of music AI will have the new form of music performance by "human-computer interaction". The "Informatics Philharmonic" system invented by Christopher Raphael at the School of Informatics and Computing, Indiana University, Bloomington can provide the complete and professional orchestral accompaniment for the soloist. In November 2018, at the "Night of AI – Music Concert of Music + AI Accompaniment System" jointly held by Central Conservatory of Music and him, in addition to the classical music performed by AI technology, The Great Wall Capriccio was also played for the first time. The "Informatics Philharmonic" system has the powerful AI learning ability and can generate a variety of different algorithms. It can be interactively changed according to the player's music rhythm and constantly adjust and improve its accompaniment ability.

(iii) Application of AI in Undergraduate Music Education

The solution to the problem of few sources of music education videos for the "3D AI music classroom – musical scenario space for primary and secondary schools" is to offer the Music Creation course in the undergraduate music education. Students can learn and master the relevant knowledge structure, acquire the ability of creation to some extent, use the AI music composition software Orb Composer, and based on the current music textbooks for primary and secondary schools, gradually create a number of music education videos with full audio atmosphere that are suitable for China's national conditions, so that the seniors of the music education major can be well prepared for their internship in primary and secondary schools and combine teaching, practice and internship together [8].

5 New Research Direction for Music Postgraduate Students

At present, the application of music AI to music education and research is just at its beginning. Artificial intelligence that is detached from music put the cart before the horse, but how can it be counted as intelligence without a powerful technical team? Hence, the interdisciplinary collaboration is the inevitable trend. Then what should music postgraduate students think over?

Intelligent interactive music teaching platforms are currently springing up like mushrooms and customize personalized teaching based on the big data analysis. Teachers teach online to reproduce the offline one-to-one and one-to-many teaching scenarios. In combination with music audio recognition of the new music AI technology, the teaching and interactions will have fun and can give answers, scoring and learning suggestions at any time, thus being efficient and low cost. The music professional groups and the technical teams should establish the long-term close cooperation, so as to guarantee not only the stability, security, advancement and ease of use of the learning platform, but also the accuracy of music expertise on the platform as well as the reasonableness, continuity and authority of music teaching. The postgraduate advisors in departments of music and technology should lead and guide their student teams for joint application of cross-disciplinary research projects, and only in this way can such research be sustainably developed [9].

At every stage of teaching, we need to make students aware that the security risks of the information system can be prevented and controlled. Operators and administrators must fully understand the possible hazards of information system risks, be familiar with information with the prevention techniques of such risks, strengthen the security operation specification of information systems, have a certain degree of foresight and the ability to deal with the emergence of risks, and be able to use the correct technology for eliminating the information system risks; and users must strengthen the security awareness of information system use in daily application, attach importance to setting passwords, anti-virus, frequent backup and careful Internet access, and strive to reduce security risks in the process of use [10].

Google's Magenta and Sony's Flow Machines are two of the world's top music AI development projects in the field of compositional AI. Sony's Flow Machines project has collected and analyzed a large number of different styles of music database, and allows users to improvise designated original music with a few simple compositional commands. A representative piece of music is the song "Daddy's Car", which is automatically generated in the database after analyzing 45 Beatles songs. The scientific achievement of the research teams of these top companies is the neural networks that can generate robots, the technological tools. It is just like the invention of the computer in the 1930 s, when it was just an arithmetic tool that needed to be constantly given new directions for learning and arithmetic. Different levels of use will give new directions to music AI. The professional music research team can develop the automatic composition and arrangement software for different music styles with the development of the artificial intelligence technology. The music AI worldwide needs to analyze and generate music of different levels, genres, ethnicities and styles to enrich itself. It will be advantageous to divide by regions, and analyze and study the ethnic and folk music databases with different geographical characteristics.

Now Google's Magenta project seems to be more technologically cutting-edge, because they disapprove of the Turing test for music AI [11], and never want robots to compose entirely according to the thinking patterns and laws of human beings. Douglas Eck, the project scientist, is also a musician. He has attempted to train the neural network of the project tool NSynth with the sound of 300,000 kinds of musical instruments, so that its computing, learning, generation and display of new sound are unique with the distinctive sound characteristics. The Magenta project team wanted the music AI to have the relatively independently thinking and innovative ability. The experimental work is not mature yet, but it may inspire a larger, freer imaginative space for young music team with the creative mind. Human needs to be so open-minded and tolerant, and the scientific and technological innovation needs to have enough N-dimensional space. Maybe one day they will be able to make music with sound effects that humans have never heard before.

6 Conclusion

Opinions vary on whether humans will be replaced by artificial intelligence, but it is an undeniable fact that humans are defeated by artificial intelligence in chess and the game of go. They cannot replace humans, but are superior in numerous aspects. Currently, robots can understand music, analyze music, create music and apply it to music teaching. With the continuous enhancement of the computing power of computers and the development and research of deep learning in the context of big data, it will be an inevitable trend to have a new music ecosystem of music AI + database + music teaching and application + social interaction.

In teaching with music AI, teachers must, on the one hand, emphasize the learning of information technology, which is a necessary skill for survival in the information age, and, on the other hand, comply with the norms of the information society because this is a precondition for a "qualified" citizen in the information age. To help students learn to survive in the digital age, teachers must lay stress on the education of the norms of the information society, so that students can consciously comply with the relevant laws and regulations, follow the moral and ethical codes in the information society, reasonably use the information technology, strive to regulate the behavior of Internet access, and become the qualified citizens in the digital age [12].

References

1. Wagman, M.: Artificial intelligence and human cognition. Q. Rev. Biol. 68.1 (2019)
2. Anagnostopoulou, C., Ferrand, M., Smaill, A.: Music and Artificial Intelligence: Second International Conference, ICMAI 2002. Music & Artificial Intelligence (2002)
3. Luo, X., Xie, L.: Research on artificial intelligence-based sharing education in the era of Internet + . In: International Conference on Intelligent Transportation IEEE Computer Society, pp. 335–338. IEEE (2018)

4. Wei, L., Zhihui, G.: Music information retrieval technology: fusion of music and artificial intelligence. Arts Exploration, **5** (2018)
5. Yonghong, G., Hengyu, G.: On the relations of education and technology in the studies of educational technology - with application of artificial intelligence in the education field as an example. Global Market Information Guide, **1** (2015)
6. Daguo, L., Weiming, L.: Prevention of risks in information system and learning to survive in the digital age. China Inf. Technol. Educ. **4**, 112–113 (2020)
7. Taoye, X.: On the application of artificial intelligence based on expert system in the education field. Science & Technology Information, **11** (2011)
8. Lulu, S.: Application of music technology in modern music education. People's Music (2012)
9. Minjie, L.: Dialogue across time and space between EMI and master composers—explorations of music language in the new media era (I). Explorations in Music, **14** (2014)
10. Daguo, L., Weiming, L.: Prevention of risks in information system and learning to survive in the digital age. China Inf. Technol. Educ. **4**, 112–113 (2020)
11. Turing, A.M.: Computing machinery and intelligence. In: Epstein, R., Roberts, G., Beber, G. (eds.) Parsing the Turing Test, pp. 23–65. Springer, Dordrecht (2009). https://doi.org/10.1007/978-1-4020-6710-5_3
12. Daguo, L., Weiming, L.: Prevention of risks in information system and learning to survive in the digital age. China Inf. Technol. Educ. **4**, 112–113 (2020)

Intra-frame Adaptive Transform Size for Video Steganography in H.265/HEVC Bitstreams

Hongguo Zhao[1], Menghua Pang[2], and Yunxia Liu[1(✉)]

[1] College of Information Science and Technology, Zhengzhou Normal University, Zhengzhou, China
liuyunxia0110@hust.edu.cn
[2] College of Mathematics and Statistics, Zhoukou Normal University, Zhoukou, China

Abstract. A video steganography method based on H.265/HEVC (High Efficiency Video Coding) video adaptive transform block size is proposed in this paper. With the adoption of new quadtree splitting structure in H.265/HEVC video compression standard, secret information could be embedded into carrier bitstreams by adaptive modifying the transform unit (TU) size during the procedures of searching best coding syntax elements set. Different from traditional video steganography methods, the proposed method could achieve at most modifying the transform unit splitting once while multiple secret information is embedded, which can achieve a high visual quality for digital carrier video. Moreover, the proposed method is only manipulated on the splitting decision process of TUs. Large volume quantity of TUs in compressed carrier video could guarantee a high embedding capacity. The experimental results have been proven the superiority of efficiency and performance about the proposed method. Moreover, the proposed method can be better applied to high-definition (HD) or higher video applications, and provides a practical tool for the protection of privacy data about digital video, as well as the contents of digital video itself.

Keywords: Video steganography · Transform unit · Splitting structure · Intra-frame prediction

1 Introduction

Digital video technology and applications has been applied in human's daily life more popular than ever before. With increasing diversity of devices, growing popularity of HD video products which are more favored in recent decades, more and more video application scenarios are developed by the continuously developing in video technology. For example, multimedia cloud [1], live broadcasting, beyond HD formats 8 K ultra-high definition (UHD) video applications [2], etc. However, information security issues which are companied by digital videos, especially for the protection of legitimate rights for owner and clients, malicious tampering, copying and broadcasting, has

D.-S. Huang and P. Premaratne (Eds.): ICIC 2020, LNAI 12465, pp. 601–610, 2020.
https://doi.org/10.1007/978-3-030-60796-8_52

brought severe challenges for the development of digital video either in applications or in technology updating.

Video steganography technology provides a powerful tool for addressing these challenges in recent years. Video steganography is a technique that utilizes the redundancy of sampled signals on human visual organ to embed secret information into carrier videos, which is transparency and unconscious for all inconclusive users except information embedding and extracting sides [3]. Similar to digital watermarking, which usually embed secret data into digital image and protect the legitimate owner interests or prevent the contents from being tampered, video steganography also embed secret information into digital video and can effectively protect the copyright information referred to the legitimate owner. In addition, different from the watermarking with limited to signal picture and low embedding capacity, video steganography is benefited for achieving larger embedding capacity due to the infinite video sequence and can be applied to more application scenarios, such as civilian or military covert communications [4]. Moreover, targeted at the illegal spread and usage of digital video, the legitimate interest of video owner can be embedded into digital video with video steganography, which can be used as the trace and declaration of the specific video. Thus, video steganography play a key role for the further security protection issues of digital videos.

The increasing demands of higher resolution digital videos and challenges traffic in networks stimulate the rapidly development in video coding technology. From the recent most popular standard of H.264/HEVC to the latest H.265/HEVC, video coding technology is going more and more outstanding in coding efficiency, as well as higher visual quality and resolutions. Compared to H.264/AVC, H.265/HEVC can achieve the reduction of approximately 50% in bit-rate while maintaining equal visual quality. The significant improvement is supported by a variety of new coding technologies and structures, such as flexible coding size, quadtree spliting structure, more angel prediction modes in spatial domain, and advanced motion vector prediction in temporal domain, etc. The problem is that these new technologies are not always compatible to its preceding H.264/AVC, especially for video steganography. So there is a strong need for efforts to explore the video steganography technologies for high resolution digital video based on H.265/HEVC.

In this paper, we examine the protection issues of video steganography in digital video. The latest and relevant researches of video steganography are elaborated and analyzed. Aimed at the challenges these methods couldn't always achieve high efficiency in the reduction of embedding error or sufficient embedding capacity, we propose to utilize transform block decision for size selection to embed secret information into carrier video. In order to minimize the embedding error, we employ the minimum processing unit-TU as the embedding carrier, and then modify at most one transform split decision by embedding multiple secret information. Experimental evaluation is proven that the proposed method can achieve high efficiency and security performance in maintaining high visual quality and sufficient embedding capacity.

The remainder of this paper is organized as follows: Sects. 2 reviews the related works on video steganography technology. Section 3 proposes the video steganography method based on the adaptive transform decision in intra-frame prediction and the experimental evaluation is presented in Sect. 4. Finally, the conclusion is shown in Sect. 5.

2 Related Works

With the continuous development of video coding technology, especially under the emergence of H.265/HEVC, digital video on network is preferred to be higher resolution along with asynchronized improvements on delay and limited traffic network. Based on this preference, video steganographyis taken into consideration for digital video security protection. Video steganography based on digital video plays a key role among various security protections (e.g., encryption) for multimedia concerning digital video occupying the majority of internet traffic.

In order to protect the secret information embedded into video carriers from attention of illegal attackers, many powerful and effective attempts such as error expansion [5], histogram shifting [6], discrete cosine transform (DCT) [7–9], etc., have been made. In addition, there are also existed some methods [10, 11] which utilize the new characteristics adopted in H.265/HEVC. Shanableh [10] utilized the splitting flag sequence of one specified coding unit (CU) to embed approximately 4-6 bits secret information. Specifically, one coding tree unit (CTU with size 64×64) can be divided into four 32×32 coding units, where each splitting flag sequence of 32×32 coding units can be represented as five binary splitting flag sequence. The front four bits of splitting flags are used for embedding secret information and the fifth bit presented whether is split into quadrants according to current 32×32 coding units as depicted in Fig. 1. It can be seen that one CTU provides sixteen bits used to embed secret information and has sufficient embedding capacity. However, the challenge is that embedding is manipulated on large scale blocks-32×32 coding units, which would introduce a large embedding error in the digital visual quality. Yiqi Tew et al. [11] proposed an improved method which mainly utilizes the splitting structure of prediction units which has rooted at CUs to embed secret information. Specifically, the splitting referred to prediction units can be divided into two categories, one is mapping to secret information 0 and the other is mapping to secret information 1 as depicted in Fig. 2. Then the true splitting referred to prediction units is constrained to this embedding mapping rule. Although this method will bring an improvement on carrier visual quality, the improved efficiency is not always rapid and significant.

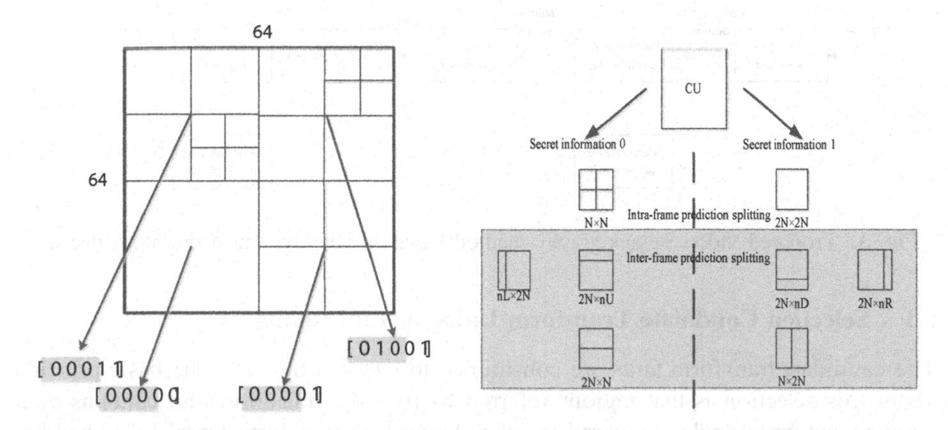

Fig. 1. Embedding into coding unit splitting. **Fig. 2.** Embedding into prediction unit splitting

Different from these previous works, we employ the smaller transform units as the actual embedding carriers to embed secret information. To decrease the embedding error, only the size of 16 × 16 transform block are utilized as embedding candidate blocks. Then, we put forward an efficiency embedding module which can embed multiple secret information while modifying only one splitting flag in most cases. Finally, we design and evaluate the high efficiency about the proposed method in terms of improvements in the reduction of embedding error for video quality and in the embedding capacity.

3 Proposed Adaptive Transform Block Decision Video Steganography Method

The proposed video steganography method based on adaptive transform block decision is illustrated in Fig. 3. The scheme is mainly composed of two components, including embedding procedures and extraction procedures. In the embedding procedures, the selection for candidate transform units is a key manipulation which is closely following the decision process of coding syntax elements determined by H.265/HEVC encoder. Then according to the pre-defined embedding mapping rules, the secret information can be embedded into TUs by modifying the TU splitting flags. After entropy encode (CABAC or CALVC), the carrier video which has been embedded secret information would be encoded to bitstream and can be transmitted on public network. The extraction procedures is an inverse loop compared to embedding, which are also included the selection of transform blocks which has been embedded secret information after video bitstream entropy decode. Then the secret information can be extracted from the transform block splitting flags with the pre-defined extraction mapping rules.

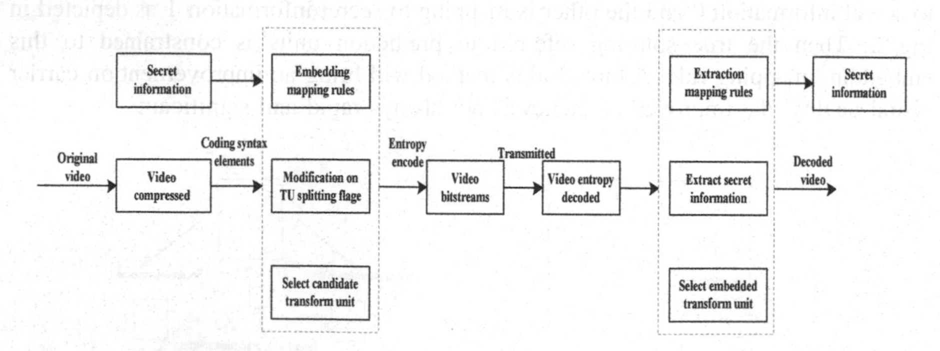

Fig. 3. Proposed video Steganography method based on adaptive transform block decision

3.1 Selection Candidate Transform Units for Embedding

The candidate transform units are constrained to size of 16 × 16. The basic principle about this selection is that regions referred to 16 × 16 transform unit contains more detailed texture signals compared to other larger size transform units, so embedding secret information into 16 × 16 transform block splitting decisions will be more unconscious of human sight and provides a higher security level than other size of transform blocks (e.g., 32 × 32).

The selection of candidate 16 × 16 transform blocks can be divided into two categories. The first category is directly 16 × 16 sizes of coding units which are determined by H.265/HEVC encoder control. In addition, conforming to proceeding constraint coding unit should also be certain that there's at least splitting into quadrants once again for transform splitting. That is, for the transform splitting structure rooted at current 16 × 16 intra-frame coding unit, the final size of transform units determined by encoder control is no more than 8 × 8 in the same covered region. One compliant typical example is depicted in Fig. 4, where the selected 16 × 16 coding unit which has been further split for the transform units contains two 8 × 8 transform units and eight 4 × 4 transform units, respectively. Another category is targeted for those coding units which size is larger than 32 × 32 (e.g. 32 × 32 or 64 × 64). When the selection of candidate blocks process meets size of 32 × 32 or 64 × 64 coding units determined by encoder control, the process will search whether there's one transform units which are confirming to the first category selection constraints with 16 × 16 coding unit substituted by current 16 × 16 transform unit. If there's one 16 × 16 transform unit inside the current coding unit is confirming to the second category constraint, the 16 × 16 transform unit is processed at the same way as the first category 16 × 16 coding units. Figure 5 depicts one typical example for the second category, where the current coding unit is 32 × 32 size and the top-right transform unit which are split from the transform splitting consists three 8 × 8 transform units and four 4 × 4 transform units in the final transform structure determined by encoder control.

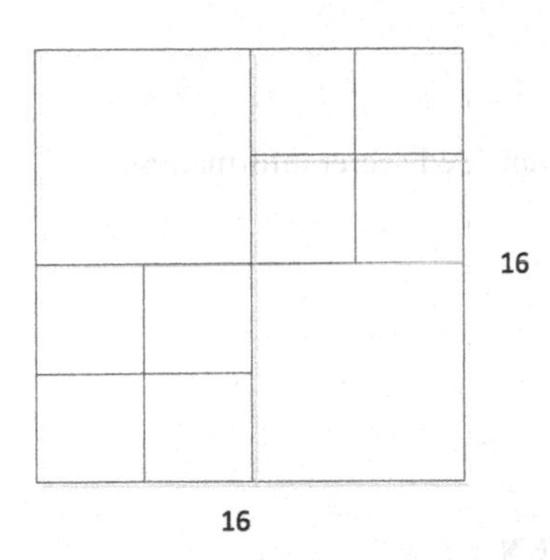

Fig. 4. First category for embedding

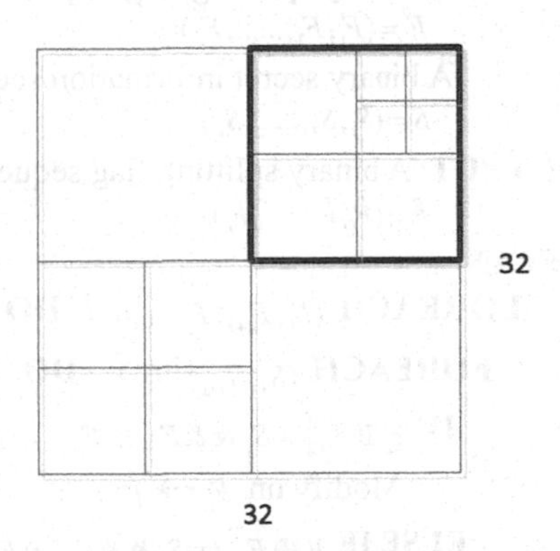

Fig. 5. Second category for embedding

After the selection of candidate coding units or transform units, we can rewrite the internal transform splitting structure as a form of binary flag sequence. The binary flag '1' implicit that based on the current splitting structure, the transform splitting should be carried out once again with quadtree format. Likewise, binary flag '0' implicit that the current splitting should be confined non-splitting. Then, the final candidate transform splitting structure determined by encode control can be represented as five flag bits as depicted in Fig. 4 and Fig. 5 with *10110* and *10100*. It is noted that the first flag bit indicates whether the candidate is split into quadrants or not, and the successive flag bit indicates whether each quadrant is further split into quadrants or not.

3.2 Embedding and Extraction Module

Embedding is manipulated on the transform splitting flag sequence. Specifically, the latter four splitting flag bit are used for embedding secret information after the complex tasks of selection for candidate codding units or transform units. One targeted goal of embedding module is that multiple secret information is embedded while only one transform splitting flag is modified. The detailed embedding and extraction module is elaborated as follows.

Embedding module depends on the characters of transform splitting flag sequence and values of secret information. Since both of splitting flag sequence and secret information is presented as the form of binary sequence, the mapping rules between them can be established as the form of matrix encoding–like procedures. The main embedding procedures are depicted in Algorithm 1. First, a binary splitting flag sequence of candidates F in single CTU can be constructed from the selection candidate transform units procedures. Then according to the binary secret information sequence

Algorithm 1. Embedding Module

INPUT: A binary splitting flag sequence of candidates:
$F = (F_1, F_2, \ldots\ldots, F_n)$
A binary secret information sequence:
$S = (S_1, S_2, \ldots\ldots, S_p)$
OUTPUT: A binary splitting flag sequence with embedded secret information:
$\hat{F} = (\hat{F}_1, \hat{F}_2, \ldots\ldots, \hat{F}_n)$
BEGIN
1. **FOREACH** (F_i, F_{i+1}, F_{i+2}) in F **DO**
2. **FOREACH** (S_j, S_{j+1}) in S **DO**
3. **IF** $F_i \oplus F_{i+2} \neq S_j \,\&\&\, F_{i+1} \oplus F_{i+2} = S_{j+1}$ **THEN**
4. Modify on $F_i \rightarrow \hat{F}_i$
5. **ELSE IF** $F_i \oplus F_{i+2} = S_j \,\&\&\, F_{i+1} \oplus F_{i+2} \neq S_{j+1}$ **THEN**
6. Modify on $F_{i+1} \rightarrow \hat{F}_{i+1}$
7. **ELSE IF** $F_i \oplus F_{i+2} \neq S_j \,\&\&\, F_{i+1} \oplus F_{i+2} \neq S_{j+1}$ **THEN**
8. Modify on $F_i \rightarrow \hat{F}_i$
9. Modify on $F_{i+1} \rightarrow \hat{F}_{i+1}$
10. **End IF**

END

and mapping rules with matrix encoding-like formats (the interval among sequence is 3bits), the selected splitting sequence can be modified as sequence $\hat{F}$. For example, if the binary splitting flag sequence is (0, 1, 1), the embedded secret information binary is (1, 0), the output of modified splitting flag sequence would be (0, 1, 1).

Extraction module is manipulated during the decoding process which is successive entropy decoding and inverse process compared to embedding module. The splitting flag sequence which is embedded secret information is obtained from the selection of transform units with the same constraints as embedding side in each CTU. We can rewrite the splitting flag sequence of transform units as array $\hat{F}$, and utilize the bit-wise *XOR* with the same interval length (e.g. 3) as the embedding module. The main extraction procedures are depicted in Algorithm 2. First, a binary splitting flag sequence of transform unit derived from transmitted bitstream $\hat{F}$ can be achieved from the selection of transform units which has the same constraints as embedding side. Then the targeted secret information binary array *S* can be generated from bitwise *XOR* in pairs. For example, if the binary splitting flag sequence is (0, 1, 1), then the output of secret information binary sequence would be (1, 0).

Algorithm 2. Extraction Module

INPUT: A binary splitting flag sequence of transform units with embedded secret information:

$$\hat{F} = (\hat{F}_1, \hat{F}_2, \ldots\ldots, \hat{F}_n)$$

OUTPUT: A binary array sequence of embedded secret information:

$$S = (S_1, S_2, \ldots\ldots, S_p)$$

BEGIN

1. **FOREACH** $(\hat{F}_i, \hat{F}_{i+1}, \hat{F}_{i+2})$ in $\hat{F}$ **DO**
2. $\quad S_j \leftarrow F'_i \oplus F'_{i+2}$
3. $\quad S_{j+1} \leftarrow F'_{i+1} \oplus F'_{i+2}$

END

3.3 Modification on Selected Transform Units

According to the modified splitting flag sequence $\hat{F}$ generated from embedding module, the initial splitting flag of selected transform units *F* should also be changed synchronously. Specifically, the modification on transform splitting is actually happened on 8 × 8 transform unit which has been spilt once in terms of 16 × 16 coding units or multiple times of splitting in terms of 32 × 32 coding units. If the modified splitting flag in $\hat{F}$ is different from the initial transform splitting flag in *F*, then the final transform splitting on this specified 8 × 8 transform unit should be confined to split again until the transform splitting structure is compatible to the corresponding modified splitting flag in $\hat{F}$. Obviously, the following transform and quantization for residuals

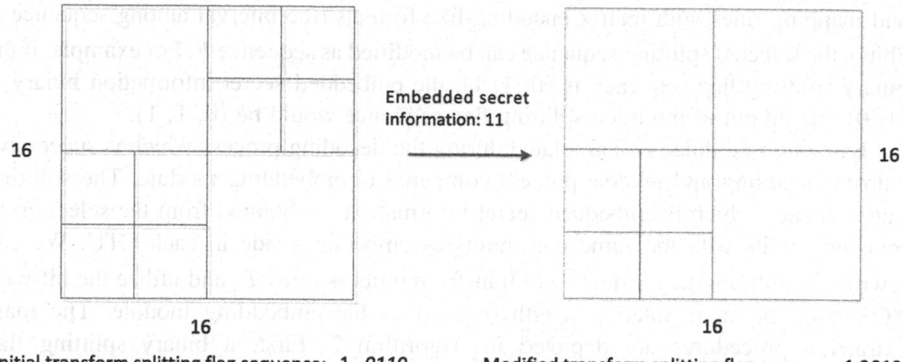

Fig. 6. An embedding example with embedding secret information (1, 1)

should also be substituted by the new residual results with new transform splitting syntax elements. Figure 6 provides an example about modification on transform units. The initial transform splitting structure on 16 × 16 coding unit is shown in left and the corresponding binary splitting sequence is (1, 0, 1, 1, 0). If we embed the secret information (1, 1), then the modified splitting flag would be (1, 0, 0, 1, 0). The actually modification on final transform splitting structure of 16 × 16 current coding unit is depicted in the right. In the end, the final transform spitting syntax elements will be encoded into bitstream by entropy coding.

4 Experimental Evaluation

The proposed adaptive transform block decision video steganography method is briefly evaluated in the reference software HM16.0 and is tested on multiple official video samples provided by JCT-VC. The tested video samples are encoded by 20 Intra-prediction frames and profile is set to main. In addition, the size of CTU, maximum and minimum size permitted in TU are configured to be 64, 32 and 4, respectively. The maximum transform splitting depth for intra is also confined to be 3. The performance of visual quality, embedding capacity and bit-rate increase for the proposed method are tested and analyzed as follows.

Figure 7 depicts the comparisons of visual quality between carrier video embedded secret information with the proposed method and video sample which just goes through the process of compressing on H.265/HEVC and with nothing to embed. Figure 8 provides values of PSNRs referred to luma, and two chroma components, respectively. The average discrepancy of PSNR among them is 0.28 dB, 0.0258 dB and 0.0075 dB in the luma, chroma_Cb and chroma_Cr, respectively. It can be seen the proposed method has maintained a high visual quality when the secret information is embedded into transform block decision.

a. Proposed Method

b. No embedding

Fig. 7. Comparisons for visual quality between proposed method and No embedding video samples.

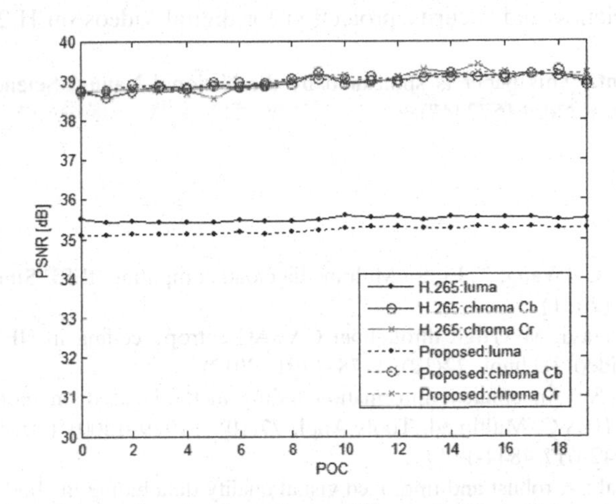

Fig. 8. Comparisons for visual quality at different POCs in the range of 0–20.

Table 1. Performance of the proposed method

Video sequence	PSNR	Proposed method			[4]		
		PSNR'	Capacity (bits)	Bit-rate increase	PSNR'	Capacity (bits)	Bit-rate increase
BasketballPass	35.50	35.22	10103	4.70%	35.07	3264	1.48%
keiba	34.76	34.48	11882	5.10%	32.55	2069	1.03%
BQSquare	42.21	39.28	9271	1.25%	38.18	10469	3.18%
BasketballDrill	41.64	40.72	38176	1.10%	40.16	5973	0.91%

Four tested video sequences are used to evaluate PSNR, embedding capacity and bit-rate increase with the proposed method and [4] in Table 1. It is noted that PSNR depicts the visual quality between the original video samples and the decoded video samples which are achieved by the encoding and decoding process. Likewise, *PSNR'* depicts the visual quality between the original video samples and the decoded video samples which carries secret information inside. It can be seen from Table 1 that the proposed method can achieve a high visual quality and sufficient embedding capacity.

5 Conclusion

In this paper, a novel and effective video steganography method based on transform block decision is proposed for video security protection. The proposed method mainly utilize the transform splitting structure of 16 × 16 blocks (coding units or transform units) to embed secret information. Effective embedding and extraction mapping rules are also established between transform splitting flag and secret information sequences. The experimental results show that the proposed method can effectively improve the embedding efficiency and security protection for digital videos on H.25/HEVC.

Acknowledgement. This paper is sponsored by the National Natural Science Foundation of China (NSFC, Grant No. 61572447).

References

1. Zhu, W., Luo, C., Wang, J., Li, S.: Multimedia cloud computing. IEEE Signal Process. Mag. **23**(3), 59–69 (2011)
2. Sze, V., Budagavi, M.: High throughput CABAC entropy coding in HEVC. IEEE Trans. Circ. Syst. Video Technol. **22**(12), 1778–1791 (2012)
3. Yang, J., Li, S.: An efficient information hiding method based on motion vector space encoding for HEVC. Multimed. Tools Appl. **77**(10), 11979–12001 (2017). https://doi.org/10.1007/s11042-017-4844-1
4. Liu, Y.X., et al.: A robust and improved visual quality data hiding method for HEVC. IEEE Access **6**, 53984–53987 (2018)
5. Kumar, M., Agrawal, S.: Reversible data hiding based on prediction error expansion using adjacent pixels. Secur. Commun. Netw. **9**(16), 3703–3712 (2016)
6. Rad, R.M., Wong, K., Guo, J.-M.: Reversible data hiding by adaptive group modification on histogram of prediction errors. Signal Process. **125**(C), 315–328 (2016)
7. Mstafa, R.J., Elleithy, K.M., Abdelfattah, E.: A robust and secure video steganography method in DWT-DCT domains based on multiple object tracking and ECC. IEEE Access **PP**(99), 1 (2017)
8. Liu, Y., Li, Z., Ma, X., Liu, J.: A robust without intra-frame distortion drift data hiding algorithm based on H.264/AVC. Multimed. Tools Appl. **72**(1), 613–636 (2013). https://doi.org/10.1007/s11042-013-1393-0
9. Swati, S., Hayat, K., Shahid, Z.: A watermarking scheme for high efficiency video coding (HEVC). PLoS ONE **9**(8), e105613 (2014). https://doi.org/10.1371/journal.Pone.0105613
10. Shanableh, T.: Data embedding in HEVC video by modifying the partitioning of coding units. IET Image Process. (2019). https://doi.org/10.1049/iet-ipr.2018.5782
11. Tew, Y.Q., Wong, K.: Information hiding in HEVC standard using adaptive coding block size decision. In: IEEE International Conference on Image Processing, pp. 5502–5506 (2015)

Towards a Universal Steganalyser Using Convolutional Neural Networks

Inas Jawad Kadhim[1,2(✉)], Prashan Premaratne[1], Peter James Vial[1], Osamah M. Al-Qershi[3], and Qasim Al-Shebani[1]

[1] School of Electrical and Computer and Telecommunications Engineering, University of Wollongong, North Wollongong, NSW 2522, Australia
ijk720@uowmail.edu.au

[2] Electrical Engineering Technical College, Middle Technical University, Baghdad, Iraq

[3] Faculty of Information Technology, Monash University, Melbourne, Australia

Abstract. A universal steganalyser has been the goal of many research leading to some good trials. Such steganalysers relied on machine learning and a wide range of features that can be extracted from images. However, increasing the dimensionality of the extracted features leads to the rapid rise in the complexity of algorithms. In recent years, some studies have indicated that well-designed convolutional neural networks (CNN) can achieve comparable performance to the two-step machine learning approaches. This paper aims to investigate different CNN architectures and diverse training strategies to propose a universal steganalysis model that can detect the presence of secret data in a colour stego-image. Since the detection of a stego-image can be considered as a classification problem, a CNN-based classifier has been proposed here. The experimental results of the proposed approach proved the efficiency in the main aspects of image steganography compared with the current state-of-the-art methods. However, a universal steganalysis is still unachievable, and more work should be done in this field.

Keywords: Steganalysis · Steganography · Convolutional neural networks · Deep learning

1 Introduction

The exponential growth in steganographic techniques in recent years has led the research community to focus on reliable steganalysis techniques. The need to hide secret information has led to Steganography where a cover image is used as the vessel to hide the information. The goal here is to leave only a minuscule change in the cover so that the stego-image is close to the cover in terms of visual quality and statistical characteristics. Steganalysis is a probing measure to extract the embedded secret information or discover the presence of confidential information from a stego-media without any prior information about the steganographic method being used [1]. Hence Kerckhoff's principle is not applicable, and it is considered as the biggest challenge in designing steganalysis techniques [1]. Generally, most of the steganographic systems

D.-S. Huang and P. Premaratne (Eds.): ICIC 2020, LNAI 12465, pp. 611–623, 2020.
https://doi.org/10.1007/978-3-030-60796-8_53

often leave some traces of secret information or distortions in the stego-image. Steganalysis makes use of these distortions to detect the presence of hidden data by identifying such distortions even though it may not be detectable by manual analysis. Steganalysis is a type of pattern recognition approach as it decides whether the given media belongs to either clean (without any secret data) or stego (with secret data). Signal processing and machine learning theory are used by steganalysis to analyse the statistical differences between cover and stego-image. Detection accuracy is enhanced by updating the number of features which leads to improved performance of the classifier. There are two main categories in steganalysis: active and passive [2]. In passive steganalysis, the primary goal is to check whether the given media is embedded with secret data or not. In active steganalysis, estimation of embedded secret data and/or its retrieval is essential. Since state-of-the-art steganographic systems use robust encryption mechanisms, the extraction of actual embedded data may be impossible. However, the latest steganalysis approaches succeeded in estimating embedding parameters, location, nature and size of the embedded secret information [3]. Steganalysis based on convolutional neural network (CNN) possesses superior accuracy while compared with traditional steganalysis approaches. CNN models can predict whether an image is embedded with some secret information or not. The goal of this paper is to propose a universal CNN-based steganalysis tool that can accurately detect stego-images. To make the model universal, the training dataset is embedded using eight different embedding techniques.

This paper is organized as follows. Section 2 presents the literature review of steganalysis. Section 3 describes the proposed methodology along with the results. Discussion of results and conclusion are given in Sect. 4.

2 Literature Review

During the past eight years, steganalysis approaches are mainly carried out by calculating a rich model (RM) [4] followed by classification by ensemble classier (EC) [5]. The first CNN-based steganalysis system was proposed by Qian *et al.* in 2015 as results of two-step approaches (EC + RM) [6]. Since then, many research articles mentioned the possibility of getting high detection performance in steganalysis approaches such as spatial, quantitative, side-informed, and JPEG steganalysis, etc. Xu *et al.* [7] got close results to the state-of-the-art conventional steganalysis using an ensemble CNN model. The Xu-Net CNN [8] was used as a base learner of the ensemble of CNNs. For JPEG steganalysis, Zeng *et al.* [9] introduced a pre-processing approach inspired from the RM and the use of a big database for learning which gives fair results while comparing the state-of-the-art approaches. Inspired from the ResNet [10], Xu proposed the Xu-Net-JPEG CNN with twenty layers and short residual connections, which resulted in higher accuracies. [11]. Later, Huang *et al.* [12] proposed the ResDer CNN, which is a variant of Xu-Net-JPEG, and obtained better results. Even though these results were highly encouraging, they were not significantly better than the results of the classical methods [13]. In most of the approaches, the design or the experimental effort was costly for minimal performance improvement in comparison to the networks such as AlexNet, VGG16, GoogleNet, ResNet, etc., that inspired those researches. By the end of 2017, and in 2018, researchers have strongly focused on spatial steganalysis as we can see in Ye-Net [14], Yedroudj-Net [15, 16], ReST-Net [17], SRNet [18]. However, Yedroudj-Net has the advantage of the small network and can learn from the small dataset, and it outperformed other related architectures. The model works well without any augmentation or transfer learning [13]. At the same time, ReST-Net uses a very large dataset with multiple sub-networks and various pre-processing filter banks. SRNet is another CNN model adapted to spatial or JPEG steganalysis. The main drawback of SRNet while comparing Yedroudj-Net is that it needs transfer learning as well as augmentation support to provide better results.

Although, CNN steganalysis models have been proposed so far, most of them optimized and trained based on one or two embedding methods. It means that those models could fail in detecting stego-images which embedded using different embedding technique. In other words, those CNN steganalysis models are not universal.

3 Proposed Method

The goal of this research is to develop a steganalysis system based on CNN that can be used to detect a wide range of embedding methods. For this purpose, two different ways to build the CNN are investigated; from scratch and from using transfer learning. The previous section demonstrated that steganalysis systems are developed based on one type of embedding techniques, which means that the steganalysis may not detect stego-images if they are embedded using a different embedding method. To avoid that issue and to make the system more universal, the training dataset is prepared by embedding the data using a wide range of embedding techniques which are the classical least significant bit (LSB), pixel value difference and modulus function (MF-PVD) [19], discrete cosine transform (DCT) with quad-tree adaptive-region approach [20], discrete wavelet transform (DWT) with minimized distortion [21], DWT with diamond encoding (DWT-DE) [22], dual-tree complex wavelet transform (DT-CWT) [23], DT-CWT with machine learning (DT-CWT-*k*-NN) [24], and DT-CWT with super-pixel (DT-CWT-SP) [25]. The reader can refer to those papers for more details. Two ways to build CNN and preparing dataset are described in the following sub-sections.

3.1 Preparing the Dataset

The dataset that has been used in this experiment is VOC2012 dataset [26] which contains 17,125 images. This dataset has been chosen because the research needs an extensive training set. So, to make it even larger, the images are divided into sub-images of size 128 × 128 and the final dataset contains 105,234 images. Only 100,000 are in the experiment. Then the 100,000-image dataset is divided randomly into eight groups where each group is embedded using one of the eight embedding methods. The final training set contains 100,000 embedded images and 100,000 original images.

3.2 Building CNN-Based Steganalysis from Scratch

To train a CNN from scratch, three different architectures are investigated. The first one is based on the network proposed in [6], the second is based on the work done by Pibre *et al.* [27], and the third is based on large convolution filters, which were proposed in [28].

3.2.1 First CNN Model

Qian *et al.* [6] proposed a network, which is illustrated in Fig. 1.

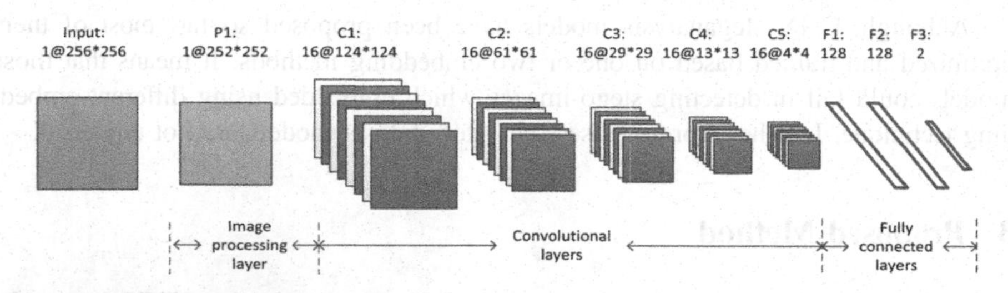

Fig. 1. Convolutional Neural Network. The form "a@b * b" means the number of feature maps a and resolution b * b of the corresponding layer [6]

The initial layer uses an image processing filtering which uses a pre-defined high-pass filter, and its coefficients remain the same during the training process. This helps to focus more on the high-frequency stego-noise (created during embedding) in the low-frequency stego image. This is a commonly adopted pre-processing tool in most of the steganalysis approaches as mentioned in the previous section. The filter that has been used is shown below.

$$k = \begin{bmatrix} -1 & 2 & -2 & 2 & -1 \\ 2 & -6 & 8 & -6 & 2 \\ -2 & 8 & -12 & 8 & -2 \\ 2 & -6 & 8 & -6 & 2 \\ -1 & 2 & -2 & 2 & -1 \end{bmatrix} \quad (1)$$

After the pre-processing step, there are five convolutional layers with 16 filters of the sizes 5×5, 3×3, 3×3, 3×3, and 5×5 respectively. The convolutional layers are followed by two fully connected layers of 128 neurons each. Qian's network is adopted in two scenarios. In the first scenario, two changes are made to Qian's network in order to train it using the prepared dataset. First, the size of the input layer is changed to 128×128 to match the size of the images. Second, the size of the filters should be changed to $5 \times 5 \times 3$, $3 \times 3 \times 3$, $3 \times 3 \times 3$, $3 \times 3 \times 3$, and $5 \times 5 \times 3$ to match the three channels of the RGB images used. In the second scenario, ensemble learning or model ensembling is used. Since RGB images have three channels; R, G and B, the proposed ensemble consists of three CNN's, and the predictions are combined using majority voting as is illustrated in Fig. 2.

3.2.2 Second CNN Model

Pibre *et al.* [27] tested different architectures trying to minimize the probabilities of error as they train the CNN using images embedded with S-UNIWARD [29] embedding method. The most efficient network they obtained uses only two convolution layers, followed by three fully connected dense layers. Figure 3 illustrates the architecture. The input images use the size of 256 × 256, and are filtered using the same filter as used in Qian et al. [6]. Since having a size of 5 × 5, the filtered feature map will possess a size of 252 × 252. In the first convolutional layer, 64 filters are used with a size of 7 × 7 and the second filter uses 16 filters of size 5 × 5. The second convolutional layer is followed by two fully connected layers of 1,000 neurons each.

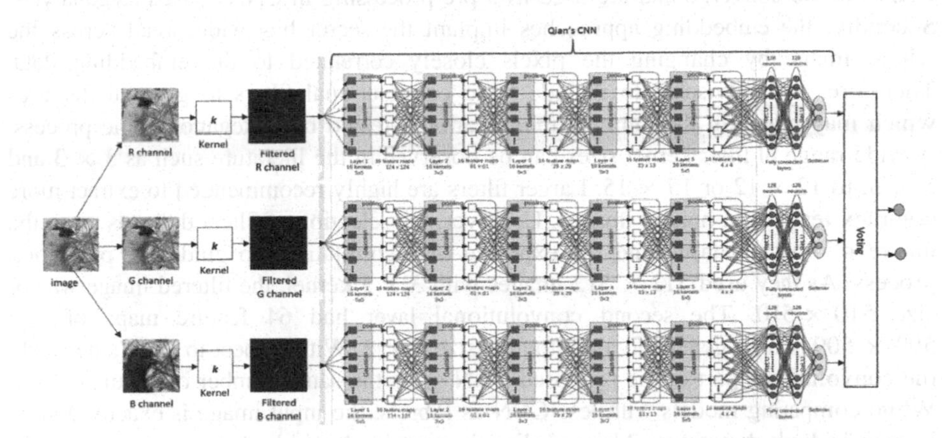

Fig. 2. Scenario 2 Ensemble of CNN's based on Qian's CNN.

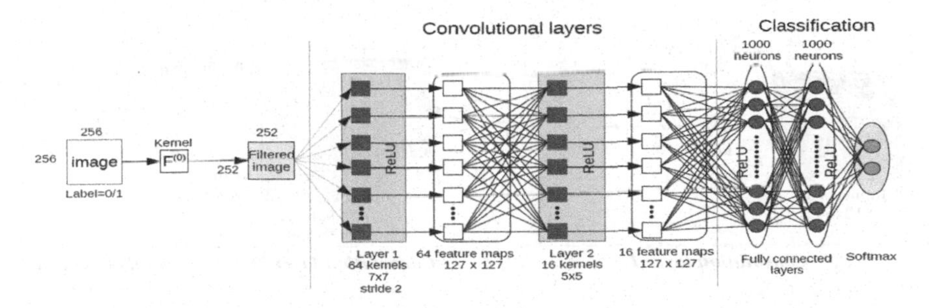

Fig. 3. The CNN architecture [27].

Again, this network is adopted in this paper for two scenarios. In the first scenario, two changes are made to the network in order to train it using the prepared dataset. First, the size of the input layer should be changed to 128 × 128 to match the size of the images. Second, the size of the filters should be changed to 7 × 7 × 3 and 5 × 5 × 3, to match the 3 channels of the RGB images used. In the second scenario, ensemble learning or model ensembling is used. Since RGB images have three channels; R, G and B, the proposed ensemble consists of three CNN's, and the predictions are combined using majority voting.

3.2.3 Third CNN Model Based on Large Convolution Filters

Salomon *et al.* [28] proposed a convolutional neural network based on the following considerations. Firstly, their proposed method suggested that CNN can learn kernel K instead of using a fixed value for K. However, in the previous two networks proposed by Qian *et al.* and Pibre et al., there was no solid proof to state the optimality of the kernel K that filter the input image was more or less the same as that of an edge detection filter. Pibre *et al.* observed it empirically that without the high-pass filter, the CNN fails to converge and are used as a pre-processing in CNNs based steganalysis. Secondly, the embedding approaches implant the secret bits widespread across the whole image by changing the pixels closely correlated to the embedding data. Therefore, it is advisable to include large convolutional filters to generate features which may highlight the minimal modifications created by a steganographic process. A wide range of filter kernel sizes can be observed in the literature such as 3×3 and 5×5, to 12×12 or 15×15. Larger filters are highly recommended to extract more complex features in natural images. However, the selection of filters depends upon the nature of image dataset, and the expected data correlations to guide the prediction process. As they used 512×512 images with 3×3 kernel, the filtered image was of size 510×510. The second convolutional layer had 64 feature maps of size 509×509, which generated an output of size 2×2. With respect to Qian's network, the convolutional part of the proposed CNN gives the same number of features (256). While comparing models with less CNN as Pibre's, the input image is exactly double in size in both directions. Also, pooling is integrated into both layers to reduce the computation cost. The final fully connected layers use classical ANN layers, and the output layer consist of two softmax neurons as shown in Fig. 4.

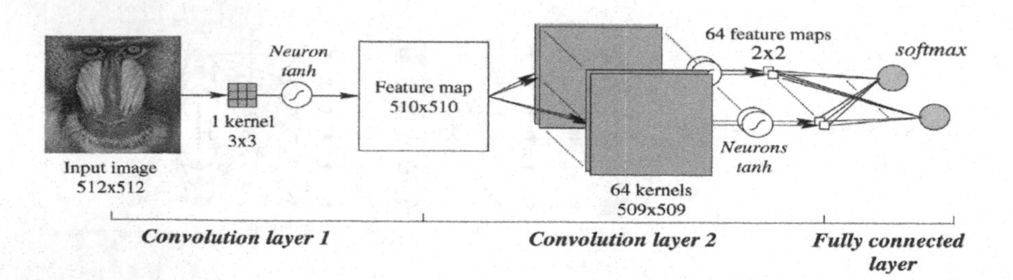

Fig. 4. CNN architecture [28].

This is the main difference in comparison with CNN models in the previous approaches of Qiam et al. and Pibre et al. Although, Salomon et al. used this shallow fully connected network without any hidden layers, their network was able to achieve the classification task and detect stego-images successfully. As described earlier, this network is adopted in this paper in two scenarios. In the first scenario, two changes should be done to the network in order to train it using the prepared dataset. First, the size of the first convolutional layers should be changed to $3 \times 3 \times 3$ as RGB images are used. Second, the second convolutional layer should be changed to $125 \times 125 \times 3$ to match the size of the images. In the second scenario, ensemble learning or model ensembling is used. Since RGB images have three channels; R, G and B, the proposed ensemble consists of three CNN's, and the predictions are combined using majority voting.

3.3 Building CNN-Based Steganalysis Using Transfer Learning

Transfer learning is used for reducing the training time, and for this purpose, the Alexnet pretrained network was used [30]. This network has outstanding performance and has been used in many applications.

3.4 Training the Networks

Networks based on all architectures have been trained using the same dataset, and the experiments are performed with MATLAB Deep Learning Toolbox on a cloud-based virtual machine with an NVIDIA Tesla P4 GPU card. After months of investigations and tuning and changing the parameters such as the strides and the learning rate, only the CNN based on the pretrained Alexnet gave outstanding results. The other networks could not converge in some cases or suffered from overfitting with low validation accuracy in other cases as shown in Fig. 5.

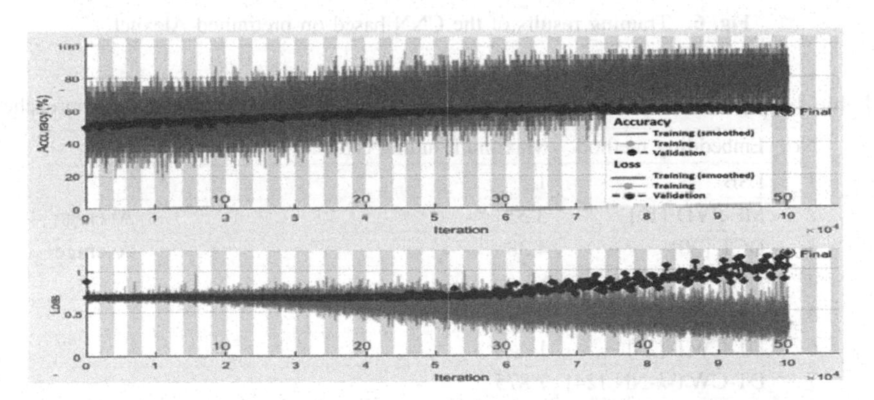

Fig. 5. Training results of the CNN based on Quin's network with overfitting.

After 200 iterations, the results of training CNN-based steganalysis using pretrained Alexnet is shown in Fig. 6 with a validation accuracy of 91.2% and a training accuracy of 94.47%. From Fig. 6, the pretrained Alexnet showed good performance; therefore, in this paper, we use the pretrained AlexNet to develop a universal steganalysis model.

3.5 CNN-Based Steganalysis Using Pretrained Alexnet Network

In order to validate the performance of the developed model for steganalysis, we ran experiments on eight state-of-the-arts spatial and transform domain steganographic approaches that are mentioned in Sect. 3. All of them are implemented and used to embed random data with different hiding capacities. To test the trained CNN model, we used a different dataset of images to the ones used for training. For this purpose, we used RGB-BMP Steganalysis Dataset [31], which has 1,500 images of size 512 × 512. The images were divided into sub-images of size 128 × 128 and the final testing dataset has 24,000. It is worth mentioning that we used the maximum available capacity during the embedding. The different hiding capacities are shown in Table 1.

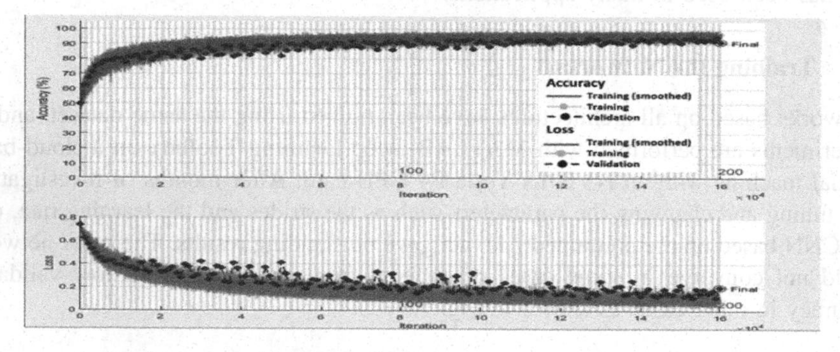

Fig. 6. Training results of the CNN based on pretrained Alexnet.

Table 1. Hiding capacities used different state-of-the-art steganography based approaches.

No.	Embedding method	Maximum hiding capacity (bit per pixel)	Notes
1	LSB	1	
2	MF-PVD [19]	3.33	Average
3	DCT [20]	4.38	Average
4	DWT [21]	8	
5	DWT-DE [22]	2.5	
6	DT-CWT [23]	1	
7	DT-CWT-*k*-NN [24]	7.875	
8	DT-CWT-SP [25]	31.2	

Since the original networks tested in this paper are not available online, and also they could not converge during training, it is impossible to compare them with Alexnet network. So, three well-known steganalysis systems are used for comparison purposes which are subtractive pixel adjacency matrix (SPAM) [32], quantitative steganalysis using rich models (QSRM) [33] and steganalysis residual network (SRNet) [18] which are available online. SPAM was developed to detect spatial steganography by merging a low-amplitude individual stego-signal, for example, LSB matching. The QSRM method is used to predict payload in the stego-image, which turns the detection process into a regression rather than classification. Since the proposed steganalysis model works as a classifier, the output of the QSRM should be converted into a classification output-like for an appropriate comparison. This is achieved by converting the output of the QSRM into 'True' or "False' based on the prediction error. In other words, if the difference between the predicted payload and the actual payload is less than 5% of the actual payload, then the prediction is correct or 'True' and vice versa. The accuracy of the QSRM then is calculated based on the number of the correct predictions. The comparison between Alexnet and three well-known steganalysis in terms of accuracy of detection is shown in Table 2.

Table 2. Accuracy of the proposed steganalysis network.

Embedding method	Accuracy of detection (%)			
	Pretrained Alexnet	SPAM [32]	QSRM [33]	SRNet [18]
LSB	72.48	89.65	63.67	**89.77**
MF-PVD	**84.94**	78.37	69.79	82.08
DCT	74.15	61.33	36.08	**89.64**
DWT	**67.84**	65.91	67.84	67.40
DWT-DE	**78.29**	76.64	37.24	76.82
DT-CWT	**40.96**	38.12	44.38	40.24
DT-CWT-*k*-NN	**35.56**	26.12	29.34	30.97
DT-CWT-SP	**30.3**	18.71	26.42	28.13

Table 3. Steganalysis error probability comparison at 0.2 and 0.4 bpp.

Steganalysis method	Error (%)			
	WOW		S-UNIWARD	
	0.2	0.4	0.2	0.4
Qian *et al.* [6]	–	29.3	–	30.9
Pibre *et al.* [27]	–	–	–	7.4
Salomon *et al.* [28]	–	4.6	–	–
SPAM [35]	36.5	25.5	36.6	24.7
Yedroudj-Net [16]	27.8	14.1	36.7	22.8
Xu-Net [8]	32.4	20.7	39.1	27.2
Ye-Net [14]	33.1	23.2	40.0	31.2
Pretrained Alexnet	49.89	49.84	49.93	49.91

In addition, for comparison with some of the other networks mentioned earlier and for a fair comparison, we used the same dataset of images from the BOSSBase database v.1.01 [34], and the same embedding methods S-UNIWARD [29], and WOW [35]. The results are shown in Table 3. All the results in the table are reported in the corresponding papers.

4 Discussion and Conclusion

In this paper, the possibility of developing a universal CNN-based image steganalysis is explored. Different CNN architectures and scenarios are investigated to try to come up with the steganalysis model that can detect stego-images regardless of the embedding technique or domain. A big dataset was used in this experiment, and the images are embedded using eight embedding methods.

The extensive experiments show that training the universal CNN-based image steganalysis from scratch is a very complex task. Sometimes the network could not converge, and sometimes it suffered from overfitting. Although 200,000 images have been used for training the networks, they are found to be insufficient for the task, and much more images are needed. However, it is worth mentioning that the original CNNs, which are investigated in this paper, are optimized for certain embedding technique. This explains the paradox between their good performance when they are used to detect a single steganographic method and their poor performance during training experiments in this paper.

In contrast, the pretrained Alexnet showed good performance with a training accuracy of 94.47% and a validation accuracy of 91.2%. However, when it was compared with other networks as in Table 3, the pretrained Alexnet network showed poor performance. The reason behind that is the fact that the network was not trained using images embedded by the S-UNIWARD and WOW methods. This emphasizes that a universal steganalysis is still hard to achieve even if it was trained using different types of embedding methods. However, when the network was tested using images embedded using the eight embedding methods; the network showed some good results, especially when images embedded using MF-PVD or DWT-DE as presented in Table 2. However, it did not perform so well with images embedded using DT-CWT-*k*-NN or DT-CWT-SP. This vast difference is to the nature of the embedding algorithms and the artefacts they created in the stego-images.

Detecting the stego-image using a classification model implies that it is quite easy to distinguish between the cover image and the stego-image. In fact, it is not that easy because the changes that were added to the image during embedding process could be random or hard to patternize using the CNN especially when the payload is small. In addition to the very small differences between cover and stego-images, utilizing different advanced embedding methods cause very different patterns of changes between cove and stego-images, which makes the detection even more difficult.

The proposed model showed the ability to outperform the SPAM and QSRM in detecting stego-images in 7 out of 8 embedding method and the ability to outperform the SRNet in detecting stego-images in 6 out of 8 embedding method. However, it has own limitations. For example, its accuracy of detection is not consistent for all embedding methods as the accuracy range from 30.3% to 84.94%. In addition, it cannot be generalized and used to detect stego-images that have been embedded using methods rather than the 8 methods used in preparing the training dataset, i.e. WOW and S-UNIWARD.

The modern blind steganalysis techniques are not universal in the sense that their performance mostly depends on the cover images as well as the embedding approaches used. As a result, further research should be concentrated on developing a real universal image steganalysis approach by considering a much wider range of hiding techniques with a much larger dataset.

References

1. Holotyak, T., Fridrich, J., Voloshynovskiy, S.: Blind statistical steganalysis of additive steganography using wavelet higher order statistics. In: Dittmann, J., Katzenbeisser, S., Uhl, A. (eds.) CMS 2005. LNCS, vol. 3677, pp. 273–274. Springer, Heidelberg (2005). https://doi.org/10.1007/11552055_31
2. Nissar, A., Mir, A.H.: Classification of steganalysis techniques: a study. Digit. Signal Process. A Rev. J. **20**(6), 1758–1770 (2010)
3. Chutani, S., Goyal, A.: A review of forensic approaches to digital image Steganalysis. Multimed. Tools Appl. **78**(13), 18169–18204 (2019). https://doi.org/10.1007/s11042-019-7217-0
4. Fridrich, J., Kodovsky, J.: Rich models for steganalysis of digital images. IEEE Trans. Inf. Forensics Secur. **7**(3), 868–882 (2012)
5. Kodovsky, J., Fridrich, J., Holub, V.: Ensemble classifiers for steganalysis of digital media. IEEE Trans. Inf. Forensics Secur. **7**(2), 432–444 (2012)
6. Qian, Y., Dong, J., Wang, W., Tan, T.: Deep learning for steganalysis via convolutional neural networks. In: Proceedings of Media Watermarking, Security, and Forensics, vol. 9409, p. 94090J (2015)
7. Xu, G., Wu, H.-Z., Shi, Y.Q.: Ensemble of CNNs for steganalysis: an empirical study. In: Proceedings of the 4th ACM Workshop on Information Hiding and Multimedia Security, pp. 103–107 (2016)
8. Xu, G., Wu, H.-Z., Shi, Y.-Q.: Structural design of convolutional neural networks for steganalysis. IEEE Signal Process. Lett. **23**(5), 708–712 (2016)
9. Zeng, J., Tan, S., Li, B., Huang, J.: Large-scale JPEG image steganalysis using hybrid deep-learning framework. IEEE Trans. Inf. Forensics Secur. **13**(5), 1200–1214 (2017)
10. He, K., Zhang, X., Ren, S., Sun, J.: Deep residual learning for image recognition. In: Proceedings of the IEEE Conference on Computer Vision and Pattern Recognition, pp. 770–778 (2016)
11. Xu, G.: Deep convolutional neural network to detect J-UNIWARD. In: Proceedings of the 5th ACM Workshop on Information Hiding and Multimedia Security, pp. 67–73 (2017)
12. Huang, X., Wang, S., Sun, T., Liu, G., Lin, X.: Steganalysis of adaptive JPEG steganography based on ResDet. In: Proceedings of the Asia-Pacific Signal and Information Processing Association Annual Summit and Conference (APSIPA 2018), pp. 549–553 (2018)
13. Chaumont, M.: Deep learning in steganography and steganalysis from 2015 to 2018. In: Digital Media Steganography: Principles, Algorithms, Advances, pp. 1–39 (2020)
14. Ye, J., Ni, J., Yi, Y.: Deep learning hierarchical representations for image steganalysis. IEEE Trans. Inf. Forensics Secur. **12**(11), 2545–2557 (2017)
15. Yedroudj, M., Chaumont, M., Comby, F.: How to augment a small learning set for improving the performances of a CNN-based steganalyzer?. Electron. Imaging **2018**(7), 317-1–317-7 (2018)

16. Yedroudj, M., Comby, F., Chaumont, M.: Yedroudj-Net: an efficient CNN for spatial steganalysis. In: Proceedings of the IEEE International Conference on Acoustics, Speech and Signal Processing (ICASSP 2018), pp. 2092–2096 (2018)
17. Li, B., Wei, W., Ferreira, A., Tan, S.: ReST-Net: diverse activation modules and parallel subnets-based CNN for spatial image steganalysis. IEEE Signal Process. Lett. **25**(5), 650–654 (2018)
18. Boroumand, M., Chen, M., Fridrich, J.: Deep residual network for steganalysis of digital images. IEEE Trans. Inf. Forensics Secur. **14**(5), 1181–1193 (2018)
19. Shen, S., Huang, L., Tian, Q.: A novel data hiding for color images based on pixel value difference and modulus function. Multimed. Tools Appl. **74**(3), 707–728 (2015). https://doi.org/10.1007/s11042-014-2016-0
20. Rabie, T., Kamel, I.: Toward optimal embedding capacity for transform domain steganography: a quad-tree adaptive-region approach. Multimed. Tools Appl. **76**(6), 8627–8650 (2017). https://doi.org/10.1007/s11042-016-3501-4
21. Kumar, V., Kumar, D.: A modified DWT-based image steganography technique. Multimed. Tools Appl. **77**(11), 13279–13308 (2018). https://doi.org/10.1007/s11042-017-4947-8
22. Atawneh, S., Almomani, A., Al Bazar, H., Sumari, P., Gupta, B.: Secure and imperceptible digital image steganographic algorithm based on diamond encoding in DWT domain. Multimed. Tools Appl. **76**(18), 18451–18472 (2017). https://doi.org/10.1007/s11042-016-3930-0
23. Kumar, S., Muttoo, S.K.: Data hiding techniques based on wavelet-like transform and complex wavelet transforms. In: Proceedings International Symposium on Intelligence Information Processing and Trusted Computing (IPTC 2010), pp. 1–4 (2010)
24. Kadhim, I.J., Premaratne, P., Vial, P.J.: High capacity adaptive image steganography with cover region selection dual-tree complex wavelet transform. Cogn. Syst. Res. **60**, 20–32 (2020)
25. Kadhim, I.J., Premaratne, P., Vial, P.J.: Improved image steganography based on super-pixel and coefficient-plane-selection. Sig. Process. **171**, 107481 (2020)
26. Everingham, M., Van Gool, L., Williams, C.K.I., Winn, J., Zisserman, A.: The pascal visual object classes (voc) challenge. Int. J. Comput. Vis. **88**(2), 303–338 (2010)
27. Pibre, L., Jérôme, P., Ienco, D., Chaumont, M.: Deep learning is a good steganalysis tool when embedding key is reused for different images, even if there is a cover source-mismatch. Electron. Imaging **2016**(8), 1–11 (2016)
28. Salomon, M., Couturier, R., Guyeux, C., Couchot, J.F., Bahi, J.M.: Steganalysis via a convolutional neural network using large convolution filters for embedding process with same stego key: a deep learning approach for telemedicine. Eur. Res. Telemed. **6**(2), 79–92 (2017)
29. Holub, V., Fridrich, J., Denemark, T.: Universal distortion function for steganography in an arbitrary domain. EURASIP J. Inf. Secur. **2014**(1), 1–13 (2014). https://doi.org/10.1186/1687-417X-2014-1
30. Krizhevsky, A., Sutskever, I., Hinton, G.E.: ImageNet classification with deep convolutional neural networks. In: Advances in Neural Information Processing Systems, pp. 1097–1105 (2012)
31. Al-Qershi, O.M., Khoo, B.E.: Evaluation of copy-move forgery detection: datasets and evaluation metrics. Multimed. Tools Appl. **77**(24), 31807–31833 (2018). https://doi.org/10.1007/s11042-018-6201-4

32. Pevny, T., Bas, P., Fridrich, J.: Steganalysis by subtractive pixel adjacency matrix. IEEE Trans. Inf. Forensics Secur. **5**(2), 215–224 (2010)
33. Kodovský, J., Fridrich, J.: Quantitative steganalysis using rich models. In: Media Watermarking, Security, and Forensics, vol. 8665, p. 866500 (2013)
34. Bas, P., Filler, T., Pevný, T.: "Break our steganographic system": the ins and outs of organizing BOSS. In: Filler, T., Pevný, T., Craver, S., Ker, A. (eds.) IH 2011. LNCS, vol. 6958, pp. 59–70. Springer, Heidelberg (2011). https://doi.org/10.1007/978-3-642-24178-9_5
35. Holub, V., Fridrich, J.: Designing steganographic distortion using directional filters. In: Proceedings of the IEEE International Workshop on Information Forensics and Security (WIFS), pp. 234–239 (2012)

A HEVC Steganography Method Based on QDCT Coefficient

Si Liu[1], Yunxia Liu[1](✉), Cong Feng[1], Hongguo Zhao[1], and Yu Huang[2]

[1] College of Information Science and Technology, Zhengzhou Normal University, Zhengzhou, China
liuyunxia0110@hust.edu.cn
[2] Peking University, Beijing, China

Abstract. This paper presents a HEVC video steganography algorithm based on QDCT coefficient without intra-frame distortion drift. We embed the secret message into the multi-coefficients of the selected 8×8 luminance QDCT blocks which meet our conditions to avert the distortion drift. With the use of three multi-coefficients in each row or column of the 8×8 luminance QDCT block, the embedded capacity is considerably larger than using single multi-coefficients. The experimental results show that this video steganography algorithm can effectively avert intra-frame distortion drift and get good visual quality.

Keywords: HEVC · Video steganography · QDCT · Multi-coefficients · Intra-frame distortion drift

1 Introduction

Video steganography is a branch of data hiding, which is a technique that embeds message into cover contents and is used in many fields such as medical systems, law enforcement, copyright protection and access control, etc. Since human visual system are less sensitive to the small changes of digital medias, especially for digital video, video steganography is a technique which hides message into a video and conceals the fact of the transmission. Video steganography techniques can be classified into compressed and uncompressed domains [1]. The steganography method in compressed domain has greater application value and has gained more attention because the video is usually transmitted or stored after compression coding. In the existing literatures, most of the steganography method based on compressed domain combining with certain aspects of the compression characteristics utilizes coding process to hide information, such as intra prediction, motion estimation, DCT/DST transform, etc., as shown in Fig. 1.

H.265/HEVC (high efficiency video coding) is the latest video coding standard published by ITUTVCEG and ISO/IEC MPEG [2]. HEVC's main achievement is its significant improvement in compression performance when compared to the previous state-of-the-art standard with at least 50% reduction in bitrate for producing video of similar perceptual quality [3], and it is well adapted for network transmission.

D.-S. Huang and P. Premaratne (Eds.): ICIC 2020, LNAI 12465, pp. 624–632, 2020.
https://doi.org/10.1007/978-3-030-60796-8_54

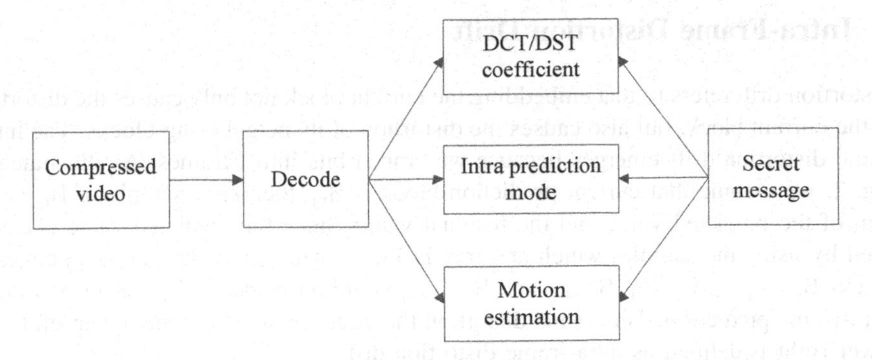

Fig. 1. Video steganography methods based on compressed domain

Since HEVC has been proposed for a short time and is very complicated, the steganography technique based on HEVC video coding standard is in the beginning stages.

Video steganography technique based on the QDCT coefficient blocks in intra-frame mode is one of the most popular compressed domain techniques adopted in H.264 [4–6]. However, distortion drift is a big problem of this steganography technique in HEVC video streams. Video steganography technique based on intra prediction [7] and motion vector [8] also have intra-frame or inter-frame distortion drift problems. Moreover, the scheme that embed the secret information into the QDCT/QDST coefficients of I frames for steganography is not suitable for HEVC video streams because of the intra-frame distortion drift. Thus, it is necessary to introduce a mechanism without intra-frame distortion drift to HEVC when data is hidden into the QDCT/QDST coefficients. Chang et al. [9] employed a three-coefficients to solve the QDST coefficient distortion drift problem for 4 4 luminance blocks. Gaj et al. [10] further proposed an improved watermarking scheme which is also robust against different noise addition and re-encoding attacks, but the embedding capacity is reduced. Liu et al. [11] proposed a robust and improved visual quality steganography method for HEVC in 4×4 luminance QDST blocks. To solve the QDCT coefficient distortion drift problem in HEVC, Chang et al. [12] proposed a paired-coefficients for 8×8 luminance QDCT blocks, But the single using of paired-coefficients will cause some obvious image hot pixels that greatly affects the visual quality of the video.

In this paper, we proposed a coefficient compensation rule for 8×8 luminance DCT blocks. According to this rule, more multi-coefficients can be obtained to compensate the intra-frame distortion drift, and with the use of multiple multi-coefficients, our steganography algorithm has better embedded capacity and visual quality.

The rest of the paper is organized as follows. Section II describes the intra-frame distortion drift. Section III describes the proposed algorithm. Experimental results are presented in Section IV and conclusions are in Section V.

2 Intra-Frame Distortion Drift

Distortion drift refers to that embedding the current block not only causes the distortion of the current block, but also causes the distortion of its neighboring blocks. The intra-frame distortion drift emerges because we embed bits into I frames. As illustrated in Fig. 2, we assume that current prediction block is $B_{i,j}$, then each sample of $B_{i,j}$ is the sum of the predicted value and the residual value. Since the predicted value is calculated by using the samples which are gray in Fig. 3. The embedding induced errors in blocks $B_{i-1,j-1}$, $B_{i,j-1}$, $B_{i-1,j}$, and $B_{i-1,j+1}$would propagate to $B_{i,j}$ because of using intra-frame prediction. This visual distortion that accumulates from the upper left to the lower right is defined as intra-frame distortion drift.

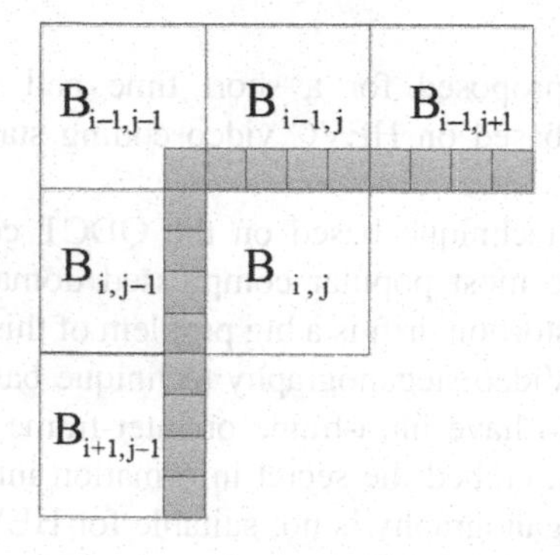

Fig. 2. The prediction block Bi,j and the adjacent encoded blocks

For convenience, we give several definitions, the 8 × 8 block on the right of the current block is defined as right-block; the 8 × 8 block under the current block is defined as under-block; the 8 × 8 block on the left of the under-block is defined as under-left-block; the 8 × 8 block on the right of the under-block is defined as under-right-block; the 8 × 8 block on the top of the right-block is defined as top-right-block, as shown in Fig. 3. The 8 × 8 block embedding induced errors transfer through the edge pixels to these five adjacent blocks.

3 Description of Algorithm Process

3.1 Embedding

According to the intra angular prediction modes of these five adjacent blocks, it can be judged that if the current block is embedded, whether the embedding error will be transmitted to the adjacent blocks by the intra-frame prediction process.

In other words, when the intra prediction mode of the five adjacent blocks satisfies certain conditions, if the embedding error just changed the other pixels of the current block instead of the edge pixels used for intra-frame angular prediction reference, then

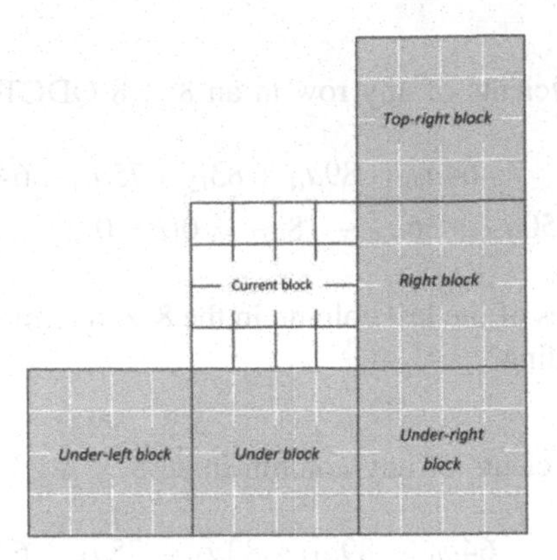

Fig. 3. Definition of adjacent blocks

the distortion drift can be avoided. We proposed two conditions to prevent the distortion drift specifically.

Condition 1: Right-mode$\in$ {2-25}, under-right-mode$\in$ {11-25}, top-right-mode $\in$ {2-9}
Condition 2: under-left-mode$\in$ {27-34}, under-mode$\in$ {11-34}.

If the current block meets Condition 1, the pixel values of the last column should not be changed in the following intra-frame prediction. If the current block meets Condition 2, the pixel values of the last row should not be changed in the following intra-frame prediction. If the current block meets the Condition 1 and 2 at the same time, the current block should not be embedded. If both the Condition 1 and 2 cannot be satisfied, the current block can be arbitrarily embedded where the induced errors won't transfer through the edge pixels to the five adjacent blocks, that means the distortion drift won't happen, but in this paper we don't discuss this situation, the current block should also not be embedded. We proposed some multi-coefficients can meet the above conditions when embedded in.

The multi-coefficients can be defined as two types. Type 1 is a four-coefficient combination (C_1, C_2, C_3, C_4), C_1 is used for bit embedding, and C_2, C_3, C_4 are used for distortion compensation. Type 2 is a paired-coefficient combination (C_1, C_2), C_1 is used for bit embedding, and C_2 is used for distortion compensation. There are three group multi-coefficients we used in this paper that applicable to 8×8 QDCT blocks, we can define them as follow:

VS(Vertical Set) = ($a_{i0} = 1$, $a_{i4} = -1$), ($a_{i6} = 1$, $a_{i7} = 2$), ($a_{i1} = 1$, $a_{i3} = -1$, $a_{i4} = 1$, $a_{i5} = 1$) ($i = 0,1,\ldots,7$)

HS(Horizontal Set) = ($a_{0j} = 1$, $a_{4j} = -1$), ($a_{6j} = 1$, $a_{7j} = 2$), ($a_{1j} = 1$, $a_{3j} = -1$, $a_{4j} = 1$, $a_{5j} = 1$) ($j = 0,1,\ldots,7$)

In fact, we found a coefficient compensation rule can create more usable multi-coefficients. We can define it as follow:

Vertical Rule:
If the embedding coefficients of any row in an 8 × 8 QDCT block meet

$$64a_{i0} - 89a_{i1} + 83_{i2} - 75a_{i3} + 64a_{i4}$$
$$-50a_{i5} + 36a_{i6} - 18a_{i7} = 0(i = 0, 1, \ldots, 7)$$

Then the pixel values of the last column in the 8 × 8 luminance block would not be changed by the embedding.

Horizontal Rule:
If the embedding coefficients of any column in an 8 × 8 DCT block meet

$$64a_{0j} - 89a_{1j} + 83a_{2j} - 75a_{3j} + 64a_{4j}$$
$$-50a_{5j} + 36a_{6j} - 18a_{7j} = 0\,(j = 0, 1, \ldots, 7)$$

Then the pixel values of the last row in the 8 × 8 luminance block would not be changed by the embedding.

As we can see, the multi-coefficients ($a_{i0} = 1, a_{i4} = -1$), ($a_{i6} = 1, a_{i7} = 2$), ($a_{i1} = 1, a_{i3} = -1, a_{i4} = 1, a_{i5} = 1$) ($i$ = 0,1,…,7) meet the Vertical Rule, the multi-coefficients ($a_{0j} = 1, a_{4j} = -1$), ($a_{6j} = 1, a_{7j} = 2$), ($a_{1j} = 1, a_{3j} = -1, a_{4j} = 1, a_{5j} = 1$) ($j$ = 0,1,…,7) meet the Horizontal Rule. According to the coefficient compensation rule, we can create many more coefficient combination to prevent the intra-frame distortion, but relatively speaking, these three multi-coefficients we used have better PSNR performance.

After the original video is entropy decoded, we get the intra-frame prediction modes and QDCT coefficients. We embed the secret data by the multi-coefficients into the 8 × 8 luminance DCT blocks of the selected frames which meet the conditions. Finally, all the QDCT coefficients are entropy encoded to get the target embedded video.

We can use ($a_{i0} = 1, a_{i4} = -1$), ($a_{i6} = 1, a_{i7} = 2$), ($a_{i1} = 1, a_{i3} = -1, a_{i4} = 1, a_{i5} = 1$) ($i$ = 0,1,…,7) to embed 3 bits in a row when current block meets Condition 1, we also can use ($a_{0j} = 1, a_{4j} = -1$), ($a_{6j} = 1, a_{7j} = 2$), ($a_{1j} = 1, a_{3j} = -1, a_{4j} = 1, a_{5j} = 1$) ($j$ = 0,1,…,7) to embed 3 bits in a column when current block meets Condition 2.

For simplicity, we refer to ($a_{i0} = 1, a_{i4} = -1$) and ($a_{0j} = 1, a_{4j} = -1$) as (1, −1) mode, refer to ($a_{i6} = 1, a_{i7} = 2$) and ($a_{6j} = 1, a_{7j} = 2$) as (1, 2) mode, refer to ($a_{i1} = 1, a_{i3} = -1, a_{i4} = 1, a_{i5} = 1$) and ($a_{1j} = 1, a_{3j} = -1, a_{4j} = 1, a_{5j} = 1$) as (1, −1, 1, 1) mode. (i, j = 0,1,…,7).

Assume (a_1, a_2, a_3, a_4) is the selected QDCT coefficients to be embedded, where a_1 is used to hide information, a_2, a_3, a_4 are used to compensate the intra-frame distortion. We take the multi-coefficients (1, −1, 1, 1) mode as an example.

(1) If the embedded bit is 1, a_1, a_2, a_3, a_4 are modified as follows:

If a_1 **mod2** = 0, then $a_1 = a_1 + 1$, $a_2 = a_2 - 1$, $a_3 = a_3 + 1$, $a_4 = a_4 + 1$. If a_1 **mod2** $\neq 0$, then $a_1 = a_1$, $a_2 = a_2$, $a_3 = a_3$, $a_4 = a_4$.

(2) If the embedded *bit* is 0, a_1, a_2, a_3, a_4 are modified as follows:

If a_1 **mod2** $\neq 0$, then $a_1 = a_1 + 1$, $a_2 = a_2 - 1$, $a_3 = a_3 + 1$, $a_4 = a_4 + 1$. If a_1 **mod2** = 0, then $a_1 = a_1$, $a_2 = a_2$, $a_3 = a_3$, $a_4 = a_4$.

The (1, −1) mode and (1, 2) mode are similar to (1, −1, 1, 1) mode, but simpler, because they have only 1 compensation coefficients.

3.2 Data Extraction and Restoration

After entropy decoding of the HEVC, we choose the embeddable blocks of one frame and decode the embedded data. Then, we extract the hidden data M as follows, $(i = 0, 1, \ldots, 7)$:

$$M = \begin{cases} 1 & \text{if} \quad \tilde{Y}_{i0} \bmod 2 = 1 \quad \text{and} \quad \text{current} \quad \text{block} \quad \text{meet} \quad \text{condition} \quad 1 \\ 0 & \text{if} \quad \tilde{Y}_{i0} \bmod 2 = 0 \quad \text{and} \quad \text{current} \quad \text{block} \quad \text{meet} \quad \text{condition} \quad 1 \end{cases}$$

$$M = \begin{cases} 1 & \text{if} \quad \tilde{Y}_{0i} \bmod 2 = 1 \quad \text{and} \quad \text{current} \quad \text{block} \quad \text{meet} \quad \text{condition} \quad 2 \\ 0 & \text{if} \quad \tilde{Y}_{0i} \bmod 2 = 0 \quad \text{and} \quad \text{current} \quad \text{block} \quad \text{meet} \quad \text{condition} \quad 2 \end{cases}$$

4 Case Study

The proposed method has been implemented in the HEVC reference software version HM16.0. In this paper we take "Keiba" (416*240), "Container" (176*144) "Akiyo" (176*144), "SlideShow" (1280*720) and "ParkScene" (1920*1080) as test video. The GOP size is set to 1 and the values of QP (Quantization Parameter) are set to be 16, 24, 32 and 40. Since we can embed 24 bits information in one 8 × 8 luminance QDCT block with the simultaneous use of multi-coefficients (1, −1) mode, (1, 2) mode and (1, −1, 1, 1) mode, The method in [12] only using (1, −1) mode is used for performance comparisons.

As shown in Table 1, the PSNR (Peak Signal to Noise Ratio) of our method is slightly lower than the method proposed in [12] in each video sequences due to the increase of embedded capacity. With the increase of QP value, the quality of the coded videos also decreases, which affects the visual performance of the embedding algorithm. Although the PSNR value of our algorithm is slightly lower than the other one, it is still acceptable.

However, the visual effect in [12] is worse than our method. When the compression quality of the videos decreases with the increase of QP value, as we can see in Fig. 4 (b), Fig. 5(b), there are some significant visual hot pixels on the pictures, and these hot pixels are becoming more and more obvious as the QP value increases. It is because the single using of the multi-coefficients (1, −1) sometimes will greatly change several fixed pixels in one 8 × 8 luminance QDCT block.

In terms of embedding capacity, as shown in Table 2, the embedding capacity of our method is nearly 3 times as much as the method in [12] of average per frame, but not exactly 3 times. This is because the method in [12] can also embed some blocks that do not satisfy Condition 1 and Condition 2, however, these blocks can be embedded only 1 bit information per block, so there is little increase in embedded capacity. It is also worth noting that when the QP value equals 24, the embedding

Table 1. PSNR(dB) of embedded frame in each video sequences

Sequences	Method	QP = 16	QP = 24	QP = 32	QP = 40
Keiba	In this paper	47.32	41.32	34.55	31.28
	In [12]	48.53	42.89	35.47	32.75
Container	In this paper	48.16	41.76	35.07	31.57
	In [12]	48.77	42.63	35.79	32.42
Akiyo	In this paper	47.27	41.34	35.87	31.55
	In [12]	47.65	42.08	36.11	32.76
SlideShow	In this paper	48.34	42.45	34.67	30.77
	In [12]	48.85	42.83	35.23	31.43
ParkScene	In this paper	47.52	41.61	35.16	31.24
	In [12]	47.99	42.45	36.13	32.07

(a)

(b)

Fig. 4. (a) Method in this paper (b) Method in [12]

(a)

(b)

Fig. 5. (a) Method in this paper (b) Method in [12]

capacity of each method is the largest, which is because the 8 × 8 luminance QDCT blocks are used most in the vicinity of this QP value.

With the increase of QP value, the visual quality of the decoded image and the embedded image are reduced. And since the single using of the multi-coefficients

Table 2. Embedding capacity (bit) of embedded frame in each video sequences

Sequences	Method	QP = 16	QP = 24	QP = 32	QP = 40
Keiba	In this paper	16440	17376	14784	10512
	In [12]	6031	6554	5878	3842
Container	In this paper	6384	7752	6576	4824
	In [12]	2203	2633	2261	1652
Akiyo	In this paper	6456	7320	5472	4152
	In [12]	2235	2507	1883	1429
SlideShow	In this paper	136392	140208	132168	84432
	In [12]	48128	49192	46248	28896
ParkScene	In this paper	236232	269688	230832	146496
	In [12]	88688	95000	78496	49864

(1, −1) sometimes will greatly change several fixed pixels in 8×8 luminance QDCT blocks, which is easy to cause significant hot pixels when the QP value is big. However, with the simultaneous use of multi-coefficients can significantly improve this visual problem, as shown in Fig. 4(a), Fig. 5(a).

5 Conclusion

This paper proposed a coefficient compensation rule to prevent the intra-frame distortion drift for 8×8 luminance QDCT blocks. As three multi-coefficients are embedded at the same time, this method can get better embedding capacity and visual effects than only using single multi-coefficients. Experimental results demonstrate the feasibility and superiority of the proposed method. In the future, we will improve the embedding coding method, such as introducing matrix coding, to further optimize the embedding efficiency

Acknowledgement. This paper is sponsored by the National Natural Science Foundation of China (NSFC, Grant 61572447).

References

1. Liu, Y., et al.: Video steganography: a review. Neurocomput. **335**, 238–250 (2019)
2. Liu, Y., Liu, S., Zhao, H., Liu, S.: A new data hiding method for H. 265/HEVC video streams without intra-frame distortion drift. Multimedia Tools Appl. **78**(6), 6459–6486 (2019)
3. Sze, V., Budagavi, M., Sullivan, G.J. (eds.): High Efficiency Video Coding (HEVC). ICS. Springer, Cham (2014). https://doi.org/10.1007/978-3-319-06895-4
4. Liu, Y.X., Li, Z.T., Ma, X.J.: Reversible data hiding scheme based on H.264/AVC without distortion drift. J. Syst. Softw. **7**(5), 1059–1065 (2012)
5. Liu, Y., et al.: A robust without intra-frame distortion drift data hiding algorithm based on H 264/AVC. Multimedia Tools Appl. **72**(1), 613–636 (2014)

6. Liu, Y., et al.: A new robust data hiding method for H. 264/AVC without intra-frame distortion drift. Neurocomput. **151**, 1076–1085 (2015)
7. Xu, J., Wang, R.D., Huang, M.L., et al.: A data hiding algorithm for HEVC based on the differences of intra prediction modes. J. Optoelectron. Laser **26**(9), 1753–1760 (2015)
8. Yang, J., Li, S.: An efficient information hiding method based on motion vector space encoding for HEVC. Multimedia Tools Appl. **2017**(1), 1–23 (2017)
9. Chang, P.C., Chung, K.L., Chen, J.J., Lin, C.H., Lin, T.J.: An error propagation free data hiding algorithm in HEVC intra-coded frames. In: Signal & Information Processing Association Summit & Conference, pp. 1–9. IEEE (2013)
10. Gaj, S., Kanetkar, A., Sur, A., Bora, P.K.: Drift compensated robust watermarking algorithm for H. 265/HEVC video stream. ACM Trans. Multimedia Comput. Commun. Appl. (TOMM). **13**(1), 1–24 (2017)
11. Liu, Y., et al.: A robust and improved visual quality data hiding method for HEVC. IEEE Access. **6**, 53984–53997 (2018)
12. Chang, P.C., Chung, K.L., Chen, J.J., Lin, C.H., Lin, T.J.: A DCT/DST-based error propagation-free data hiding algorithm for HEVC intra-coded frames. J. Visual Commun. Image Represent. **25**(2), 239–253 (2013)

Author Index

Printed in the United States
By Bookmasters